INVERSE PROBLEMS IN ENGINEERING:

THEORY AND PRACTICE

presented at
The First Conference in a Series on
Inverse Problems in Engineering
Palm Coast, Florida
June 13-18, 1993

edited by
Nicholas Zabaras
Cornell University
Keith A. Woodbury
University of Alabama
Martin Raynaud
INSA de Lyon, France

published on behalf of the
ENGINEERING FOUNDATION
BY
THE AMERICAN SOCIETY OF MECHANICAL ENGINEERS
UNITED ENGINEERING CENTER / 345 EAST 47TH STREET / NEW YORK, NEW YORK 10017

ISBN No. 0-7918-0694-4

Library of Congress Catalog Number 93-73077

FOREWORD

This volume contains the proceedings of the *First International Conference on Inverse Problems in Engineering: Theory and Practice,* held June 13–18, 1993, in Palm Coast Florida, under the auspices of the Engineering Foundation. The Conference addressed a research area that has been essentially overlooked in the engineering community despite its broad practical application — Inverse Ill-posed, Optimum Design and Parameter Estimation Problems in Engineering. The meeting emphasized a broad range of deterministic and statistical mathematical, computational, and experimental approaches that can be applied to the solution of general inverse and design problems. The program included, among others, the following sessions:

- Mathematical Aspects
- Optimal Experiment Design
- Analysis of Experimental Data
- Signal and Noise Processing
- Inverse Problems in Heat Transfer
- Inverse Heat Conduction and Convection
- Estimation of Thermomechanical Properties
- Inverse Problems in Materials Processing

The idea for the First International Conference on Inverse Problems resulted from a series of very successful Informal Annual Symposiums on Inverse Problems organized by Professor J. V. Beck at Michigan State University starting in June of 1988. Professor Beck served as the honorary chairman of this international conference and has been the driving force in the engineering community behind the increased interdisciplinary attention given to inverse problems. Even though the majority of the contributions were in the area of inverse and parameter estimation problems in heat transfer, the proceedings contain a significant number of papers in the areas of materials processing, shape optimization, and others. Also, several design problems were posed and solved as typical inverse problems. The authors are specialists from a wide range of areas, including mathematics, applied solid and fluid mechanics, heat transfer, and materials processing. The diversity of their backgrounds offered a richness of perspective and has contributed to the significance of the symposium.

As Conference Chairman, I would like to acknowledge the help of the Engineering Foundation in planning the Conference, and of ASME in publishing the proceedings. The help of my co-chairmen, Keith Woodbury and Martin Raynaud, is appreciated. The steady and significant support and contributions given by Didier Delaunay, Yvon Jarny, and Diego Murio were very essential for the success of this Conference. Finally, I would like to thank all the participants for their contributions to the success of this meeting.

Nicholas Zabaras
Cornell University
Conference Chairman

CONTENTS

Inverse Problems in Engineering: Theory and Practice
ASME 1993

INVERSE PROBLEMS AND ILL-POSEDNESS

Patricia K. Lamm
Department of Mathematics
Michigan State University
East Lansing, Michigan

ABSTRACT

We survey the historical development of the study of ill-posed problems, beginning with Hadamard. Several examples of ill-posed inverse problems are given, along with investigations of the type of operator (finitely smoothing, exponentially smoothing, etc.) governing the equation and determining its ill-posedness. More recent developments in the general area of ill-posed problems are outlined, in particular, developments in the theory of generalized regularization schemes, questions of quantification of ill-posedness, and the problem of approximation via standard numerical methods.

HISTORICAL BACKGROUND

Hadamard coined the phrase "well-posed" or "correctly set" problem in 1923, although it is clear that he was not the first to think about such problems. Cauchy, Kowalewsky, Darboux, Goursat, Holmgren, and others were already thinking about problems of this sort in the 1840's to the 1900's.

Hadamard first mentions the concept in 1902 and later in 1923, and he formalizes the notion with the following definition (Hadamard, 1923) :

Definition: A problem is *well-posed* or *correctly set* if the following three conditions hold:

1. There exists a (globally-defined) solution for all (reasonable) data;
2. The solution is unique;
3. The solution depends continuously on given data (*stability*).

A problem which is not well-posed is said to be *ill-posed*.

Hadamard gave a classic example of an ill-posed problem in his 1923 paper, an example which also clearly illustrates the distinction between a "direct problem" (which is typically well-posed) and an "inverse problem" (which is typically ill-posed). Consider the following equation which models 1-dimensional heat flow in a bar:

$$\begin{aligned} w_t &= w_{xx}, \quad 0 < x < 1, \quad t > 0, \qquad (1)\\ w(t,0) &= 0\\ w(t,1) &= 0,\\ w(0,x) &= u(x). \end{aligned}$$

A typical **forward problem** associated with this equation is as follows: **Given** (as data) the initial condition $u(x)$ describing the temperature distribution $w(0,x)$ of the bar at initial time $t = 0$, **find** the temperature distribution $f(x)$ of the bar at a later time, say $t = 1$ ($f(x) \equiv w(1,x)$). It is well-known that this problem is well-posed, with stability holding in terms of most reasonable topologies on u and f; for example, u, $f \in L_2(0,1)$.

The typical **inverse problem** associated with the above equation is as follows: **Given** (as data) $f(x) = w(1, x)$, the temperature of the bar at time $t = 1$, **find** the temperature distribution $u(x) = w(0, x)$ at $t = 0$.

Hadamard noted that inverting the heat equation solution operator (i.e., the above inverse problem) corresponds to a violation of the Second Law of Thermodynamics, which states that a transformation whose only outcome is to transfer heat from a body at a given temperature to a body of higher temperature is impossible. Using this as justification, he pronounced such inverse problems "unnatural" and concluded that they should not be of interest to mathematicians; it was Hadamard's conviction that well-posed problems (such as the above direct problem for the heat equation) are those which are physically relevant and thus are the only problems which should be studied by mathematicians.

Hadamard gave another example in his early papers to support such claims, the example of the Cauchy problem for the Laplace equation (Hadamard, 1902). To understand his example, we consider two problems, first the *Dirichlet problem for the Laplace equation*,

$$\begin{aligned} \Delta w &= 0 \quad \text{on } \Omega \subset \mathbb{R}^2 \\ w|_{\partial\Omega} &= f \end{aligned} \tag{2}$$

where $\Delta w = w_{xx} + w_{yy}$, and, second, the *Cauchy problem for the Laplace equation*,

$$\begin{aligned} \Delta w &= 0 \text{ on } \Omega, \\ &\quad \Omega \equiv \{(x, y) \in (0, \pi) \times (0, \infty)\}, \\ w|_{y=0} &= f_1, \\ \left.\frac{\partial w}{\partial y}\right|_{y=0} &= f_2. \end{aligned} \tag{3}$$

It is well known that under very general conditions on data f and domain Ω, there exists a unique solution w of (2) which depends continuously on f (in reasonable topologies). But even given *analytic* f_1, f_2, as data in (3), we generally have only local solutions of (3), global only under special circumstances. Further, there need not be continuous dependence of solutions on data $f \equiv (f_1, f_2)$, even in the case of unique solutions. As illustration, we examine the stability of the zero solution, i.e., that solution associated with data $f_1 = f_2 = 0$, as we subject that data to perturbations.

Example (Hadamard, 1902): We consider the unperturbed problem, with solution w:

$$\begin{aligned} \Delta w &= 0, \quad (x, y) \in (0, \pi) \times (0, \infty) \\ w|_{y=0} &= 0, \\ \left.\frac{\partial w}{\partial y}\right|_{y=0}(x) &= 0. \end{aligned} \tag{4}$$

and the corresponding perturbed problem, with solution w_n, for each n:

$$\begin{aligned} \Delta w_n &= 0, \quad (x, y) \in (0, \pi) \times (0, \infty) \\ w_n|_{y=0} &= 0, \\ \left.\frac{\partial w_n}{\partial y}\right|_{y=0}(x) &= \alpha_n \sin nx, \end{aligned} \tag{5}$$

where α_n is as small as desired (e.g., $\alpha_n = \frac{1}{n^p}$, e^{-cn}, etc.). It is not difficult to show (Hadamard, 1902) that $w = 0$ uniquely solves (4) and that there exists a unique, globally-defined solution w_n of (5) for any given value of α_n, where

$$w_n(x, y) = \frac{\alpha_n}{2n}(e^{ny} - e^{-ny}) \sin nx.$$

Instability is evident when we note that the perturbation in data for the second problem may be made arbitrarily small, i.e.,

$$\sup_{x \in (0,\pi)} \left| \frac{\partial w}{\partial y}\Big|_{y=0}(x) - \frac{\partial w_n}{\partial y}\Big|_{y=0}(x) \right| \le |\alpha_n|,$$

while the corresponding solutions, w of (4) and w_n of (5), may be made arbitrarily far apart for any fixed $y > 0$:

$$\sup_{x \in (0,\pi)} |w(x, y) - w_n(x, y)| = \frac{|\alpha_n|}{2n}|e^{ny} - e^{-ny}| \to \infty \quad \text{as} \quad n \to \infty.$$

Further, this result is not improved if we strengthen the topology on the data. Thus, the Cauchy problem for Laplace's equation is badly unstable. It was Hadamard's claim that this problem is "physically uninteresting", because a Cauchy problem is one which is evolving in time, while the Laplace equation is naturally used to described a steady-state process or field.

It is interesting to look more closely at one of these examples in order to understand the source of ill-posedness, in particular, *instability*, for these problems. In particular, we examine in detail the heat equation in (1). Using separation of variables, we can write the **direct problem** for (1) as

$$\mathcal{A}u = f$$

where $\mathcal{A}$ is a bounded linear operator representing the transformation of temperature $u \in U$ at $t = 0$ to temperature $\mathcal{A}u \in F$ at $t = 1$, where here $U = F = L_2(0,1)$. The eigenvalues of $\mathcal{A}$ are given by $\lambda_n = e^{-n^2\pi^2}$, associated eigenvalues given by $\varphi_n(x) = \frac{1}{\sqrt{2}} \sin n\pi x$, for $n = 1, 2, \ldots$. It is easy to show that

$$\begin{aligned} \mathcal{A}u &= \sum_{n=1}^{\infty} e^{-n^2\pi^2} \langle u, \varphi_n \rangle \varphi_n \\ \mathcal{R}(\mathcal{A}) &= \left\{ f \in F \,\middle|\, \sum_{n=1}^{\infty} |e^{n^2\pi^2} \langle f, \varphi_n \rangle|^2 < \infty \right\} \\ \mathcal{A}^{-1} f &= \sum_{n=1}^{\infty} e^{n^2\pi^2} \langle f, \varphi_n \rangle \varphi_n . \end{aligned}$$

where $\langle \cdot , \cdot \rangle$ is the $L_2(0,1)$ inner product.

To illustrate the instability of the **inverse problem** (given $f \in F$, solve $\mathcal{A}u = f$ for $u \in U$), we may take any $f \in \mathcal{R}(\mathcal{A})$ and define a perturbation $f^\delta \in \mathcal{R}(\mathcal{A})$ by $f^\delta \equiv f + \delta \sin N\pi x$, for some integer N and arbitrarily small $\delta > 0$. It follows that $\|f - f^\delta\| \le \delta$ (here $\|\cdot\|$ is the $L_2(0,1)$ norm) while $\|\mathcal{A}^{-1}(f - f^\delta)\| = e^{N^2\pi^2}\delta$; the latter may be made as large as desired by selecting sufficiently large N. It is worthwhile to note that changing the topology on $f \in F$ does *not* improve the results; i.e., requiring $\|f - f^\delta\|_{H^p} \le \delta$ (using a suitable H^p norm, $p > 0$) will still not ensure that $\|\mathcal{A}^{-1}(f - f^\delta)\|$ will remain small.

We consider now a second example of ill-posedness in order to illustrate that the *degree* of instability can vary greatly from problem to problem. To this end, we consider $\mathcal{A} : U \to F$, where $U = F = L_2(0,1)$ as before, and where $\mathcal{A}$ is bounded linear and given by

$$\mathcal{A}u(x) = \int_0^x \int_t^1 u(\tau)d\tau, \quad 0 \le x \le 1.$$

For this example, the eigenvalues of $\mathcal{A}$ are

$$\lambda_n = \left(n - \frac{1}{2}\right)^{-2} \pi^{-2}$$

and associated eigenvectors are given by

$$\varphi_n(x) = 2 \sin \left(n - \frac{1}{2}\right) \pi x, \quad n = 1, 2, \ldots .$$

It is not difficult to show, using the resolution of the identity associated with the self-adjoint operator $\mathcal{A}$,

$$\begin{aligned} \mathcal{A}u &= \sum_{n=1}^{\infty} \left(n - \frac{1}{2}\right)^{-2} \pi^{-2} \langle u, \varphi_n \rangle \varphi_n \\ \mathcal{R}(\mathcal{A}) &= \left\{ f \in F \,\middle|\, \sum_{n=1}^{\infty} | \left(n - \frac{1}{2}\right)^2 \pi^2 \langle f, \varphi_n \rangle|^2 < \infty \right\} \\ \mathcal{A}^{-1} f &= \sum_{n=1}^{\infty} \left(n - \frac{1}{2}\right)^2 \pi^2 \langle f, \varphi_n \rangle \varphi_n . \end{aligned}$$

In this example, the equation $\mathcal{A}u = f$ is not quite so ill-posed; in fact, it is easy to see that this equation is equivalent to $u = -f''$ for $f \in H^2(0,1)$, so that if $\|f - f^\delta\|_{H^2} < \delta$, the result is that the associated functions u and u^δ satisfying $\mathcal{A}u = f$ and $\mathcal{A}u^\delta = f^\delta$, respectively, are close in an L^2 sense. It follows then that the equation $\mathcal{A}u = f$ *would* be well-posed if one were willing to settle for a stronger (i.e., smoother; H^2) topology in which to measure perturbations in f. Of course, in practice, this is not acceptable as we must expect perturbations or noise in f to be, in general, quite rough. However, this does illustrate that a quantification of ill-posedness may be desirable, and in fact, the operator $\mathcal{A}$ in this problem might be considered to be "finitely smoothing" in comparison with the operator in the backwards heat example, (which might be considered "exponentially smoothing"). We will return to a more formal characterization of "degree of ill-posedness" at a later point in this paper.

It is worthwhile to note that these two examples are representative of many typical ill-posed problems in that the underlying operators are compact operators (in fact, integral operators). In the case where $\mathcal{A}$ is bounded, linear, and U, F, Banach, it follows from the Closed Graph Theorem that for $\mathcal{A}$ invertible, we have $\mathcal{A}^{-1}$ unbounded (and thus the equation $\mathcal{A}u = f$ ill-posed due to instability) if and only if $\mathcal{R}(\mathcal{A})$ is not closed in F. In the compact case, this is equivalent to $\mathcal{R}(\mathcal{A})$ being infinite-dimensional. Thus the "generic" ill-posed problem is an infinite-dimensional problem.

Because of Hadamard's statements in the 1900's – 1920's, very little was advanced regarding a general theory for ill-posed problems prior to the 1950's. What *was* done typically involved questions of existence and/or uniqueness for specialized problems (see Carleman, 1939; Holmgren, 1901; and many others) and some stability estimates for the ill-posed analytic continuation problem. For the most part, general consideration of instability of ill-posed problems was ignored until the 1940's and 1950's at which time A. N. Tikhonov began his investigations of ill-posed problems associated with geophysical exploration; these problems are essentially equivalent to the Cauchy problem for the Laplace equation (see Lavrent'ev, *et al*, 1986).

Before describing Tikhonov's contributions to the stabilized inversion of ill-posed operators, it is useful to note that questions of *existence* and *uniqueness* for linear ill-posed problems can often be handled by looking for minimum-norm least squares solutions to the original problem. That is, for U, F, Hilbert, $\mathcal{A} : U \to F$ bounded linear, and the usual problem of seeking $u \in U$ satisfying, for given data $f \in F$,

$$\mathcal{A}u = f,$$

one can restate this problem in a least squares setting by seeking $u \in U$ minimizing (over U) the least squares fit-to-data criterion,

$$J(u) = \|\mathcal{A}u - f\|^2$$

where $\|\cdot\|$ is the norm in F. All minimizers of J over U satisfy the normal equations

$$\mathcal{A}^\star \mathcal{A}u = \mathcal{A}^\star f, \tag{6}$$

and, under reasonable assumptions (Nashed, 1976a; Groetsch, 1984; Morozov, 1984) , it is not difficult to show that, for any $f \in \mathcal{R}(\mathcal{A}) + \mathcal{R}(\mathcal{A})^\perp$, there exists a unique minimum-norm solution of (6) (or more generally, there exists a unique u, solution of (6), which minimizes $\|Bu\|$, where B may be a derivative-type operator (Locker and Prenter, 1980; Morozov, 1984)). It is worth remarking that, although the least squares formulation *can* remedy problems of existence and uniqueness associated with the original equation $\mathcal{A}u = f$, in general the minimum-norm least squares solution is *less* stable with respect to perturbations in data than before. That this is true follows from the fact that, for perturbed data, least squares solutions satisfy

$$\mathcal{A}^\star \mathcal{A}u = \mathcal{A}^\star f^\delta \tag{7}$$

where the operator $\mathcal{A}^\star\mathcal{A}$ is typically more "smoothing" than the operator $\mathcal{A}$.

In any case, because the minimum-norm least squares formulation avoids problems of nonuniqueness, we will henceforth assume without loss of generality that there exists a unique solution $\overline{u}$ of the equation $\mathcal{A}u = f$.

Tikhonov's contribution to this field came with his observation that so-called ill-posed problems are *stabilizable* if sufficient *a priori* information is available about the "true solution" and provided this information is actually *used* in the construction of approximate solutions to these problems. One could claim that, in effect, Tikhonov was addressing the connection that Hadamard made between certain ill-posed problems (e.g., the backwards heat equation) and violations of the Second Law of Thermodynamics: Indeed, that law says that some "organizational effort" is required in order to actually invert the natural heat process; Tikhonov's use of *a priori* information about solutions might be considered a way of focusing such "organizational effort".

Tikhonov's work, and its many generalizations (early work in this area due to Lavrent'ev, Ivanov, Morozov, John, Miller, and many others), appeared predominantly in the 1950's and 1960's. A typical formulation of the problem as viewed by these contributors is as follows (in particular, we follow the approach here of Miller (1970)). Assume that one wishes to determine $\overline{u} \in U$ which satisfies the original equation as well as an *a priori* condition; that is, $\overline{u}$ simultaneously satisfies

$$\begin{cases} \mathcal{A}u = f \\ \|Bu\| \le \rho. \end{cases} \tag{8}$$

Here, as before, $\mathcal{A} : U \to F$ is bounded linear, and we also assume that B :dom$(B) \subset U \to Z$ is a closed linear operator, and U, F, and Z are Hilbert spaces (we will not distinguish between the use of $\|\cdot\|$ to represent the norms for these spaces, unless confusion is possible). The condition $\|B\overline{u}\| \le \rho$ is an *a priori* condition containing known information about the smoothness and boundedness of the desired solution $\overline{u}$.

In fact, however, (8) is still an idealized problem in that f^δ, rather than f, should be used, where $\|f - f^\delta\| < \delta$, for some $\delta > 0$. In this case, we accept as a reasonable solution *any* $\overline{u}^\delta \in U$ satisfying

$$\begin{cases} \|\mathcal{A}u - f^\delta\| < \delta \\ \|Bu\| \le \rho. \end{cases} \tag{9}$$

Tikhonov's proposed method of determining a solution of (9) involves solving the constrained minimization problem,

$$\min_{u \in U} \|\mathcal{A}u - f^\delta\|^2 \quad \text{subject to} \quad \|Bu\|^2 \le \rho^2, \tag{10}$$

an infinite-dimensional constrained optimization problem. The Kuhn-Tucker conditions associated with this problem are as follows: there exists $\alpha \ge 0$ such that a (sufficiently regular) solution of the constrained minimization problem (10) is also an unconstrained minimizer of

$$\|\mathcal{A}u - f^\delta\|^2 + \alpha\left(\|Bu\|^2 - \rho^2\right),$$

or, equivalently, a minimizer over U of

$$||\mathcal{A}u - f^\delta||^2 + \alpha||Bu||^2. \tag{11}$$

We note that necessary conditions for minimizers $\overline{u}_\alpha^\delta$ of (11) are

$$(\mathcal{A}^\star\mathcal{A} + \alpha B^\star B)u = \mathcal{A}^\star f^\delta. \tag{12}$$

Comparing this equation with (7), we see that the operator of interest has shifted from $\mathcal{A}^\star\mathcal{A}$ to the Tikhonov operator $(\mathcal{A}^\star\mathcal{A} + \alpha B^\star B)$ which, for $\alpha > 0$ and under suitable hypotheses on B, has its spectrum shifted away from zero. Under such conditions on B (which are easily satisfied by most differential operators; see, for example, Locker and Prenter, 1980; Morozov, 1984), we find that, unlike $\mathcal{A}^\star\mathcal{A}$ in (7), the operator $(\mathcal{A}^\star\mathcal{A} + \alpha B^\star B)$ in (12) is invertible with *continuous* linear inverse defined on all of U ($\mathcal{A}^\star f^\delta \in U$) whenever $\alpha > 0$. Thus, there exists a unique solution $\overline{u}_\alpha^\delta \in U$ of (12) which depends continuously on data f^δ, for all $\alpha > 0$ and all $f^\delta \in F$.

In general however, it is difficult to determine the correct Lagrange Multipler α which satisfies the conditions of Kuhn-Tucker theory and guarantees that solutions of (10) are also unconstrained minimizers of (11). Due to this difficulty, many authors have turned from the Kuhn-Tucker theory and have instead developed a rich theory which gives *a priori* and/or *a posteriori* methods for selecting α (see Miller, 1970; Morozov, 1984; Engl and Neubauer, 1985 ; to name just a few of the references in this area). This approach determines $\alpha = \alpha(\delta, \rho)$ such that $\alpha \to 0$ and $\overline{u}_\alpha^\delta \to \overline{u}$ the noise level $\delta \to 0$. This method of approximating $\overline{u}$ by solutions $\overline{u}_\alpha^\delta$ of (12) is called *Tikhonov regularization*, with α the corresponding *Tikhonov parameter.*

RECENT DEVELOPMENTS

In what follows, we will describe some of the developments in the general theory of ill-posed problems in recent years.

Generalized Regularization Schemes

Recent developments include a theory of generalized regularization for ill-posed operator equations. Suppose that the "true" parameter $\overline{u}$ of interest is the minimum-norm, least squares solution of $\mathcal{A}u = f$, where $\mathcal{A} : U \to F$ is bounded linear and U, F, are Hilbert spaces. Then the true solution satisfies the normal equations given above in (6), which, if $\mathcal{A}^\star\mathcal{A}$ is invertible, gives

$$\overline{u} = (\mathcal{A}^\star\mathcal{A})^{-1}\mathcal{A}^\star f;$$

in the usual case where $(\mathcal{A}^\star\mathcal{A})^{-1}$ is unbounded, $\overline{u}$ does not depend continuously on data f. We have already seen that Tikhonov regularization provides an estimate $\overline{u}_\alpha^\delta$ of $\overline{u}$, given by

$$\overline{u}_\alpha^\delta = (\mathcal{A}^\star\mathcal{A} + \alpha I)^{-1}\mathcal{A}^\star f^\delta$$

(here using in (12) the operator $B = I$, where I denotes the identity on U), where $\overline{u}_\alpha^\delta$ satisfies the following: (1) for all $\alpha > 0$, $\overline{u}_\alpha^\delta$ depends continuously on data since $(\mathcal{A}^\star\mathcal{A}+\alpha I)$ is continuously invertible (on all $\mathcal{A}f^\delta \in U$), and (2) we have $\overline{u}_\alpha^\delta \to \overline{u}$ as $\delta \to 0$ for $\alpha = \alpha(\delta)$ appropriately chosen. We may thus rewrite the Tikhonov solution as

$$\overline{u}_\alpha^\delta = R_\alpha(\mathcal{A}^\star\mathcal{A})\mathcal{A}^\star f^\delta$$

where, as above, $R_\alpha(\mathcal{A}^\star\mathcal{A})$ is (1) a continuous linear operator on all of U satisfying (2) $R_\alpha(\mathcal{A}^\star\mathcal{A}) \to (\mathcal{A}^\star\mathcal{A})^{-1}$ in some sense as $\alpha \to 0$. Here R_α is understood to mean a continuous function of the bounded linear, self-adjoint operator $\mathcal{A}^\star\mathcal{A}$ on U; the operator $R_\alpha(\mathcal{A}^\star\mathcal{A})$ is defined in usual sense from the spectral decomposition of $\mathcal{A}^\star\mathcal{A}$ and from the continuous scalar function $R_\alpha(t)$ defined on the spectrum of $\mathcal{A}^\star\mathcal{A}$. In the case of Tikhonov regularization, it is easy to see that $R_\alpha(t) = 1/(t+\alpha)$, $\quad t \in (0, ||\mathcal{A}^\star\mathcal{A}||])$, and that $R_\alpha(t) \to 1/t$ as $\alpha \to 0$.

In the work of Engl (1981) and Groetsch (1984), we find these ideas extended to more general regularization methods than simply Tikhonov regularization. For this generalized scheme it is required that solutions be defined, as above, by $\overline{u}_\alpha^\delta = R_\alpha(\mathcal{A}^\star\mathcal{A})\mathcal{A}^\star f^\delta$, for given data f^δ and given parameter $\alpha > 0$, where now the scalar function R_α is generalized and assumed to satisfy the following conditions:

1. $R_\alpha(t)$ is continuous in $t \in \sigma(\mathcal{A}^\star\mathcal{A})$.
2. $|t\, R_\alpha(t)|$ is uniformly bounded in $t, \alpha > 0$.
3. $R_\alpha(t) \to 1/t$ as $\alpha \to 0$, for each $t > 0$ in $\sigma(\mathcal{A}^\star\mathcal{A})$.

Under these assumptions it follows that $\overline{u}_\alpha^\delta \to \overline{u}$ as $\delta \to 0$ for an appropriate choice of $\alpha = \alpha(\delta)$ (Engl, 1981; Groetsch, 1984).

The non-Tikhonov regularization technique known as "spectral cut-off" (i.e., approximation via truncation of eigenfunction or singular function expansions; see Miller, 1974; Miller, 1975; Engl, 1983; Groetsch,

1984) can be easily expressed using this generalization; in this case, $R_\alpha(t) = \chi([c(\alpha), ||\mathcal{A}^\star\mathcal{A}||]) \cdot 1/t$, where $\chi([a,b]$ denotes the restriction to the interval $[a,b]$, and $c(\alpha)$ is a given positive constant, with $c(\alpha) \to 0$ as $\alpha \to 0$. Examples of other non-Tikhonov regularization methods which fit into this generalized framework may be found in (Engl, 1981, 1983) and (Groetsch, 1984).

It should be noted that this generalized theory is most naturally applied to regularization methods which are constructed based on shifting (away from zero) or otherwise altering the spectrum of $\mathcal{A}^\star\mathcal{A}$. In fact, any finite-dimensional discretization of the original equation $\mathcal{A}u = f$ (or of $\mathcal{A}^\star\mathcal{A}u = \mathcal{A}^\star f$) is a potential regularization method; unfortunately, most commonly-used discretization methods are *not* based on knowing the spectral properties of $\mathcal{A}$ or $\mathcal{A}^\star\mathcal{A}$, so such methods are not naturally described using this generalized regularization framework.

Quantifying Amounts of Ill-Posedness

The notion of quantifying the amount of ill-posedness of a problem is not a new one; Carleman, Holmgren, Franklin, John, Miller, and many others have measured the "amount of unboundedness" (in some sense) of $\mathcal{A}^{-1}$, for a number of specific bounded operators $\mathcal{A}$, and for a number of different *a priori* assumptions. For $\mathcal{A}^{-1}$ unbounded, the quantity $\sup\{||u-v|| \mid ||\mathcal{A}u-f^\delta|| \le \delta, ||\mathcal{A}v - f^\delta|| \le \delta\}$ is infinite. Thus, a reasonable way to measure the way in which the *a priori* condition $||Bu|| \le \rho$ stabilizes the solutions of $||\mathcal{A}u - f^\delta|| \le \delta$ is to measure the radius of the set of all solutions of (9); i.e., we define (Taylor, 1991)

$$\begin{aligned} \mathcal{M}_B(\delta,\rho) &\equiv \frac{1}{2}\sup\{||u-v|| \mid u,v \text{ satisfy } (9)\} \\ &= \sup\{||u|| \mid ||\mathcal{A}u|| \le \delta\,, ||Bu|| \le \rho,\}\,. \end{aligned}$$

Clearly, for the *a priori* condition to stabilize the problem $\mathcal{A}u = f$ with respect to perturbations in f, we require that $\mathcal{M}_B(\delta,\rho) \to 0$ as $\delta \to 0$. The quantity $\mathcal{M}_B(\delta,\rho)$ is often called the "modulus of continuity" for the problem (8), and, in fact, it possible to show that $\mathcal{M}_B(\delta,\rho)$ is the "best possible, worst case" rate of convergence for *any* estimates of solutions $\overline{u}^\delta$ of (9), (Tikhonov, or otherwise), the convergence to $\overline{u}$ occurring as the level δ of error converges to zero (Franklin, 1974). With regards to the modulus of continuity $\mathcal{M}_B(\delta,\rho)$, the following questions are thus of interest:

- What is the size of $\mathcal{M}_B(\delta,\rho)$ for particular $\mathcal{A}$, B, especially as $\delta \to 0$?
- Does a given method (Tikhonov, spectral cut-off, etc.) generate approximations $\hat{u}^\delta$ to $\overline{u}^\delta$ of (9) which converge to a solution $\overline{u}$ of (8) at a rate comparable to $\mathcal{M}_B(\delta,\rho)$? The optimal situation is to determine that there exists $C \ge 1$, C close to 1, such that

$$||\hat{u}^\delta - \overline{u}|| \le C\mathcal{M}_B(\delta,\rho)$$

which converges to zero, as $\delta \to 0$.

Many authors have looked at the first question for specific ill-posed problems; for example, F. John (1960) for the analytic continuation problem, Franklin (1974) and Miller (1975) for the backwards parabolic problem, and Manselli and Miller (1980) for the inverse heat conduction problem, to name just a few of the many references in this area. Recent work on more *general* problems of the form $\mathcal{A}u = f$ has appeared by Natterer (1983, 1984) and Mair (1992) ; we describe some of their findings below.

Natterer's work on quantifying ill-posedness especially facilitates a description of "finitely smoothing" operators, such as the operator $\mathcal{A}$ in the second example above, where $\mathcal{A}u(x) = \int_0^x \int_t^1 u(\tau)d\tau$, $0 \le x \le 1$. Recall that for this operator the eigenvalues were powers of $1/n$; because typical differential operators have eigenvalues which are powers of n, it is convenient to think of $\mathcal{A}$ as a function of L^{-1}, where L is a suitable differential operator.

To generalize this idea, we suppose that we are given a scale $\{U_s, s \in \mathbb{R}\}$ of Hilbert spaces generated by a self-adjoint, unbounded linear operator L ($||Lu|| \ge ||u||$), where $U_0 = U$; that is, we define $U_s \equiv \text{dom } L^s$, where powers of L are constructed in the usual way from the spectral resolution of L, and where $||u||_s \equiv ||L^s u||_0$. Suppose too that with respect to this scale $\{U_s\}$ one is able to find $a, M, m > 0$ such that

$$m||u||_{-a} \le ||\mathcal{A}u||_0 \le M||u||_{-a}. \qquad (13)$$

We note that such an inequality may always be written for any bounded linear $\mathcal{A}$ with unbounded inverse by constructing a $\mathcal{A}$-dependent Hilbert scale, e.g., let $\{U_s\}$ be generated by $L \sim (\mathcal{A}^\star\mathcal{A})^{-1}$. In the case of such an $\mathcal{A}$-dependent scale, the inequality in (13) is not particularly useful in quantifying the unboundedness of $\mathcal{A}^{-1}$. In contrast, useful information *is* obtained from (13) when $\{U_s\}$ is a standard Hilbert scale, for example, when $U_s = H^s(\Omega)$, the standard scale of Sobolev spaces for a given regular domain Ω; in this case, (13) indicates that the problem of inverting $\mathcal{A}$ is roughly equivalent to the problem of inverting L^{-a}, where L is the differential operator (with given boundary conditions) generating the Sobolev spaces. It is usual in this case to say that the operator $\mathcal{A}$ is "finitely smoothing" (Natterer, 1983; Mair, 1992) .

Under the structure of a Hilbert scale, it is possible to obtain estimates on the modulus of continuity $\mathcal{M}_B(\delta,\rho)$ in the case where $\|Bu\| \equiv \|u\|_q$. For a *finitely smoothing* $\mathcal{A}$, it is an interpolation result (Natterer, 1983) that

$$\mathcal{M}_B(\delta,\rho) = \mathcal{O}\left(\delta^{\frac{q}{a+q}} \rho^{\frac{a}{a+q}}\right). \qquad (14)$$

Thus the "best possible" rate of convergence of an estimate u^δ (of (9)) to $\overline{u}$ is $\mathcal{O}\left(\delta^{\frac{q}{a+q}}\right)$ as $\delta \to 0$.

For an *exponentially smoothing* $\mathcal{A}$ (e.g., the heat solution operator as in the first example above), which has spectral values going to zero like e^{-n^b}, one obtains (see Mair, 1992, for example)

$$\mathcal{M}_B(\delta,\rho) = \mathcal{O}\left(\left(\frac{1}{\log\frac{\rho}{\delta}}\right)^{c\frac{q}{b}}\right), \qquad (15)$$

$c > 0$, indicating a very weak form of stability, and a very slow "best possible" rate of convergence as $\delta \to 0$.

To give an indication of the size of $\mathcal{M}_B(\delta,\rho)$ as $\delta \to 0$ in these two cases, we give two examples of representative findings.

Example 1: Finitely Smoothing Example (where smoothing $\sim a$, *a priori* information $\sim q$, $\rho = 1$; see (14))

	δ	$\mathcal{M}_B(\delta,\rho)$, $a=2$	$\mathcal{M}(\delta,\rho)$, $a=3$
$q=1$	10^{-6}	$\mathcal{O}(10^{-2})$	$\mathcal{O}(10^{-1.5})$
$q=2$	10^{-6}	$\mathcal{O}(10^{-3})$	$\mathcal{O}(10^{-2.4})$
$q=1$	10^{-15}	$\mathcal{O}(10^{-5})$	$\mathcal{O}(10^{-3.75})$
$q=2$	10^{-15}	$\mathcal{O}(10^{-7.5})$	$\mathcal{O}(10^{-6})$

Example 2: Exponentially Smoothing Operator (where smoothing $\sim b$, *a priori* information $\sim q$, $\rho = 1$; see (15))

	δ	$\mathcal{M}_B(\delta,\rho)$, $b=2$	$\mathcal{M}(\delta,\rho)$, $b=3$
$q=1$	10^{-6}	$\mathcal{O}(2.6\times10^{-1})$	$\mathcal{O}(4.2\times10^{-1})$
$q=2$	10^{-6}	$\mathcal{O}(7.2\times10^{-2})$	$\mathcal{O}(1.7\times10^{-1})$
$q=1$	10^{-15}	$\mathcal{O}(1.7\times10^{-1})$	$\mathcal{O}(3.1\times10^{-1})$
$q=2$	10^{-15}	$\mathcal{O}(2.8\times10^{-2})$	$\mathcal{O}(9.4\times10^{-2})$

We see from these tables that the order of convergence of $\mathcal{M}_B(\delta,\rho) \to 0$ as $\delta \to 0$ is quite slow for exponentially smoothing operators; in practice the results are just as undesirable for the case of finitely smoothing operators with large a. In both of these two cases, since the "best possible" convergence estimate,

$$\|u^\delta - \overline{u}\| \le C\mathcal{M}_B(\delta,\rho)$$

is likely to be quite slow, it is easy to see how an improvement in the constant C can go a long way toward improving the results obtained in practice.

Natterer (1983) and Engl and Neubauer (1988) have looked at the second question above (regarding whether a given method actually converges at a rate comparable to $\mathcal{M}_B(\delta,\rho)$ as $\delta \to 0$), and have determined conditions under which standard Tikhonov regularization and Tikhonov regularization in finite-dimensional approximation spaces generate estimates for $\overline{u}$ which converge at this "best possible" rate. And, as will be seen below, Natterer has also determined that least squares solutions found in standard spline-based approximation spaces converge at a rate of order $\mathcal{M}_B(\delta,\rho)$ as $\delta \to 0$, provided the operator $\mathcal{A}$ is finitely smoothing.

Estimation of Solutions of Ill-Posed Problems using Standard Finite-Dimensional Approximations

It is an unfortunate fact of life that standard finite-dimensional approximations (methods such as projection, collocation, Galerkin, etc., all implemented in spline-type spaces) are not always applicable for the solution of ill-posed problems. We refer the reader to Chapter IV of (Groetsch, 1984) and the references therein for a lengthy discussion on the difficulties associated with employing such "standard" methods for these very nonstandard problems. There *are* ways to circumvent such difficulties, such as restricting approximations to a very narrow choice of finite-dimensional spaces (see, for example, Groetsch, 1984, and Engl, 1983a, 1983b, for L_2 projection methods; and Engl, 1983b for collocation problems), or by applying a combination projection-regularization type method, such as an implementation of Tikhonov regularization in finite-dimensional subspaces (see, for example, Marti, 1980; Wahba, 1977; Nashed, 1976b; and the references in Chapter IV of Groetsch, 1984).

Because numerical methods in standard finite-dimensional spaces are far more easily implemented than the approaches mentioned above, it is of great interest to have easy guidelines for determining when such methods are applicable for the approximation of solutions of ill-posed problems.

One desirable method for solving these problems might be to find u_h^δ solving a finite-dimensional *least squares* problem; that is, find $u_h^\delta \in S(h)$ such that

$$\min_{u \in S(h)} \|\mathcal{A}u - f^\delta\|^2 = \|\mathcal{A}u_h^\delta - f^\delta\|^2, \qquad (16)$$

where $S(h)$ is a spline space of given smoothness, defined on a grid of length h, and $\|\cdot\|$ denotes the $L_2(0,1)$ (for example) norm. Or, another desirable approach is to use a *collocation*-type method in which the goal is now to find $u_h^\delta \in S(h)$ satisfying

$$\mathcal{A}u(x_j) = f^\delta(x_j), \quad j = 1, 2, \ldots, N(h), \qquad (17)$$

where x_j, $j = 1, \ldots, N(h)$, are prescribed collocation points in, say, $[0,1]$. Each of these problems may be viewed as a projection method and expressed as the finite-dimensional problem

$$\mathcal{A}_h c_h^\delta = f_h^\delta \qquad (18)$$

for suitable matrix $\mathcal{A}_h$, vector f_h^δ, and coordinate vector c_h^δ for u_h^δ (see, for example, Natterer, 1977). As equation (18) is a finite-dimensional problem for a given h, solutions are automatically stable and depend continuously on perturbations in the finite-dimensional data f_h^δ (or f_h). But as $h \to 0$, equation (18) better approximates the original (infinite-dimensional) equation $\mathcal{A}u^\delta = f^\delta$ and thus stability can be expected to get increasingly worse. Thus it is evident that, just as α was a regularization parameter for Tikhonov regularization, so the discretization size h can be thought of as a regularization parameter; that is, the discretization size $h = h(\delta)$ should be selected such that $h(\delta) \to 0$ as $\delta \to 0$, and in order that growing instability as $h \to 0$ is offset by the decreasing noise level δ as $\delta \to 0$.

A description of the errors due to approximation and due to instability may be found in Natterer (1977) . Using this approach one may express the difference between a finite-dimensional approximation u_h^δ and a solution $\overline{u}$ of $\mathcal{A}u = f$ by writing

$$\|\overline{u} - u_h^\delta\| \leq \|\overline{u} - u_h\| + \|u_h - u_h^\delta\| \qquad (19)$$

where u_h is the approximation that would be obtained from the noise-free data f. The first term in (19) thus represents the error due to approximation only, while the second term involves the stability properties (for given h) of the finite-dimensional problem. In fact, we may write the second term as

$$\begin{aligned} \|u_h - u_h^\delta\| &\leq \|Q_h\| \, \|f - f^\delta\| \\ &\leq \|Q_h\| \, \delta \end{aligned}$$

where Q_h is the projection operator defined by the approximation method, $Q_h : F \to S(h)$. For a (reasonable) approximation method for this ill-posed problem, $\|Q_h\| \to \infty$ as $h \to 0$, so that $h = h(\delta)$ should be selected guaranteeing $h \to 0$ as $\delta \to 0$ but at a rate that still ensures $\|Q_h\| \, \delta \to 0$ as $\delta \to 0$. The task of proving convergence of approximations u_h^δ to "true" $\overline{u}$ therefore involves proving convergence of noise-free approximations (the first term in (19)) and estimating the size of $\|Q_h\|$ as $h \to 0$.

In the case of finitely smoothing operators, Natterer (1983) been able to show that spline-based least squares projection methods are convergent in the presence of noisy data, and that they are also of "best possible" order of convergence (in terms of the modulus of continuity discussions above). The key to this result is that, for least squares approximations in spline spaces, noise-free convergence is obtained at an optimal order; at the same time, the quantity $\|Q_h\|$ grows at a rate comparable to $\|(\mathcal{A}|_{S(h)})^{-1}\|$, where $\mathcal{A}|_{S(h)}$ is the restriction of $\mathcal{A}$ to $S(h)$ (an estimate that is easily computed for $\mathcal{A}$ finitely smoothing operator and $S(h)$ a spline-based space).

For infinitely smoothing operators, the situation is far more difficult. It is in fact an open question in the theory of ill-posed problems whether or not certain standard approximation methods may be used for many particular equations governed by such operators, and whether reasonable guidelines may be found for the selection of convergent methods. In any case, it is clear from the above discussion of modulus of continuity, that convergence, when obtained, is *extremely slow*; it will therefore be of practical interest to work with methods for which optimal convergence rates are obtained, and for which constant factors appearing in the convergence rates are as small as possible.

Application to Finitely Smoothing Volterra Operators

We briefly consider how the ideas presented above may be applied to the problem of solving the equation $\mathcal{A}u = f$, where $\mathcal{A}$ is the Volterra convolution integral operator given by

$$\mathcal{A}u(x) = \int_0^x k(x-s)u(s)\,ds, \quad x \in [0,1],$$

where $k \in L_2(0,1)$, $\mathcal{A} : L_2(0,1) \to L_2(0,1)$. The resulting integral equation

$$\int_0^x k(x-s)u(s)\,ds = f(x), \quad x \in [0,1], \qquad (20)$$

is a first-kind Volterra equation, a well-known ill-posed equation due to instability. In the case where k is $(\nu+1)$-times differentiable, with $\nu \geq 0$, $k^{(\nu+1)} \in L_2(0,1)$, $0 = k(0) = k'(0) = \ldots = k^{(\nu-1)}(0)$ and $k^{(\nu)}(0) \neq 0$, then, for f sufficiently smooth, the equation $\mathcal{A}u = f$ may be differentiated $\nu+1$ times yielding a second-kind equation which is well-posed (i.e., there exists a unique solution of the new equation which is continuous with respect to perturbations in $f^{(\nu+1)}$). It is then not surprising that we obtain the finitely smoothing estimate (Lamm, 1993),

$$m\|u\|_{-(\nu+1)} \leq \|\mathcal{A}u\|_0 \leq M\|u\|_{-(\nu+1)},$$

where $\|\cdot\|$ denotes the $L_2(0,1)$ norm, and the $\|\cdot\|_{-(\nu+1)}$ norm is that induced by the scale of standard Sobolev spaces, $H^s(0,1)$. It therefore follows that some of the standard numerical methods based on spline approximations (in particular, spline-based least squares and collocation methods) are applicable. Unfortunately, for ν large these methods converge at an optimal (i.e., "best possible"), albeit *very slow*, rate as $\delta \to 0$, and thus it is highly desirable to use methods for which constant factors appearing in the convergence rates are as small as possible. It appears that the "future sequential method" developed (Beck, *et al*, 1985) to treat the inverse heat conduction problem (which is also based on a Volterra convolution integral equation, with an infinitely smoothing operator), offers an improvement over sequential least squares and standard collocation methods in that it decreases the constant appearing in the convergence rate (Lamm, 1993). Whether a similar statement may be made for the inverse heat conduction problem itself is still an open question.

References

[1] Beck, J. V., Blackwell, B., and St. Clair. Jr., C. R., 1985, *Inverse Heat Conduction Problems*, Wiley, New York.

[2] Carleman, T., 1939, Sur un problème d'unicité pour les systeèmes d'Équations aux dérivées partielles à deux variables indépendantes, *Arkiv för Matematik, Astronomi och Fysik*, 26 B:1–9.

[3] Engl, H. W., 1981, Necessary and sufficient conditions for convergence of regularization methods for solving linear operator equations of the first kind, *Numer. Funct. Analy. Opt.*, 3:201–222.

[4] Engl, H. W., 1983a, On the convergence of regularization methods for ill-posed linear operator equations, In G. Hämmerlin and K. H. Hoffmann, editors, *Improperly Posed Problems and Their Numerical Treatment, Springer Int'l. Series of Num. Math.*, pages 81–96.

[5] Engl, H. W., 1983b, Regularization and least squares collocation, In P. Deuflhard and E. Hairer, editors, *Inverse Problems in Differential and Integral Equations*, Birkhäuser, Boston.

[6] Engl, H. W. and Neubauer, A., 1988, Convergence rates for Tikhonov regularization in finite-dimensional subspaces of Hilbert scales, *Proc. Amer. Math. Soc.*, 102:587–592.

[7] Engl, H. W. and Neubauer, A., 1985, Optimal discrepancy principles for the Tikhonov regularization of integral equations of the first kind, In G. Hämmerlin and K. H. Hoffmann, editors, *Constructive Methods for the Practical Treatment of Integral Equations*, pages 120–141.

[8] Franklin, J. N., 1974, On tikhonov's method for ill-posed problems, *Mathematics of Computation*, 28:889–907.

[9] Groetsch, C. W., 1984, *The Theory of Tikhonov Regularization for Fredholm Equations of the First Kind*, Pitman.

[10] Hadamard, J., 1923, *Lectures on Cauchy's Problem in Linear Partial Differential Equations*, Yale University Press.

[11] Hadamard, J., 1902, Sur les problèmes aux derivées partielles et leur signification physique, *Bull. Univ. Princeton*, 13:49–52.

[12] Holmgren, E., 1901, Über systeme von linearen partiellen differentialgleichungen, *Öfversigt af kongl. Vetenskaps-Akademiens Förhandlingar*, 58:91–103.

[13] John, F., 1960, Continuous dependence on data for solution of partial differential equations with a prescribed bound, *Comm. Pure and Appl. Math.*, 13:551–585.

[14] Lamm, P. K., 1993, Sequential regularization for the inversion of finitely smoothing volterra operators, in preparation.

[15] Lavrent'ev, M. M., Romanov, V. G., and Shishatskii, S. P., 1986, *Ill-Posed Problems of Mathematical Physics and Analysis*, Volume 64 of *AMS Transl. of Math. Monographs.*

[16] Locker, J., and Prenter, P. M., 1980, Regularization with differential operators i: general theory, *J. Math. Analy. Appl.*, 74:504–529.

[17] Mair, B. A., 1992, Tikhonov regularization for finitely and infinitely smoothing operators, preprint.

[18] Manselli, P., and Miller, K., 1980, Calculation of the surface temperature and heat flux on one side of a wall from measurements on the opposite side, *Ann. Mat. Pura Appl.*, 123:161–183.

[19] Marti, J. T., 1980, On the convergence of an algorithm for computing minimum norm solutions of ill-posed problems, *Mathematics of Computation*, 34:521–527.

[20] Miller, K., 1975, Efficient numerical methods for backward solution of parabolic problems with variable coefficients, In A. Carasso and A. Stone, editors, *Improperly Posed Boundary Value Problems*, pages 54–64.

[21] Miller, K., 1970, Least squares methods for ill-posed problems with a prescribed bound, *SIAM J. Math. Analy.*, 1:52–74.

[22] Morozov, V. A., 1984, *Methods for Solving Incorrectly Posed Problems*, Springer-Verlag, New York.

[23] Nashed, M. Z., 1976a, Editor, *Generalized Inverses and Applications*, Academic Press, New York.

[24] Nashed, M. Z., 1976b, On moment discretization and least squares solutions of linear integral equations of the first kind, *J. Math. Analy. Appl*, 53:359–366.

[25] Natterer, F., 1984, Error bounds for Tikhonov regularization in Hilbert scales, *Applicable Analysis*, 18:29–37.

[26] Natterer, F., 1983, On the order of regularization methods, In G. Hämmerlin and K. H. Hoffmann, editors, *Improperly Posed Problems and their Numerical Treatment, Springer Int'l. Series Num. Math.*, pages 189–203.

[27] Natterer, F., 1977, Regularisierung schlecht gestellter probleme durch projektionsverfahren, *Numer. Math.*, 28:329–341.

[28] Taylor, M. E., 1991, Estimates for approximate solutions to acoustic inverse scattering problems, preprint.

[29] Wahba, G., 1977, Practical approximate solutions to linear operator equations when the data are noisy, *SIAM J. Numer. Analy.*, 14:651–667.

Inverse Problems in Engineering: Theory and Practice
ASME 1993

TOWARDS A STABILITY AND ERROR ANALYSIS OF SEQUENTIAL METHODS FOR THE INVERSE HEAT CONDUCTION PROBLEM

Hans-Jürgen Reinhardt
Universität Gesamthochschule, Siegen
Siegen, Germany

ABSTRACT

Sequential approximations for the solution of the linear and nonlinear Inverse Heat Conduction Problem in one spatial dimension are analyzed. The problem is illposed and belongs to a class of Cauchy problems for parabolic initial–boundary value problems. The nonlinearity caused by a temperature dependent thermal diffusivity is linearized by freezing the temperature for certain steps in the sequential procedure known as the Beck method.

In the analysis a relation for the approximating heat fluxes showing the explicit dependency on the the previous ones is established. Moreover, a stability and error analysis provides estimates for the errors in the heat fluxes and in the temperatures when the temperature data are perturbed.

1. INTRODUCTION

In this paper, the one–dimensional nonlinear Inverse Heat Conduction Problem is studied. This problem consists in determining the transfer heat flux and temperature at one end of the spatial interval from a temperature history measured at fixed locations — inside or at the other end of the interval. The nonlinearity is due to the fact that the thermal diffusivity, thermal conductivity or specific heat depend on the temperature itself (cf. Beck et al. (1985) and the references therein).

The IHCP has many applications in the engineering sciences. It belongs to a class of boundary value problems for parabolic equations that are known to be severely illposed. Accordingly, any method for obtaining approximating solutions have to stabilize or regularize the illposed character of the problem.

Among the approximation methods, we concentrate on sequential procedures stepping forward in time. In particular, we analyze the Beck method (Beck, 1970), (Beck et al., 1982), (Beck et al., 1985) using several future times. The objectives of our analysis are twofold namely to show how the approximating heat fluxes depend explicitly on the previous ones as well as to estimate the errors in the heat fluxes and in the temperatures when the temperature data are perturbed by errors of known magnitude. The error estimates are based on energy estimates for parabolic initial–boundary value problems. Estimates of such type can be found, e. g. in Kreiss and Lorenz (1989), and Ladyzhenskaya and Ural'tseva (1968).

It should be mentioned that the method considered is not a numerical one but it is semidiscrete insofar as the time interval is subdivided similarly to the Rothe method. Clearly, the present procedure can be utilized to construct a numerical scheme which, itself, may be analyzed by appropriate techniques. Furthermore, the present analysis can be extended to two–dimensional IHCPs using an approach developed in Reinhardt (1991).

In the following Section 2, it is shown how the new heat flux q^{n+1} of the Beck method for IHCPs depends explicitly on the previous ones (cf. (13)). The representation (13) essentially utilizes appropriately defined impulse functions and is established here for linear problems only.

In Section 3, the influence of perturbed temperature data is studied. As a result, the sum of the gain coefficients is the essential factor multiplying the perturbations in the data as well as the error contribution for the previous time. This sum reflects somehow the illposed character of the problem since it is large when the time step size Δt and the number r of future times are small. At least for linear problems, a criterion can be given how to choose r and Δt related to the magnitude of the perturbations in the temperature data.

2. SEQUENTIAL APPROXIMATION BY THE BECK METHOD; BASIC RELATIONS

We consider the Inverse Heat Conduction Problem in one spatial dimension in a not necessarily linear form,

$$\begin{aligned} &\frac{\partial u}{\partial t} = \frac{\partial}{\partial x}\left(a(u)\frac{\partial u}{\partial x}\right), \quad 0 < x < L\,,\ 0 < t \le T\,, \\ &u(L,t) = g(t)\,, \quad u_x(L,t) = 0\,, \quad 0 < t \le T\,, \\ &u(.,0) = u_0\,, \qquad 0 < x < L\,. \end{aligned} \tag{1}$$

Here, the data function $g(t) = u(1,t)$, $0 < t \le T$ and the initial temperature distribution u_0 are given; the temperature and flux history at $x = 0$, $f(t) = u(0,t)$ and $q(t) = -a(u)u_x|_{x=0}$ are sought. The function $a(.)$ representing the thermal diffusivity is assumed to be continuous, bounded away from zero, $a(z) \ge \alpha_0 > 0$, and to be sufficiently smooth.

Beck (1970), and Beck et al. (1982) suggested a sequential procedure for estimating the desired heat flux at $x = 0$ where the nonlinearity is linearized by 'freezing' the thermal diffusivity for a certain number of future times. This can be understood as a certain realization of an optimal control problem with piecewise constant approximations of the heat flux.

Let $t_n = n\Delta t$, $n = 0, 1, 2, \ldots$, be equidistant (discrete) times in the time interval and Δt the time step width. With a number r of *future times* Beck's method proceeds as follows:

- Start with $n = 0$, $t_0 = 0$,
$$v_0^{(0)} = u_0\,, \quad q^0 = -\left(a^{(0)}\frac{\partial u_0}{\partial x}\right)(0).$$
- Determine $v = v^{(n+1)}$ as the solution of
$$\begin{aligned} &\frac{\partial v}{\partial t} = \frac{\partial}{\partial x}\left(a^{(n)}\frac{\partial v}{\partial x}\right), \quad 0 < x < L, \quad t_n < t \le t_{n+r}\,, \\ &-\left(a^{(n)}v_x\right)(0,t) = q^n\,, \quad v_x(L,t) = 0\,, \quad t_n < t \le t_{n+r}\,, \\ &v(.,t_n) = v_0^{(n)}\,, \quad 0 < x < L\,. \end{aligned} \tag{2}$$
- Determine q^{n+1} from:
$$\text{Minimize} \sum_{\ell=1}^{r}\left(g(t_{n+\ell}) - v^{(n+1)}(L,t_{n+\ell})\right)^2. \tag{3}$$
- Set $v_0^{(n+1)} = v^{(n+1)}(.,t_{n+1})$ and start again with $n+1$, $v_0^{(n+1)}$, q^{n+1} instead of n, $v_0^{(n)}$, q^n.

The functions $a^{(n)}$ are defined by $a^{(n)}(x) = a(v_0^{(n)}(x))$. The $v^{(n+1)}$ is thus obtained by a solution of a linear (direct) heat equation with coefficient $a^{(n)}$ determined by the solution of the previous time step. We note that $v = v^{(n+1)}$ depends on $q = q^n$.

One step of the Newton(–Raphson) method for approximating the zero of $\frac{\partial Z}{\partial q}$, with

$$Z(q) = \sum_{\ell=1}^{r}\left(g(t_{n+\ell}) - v^{(n+1)}(q;\,L,\,t_{n+\ell})\right)^2, \tag{4}$$

leads to a new estimated heat flux of the form

$$\begin{aligned} q^{n+1} &= q^n + \\ &+ \sum_{\ell=1}^{r}\gamma_\ell^{(n+1,r)}\left(g(t_{n+\ell}) - v^{(n+1)}(q^n;\,L,\,t_{n+\ell})\right) \end{aligned} \tag{5}$$

where $\Delta^{(n+1,r)} = \sum_{\ell=1}^{r} s^{(n+1)}(L,\,t_{n+\ell})^2$, $s = s^{(n+1)}$ denotes the *sensitivity function* defined by

$$\begin{aligned} &\frac{\partial s}{\partial t} = \frac{\partial}{\partial x}\left(a^{(n)}\frac{\partial s}{\partial x}\right), \quad 0 < x < L\,,\ t_n < t \le t_{n+r}\,, \\ &-\left(a^{(n)}s_x\right)(0,t) = 1\,,\ s_x(L,t) = 0\,,\ t_n < t \le t_{n+r}\,, \\ &s(x,t_n) = 0\,, \quad 0 < x < L\,, \end{aligned} \tag{6}$$

and

$$\gamma_\ell^{(n+1,r)} = s^{(n+1)}(L,t_{n+\ell})/\Delta^{(n+1,r)}$$

are called the *gain coefficients.* As in linear problems, the sensitivity function is nothing else than

$$s^{(n+1)}(x,t) = \frac{\partial v^{(n+1)}}{\partial q}(x,t)\,, \qquad 0 < x < L\,,\ t > t_n\,.$$

Analogously to Beck et al. (1985), Sec. 6.6, an *update* $\hat{v}_0^{(n+1)} = \hat{v}^{(n+1)}(.,t_{n+1})$ is obtained by solving ($\hat{v}^{(n+1)} = \hat{v}$)

$$\begin{aligned} &\frac{\partial \hat{v}}{\partial t} = \frac{\partial}{\partial x}\left(a^{(n)}\frac{\partial \hat{v}}{\partial x}\right), \quad 0 < x < L\,,\ t > t_n\,, \\ &-a^{(n)}\frac{\partial \hat{v}}{\partial x}\Big|_{x=0} = q^{n+1}\,, \quad \frac{\partial \hat{v}}{\partial x}\Big|_{x=L} = 0\,, \quad t > t_n\,, \\ &\hat{v}(.,t_n) = \hat{v}_0^{(n)}\,, \quad 0 < x < L\,, \end{aligned} \tag{7}$$

with $\hat{v}^{(0)} (= v^{(0)}) = u_0$ and q^{n+1} given by (5). The new $a^{(n+1)}$ is then determined by $\hat{v}_0^{(n+1)}$.

By superposition it is clear that

$$\hat{v}^{(n+1)} = q^{n+1}s^{(n+1)} + \hat{z}^{(n+1)}\,, \quad t > t_n\,,\ n \ge 0\,, \tag{8}$$

where $\hat{z} = \hat{z}^{(n+1)}$ is the solution of

$$\begin{aligned} &\frac{\partial \hat{z}}{\partial t} = \frac{\partial}{\partial x}\left(a^{(n)}\frac{\partial \hat{z}}{\partial x}\right), \quad 0 < x < L\,,\ t_n < t \le t_{n+r}\,, \\ &\frac{\partial \hat{z}}{\partial x}\Big|_{x=0} = \frac{\partial \hat{z}}{\partial x}\Big|_{x=L} = 0\,, \quad t_n < t \le t_{n+r}\,, \\ &\hat{z}(.,t_n) = \hat{v}_0^{(n)}\,, \quad 0 < x < L\,. \end{aligned} \tag{9}$$

By relation (8) it is obvious, that for each n a linear problem for q^{n+1} is present which is a consequence of freezing $a(.)$ in the time interval $t_n < t \le t_{n+r}$.

In our analysis, we are first interested to see in what manner the new (estimated) heat flux q^{n+1} depends on the previous ones. For notational simplicity, we only give the corresponding relation in the linear case, i. e. $a(z) = 1$. By induction, one observes that the function $\hat{z}^{(n+1)}$ defined in (9) has the representation

$$\hat{z}^{(n+1)}(x,t) = w(x,t) + \sum_{m=1}^{n} q^m y(x, t - t_{m-1}) , \quad t > t_n \quad (10)$$

where w is determined by the initial function,

$$\begin{aligned} w_t &= w_{xx} &&, t > 0 , \\ w_x|_{x=0} &= w_x|_{x=L} = 0 &&, t > 0 , \\ w|_{t=0} &= u_0 &&, 0 < x < L , \end{aligned} \quad (11)$$

and y represents a certain *impulse function*,

$$\begin{aligned} y_t &= y_{xx} &&, t > 0 , \\ y_x|_{x=0} &= 1 &&, 0 < t \leq \Delta t , \\ y_x|_{x=0} &= 0 &&, t > \Delta t , \\ y_x|_{x=L} &= 0 &&, t > 0 , \\ y|_{t=0} &= 0 &&, 0 < x < L . \end{aligned} \quad (12)$$

Insertion of (10) and (8) into (5) immediately leads to the following representation of q^{n+1} which shows the explicit dependencies on the previous heat fluxes,

$$q^{n+1} = \sum_{\ell=1}^{r} \gamma_\ell^{(r)} \left(g(t_{n+\ell}) - w(L, t_{n+\ell}) - \sum_{m=1}^{n} q^m y(L, t_{n+\ell-m+1}) \right) \quad (13)$$

with the gain coefficients $\gamma_\ell^{(r)} = s(L, \ell\Delta t)/\Delta^{(r)}$ and $\Delta^{(r)} = \sum_{\ell=1}^{r} s(L, \ell\Delta t)^2$. It should be noted that in the linear case $a(.) = 1$ the sensitivity functions do not depend on n any longer and can be expressed in the form

$$s^{(n+1)}(., t_n + \tau) = s^{(0)}(., \tau) = s(., \tau) , \quad n = 0, 1, \ldots$$

In the nonlinear case, w and y fulfill the heat equation in the different time intervals with the coefficients $a^{(n)}$ so that one has to use different $y^{(m)}$, $m = 1, \ldots, n$, and $w^{(n+1)}$ in (13). Details will be carried out elsewhere.

3. STABILITY AND ERROR ANALYSIS

For the time being, we still concentrate on the linear case and, at the end, outline the extension to nonlinear problems.

To start our analysis we like to mention an aspect which is somehow trivial — especially for an engineer — but helps to understand the question whether a chosen time step is large or small. The latter should always be related to the dimensionless form of the heat equation. Here, where the thermal diffusivity is already set equal to one, only the length L of the spatial interval determines the dimensionless time given by $t^+ = t/L^2$. In practice, L is the distance from the nearest transducer location to the surface where the temperature and heat flux should be determined. Thus, to a small dimensionless time step Δt^+ of say 0.01 there corresponds a real time step of $\Delta t = L^2 \Delta t^+ = 0.01 \times L^2$. As a consequence, the nearer the transducer is located to the surface the smaller time steps are allowed.

Our next aim is to establish *stability estimates*. A first one is immediately obtained by the definition (5) of q^{n+1},

$$|q^{n+1}| \leq |q^n| + \sum_{\ell=1}^{r} |\gamma_\ell^{(r)}| \, |g(t_{n+\ell}) - v^{(n+1)}(L, t_{n+\ell})| . \quad (14)$$

The method consists in determining q^{n+1} *and* $v^{(n+1)}$. Stability estimates for the latter are obtained by means of energy type estimates (cf., e. g. Kreiss and Lorenz (1989))

$$\|v^{(n+1)}(., t)\|_{0,2}^2 \leq C \left\{ \left\| v_0^{(n)} \right\|_{0,2}^2 + ((t - t_n)|q^n|^2 \right\} , \quad t > t_n , \quad (15)$$

$$\left\| v^{(n+1)}(., t) \right\|_{0,\infty}^2 \leq C \left\{ \left\| v_0^{(n)} \right\|_{0,2}^2 + \left\| \frac{\partial}{\partial x} v_0^{(n)} \right\|_{0,2}^2 + (1 + (t - t_n))|q^n|^2 \right\} , t > t_n . \quad (16)$$

Here, $\|.\|_{0,p}$ denotes the norm in $L^p(0, L)$ for $1 \leq p \leq \infty$. The first estimate follows from energy type inequalities for the solution of parabolic pde's with Neumann boundary conditions in connection with Gronwall's Lemma. The second one is a consequence of (15), a similar estimate for the space derivative $v_x^{(n+1)}$ and Sobolev's inequality in one spatial dimension.

The estimate for $v_x^{(n+1)}$ reads as follows,

$$\left\| \frac{\partial}{\partial x} v^{(n+1)}(., t) \right\|_{0,2}^2 \leq C \left\{ \left\| \frac{\partial}{\partial x} v_0^{(n)} \right\|_{0,2}^2 + (1 + (t - t_n)) |q^n|^2 \right\} , \quad t > t_n \quad (17)$$

Combining (14), (16) and (17) one observes that $|q^{n+1}|$, $\|v_0^{(n+1)}\|_{0,2}$, $\|\frac{\partial}{\partial x} v_0^{(n+1)}\|_{0,2}$ can be estimated by the corresponding quantities on the previous time level. This is analogous to stability estimates for time stepping algorithmus approximating solutions of parabolic pde's (cf. e. g. Reinhardt (1985), Part IV). However, here, the estimates may be too pessimistic since, due to (14), the bounds on the previous time level are always multiplied by the sum of the gain coefficients,

$$S_\Delta^{(r)} = \sum_{\ell=1}^{r} \left| \gamma_\ell^{(r)} \right| . \quad (18)$$

These numbers depend not only on r but also on the time step size Δt; they can be rather large when r or Δt are small which will be discussed later in more detail. The quantities $S_\Delta^{(r)}$ play also a major role when we study the errors in the estimted heat fluxes (cf. (5)) in the case of perturbations in the data, $g_\varepsilon^\nu = g^\nu + \varepsilon\varphi^\nu$ with $\varphi^\nu = O(1)$ and $g^\nu = g(t_\nu)$, $\nu = 1, 2, \ldots$ Perturbations of the initial function are not discussed here because corresponding analysis is standard.

TABLE 1: $S_\Delta^{(r)}$

$\Delta t \backslash r$	1	2	3	4	6	8
0.0100	9.9E+9	9.9E+9	1.5E+5	2.0E+4	1.9E+3	4.8E+2
0.0125	9.9E+9	2.2E+6	2.7E+4	4.7E+3	6.2E+2	1.9E+2
0.0200	5.6E+6	2.7E+4	1.6E+3	4.3E+2	9.9E+1	4.3E+1
0.0250	6.8E+5	5.5E+3	5.4E+2	1.8E+2	5.0E+1	2.4E+1
0.0400	1.9E+4	4.3E+2	9.1E+1	4.0E+1	1.6E+1	9.1E+0
0.0500	4.9E+3	1.7E+2	4.6E+1	2.3E+1	1.0E+1	6.2E+0
0.0800	4.8E+2	3.7E+1	1.5E+1	8.6E+0	4.6E+0	3.1E+0
0.1000	2.0E+2	2.1E+1	9.4E+0	5.9E+0	3.3E+0	2.3E+0
0.2000	2.3E+1	5.4E+0	3.1E+0	2.2E+0	1.4E+0	1.0E+0

To study $q_\epsilon^\nu - q^\nu$, one first observes that q_ϵ^{n+1} can be expressed by formula (13) with $g^{n+\ell} = g(t_{n+\ell})$ replaced by $g_\epsilon^{n+\ell}$ and q^m replaced by q_ϵ^m. The error has the following representation,

$$q_\epsilon^{n+1} - q^{n+1} = \epsilon S_\Delta^{(r)} \rho^{(n+1)}, \quad n = 0, 1, \ldots \tag{19}$$

with $\rho^{(0)} = 0$ and

$$\begin{aligned} \rho^{(n+1)} &= \sum_{\ell=1}^{r} \gamma_\ell^{(r)} \left\{ \frac{1}{S_\Delta^{(r)}} \varphi^{n+\ell} \right. \\ &\quad \left. - \sum_{m=1}^{n} \rho^{(m)} y(L, t_{n+\ell-m+1}) \right\} . \end{aligned} \tag{20}$$

The relation (19), (20) can be verified by induction. It shows that $S_\Delta^{(r)} \rho^{(n+1)}$ is nothing else than the solution of the Beck method of an IHCP with data φ^ν. We call $\rho^{(\nu)}$ the *(discrete) error function* in the following.

Another way of representing the error uses the definition (5) but $v_\epsilon^{(n+1)}$ has to be specified. Taking (19), and expressing $v_\epsilon^{(n+1)}$ as

$$v_\epsilon^{(n+1)} = v^{(n+1)} + \epsilon S_\Delta^{(r)} \tilde{v}^{(n+1)} . \tag{21}$$

$\rho^{(n+1)}$, $\tilde{v}^{(n+1)}$ are related to the previous quantities and to the data φ^ν by

$$\begin{aligned} \rho^{(n+1)} &= \rho^{(n)} + \\ &+ \sum_{\ell=1}^{r} \gamma_\ell^{(r)} \left(\frac{1}{S_\Delta^{(r)}} \varphi^{n+\ell} - \tilde{v}^{(n+1)}(L, t_{n+\ell}) \right) \end{aligned} \tag{22}$$

where $\tilde{v}^{(n+1)}$ solves the heat equation for $t > t_n$ with initial function $\tilde{v}_0^{(n)} = \tilde{v}^{(n)}(., t_n)$ at $t = t_n$ and flux boundary conditions $-\tilde{v}_x^{(n+1)}|_{x=0} = \rho^{(n)}$, $\tilde{v}_x^{(n+1)}|_{x=L} = 0$. The splitting of $v_\epsilon^{(n+1)}$ in the form (21) follows from the linear dependency on the flux boundary condition at $x = 0$ and from the representation (19) of $q_\epsilon^{(n+1)}$; then (22) is nothing else than formula (5) with $\varphi^\nu / S_\Delta^{(r)}$ as temperature data at $x = L$ instead of g^ν.

The quantities $\rho^{(\nu)}$ and $\tilde{v}^{(\nu)}$ in (22) only depend on the perturbations φ^ν in the data. If the $\rho^{(\nu)}$ are uniformly bounded — for Δt and r varying in a certain range — the errors in the heat fluxes are of magnitude $O(\epsilon S_\Delta^{(r)})$. Bounds for $\rho^{(\nu)}$ — and $\tilde{v}^{(\nu)}$ — can be obtained similarly as above (cf. (14) - (16)). The theoretical bounds, however, are rather pessimistic and include again the sum of the gain coefficients — see the discussion following up (17).

Numerical experiments in the linear case show that the $\rho^{(\nu)}$ remain bounded when Δt is not too small and r is suitably chosen. As data functions we have chosen a perturbation by a sine function and by a random number generator.

Fig. 1 displays the error functions $\rho^{(\nu)}$ with data $\varphi^\nu = \sin(\omega\nu\Delta t)$, $\omega = 20$, for a relatively large $r(= 8)$ and various step sizes. The error functions remain bounded and even reflect the periodicity of the data function. For the same φ^n and $r = 4$, Fig. 2 demonstrates how instability can occur when Δt is too small; the curve for $\Delta t = 0.0125$ and $r = 4$ cannot be seen in the figure because it is out of the indicated range. For Δt as large as 0.025 (and $r = 4$) again a bounded error function is obtained; for smaller Δt, r has to be enlarged in order to obtain stable computations. In Fig. 3, for $r = 8$ and certain small Δt's, the error functions remain also uniformly bounded similar to those in Fig. 1, where now the data function is created by a random number generator. Similar to Fig. 2, Fig. 4 shows an instable behaviour for random errors if $\Delta t = 0.0125$ and $r = 4$; with the same Δt stable computations are enforced by using $r = 6$. In all computations, we have used $L = 1$.

If we specify a tolerance τ and assume the boundedness of the error function, the requirement

$$\epsilon S_\Delta^{(r)} |\rho^{(\nu)}| \le \tau \tag{23}$$

gives a hint how to choose Δt and r according to the magnitude ϵ of the perturbations in the data. Table 1 shows the $S_\Delta^{(r)}$ for various r and Δt. In particular, if the error function is bounded by one, for simplicity, relation (23) gives a concrete criterion for choosing Δt and r.

The decrease (increase) of the $S_\Delta^{(r)}$ for increasing (decreasing) $\Delta t, r$ may allow a statement which is known since a long time namely that the Beck method performs more stably when r or Δt (or both) are increased. Then, according to our criterion (23), a smaller $S_\Delta^{(r)}$ allows perturbations

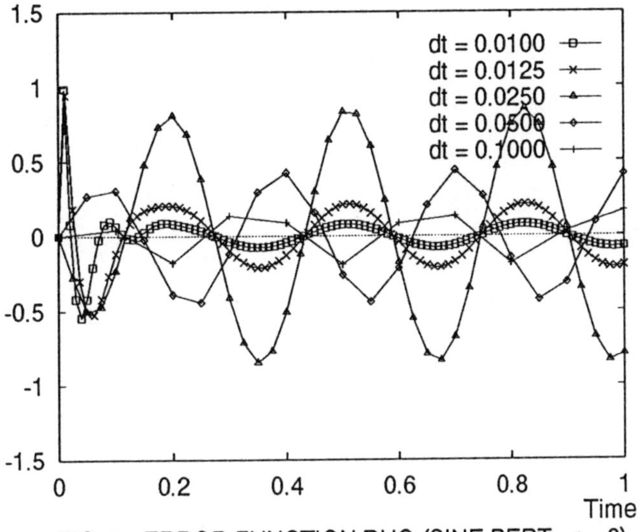

FIG. 1 : ERROR FUNCTION RHO (SINE PERT., r = 8)

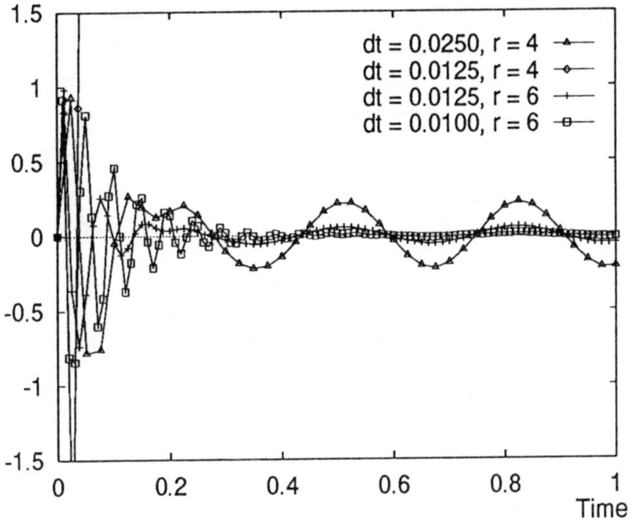

FIG. 2 : ERROR FUNCTION RHO (SINE PERT.)

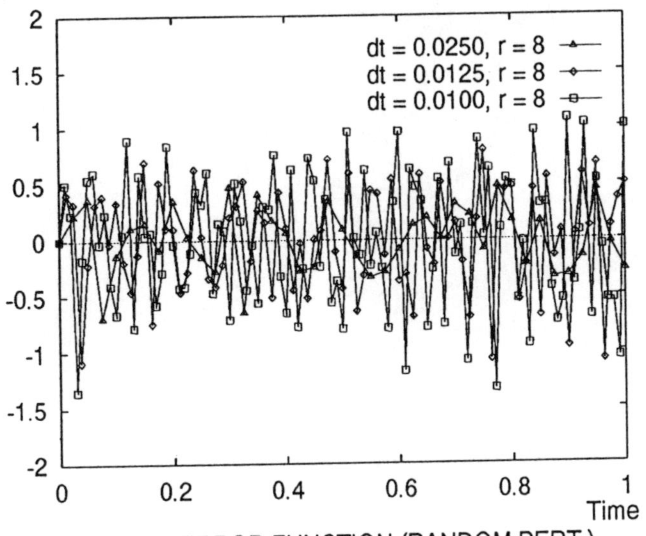

FIG. 3 : ERROR FUNCTION (RANDOM PERT.)

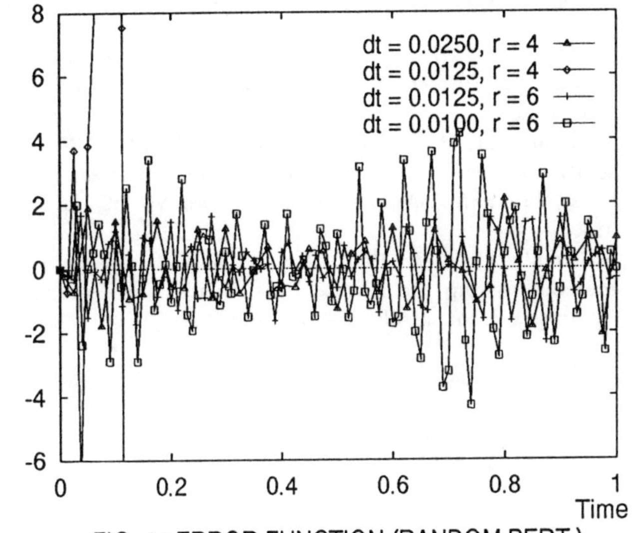

FIG. 4 : ERROR FUNCTION (RANDOM PERT.)

of larger magnitude in the data.

From the point of view of approximation methods for parabolic pde's, the quantity $S_{\Delta}^{(r)}$ plays a similar role as the amplification factors in the classical von Neumann stability analysis (see, e. g., Reinhardt (1985), 12.2). Also, the importance of the gain coefficients for the Beck method was mentioned by Beck et al. (1985), p. 127f; the decrease of the gain coefficients for increasing Δt was said to be an indicator for decreasing sensitivity to measurement errors.

The numbers in Table 1 are calculated by using the Crank–Nicolson–Galerkin method with an equidistant grid in the spatial interval of mesh size $h = 1/8$ and unit length $L = 1$.

We conclude this analysis by outlining the possible extensions to nonlinear problems of the form (1). In the basic relation (5), everything is depending on the perturbation paramater ε. Also the coefficients in the heat equation have to be modified in every time step and depend on the perturbations in the data. Estimates for the difference of solutions of parabolic pde's with and without perturbations in the initial condition, boundary conditions and in the thermal diffusivity are also provided by energy type techniques. For this, the boundedness of $a(.)$ and its derivatives up to order three has to be required. The latter is a requirement on the boundedness of $v^{(n+1)}$, $v_{\varepsilon}^{(n+1)}$ itself if one considers, for example, $a(.)$ depending linearily on u. If $a^{(n)}$, $a_{\varepsilon}^{(n)}$ (see (6), (7)) are constant in the spatial variable, e. g. if one takes average values, everthing becomes much easier and the above mentioned requirements can be omitted. Altogether, also in the nonlinear case, the essential contribution in an error estimate for $q^{n+1} - q_\epsilon^{n+1}$ is obtained by multiplying errors on the previous time level by the sum of the gain coefficients. Some additional analysis is needed to establish the relevance for practical purposes.

REFERENCES

Beck, J. V., 1970, "Nonlinear estimation applied to the nonlinear inverse heat conduction problem," *Int. J. Heat Mass Transfer*, Vol. 13, pp. 703–716.

Beck, J. V., Litkouhi, B., St. Clair, C. R. Jr., 1982, "Efficient sequential solution of the nonlinear inverse heat conduction problem," *Numerical Heat Transfer*, Vol. 5, pp. 275–286.

Beck, J. V., Blackwell, B., and St. Clair, C. R., Jr., 1985, *"Inverse heat conduction problems,"* Wiley, New York.

Kreiss, H. O., and Lorenz, J., 1989, "Initial–Boundary Value Problems and the Navier–Stokes Equations," *Pure & Appl. Math.*, Vol. 136, Academic Press, New York.

Ladyzhenskaya, O. A., and Ural'tseva, N., 1968, *"Linear and quasilinear elliptic equations."* Academic Press, New York.

Reinhardt, H.-J., 1985, *"Analysis of approximation methods for differential and integral equations,"* Springer, New York.

Reinhardt, H.-J., 1991, "A numerical method for the solution of two–dimensional inverse heat conduction problems," *Internat. J. Numer. Methods Engrg.*, Vol. 32, pp. 363–383.

Reinhardt, H.-J., 1993, "On the stability of sequential methods solving the Inverse Heat Conduction Problem," *Z. Angew. Math. Mech.*, Vol. 73, pp. T928–T930.

Inverse Problems in Engineering: Theory and Practice
ASME 1993

ON THE NUMERICAL SOLUTION OF THE TWO-DIMENSIONAL INVERSE HEAT CONDUCTION PROBLEM BY DISCRETE MOLLIFICATION

Diego A. Murio
Department of Mathematical Sciences
University of Cincinnati
Cincinnati, Ohio

ABSTRACT

A space marching implementation of the Mollification Method is introduced to numerically recover the temperature and heat flux histories at three boundary sides of a bounded two-dimensional rectangular body when the temperature and heat flux transient functions are approximately measured at the other boundary side.

NOMENCLATURE

D	*Two − dimensional domain.*
F	*Exact surface temperature.*
M	*Exact surface heat flux.*
R	*Rectangular domain.*
V	*Discrete mollified temperature*
W	*Discrete mollified heat flux.*
f	*Unknown temperature.*
h	*Space grid size.*
k	*Time grid size.*
q	*Unknown heat flux.*
s	*Space grid size.*
t	*Time variable.*
u	*Dimensionless temperature.*
v	*Mollified temperature.*
w	*Mollified heat flux.*
x	*Space variable.*
y	*Space variable.*
F_m	*Measured surface temperature.*
J_δ	*Mollified functional.*
L^2	*Square integrable functions.*
Q_m	*Measured surface heat flux.*
R^2	*Cartesian plane.*
$V^n_{i,j}$	*Discrete mollified temperature at node* (i,j,n).
$W^n_{i,j}$	*Discrete mollified heat flux at node* (i,j,n).
t_n	*Nth. time coordinate.*
x_i	*Ith. space coordinate.*
y_j	*Jth. space coordinate.*
δ	*Radius of mollification.*
ϵ	*Data error bound.*
ρ	*Gaussian kernel.*
$\epsilon_{\cdot,j,n}$	*Gaussian random variables.*
$\|\cdot\|$	L^2 *norm.*
$*$	*Convolution.*

INTRODUCTION

Analytical and computational methods for treating the inverse heat conduction problem (IHCP) have been basically restricted to one-dimensional models. The difficulties of the two-dimensional IHCP are more pronounced and very few results are available in this case.

The first analytical solution – requiring exact data – that is applicable to two-dimensional conduction systems for geometries of arbitrary shape was introduced by M. Imber (1974). Most of the literature related to the numerical treatment of the two-dimensional IHCP is based on different ways of combining finite elements realizations with the Future Temperatures Method of J.V. Beck (1970). For some of the early applications of these ideas, see B.R. Bass and L.J. Ott (1980). More numerical experimentation can be found in T. Yoshimura and K. Ituka (1985). An elaborated and comprehensive exposition of the method was presented later by J. Baumeister and H.J. Reinhart (1987), and, more recently, by N. Zabaras and J.C. Liu (1988).

In this paper we briefly review the space marching implementation of the Mollification Method for the two-dimensional inverse heat conduction problem in a slab as introduced by L. Guo and D.A. Murio (1988), and then concentrate our attention on the actual application of the algorithm to the more realistic situation related with a bounded rectangular domain in the (x,y) plane.

We assume all the functions involved to be L^2 functions in R^2 suitably extended – when necessary for the analysis – as being zero everywhere in $\{(y,t),\ t \leq 0,\ y \leq 0\}$.

TWO DIMENSIONAL IHCP IN A SLAB

We consider a two-dimensional IHCP in a semi-infinite slab, in which the temperature and heat flux histories $f(y,t)$ and $q(y,t)$ on the right-hand side ($x = 1/2$) are desired and unknown and the temperature and heat flux on the left-hand surface ($x = 0$) are approximately measurable.

The normalized linear problem can be described mathematically as follows. The unknown temperature $u(x,y,t)$ satisfies:

(1)
- (1a) $u_t = u_{xx} + u_{yy} \quad 0 < x < 1/2, \ y > 0, \ t > 0,$
- (1b) $u(0,y,t) = F(y,t), \ y > 0, \ t > 0,$ with corresponding approximate data function $F_m(y,t)$,
- (1c) $u_x(0,y,t) = Q(y,t), \ y > 0, \ t > 0,$ with corresponding approximate data function $Q_m(y,t)$,
- (1d) $u(x,y,0) = 0, \ 0 \le x \le 1/2, \ y \ge 0,$
- (1e) $u(1/2,y,t) = f(y,t), \ y > 0, \ t > 0,$ the desired but unknown temperature function,
- (1f) $u_x(1/2,y,t) = q(y,t), \ y > 0, \ t > 0,$ the desired but unknown heat flux function,
- (1g) $u(x,0,t) = 0, \ 0 \le x \le 1/2, \ t > 0.$

We hypothesize that the exact data functions $F(y,t)$ and $Q(y,t)$ and the measured data functions $F_m(y,t)$ and $Q_m(y,t)$ satisfy the L^2 data error bounds

$$\|F - F_m\| \le \epsilon \text{ and } \|Q - Q_m\| \le \epsilon. \tag{2}$$

The Fourier analysis of system (1), presented in L. Guo and D.A. Murio (1991), demonstrates that if s and w represent the Fourier transform variables associated with y and t respectively, attempting to solve problem (1) – obtaining $f(y,t)$ and $q(y,t)$ from $F(y,t)$ and $Q(y,t)$ – amplifies the error in a high frequency component by the factor

$$\exp\left[\left(\sqrt{s^4 + w^2} + s^2\right)^{1/2}\right].$$

Thus, the inverse problem is highly ill-posed in the high frequency components.

STABILIZED PROBLEM

Introducing the two-dimensional Gaussian kernel

$$\rho(y,t,\delta_1,\delta_2) = \frac{1}{\pi\delta_1\delta_2}\exp\left[-\left(\frac{y^2}{\delta_1^2} + \frac{t^2}{\delta_2^2}\right)\right]$$

and denoting $\rho(y,t,\delta_1,\delta_2)$ by $\rho_\delta(y,t)$ if $\delta_1 = \delta_2 = \delta$, the two-dimensional convolution of any locally integrable function $g(y,t)$ with the gaussian kernel $\rho_\delta(y,t)$ – the mollification of the function $g(y,t)$ – is written as

$$J_\delta g(y,t) = (\rho_\delta * g)(y,t) = \int_{-\infty}^{\infty}\int_{-\infty}^{\infty} \rho_\delta(y',t')g(y-y',t-t')dy'dt'.$$

Mollifying system (1), we obtain the following associated problem: attempt to find $J_\delta f_m(y,t) = J_\delta u(1/2,y,t)$ and $J_\delta q_m(y,t) = J_\delta u_x(1/2,y,t)$ at some point (y,t) of interest and for some radius $\delta > 0$ given that $J_\delta u(x,y,t)$ satisfies

$$(3)\quad \begin{array}{ll} (J_\delta u)_t = (J_\delta u)_{xx} + (J_\delta u)_{yy}, & 0 < x < 1/2, \ y > 0, \ t > 0, \\ J_\delta u(0,y,t) = J_\delta F_m(y,t), & y > 0, \ t > 0, \\ J_\delta u_x(0,y,t) = J_\delta Q_m(y,t), & y > 0, \ t > 0, \\ J_\delta u(x,y,0) = 0, & 0 \le x \le 1/2, \ y > 0, \\ J_\delta u(x,0,t) = 0, & 0 \le x \le 1/2, \ t > 0. \end{array}$$

This problem and its solutions satisfy the following theorem.

Theorem 1. Suppose that $\|F - F_m\| \le \epsilon$ and $\|Q - Q_m\| \le \epsilon$. Then,

1. Problem (3) is a formally stable problem with respect to perturbations in the data.
2. If the exact boundary temperature $f(y,t)$ and heat flux $q(y,t)$ have uniformly bounded first order partial derivatives on the bounded domain $D = [0,Y] \times [0,T]$, then $J_\delta f_m$ and $J_\delta q_m$ verify

$$\|f - J_\delta f_m\|_D \le O(\delta) + 2\epsilon \exp[7\delta^{-2}] \tag{4}$$

and

$$\|q - J_\delta q_m\|_D \le O(\delta) + 2\epsilon \exp[7\delta^{-2}]. \tag{5}$$

The proof of this statement can be found in L. Guo and D.A. Murio (1991).

NUMERICAL METHOD AND ERROR ANALYSIS

With $v = J_\delta u$ and $w = v_{yy}$, system (1) is equivalent to

$$(6)\quad \begin{array}{ll} v_t = w_x + v_{yy}, & 0 < x < 1/2, \ y > 0, \ t > 0, \\ w = v_x, & 0 < x < 1/2, \ y > 0, \ t > 0, \\ v(0,y,t) = J_\delta F_m(y,t), & y > 0, \ t > 0, \\ w(0,y,t) = J_\delta Q_m(y,t), & y > 0, \ t > 0, \\ v(x,y,0) = 0, & 0 \le x \le 1/2, \ y \ge 0, \\ v(1/2,y,t) = J_\delta f_m(y,t), & y > 0, \ t > 0, \ \text{unknown}, \\ w(1/2,y,t) = J_\delta q_m(y,t), & y > 0, \ t > 0, \ \text{unknown}, \\ v(x,0,t) = 0, & 0 \le x \le 1/2, \ t > 0. \end{array}$$

Without loss of generality, we will seek to reconstruct the unknown mollified boundary temperature function $J_\delta f_m$ or mollified boundary heat flux function $J_\delta q_m$ in the unit square $D = [0,1] \times [0,1]$ of the (y,t) plane $x = 1/2$. Consider a uniform grid in the (x,y,t) space:

$$\{(x_i = ih, \ y_j = js, \ t_n = nk), \ i = 0,1,\ldots,N, \ Nh = 1/2;$$
$$j = 0,1,\ldots,M, \ Ms = L; \ n = 0,1,\ldots,P, \ Pk = C\}$$

where L and C depend on h, s and k in a way to be specified later, $L, C > 1$. Let the grid functions V and W be defined by

$$V_{i,j}^n = v(x_i,y_j,t_n), \ W_{i,j}^n = (x_i,y_j,t_n),$$
$$0 \le i \le N, \ 0 \le j \le M, \ 0 \le n \le P.$$

We notice that

$$\begin{array}{ll} V_{0,j}^n = J_\delta F_m(y_j,t_n), & 0 \le j \le M, \ 0 \le n \le P, \\ W_{0,j}^n = J_\delta Q_m(y_j,t_n), & 0 \le j \le M, \ 0 \le n \le P, \\ V_{i,0}^n = 0 & 0 \le i \le N, \ 0 \le n \le P, \\ V_{i,j}^0 = 0 & 0 \le i \le N, \ 0 \le j \le M. \end{array}$$

We approximate the system of partial differential equations (6) with the consistent finite difference schemes

$$(7)\quad \begin{array}{rcl} W_{i+1,j}^n &=& W_{i,j}^n + \frac{h}{2k}\left(V_{i,j}^{n+1} - V_{i,j}^{n-1}\right) \\ && -\frac{h}{s^2}\left(V_{i,j-1}^n - 2V_{i,j}^n + V_{i,j+1}^n\right), \\ V_{i+1,j}^n &=& V_{i,j}^n + hW_{i+1,j}^n, \\ && 0 \le i \le N-1, 1 \le j \le M-i-1, \\ && 1 \le n \le P-i-1, \\ V_{0,j}^n &=& (J_\delta F_m)_j^n, \\ && 1 \le j \le M-1, 1 \le n \le P-1, \\ W_{0,j}^n &=& (J_\delta Q_m)_j^n, \\ && 1 \le j \le M-1, 1 \le n \le P-1, \\ V_{i,0}^n &=& 0, \\ && 0 \le i \le N-1, 1 \le n \le P-i, \\ V_{i,j}^0 &=& 0, \end{array}$$

$$1 \le i \le N, 1 \le j \le M.$$

Notice that as we march forward in the x-direction in space, we must drop the estimation of the interior temperature from the highest previous point in time and the associated right-most point in the y-direction. Since we want to evaluate $\{V^n_{N,j}\}$ and $\{W^n_{N,j}\}$ at the grid points of the unit square $D = [0,1] \times [0,1]$ – in the (y,t) plane at $x = 1/2$ – after N iterations, the minimum initial length C of the data sample interval in the time axis needs to satisfy the condition $C = Pm = 1 - k + k/h$. Similarly, the minimum initial length L of the data sample interval in the y-direction satisfies $L = Ms = 1 - s + s/h$.

The stability of the finite difference scheme (7) and the convergence of the numerical solution to the solution of the mollified problem (3) are shown in L. Guo and D.A. Murio (1991).

Remarks:

1. The radius of mollification, δ, can be selected automatically as a function of the level of noise in the data. In fact, for a given $\epsilon > 0$, there is a unique $\delta > 0$, such that

   (8) $$\|J_\delta F_m - F_m\|_D = \epsilon.$$

2. It is also possible to replace the discrete two-dimensional mollification of the data functions by two successive one-dimensional mollifications in time and y-space. In this manner, the data filtering task can be executed as a parallel process.

NUMERICAL RESULTS

We investigate now the two-dimensional IHCP when the space domain in the (x,y) plane is restricted to the bounded prototype rectangle $R = [0,1/2] \times [0,1]$. We must add the "boundary" condition – from the point of view of the IHCP and the x-space marching scheme –

$$u(x,1,t) = h_1(x,y), \quad 0 \le x \le 1/2, \quad t > 0,$$

and, at the same time, we shall not require the homogeneity of the boundary condition (1g) that now should read

$$u(x,0,t) = h_0(x,t), \quad 0 \le x \le 1/2, \quad t > 0.$$

However, since the functions f, h_0, h_1 and F uniquely determine the solution of the direct problem – including the heat fluxes Q and q at $x = 0$ and $x = 1/2$ respectively – it is clear that, for the inverse problem, the data functions F and Q possess all the necessary information for the recovery of the temperature and heat flux functions f and q at the surface $x = 0$ and also the boundary conditions h_0 and h_1 at $y = 0$ and $y = 1$ respectively. Consequently, the functions $u(x,0,t) = h_0(x,t)$ and $u(x,1,t) = h_1(x,t)$ will be treated as <u>unknowns</u> and their recovery becomes a natural task for the two-dimensional IHCP. Mathematically, the new inverse problem can be stated as follows:

(9)

(9a) $u_t = u_{xx} + u_{yy} \quad 0 < x < 1/2, \quad 0 < y < 1 \quad t > 0,$

(9b) $u(0,y,t) = F(y,t), \; 0 < y < 1, \; t > 0,$
with corresponding approximate data function $F_m(y,t)$,

(9c) $u_x(0,y,t) = Q(y,t), \; 0 < y < 1, \; t > 0,$
with corresponding approximate data function $Q_m(y,t)$,

(9d) $u(x,y,0) = 0, \; 0 \le x \le 1/2, \; 0 < y < 1,$

(9e) $u(1/2,y,t) = f(y,t), \; 0 < y < 1, \; t > 0,$
the desired but unknown temperature function,

(9f) $u_x(1/2,y,t) = q(y,t), \; 0 < y < 1, \; t > 0,$
the desired but unknown heat flux function,

(9g) $u(x,0,t) = h_0(x,t), \; 0 \le x \le 1/2, \; t > 0,$
unknown boundary temperature function,

(9h) $u(x,1,t) = h_1(x,t), \; 0 \le x \le 1/2, \; t > 0,$
unknown boundary temperature function.

A few words about the uniqueness of the solution of system (9) are appropriate.

Consider the direct problem for the two-dimensional heat conduction system above, with boundary data temperature $\bar{F}(y,t) = u(0,y,t)$, $\bar{h}_0(x,t) = u(x,0,t)$, $\bar{h}_1(x,t) = u(x,1,t)$ and $\bar{f}(y,t) = u(1/2,y,t)$. This problem has a unique temperature solution and, in particular, an induced heat flux solution at $x = 0$ that we denote $\bar{Q}(y,t) = u_x(0,y,t)$. On the other hand, the same direct problem with data temperature $F(y,t) = v(0,y,t)$, $h_0(x,t) = v(x,0,t)$, $h_1(x,t) = v(x,1,t)$ and $f(y,t) = v(1/2,y,t)$ also has a unique solution and the induced heat flux at $x = 0$ will be denoted $Q(y,t) = v_x(0,y,t)$.

Is it possible to have $Q = \bar{Q}$ if $f \ne \bar{f}$ and/or $h_0 \ne \bar{h}_0$ and/or $h_1 \ne \bar{h}_1$? By superposition, the solution of the direct problem with boundary data temperature $\Delta\tilde{F} = F - \bar{F} = 0$, $\Delta h_0 = h_0 - \bar{h}_0$, $\Delta h_1 = h_1 - \bar{h}_1$ and $\Delta f = f - \bar{f}$ is such that the heat flux at $x = 1/2$, $\Delta\tilde{Q} = Q - \bar{Q}$, is identically zero. We can now imbed the original rectangular domain symmetrically into the square $[1/2, 1/2] \times [0,1]$ in the (x,y) plane – for every time t – and ask for some relationship (if any) among the boundary data functions Δf, Δh_0, Δh_1, ΔF, ΔH_0 and ΔH_1 such that $\Delta\tilde{F} = \Delta\tilde{Q} = 0$. Pictorially, with $T = u - v$, the situation looks like this:

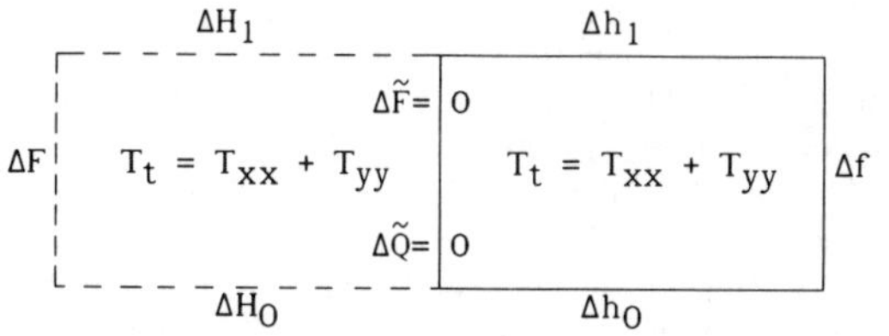

Figure 1. Imbedded domain.

Given that we do not consider interior sources in our heat transfer model, the condition $\Delta\tilde{f} = 0$ is achieved if and only if "the effect of each of the two symmetric boundary data temperatures annihilates each other at every point (x,y,t)", i.e., if and only if

(10) $$\Delta f = -\Delta f, \Delta h_1 = -\Delta H_1 \text{ and } \Delta h_0 = -\Delta H_0.$$

The same argument implies that for the heat flux condition $\Delta\tilde{Q} = 0$ to occur, we must have

(11) $$\Delta f = \Delta F, \Delta h_1 = \Delta H_1 \text{ and } \Delta h_0 = \Delta H_0.$$

From equalities (10) and (11), it follows that

$$\Delta f = \Delta h_1 = \Delta h_0 = 0,$$

and we conclude that $\Delta\tilde{F} = \Delta\tilde{Q} = 0$ if and only if the temperature data on the boundary is identically zero.

This argument shows that the IHCP (9) has a unique solution.

We assume the data functions F_m and Q_m to be discrete functions measured at equally spaced points in the (y,t) domain $[0,L] \times [0,C]$, where $L = 1 - s + s/h$, $C = 1 - k + k/h$, $Nh = 1/2$, $h = \Delta x$, $s = \Delta y$ and $k = \Delta t$. The $(M+1) \times (P+1)$ sample points in this rectangle have coordinates $(y_j, t_n) = (js, nk)$, $0 \leq j \leq M$, $0 \leq n \leq P$, $Ms = L$, $Pn = C$. After extending the data functions in such a way that they decay smoothly to zero outside $[0, L] \times [0, C]$, we consider the data functions F_m and Q_m defined at equally spaced sample points on any rectangular domain of interest in the (y, t) plane.

Once the radii of mollification δ_F and δ_Q, associated with the data functions F_m and Q_m respectively, have been obtained – after solving the discrete version of equation (8) – and the discrete filtered data functions

$$\begin{aligned} J_{\delta_F} F_m(y_j, t_n) &= V^n_{0,j}, \\ J_{\delta_Q} F_m(y_j, t_n) &= W^n_{0,j}, \quad 0 \leq j \leq M, 0 \leq n \leq P, \end{aligned}$$

are determined, we approximate system (9) with the consistent finite difference scheme (7).

As we march forward in the x-direction in space, we drop the estimation of the interior temperature from the highest previous point in time but the discrete heat fluxes $W^n_{i+1,0}$ and $W^n_{i+1,M}$ are computed by linear extrapolation from the already estimated values $W^n_{i+1,1}$, $W^n_{i+1,2}$ and $W^n_{i+1,M-2}$, $W^n_{i+1,M-1}$, introducing an extra local error of order $O(s)$:

$$\begin{aligned} W^n_{i+1,0} &= 2W^n_{i+1,1} - W^n_{i+1,2} \\ W^n_{i+1,M} &= 2W^n_{i+1,M-1} - W^n_{i+1,M-2} \end{aligned}$$

The corresponding unknown boundary temperatures are then calculated directly using the usual formulas

$$V^n_{i+1,0} = V^n_{i,0} + hW^n_{i+1,0}$$

and

$$V^n_{i+1,M} = V^n_{i,M} + hW^n_{i+1,M}.$$

This process is continued for $n = 1, 2, \ldots, P-i-1$ until the approximate solution is computed at all the grid points of the domain in the plane x_{i+1}. Then the entire cycle is repeated for the discrete points of the plane x_{i+2}, etc.

The values $V^n_{N,j}$ and $W^n_{N,j}$, $0 \leq j \leq M - N$, $0 \leq n \leq P - N$, so obtained, are then taken as the accepted approximations for the boundary temperature and heat flux histories respectively at the different y-locations in the domain $D = [0, 1] \times [0, 1]$ of the (y, t) plane at $x = 1/2$.

If the exact data temperature (heat flux) is denoted by $F(y, t)$ $(Q(y, t))$, the noisy data $F_m(y_j, t_n)$ $(Q_m(y_j, t_n))$ is obtained by adding a random error to F_m (Q_m); i.e., for every grid point (y_j, t_n),

$$\begin{aligned} F_m(y_j, t_n) &= F(y_j, t_n) + \epsilon_{1,j,n}, \\ Q_m(y_j, t_n) &= Q(y_j, t_n) + \epsilon_{2,j,n}, \end{aligned}$$

where $\epsilon_{1,j,n}$ and $\epsilon_{2,j,n}$ are Gaussian random variables of variance ϵ^2.

Numerical Example:

Figures 2 and 3 show the computed temperature solution – obtained with the numerical scheme introduced above with $h = 0.1$, $s = k = 0.01$, $\delta = 0.04$, $\epsilon = 0.005$ – and the corresponding discrete error surface at $x = 1/2$ and $0 \leq t \leq 0.8$, respectively, for the IHCP described below.

$$(12) \quad \begin{array}{ll} u_t = u_{xx} + u_{yy}, & 0 < x < 1/2, \\ & 0 < y < 1, t > 0, \\ u(0, y, t) = 0, & 0 < y < 1, t > 0, \\ u_x(0, y, t) = \frac{1}{\sqrt{2}} e^{-t} \sin \frac{y}{\sqrt{2}}, & 0 < y < 1, t > 0, \\ u(0, y, 0) = \sin \frac{y}{\sqrt{2}} \sin \frac{x}{\sqrt{2}}, & 0 \leq y \leq 1/2, \\ & 0 < y < 1, \\ u(1/2, y, t) = f(y, t), \text{ unknown}, & 0 < y < 1, t > 0, \\ u_x(1/2, y, t) = f(y, t), \text{ unknown}, & 0 < y < 1, t > 0, \\ u(x, 0, t) = h_0(x, t), \text{ unknown}, & 0 \leq x \leq 1/2, t > 0, \\ u(x, 1, t) = h_1(x, t), \text{ unknown}, & 0 \leq x \leq 1/2, t > 0. \end{array}$$

The unique exact temperature solution for problem (11) is

$$u(x, y, t) = e^{-t} \sin \frac{y}{\sqrt{2}} \sin \frac{x}{\sqrt{2}}, \quad 0 \leq x \leq 1/2, \ 0 \leq y \leq 1, \ t > 0,$$

and, consequently, the unknown functions for the IHCP are given by

$$\begin{array}{ll} f(y, t) = e^{-t} \sin \frac{y}{\sqrt{2}} \sin \frac{1}{2\sqrt{2}}, & 0 \leq y \leq 1, t > 0, \\ q(y, t) = \frac{1}{\sqrt{2}} e^{-t} \sin \frac{y}{\sqrt{2}} \cos \frac{1}{2\sqrt{2}}, & 0 \leq y \leq 1, t > 0, \\ h_0(x, t) = 0, & 0 \leq x \leq 1/2, t > 0, \\ h_1(x, t) = e^{-t} \sin \frac{1}{\sqrt{2}} \sin \frac{x}{\sqrt{2}}, & 0 \leq x \leq 1/2, t > 0. \end{array}$$

To study the numerical stability of the algorithm, we use different average perturbations for $\epsilon = 0$, 0.001, 0.002, 0.003, 0.004 and 0.005. The solution errors for the discrete temperature and heat flux functions at $x = 1/2$ in the discretized time interval $I = [0, 1]$ at the $y = 1/2$ location, are respectively given by

$$\|V_{N,j} - f_j\|_I = \left\{ \frac{1}{P - N + 1} \sum_{n=1}^{P-N+1} (V^n_{N,j} - f^n_j)^2 \right\}^{1/2},$$

and

$$\|Q_{N,j} - q_j\|_I = \left\{ \frac{1}{P - N + 1} \sum_{n=1}^{P-N+1} (W^n_{N,j} - q^n_j)^2 \right\}^{1/2}, \quad js = \frac{1}{2}.$$

Table 1 shows the results of the numerical experiments associated with Problem 1.

Temperature			Heat Flux		
ϵ	δ_F	Error Norm	ϵ	δ_Q	Error Norm
0.000	0.01	0.003580	0.000	0.01	0.017352
0.001	0.04	0.003580	0.001	0.04	0.017356
0.002	0.04	0.003581	0.002	0.04	0.017364
0.003	0.04	0.003583	0.003	0.04	0.017374
0.004	0.04	0.003585	0.004	0.04	0.017388
0.005	0.04	0.003588	0.005	0.04	0.017404

Table 1. Error norm as a function of the amount of noise in the data for the surface temperature and surface heat flux in Problem 1 at $x = y = 1/2$.

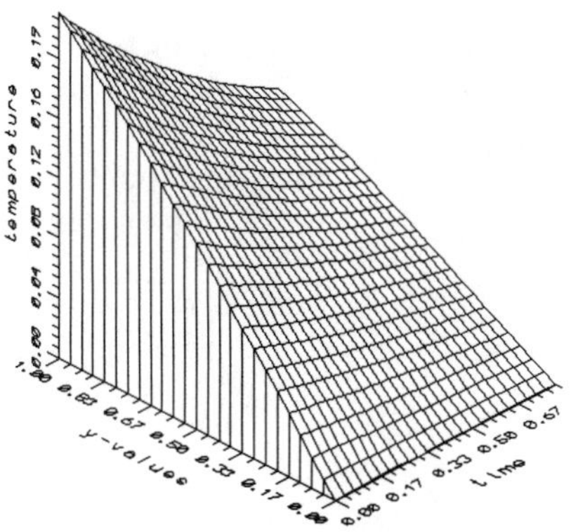

Figure 2. Reconstructed temperature function at $x = 1/2$.

$h = 0.1,\ s = k = 0.01,\ \delta = 0.04,\ \epsilon = 0.005.$

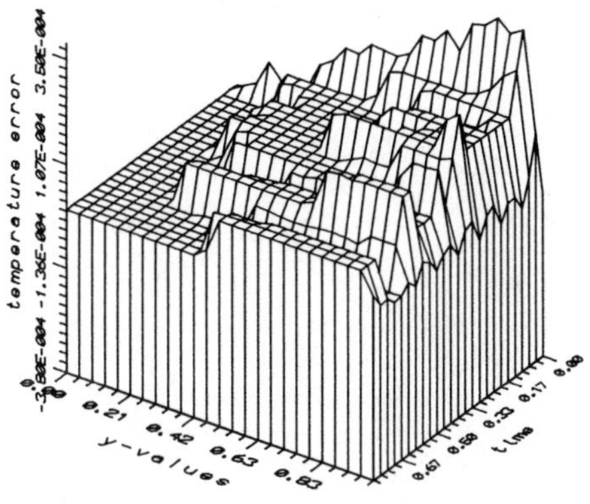

Figure 3. Temperature error function at $x = 1/2$.

$h = 0.1,\ s = k = 0.01,\ \delta = 0.04,\ \epsilon = 0.005.$

REFERENCES

B.R. Bass and L.J. Ott, 1980, "A finite element formulation of the two-dimensional inverse heat conduction problem", Adv. Comput. Technol., Vol. 2, pp. 238-248.

J. Baumeister and H.J. Reinhart, 1987, "On the approximate solution of a two-dimensional inverse heat conduction problem", Inverse and Ill-posed Problems, H.G. Engl and C.W. Groetsch, (Eds.), Academic Press, Orlando, Florida, pp. 325-344.

J.V. Beck, 1970, "Nonlinear estimation applied to the nonlinear inverse heat conduction problem", Int. J. Heat Mass Transfer, Vol. 13, pp. 703-716.

L. Guo and D.A. Murio, 1991, "A mollified space-marching finite difference algorithm for the two-dimensional inverse heat conduction problem with slab symmetry", Inverse Problems, Vol. 7, pp. 247-259.

M. Imber, 1974, "Temperature extrapolation mechanism for two-dimensional heat flow", AIAA J., Vol. 12, pp. 1089-1093.

T. Yoshimura and K. Ituka, 1985, "Inverse heat conduction problem by finite element formulation", Int. J. Systems Sci., Vol. 16, pp. 1365-1376.

N. Zabaras and J.C. Liu, 1988, "An analysis of two-dimensional linear inverse heat transfer problems using an integral method", Numer. Heat Transfer, Vol. 13, pp. 527-533.

Inverse Problems in Engineering: Theory and Practice
ASME 1993

COMPARISON OF THE ITERATIVE REGULARIZATION AND FUNCTION SPECIFICATION ALGORITHMS FOR THE INVERSE HEAT CONDUCTION PROBLEM

James V. Beck
Department of Mechanical Engineering
Michigan State University
East Lansing, Michigan

ABSTRACT

One method for solving the inverse heat conduction problem is the function specification method. Another method is the iterative regularization algorithm which was originally developed by Alifanov in Russia.

There are several variations of both methods. The purpose of this paper is to give a detailed comparison of a one-dimensional form of both methods using the same choice of approximating functions.

The methods generally give similar answers but the iterative regularization gives slightly better results for smooth heat fluxes which start and stop at zero. The function specification method is better when the heat flux varies more abruptly and the heat flux does not start and stop with zero values.

INTRODUCTION

The inverse heat conduction problem (IHCP) is the determination of the surface heat flux from one or more measured interior, transient temperature histories in a heat conducting solid. A review of the literature and a presentation of various methods is given in Beck, Blackwell and St. Clair (1985). One important method which has not been discussed in this reference is the iterative regularization algorithm developed by Alifanov (1979) in Russia. Artyukhin and Rumyantsev (1980) and others in Russia have also used and extended this method. It is also being used by others as well (Jarny, Ozisik, and Bardon, 1991).

The iterative regularization method (IRM) can estimate both the surface heat flux as a function of time (IHCP) and thermal properties as functions of temperature and position; other parameters can also be estimated using the method (Jarny et al., 1991). There are several variations of the IRM and the function specification method (FSM), depending, for example, which of the following are used: a) the steepest descent, conjugate gradient, ordinary least squares or other minimization methods, b) numerical convolution, finite difference, finite element, boundary element or other numerical methods, and c) constant, linear, parabolic or cubic spline surface heat flux approximations.

The purpose of this paper is to give the basic equations of a form of the IRM, give an algorithmic form of the IRM, and give some detailed examples. The comparable function specification method is also given. In order to reduce the dependence on the spatial grid in the FD and FE calculations, the convolution (i.e., Duhamel's) integral is used (Beck et al., 1985) for both methods.

The IHCP is solved for constant properties and can be mathematically described by

$$k\,\frac{\partial^2 T}{\partial x^2} = \rho c\,\frac{\partial T}{\partial t},\quad 0 < x < L,\ 0 < t < t_f \tag{1}$$

$$-k\,\frac{\partial T(0,t)}{\partial x} = q_0(t) = ?,\quad -k\,\frac{\partial T(L,t)}{\partial x} = q_L(t),\ known \tag{2,3}$$

$$T(x,0) = T_0(x),\quad T(d,t) = Y(t) \tag{4,5}$$

The body is a plate and has thickness L and the time domain is zero to t_f. (In some of the examples to be given, the time domain actually starts at some negative time, t_0, and continues to the final time. The theory is the same for both cases; however, Duhamel integrals must take this into account.) The temperature is measured at location $x = d$ and is denoted $Y(t)$. For simplicity, the known heat flux at $x = L$ is zero and $T_0(x)$ is set equal to zero. In other words, $T(x,t)$ henceforth represents the temperature rise.

ITERATIVE REGULARIZATION METHOD (IRM)

In the IRM, three basic problems are solved at each iteration. First, eqs. (1) to (4) are solved with $q_0(t)$ replaced with its estimated function $q^{(n)}(t)$ for the nth iteration. Then the adjoint and sensitivity problems for the nth iteration must be solved. The adjoint problem is

$$k\frac{\partial^2 \psi}{\partial x^2} = -\rho c\frac{\partial \psi}{\partial t} - \delta(x-d)[T(x,t;q(t)) - Y(t)],\quad t = t_f \text{ to } 0 \tag{6}$$

$$\frac{\partial \psi(0,t)}{\partial x} = \frac{\partial \psi(L,t)}{\partial x} = 0,\quad \psi(x,t_f) = 0 \tag{7,8,9}$$

Notice that this adjoint problem goes backward in time, starting at time t_f, the final time. The driving term in the $\psi(x,t)$ problem is the difference between the temperature calculated in the eq. (1) to (5) problem and the measured temperature, $Y(t)$; $\delta(x-d)$ is the Dirac delta function and is zero except when $x = d$. The units of the adjoint variable, ψ, are $m^2 - K^2/W$.

The sensitivity problem is the solution of

$$k\frac{\partial^2\theta}{\partial x^2} = \rho c\frac{\partial\theta}{\partial t},\ 0 < x < L,\ 0 < t < t_f \tag{10}$$

$$-k\frac{\partial\theta(0,t)}{\partial x} = p^{(n)}(t),\ \frac{\partial\theta(L,t)}{\partial x} = 0,\ \theta(x,0) = 0 \tag{11,a,b,c}$$

where $p^{(n)}(t)$ comes from the solution of the adjoint problem (see eqs. (19) and (20) below). The units of $p(t)$ are the same as ψ. The units of θ are $K^3 - m^4/W^2$.

The above quantities are used in the approach to a minimum of the function

$$S(q(t)) = \int_0^{t_f}[Y(t) - T(d,t)]^2 dt \tag{12}$$

Implicit in the iterative procedure is that $S(\cdot)$ is not precisely minimized but is reduced to the level where it is just less than δ^2, or

$$S^{(n)} \leq \delta^2 \tag{13}$$

where the superscript n refers to the iteration and δ^2 is a measure of the errors in the temperature measurements, $Y(t)$. Other stopping criteria are suggested in Alifanov and Balashova (1985). The regularization for finding $q(t)$ is incorporated in the natural "viscosity" or slowness in the approach to the minimum provided by the methods of steepest descent or conjugate gradient. The latter method converges much more rapidly than the steepest descent method.

ITERATIVE REGULARIZATION ALGORITHM

The procedure is now outlined in the form of an algorithm.

Set $n = 0$ and start with an estimate of $q(t)$. Usually $q^{(0)}(t) = 0$ is chosen.

1. Set $n = n + 1$. Solve the temperature problem using eqs. (1) to (4) with $q_o(t)$ replaced by the estimated function $q^{(n)}(t)$.

Calculate

$$S^{(n)}(t_f) = \int_0^{t_f}[Y(t) - T^{(n)}(d,t)]^2 dt \tag{14a}$$

If

$$S^{(n)} < \delta^2 \tag{14b}$$

terminate the computations.

2. Solve the adjoint problem for $\psi(x,t)$. Use eqs. (6) to (9). The derivative of S with respect to the $q^{(n)}(t)$ function is denoted $\nabla S^{(n)}(t)$ and happens to be equal to the adjoint variable evaluated at $x = 0$,

$$\nabla S^{(n)}(t) = \psi(0,t;q^{(n)}(t)) \tag{15}$$

3a. If n= 1, set

$$\gamma^{(1)} = 0 \tag{16}$$

and go to step 5.

3b. If n ≥ 2, calculate

$$\gamma_N^{(n)} = \int_0^{t_f}\psi^{(n)}(t)\,[\psi^{(n-1)}(t) - \psi^{(n)}(t)]dt \tag{17}$$

where all the auguments of ψ are omitted for convenience and where the N subscript denotes numerator.

4. If n ≥ 2, calculate

$$\gamma^{(n)} = -\gamma_N^{(n)} / \gamma_D^{(n-1)} \tag{18}$$

where $\gamma_D^{(n-1)}$ is obtained from step 10 of the previous iteration. (The D denotes denominator.)

5a. For $n = 1$ or for the steepest descent method, use

$$p^{(n)}(t) = \psi(0,t;q^{(n)}(t)) \tag{19}$$

5b. For $n \geq 2$ and for the conjugate gradient method, use

$$p^{(n)}(t) = \psi(0,t;q^{(n)}(t)) + \gamma^{(n)}p^{(n-1)}(t) \tag{20}$$

6. Calculate $\theta^{(n)}(x,t)$ using eqs. (10) to (12).

7. Calculate

$$\beta_D^{(n)} = \int_0^{t_f}[\theta^{(n)}(d,t)]^2 dt \tag{21}$$

8. Calculate

$$\beta_N^{(n)} = \int_0^{t_f}\psi(0,t;q^{(n)}(t))p^{(n)}(t)dt \tag{22}$$

9. Calculate

$$\beta^{(n)} = \beta_N^{(n)} / \beta_D^{(n)} \tag{23}$$

10. Calculate

$$\gamma_D^{(n)} = \int_0^{t_f} [\psi(0,t;q^{(n)}(t))]^2 dt \tag{24}$$

11. Calculate

$$q^{(n+1)}(t) = q^{(n)}(t) - \beta^{(n)} p^{(n)}(t) \tag{25}$$

Go to step 1.

CONVOLUTION INTEGRAL

One relatively simple way to investigate the above algorithm is to use the convolution integral to calculate the T's, ψ's, and θ's. For the case of $T_0(x) = T_0$ and $q_L(t) = 0$, the temperature at any time t_M can be calculated using

$$T_M^{(n)} = T^{(n)}(d,t_M) = \int_0^{t_M} q^{(n)}(\lambda) \frac{\partial\phi(d,t_M-\lambda)}{\partial t} d\lambda + T_0 \tag{26}$$

where $\phi(d,t)$ is the temperature at $x = d$ caused by a heat flux q of 1 starting at $t = 0$, assuming $q(t)$ is zero for $t > 0$. Eq. (28a) can be approximated as

$$T_M^{(n)} = \sum_{i=1}^{M} q_i \nabla\phi_{M-i+1} + T_0, \; M=1,2,\ldots,M_f \tag{27}$$

$$\nabla\phi_{M-i+1} = \phi_{M-i+1} - \phi_{M-i}, \; M_f \Delta t = t_f \tag{28}$$

and ϕ_{M-i} is ϕ evaluated at $x = d$ and time $t_{M-i} = (M-i)\Delta t$, with Δt being the time step. The heat flux q_i is for the time period t_{i-1} to t_i and is best plotted at time $t_{i-1/2}$.

A simpler case than arbitrary d is for d being equal to L. For this case $\psi_k^{(n)}$ and $\theta_M^{(n)}$ are approximated by

$$\psi_k^{(n)} = \sum_{i=1}^{M_f-k+1} [T_k^{(n)} + i - 1 - Y_{k+i-1}] \nabla\phi_i, \; k = 1,2,\ldots,M_f \tag{29}$$

$$\theta_M^{(n)} = \sum_{i=1}^{M} p_i^{(n)} \nabla\phi_{M-i+1} \tag{30}$$

$$p_i^{(n)} = \psi_{M_f}^{(n)} - i + 1 + \gamma^{(n)} p_i^{(n-1)} \tag{31}$$

An integral such as $\beta_D^{(n)}$ in eq. (21) can be approximated by

$$\beta_D^{(n)} = \sum_{i=1}^{M_f} [\theta_i^{(n)}]^2 \Delta t \tag{32}$$

FUNCTION SPECIFICATION ALGORITHM

The FS algorithm can be expressed as eq. (4.4.24) of Beck et al. (1985),

$$\hat{q}_M = \frac{\sum_{i=1}^{r} [Y_{M+i-1} - \hat{T}_{M+i-1}|_{q_M=q_{M+1}=q_{M+r-1}=0}]\phi_i}{\sum_{i=1}^{r} \phi_i^2} \tag{33}$$

where r is the number of "future" time steps and $r = 1$ is for "exact" matching. EQ. (33) is much simpler than the IRM. It is necessary to compare the results of the two methods. In the comparison, r is selected using the same criterion used in the IRM, namely, eq. (13). Expressions for ϕ, θ and ψ are given below.

For the case of a plate heated at $x = 0$ with an unknown heat flux and insulated at $x = L$, the ϕ function is obtained from the X22B10T0 solution (Beck et al, 1992, p. 166)

$$T(x^+,t^+) = \frac{q_0 L}{k}[t^+ + \frac{1}{3} + \frac{1}{2}(x^+)^2 - \frac{2}{\pi^2}\sum_{n=1}^{\infty}\frac{1}{n^2}e^{-n^2\pi^2 t^+}\cos(n\pi x^+)] \tag{34}$$

where

$$t^+ = \frac{\alpha t}{L^2}, \; x^+ = \frac{x}{L}, \; q_0 = 1 \tag{35}$$

Hence for the sensor at $x^+ = d^+ = d/L$, the ϕ function is given by

$$\phi_i = T(d^+, t_i^+) = T(d^+, i\Delta T) \tag{36}$$

The ψ function is the temperature at $x = 0$ for a planar heat source at $x = d$. For the present geometry, the solution for ψ uses eq. (33) as a building block with x^+ set equal to d^+.

The θ solution given by eqs. (10) to (12) uses the basic solution for q_0 at $x = 0$ and the response at $x = d$. This means that θ uses the same ϕ as defined by eq. (34).

Linear with Time q Test Case

The case of the heat flux linearly increasing with time case (X22B20T0, Beck et al., 1992, p. 169) is an important one for test cases and also to test the constant heat flux assumption over a time interval. It is given by

$$T(x^+,t^+) = \frac{q_N L}{k}[T_t^+(x^+,t^+) + T_s(x^+,t^+)] \tag{37}$$

$$T_t^+(x^+,t^+) = \frac{2}{\pi^4}\sum_{n=1}^{\infty}\frac{1}{n^4}e^{-n^2\pi^2 t^+}\cos(n\pi x^+) \tag{38}$$

$$T_s^+(x^+,t^+) = \frac{t^{+2}}{2} + [\frac{1}{3} - x^+ + \frac{1}{2}x^{+2}]t^+ - \frac{1}{360}[8 - 60x^+ + 60x^{+3} - 15x^{+4}] \tag{39}$$

where eq. (38) gives a transient, decaying term and eq. (39) gives a quasi-steady state term, one part of which is time-independent.

For dimensionless times greater than 0.5, the contribution of eq. (38) is very small, less than about 0.00015; at this dimensionless time, the temperature at the heated surface $(x = 0)$ is about 0.25 and it gets larger rapidly.

The dimensionless heat flux, $q^+(t^+)$, is related to q_N, used in eq. (37), by

$$q^+(t^+) = \frac{q(t)}{q_N} = t^+ \tag{40}$$

The heat flux q_N is the nominal heat flux corresponding to $q(t)$ at time $t^+ = 1$.

COMPARISON OF ITERATIVE REGULARIZATION AND FUNCTION SPECIFICATION METHODS

Comparison of the IRM with other methods such as the FSM is difficult for a number of reasons. One of these is that the IRM with conjugate gradients is a nonlinear method even for linear problems, while the FSM is linear for linear problems. Hence, the comparison methods suggested in Beck et al. (1985) and in Raynaud and Beck (1988) cannot be used because they assume that the methods are linear for linear IHCPs. Yet another problem in the comparison is that the IRM sometimes uses many more measurement times than the number of unknown constants to find q; in contrast the FSM typically uses the same number of constants as the measurement times, but it can be readily modified to treat multiple T's for each q. (See also the second paragraph in the introduction.) Hence the comparison is made with the same q approximation (or basis functions) and the same number of q components as measurement times.

Iterative regularization results for linear-with-time q

An example of linear-with-time q is now discussed for t^+ from -0.24 to 0.96. For negative t^+, the temperature is zero and q increases linearly with time thereafter. Simulated experimental temperatures have been generated for dimensionless times of 0.06, a total of 20 times. Additive, zero mean, constant variance, uncorrelated, and gaussian errors have been included with dimensionless errors, σ_Y, of 0.00, 0.01 and 0.02.

Fig. 1 shows the results of the iterative regularization method for zero simulated measurement errors. (Actually only 7 significant figures are used, but that is sufficient to approximate "zero errors" in this case). Fig. 2 depicts the estimated q's for simulated errors which are additive, zero mean, constant variance $\sigma_Y =$ 0.01 and 0.02, uncorrelated and gaussian. One stopping criterion uses the number of iterations which reduces the standard deviation of the calculated errors in the simulated temperature, s_Y, to the expected value, which is zero in Fig. 1. Some values of this estimated standard deviation are shown in Table 1; these values decrease with the iterations but zero is not attained by 20 iterations. The second column in Table 1 gives the root mean square of the q estimates. It is given by

$$s_q = \left[\frac{1}{n}\sum_{i=1}^{n} (\hat{q}_i - q_i)^2\right]^{1/2} \tag{41}$$

where $\hat{q}_i$ = *estimated* q *at time,* t_i

and q_i is the true value at the same time. Also note that $ns_q^2 = \delta^2$, where δ^2 is the value given in Eq. (14b). In both Tables 1 and 2, s_q is about 0.3 or larger which is quite large since q varies from zero to a maximum value of 0.92. A most interesting point is that the s_q values in Table 2 (for $\sigma_Y =$ 0.01) have a minimum at four iterations. Notice in Table 2 that the s_Y (an estimate of σ_y) values decrease monotonically, as they do for zero errors (Table 1). Five iterations are required for s_Y to just go below $\sigma_Y =$ 0.01 in Table 2. Fig. 2 also shows estimated q's for $\sigma_Y =$ 0.02 and Table 3 shows more details. Doubling the random measurement errors in Fig. 2 does not double the errors in q, with the errors being 50% larger at the most; this is a desirable characteristic which the FSM does not share. The iterative regularization algorithm has the desirable feature of requiring fewer iterations as the measurement errors are increased; notice that only four rather than five iterations are now required if the criterion of s_Y going just below σ_Y is used.

COMPARISON OF IRM AND FSM FOR 80 TIME STEPS AND LINEAR q

One of the purposes of this paper is to provide a comparison with FSM. Fig. 3 displays some results using FSM for the same data used in Fig. 2. The exact values for the q's are indicated by triangles, and the estimates of q are denoted by stars, circles and crosses for $\sigma_Y =$ 0, 0.01 and 0.02, respectively. The number of future time steps, r, was chosen to be equal to the value so that s_Y was just greater than σ_Y (because the s_Y values start small and increase with r). This criterion is consistent with what was chosen for the iterative regularization method and works quite well. The needed value of r increases with the magnitude of the measurement errors, but only slowly.

A comparison of Figures 2 and 3 indicates that the FSM is superior to the IRM. This is noted for all times but is particularly true at the final times even with negligible measurement errors.

In the IRM, the usual initial estimates of the q components being equal to zero was chosen. Fig. 4 shows some results for the $q^{(0)}(t)$ chosen to be equal to $q(t_n)$. Although the values near the final time, t_n, of the estimated q are improved in this case, the estimated values near $t = 0$ are less accurate.

The value of 20 time steps is convenient for insight and plotting.

However, much larger numbers of time steps are encountered in practice. Hence an example with 80 time steps of the same value as above and $\sigma_Y =$ 0.01 are considered. Following the same rule for selecting the number of iterations, 15 iterations are needed. See Table 4.

Two important comparisons should be made with the 20 time step cases. First, there is less variability in the estimated q values for the 80 time step IRM case than there is in the 20 time step IRM case. Hence, the 80 time step case produces better estimates than a smaller number of time steps. Also, the estimates of q near $t = 0$ depend upon the number of time steps in the complete domain in the IRM; the FSM, however, is independent of the total time domain, since only the r future steps are relevant for each q. Second, the number of required iterations increases with the number of time steps. (The number of iterations for 20 time steps was 5 and that for 80 steps was 15.) The FSM (Beck et al., 1985) is independent of the number of time steps.

COMPARISON OF RESULTS FOR THE IR AND FS METHODS FOR A TRIANGULAR HEAT FLUX

One of the standard test cases for comparing results of inverse heat conduction algorithms is for a triangular heat flux. The geometry is a flat plate and the sensor is located at $x = L$, which is insulated. See Beck et al. (1985), chap. 5. The heat flux is zero before time zero and is again zero after time 1.2, with a triangular flux in between and a maximum occurring at time 0.6. See Figures 5 and 6. The time steps are again the dimensionless time of 0.06. The values of temperature are used starting at the dimensionless time equal to -0.24 and end at time 1.56. There are then 30 time steps.

Fig. 5 is for iterative regularization. The curve with the stars is for zero simulated measurement errors (or actually accurate only to seven significant figures). Twenty iterations give excellent results. For errors equal to $\sigma_Y =$ 0.0017 (about 0.5% of the maximum temperature rise of the measured temperature) the results shown in Fig. 5 are very good but not as good as for no measurement errors. Only 5 iterations were needed. The locations of greatest inaccuracy tended to be where the heat flux changed abruptly. The root mean squared error in q for $\sigma_Y =$ 0.0017 is 0.01814 which is 3.02% of the maximum true q of 0.6. Fig. 6 is the function specification method and can be compared with Fig. 5. For zero measurement errors, two future temperatures ($r = 2$) were used and again excellent results were obtained. For the case of $\sigma_Y =$ 0.0017, the results look good but not as smooth as those for the IR method. Instead of relying on visual comparisons, we again give the root mean squared error in q for $\sigma_Y =$ 0.0017 which is 0.0244, or 4.07% of the maximum true q of 0.6.

It is clear that for this particular case the IRM method is better than the function specification method. However, the difference between 3 and 4% is not large.

COMPARISON USING 60 TIME STEPS FOR TRIANGULAR HEAT FLUX EXAMPLE

The above example used 30 time steps with a dimensionless time step of 0.06. The results of the same triangular heat flux with 60 time steps and time steps of 0.03 are now given. The same time period is covered. The results are shown in Fig. 7. The IR solution needed 8 iterations and the FS solution needed $r = 6$ future temperatures to satisfy the stopping criterion given above. Again the IR method produces better results than the FS, but the difference is not large (4.0 and 5.1% of the maximum true q for IRM and FSM, respectively).

SQUARE HEAT FLUX TEST CASE

Another test case is that of a square heat flux, one which starts at zero at $t^+ = -0.24$, jumps to 1.0 at $t^+ = 0.0$, drops to zero at $t^+ = 0.96$ and remains at zero until $t^+ = 1.2$. The time steps are 0.06. By allowing the heat flux to start at zero and return to zero, the IRM is favored in this example. Fig. 8 shows the results which are very similar. The main difference is that 20 iterations were required for IRM versus only $r = 2$ for FSM. IRM has $s_q =$ 0.0575 while the corresponding FSM value is 0.0494, which favors FSM.

This example can be continued with random errors introduced. For the same set of random errors with $s_Y =$ 0.01 and the usual stopping criterion for both methods, the IRM requires seven iterations and the FSM requires $r = 4$. The results are shown in Fig. 9 which indicates that the FSM is better for this example. The s_q value for IRM is 0.318 and the FSM s_q value is 0.189, or the IRM method is 68% larger, again indicating that FSM is better for this example which involves rapidly changing heat fluxes.

COMPUTING TIME

In terms of computing time, the FSM takes much less time than the IRM, particularly as more time steps are considered. For the present examples involving one dimension, the computing time is so small that it is not an important consideration. However, for two and three dimensional problems, the computation time could be quite significant and important. It is possible that the FSM might be attractive for the time domain but the IRM be best for the space domain (such as in 2D and 3D problems). In other words, a hybrid technique might be superior to either method alone.

CONCLUSIONS

The iterative regularization and the function specification methods are compared for the same heat flux approximations and the same cases. The heat flux approximation (or basis function) is simply a constant for each time step. Other approximations may show different results.

The results are quite similar in the values, provided the heat flux starts and stops at zero. However, for smoothly varying q's (starting and stopping at zero), the IRM gives more accurate results, such as 3% compared to 4% errors for the IRM and FSM, respectively. For heat fluxes which vary abruptly with time, the FSM had smaller errors, over 40% smaller in one case.

The computer time for the FSM is smaller than that for the IRM and the FSM algorithm is much less complex than that for the IRM. For two and three dimensional problems, a hybrid method might be better than either one alone. The IRM has the advantage that it can also be used for finding other functions, such as the thermal conductivity as a function of temperature; in that case, the adjoint and sensitivity equations in this paper would have to be modified. The parameter estimation techniques in Beck and Arnold (1977) would be in competition with the IRM for k(T).

The iterative regularization method is very powerful and adaptable but it is not better than the FSM for all IHCP cases considered herein.

ACKNOWLEDGEMENT

The help and insights of Prof. Patricia Lamm, Dean O.M. Alifanov, and Prof. E.A. Artyukhin are greatly appreciated. This work was partially supported by Sandia National Laboratories.

REFERENCES

Alifanov, O.M., 1979, **Identification of Heat Transfer Processes in Flying Vehicles. Introduction to the Theory of Inverse Heat Transfer Problems**, Machinostroenie, Moscow, in Russian.

Alifanov, O.M. and Balashova, I.E., 1985, "Choice of Approximate Inverse Heat Conduction Problem Solution, "**Inzhenerno-Fiz. Zhurn.**, Vol. 48, pp. 851 - 860.

Artyukhin, E.A. and Rumyantsev, S.V., 1980, "The Gradient Method for Smooth Solutions of Inverse Boundary Heat Conduction Problems, "**Inzhenerno-Fiz. Zhurn.**, Vol. 39, pp. 259 - 263.

Beck, J.V. and Arnold, K.J., 1977, **Parameter Estimation in Engineering and Science**, Wiley, NY, NY.

Beck, J.V., Blackwell, B. and St. Clair, C.R.,Jr., 1985, **Inverse Heat Conduction: Ill-Posed Problems**, Wiley-Interscience, NY, NY.

Beck, J.V., Cole, K., Haji-Sheikh, A. and Litkouhi, B., 1992, **Heat Conduction Using Green's Functions**, Hemisphere, Wash., D.C.

Jarny, Y., Ozisik and Bardon, J.P., 1991, "A General Optimization Method Using Adjoint Equation for Solving Multidimensional Inverse Heat Conduction", **Int. J. Heat and Mass Transfer**, vol. 34, pp.2911-2919.

Raynaud, M. and Beck, J.V., 1988, "Methodology for Comparison of Inverse Heat Conduction Methods," **ASME J. of Heat Transfer**, Vol. 110, pp. 30 -37.

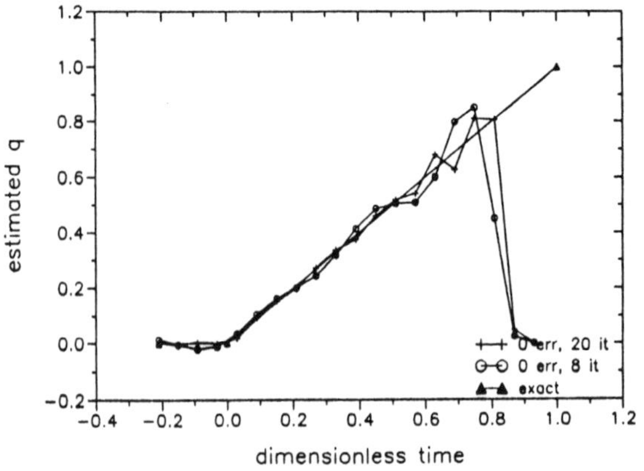

Fig. 1 Iteration regularization with no errors

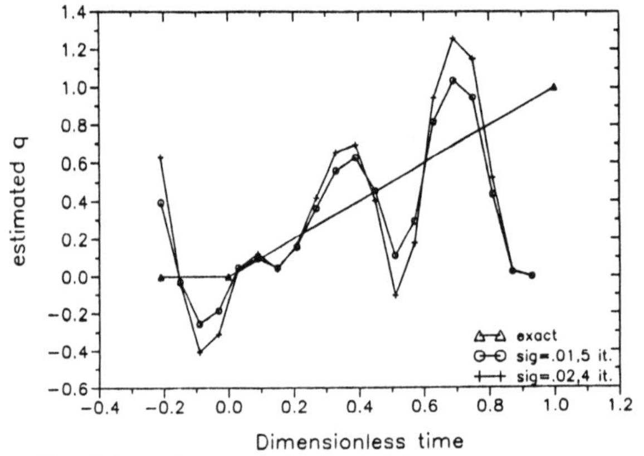

Fig. 2 Iterative regularization with errors

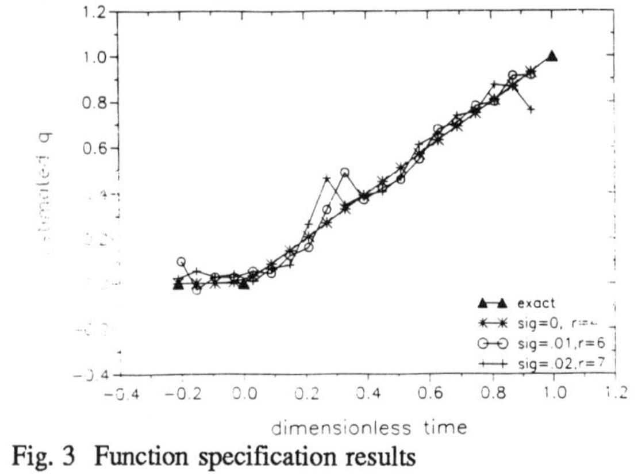

Fig. 3 Function specification results

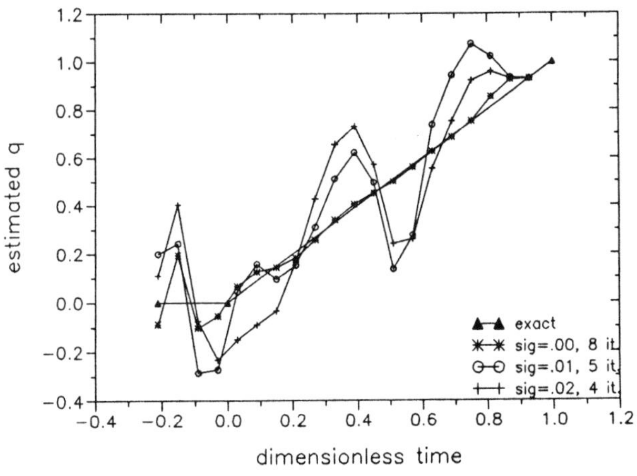

Fig. 4 Iterative regulation with errors and initial q = 0.93

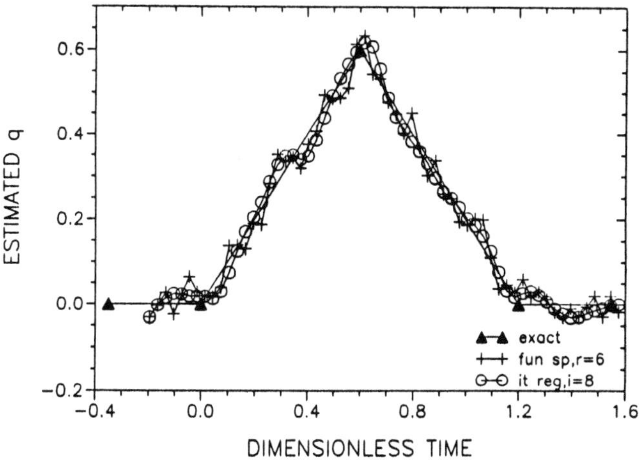

Fig. 7 Iterative regularization and function specification results for 0.03 time step and 80 time steps, s = 0.0017. Triangular q.

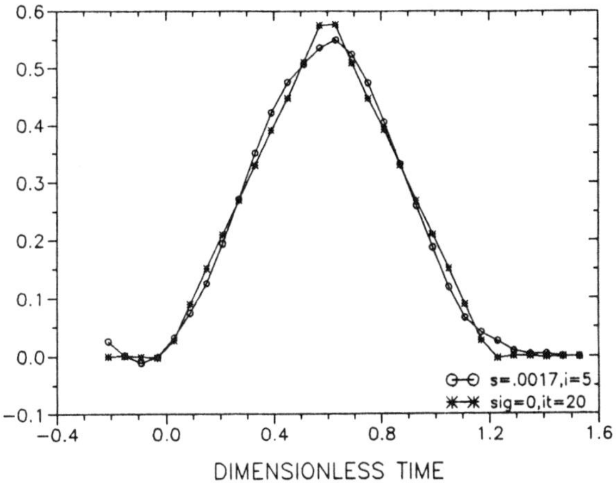

Fig. 5 Iterative regularization, triangular heat flux case

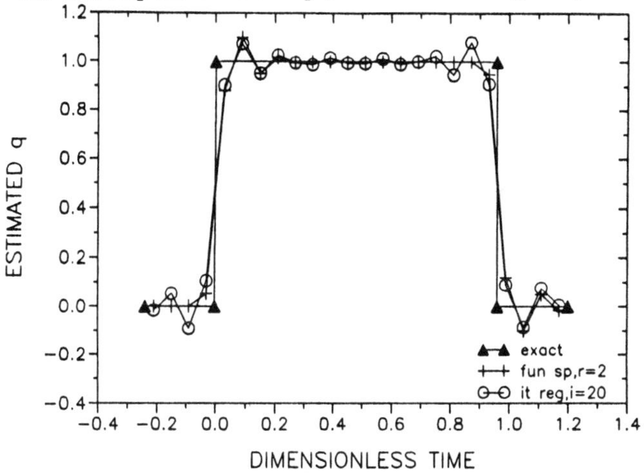

Fig. 8 Iterative regularization and function specification results for 0.06 time step and 24 time steps, s = 0. Step heat flux.

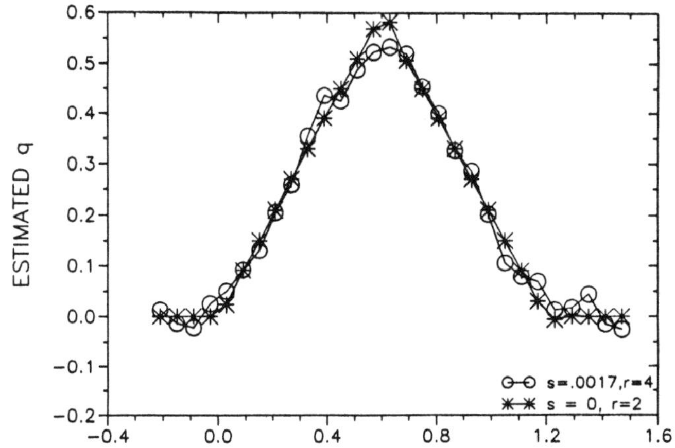

Fig. 6 Function specification results, triangular heat flux

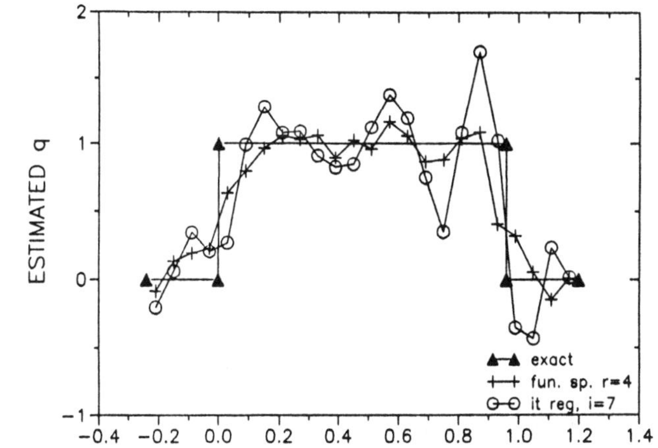

Fig. 9 Iterative regularization and function specification results for 0.06 time step and 24 time steps, s = 0.01. Step heat flux.

Table 1

Results for linear q(t) using IRM

$\sigma_Y = 0$, 20 time steps

#It	s_q	s_Y
1	.458	.107
4	.323	.00584
8	.295	.000600
12	.283	.000236
20	.278	.000043

Table 2

Results for linear q(t) using IRM

$\sigma_Y = 0.01$, 20 time steps

#It	s_q	s_Y
1	.458	.1097
4	.319	.0115
5	.356	.0089
6	.424	.0076
8	.462	.0062
10	1.134	.0050

Table 3

Results for linear q(t) using IRM,

$\sigma_Y = 0.02$, 20 time steps

#It	s_q	s_Y
1	.458	.1133
3	.346	.0249
4	.428	.0198
5	.586	.0157
16	4.203	.0053

Table 4

Results for linear q(t) using the IRM,

$\sigma_Y = 0.01$, 80 time steps

#It	s_q	s_Y
1	2.073	4.061
8	.733	.0253
12	.663	.0123
14	.649	.0102
15	.641	.0093
16	.640	.0086

Inverse Problems in Engineering: Theory and Practice
ASME 1993

BOUNDARY INVERSE HEAT CONDUCTION PROBLEM IN EXTREME FORMULATION

O. M. Alifanov and A. V. Nenarokomov
Moscow Aviation Institute
Moscow, Russia

ABSTRACT

Methods based on solving boundary inverse heat conduction problems are widely used at present in experimental investigations of thermal processes between solids and the environment. To solve three-dimensional ill-posed boundary inverse problems an iterative regularization method is used. The method is based on minimizing the residual functional by means of gradient methods of the first kind. The exactness of the inverse problem solution obtained by the suggested algorithms are analyzed.

NOMENCLATURE

Roman

C	Volumetric heat capacity
f	Experimental measurements
g	Increment of unknown function
J	Residual fanctional
L	Number of layers
$q_1 - q_6$	Heat fluxes
r	Spatial coordinate
R	Thermal contact resistance
T	Temperature
T_0	Initial temperature
u	Unknown function
x,y,z	Spatial coordinates

Greek

β	Parameter of minimization method
γ	Descent step
λ	Thermal condactivity
σ	Diviation of measurement
τ	Time
φ,θ	Angle coordinates
ψ	Ajoint variable

INTRODUCTION

Experimental-and-computational methods, based on solving the boundary inverse heat conduction problem, form an intensively developing direction in the field of unsteady heat transfer processes investigation. When analysing high-temperature processes it is usually necessary to use nonlinear mathematical models of heat transfer with thermal properties depending on temperature. The above circumstances leads to an additional difficulties, when working out algorithms for the solution of corresponding inverse problems. During last years different algorithms were devveloped to solve multi-dimensional boundary inverse heat conduction problems. Most of them can be applied only to linear inverse problems. With others it is possible to consider nonlinear inverse heat transfer problems. Still others are universal and for them a mathematical model of the considered heat transfer processes is not essential. Iterative methods for inverse problems solution [1] refer to the last group of methods. In the authors opinion such methods are the most universal and they allow to investigate a number of practical problems. In this paper problem of reconstruction of time- and coordinate of surface- depended heat flux for simple three-dimencional shape (slab, cylinder and spherical segment), consisted from L layers with boundaries r_l, $l=\overline{1,L+1}$, is under consideration (Fig.1).

ALGORITHM

Let us suppose that in the body analysed the heat transfer process is covered by the boundary-value problem for the three-dimensional quazi-linear heat conduction equation. Coefficients of parabolic equations are the function of temperature. In real situations, there is contact heat transfer between the layers at the boundaries, determined by the values of thermal contact resistance R $l=\overline{1,L-1}$. On the internal boundary ($r=r_1$) boundary condition of second or third kind are prescribed. On the side boundaries arbitrary boundary conditions can be considered.

Then the mathematical formulation of the problem of heat conduction for slab takes the following form

$$C_l(T)\frac{\partial T_l}{\partial \tau}=\frac{\partial}{\partial r}\left(\lambda_l(T)\frac{\partial T_l}{\partial r}\right)+\frac{\partial}{\partial x}\left(\lambda_l(T)\frac{\partial T_l}{\partial x}\right)+ + \frac{\partial}{\partial y}\left(\lambda_l(T)\frac{\partial T_l}{\partial y}\right) \quad (1a)$$

$$T_l = T_l(\tau,r,x,y), \qquad r \in (r_l,r_{l+1}), \quad l=\overline{1,L},$$

$$x \in (x_1,x_2), \quad y \in (y_1,y_2), \quad \tau \in (\tau_{min},\tau_{max}]$$

$$T_l(0,r,x,y) = T_{0l}(r,x,y), \quad r \in (r_l,r_{l+1}), \tag{2a}$$

$$l = \overline{1,L}, \quad x \in (x_1,x_2), \quad y \in (y_1,y_2)$$

$$-\lambda_l(T)\frac{\partial T_l}{\partial r}(\tau,r_l,x,y) = q_l(\tau,x,y), \tag{3a}$$

$$x \in (x_1,x_2), \quad y \in (y_1,y_2), \quad \tau \in (\tau_{min},\tau_{max}]$$

$$-\lambda_L(T)\frac{\partial T_L}{\partial r}(\tau,r_{L+1},x,y) = u(\tau,x,y) \tag{4a}$$

$$x \in (x_1,x_2), \quad y \in (y_1,y_2), \quad \tau \in (\tau_{min},\tau_{max}]$$

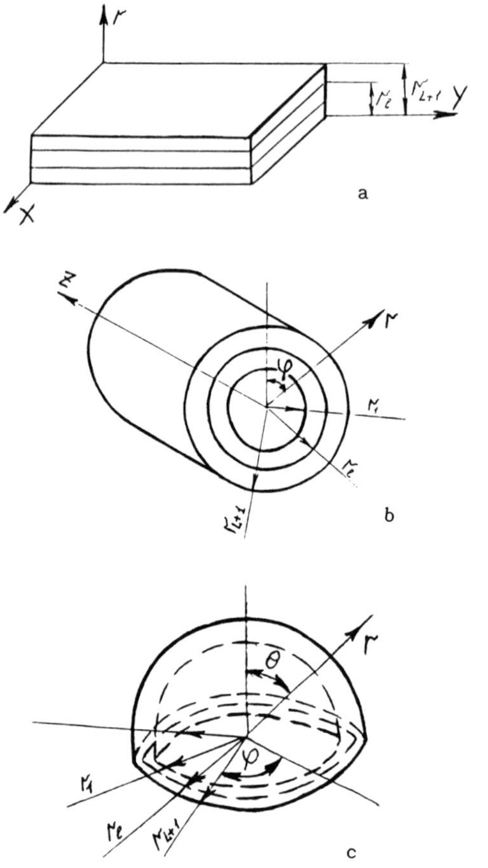

Figure 1: Coordinate systems. a - rectangular, b - cylindrical, c - spherical.

$$-\lambda_l(T)\frac{\partial T_l}{\partial y}(\tau,r,x,y_1) = q_3(\tau,r_l,x), \tag{5a}$$

$$x \in (x_1,x_2), \quad r \in (r_l,r_{l+1}), \; l = \overline{1,L}, \quad \tau \in (\tau_{min},\tau_{max}]$$

$$-\lambda_l(T)\frac{\partial T_l}{\partial y}(\tau,r,x,y_2) = q_4(\tau,r_l,x), \tag{6a}$$

$$x \in (x_1,x_2), \quad r \in (r_l,r_{l+1}), \; l = \overline{1,L}, \quad \tau \in (\tau_{min},\tau_{max}]$$

$$-\lambda_l(T)\frac{\partial T_l}{\partial x}(\tau,r,x_1,y) = q_5(\tau,r_l,y), \tag{7a}$$

$$y \in (y_1,y_2), \quad r \in (r_l,r_{l+1}), \; l = \overline{1,L}, \quad \tau \in (\tau_{min},\tau_{max}]$$

$$-\lambda_l(T)\frac{\partial T_l}{\partial x}(\tau,r,x_2,y) = q_6(\tau,r_l,y), \tag{8a}$$

$$y \in (y_1,y_2), \quad r \in (r_l,r_{l+1}), \; l = \overline{1,L}, \quad \tau \in (\tau_{min},\tau_{max}]$$

$$-\lambda_l(T)\frac{\partial T_l}{\partial r}(\tau,r_{l+1},x,y)\; R_l(T) =$$

$$= T_l(\tau,r_{l+1},x,y) - T_{l+1}(\tau,r_{l+1},x,y) \tag{9a}$$

$$x \in (x_1,x_2), \quad y \in (y_1,y_2), \quad l = \overline{2,L-1} \quad \tau \in (\tau_{min},\tau_{max}]$$

$$\lambda_l(T)\frac{\partial T_l}{\partial r}(\tau,r_{l+1},x,y) = \lambda_{l+1}(T)\frac{\partial T_{l+1}}{\partial r}(\tau,r_{l+1},x,y) \tag{10a}$$

$$x \in (x_1,x_2), \quad y \in (y_1,y_2), \quad l = \overline{1,L-1} \quad \tau \in (\tau_{min},\tau_{max}]$$

Then for cylinder

$$r\,C_l(T)\frac{\partial T_l}{\partial \tau} = \frac{\partial}{\partial r}\left(r\,\lambda_l(T)\frac{\partial T_l}{\partial r}\right) + \frac{\partial}{\partial \varphi}\left(\frac{1}{r}\lambda_l(T)\frac{\partial T_l}{\partial \varphi}\right) +$$

$$+\frac{\partial}{\partial z}\left(r\,\lambda_l(T)\frac{\partial T_l}{\partial z}\right) \tag{1b}$$

$$T_l = T_l(\tau,r,\varphi,z), \quad r \in (r_l,r_{l+1}), \; l = \overline{1,L}, \quad \varphi \in (0,2\pi),$$

$$z \in (z_1,z_2), \quad \tau \in (\tau_{min},\tau_{max}]$$

$$T_l(0,r,\varphi,z) = T_{0l}(r,\varphi,z), \quad r \in (r_l,r_{l+1}), \tag{2b}$$

$$l = \overline{1,L}, \quad \varphi \in (0,\ 2\pi), \quad z \in (z_1,z_2)$$

$$-\lambda_l(T)\frac{\partial T_l}{\partial r}(\tau,r_l,\varphi,z) = q_l(\tau,\varphi,z), \tag{3b}$$

$$\varphi \in (0,\ 2\pi), \quad z \in (z_1,z_2), \quad \tau \in (\tau_{min},\tau_{max}]$$

$$-\lambda_L(T)\frac{\partial T_L}{\partial r}(\tau,r_{L+1},\varphi,z) = u(\tau,\varphi,z) \tag{4b}$$

$$\varphi \in (0,\ 2\pi), \quad z \in (z_1,z_2), \quad \tau \in (\tau_{min},\tau_{max}]$$

$$-\lambda_l(T)\frac{\partial T_l}{\partial z}(\tau,r,\varphi,z_1) = q_3(\tau,r_l,\varphi), \tag{5b}$$

$$\varphi \in (0,2\pi), \quad r \in (r_l,r_{l+1}), \; l = \overline{1,L}, \quad \tau \in (\tau_{min},\tau_{max}]$$

$$- \lambda_l(T) \frac{\partial T_l}{\partial z}(\tau,r,x,z_2) = q_4(\tau,r_l,\varphi), \qquad (6b)$$

$$\varphi \in (0,2\pi), \quad r \in (r_l,r_{l+1}), \; l = \overline{1,L}, \quad \tau \in (\tau_{min},\tau_{max}]$$

$$-\lambda_l(T) \frac{\partial T_l}{\partial r}(\tau,r_{l+1},\varphi,z) \; R_l(T) =$$

$$= T_l(\tau,r_{l+1},\varphi,z) - T_{l+1}(\tau,r_{l+1},\varphi,z) \qquad (9b)$$

$$\varphi \in (0,2\pi), \quad z \in (z_1,z_2), \; l = \overline{2,L}, \quad \tau \in (\tau_{min},\tau_{max}]$$

$$\lambda_l(T) \frac{\partial T_l}{\partial r}(\tau,r_{l+1},\varphi,z) = \lambda_{l+1}(T) \frac{\partial T_{l+1}}{\partial r}(\tau,r_{l+1},\varphi,z) \qquad (10b)$$

$$\varphi \in (0,2\pi), \quad z \in (z_1,z_2), \; l = \overline{2,L}, \quad \tau \in (\tau_{min},\tau_{max}]$$

And then for spherical segment

$$r^2 \sin\theta \; C_l(T) \frac{\partial T_l}{\partial \tau} = \frac{\partial}{\partial r}\left(r^2 \sin\theta \; \lambda_l(T) \frac{\partial T_l}{\partial r}\right) +$$

$$+ \frac{\partial}{\partial \varphi}\left(\frac{\lambda_l}{\sin\theta}(T) \frac{\partial T_l}{\partial \varphi}\right) + \frac{\partial}{\partial \theta}\left(\sin\theta \; \lambda_l(T) \frac{\partial T_l}{\partial \theta}\right) \qquad (1c)$$

$$T_l = T_l(\tau,r,\varphi,z), \quad r \in (r_l,r_{l+1}), \; l = \overline{1,L}, \quad \varphi \in (0,2\pi),$$

$$\theta \in (\theta_1,\theta_2), \quad \tau \in (\tau_{min},\tau_{max}]$$

$$T_l(0,r,\varphi,\theta) = T_{0l}(r,\varphi,\theta), \quad r \in (r_l,r_{l+1}), \qquad (2c)$$

$$l = \overline{1,L}, \quad \varphi \in (0, 2\pi), \quad \theta \in (\theta_1,\theta_2)$$

$$- \lambda_l(T) \frac{\partial T_l}{\partial r}(\tau,r_l,\varphi,\theta) = q_1(\tau,\varphi,\theta), \qquad (3c)$$

$$\varphi \in (0, 2\pi), \quad \theta \in (\theta_1,\theta_2), \quad \tau \in (\tau_{min},\tau_{max}]$$

$$- \lambda_L(T) \frac{\partial T_L}{\partial r}(\tau,r_{L+1},\varphi,\theta) = u(\tau,\varphi,\theta) \qquad (4c)$$

$$\varphi \in (0, 2\pi), \quad \theta \in (\theta_1,\theta_2), \quad \tau \in (\tau_{min},\tau_{max}]$$

$$-\lambda_l(T) \; \frac{1}{r} \frac{\partial T_l}{\partial \theta}(\tau,r,\varphi,\theta_1) = q_3(\tau,r_l,\varphi), \qquad (5c)$$

$$\varphi \in (0,2\pi), \quad r \in (r_l,r_{l+1}), \; l = \overline{1,L}, \quad \tau \in (\tau_{min},\tau_{max}]$$

$$- \lambda_l(T) \; \frac{1}{r} \frac{\partial T_l}{\partial \theta}(\tau,r,x,\theta_2) = q_4(\tau,r_l,\varphi), \qquad (6c)$$

$$\varphi \in (0,2\pi), \quad r \in (r_l,r_{l+1}), \; l = \overline{1,L}, \quad \tau \in (\tau_{min},\tau_{max}]$$

$$-\lambda_l(T) \frac{\partial T_l}{\partial r}(\tau,r_{l+1},\varphi,\theta) \; R_l(T) =$$

$$= T_l(\tau,r_{l+1},\varphi,\theta) - T_{l+1}(\tau,r_{l+1},\varphi,\theta) \qquad (9c)$$

$$\varphi \in (0,2\pi), \quad \theta \in (\theta_1,\theta_2), \; l = \overline{2,L}, \quad \tau \in (\tau_{min},\tau_{max}]$$

$$\lambda_l(T) \frac{\partial T_l}{\partial r}(\tau,r_{l+1},\varphi,\theta) = \lambda_{l+1}(T) \frac{\partial T_{l+1}}{\partial r}(\tau,r_{l+1},\varphi,\theta) \qquad (10c)$$

$$\varphi \in (0,2\pi), \quad \theta \in (\theta_1,\theta_2), \; l = \overline{2,L}, \quad \tau \in (\tau_{min},\tau_{max}]$$

where $u(\tau)$ - unknown boundary condition. In addition the results of temperature measurements on the internal boundary are available

$$T_{exp}(\tau,r_1,x,y) = f(\tau,x,y), \quad x \in [x_1,x_2], \qquad (11a)$$

$$y \in [y_1, y_2]$$

or

$$T_{exp}(\tau,r_1,\varphi,z) = f(\tau,\varphi,z), \quad \varphi \in [0 ,2\pi], \qquad (11b)$$

$$z \in [z_1, z_2]$$

or

$$T_{exp}(\tau,r_1,\varphi,\theta) = f(\tau,\varphi,\theta), \quad \varphi \in [0 ,2\pi], \qquad (11c)$$

$$\theta \in [\theta_1, \theta_2]$$

The problem of determining function $u(\tau)$ is solved by means of minimization of the residual functional which is the mean-square deviation of the temperatures calculated at the internal boundary the mathematical model (1) - (4) from experimental temperatures

$$u = \arg \min_{u \in L_2} J(u), \qquad (12)$$

where $J(u) = \int_{\tau_{min}}^{\tau_{max}} \int_{x_1}^{x_2} \int_{y_1}^{y_2} (T(\tau,r_1,x,y) - f(\tau,x,y))^2 dy dx d\tau$

or $J(u) = \int_{\tau_{min}}^{\tau_{max}} \int_{0}^{2\pi} \int_{z_1}^{z_2} (T(\tau,r_1,\varphi,z) - f(\tau,\varphi,z))^2 dz d\varphi d\tau$

or $J(u) = \int_{\tau_{min}}^{\tau_{max}} \int_{0}^{2\pi} \int_{\theta_1}^{\theta_2} (T(\tau,r_1,\varphi,\theta) - f(\tau,\varphi,\theta))^2 d\theta d\varphi d\tau$

The solution of the problem (6) in parametric form is carried out by means of gradient methods of unconstrained minimization. Iterations are buit in the following manner

$$u_k^s = u_k^{s-1} + \gamma^s g_k^s, \qquad s = 1,\ldots,s^* \qquad (13)$$

$$g^s = - (J_u'(u))^s + \beta^s g^{s-1} ,$$

$$\beta^s = \langle \overline{(J')^s} - \overline{(J')^{s-1}} \rangle_{L_2} / \| \overline{(J')^s} \|_{L_2}, \quad \beta^0 = 0,$$

The number of the last iteration s^* is chosen according to the iterative regularization principle

$$J(u^{s^*}) \le \delta_f^2, \qquad (14)$$

where δ_f^2 is the mean-square temperature measurement error

$$\delta_f^2 = \int_{\tau_{min}}^{\tau_{max}}\int_{x_1}^{x_2}\int_{y_1}^{y_2} \sigma^2(\tau,x,y)\, dydxd\tau$$

or $$\delta_f^2 = \int_{\tau_{min}}^{\tau_{max}}\int_{0}^{2\pi}\int_{z_1}^{z_2} \sigma^2(\tau,\varphi,z)\, dzd\varphi d\tau$$

or $$\delta_f^2 = \int_{\tau_{min}}^{\tau_{max}}\int_{0}^{2\pi}\int_{\theta_1}^{\theta_2} \sigma^2(\tau,\varphi,\theta)\, d\theta d\varphi d\tau$$

The descent parameter γ^s is determined from the condition

$$\gamma = \arg\min_{\gamma \in R^+} \left(J\,(u^s + \gamma^s g^s) \right) \qquad (15)$$

The gradient of the functional minimized is computed by using the solution of a boundary-value problem for a conjugate variable. The expression for the gradient is

$$J'_u = \psi(\tau,r_1,x,y) \qquad (16)$$

where $\psi(\tau,r,x,y)$ (or $\psi(\tau,r,\varphi,z)$ or $\psi(\tau,r,\varphi,\theta)$) is the solution of the following conjugate boundary-value problem:

$$C_l(T)\frac{\partial\psi_l}{\partial\tau} = \frac{\partial}{\partial r}\left(\lambda_l(T)\frac{\partial\psi_l}{\partial r}\right) + \frac{\partial}{\partial x}\left(\lambda_l(T)\frac{\partial\psi_l}{\partial x}\right) + \frac{\partial}{\partial y}\left(\lambda_l(T)\frac{\partial\psi_l}{\partial y}\right) - \frac{d\lambda}{dT}\frac{\partial T}{\partial r}\frac{\partial\psi_l}{\partial r} - \frac{d\lambda}{dT}\frac{\partial T}{\partial x}\frac{\partial\psi_l}{\partial x} - \frac{d\lambda}{dT}\frac{\partial T}{\partial y}\frac{\partial\psi_l}{\partial y} \qquad (17a)$$

$\psi_l = \psi_l(\tau,r,x,y)$, $r \in (r_l,r_{l+1})$, $l = \overline{1,L}$,

$x \in (x_1,x_2)$, $y \in (y_1,y_2)$, $\tau \in (\tau_{min},\tau_{max}]$

$$\psi_l(\tau_{max},r,x,y) = 0, \quad r \in (r_l,r_{l+1}), \qquad (18a)$$

$l = \overline{1,L}$, $x \in (x_1,x_2)$, $y \in (y_1,y_2)$

$$-\lambda_1(T)\frac{\partial\psi_1}{\partial r}(\tau,r_1,x,y) = 2\,(T(\tau,r_1,x,y) - f(\tau,x,y)) \qquad (19a)$$

4 $x \in (x_1,x_2)$, $y \in (y_1,y_2)$, $\tau \in (\tau_{min},\tau_{max}]$

$$-\lambda_L(T)\frac{\partial\psi_L}{\partial r}(\tau,r_{L+1},x,y) = 0 \qquad (20a)$$

$x \in (x_1,x_2)$, $y \in (y_1,y_2)$, $\tau \in (\tau_{min},\tau_{max}]$

$$-\lambda_l(T)\frac{\partial\psi_l}{\partial y}(\tau,r,x,y_1) = 0 \qquad (21a)$$

$x \in (x_1,x_2)$, $r \in (r_l,r_{l+1})$, $l = \overline{1,L}$, $\tau \in (\tau_{min},\tau_{max}]$

$$-\lambda_l(T)\frac{\partial\psi_l}{\partial y}(\tau,r,x,y_2) = 0 \qquad (22a)$$

$x \in (x_1,x_2)$, $r \in (r_l,r_{l+1})$, $l = \overline{1,L}$, $\tau \in (\tau_{min},\tau_{max}]$

$$-\lambda_l(T)\frac{\partial\psi_l}{\partial x}(\tau,r,x_1,y) = 0 \qquad (23a)$$

$y \in (y_1,y_2)$, $r \in (r_l,r_{l+1})$, $l = \overline{1,L}$, $\tau \in (\tau_{min},\tau_{max}]$

$$-\lambda_l(T)\frac{\partial\psi_l}{\partial x}(\tau,r,x_2,y) = 0 \qquad (24a)$$

$y \in (y_1,y_2)$, $r \in (r_l,r_{l+1})$, $l = \overline{1,L}$, $\tau \in (\tau_{min},\tau_{max}]$

$$-\lambda_l(T)\frac{\partial\psi_l}{\partial r}(\tau,r_{l+1},x,y)\,R_l(T) = \left(1 + \lambda_l\frac{dR_l}{dT}\frac{\partial T_l}{\partial r}\right)\left(\psi_l(\tau,r_{l+1},x,y) - \psi_{l+1}(\tau,r_{l+1},x,y)\right) \qquad (25a)$$

$x \in (x_1,x_2)$, $y \in (y_1,y_2)$, $l = \overline{1,L-1}$, $\tau \in (\tau_{min},\tau_{max}]$

$$\lambda_l(T)\frac{\partial\psi_l}{\partial r}(\tau,r_{l+1},x,y) = \lambda_{l+1}(T)\frac{\partial\psi_{l+1}}{\partial r}(\tau,r_{l+1},x,y)\left(1 + \lambda_l\frac{dR_l}{dT}\frac{\partial T_l}{\partial r}\right) \qquad (26a)$$

$x \in (x_1,x_2)$, $y \in (y_1,y_2)$, $l = \overline{1,L-1}$, $\tau \in (\tau_{min},\tau_{max}]$

or for cylinder

$$r\,C_l(T)\frac{\partial\psi_l}{\partial\tau} = \frac{\partial}{\partial r}\left(r\,\lambda_l(T)\frac{\partial\psi_l}{\partial r}\right) + \frac{\partial}{\partial\varphi}\left(\frac{1}{r}\lambda_l(T)\frac{\partial\psi_l}{\partial\varphi}\right) + \frac{\partial}{\partial z}\left(r\,\lambda_l(T)\frac{\partial\psi_l}{\partial z}\right) - r\frac{d\lambda}{dT}\frac{\partial T}{\partial r}\frac{\partial\psi_l}{\partial r} - \frac{1}{r}\frac{d\lambda}{dT}\frac{\partial T}{\partial\varphi}\frac{\partial\psi_l}{\partial\varphi} - r\frac{d\lambda}{dT}\frac{\partial T}{\partial z}\frac{\partial\psi_l}{\partial z} + \lambda(T)\,\psi_l \qquad (17b)$$

$\psi_l = \psi_l(\tau,r,\varphi,z)$, $r \in (r_l,r_{l+1})$, $l = \overline{1,L}$, $\varphi \in (0,2\pi)$,

$z \in (z_1,z_2)$, $\tau \in (\tau_{min},\tau_{max}]$

$$\psi_l(\tau_{max},r,\varphi,z) = 0 \quad r \in (r_l,r_{l+1}), \qquad (18b)$$

$l = \overline{1,L}$, $\varphi \in (0,\ 2\pi)$, $z \in (z_1,z_2)$

$$-\lambda_1(T)\frac{\partial\psi_1}{\partial r}(\tau,r_1,\varphi,z) = 2\,(T(\tau,r_1,x,y) - f(\tau,x,y)) \qquad (19b)$$

$\varphi \in (0,\ 2\pi)$, $z \in (z_1,z_2)$, $\tau \in (\tau_{min},\tau_{max}]$

$$-\lambda_L(T)\frac{\partial\psi_L}{\partial r}(\tau,r_{L+1},\varphi,z) = 0 \qquad (20b)$$

$\varphi \in (0, 2\pi), \quad z \in (z_1,z_2), \quad \tau \in (\tau_{min},\tau_{max}]$

$$-\lambda_1(T)\frac{\partial\psi_1}{\partial z}(\tau,r,\varphi,z_1) = q_3(\tau,r_1,\varphi), \qquad (21b)$$

$\varphi \in (0,2\pi), \quad r \in (r_1,r_{1+1}), \; 1 = \overline{1,L}, \quad \tau \in (\tau_{min},\tau_{max}]$

$$-\lambda_1(T)\frac{\partial\psi_1}{\partial z}(\tau,r,x,z_2) = 0 \qquad (22b)$$

$\varphi \in (0,2\pi), \quad r \in (r_1,r_{1+1}), \; 1 = \overline{1,L}, \quad \tau \in (\tau_{min},\tau_{max}]$

$$-\lambda_1(T)(\frac{\partial\psi_1}{\partial r}(\tau,r_{1+1},\varphi,z) - \frac{1}{r}\psi_1(\tau,r_{1+1},\varphi,z))R_1(T) =$$
$$= (1 + \lambda_1\frac{dR_1}{dT}\frac{\partial T_1}{\partial r})(\psi_1(\tau,r_{1+1},\varphi,z) - \psi_{1+1}(\tau,r_{1+1},\varphi,z)) \qquad (25b)$$

$\varphi \in (0,2\pi), \quad z \in (z_1,z_2), \quad 1 = \overline{2,L}, \quad \tau \in (\tau_{min},\tau_{max}]$

$$\lambda_1(T)\frac{\partial\psi_1}{\partial r}(\tau,r_{1+1},\varphi,z) - \frac{\lambda_1}{r}\psi_1(\tau,r_{1+1},\varphi,z) =$$
$$= (\lambda_{1+1}(T)\frac{\partial\psi_{1+1}}{\partial r}(\tau,r_{1+1},\varphi,z)$$
$$- \frac{\lambda_1}{r}\psi_1(\tau,r_{1+1},\varphi,z))(1 + \lambda_1\frac{dR_1}{dT}\frac{\partial T_1}{\partial r}) \qquad (26b)$$

$\varphi \in (0,2\pi), \quad z \in (z_1,z_2), \quad 1 = \overline{2,L}, \quad \tau \in (\tau_{min},\tau_{max}]$

or for spherical segment

$$r^2\sin\theta\, C_1(T)\frac{\partial\psi_1}{\partial\tau} = \frac{\partial}{\partial r}\left(r^2\sin\theta\,\lambda_1(T)\frac{\partial\psi_1}{\partial r}\right) +$$
$$+ \frac{\partial}{\partial\varphi}\left(\frac{\lambda_1}{\sin\theta}(T)\frac{\partial\psi_1}{\partial\varphi}\right) + \frac{\partial}{\partial\theta}\left(\sin\theta\,\lambda_1(T)\frac{\partial\psi_1}{\partial\theta}\right) -$$
$$- r^2\sin\theta\frac{d\lambda}{dT}\frac{\partial T}{\partial r}\frac{\partial\psi_1}{\partial r} - \frac{1}{\sin\theta}\frac{d\lambda}{dT}\frac{\partial T}{\partial\varphi}\frac{\partial\psi_1}{\partial\varphi} -$$
$$- \sin\theta\frac{d\lambda}{dT}\frac{\partial T}{\partial\theta}\frac{\partial\psi_1}{\partial\theta} + (2\sin\theta + \frac{1}{\sin\theta})\lambda(T)\psi_1 \qquad (17c)$$

$\psi_1 = \psi_1(\tau,r,\varphi,z), \quad r \in (r_1,r_{1+1}), \; 1 = \overline{1,L}, \quad \varphi \in (0,2\pi),$

$\theta \in (\theta_1,\theta_2), \quad \tau \in (\tau_{min},\tau_{max}]$

$$\psi_1(\tau_{max},r,\varphi,\theta) = 0 \qquad r \in (r_1,r_{1+1}), \qquad (18c)$$

$1 = \overline{1,L}, \quad \varphi \in (0, 2\pi), \quad \theta \in (\theta_1,\theta_2)$

$$-\lambda_1(T)\frac{\partial\psi_1}{\partial r}(\tau,r_1,\varphi,\theta) = 2\,(T(\tau,r_1,x,y) - f(\tau,x,y)) \qquad (19c)$$

$\varphi \in (0, 2\pi), \quad \theta \in (\theta_1,\theta_2), \quad \tau \in (\tau_{min},\tau_{max}]$

$$-\lambda_L(T)\frac{\partial\psi_L}{\partial r}(\tau,r_{L+1},\varphi,\theta) = 0 \qquad (20c)$$

$\varphi \in (0, 2\pi), \quad \theta \in (\theta_1,\theta_2), \quad \tau \in (\tau_{min},\tau_{max}]$

$$-\lambda_1(T)\frac{1}{r}\frac{\partial\psi_1}{\partial\theta}(\tau,r,\varphi,\theta_1) = 0 \qquad (21c)$$

$\varphi \in (0,2\pi), \quad r \in (r_1,r_{1+1}), \; 1 = \overline{1,L}, \quad \tau \in (\tau_{min},\tau_{max}]$

$$-\lambda_1(T)\frac{1}{r}\frac{\partial\psi_1}{\partial\theta}(\tau,r,x,\theta_2) = 0 \qquad (22c)$$

$\varphi \in (0,2\pi), \quad r \in (r_1,r_{1+1}), \; 1 = \overline{1,L}, \quad \tau \in (\tau_{min},\tau_{max}]$

$$-\lambda_1(T)(\frac{\partial\psi_1}{\partial r}(\tau,r_{1+1},\varphi,\theta) - \frac{2}{r}\psi_1(\tau,r_{1+1},\varphi,\theta))R_1(T) =$$
$$= (1 + \lambda_1\frac{dR_1}{dT}\frac{\partial T_1}{\partial r})(\psi_1(\tau,r_{1+1},\varphi,\theta) - \psi_{1+1}(\tau,r_{1+1},\varphi,\theta)) \qquad (25c)$$

$\varphi \in (0,2\pi), \quad \theta \in (\theta_1,\theta_2), \quad 1 = \overline{2,L}, \quad \tau \in (\tau_{min},\tau_{max}]$

$$\lambda_1(T)\frac{\partial\psi_1}{\partial r}(\tau,r_{1+1},\varphi,\theta) - \frac{\lambda_1}{r}\psi_1(\tau,r_{1+1},\varphi,\theta) =$$
$$= (\lambda_{1+1}(T)\frac{\partial\psi_{1+1}}{\partial r}(\tau,r_{1+1},\varphi,\theta)$$
$$- \frac{\lambda_1}{r}\psi_1(\tau,r_{1+1},\varphi,\theta))(1 + \lambda_1\frac{dR_1}{dT}\frac{\partial T_1}{\partial r}) \qquad (26c)$$

$\varphi \in (0,2\pi), \quad \theta \in (\theta_1,\theta_2), \quad 1 = \overline{2,L}, \quad \tau \in (\tau_{min},\tau_{max}]$

Linear estimation was used for determination descent step. As example, for spherical segment it can be calculated as

$$\gamma^s = \int_{\tau_{min}}^{\tau_{max}}\int_0^{2\pi}\int_{\theta_1}^{\theta_2}\vartheta_1(\tau,r_1,\varphi,\theta,)(T(\tau,r_1,\varphi,\theta) - f(\tau,\varphi,\theta))^2 d\theta d\varphi d\tau \Big/ \int_{\tau_{min}}^{\tau_{max}}\int_0^{2\pi}\int_{\theta_1}^{\theta_2}\vartheta_1^2(\tau,r_1,\varphi,\theta,)\, d\theta d\varphi d\tau \qquad (27c)$$

where $\vartheta(\tau,r,\varphi,\theta)$ is the Frechet derivative of $T(\tau,r,\varphi,\theta)$ by $u(\tau,\varphi,\theta)$ and is the solution of the following boundary-value problem:

$$r^2\sin\theta\, C_1(T)\frac{\partial\vartheta_1}{\partial\tau} = \frac{\partial}{\partial r}\left(r^2\sin\theta\,\lambda_1(T)\frac{\partial\vartheta_1}{\partial r}\right) +$$
$$+ \frac{\partial}{\partial\varphi}\left(\frac{\lambda_1}{\sin\theta}(T)\frac{\partial\vartheta_1}{\partial\varphi}\right) + \frac{\partial}{\partial\theta}\left(\sin\theta\,\lambda_1(T)\frac{\partial\vartheta_1}{\partial\theta}\right) -$$
$$- r^2\sin\theta\frac{d\lambda}{dT}\frac{\partial T}{\partial r}\frac{\partial\vartheta_1}{\partial r} - \frac{1}{\sin\theta}\frac{d\lambda}{dT}\frac{\partial T}{\partial\varphi}\frac{\partial\vartheta_1}{\partial\varphi} -$$
$$- \sin\theta\frac{d\lambda}{dT}\frac{\partial T}{\partial\theta}\frac{\partial\vartheta_1}{\partial\theta} + \left(\frac{d\lambda}{dT}\left(r^2\sin\theta\frac{\partial T^2}{\partial r^2} + \frac{1}{\sin\theta}\frac{\partial^2 T}{\partial\varphi^2} + \right.\right.$$

$$+ \sin\theta \frac{\partial^2 T}{\partial\theta^2}\Bigg) + \frac{d^2\lambda}{dT^2}\Bigg(r^2\sin\theta\left(\frac{\partial T}{\partial r}\right)^2 + \frac{1}{\sin\theta}\left(\frac{\partial T}{\partial\varphi}\right)^2 +$$

$$+ \sin\theta\left(\frac{\partial T}{\partial\theta}\right)^2\Bigg) + 2r\sin\theta\frac{d\lambda}{dT}\frac{\partial T}{\partial r} + \cos\theta\frac{d\lambda}{dT}\frac{\partial T}{\partial\theta} -$$

$$- r^2\sin\theta\frac{dC}{dC}\frac{\partial T}{\partial\tau}\Bigg)\vartheta_l \quad (28c)$$

$\vartheta_l = \vartheta_l(\tau,r,\varphi,z)$, $r \in (r_l,r_{l+1})$, $l = \overline{1,L}$, $\varphi \in (0,2\pi)$,

$\theta \in (\theta_1,\theta_2)$, $\tau \in (\tau_{min},\tau_{max}]$

$$\vartheta_l(\tau_{min},r,\varphi,\theta) = 0 \qquad r \in (r_l,r_{l+1}), \quad (29c)$$

$l = \overline{1,L}$, $\varphi \in (0, 2\pi)$, $\theta \in (\theta_1,\theta_2)$

$$- \lambda_1(T)\frac{\partial\vartheta_1}{\partial r}(\tau,r_1,\varphi,\theta) - \frac{d\lambda_1}{dT}\frac{\partial T}{\partial r}\,\vartheta_1(\tau,r_1,\varphi,\theta) = 0 \quad (30c)$$

$\varphi \in (0, 2\pi)$, $\theta \in (\theta_1,\theta_2)$, $\tau \in (\tau_{min},\tau_{max}]$

$$- \lambda_L(T)\frac{\partial\vartheta_L}{\partial r}(\tau,r_{L+1},\varphi,\theta) - \frac{d\lambda_l}{dT}\frac{\partial T}{\partial r}\,\vartheta_l(\tau,r_{L+1},\varphi,\theta) =$$

$$= g^s \quad (31c)$$

$\varphi \in (0, 2\pi)$, $\theta \in (\theta_1,\theta_2)$, $\tau \in (\tau_{min},\tau_{max}]$

$$-\lambda_l(T)\,\frac{1}{r}\,\frac{\partial\vartheta_l}{\partial\theta}(\tau,r,\varphi,\theta_1) - \frac{1}{r}\frac{d\lambda_l}{dT}\frac{\partial T}{\partial\theta}\,\vartheta_l(\tau,r,\varphi,\theta_1) =$$

$$= 0 \quad (32c)$$

$\varphi \in (0,2\pi)$, $r \in (r_l,r_{l+1})$, $l = \overline{1,L}$, $\tau \in (\tau_{min},\tau_{max}]$

$$- \lambda_l(T)\,\frac{1}{r}\,\frac{\partial\vartheta_l}{\partial\theta}(\tau,r,x,\theta_2) - \frac{1}{r}\frac{d\lambda_l}{dT}\frac{\partial T}{\partial\theta}\,\vartheta_l(\tau,r,\varphi,\theta_2) =$$

$$= 0 \quad (33c)$$

$\varphi \in (0,2\pi)$, $r \in (r_l,r_{l+1})$, $l = \overline{1,L}$, $\tau \in (\tau_{min},\tau_{max}]$

$$- \lambda_l(T)\frac{\partial\vartheta_l}{\partial r}(\tau,r_{l+1},\varphi,\theta)\;R_l(T) -$$

$$- \frac{d\lambda_l}{dT}\frac{\partial T}{\partial r}\,\vartheta_l(\tau,r_{l+1},\varphi,\theta)\;R_l(T) -$$

$$- \lambda_l(T)\,\frac{dR_l}{dT}\frac{\partial T}{\partial r}\,\vartheta_l(\tau,r_{l+1},\varphi,\theta) =$$

$$= \vartheta_l(\tau,r_{l+1},\varphi,\theta) - \vartheta_{l+1}(\tau,r_{l+1},\varphi,\theta) \quad (34c)$$

$\varphi \in (0,2\pi)$, $\theta \in (\theta_1,\theta_2)$, $l = \overline{2,L}$, $\tau \in (\tau_{min},\tau_{max}]$

$$\lambda_l(T)\frac{\partial\vartheta_l}{\partial r}(\tau,r_{l+1},\varphi,\theta) + \frac{d\lambda_l}{dT}\frac{\partial T}{\partial r}\,\vartheta_l(\tau,r_{l+1},\varphi,\theta) =$$

$$= \lambda_l(T)\frac{\partial\vartheta_{l+1}}{\partial r}(\tau,r_{l+1},\varphi,\theta) +$$

$$+ \frac{d\lambda_{l+1}}{dT}\frac{\partial T}{\partial r}\,\vartheta_{l+1}(\tau,r_{l+1},\varphi,\theta) \quad (35c)$$

$\varphi \in (0,2\pi)$, $\theta \in (\theta_1,\theta_2)$, $l = \overline{2,L}$, $\tau \in (\tau_{min},\tau_{max}]$

NUMERICAL SIMULATION

On the basis of the suggested algorithm were developed a computer program. The boundary problems in Eqs. (1)-(10), (17)-(26) , and (28)-(35) are solved using the finite-difference method (implicit fractional-step scheme); all the differential operators are approximated in the same different grid, with a constant time step and a constant step over the spatial coordinate inside each layer. The construction of finite-difference analog of the differential problem is undertaken in individual layers, and the solutions in adjacent layers are matched using a finite -difference representation of the energy-matching conditions. This approach disrupts the homogenity of the difference scheme, but allows the accuracy of numerical solution to be increased in the presence of discontinuiti of the firs kind in the solution at the boundary between the layers. The initial direct boundary problem in Eqs. (1)-(10) is nonlinear, and therefore its solution is successively refined in each time step by the method of simple iteration, until the solution in two adjacent iterations coincide with an a priori specified relative accuracy of 0.001.

Different computational experiments were carried out to analyse the efficiency of the algorithm described. Calculations were done under the conditions that L=1, r_1 = .3 m, r_2= .31 m, $x_1 = y_1 = z_1 = \theta_1 = 0$, $x_2 = y_2 = z_2$ = .3, θ_2 = .075035 rad, $q_1 = 0$, $q_3 = q_4 = q_5 = q_6$ = 100 W/m^2, unknown heat flux was uniform by the surface ($u(\tau,x,y) = u(\tau,\varphi,z) = u(\tau,\varphi,\theta) = q_2(\tau)$), λ = 75 W/m/K, C = .2 10^7J/m^{3}K.

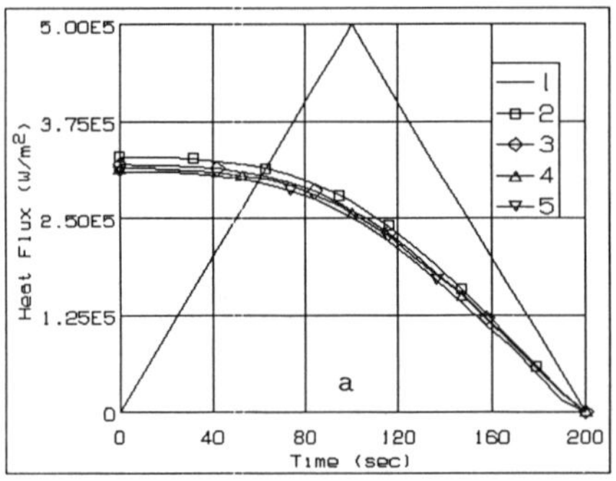

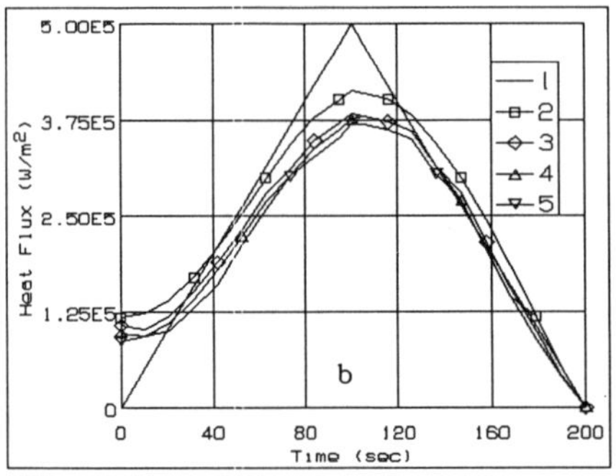

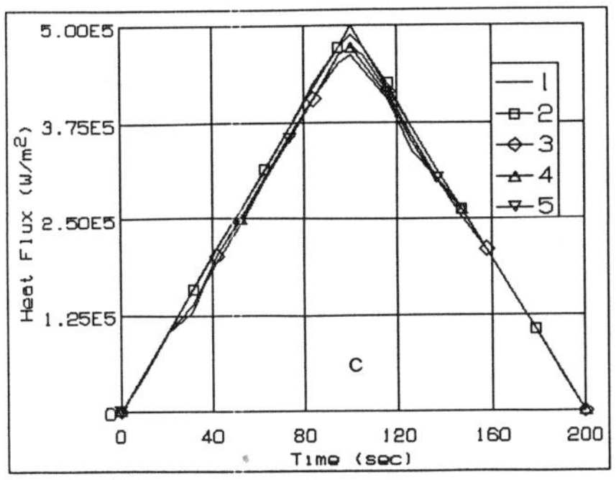

Figure 2: Heat flux u(τ) determination, a - 1st iteration, b - 3rd iteration , c - 20th iteration; 1 - exact "known" function, 2 - one dimensional case, 3 - rectangular case, 4 - cylindrical case, 5 - spherical case.

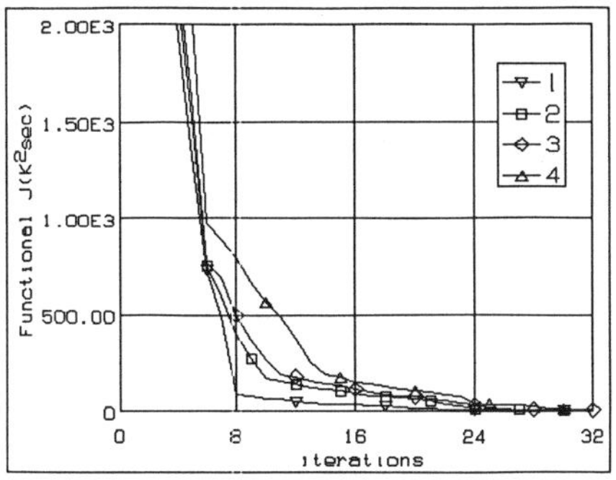

Figure 3: Convergence for heat flux u(τ) determination, 1 - one dimensional case, 2 - rectangular case, 3 - cylindrical case, 4 - spherical case.

Fig. 2 and 3 displays some results of reconstructing function q (τ) for different coordinate system.

The results of solving the boundary inverse heat conduction problem, when initial data are known with errors, are shown in Fig. 4 (50th iteration). The following expression was used to simulate measurement errors

$$f(\tau,x,y) = \overline{f(\tau,x,y)}\,(1 + \gamma\omega), \qquad (36)$$

where $f(\tau,x,y)$ is noised temperature measurements", $\overline{f(\tau,x,y)}$ is the exact function calculated from the numerical solution of the boundary-value problem (1) - (10), ω is random numbers distributed according to the normal law with a zero mean, γ is the maximum of a relative random temperature error (5%). Heat flux $u(\tau,x,y)$ is localyzed in the one point (x,y) = (.05,.05). ($r_1 = x_1 = y_1 =$ 0.0, $x_2 = y_2 = .1$, $r_2 = .01$)

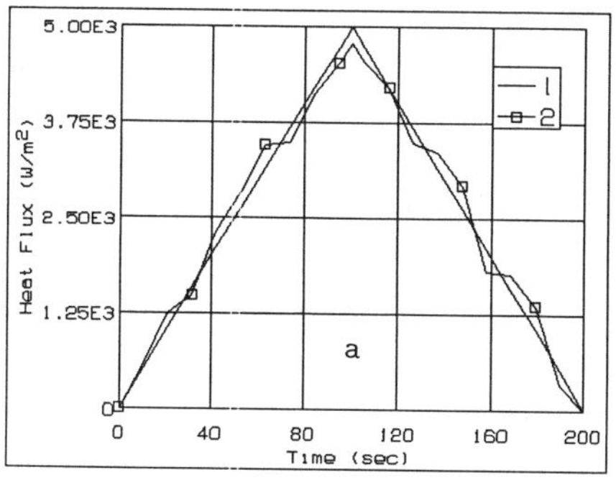

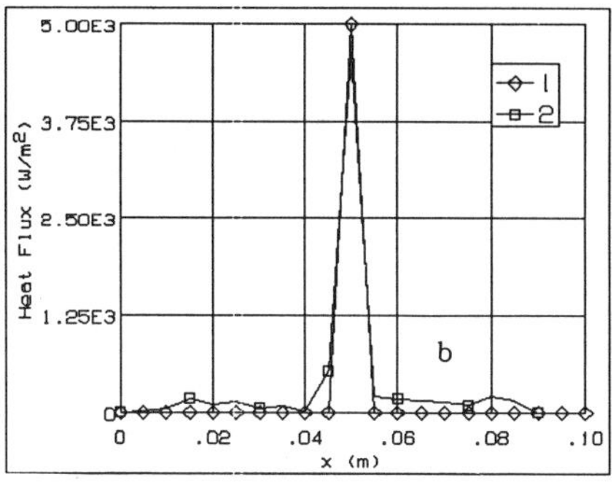

Figure 4: Heat flux $u(\tau,x,y)$ determination. a - $u(\tau,x,y)|_{x=.5,\ y=.5}$, b - $u(\tau,x,y)|_{\tau=10.,y=.05}$ 1 - exact "known" function, 2 - solution of inverse priblem.

CONCLUSIONS

The purpose of this papers is to describe the developed algorithm by means of which to analyze and process the data of the transient thermal experiments in the determination of the external heating of bodies.

The approach has been suggested for determination the unknown boundary conditions on the surface of multilayer simple-shape bodies as the solution of nonlinear inverse heat conduction problem in extreme form. Such method made it possible to include a priori information about unknown functions [6], and to use parametrization of unknown functions[5]. The algorithm has been presented for rectangular, cylindrical and spherical coordinate systems.

Ullustrative examples have been presented for the comparison of the rate of convergence for different coordinate system. Exept this results of determination of magnitude and position of concentrated heat flux has been presented too. The accuracy of the solution obtained in these cases for inverse problem corresponds to the errors of the simulated "experimental measurements".

REFERENCES

1. Alifanov, O.M., 1988, *Inverse Heat Transfer Problems*, Mashinostroenie, Moscow (in Russian).

2. Alifanov, O.M., Artyukhin, E.A. and Rumyantsev, S.V.,1988, *Extreme methods of ill-posed problems solving*, Nauka, Moscow (in Russian).

3. Jarny, Y., Ozisik, M.N. andBardon, J.P., "A General Optimization Method Using Adjoint Equation for Solving Multidimensional Inverse Heat Conduction", *International Journal of Heat and Mass Transfer*, Vol.34, pp.2911-2919.

4. Alifanov, O.M. E.A. and Nenarokomov, A.V. "Three Dimensional Inverse Heat Conduction Problem in Extreme Form", *Soviet Physics. Doklady*, Vol.325, pp.950-954.

5. Alifanov, O.M. E.A. and Nenarokomov, A.V. "Influence of Different Factors on the Accuracy of the Solution of the Parametrized Inverse Heat Conduction Problem", *Journal of Engineering Physics*, Vol.56, pp.308-312.

6. Alifanov, O.M., Artyukhin, E.A. and Nenarokomov, A.V., "Spline Approximation of the Solution of the Inverse Heat Conduction Problem, Taking Account of the Smoothness of the Desired Function", *High Temperature*, Vol.25, pp.520-526.

INVERSE PROBLEM RESEARCH AT ENERGY ENGINEERING DEPARTMENT OF GENOVA UNIVERSITY (ITALY)

G. Milano and F. Scarpa
Dipartimento di Ingegneria Energetica

D. Pescetti
Istituto di Fisica di Ingegneria

Università di Genova
Genova, Italy

ABSTRACT

In this paper an outline is given of some inverse problems developed at the Energy Engineering Department of Genova University. In particular the following topics are reported and discussed: the identification of temperature dependent thermophysical properties of materials subjected to one-dimensional transient heat conduction, from a set of thermal measurements at the boundaries and/or inside the sample; a problem of magnetic field synthesis related to the determination of the characteristics of a given number of electric coils, independently fed, able to produce in a certain region of space a desired spatial distribution of magnetic field; the two-dimensional temperature profile reconstruction in a fluid during a heat transfer process by convection, from the knowledge of the iso-deflection lines given by quantitative Schlieren images. For each topic a brief introduction of the physical aspect of the problem is given and the main results obtained by using different solution algorithms (OLS, MAP or Kalman techniques), are reported. When necessary, a regularization method for linear and non linear problems is also introduced in the solution, whose effect is discussed too.

IDENTIFICATION OF TEMPERATURE DEPENDENT THERMOPHYSICAL PROPERTIES FROM TRANSIENT DATA

The research activity in the field of inverse problems has been mainly addressed to the estimation of thermophysical properties of materials from transient thermal measurements . In particular a technique has been developed by which it is possible to identify thermal conductivity and volumetric heat capacity as a function of temperature from a single experiment of transient heat conduction. The identification algorithm is formulated in terms of a non linear, time varying, inverse heat conduction problem in which the temperature dependent thermophysical properties are the unknown functions and the temperature or heat flux histories at the boundaries and inside the specimen, at known positions, are the measured quantities.

As heat transfer model the one-dimensional heat conduction equation for homogeneous and isotropic material is assumed :

$$C(T)\frac{\partial T}{\partial t} = \frac{\partial}{\partial x}\left[k(T)\frac{\partial T}{\partial x}\right] , \quad T=T(x,t) \tag{1}$$

$$0<x<b \qquad 0<t\le t_m \tag{2}$$

where $k(T)$, and $C(T)$ are the temperature dependent thermal conductivity, and volumetric heat capacity; x the spatial coordinate, t the time, T the temperature, b the slab thickness and t_m the total test duration.

Equation (1) is coupled with the following initial and boundary conditions:

$$T(x,0)=\Phi(x) \ ; \quad 0<x<b \tag{3}$$

$$T(0,t)=U_1(t) \ \text{ or } \ q(0,t)=-k(T)\left.\frac{\partial T}{\partial x}\right|_{x=0} = u_1(t) \tag{4}$$

$$T(b,t)=U_2(t) \ \text{ or } \ q(b,t)=-k(T)\left.\frac{\partial T}{\partial x}\right|_{x=b} = u_2(t) \tag{5}$$

$$0<t\le t_m$$

Two different kinds of parameterization for thermal conductivity and heat capacity have been developed: polynomial or cubic B-splines approximation. The first one is a very simple and practical representation for the temperature dependence effect and it is easy to implement. The cubic B-splines require a greater volume of calculations but they are able to fit in more flexible way the local variation of the temperature dependent unknown functions.

Following the polynomial approximation, thermal conductivity and heat capacity can be parameterized :

$$k(T)= \Sigma_j k_j (T-T_{ref})^j \; ; \; C(T)= \Sigma_j C_j (T-T_{ref})^j \quad (6a)$$

where T_{ref} is a reference temperature.

With the cubic B-splines approximation we have:

$$k(T)= \Sigma_j k_j B_j(T) \; ; \; C(T)= \Sigma_j C_j B_j(T) \quad (6b)$$

where $T=[T_{min}, T_{max}]$; $B_j(u)$ are the well known basis functions for uniform cubic B-splines; k_j and C_j are the set of the unknown control vertices.

The vector β of unknown parameters can be defined for both kinds of approximations:

$$\beta = \{ k_0, k_1, k_2, .. , C_0, C_1, C_2, .. \}$$

If only temperature time histories are processed (i.e. no heat flux measurements are made during experiments) the original unknown functions $C(T)$ and $k(T)$ in Eq.(1) are both identified with reference to a constant parameter $C_0=C(T_{ref})$, whose value cannot be estimated separately. In this case the thermophysical properties to be determined are defined :

$$k'(T)=k(T)/C_0; \qquad C'(T)=C(T)/C_0 \quad (7)$$

where $k'(T)$ has the meaning (unit) of a temperature dependent thermal diffusivity,and $C'(T)$ is a normalized volumetric heat capacity. The vector of unknown parameters β results in this case:

$$\beta = \{ k'_0, k'_1, k'_2, .. , C'_1, C'_2, C'_3, .. \}$$

Different methodologies were developed for solving the inverse conduction problem, namely the stochastic Kalman filter technique and deterministic methods such as MAP (Maximum A Posteriori) or OLS (Ordinary Least Squares). The Kalman filtering technique was described in (Scarpa et al., 1991) and, with more detail, in (Scarpa and Milano, 1993) and in the companion paper of the present analysis (Scarpa et al., 1993). The OLS and MAP algorithms were developed and used in (Milano and Scarpa, 1991) and in (Bartolini et al., 1991).
A large number of simulated experiments were made and the main results obtained have put in evidence some issues. The Kalman filter is a more complete technique: its statistical formulation provides the means to take into account the uncertainty relative to the initial temperature distribution and to the measured boundary conditions. It always provides accurate estimates and reliable confidence bounds. The other side of the coin is the great computational effort required by this sophisticated technique; if the initial condition is known with a good accuracy and only a small quantity of noise affects the control variables, then the good behavior (Scarpa et al., 1993) of more conventional least squares techniques makes advisable the use of OLS or MAP methods. The implementation of these algorithms is a straightforward task and no problems arise even during the elaboration of true experimental data.

As minimization algorithm both the gradient method and the Gauss technique have been developed and tested. The gradient method is perhaps more accurate in the parameter estimation but it requires more iterations in the case of non linear problems (Bartolini et al., 1991). Moreover if the number of unknown parameters is more than 2 or 3 the number of iterations increases dramatically.
With the use of the Gauss method, together with a large amount of data, the reconstruction process takes place with a limited number of iterations, smoothly and , from the authors' experience, never requires the presence of a regularization algorithm.

Besides the numerical simulations, also an experimental apparatus has been set up for measuring the transient response of materials described in detail in (Bartolini et al.,1991). In that follows, a short description of the experimental facility is reported together with an example of the reconstruction of temperature dependent thermophysical properties of insulating materials.

The test section which includes the heater the specimen and the cold plate is placed inside a thermostatted bell-shaped vessel, provided with vacuum pumps and with an air drier system. The combination of the air drier and the vacuum system is utilized to obtain an efficient "in-place" drying for open-cell materials (e.g. fibrous insulators). In fact the presence of moisture greatly influences the transient conduction and it can be one of the major source of distortion in the reconstructed parameters.

The measuring section, schematically shown in Fig. 1, is made of an electrically heated, disk-shaped,copper plate 24 cm in diameter and 0.7 cm thick, and of a refrigerated square plate 26 cm x 26 cm x 3 cm ,aluminium made. The heated and the cold plate are kept in fixed position by means of some spacers whose adjustable height determines the thickness of the specimen in case of soft material. The temperature of the hot and cold surface is measured by stainless steel, sheathed thermocouples (sheath diameter 0.05 cm , wire diameter 0.008 cm) arranged in proper grooves of the heated and of the refrigerated plate. The temperature time response of the sample is measured inserting inside the material, at known position, a number (normally 6) of sheathed thermocouples of the same kind as above. After the manual insertion, the no-

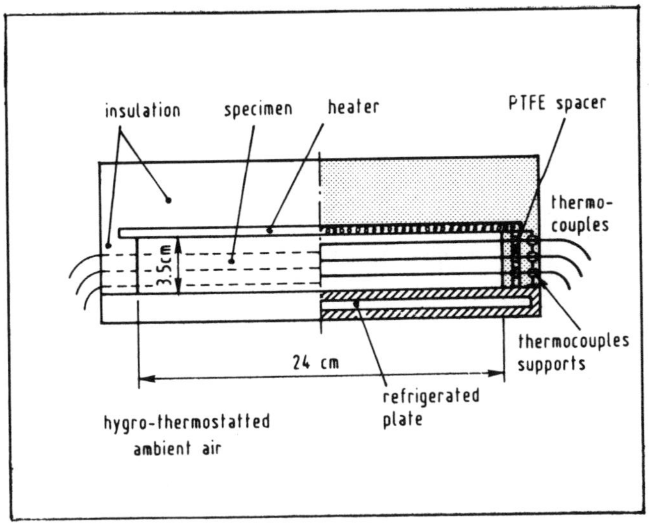

Fig. 1 Schematic view of the test section of the experimental apparatus

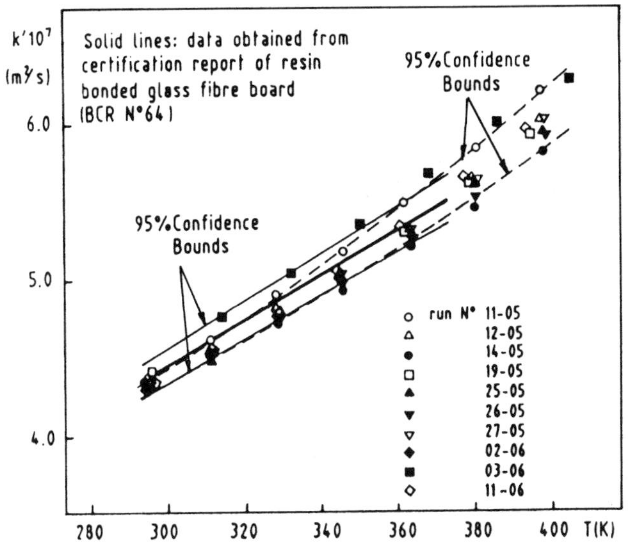

Fig. 2 Comparison between experimental results given by the inverse technique and literature data

minal position of the inner TC junctions is usually known within $\pm$ 0.1 cm. Uncertainties of the above order of magnitude can produce strongly biased estimates.

To greatly reduce this kind of error the position of each internal sensor is considered as further parameter to be determined . The vector $\mathbf{\beta}$ is properly augmented and an interpolation algorithm is inserted in the measurement model. Details of the above procedure as far as the results of several simulated and experimental tests are reported in (Milano and Scarpa, 1991) and (Milano et al., 1991).

The total test duration is typically 1 hour and all temperatures are measured with an A/D converter unit every 4 seconds. Each temperature value is obtained in turn as the average of 1000 samples (sampling rate 13000/s, A/D resolution 12 bit). A correction for the delay due to the thermocouple time response and to the data acquisition system is also introduced in the algorithm. In Fig. 2 some experimental results are shown concerning the european reference material for the thermal conductivity of insulators (resin bonded glass fibre board BCR 64).

The specimen, 3.5 cm thick ,is subjected to an exponential increase of temperature from ambient up to 400 K from one side, while the opposite surface is kept at the initial ambient temperature. The maximum temperature is limited in this case by the chemical stability of the resin bonded glass fibers. The temperature dependent parameter $k'(T)= k(T)/C_0$ estimated with the transient technique is compared with the corresponding quantity obtainable from distinct measurements of thermal conductivity $k(T)$ (hot guarded plate in steady state) and of volumetric heat capacity C_0 (differential scanning calorimeter). The comparison shows a good agreement between the two different methodologies. The functions $k'(T)$ and $C'(T)$ (see Eq.7) are parameterized by uniform cubic B-splines with 7 and 3 control vertices respectively. The parameter $C'(T)$, not reported here, shows a temperature dependence less pronounced than that of $k'(T)$ and the temperature excursion imposed during the experiment is not sufficient for its reliable identification. The standard deviation of the temperature residuals, averaged for the 6 inner measuring points results less than 0.04 K . The 95% confidence bound given by the inverse solution and averaged for the 10 runs reported in Fig. 2 varies from $\pm$ 1.2% at the initial ambient temperature up to $\pm$ 2.5% at the maximum temperature. As it can be seen from Fig. 2 , the experimental results are encouraging in the use of transient techniques.

Among the future developments of the research, the following topics will be considered: the application of the previously described techniques also to composite materials; the real time identification of the boundary conditions during a transient heat conduction process; the reconstruction of the initial temperature distribution, within a sample, by means of smoothing techniques.

A PROBLEM OF MAGNETIC FIELD SYNTHESIS

Consider the synthesis (design) problem related to the determination of the characteristics of a limited number of electric coils, independently fed, able to produce in a certain region of space a wanted spatial distribution of magnetic field. It is well known that this kind of problem, which has a lot of practical applications in mass spectrography, in electron paramagnetic resonance imaging, in superconducting devices, etc., is highly "improperly posed". In order to obtain stable and acceptable solutions, special cures are required such as the use of regularization techniques. In that follows the Ordinary Least Squares (OLS) approach, combined to the Levenberg-

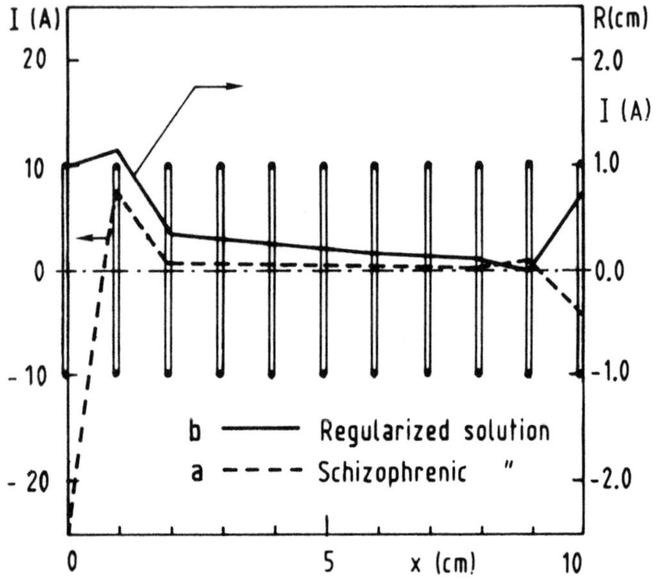

Fig. 3 Current intensity distribution for (a) schizophrenic solution and (b) regularized solution

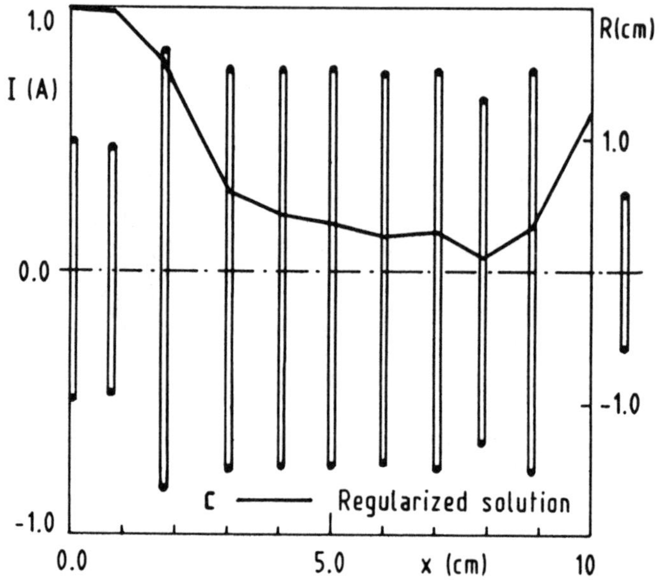

Fig. 4 An improved solution obtained by adding radius and position of the coils to the sought-for parameters

Marquardt regularization method, is proposed for solving the aforementioned synthesis problem and some very simple examples of application are presented.

In particular we consider the magnetic field generated by N coaxial, electric coils in a generic point along the axis. Each kth coil ($k=1,..,N$), having a radius R_k, is located at a distance x_k with respect to an arbitrary reference position and it can be independently fed by a current of intensity I_k. The intensity $\boldsymbol{H}_c$ of the calculated magnetic field at the generic position x_j, along the axis, is given by the well known formula:

$$H_c(x_j) = \sum_{1}^{N} {}_k \frac{R_k^2}{2\cdot[R_k^2+(x_j-x_k)^2]^{3/2}} \cdot I_k \tag{1}$$

The synthesis problem here considered consists in finding the set of parameters I_k, R_k and x_k which produce a wanted function $\boldsymbol{H}_w(x)$ within a given interval $a \leq x \leq b$.

Following the OLS approach the problem can be solved by minimizing the least squares residuals:

$$S= [\, \boldsymbol{H}_w\text{-}\boldsymbol{H}_c(\boldsymbol{\beta})\,]^t \cdot [\, \boldsymbol{H}_w\text{-}\boldsymbol{H}_c(\boldsymbol{\beta})\,] \longrightarrow \min \tag{2}$$

where the summation in Eq. 2 is carried out to a certain number of discrete points xj (j=1,..,M) within the interval (a,b) and the vector $\boldsymbol{\beta}$ of the unknown parameters results in this case:

$$\boldsymbol{\beta} = \{\, I_1, I_2,.., I_N, R_1, R_2,.., R_N, x_1, x_2,..,x_N \,\} \tag{3}$$

Due to the general non linearity of the problem the minimization process can be constructed with successive approximations using recursively the Gauss method. The solution vector β_{i+1} at the iteration $i+1$ is obtained moving the previous solution at the iteration i along a quasi-Newton direction. In vectorial form the iterative process can be written:

$$\boldsymbol{\beta}_{i+1} = \boldsymbol{\beta}_i - 1/2\, \mathbf{P}_i \cdot \boldsymbol{grad}\{\, S(\boldsymbol{\beta})_i \,\} \tag{4}$$

The matrix **P** at the iteration i is defined as:

$$\mathbf{P}_i = [\mathbf{X}_i^t \cdot \mathbf{X}_i]^{-1} \tag{5}$$

where **X** is the Jacobian or sensitivity matrix of $\boldsymbol{H}_c$ in respect to $\boldsymbol{\beta}$ calculated at the discrete points x_j ($j=1,..,M$).

The gradient of S in Eq. (4) as well as the sensitivity coefficients $[\mathbf{X}]_{j,k}$ can be directly determined by performing the derivatives of Eq.(1):

$$[\mathbf{X}]_{j,k} = \frac{\partial H_j}{\partial \beta_k} \tag{6}$$

In order to obtain stable solutions, the Levenberg-Marquardt regularization algorithm has been introduced by replacing the original matrix **P** with the modified:

$$\mathbf{P}_i = [\, \mathbf{X}_i^t \cdot \mathbf{X}_i + \mu_i \boldsymbol{\Omega} \,]^{-1} \tag{7}$$

where $\boldsymbol{\Omega}$ is a matrix set to the diagonal terms of $\mathbf{X}_i^t \cdot \mathbf{X}_i$ and μ_i is a properly assigned coefficient which decreases when the iteration number increases.

Incidentally we can remark that the Levemberg regularization algorithm (Levenberg,1944) based on the simultaneous minimization of both the squares of the residuals and of the parameter increments shows for this application an interesting physical meaning. In fact the minimization of the norm $\|\, \boldsymbol{\beta} \,\|^2$ includes the minimization of

Table 1 Influence of the regularization and of the degree of freedom

case	M	$Q \propto \Sigma_k I_k^2 R_k$	e%
a	11	100	$1.5 \cdot 10^{-14}$
b	11	0.434	0.113
c	101	0.437	0.052

the squares of the current intensity, which are directly proportional to the ohmic losses, and therefore to the dissipated energy. From this point of view the Levenberg regularization makes "acceptable" the solution in the sense that the energy required to realize the wanted magnetic field, with a certain degree of approximation, is minimized.

In Fig. 3 an example of regularized and non regularized solution is shown. In this case we want to produce, along the axis of the coils, a magnetic field whose intensity is given by the function:

$$H(x) = \frac{1}{x} \quad \text{within the interval } 1.5 \leq x \leq 8.5 \text{ cm}$$

having at disposal eleven equal rings, uniformly distributed within the interval $0.0 \leq x \leq 10.0$ cm. The radius of each ring is $R = 1.0$ cm and both the radius and the position are fixed. The magnetic field residuals are evaluated at the eleven equispaced points ($M=11$) within the sub-interval $1.5 \leq x \leq 8.5$ cm. In Fig. 3 the dashed line represents the non regularized solution (in this case the exact one) while the continuous line is the regularized pseudo-solution . As it can be seen, the non regularized solution shows the typical "schizophrenic" behavior with sudden jump from large negative to positive value. The physical meaning of the solution is preserved even with negative values of I_k because the current may circulate in the rings clockwise or not, but the large current intensity makes the solution non practical. Assuming as $Q= 100$ the energy required for the schizophrenic solution (case a) the regularization effect (case b) reduces Q by a factor more than 200, as one can see in Table 1. By introducing as a measure of the accuracy of the solution the average percentage error $e\%$ defined as :

$$e\% = \frac{1}{M} \Sigma_{1}^{M} {}_{j} \frac{| H_w(x_j)-H_c(x_j) |}{H_w(x_j)} \cdot 100$$

the case (b) is affected , on average, by a percentage error $e= 0.11\%$. If we increase the degree of freedom of the system by allowing a variation of the radius and position of each coil, the accuracy of the pseudo-solution increases at the same energy required as shown in Fig.4 and in Table 1 (case c). The number of points, in which the residuals are evaluated, is one order of magnitude greater and the error e% is halved. On the basis of the above procedure but for a more complicated situation a central force apparatus was designed and manufactured (Pescetti et al., 1985).

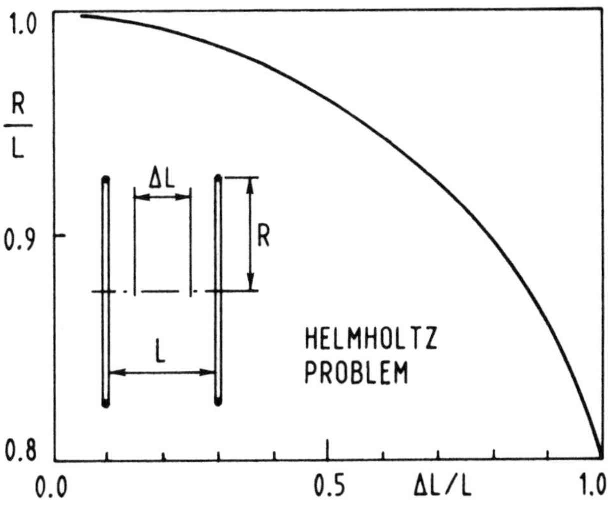

Fig. 5 Inverse solution of the Helmholtz pair

One other problem which is returned of great interest, from the pioneering work of Helmholtz ,is the following: "find the ratio between the radius and the distance of two coils in order to have a uniform magnetic field on the axis (Helmholtz problem) or a magnetic field with constant gradient (anti-Helmholtz problem) (Quine and Rinard, 1992) or more complicated functions". This kind of problem can be easily solved by using the above regularized procedure as shown in Fig. 5 in which the Helmholtz pair is considered. A uniform magnetic field is desired on the axis between the two coils ; the classic Helmholtz solution, obtained by putting to zero the first second and third derivative of H at the middle of the distance L, gives $R/L \approx 1$. The results of Fig.5 clearly show that if the sub-interval L in which the residuals are evaluated ,is progressively reduced the ratio R/L approaches the Helmholtz solution.

TEMPERATURE PROFILE RECONSTRUCTION FROM SCHLIEREN IMAGES

Among the optical methods available for measuring temperature profiles in fluids, the quantitative Schlieren technique is widely used as effective experimental support in thermofluid dynamic researches.

As well known, the Schlieren method allows one to visualize the deflection of a light ray passing through a transparent medium in which a gradient of the refractive index is present. Since in a thermal boundary layer of a fluid the temperature gradient is related to the density gradient and therefore to the refractive index gradient, the Schlieren images can be processed in order to evaluate the fluid temperature distribution.

Many particular versions of the quantitative Schlieren technique were developed; among these the so called "focal filament method" is simple to realize, easy to operate and inexpensive. With the focal filament method (Vasil'ev, 1971) the measurement of the light ray deflection is made by shifting an opaque filament along the focal plane of the Schlieren head.

In the assumption of two-dimensional temperature field ($T=f(x,y)$) the amount of light deflection D can be related (Eckert and Goldstein, 1976) to the local temperature and temperature gradient of the fluid by the equation :

$$D = \Gamma \cdot \frac{1}{T^2} \cdot \frac{\partial T}{\partial y} \qquad (1)$$

where T is the local absolute temperature, y the spatial coordinate along which the temperature gradient is measured and Γ (m^2 K) is a global coefficient which includes the Gladstone-Dale constant, the ideal gas constant, the fluid refractive index, the ambient pressure, the length of the disturbed optical path and the focal length of the Schlieren head. For each experiment performed with a given apparatus the constant Γ assumes a proper value which can be easily calculated. Therefore, having at disposal, at a certain number of elevations x_k ($k=1,2,..,M$) of a plate subjected to a convective process, the set of discrete values of displacements $D_{k,i}$ ($i=1,2,..,N$), the problem consists in finding the family of temperature distributions $T_k = T_k(y)$ ($k=1,2,..,M$) which satisfy Eq. (1).

It can be noted that equation (1) can be immediately integrated and the temperature distribution can be obtained at the generic elevation x_k as follows:

$$T_k(y) = \frac{1}{\dfrac{1}{T_k(0)} - \dfrac{1}{\Gamma}\displaystyle\int_0^y D_k(y)\, dy} \qquad (2)$$

If the function $D_k(y)$ is known with a good precision and at several discrete points close to each other, Eq.(2) gives the solution of the problem with the best accuracy. In fact the only error affecting the results is the approximation of the formula utilized for the numerical integration. In the above assumptions this kind of error can be reduced to a negligible quantity. The direct integration technique possesses also a natural capability in damping the oscillations of the solution caused by possible noise present in input data. However, the direct integrating technique has the following disadvantages: it requires the knowledge of the temperature at the starting point of the integration ($y=0$) and it cannot use other boundary conditions such as for example the knowledge of the temperature far from the plate in the undisturbed region and/or some symmetry condition. If the input data are corrupted by noise, as always occurs in experimental measurements, the above technique is stable but unable to estimate the quality (variance) of the solution which has the same importance as the solution itself. Taking into account the above considerations the solution of the problem based on an iterative, inverse technique appears appropriate and it may be considered a valid alternative to the direct integration method.

Following the OLS formulation and denoting by $\boldsymbol{D}_m$ and $\boldsymbol{D}_c$ the measured and the calculated light rays displacement at a generic elevation x_k, the solution of the inverse problem requires the minimization of the functional :

$$S = [\, \boldsymbol{D}_m - \boldsymbol{D}_c(\boldsymbol{T})]^{\mathrm{T}} \cdot [\, \boldsymbol{D}_m - \boldsymbol{D}_c(\boldsymbol{T})] \longrightarrow \min \qquad (3)$$

where $\boldsymbol{T} = \{T_i\}$ ($i=1,2,..,N$) is the vector of the sought for temperatures which must satisfy Eq.(1). Depending on the kind of the convective experiment some constraints must be imposed to the vector $\boldsymbol{T}$; for example in the case of a plate at uniform temperature T_w, surrounded by a fluid at constant temperature T_∞ we have:

$$y = 0 \quad T = T_w$$
$$y > L \quad T = T_\infty \qquad (4)$$

where L is a point in the fluid sufficiently far from the plate (note that an optimum value for L can be selected by considering L as a further parameter to estimate).

Owing to the non linearity of Eq.(1), the minimization process is iterative and the successive approximations can be obtained with the same Gauss technique described in the previous section. The more delicate problem here is the choice of the parameterization for the unknown temperature function $T(y)$. In fact the selected approximation must be able to fit, with the greatest degree of accuracy, the typical temperature variation of the fluid within the thermal boundary layer.

Two kinds of parameterization are developed and tested: a local, second order, three points, polynomial approximation (moving parabola) and the cubic B-Splines approximation. Following the first kind of parameterization the temperature function is approximated by the moving parabola:

$$T = T_i + a(y-y_i) + b(y-y_i)^2 \qquad i = 2,3,..,N\text{-}1 \qquad (5)$$

where:

$$a = (T_{i+1}\, d_i^2 + T_i\,(d_{i+1}^2 - d_i^2) - T_{i-1}\, d_{i+1}^2)/c$$

$$b = (T_{i+1}\, d_i - T_i\,(d_{i+1} + d_i) + T_{i-1}\, d_{i+1})/c$$

$$c = d_i\, d_{i+1}(d_{i+1} + d_i)$$

$$d_i = y_i - y_{i-1}\,;\; d_{i+1} = y_{i+1} - y_i$$

Table 2 Comparison between calculated and exact temperature for laminar free convection; unperturbed data

η	Θ_{exact}	Θ_{spl}	Θ_{par}	Θ_{int}
0.00	1.0000	1.0000	1.0000	1.0000
0.40	0.7742	0.7749	0.7756	0.7751
0.80	0.5602	0.5606	0.5575	0.5629
1.30	0.3381	0.3343	0.3276	0.3376
1.60	0.2379	0.2308	0.2207	0.2341
2.00	0.1422	0.1352	0.1365	0.1381
2.40	0.0817	0.0762	0.0711	0.0769
2.80	0.0457	0.0400	0.0415	0.0402
3.20	0.0250	0.0204	0.0195	0.0192
5.50	0.0006	0.0000	0.0000	0.0000
$\Theta'(0)$	-0.5671	-0.5649	-0.5687	-0.5624

Proper expressions can be easily derived for the initial $i=1$ and last point $i=N$. The derivatives $T/\ y$ can be calculated from Eq.(5) with a central, second order accuracy formula and introduced in Eq.(1) in order to evaluate the light ray displacements D_c. Furthermore Eq.(5) allows one to easily calculate the local sensitivity coefficients and the Jacobian matrix. With the use of the cubic B-Splines approximation the temperature can be parameterized by the following transformations:

for $0 \leq y < 2y/6$

$$T = T_1+(T_2-T_1)u^3 \tag{6}$$

for $2y/6 \leq y < (4y_2 + y_3)/6$

$$T = [T_1(5\text{-}3u\text{-}3u^2+2u^3)+ + T_2(1+3u+3u^2\text{-}3u^3)+T_3u^3]/6 \tag{7}$$

for $(y_{i\text{-}3}+4y_{i\text{-}2}+y_{i\text{-}1})/6 \leq y < (y_{i\text{-}2}+4y_{i\text{-}1}+y_i)/6$

$$T = [T_{i\text{-}3}(1\text{-}3u+3u^2\text{-}u^3)+T_{i\text{-}2}(4\text{-}6u^2+3u^3)+ + T_{i\text{-}1}(1+3u+3u^2\text{-}3u^3)+T_iu^3]/6 \tag{8}$$

for $(y_{N\text{-}2}+4y_{N\text{-}1}+y_N)/6 \leq y < (y_{N\text{-}1}+5y_N)/6$

$$T = [T_{N\text{-}2}(1\text{-}3u+3u^2\text{-}u^3)+T_{N\text{-}1}(4\text{-}6u^2+3u^3)+ + T_N(1+3u+3u^2\text{-}2u^3)]/6 \tag{9}$$

for $(y_{N\text{-}1} + 5y_N)/6 \leq y < y_N/6$

$$T = [T_{N\text{-}1}(1\text{-}3u+3u^2\text{-}u^3)+T_N(5+3u\text{-}3u^2+u^3)]/6 \tag{10}$$

where $y_1 = 0$; $y_N = L$; $T_1 = T_w$; $T_N = T_\infty$; and u is the local variable $0 \leq u \leq 1$. Equations (6),(10) and (7),(9) are derived by imposing a triple vertices and a double vertices end condition respectively. As test case two typical laminar convective flows are considered: the natural convection along a vertical flat plate at uniform temperature,and the forced convection along a flat plate at zero incidence. For both kinds of convective flows the exact (semi-analytical) solution is available, the corresponding iso-deflection lines can be evaluated with Eq. (1) and used as simulated input data for verifying the inverse solution. In Tab.1 the results for free laminar convection along a vertical flat plate are reported while the sketch of the thermal boundary layer and the corresponding iso-deflection lines is shown in Fig.6. The exact solution Θ_{exact} (Ostrach, 1953) is compared with the reconstructed ones by using the cubic B-Splines approximation Θ_{spl} and the moving parabola Θ_{par}. The solution obtained by direct integration (Eq. (2)) Θ_{int} is reported too. Only a limited number of iso-deflection lines are utilized as it happens in real experiments.

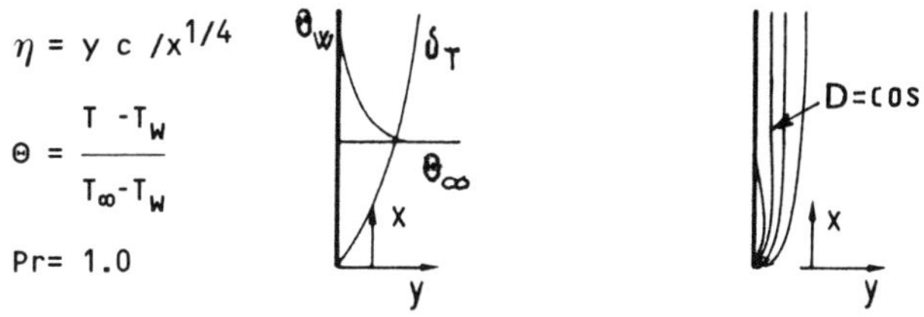

Fig. 6 Laminar free convection along a vertical plate; (a) thermal boundary layer, (b) Schlieren images

In Tab.2 and in Fig.7 the comparison is done for the laminar forced convection; the exact solution Θ_{exact} (Howart,1938) is compared, as in the previous case, with Θ_{spl} Θ_{par} and Θ_{int}. The results of Tab.1 and 2 show, as a whole, a good capability of the inverse technique in reconstructing the temperature field and the temperature gradient at the wall ($\Theta'(0)$) for both kinds of parameterization. The solution Θ_{spl} shows, on average, the better accuracy, comparable to that given by direct integration, but it requires a greater volume of calculation.

When the input data are perturbed by random errors the estimates of the inverse solution maintain an accuracy comparable, on average, to that of the direct integration as shown in Tab.3. The better accuracy for Θ_{spl} occurs far from the wall because, in this region, the inverse technique may take advantage of the second boundary condition. At the same time the confidence bounds give an estimate of the quality of the solution which is a vital information in the analysis of experimental data.

The inverse technique here described was proven effective in the reconstruction of temperature field from Schlieren images in buoyancy induced flows for vertical flat plate and for more complicated geometries such as vertical plate arrays (Devia, 1993).

An interesting development for future applications is the extension of the inverse technique to the reconstruction of a three-dimensional temperature field.

Table 3 Comparison between calculated and exact temperature for laminar forced convection; unperturbed data

η	Θ_{exact}	Θ_{spl}	Θ_{par}	Θ_{int}
0.00	0.0	0.0	0.0	0.0
0.80	0.2647	0.2644	0.2640	0.2639
1.60	0.5168	0.5168	0.5242	0.5134
2.40	0.7290	0.7228	0.7387	0.7228
3.20	0.8761	0.8763	0.8880	0.8692
4.00	0.9555	0.9554	0.9615	0.9503
4.80	0.9878	0.9877	0.9900	0.9847
5.60	0.9975	0.9975	0.9974	0.9957
7.20	0.9999	1.0000	1.0000	1.0000
$\Theta'(0)$	0.3321	0.3327	0.3329	0.3299

Table 4 Comparison between calculated and exact temperature for laminar forced convection; perturbed data $\sigma_p = 5\%\ D_{max}$

η	Θ_{exact}	Θ_{spl}	C.B.99%	Θ_{int}
0.00	0.0	0.0	0.0	0.0
0.80	0.2647	0.2615	±0.0115	0.2604
1.60	0.5168	0.5212	±0.0191	0.5138
2.40	0.7290	0.7409	±0.0190	0.7312
3.20	0.8761	0.8728	±0.0197	0.8742
4.00	0.9555	0.9443	±0.0189	0.9353
4.80	0.9878	0.9688	±0.0189	0.9575
5.60	0.9975	0.9928	±0.0159	0.9789
7.20	0.9999	1.0000	±0.0000	0.9828
$\Theta'(0)$	0.3321	0.3253	±0.0210	0.3247

ACKNOWLEDGEMENT

This work was supported by Italian MURST (40%)

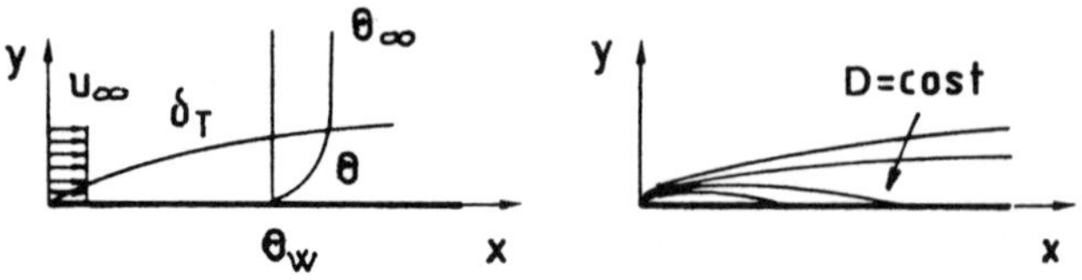

$\eta = y\ (u_\infty/\nu x)^{1/2}$; $\Theta = (T-T_w)/(T_\infty-T_w)$; Pr=1.0

Fig. 7 Laminar forced convection on a flat plate at zero incidence; (a) thermal boundary layer, (b) Schlieren images (iso-deflection lines)

REFERENCES

Bartolini, R., Scarpa, F., Milano, G., 1991, "Determination of Thermal Diffusivity of Fibrous Insulating Materials", *High Temperatures-High Pressures*, Pion Limited, London, Vol. 23, pp. 659-673.

Devia, F., Milano, G., and Tanda, G., 1993, "*Evaluation of Thermal Field in Buoyancy-Induced Flows By a Schlieren Method*" Paper accepted at the 3rd Worl Conference on Experimental Heat Transfer, Fluid Mechanics and Thermodynamics, Honolulu, November 1993.

Eckert, E.R.G., and Goldstein, R.J., 1976, *Measurements in Heat Transfer* , 2nd ed., McGraw-Hill, New York.

Howarth,L., 1938, "*On the Solution of the Laminar Boundary Layer Equations* ", Proc. Roy. Soc. London, 164A, pp.547-579.

Levenberg, K., 1944, "A Method for the Solution of Certain Non-Linear Problems in Least Squares", Quart. Appl. Math., Vol 2, N. 2, pp. 164-168.

Milano, G., and Scarpa, F., 1991, "Numerical Experiments on Thermophysical Properties Identification from Transient Temperature Data ", *Proceedings of the 4th Annual Inverse Problems in Engineering Seminar*, Michigan State University, Lansing.

Milano, G., Scarpa, F., and Bartolini, R., 1991, "Identification of Temperature Dependent Thermophysical Properties of Insulating Materials", *Proceedings of the 18th Int. Congress of Refrigeration* , International Institute of Refrigeration, Montreal, Vol. 2, pp.712-716.

Ostrach, S., 1953, "*An Analysis of Laminar Free-Convection Flow and Heat Transfer About a Plate Parallel to the Direction of the Generating Body Force*" NACA Report 1111.

Pescetti, D., Castelli,P. ,and Parodi, M., 1985, "Computer Simulation of a Central Force Apparatus" *Conference on Physics Education*, GIREP/SVO/UNESCO, W.C.C.-Utrecht.

Quine,R.W., Rinard, G., 1992, " Design of magnetic-field gradient coils for imaging", *Computers in Physics*, Vol.6, N.6.

Scarpa, F., Bartolini R., and Milano, G., 1991, "State Space (Kalman) Estimator in the Reconstruction of Thermal Diffusivity From Noisy Temperature Measurements", *High Temperatures-High Pressures*, Pion Limited, London, Vol.23, pp. 633-642.

Scarpa, F., Milano, G., and Pescetti, D., 1993, "Thermophysical Properties Estimation From Transient Data: Kalman Versus Gauss Approach", *First International Conference On Inverse Problems in Engineering: Theory and Practice*, Palm Coast, Florida.

Scarpa, F., and Milano, G., 1993, "Identification of Temperature Dependent Thermophysical Properties from Transient Data: a Kalman Filtering Approach", Internal report, Dipartimento di Ingegneria Energetica, Università di Genova, Genova, pp. 1-16.

Vasil'ev, L.A., 1971, *Schlieren Methods*, Keter Inc., New York.

Inverse Problems in Engineering: Theory and Practice
ASME 1993

SOLUTION OF AN INVERSE HEAT CONDUCTION PROBLEM USING STRAIN GAGE MEASUREMENTS

M. Raynaud and G. Blanc
Centre de Thermique (CETHIL)
Institut National de Sciences Appliquées de Lyon
Villeurbanne, France

T. H. Chau
Direction des Etudes et Recherches – REME
Electricité de France
Saint-Denis, France

ABSTRACT

The inverse heat conduction problem (IHCP) is commonly referred as the problem of determining unknown thermal boundary conditions from remote temperature measurements. However other types of measurements may be used provided their sensitivity to the thermal field is large enough. This work focuses on the utilisation of thermal stress measurements for solving the inverse heat conduction problem. The objective being to compare the quality of the information given by thermal and by strain sensors.

The case of a cylindrical tube is considered. The internal boundary condition is unknown and must be estimated from sensors placed at the outer surface only. The method used for solving the IHCP from stress measurements is developped first. The method is then applied to experimental measurements. The data are examined in details to distinguish the noise from the true signal. Results given by thermal or/and strain sensors are compared. The comparison of the estimated and measured internal temperature show that, for the case of interest, much better estimates can be obtained with strain gages than with thermocouples.

NOMENCLATURE

E	Young modulus
N	total number of spatial node
Q	sensitivity coefficient
T	temperature
a	inner tube radius
b	outer tube radius
q	inner surface heat flux density
r	radius
t	time
w	weight

subscripts

i	spatial node (1 for inner surface, N for outer surface)
m	measured value
s	thermal stress
t	temperature

superscripts

n	temporal node

greeks

α	coefficient of thermal expansion
λ	thermal conductivity
ν	Poisson coefficient
σ	thermal stress
ρ	density

INTRODUCTION

The temperature fluctuations at the cold-hot interface of a stratified flow can induce in the long run a thermal striping of the surface materials. Such problems are encountered on some PWR pipes. The prevision of the thermal fatigue requires the knowledge of the tube temperature field at any time which is usually calculated from the temperature variations at the inner and outer radius. Since it is often difficult to predict the fluid temperature variations, and it is, for security reason, impossible to place sensors on the tube inner face, the thermal stresses cannot be calculated directly. Thus, the pipe thermal field can only be determined through the solution of an inverse heat conduction problem (IHCP).

The IHCP is commonly referred as the problem of determining unknown thermal boundary conditions from remote temperature measurements. However, it is now well known (Beck et al., 1985, Hensel 1990) that there are cases where it is impossible to infer the boundary conditions. This is due to the lack of sensitivity of the measurements to the parameter or function to be determined. In such cases, it is necessary to find other propeties that can be measured and which depend on the thermal field. These measurements may be used to determined the unknown thermal condition provided their variations with the thermal field are large enough.

One can imagine for instance to use ultrasonic measurements since the speed of sound varies with temperature. This work focuses on the utilization of thermal strain measurements for solving the inverse heat conduction problem. Such an approach has been used by several researchers. Cialkowski and Grysa (1980) and Grysa et al. (1981) studied a one-dimensional uniaxial strain problem. They used an analytical approximate method for solving the IHCP. The case of a long cylinder has been adressed by Noda (1989) and by Morilhat et al. (1992). Their solution procedures are based on the Laplace transform and on the Fourier transform, respectively. All of these methods are limited to linear problems.

In this paper a method that can be applied to nonlinear problem is developped. The data measured on an experimental set-up (Morilhat et al., 1992) which was designed to study the faisability of such an approach are used to compare the quality of the information given by thermal and by strain sensors.

The inverse methods is briefly described. Then the data are analysed and used for the inversion. The respective advantages/disavantages of the two measurement types are discussed and condusions are given.

DESCRIPTION OF THE PROBLEM

A 0.3636 m inner diameter and 21.4 mm thick stainless steel round tube is studied, figure 1. The thermal and mechanical properties of the material are known : $\lambda = 50.3$ $W.m^{-1}.K^{-1}$, $\rho c = 3976285$ $J.m^{-3}$, $E = 210\,000$ MPa , $\alpha = 11{,}87\ 10^{-6}\ K^{-1}$ and $\nu = 0.3$. The tube is isolated on its outer face with 50 mm of mineral fiber. Thermocouples have been placed on the inner and outer tube surface as well as a gage for measuring the axial stress, σ_{zz}, on the external face.

DESCRIPTION OF THE METHOD

The objective being to use as much information as possible to estimate the internal temperature, a method which could treat the thermocouple and strain gage measurements simultaneoulsly has been sought. Although the method is herein applied to the determination of surface temperature for a linear axysimmetric problem, it should not be limited to such configuration. These considerations have leaded us to develop a procedure derived from Beck's method.

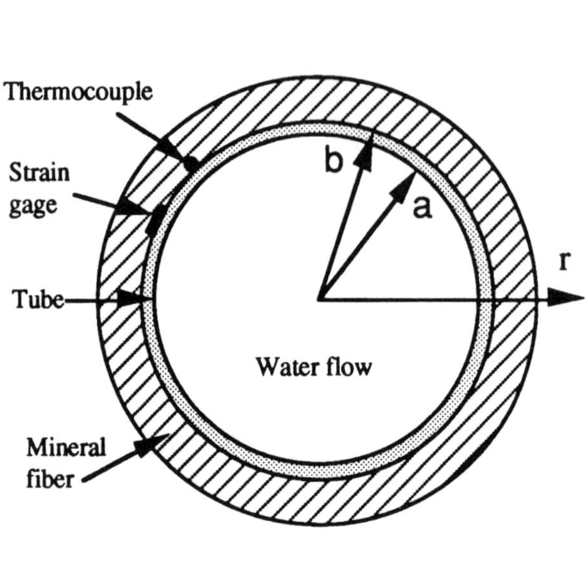

FIGURE 1 : PROBLEM GEOMETRY

Beck's method for the solution of the IHCP is a sequential procedure in which the unknown heat flux is determined step by step (since it is more difficult to is sought directly) . For q^n known, the variation Δq from q^{n+1} to q^n is obtain by minimizing with respect to q^{n+1} the square of the difference between the measured and the calculated temperatures over nt time steps :

$$F_t(q^{n+1}) = \sum_{j=1}^{nt} \left(T_m^{n+j} - T_N^{n+j}\right)^2 \quad (1)$$

which leads (Beck, 1985) to :

$$\Delta q = \frac{\sum_{j=1}^{nt} \left(T_m^{n+j} - T_N^{n+j}\right) Q_t^{n+j}}{\sum_{j=1}^{nt} \left[Q_t^{n+j}\right]^2} \quad (2)$$

with

$$Q_t^{n+j} = \frac{\partial T_N^{n+j}}{\partial q^{n+1}} \quad (3)$$

Similarly the new functionnal :

$$F_s(q^{n+1}) = \sum_{j=1}^{ns} \left(\sigma_m^{n+j} - \sigma_N^{n+j}\right)^2 \qquad (4)$$

is defined. The thermal axial stress, on the external face, due to a radial thermal gradient is :

$$\sigma_N^n = P\left[I^n - T_N^n \right] \qquad (5)$$

with

$$I^n = I_o \int_a^b r\, T(r,t)\, dr \qquad (6)$$

$$I_o = \frac{2\nu}{b^2 - a^2} \qquad (7)$$

$$P = \frac{E\alpha}{1-\nu} \qquad (8)$$

The integral of Eq. (6) can be approximated with the trapezoïdal rule :

$$\int_a^b rT(r,t)dr = \Delta r \sum_{i=2}^{N-1} (a+(i-1)\Delta r)T_i + \frac{\Delta r}{2}\left[aT_1 + (a+(N-1)\Delta r)T_N\right] \qquad (9)$$

By subtituing Eqs. (6) to (9) into Eq. (4), a functionnal that can be derived with respect to q^{n+1} is obtained. Then Beck's procedure is used to get the expression of the desired correction for the heat flux :

$$\Delta q = \frac{\sum_{j=1}^{ns} \left(\frac{\sigma_m^{n+j}}{P} - \sigma_N^{n+j}\right) Q_s^{n+j}}{\sum_{j=1}^{ns} \left[Q_s^{n+j}\right]^2} \qquad (10)$$

with

$$\sigma_N^{n+j} = T_N^{n+j} - 2I_o\Delta r \sum_{i=2}^{N-1} (a+(i-1)\Delta r)T_i^{n+j} - I_o\Delta r\left[aT_1^{n+j} + (a+(N-1)\Delta r)T_N^{n+j}\right] \qquad (11)$$

and

$$Q_s^{n+j} = \frac{\partial \sigma_N^{n+j}}{\partial q^{n+1}} \qquad (12)$$

Equation (5) along with the trapezoidal rule can be used to relate the stress sensitivity coefficient to the temperature sensitivity coefficient :

$$Q_s^{n+j} = Q_N^{n+j} - 2I_o\Delta r \sum_{i=2}^{N-1} (a+(i-1)\Delta r)Q_i^{n+j} - I_o\Delta r\left[aQ_1^{n+j} + (a+(N-1)\Delta r)Q_N^{n+j}\right] \qquad (13)$$

For a linear problem, the sensitivity coefficients, Eq. (12), are constant and, like for the classical IHCP, no iteration are needed. Once q^{n+1} is determined, the whole temperature field at t^{n+1} can be calculated and the procedure is repeated for the next time step.

Since the functionnals F_t and F_s have different units (the latter being larger by several order of magnitude), a weighted functionnal is used for taking simultaneously into account the information given by the two sensor types:

$$F_{st}(q^{n+1}) = w_t F_t + \frac{w_s}{P^2} F_s \qquad (14)$$

The minimization procedure leads to :

$$\Delta q = \frac{w_t N_2 + \frac{w_s}{P^2} N_{10}}{w_t D_2 + w_s D_{10}} \qquad (15)$$

where N_i and D_i refer to the numerator and denominator of equations 2 and 10, respectively.

The thermal and stress sensitivity coefficients are plotted figure 2 for ease of comparison. It shows that at the measurement location the stress is more sensitive to the heat flux than the temperature. Thus the estimation of the internal surface temperature will be easier (provided that both type of measurements are accurate) with the strain gage. The radial stress $\sigma_{\theta\theta}$ is calculated from the equations (5, 6 and 8) with $I_o = \nu/(b^2 - a^2)$. Its sensitivity coefficients being the largest, figure 2, it would have been preferable to measure the radial stress instead of the axial one. The difference between the sensitivity coefficients is so large that $N_2 << N_{10}$ and for equal functional weights, the influence of the temperature measurements is negligible. If the confidence in both types of measurement is similar, then the weights ratio can be chosen so that :

EXPERIMENTAL RESULTS

Inversion from the temperatures

The tube temperature field is calculated with a pure implicit finite differences method, the thickness being divided into 30 equal spaces (N=31). The internal temperature determined from the filtered external temperature measurements are shown figure 6. The dimensionless time step characteristic of 1-D IHCP (Beck, 1984) problem being small (0.007), a large number of future time temperatures, nt=25, must be used to obtained appropriate results. The agreement between the estimated and measured internal temperatures is good excepted for the early times. The difference for t<15 s is due to the unknown initial temperature distribution. The fluctuations around 20 and 44 s are caused by the amplification of the changes of slope (which were not completely removed by the filtering procedure) of the external temperature around 24 and 48 s.

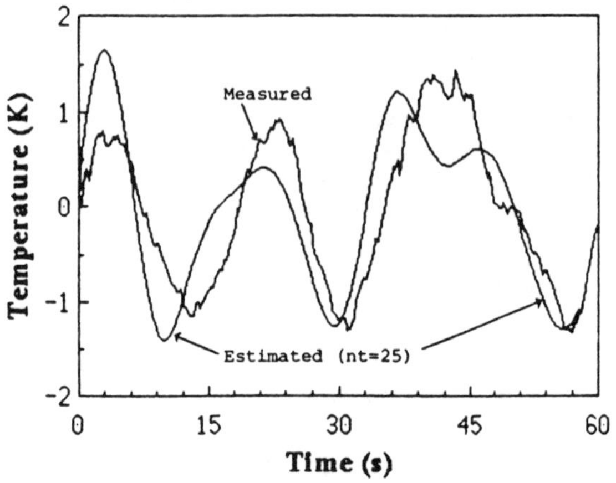

FIGURE 6 : INVERSION WITH THE TEMPERATURE MEASUREMENTS

Inversion from the thermal stresses

A finer spatial mesh being necessary to ensure a good approximation of the integral of Eq. (9) with the trapezoidal rule, the direct problem is solved with 51 nodes. The estimated internal temperature is plotted figure 7. The results are worse, even then the sensitivity coefficients are larger, than the one obtained from the temperature. This is due to the noise level which is too large and which cannot be completely eliminated. However the temperature measurement have also been used to reset the mean tube temperature. As a matter of fact, the thermal stress for linear problem depends only on the temperature difference, Eq. (5), and not on the temperature level. Consequently, the temperature level cannot be assesed with the strain gage measurements only.

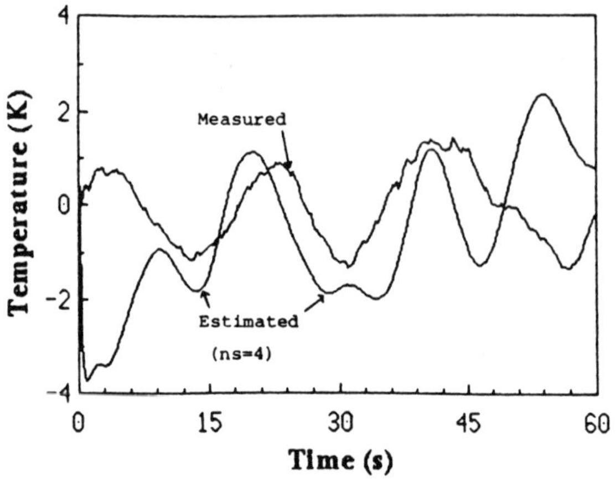

FIGURE 7 : INVERSION WITH THE THERMAL STRESS MEASUREMENTS

Inversion from the temperatures and thermal stresses

In order to show the faisability of the method, the internal temperature variations have been estimated by using simultaneously the temperature and stress measurements. For this case, the signal delivered by the strain gage has been treated as if it was the radial stress and not the axial stress and has not been filtered. The results, presented in figure 8, show that the inversion is much more stable and that the computationnal time is smaller than when the temperature measurements are used alone since nt=ns=4 are small compared to 25. The weights ratio, Eq. 16, is equal to 310. The agreement with the internal measurements is poor since, as figure 5 showed, the two stresses differ in both amplitude and phase.

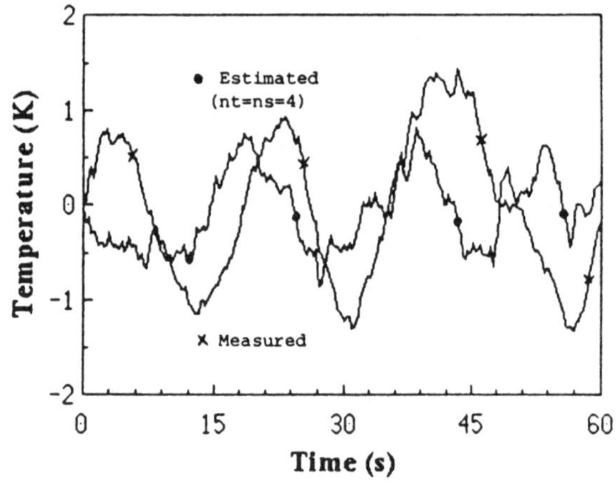

FIGURE 8 : INVERSION WITH THE THERMAL STRESS MEASUREMENTS

$$\frac{W_t}{W_s} = \frac{Q_s^{n+ns}}{Q_t^{n+nt}} \qquad (16)$$

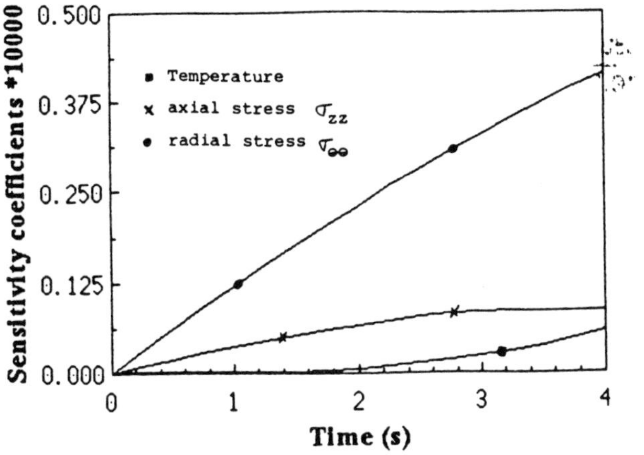

FIGURE 2 : SENSITIVITY COEFFICIENTS

EXPERIMENTAL DATA

During the experiment the mean tube temperature decreased from 160 to 60°C over a one hour period. Since the study focuses on the frequency fluctuations of the order of an Hertz, the temperatures variations - around the mean temperature - at the inner and outer radius are shown figure 3 over a 64 s interval instead of the variations over the whole time. The temperatures (and stress) were measured with a 4 Hz frequency. It can be seen that the signals are quite noisy. If the external temperature is used, as it is, in the inverse method then it leads to enormous and unphysical oscillations.

The direct problem with the internal temperature measurements as inner boundary condition and an adiabatic condition at the outer radius has been solved. The temperature calculated at the outer radius is drawn figure 4 and clearly shows the measurement noise. A gaussian and a square smoothing filters (Raynaud, 1986) have then been applied on the data to remove the higher frequency fluctuations. The smoothed data fits much better with the calculated one, figure 3. The difference between the two at the early time is caused by the initial temperature distribution which is unknown and has arbitrarily been set to zero.

The axial stress measured on the external face is plotted figure 5 along with the axial and radial stresses calculated from the solution of the direct problem. The amount of noise is very large. It may be caused by mechanical stress on the tube assembly. This higher frequency noise can also be partially removed with a smoothing filter. This figure shows that the amplitude of the radial stress fluctuations are larger than the ones of the axial stress and that there are lagged in time.

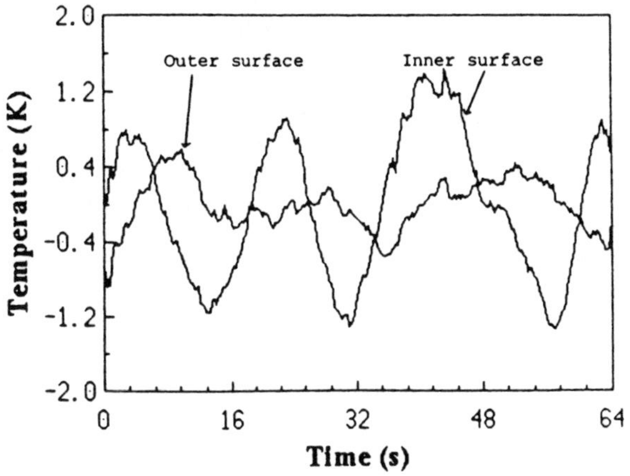

FIGURE 3 : MEASURED TEMPERATURES

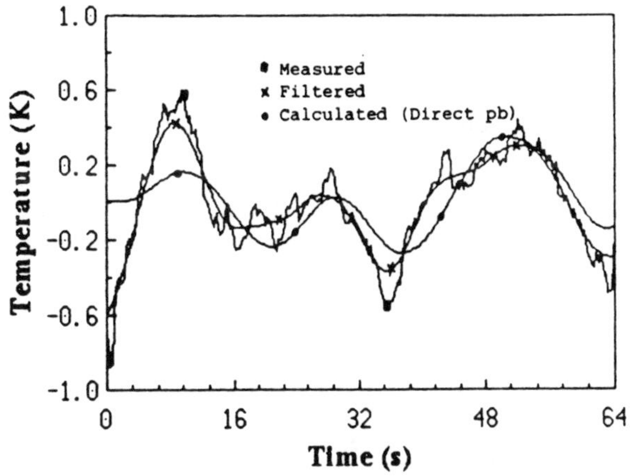

FIGURE 4 : SMOOTHED TEMPERATURES

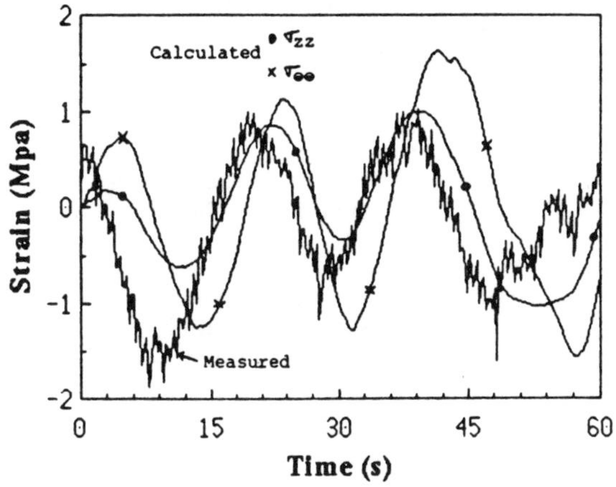

FIGURE 5 : EXTERNAL THERMAL STRESSES

CONCLUSIONS

This study has shown that the external thermal stress of a long tube is more sensitive to the internal temperature variations than the external temperature. But the mechanical stress on the tube induce a "noise" in the thermal stress measurements and lead to worse results when the inversion is performed from the axial stress measurements than from the temperature measurements. On the other hand, it has been shown that the internal temperature could have been better determined if the radial stress had been measured instead of the axial one. However, it has been shown that the temperature measurements are necessary for determining the mean tube temperature.

The experimental set-up has been designed to obtain an axisymmetric problem. The next step of this work is to consider a two-dimensionnal problem, T(r,q,t). The number of thermal and strain sensors and their locations will have to be determined as well as the conditions under which the thermal stresses can be detected from the mechanical ones.

REFERENCES

Beck, J.V., Blackwell, B., St. Clair, C.R., 1985, "Inverse heat conduction, ill-posed problem," *Wiley Interscience*, New-York.

Cialkowski, M.J., Grysa, K.W., 1980, "On a certain inverse problem of temperature and thermal stress fields," *Acta Mechanica*, Vol. 36, pp. 169-185.

Grysa, K.W.Cialkowski, M.J., Kaminski, H., 1981, "An inverse temperature field problem of the theory of thermal stresses," *Nuclear Eng. Design*, Vol. 64, pp. 169-184.

Hensel, E., 1990, "Inverse theory and applications for engineers," *Prentice Hall,* Englewood Cliffs.

Morilhat, P., Maye, J.P., Brendle, E., Hay, B., 1992, "Résolution d'un problème thermique inverse par méthodes analytique et impulsionnelle. Application à la détermination des fluctuations thermiques parétiales à l'intérieur de la conduite," *Int. J. Heat Mass Transfer*, Vol. 35, pp. 1377-1383.

Noda, N., 1989, "An inverse problem of coupled thermal stress in a long circular cylinder," JSME Int. Journal, Vol. 32, pp. 348-354.

Raynaud, M., 1986, "Combination of methods for the inverse heat conduction problem with smoothing filters ," *AIAA/ASME paper, 4th Thermophysics and Heat Transfer Conf.*, Boston.

Inverse Problems in Engineering: Theory and Practice
ASME 1993

A GENERAL OPTIMIZATION ALGORITHM TO SOLVE 2-D BOUNDARY INVERSE HEAT CONDUCTION PROBLEMS USING FINITE ELEMENTS

B. Truffart, Y. Jarny, and D. Delaunay
Laboratoire de Thermocinétique
ISITEM
Université de Nantes
Nantes, France

ABSTRACT

A general optimization algorithm to solve two-dimensionnal Boundary Inverse Heat Conduction Problems is presented. The solution of the direct problem is approximated using Finite Elements over any complex domain and for any configuration of materials. Different formulations are considered to determine the unknown functions on some parts of the boundary, but an unified approach based on gradient methods is used to minimize a regularized residual functionnal. The numerical code is built using standard modules of MODULEF library, for mesh building and for graphic data processing.

INTRODUCTION

Importance of Inverse Methods in Heat Transfer for mathematical modeling, thermal design and optimization of engineering systems has been largely analyzed in a recent american-russian workshop [1]. Development of these methods is based on progresss in mathematical theory of ill-posed problems and on advanced computional methods and computer technology. One of the recommendations of the workshop concerns the development of methods and algorithms for multi-dimensional Inverse Heat Conduction Problems. Many applications require estimation of thermophysical properties or boundary functions in two or three-dimensional spaces and cannot be studied with the 1-D approach. In this paper we are interested to show how well developped methods for one-dimensional problems can be extended to higher dimensions. Due to the complexity of these problems we consider here only stationary cases.

Hsu [2] (1992) has developped a 2-D finite element code based on the Beck's method, boundary conditions are computed with time sequential algorithm using sensitivity equations. Hensel & Hills [3] (1989), Pasquetti & Le Niliot [4] (1991), Maillet et al. [5] (1991) follow the same approach but use different ways to solve the direct problem: finite differences in [3] and boundary elements in [4], [5].Thanks to linearity of the model, the determination of the unknown boundary conditions is carried on by using explicit relationship between these unknowns and temperature measurements, regularization is needed to deal with ill-posedness.

The algorithm presented here is based on the extension for two-dimensional geometries of a well known methodology used for one-dimensional cases. The main features of the method are described in Jarny [6], (1991). Unknown boundary conditions are determined by an iterative method which consists in minimizing a regularized functional. At each step, direct problem is solved with finite element approximation, and the use of adjoint equations allows conjugate gradient minimization of the criterion.

The Finite Element Library MODULEF [7] has been chosen to deal with general shape of the geometrical domain and with any complex configuration of materials. This choice allows the use of several standard MODULEF programs developped for two- (or three-) dimensional data processing purposes. Numerical results are presented.Influence of sensor locations, measurement accuracy and regularization term are evaluated on an example studied by Maillet [5].

THE DIRECT HEAT CONDUCTION PROBLEM

The continuous model

We consider a domain Ω, and in order to describe different kinds of boundary conditions, the boundary Γ is divided in three parts Γ_s, each of them can be subdivided in several parts $\Gamma_{s,j}$, $j = 1, \dots, N_s$, $s = 1, 2, 3$ as it is shown on the figure 1. To simplify the presentation we assume $N_s = 1, \forall s$.

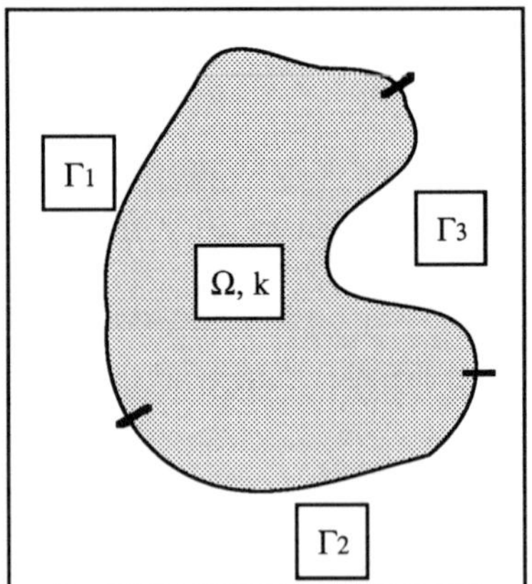

Fig.1 Direct Problem

At any point x of $\Omega \cup \Gamma$ the temperature field T(x) is solution of the elliptic equation governing the steady-state heat conduction in an anisotropic solid:

$$\nabla(\mathbf{k}\ \nabla T) + f_\Omega = 0 \text{ in } \Omega \qquad (1)$$

$$T = T_p \quad \text{on } \Gamma_1 \qquad (2.1)$$

$$\mathbf{k}\ \nabla T \cdot \mathbf{n} + q = 0 \quad \text{on } \Gamma_2 \qquad (2.2)$$

$$\mathbf{k}\ \nabla T \cdot \mathbf{n} + h\,T = h \cdot T_e \quad \text{on } \Gamma_3 \qquad (2.3)$$

Where

k is the symetric thermal conductivity tensor,
f_Ω is a general source term,
T_p is the temperature imposed on Γ_1,
q is the surface heat flux density leaving Ω through Γ_2,
h is the heat transfer coefficient on Γ_3,
(All this parameters are spatially varying.),
n is the outward normal to the boundary Γ.

The variational formulation

Let us introduce V, a space of real functions defined on $\Omega \cup \Gamma$ and consider the subspace V_1 of V, such as:

$$V_1 = \left\{ v \in V \ / \ v = 0 \text{ on } \Gamma_1 \right\} \qquad (3)$$

Introduce the bilinear form a(T,v) and the linear form L(v) :

$$a(T,v) = \int_\Omega \mathbf{k}\ \nabla T\ \nabla v\ dx + \int_{\Gamma_3} h\ T\ v\ ds \qquad (4.1)$$

$$L(v) = \int_\Omega f_\Omega\ v\ dx - \int_{\Gamma_2} q\ v\ ds + \int_{\Gamma_3} h\ T_e\ v\ ds \qquad (4.2)$$

Then the direct heat conduction problem consists to find $T \in V$ such as :

$$a(T,v) = L(v) \ \text{ in } \Omega \cup \Gamma, \forall v \in V_1 \qquad (5.1)$$

$$T = T_p \quad \text{on } \Gamma_1 \qquad (5.2)$$

In this formulation two kinds of boundary conditions are clearly distinguished: "natural" conditions on Γ_2 and Γ_3 which are included in the bilinear form a(.,.) and the linear form L(.), and "fixed" or Dirichlet conditions on Γ_1.

The Finite Element approximation

Finite element methods can be derived from this variational formulation. They consist in searching for an approximated solution denoted T_h in a finite dimensional space V.In this paper we have considered Lagrange Finite Elements P_1 and triangular geometric elements K.So the space in which the approximated solution T_h is searched is defined by :

$$V_h = \left\{ v_h \in C^\circ(\Omega_h),\ \forall\ K,\ v_{h|K} \in P^1(K) \right\} \qquad (6)$$

Then the discret heat conduction problem is :

$$\left[\begin{array}{l} \text{Find } T_h \in V_h, \text{ solution of} \\ a(T_h, v_h) = L(v_h) \qquad \forall\ v_h \in V_{1h} \end{array} \right. \qquad (7)$$

$$\text{with } V_{1h} = \left\{ v_h \in V_h \ / \ v_{h|\Gamma_1} = 0 \right\} \qquad (8)$$

Note: The subscript "h" is relative to a "length" of the geometric elements K. The main result of the finite element approximation of T is the uniqueness of the solution T_h and convergence, that is $\| T_h - T \|_V \to 0$ when $h \to 0$.

Let us denote:
NSG = the set of the nodes of the mesh on Ω
$NS\Gamma_i$ = the set of the nodes of the mesh on Γ_i, i = 1 to 3
NSO = NSG - $NS\Gamma_1$ = the set of free nodes
$\{ p_i \in V_h\ ,\ i \in NSG \}$ a polynomial basis of V_h

Then any function $v \in V$ is approximated by $v_h \in V_h$,and

$$v_h(x) = \sum_{i=1}^{NSG} v_{ih}\ p_i(x) \qquad (9)$$

In such a way, the direct problem can be written in a matrix form as follow:

Find $T_h \in V_h$ such as, $\forall m \in NSO$:

$$\sum_{n\in NSO}\left(A_{mn}+B_{mn}\right)T_{hn} = - \sum_{n\in NSG\text{-}NSO} A_{mn}T_{hn}+C_m+D_m+E_m \quad (10)$$

With: $$A_{mn}=\int_{\Omega} \mathbf{k}(X)\,\nabla.p_m\,.\,\nabla.p_n\,d\Omega \quad (11.1)$$

$$B_{mn}=\int_{\Gamma_3} h(\Gamma)\,p_m\,p_n d\Gamma \quad (11.2)$$

$$C_m=\int_{\Omega} f_\Omega(X)\,p_m\,d\Omega \quad (11.3)$$

$$D_m=-\int_{\Gamma_2} q(\Gamma)\,p_m\,d\Gamma \quad (11.4)$$

$$E_m=T_e\int_{\Gamma_3} h(\Gamma)\,p_m\,d\Gamma \quad (11.5)$$

To simplify our notation in the following, the subscript "h" used to denote the elements of the finite space V_h is dropped , so the approximated solution will be denoted T(x) instead of $T_h(x)$.

THE BOUNDARY INVERSE HEAT CONDUCTION PROBLEM

Statement of the Inverse Problem

The goal of the inverse analysis is the determination of unknown functions defined on a new part Γ_u of the boundary Γ as it is shown on the Fig.2. Depending on the model chosen to describe the boundary condition on this part, the unknown to be determined is temperature, heat flux or heat transfer coefficient. The determination is based on the following assumptions:

- the geometry of the boundary Γ_u is known
- additional temperature measurement Y_i are available in the domain at specified locations $X_i \in \Omega$, i= 1, ..., N_t .

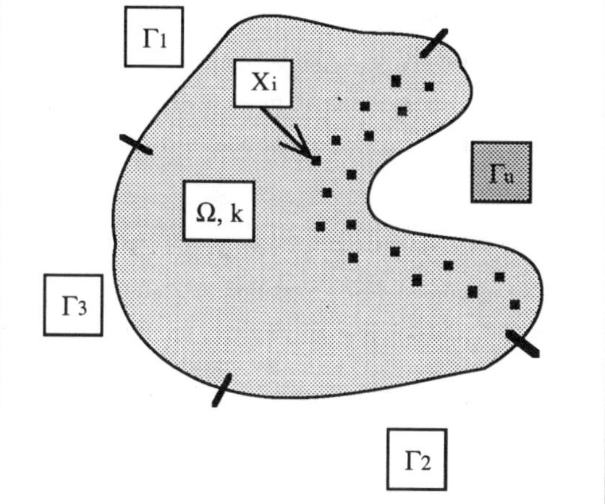

Fig.2 Inverse Problem

Three cases are considered:

Cases 1, 2 & 3 :

$$\nabla(\mathbf{k}\,\nabla T) + f_\Omega = 0 \text{ in } \Omega \quad (12)$$

$$T = T_p \quad \text{on } \Gamma_1 \quad (13.1)$$

$$\mathbf{k}\,\nabla T\,.\,\mathbf{n} + q = 0 \quad \text{on } \Gamma_2 \quad (13.2)$$

$$\mathbf{k}\,\nabla T\,.\,\mathbf{n} + h\,T = h\,.\,T_e \quad \text{on } \Gamma_3 \quad (13.3)$$

Case 1: $$T = T_u \text{ unknown on } \Gamma_u \quad (14.1)$$

Case 2: $$q_u = -\mathbf{k}\,\nabla T\,.\,\mathbf{n} \text{ unknown on } \Gamma_u \quad (14.2)$$

Case 3: $$\mathbf{k}\,\nabla T\,.\,\mathbf{n} + h_u\left(T - T_e\right) = 0\,,$$

$$h_u \text{ unknown on } \Gamma_u \quad (14.3)$$

Let us denote $Z_u(s)$ the unknown function to be determined on the boundary Γ_u , U = a set of real functions defined on Γ_u , U_{ad} a subset of admissible functions.

We assume that for any $Z \in U_{ad}$, the direct problem has an unique solution denoted $T(Z) \in V$ which is calculated by (10). Then the inverse problem aims to get $Z_u \in U_{ad}$ such as:

$$T(X_i;Z_u) = Y_i \qquad \text{for } i=1,..,N_t \quad (15)$$

Because of measurement and modeling errors, the problem is solved in the least-square sense and consists to approximate Z_u by searching for $Z^*_\alpha \in U_{ad}$ which minimizes the criterion J(Z):

$$J(Z) = \sum_{i=1}^{N_t} \int_{\Omega} \omega_i\left(T(x;Z)-Y_i\right)^2 dx + \alpha \int_{\Gamma_u}\left(Z(s)-Z_{hest}(s)\right)^2 ds \quad (16)$$

Where Z_{est} is an apriori estimation of Z_u

α is a regularization parameter

$$\omega_i(x) - \begin{cases} \neq 0 \text{ for x at neighbourrhoud of } X_i \\ = 0 \text{ elsewhere} \end{cases}$$

$$\text{with } \int_{\Omega}\omega_i(x)dx = 1$$

The left part of J is called the residual measurement and the right part the regularization term. Depending on the desired regularity of the function Z to determine, additionnal regularization term can be added.

Correctness of the optimization problem

Let us consider separately cases 1&2 from case 3.

If the direct problem is linear, then $\overline{T}(Z)$= [T(Z)-T(0)] is a linear application from the space U to the space V, and the criterion J can be written as:

$$J(Z) = \Pi(Z,Z) + 2\,l\,(Z) + C \quad (17)$$

$$\Pi(Z,Z)= \left[\sum_{i=1}^{N_t} \int_{\Omega} \omega_i \; \overline{T}^2(Z)dx + \alpha \int_{\Gamma_u} Z^2 ds\right] \qquad (18.1)$$

$$l(Z) = \left[\sum_{i=1}^{N_t} \int_{\Omega} \omega_i \; \overline{T}(Z) \left[T(0)-Y_i\right]dx + \alpha \int_{\Gamma_u} Z\text{-}Z_{est} ds\right] \qquad (18.2)$$

$$C = \sum_{i=1}^{N_t} \int_{\Omega} \omega_i [T(0)-Y_i]^2 dx + \alpha \int_{\Gamma_u} Z^2_{est} ds \qquad (18.3)$$

From the linearity of $\overline{T}$ with respect to Z, J(Z) is a quadratic functional, $\Pi(.,.)$ is a bilinear form, l(.) is linear and C is constant.

By taking α strickly positive, then J(Z) is strictly convex and there exists an unique solution Z^*_α to the optimization problem; this solution is characterized (for $U_{ad} = U$) by

$$\delta J = \left(\nabla J\,(Z^*_\alpha)\,,\, \delta Z\right)_U = 0 \qquad \forall\, \delta Z \qquad (19)$$

Where ∇J is the gradient of J.

In practice the difficulty will be to choose α in the regularization term. As it will be shown, the choice depends on the "quality" of the available data : number and location of sensors, accuracy measurement and on the desired quality of the unknown function Z_u to determine. In practice the space U is finite dimensional and any function Z is approximated by:

$$Z(s) = \sum_{i=1}^{NR} Z_i\; \sigma_i(s) \qquad (20)$$

Where $\left\{\sigma_i\,,\, i = 1 \text{ to } NR\right\}$ is a set of basis functions defined on Γ_u. The "quality" of Z can be specified by the dimension NR and by the "regularity" of the chosen basis.

For case 3, $\overline{T}(Z)$ is not linear, J(Z) is not quadratic and uniqueness of Z^*_α is not ensured.

Some properties of the solution Z^*_α can be underlined:

- by taking $\alpha > 0$, the criterion J is not identical to the residual measurement, so even without error a bias will be introduced in the estimation of the unknown function Z_u by Z^*_α.

- depending on the formulation of the optimization problem, different estimation of the unknown funtion can be obtained:

For case 1, when $T_u = Z^*_\alpha$ has been determined, resolution of the direct problem gives the heat flux q_{u1} on the boundary Γ_u, and the heat transfer coefficient h_{u1} can be derived from the relation ship: $q_{u1} = h_{u1}\,(Z^*_\alpha - T_e)$

For case 2 the knowledge of $q_u = Z^*_\alpha$ and resolution of the direct problem will give the temperature T_{u2} and the coefficient h_{u2} on Γ_u.

For case 3 , knowing $h_u = Z^*_\alpha$, the direct problem allows to calculate temperature T_{u3} and heat flux q_{u3} on Γ_u.

Some numerical results are presented to illustrate these different formulations. The question of knowing if there is a formulation better than the others has not been answered. But two advantages of case 2 can be recalled: for linear problem, the criterion J(Z) is quadratic and in the variationnal formulation, boundary condition of the second kind is "natural", which makes the computation of the gradient much easier.

MINIMIZATION OF THE CRITERION

The Descent method

To reach the optimal solution $Z^*_\alpha \in U_{ad}$, an iterative descent method is used:

$$\underline{Z}^{n+1} = \underline{Z}^n + \rho^n.\underline{d}^n \qquad (21)$$

Where $\underline{Z}^n = \left\{Z^n_1, \ldots, Z^n_{NR}\right\}$ is defined according to (20)

$\underline{d}^n$ is the descent direction relative to the same basis

$\rho^n > 0$ is the descent depth

n is the current step of iteration

Each step consists to compute $\underline{d}^n$ and ρ^n in order to have:

$$J(\underline{Z}^{n+1}) < J(\underline{Z}^n)\,,\; \forall\; n \qquad (22.1)$$

$$\| \underline{Z}^n - \underline{Z}^*_\alpha \|_U \to 0 \text{ when } n \to \infty \qquad (22.2)$$

Computation of the direction $\underline{d}^n$ is based on the gradient ∇J of the criterion which can be derived from an adjoint equation coupled to the direct heat conduction problem.

Computation of the depth ρ^n depends on the formulation of the optimization problem:

- for quadratic cases (cases 1&2), a linear search along the descent direction gives an optimal value ρ^{n*};

- for case 3 which is non-quadratic with respect to Z, linearization of the criterion can be achieved.

Computation of the first variation δT

Let us consider δZ a variation of Z along the boundary Γ_u and denote T_ε the solution of the direct problem corresponding to $(Z+\delta Z) \in U_{ad}$. Then $\delta T = \lim\limits_{n\to\infty} \dfrac{T_\varepsilon - T}{\varepsilon}$ is called the first variation of T with respect to Z.

The function $\delta T(x)$is solution of the sensitivity equations:

Cases 1, 2 & 3 :

$$\nabla(\mathbf{k}\ \nabla\delta T) = 0 \quad \text{in } \Omega \quad (23)$$

$$\delta T = 0 \quad \text{on } \Gamma_1 \quad (24.1)$$

$$\mathbf{k}\ \nabla\delta T.\ \mathbf{n} = 0 \quad \text{on } \Gamma_2 \quad (24.2)$$

$$\mathbf{k}\ \nabla\delta T.\ \mathbf{n} + h\ \delta T = 0 \quad \text{on } \Gamma_3 \quad (24.3)$$

Case 1: $\delta T = \delta Z$ on Γ_u (25.1)

Case 2: $\mathbf{k}\ \nabla\delta T.\ \mathbf{n} + \delta Z = 0$ on Γ_u (25.2)

Case 3: $\mathbf{k}\ \nabla\delta T.\ \mathbf{n} + h_u\ \delta T = -\delta Z\,(T-T_e)$ on Γ_u (25.3)

Using the finite element approximation of the direct problem as described in section II, we can compute δT due to any variation δZ .Then the corresponding variation δJ of the criterion is given by:

$$\delta J(Z)=2\sum_{i=1}^{N_t}\int_\Omega \omega_i\big(T(x;Z)-Y_i\big)\delta T dx+2\alpha\int_{\Gamma_u}\big(Z(s)-Z_{hest}(s)\big)\delta Z ds \quad (26)$$

Computation of the gradient ∇J

We consider the case $U= L^2(\Gamma_u)$; in order define ∇J the gradient of J such as : $\delta J = \int_{\Gamma_u}\nabla J\ \delta Z\ d\Gamma$ (27),

we introduce an adjoint variable $\Psi(x)$ defined on $\Omega\cup\Gamma$ and solution of the adjoint equations :

Cases 1, 2 & 3 :

$$\nabla(\mathbf{k}\ \nabla\Psi) = \sum_{i=1}^{N_t}\omega_i\big(T(x)-Y_i\big) \quad \text{in } \Omega \quad (28)$$

$$\Psi = 0 \quad \text{on } \Gamma_1 \quad (29.1)$$

$$\mathbf{k}\ \nabla\Psi.\ \mathbf{n} = 0 \quad \text{on } \Gamma_2 \quad (29.2)$$

$$\mathbf{k}\ \nabla\Psi.\ \mathbf{n} + h\ \Psi = 0 \quad \text{on } \Gamma_3 \quad (29.3)$$

Case 1: $\Psi = 0$ on Γ_u (30.1)

Case 2: $\mathbf{k}\ \nabla\Psi.\ \mathbf{n} = 0$ on Γ_u (30.2)

Case 3: $\mathbf{k}\ \nabla\Psi.\ \mathbf{n} + h_u\ \Psi = 0$ on Γ_u (30.3)

Consequently, we have:

$$\delta J(Z) = 2\int_\Omega \nabla(\mathbf{k}\nabla\Psi)\delta T dx+2\alpha\int_{\Gamma_u}\big(Z(s)-Z_{hest}(s)\big)\delta Z ds \quad (31)$$

Using Green Formula, we get:

$$\int_\Omega \nabla(\mathbf{k}\nabla\Psi)\delta T dx = -\int_\Omega \mathbf{k}\nabla\Psi\nabla\delta T dx + \int_\Gamma \mathbf{k}\nabla\Psi\mathbf{n}\delta T ds \quad (32)$$

Using the sensitivity equations and the equations (27) to (31), it comes the general gradient formulas:

Case 1: $\nabla J\ (s) = 2\left[\mathbf{k}\ \nabla\Psi.\ \mathbf{n} + \alpha\big(Z-Z_{hest}\big)\right]$ on Γ_u (33.1)

Case 2: $\nabla J\ (s) = 2\left[\Psi + \alpha\big(Z-Z_{hest}\big)\right]$ on Γ_u (33.2)

Case 3: $\nabla J\ (s) = 2\left[\Psi\ (T - T_e) + \alpha\big(Z-Z_{hest}\big)\right]$ on Γ_u (33.3)

For any function $\nabla J \in U$, we have: $\nabla J\ (s) = \sum_{i=1}^{NR}\nabla J_i\ \sigma_i(s)$ (34)

then the components ∇J_i of the gradient in this basis satisfy:

$$\nabla J_i = \int_{\Gamma_u}\nabla J(s)\ \sigma_i(s)\, d\Gamma \qquad i = 1, ..., NR \quad (35)$$

Descent Direction Research

Descent directions $\underline{d}^n$ are built using the gradient ∇J of the criterion as follows:

$$\underline{d}^n = a^n\ \underline{\nabla J}^n + \sum_{j=1}^{n-1} b^j\ \underline{d}^j \quad (36)$$

several techniques are used to compute the real numbers a^n and b^j [8] :

The *Steepest Descent Method* (SDM) consists in choosing the gradient ∇J as the descent direction:

$$\underline{d}^n = -\ \underline{\nabla J}^n \quad (37)$$

The *Polack-Ribière Method* (PRM) conjugates the directions $\underline{\nabla J}^n$ and $\underline{d}^{n-1}$:

$$\underline{d}^n = -\ \underline{\nabla J}^n + b^{n-1}\ \underline{d}^{n-1} \quad (38.1)$$

$$\text{With} \quad b^{n-1} = \frac{\left(\underline{\nabla J}^n\ ,\ \underline{\nabla J}^n - \underline{\nabla J}^{n-1}\right)_E}{\left(\underline{d}^{n-1}\ ,\ \underline{\nabla J}^n - \underline{\nabla J}^{n-1}\right)_E}\ , \qquad E = \mathbb{R}^{NR} \quad (38.2)$$

The *Beale-Powell Method* (BPM) uses previous directions $\underline{d}^{n-1}$ and $\underline{d}^q$ ($q < n-1$), where q is an iteration number for which the Hessian $\nabla^2 J$ is quasi-constant:

$$\underline{d}^n = -\ \underline{\nabla J}^n + b^{n-1}\ \underline{d}^{n-1} + b^q\ \underline{d}^q \quad (39.1)$$

$$\text{With} \quad b^q = \frac{\left(\underline{\nabla J}^n\ ,\ \underline{\nabla J}^{q+1} - \underline{\nabla J}^q\right)_E}{\left(\underline{d}^q\ ,\ \underline{\nabla J}^{q+1} - \underline{\nabla J}^q\right)_E}\ , \qquad E = \mathbb{R}^{NR}$$

(39.2)

At each step "n", the choice of $\underline{d}^n$ among these three methods is governed by the following motives :

- if n=1 or q=n then use SDM (40.1)
- if n=2 or q=n-1 then use PRM (40.2)
- if n>2 or q<(n-1) then use BPM (40.3)
- if q<(n-1) and $\left(\underline{d}^n, \underline{\nabla J}^n\right)_E$ is to small (it means that the Hessian becomes quasi-constant) , then q=n-1 (40.4)
- if q<n and $\left(\underline{\nabla J}^{n+1}, \underline{\nabla J}^n\right)_E$ is not enough small (it means that gradients are not enough orthogonal) , then q=n (40.5)

Descent Depth Research

When $\underline{d}^n$ has been determined, the optimal descent depth ρ^{n*} is searched for, according to the equation:

$$\rho^{n*} = \underset{\rho>0}{\text{Arg min}}\ J(\underline{Z}^n + \rho^n.\underline{d}^n) \quad (41)$$

Depending on the linearity of the direct heat conduction solution $\overline{T}(Z)$ with respect to the unknown function Z, two methods are considered.

For cases 1&2 , the direct problem is linear, so

$$\overline{T}(\underline{Z}^{n+1}) = \overline{T}(\underline{Z}^n) + \rho^n \delta T(\underline{Z}^n) \quad (42)$$

where $\delta T(\underline{Z}^n)$ is the variation of temperature corresponding to the variation $\underline{\delta Z}^n$ defined by $\underline{d}^n$.

Using the notation of the quadratic criterion J, and linearity of the forms $\Pi(.,.)$, $l(.)$ (eq.18)

$$J(\underline{Z}^{n+1}) = \rho^{n^2}\ \Pi(\underline{d}^n,\underline{d}^n) + 2\rho\left[\Pi(\underline{Z}^n,\underline{d}^n) + l(\underline{d}^n)\right] + J(\underline{Z}^n) \quad (43)$$

The optimal value ρ^{n*} satisfy : $\dfrac{dJ}{d\rho} = 0$ (44)

As a result,

$$\rho^{n*} = -\frac{\Pi(\underline{Z}^n,\underline{d}^n) + l(\underline{d}^n)}{\Pi(\underline{d}^n,\underline{d}^n)} \quad (45)$$

For the case 3, there is no linearity between T and Z, so the value ρ^{n*} computed by linearization according to the previous equation is not optimal. Following Flechter [9], a polynomial interpolation $q(\rho) = J(\underline{Z}^n + \rho.\underline{d}^n)$ can be used to improve the value obtained by linearization.

Stopping Criterion of the algorithm

For ideal case whithout any measurement errors, the checked condition for the minimization of the criterion is:

$$J(\underline{Z}^n) < \varepsilon_1 \quad (46)$$

where ε_1 is related to the accuracy of the direct problem solution.

For measurements Y_i perturbed by an additive random error with zero mean value with standard deviation σ_Y, the stopping criterion is related to σ_Y by:

$$\varepsilon_1 = 4\, N_t\, \sigma_Y^2 \quad (47)$$

Computional Algorithm

In general case, Z_{est} is not available , so by taking $\alpha = 0$ and assuming Z(s) = constant on Γ_u, a first minimization of J(Z) gives Z_0^* which is taken as Z_{est} .

At iteration "n", $\underline{Z}^n$, $\underline{\nabla J}^n$, $\underline{d}^j$ (j < n), are known. The algorithm computes the new iterate $\underline{Z}^{n+1}$ as follows:

step 1: compute the new direction $\underline{d}^n$ (37-38-39) by taking into account the observations (40.1-2-3) and check condition(40.4)

step 2: compute the optimal descent depth ρ^{n*} according to (46) and [9]

step 3: change $\underline{Z}^n$ into $\underline{Z}^\#$ according to (21), resolve the direct problem (10), and compute the criterion

step 4: stopping criterion (46-47)

step 5: solve the adjoint equation (29-30) and compute the gradient $\nabla \underline{J}^\#$ thanks to (38,39,40)

step 6: check the orthogonality of the gradients $\nabla \underline{J}^\#$ and $\underline{J}^n$ according to (40.5); If (40.5) is not ensured , goto step 1

step 7: reinitialisation: n+1←#; n - 1←n; n ←n+1

NUMERICAL RESULTS

A case studied by Maillet [5] is considered; it consists to determine the heat transfer at the external surface of a tube submitted to an external transversal flow (fig.3). Due to the symetry of the problem , only one semo-cylinder is considered.

Fig.3 Mesh of the domain Ω

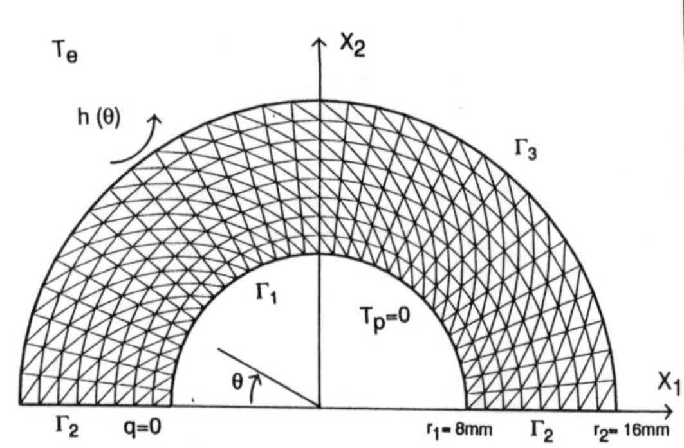

The direct Problem

Temperature inside the tube is solution of the heat conduction system (1-2) on which:

- on Γ_1 defined by (y=0, $x \in [-r_2,-r_1] \cup [r_1,r_2]$), $\nabla T.\mathbf{n} = 0$
- on Γ_2 defined by ($x^2+y^2=r_1$), $T=T_p=0$
- on Γ_3 defined by ($x^2+y^2=r_2$), $\mathbf{k}\ \nabla T.\mathbf{n} + h(s)T = h(s)T_e$

The spatial distribution of h along Γ_3 is defined on fig.5. T_e is equal to 40°C. Thermal conductivity is equal to 0.26 $W.m^{-1}.°C^{-1}$, and source term f_Ω is equal to zero.

Influence of the mesh on the accuracy of the temperature field has led to take NK = 512 elements with NSG = 297 nodes (fig.3). The thermal isovalues are given by fig.4.

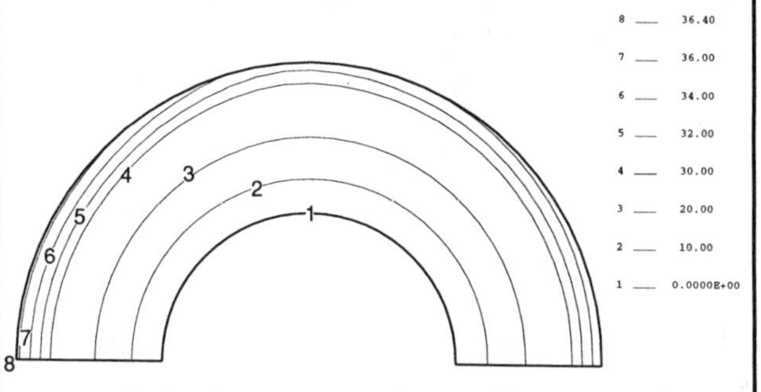

Fig.4 Temperature field (Direct problem)

The Inverse Problem

Influence of formulation, of error measurements, of number and locations of sensors are evaluated with some examples to illustrate the possibilities of the algorithm.

A standard problem is defined by the following parameters:

- $\Gamma_u = \Gamma_3$,
- Nt = 17 sensors regularly distributed on a circumference with radius r = 14 mm.
- The unknown Z_u is approximated by piecewise constant function, with NR = 16,
- No error measurement,
- No regularization term.

Influence of formulation. The standard problem has been solved by considering $Z_u = q_u$ (case 2) and $Z_u = h_u$ (case 3). Figure 5 shows that results obtained with theses two formulations are quite similar. The main difference is computing time which is about twenty time longer for case 3. For this problem, the algorithm is more efficient to minimize quadratic criterion (case 2) than non quadratic one (case 3). As a result, in the following we consider only the case $Z_u = q_u$.

Influence of error measurement. The temperature Y_i (i = 1, .., Nt) of the standard problem are perturbed by a random noise ($\sigma_Y = 0{,}05$ °C). On fig.6, it can be observed that without regularization ($\alpha = 0$), the computed solution is unstable even with a stopping criteria chosen according to (47). Goods results are obtained with $\alpha = 0.1$. With $\alpha = 0.01$ regularization is not sufficient and with $\alpha = 1$ the bias on the approximated solution becomes too important.

Fig.5 Estimation of h (case 2 and 3)
r=14mm Nt=17 NR=16 err=0 eups=0 alpha=0

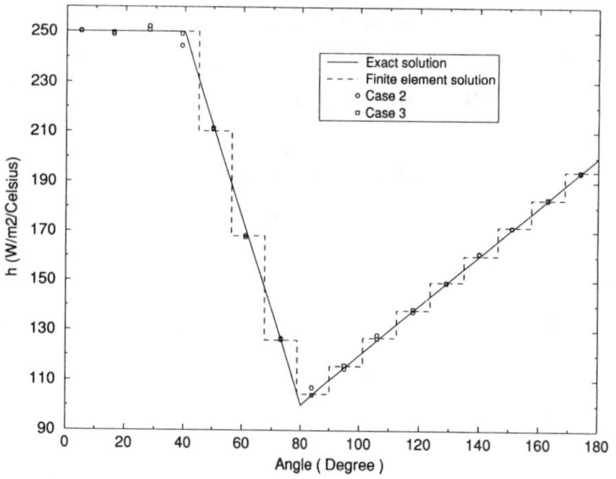

Fig.6 Estimation of h with Measurement Data Errors
r=14mm Nt=17 NR=16 err=+/-0.05 Celsius

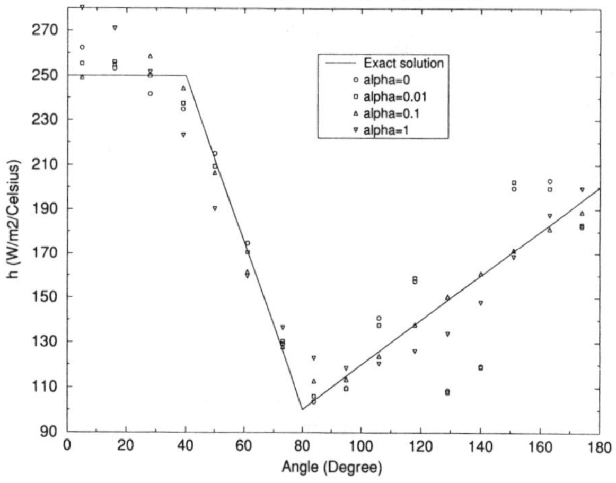

Fig.7 Influence of Number of Sensors
r=14mm NR=16

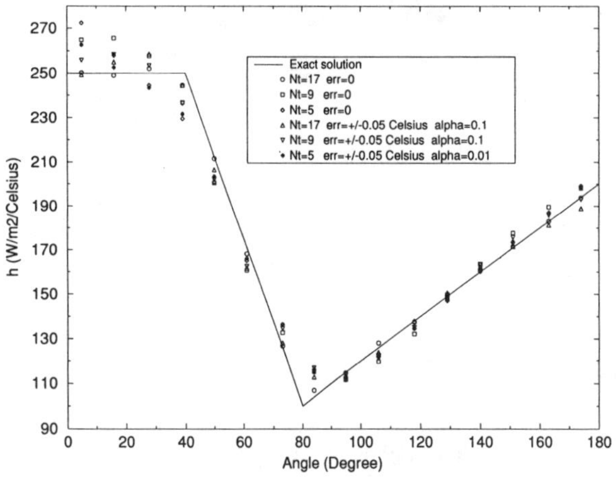

Influence of number of sensors. Three cases Nt=17, 9 and 5 , with and without noise on the measurements, are presented on fig.7. Even with 5 sensors the algorithm gives a satisfying solution. Obviously the more sensors there are, the better the unknown function determination is.

Influence of sensor locations. Three cases with Nt=17, with and without noise measurement, are compared in fig.8. The radial positions of the sensors are r = 13, 14 , 15 mm. Without any noise, results are quite similar; with error measurement, better results are obtained with sensors nearest the boundary.

Fig.8 Influence of Sensor Location
Nt=17 NR=16

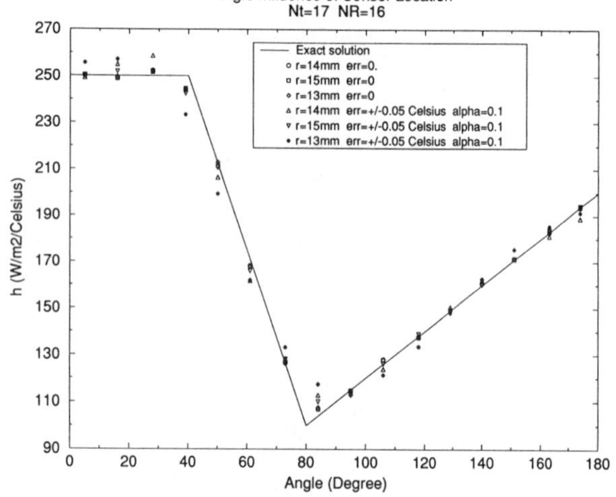

Fig.9 Adjoint field Y(x) at the final step

r=14mm Nt=17 NR=16 Err=0 alpha=0

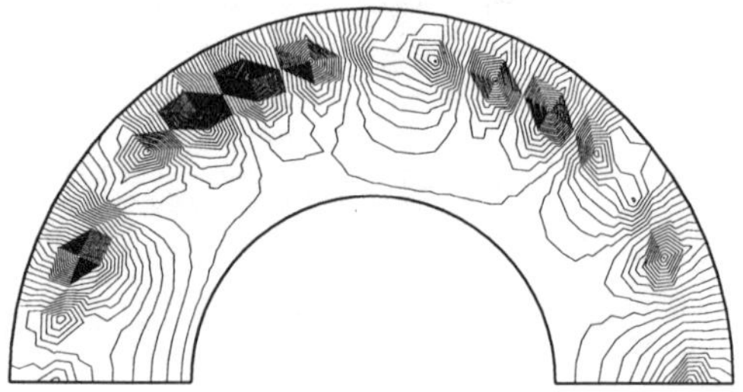

CONCLUSION

A general optimization algorithm has been described to solve boundary inverse heat conduction problem for general two-dimensional domains.

A unified approach using descent method has been presented to determine surface temperature, surface heat flux or heat transfer coefficient . Three different formulation have been considered but the same approach is used to compute the gradient of the criterion by introducing an adjoint variable with convenient boundary conditions . The formulation with heat flux unknown has given better results for the studied problem.

The algorithm has been developped using general modules of the FORTRAN finite element library MODULEF. As a result, the code has a great flexibility to deal with complex geometry. It can be used also for sensitivity analysis, experimental design; for example, fig.9 shows the influence of sensor location on the residual adjoint field.

Using the same approach, the extension of this inverse conduction code for non stationnarry cases is in progress.

References

[1] *Joint American-Russian NSF Workshop on Inverse Problem of Heat Transfer* - Final Report, Michigan State University, June 13-15, 1992.

[2] T.R. Hsu, N.S. Sun, G.G. Chen, Z.L. Gong, 1992, "Finite Element Formulation for Two-Dimensional Inverse Heat Conduction ", *ASME Journal of Heat Transfer* , vol. 114, pp 553-557.

[3] E. Hensel & R. Hills, 1989, "Steady State Two-Dimensional Inverse Heat Conduction", *Numerical Heat Transfer*, Part B, vol. 15, pp 227-240.

[4] R.Pasquetti, C. Le Niliot, 1991, "Boundary Element Approach For Inverse Heat Conduction Problems: Application to a bidimensional transient numerical experiment", *Numerical Heat Transfer*, Part B, vol. 20, pp 169-189.

[5] D.Maillet, A. Degiovanni & R.Pasquetti, 1991, "Inverse Heat Conduction Applied to the Measurement of Heat Transfer Coefficient on a Cylinder: Comparaison between an Analytical and a Boundary Element Technique", *ASME Journal of Heat transfer*, vol. 113, pp 549-557.

[6] Y. Jarny, M.N. Ozisik & J.P. Bardon, 1991, "A General Optimization Method using Adjoint Equation for Solving Multidimensional Inverse Heat Conduction", *Int. J. Heat Mass Transfer*, vol. 34, n° 11, pp 2911-2919.

[7] M. Bernadou & al., 1988, *Modulef ,une bibliothèque modulaire d' éléments finis*, INRIA, France.

[8] C.Lemarechal,1989,*Methodes Numériques d'Optimisation*, Collection didactique, INRIA, France.

[9] R. Flechter, 1980-1981, *Practical Methods of Optimization*, vol.1-2,Wiley-Interscience Publication.

Inverse Problems in Engineering: Theory and Practice
ASME 1993

THE METHOD OF IDENTIFICATION OF A HEAT FLOW ON THE SURFACE OF A SEMI-BOUNDED SOLID BODY

Yuri M. Matsevity
Department of Modelling Thermal
and Mechanical Processes
Institute for Problems in Machinery
Ukrainian Academy of Sciences
Kharkov, Ukraine

ABSTRACT

The method of solving the heat conduction inverse problem based on the lumped-capacitance concept and thermophysical smoothening of the input information is proposed. The transition to the lumped-capacitance model narrows the domain of feasible solutions, which has a regularizing effect on the process of identification of the heat exchange boundary conditions.

The problem of identification of a heat flow on the surface of a semi-bounded solid body by its temperature belongs to the class of incorrect inverse problems in mathematical physics. The computational problems arising during its solution are due to the absence of a continuous dependence of the heat flow being identified on the temperature measured with some error. From the physical point of view this means that even slight temperature variations can cause any, even infinitely large variations of the sought for heat flow.

Under these conditions the measurement errors have an essentially increasing influence on the accuracy and stability of the identification process, which makes it necessary to smoothen the initial data. However, conventional smoothening based on the least-squares method and effected by means of polynomials or splines can lead to significant distortion of the initial information.

The method of identification of non-stationary heat flows proposed herein is based on the lumped-capacitance concept and thermophysical smoothening of the initial information about the temperature to be measured. The essence of the method consists in the following

The non-stationary temperature field $T(x,t)$ of a semi-bounded body (Fig. 1) heated by a heat flow with the specific rate of flow $q(t)$ can be represented by the solution of the heat conduction problem:

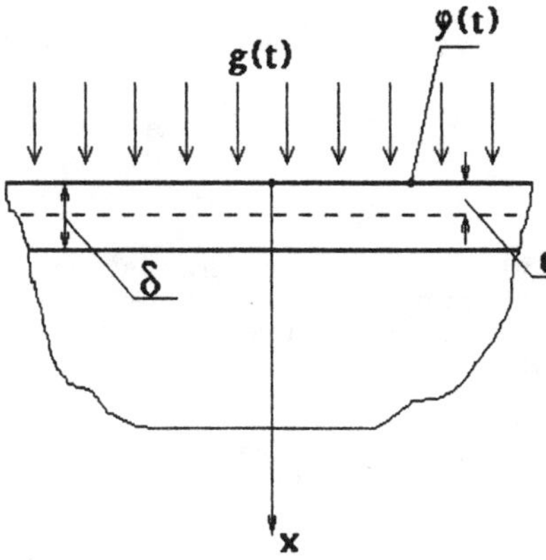

FIG. 1. SEMI-BOUNDED BODY

$$\frac{\partial T}{\partial t} = a\,\frac{\partial^2 T}{\partial x^2}\,,\quad 0<x<\infty,\ \ t>t_o, \tag{1}$$

$$-\lambda\,\frac{\partial T}{\partial x} = q(t),\ \ x=0,\ \ t>t_o, \tag{2}$$

$$T = T_f,\ \ 0\le x<\infty,\ \ t=t_o \tag{3}$$

where a is the thermal diffusivity ($a=\frac{\lambda}{c}$),

λ is the heat conduction, c is the specific heat capacity per unit volume, t_o is the zero time, T_f is the initial temperature (ambient temperature).

Problem (1) - (3) is stated correctly. Its solution given by the formula :

$$T(x,t)=T_f+\frac{1}{\lambda}\sqrt{\frac{a}{\pi}}\int_{t_o}^{t} q(\tau)\,\frac{\exp[-\frac{x^2}{4a(t-\tau)}]}{\sqrt{t-\tau}}\,d\tau \qquad (4)$$

(Beck et al, 1989) exists, is unique and continuously depends on the value of the heat flow $q(t)$. The satisfaction of these three conditions is sufficient for a mathematical statement of the problem correct by Hadamard.

Conversely, the solution of the inverse problem of determining the heat flow $q(t)$ by the measured surface temperature $T(0,t)=\varphi(t)$ found from the formula

$$\tilde{q}(t)=\sqrt{\frac{ac}{\pi}}\int_{t_o}^{t}\frac{\partial\varphi}{\partial t}\cdot(t-\tau)^{-0.5}d\tau \qquad (5)$$

is not continuously dependent on the measured temperature. Indeed, if the temperature in the interval between two measurement points t_o and t_1 changes linearly, then we have:

$$\tilde{q}(t_1)=2\sqrt{\frac{ac}{\pi}}\left[\frac{\varphi_1-\varphi_2}{\sqrt{\tau}}\right], \qquad (6)$$

where $\tau=t_1-t_o$. If thereat the temperature measurement error is equal to ΔT, then the identification error Δq found from (6) is equal to:

$$\Delta q=4\sqrt{\frac{ac}{\pi\tau}}\,\Delta T. \qquad (7)$$

Analysis of relationship (7) shows that at a fixed value of ΔT and at τ tending to zero, the identification error Δq can reach any, however great value. Here the third condition of problem correctness by Hadamard is violated. Therefore its solution requires the development of special methods allowing to obtain solutions which would be stable with respect to the temperature measurement errors.

We shall proceed as follows. Let the heat transfer in a semi-bounded body be thought of as a thermal interaction of a thermally thin skin layer of thickness δ with the remaining part of the body (Fig. 1). For this let us integrate the heat conduction equation (1) over the δ-layer thickness. We obtain:

$$c\delta\frac{\partial T}{\partial t}=\lambda\frac{\partial T}{\partial x}\Big|_{x=\delta}-\lambda\frac{\partial T}{\partial x}\Big|_{x=0} \qquad (8)$$

Using the condition of an ideal thermal contact on the formal boundary of the δ-layer

$$\lambda\frac{\partial T}{\partial x}\Big|_{x=\delta-0}=\lambda\frac{\partial T}{\partial x}\Big|_{x=\delta+0},\quad T\Big|_{x=\delta-0}=T\Big|_{x=\delta+0} \qquad (9)$$

and the boundary condition (2), from (8) we obtain:

$$c\delta\frac{\partial T}{\partial t}-\lambda\frac{\partial T}{\partial x}=q(t),\quad x=0,\quad t>t_o \qquad (10)$$

and then we pass from the boundary problem (1) - (3) to the mathematical model of heat transfer with a lumped capacitance:

$$\frac{\partial T}{\partial t}=a\frac{\partial^2 T}{\partial x^2},\quad 0<x<\infty,\quad t>t_o, \qquad (11)$$

$$c\delta\frac{\partial T}{\partial t}-\lambda\frac{\partial T}{\partial x}=q(t),\quad x=0,\quad t>t_o, \qquad (12)$$

$$T=T_f,\quad 0\le x<\infty,\quad t=t_o. \qquad (13)$$

The boundary problem (11) - (13) is a somewhat rough approximation to the problem (1) - (3), since in it the surface temperature of a solid body is identified with the mean integral temperature of the skin δ-layer. This temperature is calculated as follows. The temperature gradient $\partial T/\partial x|_{x=0}$ is expanded into a Taylor's series in terms of the powers of the small parameter $\varepsilon\in[0,\delta]$;

$$\frac{\partial T}{\partial x}\Big|_{x=0}=\frac{\partial T}{\partial x}\Big|_{x=\varepsilon}+\varepsilon\frac{\partial^2 T}{\partial x^2}\Big|_{x=\varepsilon}+O(\varepsilon^2)$$

and the value $\partial^2T/\partial x^2$ is expressed in terms of $\partial T/\partial t$ taken from the heat conduction equation (11). We obtain

$$\frac{\varepsilon}{a}\frac{\partial T}{\partial t}=\frac{\partial T}{\partial x}\Big|_{x=\varepsilon}-\frac{\partial T}{\partial x}\Big|_{x=0}. \qquad (14)$$

By replacing the derivatives with respect to x in (14) with the finite-difference ratios, we obtain

$$\frac{\varepsilon}{a}\frac{dT_\varepsilon}{dt}=\frac{T_o-T_1}{h}-\frac{T_\varepsilon-T_1}{h-\varepsilon}, \qquad (15)$$

where T_o is the surface temperature; T_ε and T_1 are the temperatures at the distances ε and h from the surface; h is the finite-difference mesh width.

From (15) it follows that

$$\frac{dT_\varepsilon}{dt} + \beta \cdot T_\varepsilon = \psi(t), \quad T_\varepsilon(0)=T_f, \qquad (16)$$

where

$$\beta=\frac{a}{\varepsilon(h-\varepsilon)}; \quad \psi(t) = \frac{a}{\varepsilon h} T_o + \frac{a}{h(h-\varepsilon)} T_1;$$

$$T_o = \varphi(t).$$

Assuming that $\psi(t)$ is a piecewise-constant function of time, the solution of problem (16) can be presented as follows:

$$T_\varepsilon(t_{k+1})=T_\varepsilon(t_k)e^{-\beta\tau}+(1-e^{-\beta\tau})\cdot\psi(t_{k+1}), \qquad (17)$$

where $\tau= t_{k+1}- t_k$ is the interval of constant values of the function $\psi(t)$.

The result of calculations by formula (17) depends on three parameters: spatial ones h, ε and the temporal one τ. Choosing ε on the condition that $\varepsilon \ll h$, the form of the solution of (17) can be simplified:

$$T_\varepsilon(t_{k+1})=T_\varepsilon(t_k)e^{-\beta\tau}+(1-e^{-\beta\tau})\cdot\varphi(t_{k+1}) \qquad (18)$$

Expression (18) formalizes the process of artificial approximation of measurements, which due to the inertiality of the thermal system smoothens the measurements taken on the surface of the body being investigated. Since such smoothening is carried out in strict correspondence with the physical process taking place in the body, it is correct in contrast to other kinds of smoothening (for instance, the least-squares method), and it is called by us as thermophysical smoothening.

From the mathematical point of view the thermophysically smoothened surface temperature is a bounded function with a limited local variation on some finite time interval $[0,\tau]$. If for computing $q(t)$ we use relationship (12) at $x=\varepsilon$, then

$$c\delta \frac{\partial T}{\partial t} - \lambda \frac{\partial T}{\partial x}\bigg|_{x=\varepsilon} = \tilde{q}(t). \qquad (19)$$

Here $\partial T_\varepsilon/\partial t$ is a continuous bounded time function having a bounded partial derivative $\partial^2 T_\varepsilon/\partial t^2$. Since $\partial T/\partial x$ is continuous also and has a bounded partial derivative (Rektoris, Sultangazin, 1976), then function $\tilde{q}$ is continuous and a continuously differentiated function, i.e. it belongs to some compact on $[0,\tau]$.

Thus, the transition from the initial mathematical statement of the problem in the form of a system of equations (1) - (3) to a model with a lumped-capacitance (11) - (13) constricts the domain of feasible solutions of the inverse problem to a compact set. This, in combination with thermophysical smoothening of the initial information, ensures correctness by Tikhonov of the mathematical statement of the problem of identification of a heat flow by the surface temperature of the body being investigated.

REFERENCES

Beck, J., Blackwell, B., and Saint-Claire, Ch. (Jr.), 1989, "Non-correct Inverse Problems of Heat Conduction," B. Artyukhin, ed., Mir Publishers, Moscow, 312 pp.

Rektoris, K., and Sultangazin, U., 1976, "On One Model Problem for the Heat Conduction Equation," Vesti AN Kaz. SSR, N 5, pp. 63-68.

Inverse Problems in Engineering: Theory and Practice
ASME 1993

INVERSE PROBLEMS AND TECHNIQUES IN METAL FORMING PROCESSES

Nicholas Zabaras and Seshadri Badrinarayanan
Sibley School of Mechanical and Aerospace Engineering
Cornell University
Ithaca, New York

ABSTRACT

Processing of materials plays a very important role in every manufacturing industry today. To achieve products with required characteristics, the process should be carefully designed. Many of these issues can be posed as inverse problems wherein the final desired output is known and the system input is to be determined along with the complete process conditions. In this paper, we have tried to classify some of the problems of Materials Processing that can be formulated as inverse problems and discuss some ways of attacking these problems. Some new results on ideal forming paths of rate-dependent and rate-independent solids are presented together with algorithms for the design of optimum processes.

INTRODUCTION

In engineering science, a physical process is typically analyzed using a mathematical model in which the actual system is represented by a set of equations containing parameters. The classical direct problem is to find the output of the system given the input and the system parameters. Inverse problems, on the other hand, involve determining the unknown causes of known consequences. There are two main types: 1) the reconstruction problem, where the system input is determined given the parameters and output, and 2) the identification or parameter estimation problem, where the parameters are found given the input and output. Note that the given output can either be a measured response of the system (inverse problem) or a desired response (inverse design problem). Nearly all inverse problems are classified as ill-posed (Hadamard [1]) in that their solutions do not necessarily satisfy conditions of existence, uniqueness, and stability.

In metal forming processes, a workpiece is assumed to be deformed as it passes through a sequence of dies in cold or hot environment. A typical direct problem includes the evaluation of the material deformation and state (i.e. the stresses and state variables) during the process, given the constitutive behavior and proper thermomechanical boundary conditions that are fixed or usually deformation dependent. These problems are highly non-linear and must be solved following the history of deformation. Their solution usually consists of two coupled problems: the kinematic problem and the constitutive problem. In the kinematic problem, one calculates the incremental deformation given the material state, while in the constitutive problem, the material state is evaluated given the material deformation.

There are still several open issues related with the direct analysis of forming processes, as for example is the problem of specification of an accurate constitutive model that is valid in the range of stress and temperature that are involved in the particular process of interest or the problem of the modeling of frictional conditions.

Direct FEM models are being used extensively as a tool for the design of processes, selection of materials, design of dies and control of deformations. These design techniques are mostly trial and error algorithms where one is solving the direct problem for several die geometries, material behavior, boundary conditions, etc. This is a rather time consuming and tedious way for analyzing the sensitivity of a process to several of the parameters involved and a need for a more mathematically structured methodology exists.

A typical parameter estimation problem that can be posed as inverse ill-posed problem is the calculation of the functional form of the constitutive model and the evaluation of the related parameters. Other inverse deformation problems are the identification problems of specification of thermomechanical boundary conditions from given deformation and temperature transient data inside the workpiece during a deformation process. Inverse design forming problems include the optimum design of dies to achieve a final product with desired properties (for example given geometry, minimum residual stresses, minimum required work, etc.), the optimum design of initial preforms that for a given process result in a desired final state of the workpiece with a minimum work expenditure, etc.

The main differences of these problems from other traditional inverse problems (as for example the inverse heat conduction problem, IHCP [2]) are the following:

- The parameters to be optimized are functions of position and time and the cost functional may be a tensor field, such as the distribution of residual stresses in the final product.
- The cost functional depends on the solution of a coupled non linear set of partial differential equations and it cannot be written explicitly in terms of the unknown initial conditions, boundary conditions or forcing terms.
- The associated direct problems are coupled thermomechanical problems with a strong history dependence and they are defined in complicated geometries.
- The number of material parameters involved is significantly higher than the one in typical parameter estimation problems in heat transfer and their calculation is not straight forward.
- The functional form of the constitutive problems is not unique and not well defined.
- Experimentation for the calculation of the material state is difficult and expensive.
- The finite element analysis of the direct problems is a very time consuming computational process.

In a sense, several classical ill-posed inverse problems of interest to the inverse community at large are related to materials processing as for example the IHCP, the backward heat conduction problem, the thermal control of the final state problem, and other. In the remaining of this paper, the basic theory of large deformations in forming processes will be reviewed and some typical inverse forming problems will be defined. Ideal forming is then presented and possible methods of solution are reviewed. The present paper does not claim to provide a comprehensive literature review on the subject and it is restricted to a small portion of the inverse and design problems that could be addressed in the area of forming processes.

DIRECT METAL FORMING PROBLEMS

In a metal forming process, an initially simple workpiece is plastically deformed in cold or hot conditions to produce a relatively complex final configuration. The "computational" *direct* problem in an updated Lagrangian formulation is usually divided into two parts [3-5].

1. The first part involves the calculation of the material state at the end of a time step, given the material state in the beginning of the step and the deformation gradient $\mathbf{F}_u$ of the configuration at the end of the step with respect to that in the beginning of the step.
2. The second part, which computes the deformation field, involves the development of a Newton-Raphson scheme from a linearized form of the principle of virtual work and includes the calculation of the linearized material moduli.

The stress is calculated from a hyperelastic model and an appropriate radial return mapping. A consistency is maintained between the tangent operator and the integration scheme employed.

The constitutive equation for the stress is here given as (see reference [3]):

$$\bar{\mathbf{T}} = \mathbf{L}^e [\bar{\mathbf{E}}^e] \tag{1}$$

where the strain measure $\bar{\mathbf{E}}^e$ is defined with respect to the intermediate (unstressed) configuration as

$$\bar{\mathbf{E}}^e = \ln \mathbf{U}^e \tag{2}$$

while the conjugate stress measure $\bar{\mathbf{T}}$ is the pull back of the Kirchoff stress with respect to $\mathbf{R}^e$, i.e.

$$\bar{\mathbf{T}} = \mathbf{R}^{eT} \det(\mathbf{U}^e)\, \mathbf{T}\, \mathbf{R}^e \tag{3}$$

Here $\mathbf{U}^e$ and $\mathbf{R}^e$ are calculated from the polar decomposition of $\mathbf{F}^e$, i.e.

$$\mathbf{F}^e = \mathbf{R}^e \mathbf{U}^e \tag{4}$$

where [6]:

$$\mathbf{F} = \mathbf{F}^e \bar{\mathbf{F}}^p \mathbf{F}^\theta \tag{5}$$

and $\mathbf{F}$ is the total deformation gradient in the current configuration where the stresses must be calculated and the multiplicative decomposition of $\mathbf{F}$ is defined in Fig. 1.

The elastic isotropic moduli $\mathbf{L}^e$ are defined as:

$$\mathbf{L}^e = 2G\,\mathfrak{I} + \left(K - \frac{2}{3} G \right) \mathbf{I} \otimes \mathbf{I} \tag{6}$$

and if the thermal deformation is isotropic, then the evolution of the thermal deformation gradient is given as:

$$\dot{\mathbf{F}}^{\theta}\,\mathbf{F}^{\theta\,-1} = \dot{\theta}\,\beta\,\mathbf{I} \tag{7}$$

where θ is the temperature field to be calculated from the solution of an appropriate heat transfer initial value problem.

A flow rule is given in the form of the evolution of $\bar{\mathbf{F}}^p$ with zero spin of the intermediate configuration:

$$\bar{\mathbf{L}}^p = \bar{\mathbf{D}}^p = \dot{\bar{\mathbf{F}}}^p\,\bar{\mathbf{F}}^{p^{-1}} = \sqrt{\frac{3}{2}}\,\dot{\bar{\varepsilon}}^p\,\bar{\mathbf{N}}^p(\bar{\mathbf{T}}', \tilde{\sigma}) \tag{8a}$$

$$\bar{\mathbf{W}}^p = \mathbf{0} \tag{8b}$$

with $$\bar{\mathbf{N}}^p(\bar{\mathbf{T}}', \tilde{\sigma}) = \sqrt{\frac{3}{2}}\,\frac{\bar{\mathbf{T}}'}{\tilde{\sigma}} \tag{8c}$$

where $$\tilde{\sigma} = \sqrt{\frac{3}{2}\,\bar{\mathbf{T}}'\cdot\bar{\mathbf{T}}'} \tag{8d}$$

and $$\bar{\mathbf{T}}' = \bar{\mathbf{T}} - \frac{\mathrm{tr}\,\bar{\mathbf{T}}}{3}\,\mathbf{I} \tag{8e}$$

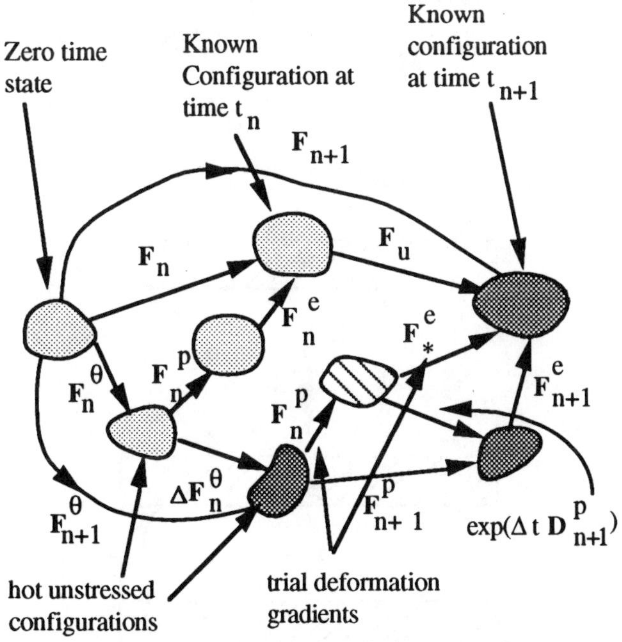

Fig. 1. The constitutive problem showing the multiplicative decomposition of **F** and the evolution of the actual, stress free and "thermal" configurations.

For rate independent plasticity, the evolution of the equivalent plastic strain $\dot{\bar{\varepsilon}}^p$ is specified from the consistency conditions:

$$\tilde{\sigma} = s \tag{9a}$$

$$\dot{\tilde{\sigma}} = \dot{s} \tag{9b}$$

while for rate dependent plasticity, $\dot{\bar{\varepsilon}}^p$ is specified via uniaxial experiments as:

$$\dot{\bar{\varepsilon}}^p = f(\tilde{\sigma}, s, \theta) \tag{10}$$

while the evolution of the isotropic scalar resistance s is defined as:

$$\dot{s} = g(\tilde{\sigma}, s, \theta) = h(s)\,\dot{\bar{\varepsilon}}^p \tag{11}$$

where h(s) is the hardening function. The functional form of f and g varies for different models in the literature. In Fig. 1, the configurations at time t_n and t_{n+1} are shown, as well as the evolution of the intermediate unstressed configurations. Let $\mathbf{F}_u$ be the known relative deformation gradient between the two configurations, i.e.

$$\mathbf{F}_u = \mathbf{F}_{n+1}\,\mathbf{F}_n^{-1} \tag{12}$$

Let us now express the equilibrium equation in an updated Lagrangian framework. Let the configuration $\mathbf{B}_n$ of the body at time $t = t_n$ be known and under equilibrium. Then the incremental quasi-static boundary value problem at time $t = t_{n+1}$ is to find the incremental (with respect to configuration $\mathbf{B}_n$) displacement field $\mathbf{u}(\mathbf{x}_n, t_{n+1}) \equiv \mathbf{u}_{n+1}$ such that:

$$G(\mathbf{u}_{n+1}, \tilde{\mathbf{u}}(\mathbf{x}_n)) =$$

$$\int_{\mathbf{B}_n} \mathbf{P}_u \cdot \frac{\partial \tilde{\mathbf{u}}}{\partial \mathbf{x}_n}\,dV - \left(\int_{\partial \mathbf{B}_{n+1}} \mathbf{t}\cdot\tilde{\mathbf{u}}\,ds + \int_{\mathbf{B}_{n+1}} \mathbf{b}\cdot\tilde{\mathbf{u}}\,dv \right) = 0 \tag{13}$$

for each test vector field $\tilde{\mathbf{u}}(\mathbf{x}_n)$, which is zero on the portion of the boundary where kinematic boundary conditions are applied. The above equation is a mixed form of the principle of virtual work. The internal work is expressed in the reference configuration B_n using the Piola-Kirchhoff I stress, $\mathbf{P}_u$, while the external work is expressed in the current configuration where the applied surface tractions, $\hat{\mathbf{t}}$, and body forces, $\hat{\mathbf{b}}$, are given. Part of the plastic work rate $\int_{B_n} \mathbf{T} \cdot \bar{\mathbf{D}}^p dV$ is used as a heat source in the appropriate heat transfer equations.

In order to solve the above set of non-linear equations for the incremental displacement field $\mathbf{u}(\mathbf{x}_n, t_{n+1})$, an iterative scheme must be used (see Fig. 2). A Newton-Raphson scheme is usually adopted, which requires linearization of equ. (13) about the last obtained solution for $\mathbf{U}_{n+1}$. The incremental strain measures must be calculated in terms of incremental displacements using objective kinematic approximations that are consistent with the integration scheme used for the solution of the constitutive model [3-6].

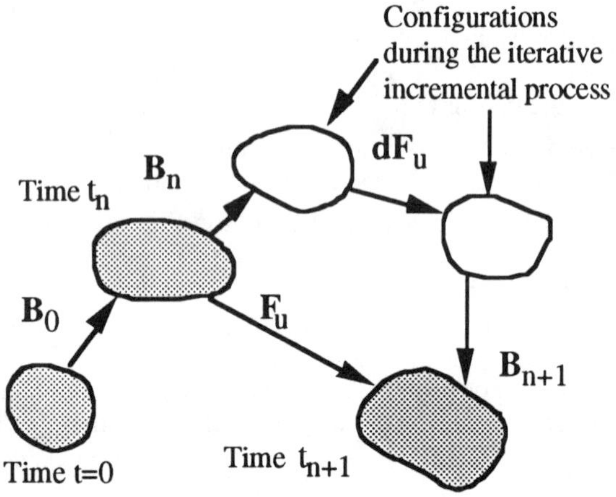

Fig. 2. The kinematic problem with a Newton-Raphson technique.

PREFORM DESIGN VIA BACKWARD TRACING

In the design of forming processes, the most important objective is, with a minimum effort and cost, to obtain a product with a desired state and geometry. When the final configuration required is known along with the required material properties, the forming process has to be designed so that the initial billet is successfully transformed to the final product with the specified shape and microstructure. In some processes like simple disk forging, the final shape may not be complex but a desired material property may be required.

The design of forming processes can be considered as the design of the initial workpiece and of the subsequent shapes or otherwise known as preforms, the design of the die and loading conditions and temperature and friction interface conditions, the number of steps involved in the forming operation (or sequence of operations), etc. Deformation during the process is purely controlled by the contact between the workpiece and the die. This along with limited workability of the material imposes severe restrictions on the design of the process and makes the choice of preforms and die a very important decision.

The earlier work done in preform design in metal forming is reviewed briefly in [7,8]. The main idea has been to computerize the design calculations based on qualitative guidelines derived from experience and experimental studies. Kobayashi and co-researchers [7-11] came up with a new method of preform designing. They used backward tracing - which is "*to trace backward the loading path in the actual forming process from a given final configuration by the finite element method* [8]".

We briefly review some important preform problems.

Disk Forging

In forging, preform design involves the determination of a number of preforms and the design of the shapes and dimensions of each preform. *The requirement here is to produce a flat disk with a uniform strain of a required amount from a cylindrical bar stock* [7]. The forging of a simple cylindrical stock produces a barreled piece because of the friction between the die and the workpiece. Also the regions close to the axis do not deform much compared to the periphery (Fig. 3).

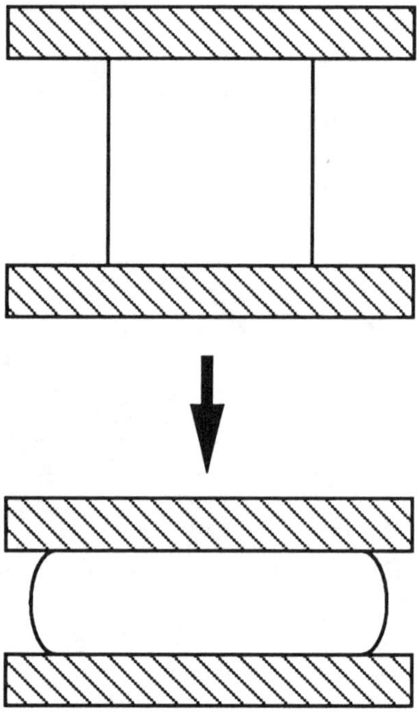

Fig. 3. Axisymmetric disk forging.

Closed Die Forging

In conventional processes, the formation of flash restricts the lateral flow of material which results in the workpiece filling the die cavity. The excess material is then trimmed at the completion of the process [7]. The cost of excess material amounts to 15% of the total forging cost and the trimming-off process needs additional machining cost [7,8]. The forging load due to flash will result in die wear [7]. *The aim here is to design the die and the preforms so that the die cavity is filled without forming any flash* (see sequence of operations in Fig. 4).

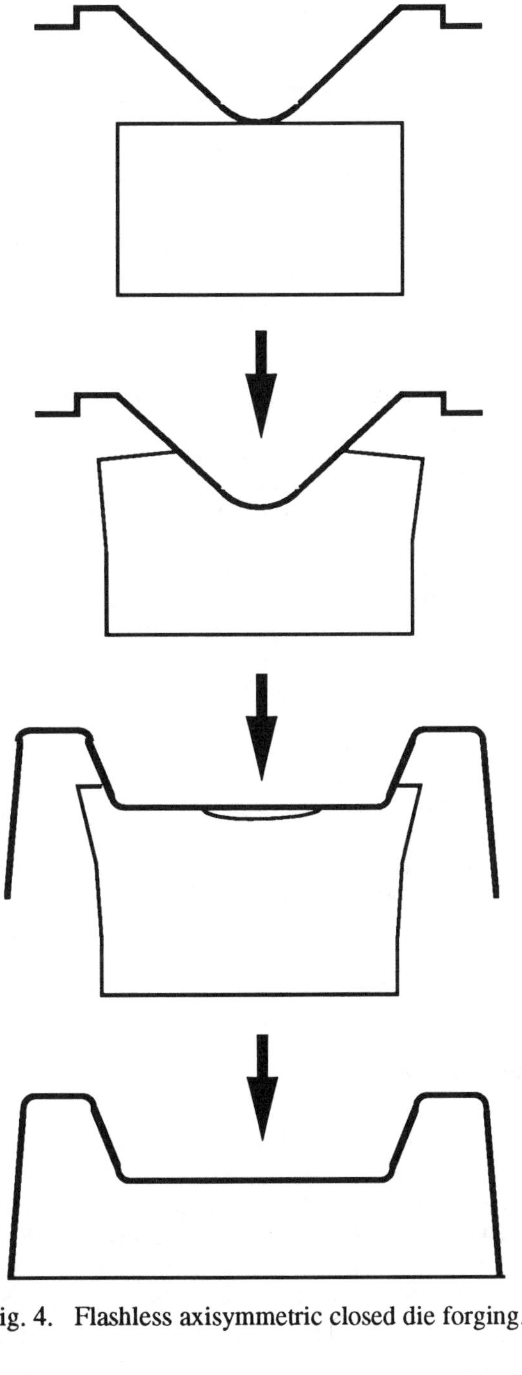

Fig. 4. Flashless axisymmetric closed die forging.

Rolling

When ingots with flat end are rolled, the end shape obtained may be defective. *The aim here is to design the preform end shapes which will result in flat ends after rolling, thereby eliminating the crop loss.* This can improve the yield by about 4% [7,10] (see Fig. 5).

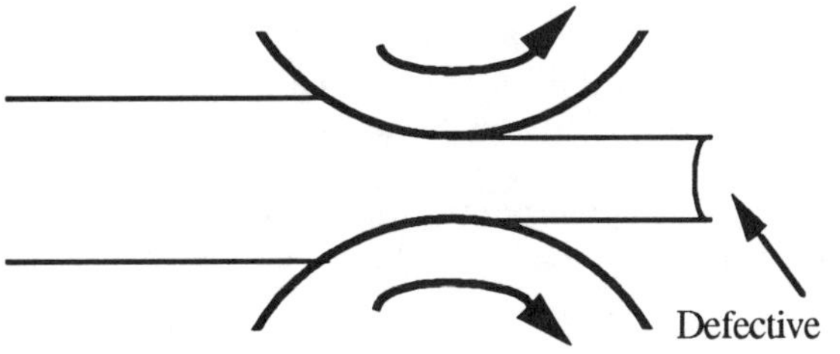

Fig 5. Defective front end shape in flat end ingot rolling.

Shell Nosing

In shell nosing, an ogive nose is formed at a open end of a tubular part by pressing the tube into a contoured die [8,9]. *The idea here is to design a preform so that after nosing, the wall thickness is uniform.* Shell nosing can be done at room temperature (cold nosing) or at elevated temperature (hot nosing).

To address these preform design problems, Park et al. [8] introduced the method of "*backward tracing by FEM*," and included work-hardening, strain rate and temperature effects as well as changing boundary conditions during the process and the possibility of multiple designs.

However, none of these issues was addressed in depth from an inverse problem solving, constitutive representation or modeling point of view. The material modeling most of the time was a very simple - rigid plastic model. Hardening and temperature effects were included in selected problems. In shell nosing problems, backward tracing was done for both the temperature and strain fields.

The boundary conditions are controlled in a heuristic (ad-hoc) manner and the qualitative solution for most of the problems were a-priori known. Hence a solution is expected with at most some perturbations. Based on this, the contact conditions are manipulated so as to get a preform with some desired characteristics.

The general idea seems to be that of developing a design procedure - one that is better than existing methods. There does not seem to be any mathematical validity for doing "backward tracing." The question of whether a stable solution is obtained every time the backward tracing is performed was not at all addressed.

There are genuine difficulties as such in any kind of inverse problem involving preform design by FEM. The boundary conditions very much depend upon the shape of the preform. This boundary condition history changes abruptly. A node in contact with die in one step might have been free the previous step. We must emphasize that in the work just reviewed, *the backward path was in some sense forced to be very similar with paths known from a-priori information based on the solution of the direct problem. Without these a-priori information there is no unique design. Also, as a result of the specification of the backward path, the problem cannot be stated as an optimization problem.* These issues will be further explored in the Ideal Forming section of this paper.

PARAMETER ESTIMATION PROBLEMS IN CONSTITUTIVE MODELING

Of importance here, is the determination of the scalar or tensor parameters that define the functions $f(\tilde{\sigma},s,\theta)$ and $g(\tilde{\sigma},s,\theta)$. We assume that appropriate state variables have already been selected and that the functional dependence of the functions f and g upon $\tilde{\sigma}$ and s has been decided based on physical arguments. The number of measurements of deformation inside a specimen (strain or displacement), the location of the sensors, the optimum specimen geometry for maximum sensitivity of the measured quantities on the unknown parameters, the selection of the type and number of tests (tensile, shear, biaxial, etc.) are important parameter estimation problems that must be addressed as ill-posed inverse problems. The deterministic mathematical structure of these models is similar to the one for inverse boundary problems that are discussed next.

INVERSE BOUNDARY PROBLEMS

Here, the calculation of the boundary thermomechanical conditions during a forming process are of main interest. An example is the problem of calculating the frictional conditions in a flat rolling process using experimental measurements of strain or displacement inside the rolls, which was addressed by Schnur and Zabaras [12] and is given graphically in Fig. 6.

Similarly, one may address inverse problems where thermal boundary conditions must be calculated (e.g. boundary flux between the die and the workpiece, thermal contact resistances, etc.).

The above problems and the parameter estimation problems mentioned earlier, share the same mathematical structure. Specifically, they take the form of iteratively looking for a solution that minimizes the error between the calculated quantities $\{u^{*r}\}$ at the r-iteration step and the measured or desired quantities $\{\hat{u}\}$ at all times:

$$E = \frac{1}{2}\{u^{*r}-\hat{u}\}^T\{u^{*r}-\hat{u}\} \qquad (14)$$

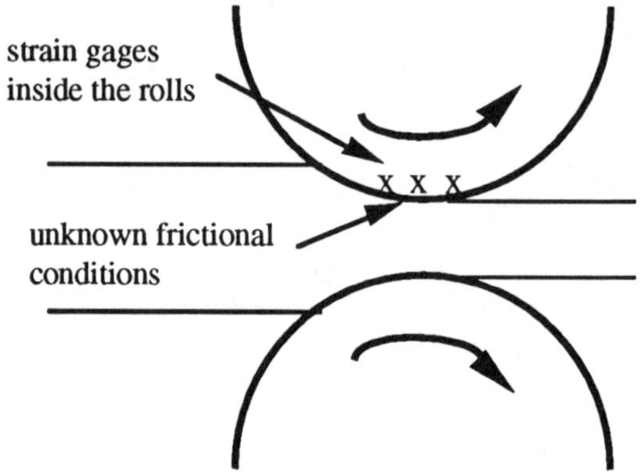

Fig. 6. Determination of frictional tractions using measurements of strain inside the rolls.

The error is minimized with respect to the unknown parameters $\{\beta\}$. Here $\{\beta\}$ is used to denote the boundary nodal traction and/or fluxes, or the parameters in the constitutive model, or the shape of the specimen in a given test for the evaluation of the constitutive model. Also, $\{u\}$ is used to denote discrete measurements of the deformation or temperature field during the deformation process or the testing necessary for the evaluation of the constitutive model. To stabilize the solution, the error E is usually augmented with spatial or temporal regularization. Some details on this topic will be given later.

To minimize E with respect to $\{\beta\}$ a sensitivity analysis can be used (Zabaras et al. [13-16]). The sensitivity coefficients, $[S_\beta]$, represent the change in $\{u\}$ at the measurement locations with respect to the change in $\{\beta\}$. An iterative scheme is required. The calculated $\{u\}$ at each iteration is expressed as

$$\{u^{*r}\}=\{u^{*r-1}\}+[S_\beta]\{\beta^r-\beta^{r-1}\} \qquad (15)$$

The equation for updating $\{\beta\}$ is obtained by substituting equation (15) in (14) and minimizing with respect to $\{\beta^r\}$:

$$[S_\beta]^T[S_\beta]\{\beta^r-\beta^{r-1}\}=[S_\beta]^T\{\hat{u}-u^{*r-1}\} \qquad (16)$$

The iteration process stops when the difference between parameter values of consecutive iterations is sufficiently small. Each evaluation of the error in equation (16) requires solving the direct (thermomechanical) problem for $\{u^*\}$. The gradient must be approximated by finite differences, requiring more function evaluations. Quasi-Newton, conjugate gradient, and modified Levenberg-Marquardt methods can be employed [17].

Separate approximations are used for the state and parameter spaces. The approximation of the state space is linked to the numerical solution of the related boundary value problem. In the inverse and design problems, the "parameter space" applies only to the spatially varying boundary tractions or flux or the function defining the unknown die surface. To obtain accurate solutions of the direct problem, an adequate discretization of the state space is needed. In addition, to stabilize the solution of the inverse problems, the discretization of the state space should be much higher than that of the parameter space. The advantage of imposing a functional form on the unknown variables is clearly shown in references [13-16]. Over large intervals, polynomial approximations of higher order, tend to oscillate. However, lower order splines exhibit flexibility without oscillation and the B-spline basis is preferred with the benefit that part of the B-splines can be modified without greatly affecting the rest [18].

The stability of the least squares minimization process primarily depends on the approximation and discretization of the parameter space. The regularization method of Tikhonov [19] increases stability by ensuring that the minimum lies in a compact set and by imposing a penalty against excessive oscillations. In the context of smoothing the solution, regularization only makes sense for problems with spatially varying unknown parameters. For example, for the problem of calculating the boundary traction τ over $\partial\Omega_0$, the regularization method involves the minimization of a smoothing functional, M:

$$M = E + \alpha R \tag{17}$$

where E is the error defined in equation (14), R is the stabilizing functional, and α is the regularization parameter with $\alpha > 0$. The stabilizing functional of order p is expressed in terms of the boundary traction components, τ_i, and their derivatives over the entire boundary $\partial\Omega_0$ as:

$$R = \sum_{i=1}^{3} \sum_{m=0}^{p} \zeta_{mi} \int_{\partial\Omega_0} \left(\frac{\partial^m \tau_i(x,y)}{\partial s^m} \right)^2 ds \tag{18}$$

where ds is a differential segment on $\partial\Omega_0$. The weighting factors, ξ_m, ($\xi_m \geq 0$) control the relative importance of each derivative in the stabilizing functional. Lee and Seinfeld [20] recommend dimensionless variables instead of s and taking the weighting factors as 1.

The regularization parameter, α, determines the weight given to smoothing relative to matching the given data. The most well known of the criteria for the calculation of α, are the discrepancy principle [21] and the modified order of magnitude rule [20]. In deformation processing, certain measures of the energy stored or dissipated within the material can also be used as the regularizing functional R.

OPTIMUM DIE DESIGN

The problems in this category are optimization problems that can be characterized as *inverse retrospective optimum design problems*. The major objective is to select the geometry of the die in such a way that, for a given initial workpiece, a desired state (stress and state variables) is obtained in the final product. For example, one may desire a uniform residual stress distribution or a stress distribution that does not violate certain design/failure criteria, etc. Mathematically, one can define these problems as follows:
Minimize with respect to the die surface d(x,y) = 0, the L_2 norm of the difference between the equivalent stress $\tilde{\sigma}$ and a desired distribution $\hat{\sigma}(x,y)$, over the body :

$$\min_{d(x,y)} \int_{\Omega_0} \left(\tilde{\sigma}(x,y,t_f) - \hat{\sigma}(x,y) \right)^2 d\Omega \tag{19}$$

where t_f is defined as the time at which a steady stress, state variable and temperature fields ($T=T_a$) have been achieved in the body. The desired residual stress distribution must be an achievable one. Another possibility is to obtain a least standard deviation of the final stress distribution. This can be achieved by taking the desired residual distribution as the average of the distribution at time t_f. Similar criteria can be written in terms of the hydrostatic pressure p. Constraint optimization versions of the above problems are also useful. Similar to the above criteria have already been applied by Kang and Zabaras [22], who calculated the optimum cooling history in unidirectionally solidifying bodies that leads to minimum or uniform residual stresses at the end of the process.

Here, a p-th dimensional approximation of d(x,y) could be introduced, e.g.

$$d(x,y) = y - w(x) = y - \sum_{i=1}^{p} \alpha_i \phi_i(x) = 0 \tag{20}$$

where p is the number of a priori defined basis functions and $\phi_i(x)$ belongs to the approximation function space.

A finite dimensional form of the functional of equ. (19) can now be written where the main unknowns are the coefficients α_j, $j = 1, \ldots, p$.

IDEAL FORMING

A topic of high interest in metal forming is that of designing efficient process conditions so that the total work done is minimum. The problem is defined as follows:

Design the initial preform and all the intermediate shapes as well as the required load history and die surfaces, such that a final desired shape is achieved with the least effort. Actually this problem is one of finding an extremum path between two points in the deformation space but *with none of the end points fixed.* The deformation gradient distribution in the final configuration is not known, however, the final configuration is known. Also, the deformation gradient distribution in the initial configuration is known, while the initial configuration itself is unknown.

For demonstration of ideas, let us consider the following simple problem. Assume that the initial and final configurations and the state variables in these configurations are well known. Then the problem reduces to *finding a deformation path that results in the least work.* Under assumptions of homogeneous deformations, this problem has been solved in [23,24]. The material was considered as time-independent, rigid plastic and isotropic with a convex yield surface. Hardening was also taken into account in [24]. Using small deformation plasticity theory, the ideal deformation path under isotropic hardening turns out to be a radial stress path in the stress space or in other words (1) *the principal material lines are fixed with respect to the material during deformation* and (2) *the ratio of the principal true strain rates is constant.*

In the remaining of this paper, we will employ the large deformation theory discussed earlier, to calculate the optimum deformation paths (minimum plastic work paths) for both rate-independent and rate-dependent solids during forming processes.

Rate-Independent Model

Consider the Hyperelastic Rate-Independent plasticity model with a single internal variable as described earlier. Assume that all processes are isothermal. *Supposing that we know the initial and final configurations and the initial and final values of* ***F*** *at all points in the body, our aim is to obtain the complete time history of* ***F*** *such that the total work done is minimum.* The total work done can be written as

$$W = \int_0^{t_f} \int_{\mathbf{B}_0} \mathbf{P} \cdot \dot{\mathbf{F}} \, dV \, dt \qquad (21)$$

Considering a multiplicative decomposition $\mathbf{F} = \mathbf{F}^e \mathbf{F}^p$, and equ. (4), equ. (21) can be written as

$$W = \int_0^{t_f} \int_{\mathbf{B}_0} \bar{\mathbf{T}}' \cdot \dot{\mathbf{U}}^e \mathbf{U}^{e-1} \, dV \, dt + \int_0^{t_f} \int_{\mathbf{B}_0} \bar{\mathbf{T}}' \cdot \mathbf{D}^p \, dV \, dt \qquad (22)$$

For a hyperelastic model, clearly, the first term of equ. (22) depends only on the initial and the final states and it is only the second term that is path dependent. Note that the integrand is a function of the position in *the initial configuration* and time. Hence, our task reduces to optimizing w^p for each material point, where w^p is the plastic work per unit volume in the initial configuration.

$$w^p = \int_0^{t_f} \bar{\mathbf{T}}' \cdot \mathbf{D}^p \, dt \qquad (23)$$

The yield surface is represented by equ. (9a) as

$$f(\bar{\mathbf{T}}', s) = \tilde{\sigma} - s = 0 \qquad (24)$$

and s follows the evolution equ. (11). In order to eliminate this constraint equation, consider the following assumptions:

(1) There is no unloading or neutral loading during the process.

(2) The hardening function h(s) is positive.

Now, the consistency condition implies:

$$\dot{\tilde{\sigma}} = \dot{s} = h(s)\, \dot{\tilde{\varepsilon}}^p$$

hence

$$\dot{\tilde{\varepsilon}}^p = \frac{1}{h(s)} \dot{\tilde{\sigma}} = \frac{1}{h(s)} \frac{3}{2\tilde{\sigma}} \left(\dot{\bar{\mathbf{T}}}' \cdot \bar{\mathbf{T}}' \right) \qquad (25)$$

From equ. (23), we can derive that:

$$\dot{w}^p = \bar{\mathbf{T}}' \cdot \mathbf{D}^p = \tilde{\sigma}\, \dot{\tilde{\varepsilon}}^p \qquad (26)$$

Using the consistency condition of equ. (25), we conclude that:

$$\frac{s}{h(s)} \dot{s} = \dot{w}^p$$

or with integration of both sides, that

$$w^p = \hat{w}(s) \qquad (27)$$

Since h(s) is a positive function, $\hat{w}(s)$ is a strictly increasing function of s. Applying the inverse function theorem, we obtain

$$s = g(w^p) \quad (28a)$$

and

$$\frac{d\hat{w}}{ds} = \left(\frac{dg}{dw^p}\right)^{-1} \quad (28b)$$

Thus the equation of the yield surface becomes

$$f(\bar{T}', w^p) = \tilde{\sigma} - g(w^p) = 0 \quad (29)$$

and the flow rule takes the form (note that the condition is loading all the time):

$$D^p = \left\{\frac{dg}{dw^p}\tilde{\sigma}\right\}^{-1} \left(\frac{3}{2\tilde{\sigma}}\right)^2 \left(\dot{\bar{T}}'\cdot\bar{T}'\right) \bar{T}' \quad (30)$$

Let us now define the plastic strain E^p such that

$$\dot{E}^p = D^p \quad (31)$$

We are looking for stationary values of the total plastic work subject to fixed initial and final plastic strains at times t=0 and t=tf, respectively. This implies that:

$$\delta w^p = \int_0^{t_f} \left(\delta\bar{T}'\cdot D^p + \bar{T}'\cdot \delta D^p\right) dt = 0$$

or with integration by parts that:

$$\delta w^p = \int_0^{t_f} \left(\delta\bar{T}'\cdot D^p - \dot{\bar{T}}'\cdot\delta E^p\right) dt = 0 \quad (32)$$

From the consistency condition, we have:

$$\frac{3}{2\tilde{\sigma}}\left(\bar{T}'\cdot\delta\bar{T}'\right) - \frac{dg}{dw^p}\left(\bar{T}'\cdot\delta E^p\right) = 0 \quad (33)$$

Combining equs. (32) and (33) and using the flow rule [equ. (30)], gives:

$$\delta w^p = \int_0^{t}\left(\frac{3}{2\tilde{\sigma}^2}\left(\dot{\bar{T}}'\cdot\bar{T}'\right)\bar{T}' - \dot{\bar{T}}'\right)\cdot\delta E^p\, dt$$

Since δE^p is arbitrary, we conclude that:

$$\frac{3}{2\tilde{\sigma}^2}\left(\dot{\bar{T}}'\cdot\bar{T}'\right)\bar{T}' = \dot{\bar{T}}' \quad (34)$$

This implies that *for an optimum deformation path and at any time instant, the deviatoric rotation neutralized Kirchoff stress rate must have the same direction as the deviatoric rotation neutralized Kirchoff stress.*

Rate-Dependent Model

Here we assume that the model is rate-dependent and that the initial and final configurations as well as the material state in both configurations are known. Now the problem becomes

$$\text{minimize } w^p = \int_0^{t_f} \tilde{\sigma}\, f(\tilde{\sigma},s)\, dt \quad (35a)$$

$$\text{subject to } \dot{s} = g(\tilde{\sigma}, s) = h(s)\, f(\tilde{\sigma},s) \quad (35b)$$

Using a Lagrange multiplier method, this is written as

$$\text{minimize } J = \int_0^{t_f} \left[\tilde{\sigma}\, f(\tilde{\sigma},s) + \lambda\,(\dot{s} - g(\tilde{\sigma}, s))\right] dt$$

For brevity, functional dependencies are dropped and subscripts are used to denote partial derivatives. The Euler-Lagrange equations for the above minimization problem, can be shown to be as follows:

$$f + \tilde{\sigma} f_{\tilde{\sigma}} - \lambda g_{\tilde{\sigma}} = 0 \quad (36a)$$

$$\tilde{\sigma} f_s - \lambda g_s - \dot{\lambda} = 0 \quad (36b)$$

$$\dot{s} - h f = 0 \quad (36c)$$

Simplifying these equations, we finally arrive at the following optimality condition:

$$\dot{\tilde{\sigma}} = -h f \frac{\left(2 f_s f_{\tilde{\sigma}} - f f_{\tilde{\sigma}s}\right)}{\left(2 f_{\tilde{\sigma}}^2 - f f_{\tilde{\sigma}\tilde{\sigma}}\right)} \quad (37)$$

Notice that the above equation as well as the corresponding equation for the rate independent case [equ. (34)] cannot be used to march forward in time starting at t=0. However, any extremal solution must satisfy these equations.

For a small deformation theory with a yield surface rate-independent model, radial paths in the stress space are the optimal paths. Under these conditions,

Chung and Richmond [25-27] solved an ideal sheet metal forming process and calculated the preform and intermediate shapes. They assumed that *the process is ideal, if each of the material points undergoes deformation along its minimum plastic work path. Hence, once a mapping of elements in the initial undeformed configuration to elements in the desired deformed final configuration has been defined, the strain evolution is pre-determined for each material element in the whole body.*

However, in the much more realistic large deformation theory involving elasticity that was presented earlier, the strain evolution is not explicitly predetermined. *For a given deformation gradient history, we can only check whether it constitutes an extremal path or not.* Hence, we resort to the following method to determine the optimal path.

Determination of the Ideal Deformation Path

Suppose the initial and the final configurations of the workpiece are given and that the deformation gradient in the final configuration is also provided. Rather than determining the complete deformation history, supposing we are simply looking for 'n' intermediate stages. We can make an initial guess of the deformation history by means of a spline approximation (Fig. 7).

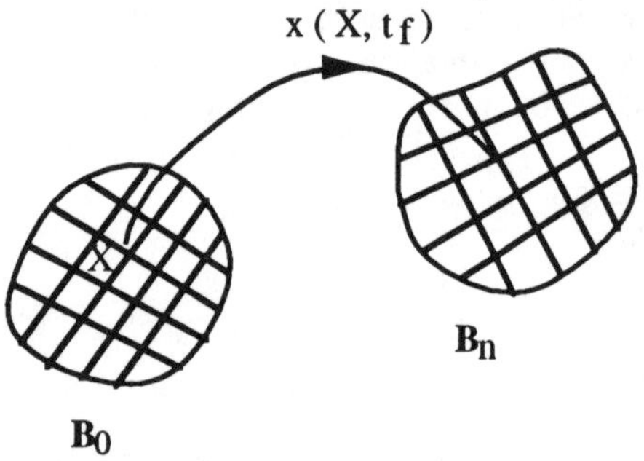

Fig. 7. Mapping of material points from the initial configuration to the final configuration.

The time domain is split into 'n' equal parts and we know the value of the function x(X,t) at t=0 and $t=t_f$. The intermediate n-1 values are initially guessed and a smooth B-spline fitting is done. The gradient of this function with respect to X gives the initial trial deformation gradient. If the process is a smooth deformation in a die, this amounts to guessing the equation of the die surface. Once an initial guess is formed, the constitutive problem is solved and the optimality condition is checked. A sketch of the algorithm is given below.

(1): Discretize the given domain spatially. Then discretize the time domain into n parts.

(2): Make an initial guess for the position values for all the nodes in the domain for all the intermediate time steps.

(3): Solve the constitutive problem to get $\bar{\mathbf{T}}$, $\tilde{\sigma}$, $\dot{\tilde{\sigma}}$, s

(4): Construct the error function

(a) If the model is rate independent, then

$$E = \int_0^{t_f} \| \mathbf{N}_{\mathbf{T}} - \mathbf{N}_{\dot{\mathbf{T}}} \| \, dt \qquad (38)$$

where $\mathbf{N}_{\mathbf{T}}$ is the direction of $\bar{\mathbf{T}}'$ and $\mathbf{N}_{\dot{\mathbf{T}}}$ is the direction of $\dot{\mathbf{T}}'$ and the norm is the L_2 norm.

(b) If the model is rate dependent, then

$$E = \int_0^{t_f} \left(\dot{\tilde{\sigma}} + h\, f \, \frac{\left(2\, f_s\, f_{\tilde{\sigma}} - f\, f_{\tilde{\sigma}s} \right)}{\left(2\, f_{\tilde{\sigma}}^{2} - f\, f_{\tilde{\sigma}\tilde{\sigma}} \right)} \right)^2 dt \qquad (39)$$

(5): Linearize the error E with respect to the unknown parameters in the B-spline space and use the Newton-Raphson method to find the increments on the unknowns.

(6): Update the unknown parameters, check for convergence and if not converged go to step (3).

Preform Design with Ideal Forming

Once the optimal path is calculated knowing both the initial and the final configurations, we can try to design the preform so as to obtain the final product with the least effort. *We assume that we are given only the final required configuration and we have to design the preform, intermediate shapes and the process conditions.*

(1): Discretize the final configuration.

(2): Guess an initial configuration such that each node in it corresponds to a particular node in the final configuration.

(3): Obtain the optimal path for this initial configuration that results in the required final configuration (refer to the previous algorithm).

(4): Evaluate w^p as a function of the initial configuration.

(5): Perform a Newton-Raphson iteration with main variables as the nodal locations of the initial configuration, until convergence is achieved. This step requires the linearization of w^p with respect to the deformation gradient. If not converged, go to step (3).

At the end of this step the preform has been designed along with all the intermediate shapes and the deformation gradient history. Knowing the deformation gradient history, one can go ahead and calculate the thermo-mechanical interface boundary conditions that result in the required deformation gradient history. The deformation history which we have calculated may not be achievable all the time. Hence, we may have to consider a family of deformation gradients that are achievable and find the optimum one from this space. Also, while designing the preform shape, there may be some constraints imposed on the preform. That should also be taken into account while applying the Newton-Raphson method.

CONCLUSIONS

A few inverse, design and parameter estimation problems of great interest to the materials industry have been highlighted. Such problems present a formidable challenge in the areas of computational mechanics, applied mathematics, applied forming processing and experimentation.

The objective of these problems is to design processes with a minimum cost that lead to products with desired thermomechanical properties. The main variables to be controlled in the proposed analysis, include the die surface, the preforms, the history of thermomechanical loading during the process and other. The non-uniqueness of the solution of these problems can be avoided by considering constrained optimization problems and spatial and/or temporal regularization.

This work could have two major applications in materials processing.

(1) It will provide a computational methodology based on Applied Mechanics and Applied Mathematics techniques that can be used to design two- and three-dimensional forming processes that achieve a product with desired structure or result in minimum deficiencies in the finished product due to the generation of residual stresses and distortions; and

(2) It will improve our knowledge about the process by applying inverse techniques to assist in the calculation of the applied load conditions and optimum die surfaces. Guidance will be provided on the type, number and location of measurements required to achieve the best possible knowledge about the thermomechanical material state during the forming process. This will further enhance our understanding of the process and will lead to better designing and evaluating of forming processes.

ACKNOWLEDGEMENTS

This work was funded by NSF grant DDM-9157189 to Cornell University. The computing for this project was supported by the Cornell National Supercomputer facility, which receives major funding by the NSF and IBM Corporation, with additional support from New York State.

REFERENCES

1. Hadamard, J. (1902), "Sur les problèmes aux dérivées partielles et leur signification physique", *Bull. Univ. Princeton*, **13**.

2. Beck, J. V., Blackwell, B., and St. Clair, C. R. (1985), *Inverse Heat Conduction: Ill-Posed Problems*, Wiley-Interscience, New York.

3. Weber, G. and Anand, L. (1990), "Finite Deformation Constitutive Equations and a Time Integration Procedure for Isotropic, Hyperelastic-Viscoplastic Solids", *Comp. Meth. in Appl. Mech. and Eng.*, **79**, 173-202.

4. Zabaras, N. and Arif, A. F. M. (1992), " A Family of Integration Algorithms for Constitutive Equations in Finite Deformation Elasto-Viscoplasticity", *Int. j. numer. methods eng.*, **33**, 59-84.

5. Arif, A. F. M. and Zabaras, N. (1992), " On the Performance of Two Tangent Operators for Finite Element Analysis of Large Deformation Inelastic Problems", *Int. j. numer. methods eng.*, **35**, 369-389.

6. Badrinarayanan, S. and Zabaras, N. (1993), " A Hyperelastic-ThermoViscoplastic FEM model for Hot Forming Processes", in preparation.

7. Kobayashi, S., Oh, S., and Altan, T. (1989), *Metal Forming and the Finite-Element Method*, Oxford University Press, New York.

8. Park, J.,. Rebelo, N., and Kobayashi, S. (1983), "A New Approach to Preform Design in Metal Forming with the Finite Element Method," *Int. J. Mech Tool Des. Res.* **23**, 71-79.

9. Hwang, S.M. and Kobayashi, S. (1987), "Preform Design in Shell Nosing at Elevated Temperatures," *Int. J. Mech. Tool Manufacture*, **27**, 1-14.

10. Hwang, S.M. and Kobayashi, S. (1984), "Preform Design in Plane-Strain Rolling by the Finite-Element Method," *Int. J. Mech. Tool Des. Res.*, **24**, 253-266.

11 Hwang, S.M. and Kobayashi, S. (1986), "Preform Design in Disk Forging," *Int. J. Mech. Tool Des. Res.*, **26**, 231-243.

12. Schnur, D. and Zabaras, N. (1992), " An Inverse Method for Determining Elastic Material Properties and a Material Interface", *Int. j. numer. methods eng.*, **33**, 2039-2057.

13. Schnur, D. S. and Zabaras, N. (1990), "Finite element solution of two-dimensional inverse elastic problems using spatial smoothing", *Int. j. numer. methods eng.*, **30**, 57-75.

14. Zabaras, N. (1990), "Inverse Modeling of Solidification and Welding Processes", Modeling of Casting, Welding and Advanced Solidification Processes, (Proceedings of the Fifth International Conference on Modeling of Casting and Welding Processes, Davos, Switzerland, September 16-21, 1990) ed. M. Rappaz et al., 523-530.

15. Zabaras, N., Ruan, Y., and Richmond, O. (1992), " On the Design of Two-Dimensional Stefan Processes with Desired Freezing Front Motion", *Numerical Heat Transfer*, Part B, **21** (1992), 307-325.

16. Zabaras, N. and Kang S. (1993), " On the Solution of an Ill-Posed Inverse Design Solidification Problem Using Minimization Techniques in Finite and Infinite Dimensional Spaces", *Int. j. numer. methods eng.*, in press.

17. Fletcher, R. (1987), *Practical Methods of Optimization*, John Wiley & Sons, New York.

18. Schumaker, L. L. (1981), *Spline Functions: Basic Theory*, John Wiley & Sons, New York.

19. Tikhonov, A. N. and Arsenin, V. Y. (1977), *Solution of Ill-Posed Problems*, V. H. Winston, Washington, D.C.

20. Lee, T.-Y. and Seinfeld, J. H. (1987), "Estimation of Petroleum Reservoir Properties", Proc. 26th IEEE Conf. on Decision and Control, Los Angeles, CA, December 9-11, 1386-1390.

21. Morozov, V. A. (1984), *Methods of Solving Incorrectly Posed Problems*, Springer-Verlag, New York.

22. Kang, S. and Zabaras, N. (1993), "On the Optimization of Residual Stresses in a Unidirectional Casting Process", Modeling of Casting, Welding and Advanced Solidification Processes VI (proceedings of the Sixth International Conference on Modeling of Casting and Welding Processes, held at Palm Coast, Florida, March 21-26, 1993), edt. T. S. Piwonka, 655-662.

23. Hill, R. (1986) "Extremal Paths of Plastic Work and Deformation," *J. Mech. Phys. Solids*, **34**, No. 5, 511-523.

24. Ponter, A.R.S. and Martin, J.B. (1972) "Some Extremal Properties and Energy Theorems for Inelastic Materials and their Relationship to the Deformation Theory of Plasticity," *J. Mech. Phys. Solids*, **20**, 281-300.

25. Chung, K. and Richmond, O. (1992) "Sheet Forming Process Design Based on Ideal Forming Theory," NUMIFORM.1992, eds. Chenot, Wood and Zienkiewicz, 455-460.

26. Chung, K. and Richmond, O. (1992), "Ideal Forming - I. Homogeneous Deformation with Minimum Plastic Work," *Int. J. Mech. Sci.*, **34**, No. 7, 575-591.

27. Chung, K. and Richmond, O. (1992), "Ideal Forming - II. Sheet Forming with Optimum Deformation," *Int. J. Mech. Sci.*, **34,** No. 8, 617-633.

Inverse Problems in Engineering: Theory and Practice
ASME 1993

A NUMERICAL INVESTIGATION OF THE ELASTIC MODULI IN AN INHOMOGENEOUS BODY

Andrei Constantinescu
Laboratoire de Mécanique des Solides
École Polytechnique, Mines, Ponts et Chaussées
Palaiseau, France

Abstract

In this paper we present a numerical method for the identification of the interior distribution of the elastic moduli in an inhomogeneous body from boundary measurements. We suppose that the body is isotropic and that we dispose of a number of simultaneous displacement and force measurements on the hole boundary. The proposed variation-al method minimizes the error on constitutive law. Numerical results are presented for the determination of a continuous and a discontinuous interior distribution of the Young modulus.The Poisson coefficient was taken constant. The boundary data was provided by numerical simulation and introduction of a uniform noise. The method is found to be generally robust reproducing the general characteristics of the interior distribution.

1 Introduction

Traditional problems in mechanics are looking for the response of a structure to a known force (or displacement), were the parameters of the structure are supposed to be known. The inverse problem studied in this paper seeks to determine the unknown parameters of the structure, from the known force and the known response of the structure.

Our structure is an elastic body of known shape, with the interior distribution of the elastic moduli as the unknown parameters. The response of this structure to a known force is given by a measurable boundary displacement. The identification problem we are posing here, seeks to recover the interior distribution of elastic moduli from displacement and force measurements on the boundary.

Inverse problems of this kind have been treated in exploration seismology and acoustic non-destructive testing. Contrary to these techniques we are interested in the static response of the structure.

The focus of this article is to explore a numerical method to solve the identification problem in linear isotropic elastostatics. The method is based on the minimization of the error on constitutive law on statically and kinematic ally admissible fields. The error on constitutive law was introduced in elasticity by Ladevèze and Leguillon [5] to study an expression of the finite element error, leading to a distribution of the calculation accuracy. Later, Ladevèze, Reynier and Nedjar [6] have studied an inverse problem for free elastic vibrations, using the error on constitutive law as a criterion to change the rigidity or the mass distribution of the finite element model to find the given modal values and frequencies.

An approach similar to the one presented here, also based on the error on constitutive law, has been used by Kohn and McKenney to determine the interior distribution of electric conductivities from boundary measurements. The elastic and electric problems are closely related to each other as both are looking for the unknown coefficient of an elliptic equation. The electric problem received much attention in the recent time both from mathematical and technical point of view; a survey of the existing bibliography on the subject can be found in [2] and [3].

The key point of this paper is a new expression of the error on constitutive law. This expression is obtained by using the decomposition of tensors in spherical and deviatorical parts, and the bulk and shear modulus as elastic moduli. This expression permits the application of minimization methods similar to those used by

Kohn and McKenney [3] for the electric identification problem. The algorithms are an alternating direction implicate (ADI) method and a modified newton (MN) method.

In order to demonstrate the practical applicability of this technique, we give some numerical examples. Using a direct finite element calculation we provide a set of boundary data (displacement and force measurements). From this data, we then reconstruct the distribution of the elastic moduli using the ADI method.

We have studied continuous and discontinuous distributions of elastic moduli, and influence of uniform noise on the measurements. The technique could be qualified as generally robust, as principal characteristics are recovered even from low noise measurements.

2 The identification problem

We consider a linear elastic body occupying in the reference configuration a finite regular domain Ω, with the boundary $\partial\Omega$. Let $\mathbf{u}, \boldsymbol{E}, \boldsymbol{T}$ stand for the vector field of displacements, and the related tensor fields of strain and stress respectively.

We also assume that the general behavior of the body is governed by the following system of equations:

$$\boldsymbol{E} = \frac{1}{2}(\nabla \boldsymbol{u} + \nabla^T \boldsymbol{u}), \quad \boldsymbol{T} = \mathrm{C}(x)\boldsymbol{E}, \quad div\, \boldsymbol{T} = 0, \qquad (1)$$

were C denotes the forth order elasticity tensor, and $\boldsymbol{x}$ the current point of Ω. We have written $\mathrm{C}(\boldsymbol{x})$ to stress that we suppose the body inhomogeneous. For an elastic isotropic body the constitutive law takes the following form:

$$\boldsymbol{T} = \mathrm{C}_\gamma(\boldsymbol{x})\boldsymbol{E} = \lambda(\boldsymbol{x})(tr\boldsymbol{E})1 + 2\mu(\boldsymbol{x})\boldsymbol{E} \qquad (2)$$

where $\gamma(\boldsymbol{x}) = (\lambda(\boldsymbol{x}), \mu(\boldsymbol{x}))$ are the Lamé coefficients.

If $\mathrm{C}(\boldsymbol{x})$ is positive definite for each $\boldsymbol{x} \in \Omega$, then the equation (1) are equivalent with an elliptic equation for the displacement $\boldsymbol{u}$ in Ω:

$$div\,(\mathrm{C}(\boldsymbol{x})\nabla \boldsymbol{u}(\boldsymbol{x})) = 0. \qquad (3)$$

In *direct problems* the distribution of $\mathrm{C}(\boldsymbol{x})$, is known for each $\boldsymbol{x} \in \Omega$, and the displacement $\boldsymbol{u}$ solution of (1), is determined by one of the following boundary conditions:

- imposed displacements: $\boldsymbol{u}|_{\partial\Omega} = \boldsymbol{\xi}$,
- imposed forces: $\boldsymbol{Tn}|_{\partial\Omega} = \boldsymbol{\varphi}$.

The solution of these problems is unique in the first case and unique modulo a rigid displacement in the second one.

The *inverse problem* presented here seeks to determine the distribution of $\mathrm{C}(\boldsymbol{x})$ in Ω from a super abundant boundary conditions. We assume that we *simultaneously* known the displacement $\boldsymbol{u}|_{\partial\Omega}$ and the force $\boldsymbol{Tn}|_{\partial\Omega}$ on the boundary for various independent solutions $\boldsymbol{u}$ of the elastic displacement equation (1). *Perfect* knowledge of *all* possible measurements is equivalent with the knowledge of the 'Dirichlet-to-Neumann' data map:

$$\Lambda_{\mathrm{C}} : \boldsymbol{u}|_{\partial\Omega} \longrightarrow \boldsymbol{Tn}|_{\partial\Omega}$$

associating to each boundary displacement, the corresponding boundary force. It is the map Λ_{C} that determines C. In effect, Nakamura and Uhlmann [7] have proven for a two dimensional isotropic body that C, is uniquely determined by Λ_{C}. Moreover, Ikehata [4] has proven a constructive result for the linearized problem in isotropic elasticity. For Lamè moduli of the form: $\lambda(\boldsymbol{x}) = \lambda_0 + \delta\lambda(\boldsymbol{x})$ and $\mu(\boldsymbol{x}) = \mu_0 + \delta\mu(\boldsymbol{x})$, with λ_0 and μ_0 constants, and $\lambda \gg \delta\lambda$ and $\mu \gg \delta\mu$, we can express the spatial Fourier transform of $\delta\lambda$ and $\delta\mu$ in terms of the work $\boldsymbol{u_k} \cdot \Lambda_{\mathrm{C}}(\boldsymbol{u_k})$ on the boundary. For each point $\boldsymbol{k}$ of the transformed space , $\boldsymbol{u_k}$ is a known displacement solution of the elastic problem of moduli λ_0 and μ_0. Even if constructive, the result of Ikehata is of limited numerical interest. It is well known that the inversion of spatial Fourier transform is not a well posed problem, being very sensitive to sampling.

Both results cited before consider an isotropic body. To our knowledge no result has been proven for the anisotropic body.

In practice, one cannot expect to measure Λ_{C} for *all* possible boundary displacements. We could only hope to know $\boldsymbol{\varphi} = \Lambda_{\mathrm{C}}(\boldsymbol{\xi})$ for a number of boundary displacement distributions $\boldsymbol{\xi}$. But even this is practically impossible, as we measure in a finite number of points of the boundary and within the accuracy of our instruments. That signifies, we are obliged to interpolate after-words the measured values to the hole boundary.

In view of this considerations we shall assume that the body is isotropic and that we dispose of N "measurements" on the hole boundary $\partial\Omega$:

$$(\boldsymbol{\xi}_i, \boldsymbol{\varphi}_i)_{i=1}^{N}, \qquad \boldsymbol{\varphi}_i \approx \Lambda_\gamma(\boldsymbol{\xi}_i),$$

The measurement $\boldsymbol{i}$ consists of a pair $(\boldsymbol{\xi}_i, \boldsymbol{\varphi}_i)$, where $\boldsymbol{\xi}_i$ of a boundary displacement $\boldsymbol{\varphi}_i$ and the related boundary force.

3 The error on constitutive law

Our problem can be stated as follows:

For N given measurements (in other words, pairs $(\xi_i, \varphi_i)_{i=1}^N$, such that $\varphi_i = \Lambda_\gamma(\xi_i)$), we are looking for a solution:

$$(\gamma, u_1, ..., u_N, T_1, ..., T_N)$$

of the following system of equations:

$$E_i = \frac{1}{2}(\nabla u_i + \nabla^T u_i), \tag{4}$$

$$T_i = C_\gamma E_i, \tag{5}$$

$$div\, T_i = 0 \qquad i = 1, N \tag{6}$$

with the corresponding Boundary conditions:

$$u_i|_{\partial\Omega} = \xi i, \qquad i = 1, N \tag{7}$$

$$T_i n|_{\partial\Omega} = \varphi_i \qquad i = 1, N \tag{8}$$

If we take the constitutive law apart, we can group the equations and boundary conditions to obtain statically ((6) and (8)) and kinematically ((4) and (7)) admissible fields. It is now natural, to look for elastic moduli, statically and kinnematically admissible fields, such that these fields fit best in the constitutive equation. That means we want to minimize a norm (defined by the constitutive law), over elastic moduli, statically and kinnematically admissible fields. We have chosen the norm [1] $|C_\gamma^{-\frac{1}{2}}T - C_\gamma^{\frac{1}{2}}E|$, rather than $|C_\gamma^{-1}T - E|$ or $|T - C_\gamma E|$, because this norm has the physical dimension of the energy. This choice will later show up as very advantageous in our reasoning. We are thus looking for the minimum of the error on constitutive law:

$$I(\gamma, u_1, ..., u_N, T_1, ..., T_N) = \tag{9}$$

$$\sum_{i=1}^{N} \frac{1}{2} \int_\Omega |C_\gamma^{-\frac{1}{2}} T_i - C_\gamma^{\frac{1}{2}} E_i|^2 \, dx \tag{10}$$

over all arguments subject to the constraints:

$$u_i|_{\partial\Omega} = \xi_i, \quad T_i n|_{\partial\Omega} = \varphi_i, \qquad (i = 1, N) \tag{11}$$

The minimum value of I is exactly 0 , and it is achieved exactly for the solution of the identification problem.

The definition of I, doesn't use the assumption of isotropy.

At this point we make use of the assumption of isotropy. We shall rewrite I, splitting tensors in spherical and deviatorical parts, and expressing the elastic moduli in terms of the bulk modulus $\eta = (3\lambda + 2\mu)$, and the shear modulus $\omega = 2\mu$. This means:

$$I(\eta, \omega, u_1, ..., u_N, T_1, ..., T_N) =$$

$$\sum_{i=1}^{N} \frac{1}{2} \int_\Omega [\omega^{-\frac{1}{2}} \overset{o}{T}_i - \omega^{\frac{1}{2}} \overset{o}{E}_i]^2 \, dx +$$

$$+\sum_{i=1}^{N} \frac{1}{2} \int_{\partial\Omega} \frac{1}{3} [\eta^{-\frac{1}{2}}(trT_i) - \eta^{\frac{1}{2}}(trE_i)]^2 \, dx \tag{12}$$

The advantage of the previous choice is the splitting of I in two terms each containing only one elastic coefficient. By developing the squares we obtain: [2]

$$I(\eta, \omega, u_1, ..., u_N, T_1, ..., T_N) = \tag{13}$$

$$\sum_{i=1}^{N} \frac{1}{2} \int_\Omega [\omega^{-1} \overset{o}{T}_i \cdot \overset{o}{T}_i + \omega \overset{o}{E}_i \cdot \overset{o}{E}_i] \, dx$$

$$+\sum_{i=1}^{N} \frac{1}{6} \int_\Omega [\eta^{-1}(trT_i)^2 + \eta(trE_i)^2] \, dx$$

$$-\sum_{i=1}^{N} \frac{1}{2} \int_\Omega T_i \cdot E_i \, dx$$

Applying the theorem of work and energy for the last term we have:

$$\sum_{i=1}^{N} \frac{1}{2} \int_\Omega T_i \cdot E_i \, dx = \sum_{i=1}^{N} \int_\Omega \xi_i \cdot \varphi_i \, dx$$

He is thus constant for a given boundary data set $(\xi_i, \varphi_i)_{i=1}^N$, and is irrelevant to the process of minimization.

This fact is a direct consequence of the initial choice of the norm in the error on constitutive law.

4 The minimization algorithms

In this section we shall give two minimization algorithms for the error on constitutive law using the previous developments of I. The minimization schemes are similar to the ones proposed by Kohn and McKenney [3] in the electric problem.

The first algorithm uses an *alternating direction implicit* (ADI) method for the minimization. A single iteration consists of the following steps:

1. with fixed [3] (η, ω) compute $(E_i)_{i=1,N}^N$ from the solutions u_i of (3) with the boundary conditions: $u_i|_{\partial\Omega} = \varphi_i$, $i = 1, N$;

[1] C is symmetric et positive definite so $C^{\frac{1}{2}}$ and $C^{-\frac{1}{2}}$ are well defined.

[2] Let $\cdot$ denote the double contracted tensor product and the simple contacted vector product.

[3] by an initial guess or by the previous iteration

2. with the same (η, ω) as before, compute $(\boldsymbol{T}_i)_{i=1,N}^{N}$ from the solutions $\boldsymbol{v}_i$ of (3) with the boundary conditions: $\boldsymbol{T}_i \boldsymbol{n}|_{\partial\Omega} = \boldsymbol{\xi}_i$, $i = 1, N$;
3. update $(\eta(\boldsymbol{x}), \omega(\boldsymbol{x}))$ for every $\boldsymbol{x} \in \Omega$ minimizing: $J(\eta, \omega) = I(\eta, \omega, \boldsymbol{E}_1, ..., \boldsymbol{E}_N, \boldsymbol{T}_1, ..., \boldsymbol{T}_N)$ where $(\boldsymbol{E}_i)_{i=1}^{N}$, $(\boldsymbol{T}_i)_{i=1}^{N}$ are the last computed values

The last step has the advantage of being trivial. The optimal choice of $(\eta(\boldsymbol{x}), \omega(\boldsymbol{x}))$ is obtained by minimizing pointwise the integrand of I:

$$\begin{aligned} \omega(\boldsymbol{x}) &= [\sum_{i=1}^{N} \mathring{\boldsymbol{T}}_i(\boldsymbol{x}) \, \mathring{\boldsymbol{T}}_i(\boldsymbol{x})]^{\frac{1}{2}} [\sum_{i=1}^{N} \mathring{\boldsymbol{E}}_i(\boldsymbol{x}) \, \mathring{\boldsymbol{E}}_i(\boldsymbol{x})]^{-\frac{1}{2}} \\ \eta(\boldsymbol{x}) &= [\sum_{i=1}^{N} (tr\boldsymbol{T}_i(\boldsymbol{x}))^2]^{\frac{1}{2}} [\sum_{i=1}^{N} (tr\boldsymbol{E}_i(\boldsymbol{x}))^2]^{-\frac{1}{2}}. \end{aligned} \quad (14)$$

This method has the advantage of decreasing the value of I at every step. Its disadvantage is that it converges very slowly once I get near an minimum. That is also a reason why in some practical cases (for example noise data) we effectively increase the distance between the real moduli and the computed ones.

The second scheme is a *modified Newton* (MN) method. At every step we approximate I by a quadratic form, which is minimized, and we proceed for the next approximation.

To simplify the writing, we introduce the following notations:

$$\boldsymbol{u} = (\boldsymbol{u}_1, ..., \boldsymbol{u}_N), \boldsymbol{E} = (\boldsymbol{E}_1, ..., \boldsymbol{E}_N), \boldsymbol{T} = (\boldsymbol{T}_1, ..., \boldsymbol{T}_N)$$

Furthermore we shall completely neglect the details of calculation and present only the principal steps of the reasoning. The final expressions of the approximations are given in the appendix. We begin by computing a quadratic approximation $I_0 + \delta I_0 + \frac{1}{2}\delta^2 I_0$ of I . The computation evaluates I at the point $\eta = \eta_0 + \delta\eta$, $\omega = \omega_0 + \delta\omega$, $\boldsymbol{E} = \boldsymbol{E}_0 + \delta\boldsymbol{E}$, $\boldsymbol{T} = \boldsymbol{T}_0 + \delta\boldsymbol{T}$ and expands to second order. In order to fulfill the boundary conditions, we impose to the perturbations $\delta\boldsymbol{u}$ and $\delta\boldsymbol{T}$ the constraints:

$$\delta\boldsymbol{u}_i|_{\partial\Omega} = 0, \quad \delta\boldsymbol{T}_i\boldsymbol{n}|_{\partial\Omega} = 0, \quad div\,\delta\boldsymbol{T}_i = 0 \;\; i = 1, N. \quad (15)$$

By using the expansion to first order in δE, δT of the "optimal choice" (14), we can eliminate the terms $\delta\omega$ and $\delta\eta$ from $I_0 + \delta I_0 + \frac{1}{2}\delta^2 I_0$. This gives a new quadratic approximation $I_0 + \delta I_0' + \frac{1}{2}\delta^2 I_0'$ of I, with I_0 as before. Unfortunately this approximation is not convex, as the Hessian $\frac{1}{2}\delta^2 I_0'$ is not positive definite. Its then interesting to introduce a regularization parameter $0 \leq \epsilon \ll 1$ such that $\frac{1}{2}\delta^2 I_0''(\epsilon)$ is positive definite. We shall then use the convex approximation $I_0 + \delta I_0' + \frac{1}{2}\delta^2 I_0''(\epsilon)$ in the minimization scheme.

The steps of a single iteration of the NM scheme can be expressed as:

1. update (η^0, ω^0) using the last computed value of $(\boldsymbol{E}_i)_{i=1,N}^{N}$, $(\boldsymbol{T}_i)_{i=1,N}^{N}$ and the "optimal choice" (14)
2. minimization of $I_0 + \delta I_0' + \frac{1}{2}\delta^2 I_0''$ as a function of $\delta\boldsymbol{T}$, $\delta\boldsymbol{E}$ under the constraints (15).

5 Test calculations using the IDA method

In this part we present some numerical results using the IDA method for the two dimensional identification problem.

The solution of direct elastic problems, needed for the construction of the synthetic data or for the process of identification, were obtained by the finite element method, using FEM code CASTEM2000 (CEA, France).

All our numerical test were done on a square domain, 1 unit by 1 unit. It was divided into $n \times n$ quadrangular elements. The finest mesh had 48 x 48 elements.

The displacement and the force measurements on the boundary were created by solving direct elastic problems with the "real" distribution of elastic moduli. We imposed the boundary forces $(\varphi_i)_{i=1}^{N}$ in order to solve the following N Neumann problems:

$$div\,(\mathbf{C}_{reel}\boldsymbol{E}(\boldsymbol{u}_i)) = 0$$
$$\mathbf{C}_{reel}\boldsymbol{E}(\boldsymbol{u}_i)\boldsymbol{n}|\partial\Omega = \varphi_i$$

and to extract the boundary displacements $\boldsymbol{\xi}_i = \boldsymbol{u}_i|_{\partial\Omega}$.

As imposed forces we have chosen a parabolic pressure distribution over a number of nodes. Only the application point of this pressure distribution changed from one measurement to the other. The choice of a parabolic pressure distribution, was inspired by the Hertz pressure solution of the contact problem of two elastic bodies.

The results obtained in electricity showed finer reconstruction-s when concentrated moments were used as a force distribution. This fact could be related to the intimate mathematical structure of the identification problem. We did not consider concentrated moments, because they are difficult from the experimental point of view.

The noise was introduced at this point in the tests . The formulas for the noisy data are:

$$\boldsymbol{\xi}_i^n = \boldsymbol{\xi}_i + a\; r(\boldsymbol{x})\; max\boldsymbol{\xi} \quad (16)$$

$$\varphi_i^n = \varphi_i + +a\ r(x)\ max\xi\ , \qquad (17)$$

were n is a the amplitude of the perturbation given in percent, $r(x)$ is a random number in $[-1, 1]$. The proportionality of noise to the maximum of measured values is very penalisating choice, we generally accept a linear dependence of noise to the measures value. This time we made this choice to test the robustness of the method.

In the test presented here we took as real distributions of elastic moduli, a varying Young modulus and a constant Poisson coefficient. The Young moduli varied continuously after an exponential law, or discontinuously representing square inclusions. The exact form of distributions is:

- exponential distribution (fig. 7):
  $E_{real}(x) = E_0 + \delta E_1\ exp(-x^2/l)$

- square inclusion:

$$E_{real} = \begin{cases} E_0 & \text{si } x \in \Delta_1 \\ E_0 + \delta E_1 & \text{si } x \in \Omega - \Delta_1 \end{cases}$$

- sandwich inclusion (one square inclusion into another, fig. 3):

$$E_{real} = \begin{cases} E_0 & \text{si } x \in \Delta_1 \\ E_0 + \delta E_1 & \text{si } x \in \Delta_2 - \Delta_1 \\ E_0 + \delta E_2 & \text{si } x \in \Omega - \Delta_2 \end{cases}$$

Δ_1 and Δ_2 are squares such that: $\Delta_1 \subset \Delta_2 \subset \Omega$. As

a result of the tests, we can state the following remarks:

- The calculated moduli are after 5-10 iterations at 10-20% distance in relative error from the real moduli. The relative error depends of the moduli distribution, and the noise level. For high noise levels ($n > 5\%$) we rest sometimes even at more than 20% distance in relative error (fig. 5)

  If the iterations are continued, the relative error decreases slowly (fig. 4), and in some cases we arrive even to an increase of the relative error (fig 5).

- The error on constitutive law has a similar behavior. After a great descent in the first iterations we asymptotically tend to a limit value, with little decrease (fig. 12).

  However, it never happened that the error on constitutive law increases.

  Noise can influence the limiting value. It changes also the descent velocity of the first iterations (fig. 10).

- The general characteristics (location of inclusions or maxima, level) of the distribution of elastic moduli are found in the first 5-15 iterations (fig. 1, 2, 11).

  We were not capable of reconstructing sharp corners (fig. 1, 2), or to reproduce a soft inclusion around-ed by a harder kernel (fig. 3, 6).

- The noise affects essentially the values of the distribution on a strip near the boundary (fig. 2, 8, 9).

- For great $\delta E/E$ (> 1.5) we remark oscillations in the reconstructed distribution of the moduli, increasing with the number of oscillations.

- An important stabilization effect had been obtained by applying measurements near the corners of the domain.

Similar remarks were also made for the electric identification problem.

6 Conclusions

Presented above is a numerical method for determining the interior distribution of elastic moduli by boundary measurements. It was seen that the solution was still found in the presence of noise, even if with less accuracy than if noise was absent. A general enhancement of the solution accuracy could be obtained, as in most inverse problems by introducing some apriori information in the process. This would essentially decrease the number of unknown of the problem, conducting thus to a better result.

It may be concluded that the method could be extended to solve identification problems in linear elastostatics of a more general nature.

References

[1] Bonnet M., Bui H.D., Maigre H., Planchard J. - *Identification of heat conduction coefficient: application to nondestructive testing* IUTAM Symposium on Inverse problems in engineering mechanics (11-15 May 1992, Tokyo, Japan), ed M.Tanaka and H.D. Bui, Springer Verlag, 1993

[2] Bonnet M., Bui H.D., Maigre H., Planchard J. - Identification of heat conduction coefficient: application to nondestructive testing. In *IUTAM Symposium on Inverse problems in engineering mechanics (11-15 May 1992, Tokyo, Japan).*, ed. M.Tanaka and H.D. Bui, Springer Verlag, 1993.

[3] R.Kohn et A.McKenney - Numerical implementation of a variational method for electric impedance

tomography *Inverse problems*, No. 6 , 1990, p. 389-414

[4] M. Ikehata - *Inversion for the linearized problem for an inverse boundary value problem in elastic prospection SIAM J.Appl.Math.*, Vol. 50 , No. 6 , dec. 1990 , p. 1635-1644

[5] P.Ladevèze and D.Leguillon - Error estimates procedures in the finite element method and applications *SIAM J.Numer.Anal.*, Vol.20, No.3, June 1983

[6] P.Ladevèze, M.Reynier and D.Nedjar - Parametric Correction of Finite Element Models using Modal Tests *IUTAM Symposium on Inverse problems in engineering mechanics (11-15 May 1992, Tokyo, Japan).*, ed. M.Tanaka and H.D. Bui, Springer Verlag, 1993.

[7] G.Nakamura et G.Uhlmann - Uniqueness for identifying Lamé moduli by Dirichlet to Neumann map *Inverse Problems in Engineering Sciences* - ICM-90 Satellite Conference Proceedings, ed. M.Yamaguti, Springer Verlag, Tokio 1991

[8] Z.Sun et G.Uhlmann *Generic uniqueness for determined inverse problems in 2 dimensions Inverse Problems in Engineering Sciences* - ICM-90 Satellite Conference Proceedings, ed. M.Yamaguti, Springer Verlag, Tokio 1991

The second order approximations of I

We give here the detailed expressions for the approximations of I , which have been used in the description of the MN minimization method.

The terms of the expansion of I at second order are:

$$\begin{aligned} I_0 &= I(\eta_0, \omega_0, \boldsymbol{E}_0, \boldsymbol{T}_0) \\ &= \frac{1}{2}\int_\Omega [\omega_0^{\frac{1}{2}} \mathring{\boldsymbol{E}}_0 - \omega_0^{-\frac{1}{2}} \mathring{\boldsymbol{T}}_0]^2\, dx \\ &+ \frac{1}{6}\int_\Omega [\eta_0^{\frac{1}{2}}(tr\boldsymbol{E}_0) - \eta_0^{-\frac{1}{2}}(tr\boldsymbol{T}_0)]^2\, dx \end{aligned}$$

$$\begin{aligned} \delta I_0 = \frac{1}{2}\int_\Omega [(| \mathring{\boldsymbol{E}}_0 |^2 - \frac{1}{\omega^2}| \mathring{\boldsymbol{T}}_0 |^2)\,\delta\omega\, + \\ 2\omega_0\, \mathring{\boldsymbol{E}}_0 \cdot\delta\, \mathring{\boldsymbol{E}} + \frac{2}{\omega_0}\, \mathring{\boldsymbol{T}}_0 \cdot\delta\, \mathring{\boldsymbol{T}}]\, dx \\ + \frac{1}{6}\int_\Omega [((tr\, \boldsymbol{E}_0)^2 - \frac{1}{\eta^2}(tr\boldsymbol{T}_0)^2)\,\delta\eta\, + \\ 2\eta_0(tr\, \boldsymbol{E}_0)(tr\,\delta\boldsymbol{E}) \\ + \frac{2}{\eta_0}(tr\, \boldsymbol{T}_0)(tr\,\delta\, \mathring{\boldsymbol{T}})]\, dx \end{aligned}$$

$$\begin{aligned} \delta^2 I_0 = \int_\Omega [\frac{1}{\omega^3}| \mathring{\boldsymbol{T}}_0 |^2(\delta\omega)^2 + \\ 2(\mathring{\boldsymbol{E}}_0 \cdot\delta\, \mathring{\boldsymbol{E}} - \frac{1}{\omega^2}\, \mathring{\boldsymbol{T}}_0 \cdot\delta\, \mathring{\boldsymbol{T}})\,\delta\omega + \omega_0|\delta\boldsymbol{E}|^2 + \frac{1}{\omega_0}|\delta\boldsymbol{T}|^2]\, dx \\ + \frac{1}{3}\int_\Omega [\frac{1}{\eta^3}|tr\,\boldsymbol{T}_0|^2(tr\,\delta\boldsymbol{T}_0)^2 + 2(tr\,\boldsymbol{E}_0)(tr\,\delta\boldsymbol{E}_0)]\, dx \\ + \int_\Omega [-\frac{1}{\eta^2}(tr\,\boldsymbol{T}_0)(tr\,\delta\boldsymbol{T})\,\delta\eta + \eta_0(tr\,\delta\boldsymbol{E})^2 \\ + \frac{1}{\eta_0}(tr\,\delta\boldsymbol{T})^2]\, dx \end{aligned}$$

The second ordre expansion in $\delta\boldsymbol{T}$, $\delta\boldsymbol{E}$ of the "optimal choice" (14) is:

$$\begin{aligned} \delta\omega = | \mathring{\boldsymbol{T}}_0 |^{-1} | \mathring{\boldsymbol{E}}_0 |^{-1} (\mathring{\boldsymbol{T}}_0 \cdot\delta\, \mathring{\boldsymbol{T}} \\ - \omega_0^2\, \mathring{\boldsymbol{E}}_0 \cdot\delta\, \mathring{\boldsymbol{E}}) \\ + O(|\delta\boldsymbol{T}|^2 + |\delta\boldsymbol{E}|^2) \\ \delta\eta = (tr\,\boldsymbol{T}_0)^{-1}(tr\,\boldsymbol{E}_0)^{-1}((tr\,\boldsymbol{T}_0)(tr\,\delta\boldsymbol{T}) \\ - \eta_0^2(tr\,\boldsymbol{E}_0)(tr\,\delta\boldsymbol{E})) + O(\delta\boldsymbol{T}|^2 + |\delta\boldsymbol{E}|^2) \end{aligned}$$

With the previous expressions of $\delta\omega$ and $\delta\eta$ the the terms of the approximations of I became:
$I = I_0 + \delta I_0' + \frac{1}{2}\delta^2 I_0' + O(|\delta\boldsymbol{T}|^2 + |\delta\boldsymbol{E}|^2)$:

$$\begin{aligned} \delta I_0' = \int_\Omega [\omega_0\, \mathring{\boldsymbol{E}}_0 \cdot\delta\, \mathring{\boldsymbol{E}} + \frac{1}{2\omega_0}\, \mathring{\boldsymbol{T}}_0 \cdot\delta\, \mathring{\boldsymbol{T}}]\, dx \\ + \int_\Omega [\eta_0(tr\,\boldsymbol{E}_0)(tr\,\delta\boldsymbol{E}) + \frac{1}{2\eta_0}(tr\,\delta\boldsymbol{T})(tr\,\boldsymbol{T})]\, dx \\ \delta^2 I_0' = \int_\Omega [\omega_0|\delta\, \mathring{\boldsymbol{E}} |^2 + \frac{1}{\omega_0}|\delta\, \mathring{\boldsymbol{T}} |^2]\, dx\, + \\ - \int_\Omega \frac{\omega_0}{| \mathring{\boldsymbol{E}}_0 |^2}(\mathring{\boldsymbol{E}}_0 \cdot\delta\, \mathring{\boldsymbol{E}} - \omega_0^2\, \mathring{\boldsymbol{T}}_0 \cdot\delta\, \mathring{\boldsymbol{T}})^2\, dx \\ + \int_\Omega [\eta_0|(tr\,\delta\boldsymbol{E})|^2 + \frac{1}{\eta_0}|(tr\,\delta\boldsymbol{T})|^2]\, dx \\ - \int_\Omega \frac{\eta_0}{|(tr\,\boldsymbol{E}_0)|^2}((tr\,\boldsymbol{E}_0)(tr\,\delta\boldsymbol{E}) \\ - \eta_0^2(tr\,\boldsymbol{T}_0)(tr\,\delta\boldsymbol{T}))^2\, dx \end{aligned}$$

The modified Hessian with the regularisation parameter $0 \leq \epsilon \ll 1$ is:

$$\begin{aligned} \delta^2 I'' = \int_\Omega [\omega_0|\delta\, \mathring{\boldsymbol{E}}_0 |^2 + \frac{1}{\omega_0}|\delta\, \mathring{\boldsymbol{T}}_0 |^2 \\ - \frac{1}{1+\epsilon}\frac{\omega_0}{| \mathring{\boldsymbol{E}}_0 |^2}(\mathring{\boldsymbol{E}}_0 \cdot[\delta\, \mathring{\boldsymbol{E}} - \frac{1}{\omega_0}\delta\, \mathring{\boldsymbol{T}}])^2]\, dx \\ + \int_\Omega \eta_0|tr\,\delta\boldsymbol{E}_0|^2 + \frac{1}{\eta_0}|tr\,\delta\boldsymbol{T}_0|^2 \\ - \frac{1}{1+\epsilon}\frac{\eta_0}{|tr\boldsymbol{E}_0|^2}((tr\boldsymbol{E}_0)[(tr\,\delta\boldsymbol{E}) - \frac{1}{\eta_0}(tr\,\delta\boldsymbol{T})])^2\, dx \end{aligned}$$

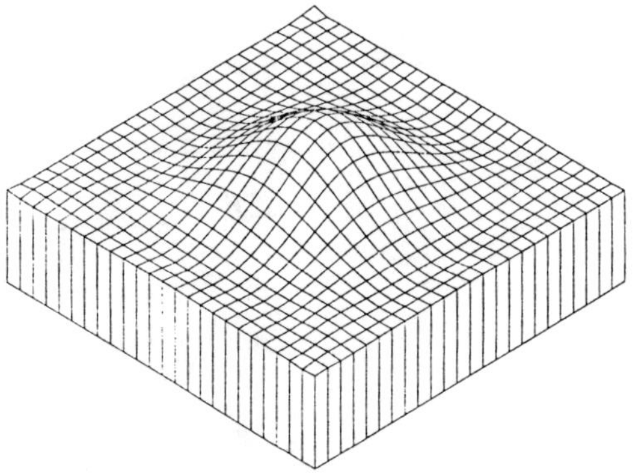

Figure 1: Computed distribution of the Young modulus (5 iterations) for a centred square inclusion $[1/3, 2/3] \times [1/3, 2/3]$ $\delta E/E_0 = 1$ $n = 0\%$

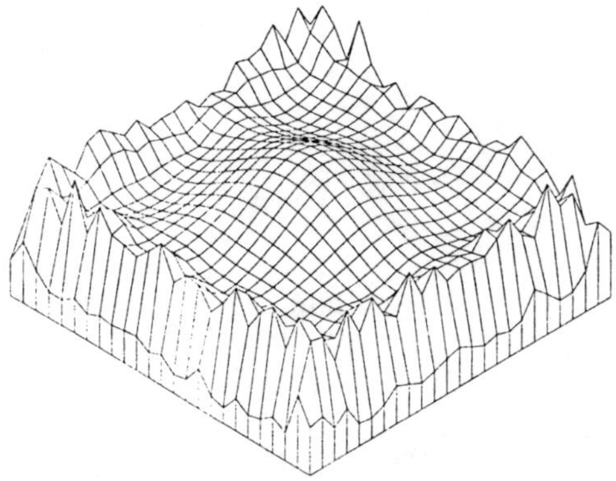

Figure 2: Computed distribution of the Young modulus (5 iterations) for a centered square inclusion $[1/3, 2/3] \times [1/3, 2/3]$ $\delta E_1/E_0 = 1$ $n = 5\%$

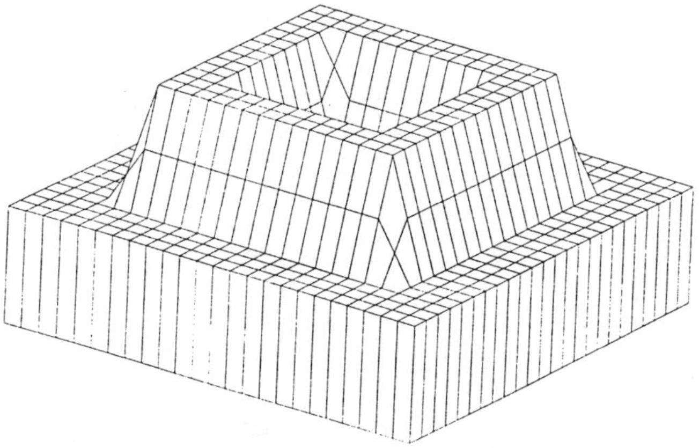

Figure 3: Real distribution of the Young modulus for a centered sandwich inclusion $\Delta_1 = [2/6, 4/6] \times [2/6, 4/6]$ $\Delta 21 = [1/6, 5/6] \times [1/6, 5/6]$ $\delta E_1/E_0 = 1$ $\delta E_2 = 0$ $n = 0\%$

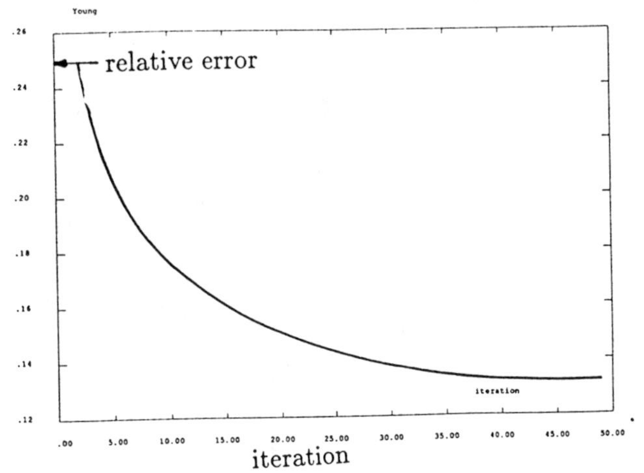

Figure 4: Relative error for the computed distribution of the Young modulus for a centered square inclusion $[1/3, 2/3] \times [1/3, 2/3]$ $\delta E_1/E_0 = 1$ $n = 0\%$

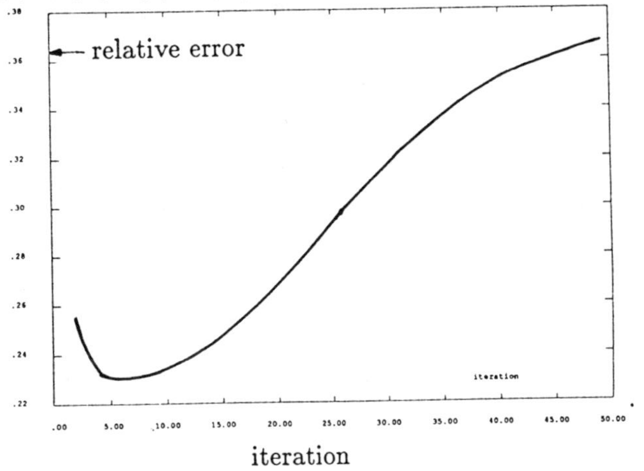

Figure 5: Relative error for the computed distribution of the Young modulus for a centered square inclusion $[1/3, 2/3] \times [1/3, 2/3]$ $\delta E_1/E_0 = 1$ $n = 5\%$

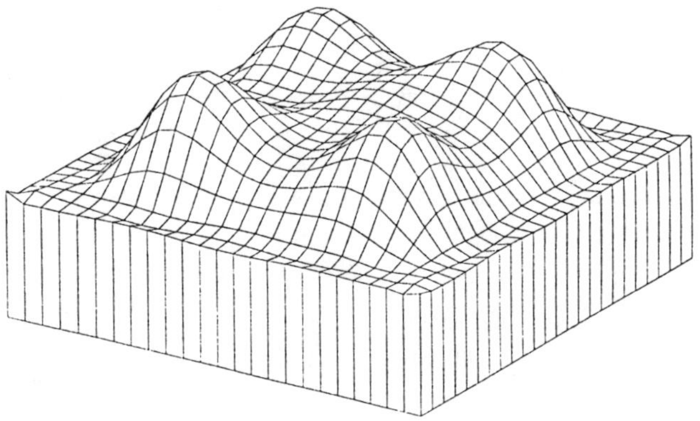

Figure 6: Computed distribution of the Young modulus (15 iterations) for a centered sandwich inclusion $\Delta_1 = [2/6, 4/6] \times [2/6, 4/6]$ $\Delta 21 = [1/6, 5/6] \times [1/6, 5/6]$ $\delta E_1/E_0 = 1$ $\delta E_2 = 0$ $n = 0\%$

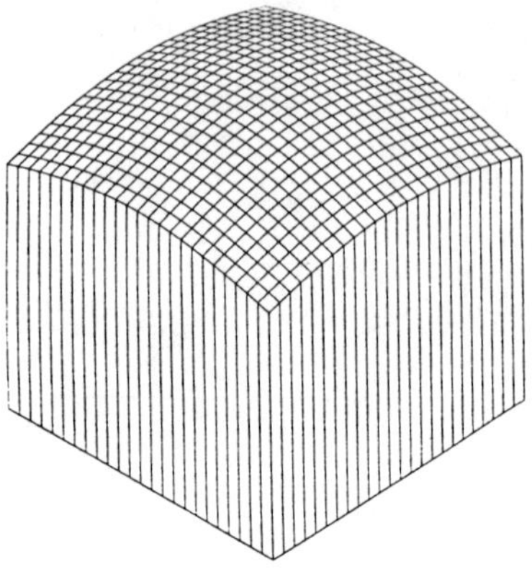

Figure 7: Real exponential distribution of the Young modulus $\delta E_1/E_0 = 0.5$

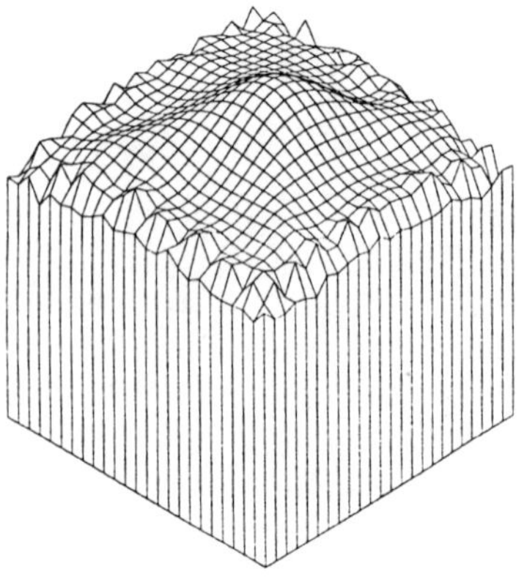

Figure 8: Computed Young modulus for the exponential distribution $\delta E_1/E_0 = 0.5$ $n = 1\%$

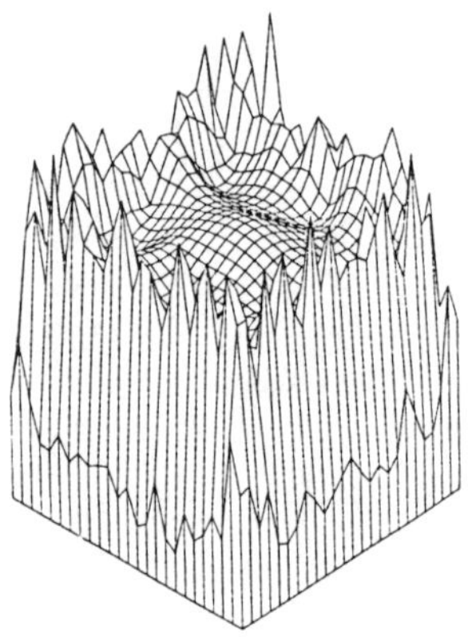

Figure 9: Computed Young modulus for the exponential distribution $\delta E_1/E_0 = 0.5$ $n = 5\%$

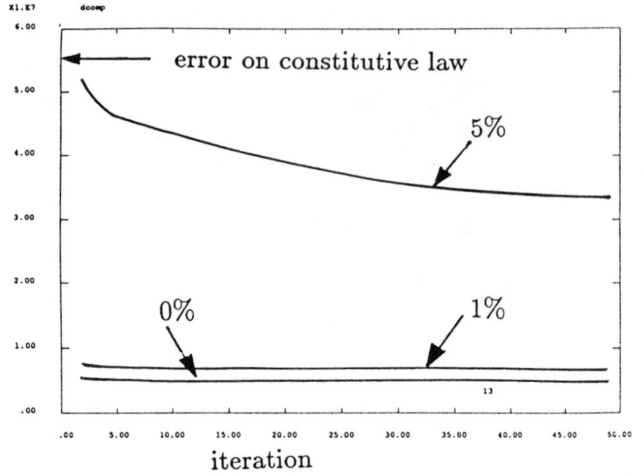

Figure 10: Decreasing error on constitutive law for different noise levels (exponential distribution $\delta E_1/E_0 = 0.5$)

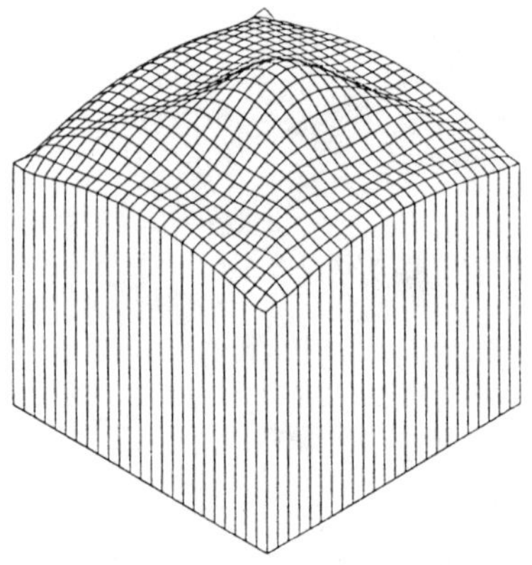

Figure 11: Computed Young modulus for the exponential distribution $\delta E_1/E_0 = 0.5$ $n = 0\%$

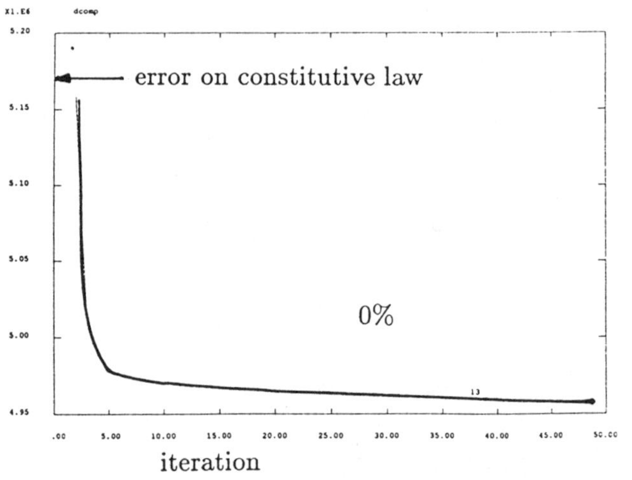

Figure 12: Decreasing error on constitutive law for $n -$ 0% (exponential distribution $\delta E_1/E_0 = 0.5$)

Inverse Problems in Engineering: Theory and Practice
ASME 1993

IDENTIFICATION OF UNKNOWN OBSTACLES USING BOUNDARY ELEMENTS AND SHAPE DIFFERENTIATION

Marc Bonnet
Laboratoire de Mécanique des Solides
École Polytechnique
Palaiseau, France

Abstract.

In this paper, we consider the application of shape differentiation and boundary elements to shape identification problems in infinite acoustic media using gradient minimization methods. An analytical expression of the first derivative of an integral functional with respect to a moving surface (the boundary of the unknown scatterer) is established in the case of a penetrable bounded obstacle illuminated by a known incident pressure wave. This formulation is incorporated in an unconstrained minimization algorithm using gradients, (namely BFGS quasi-Newton) in order to solve numerically the inverse problem.

Numerical results are presented for the search of a rigid bounded obstacle embedded in an infinite 3D acoustic medium, where the measurements are taken to be values of the pressure field on a remote measurement surface. They demonstrate the efficiency of the proposed method. Some computational issues (accuracy, CPU time, influence of measurements errors) are discussed.

Formulation of the direct problem

In this paper, a formulation for the numerical solution to the problem of identifying an unknown domain from external measurements, based on boundary integral equations (BIE) and shape differentiation, is considered. Such identification problems can be stated in various contexts: acoustics, elastodynamics, thermal sciences and can lead to applications e.g. to NDT techniques and other identification problems. The use of BIE formulations is suggested by the basic nature of the inverse problem under consideration (to search an unknown domain, thus an unknown surface), because domain discretization is avoided.

Let us consider, as a model situation, the problem of identifying a 3D homogeneous scatterer of finite extent Ω^- (wave velocity c^-, mass density ρ^-) imbedded in an infinite acoustical medium $\Omega^+ = \mathbf{R}^3 - \Omega^-$ (wave velocity c_+, mass density ρ^+). The scatterer Ω^- is illuminated with a known incident time-harmonic pressure $p^I(\boldsymbol{x})\exp(-i\omega t)$, which satisfies Helmholtz' equation $\Delta p^I + k_+^2 p^I = 0$ everywhere in $\mathbf{R}^3$ (with $k_+ = \omega/c_+$). Hence every acoustic field quantity is time-harmonic of pulsation ω and the factor $\exp(-i\omega t)$ will be discarded throughout, as is usually done.

The scattered pressure field $p_\Gamma(\boldsymbol{x})$ induced by the presence of the penetrable obstacle Ω^- (ie the solution of the *direct problem*) is governed by the following 'state equation':

$$\left\{\begin{array}{ll} \Delta p + k_+^2 p = 0 & \text{in } \Omega^+ \\ \Delta p + k_-^2 p = 0 & \text{in } \Omega^- \\ [\![p + p^I]\!] = 0 & \text{on } \Gamma \\ [\![\frac{1}{\rho\omega^2}(p_{,n} + p^I_{,n})]\!] = 0 & \text{on } \Gamma \\ \text{(radiation condition)} & \end{array}\right. \tag{1}$$

where $\Gamma \equiv \partial\Omega^+ = \partial\Omega^-$. The unit normal $\boldsymbol{n}$ is directed outside Ω^+, i.e. is interior to Γ and $(\cdot)_{,n}$ denotes the normal derivative $\nabla(\cdot).\boldsymbol{n}$. The operator $[\![f]\!]$ denotes the jump accross Γ:

$$[\![f]\!](y) = f^-(y) - f^+(y) = \lim_{\epsilon \searrow 0} f(y + \epsilon\boldsymbol{n}) - f(y - \epsilon\boldsymbol{n})$$

p defined by (1) depends on Γ; this is symbolically emphasized using the notation p_Γ for the solution of (1).

The direct problem (1) can be stated in terms of two coupled regularized [1] BIEs on the two independent boundary unknowns $p = p^+ = p^-, p^+_{,n}$ which remain after accounting for the jump conditions which appear in (1), as follows:

$$p(\boldsymbol{x}) + \int_\Gamma p(\boldsymbol{y})[G^+_{,n}(\boldsymbol{x},\boldsymbol{y}) - G^0_{,n}(\boldsymbol{x},\boldsymbol{y})]\, dS_y$$

$$+\int_\Gamma [p(y)-p(x)]G^0_{,n}(x,y)\,dS_y$$
$$-\int_\Gamma p^+_{,n}(y)G^+(x,y)\,dS_y \quad = \quad 0 \quad (2)$$
$$\int_\Gamma p(y)[G^-_{,n}(x,y)-G^0_{,n}(x,y)]\,dS_y$$
$$+\int_\Gamma [p(y)-p(x)]G^0_{,n}(x,y)\,dS_y$$
$$-\int_\Gamma \left(ap^+_{,n}+(a-1)p^I_{,n}\right)G^-(x,y)\,dS_y \quad = \quad 0 \quad (3)$$

where a denotes the ratio ρ^-/ρ^+, $G^\pm(x,y) = e^{ik_\pm r}/(4\pi r)$ and $G^0(x,y) = 1/(4\pi r)$ are the dynamic and static Green functions, while $r = \| x - y \|$, $(\cdot)_{,n} \equiv n(y)\cdot\nabla_y$. The non-integral term in (2) stems only from the unboundedness of the exterior domain Ω^+, and is therefore not to be mistaken with the conventional free-term coefficient of usual Cauchy principal value (CPV) BIEs. Eqns. (2), (3) hold for any $x \in \mathbf{R}^3$. The integrands in eqn. (2), (3) are weakly singular when $x \in \Gamma$, provided $p(x) \in C^{0,\alpha}(\Gamma)$, thanks to the regularizing effect of the term $[p(y) - p(x)]$, and contain no CPV integral. This allows a proper handling of the singular integrals at the implementation stage.

Statement of the inverse problem

The inverse problem under consideration is the reconstruction of the shape of Ω^+. Hence the boundary Γ is the primary unknown. Any attempt to reconstruct Γ relies upon the knowledge of supplementary date. Accordingly, ssume that the values of the pressure field scattered by the true obstacle are known on a measurement surface $C \subset \Omega^+$, exterior to Γ:

$$p(x) = \hat{p}(x) \qquad \text{on } C \qquad (4)$$

Thus Γ can be sought as a minimizer of some distance between computed and measured p's.

$$\min_\Gamma J(p_\Gamma) \quad \text{where} \quad J(p) = \int_C j(p-\hat{p})(y)\,dC_y \quad (5)$$

j being a positive function of its argument (e.g. $j(p-\hat{p}) = \frac{1}{2} \mid p - \hat{p} \mid^2$ for the least-squares distance).

In the cases where j is differentiable with respect to Γ, the nonlinear minimization problem (5) is best solved (in terms of both computational efficiency and accuracy) using gradient methods, such as Quasi Newton or conjugate gradient [5].

These algorithms need repeated computations of the derivative of $J(p_\Gamma)$ with respect to Γ (or, in practice, the design parameters which define the current location of Γ). They may be computed using finite-difference methods. However, this is computationally expensive (because the evaluation of each partial derivative needs a complete solution of (1) on a perturbed geometry $\Gamma + \delta G$) and may be poor in terms of accuracy.

The main task of the present paper is the derivation of an analytical formulation for the evaluation of the gradient of $J(p_\Gamma)$ with respect to Γ. This operation relies upon the ***shape differentiation*** concept and the adjoint problem approach. The result can then be discretized, using boundary elements in the present case. This is expected to improve the use of gradient minimization methods for the solution of shape identification problems such as (5), in terms of both accuracy and computational efficiency.

The shape differentiation approach.

The shape differentiation approach deals with derivatives of functionals with respect of variable domains or boundaries (e.g. $J(p_\Gamma)$) involving fields which themselves depend on the geometry, notably the solutions of boundary-value problems like p_Γ.

Let Γ denote the current location of the unknown boundary during the minimization process, and consider a further (small) evolution of the surface Γ defined by means of a normal 'velocity' field $\theta(y)$:

$$y \in \Gamma \rightarrow y + \theta(y)n(y)\tau \qquad (\tau \geq 0) \qquad (6)$$

while the measurement surface C is kept fixed ($\forall y \in C \;\; \theta(y) = 0$). Definition (6) is considered only for small values of τ, in order to define derivatives at $\tau = 0$, i.e. at the current Γ.

Various formulas are given in the literature (see e.g. [8]) for the derivative of integrals with respect to variable volumes Ω or surfaces Γ. The following ones are used here:

$$\frac{d}{d\tau}\int_\Omega u\,dV_y \quad = \quad \int_\Omega u_{,\tau}\,dV_y + \int_{\partial\Omega} u\theta\,dS_y \qquad (7)$$
$$\frac{d}{d\tau}\int_\Gamma u\,dS_y \quad = \quad \int_\Gamma \{u_{,\tau} + (u_{,n} - 2Ku)\,\theta\}\,dS_y \quad (8)$$

where $u_{,\tau}(y,\tau)$ denotes the ordinary partial derivative of a field $u(y,\tau)$ with respect to τ ('eulerian' derivative):

$$u_{,\tau}(y,\tau) \equiv \frac{\partial}{\partial\tau}u(y,\tau) \qquad (9)$$

which commutes with space derivatives:

$$(\nabla u)_{,\tau}(y,\tau) = \nabla u_{,\tau}(y,\tau) \qquad (10)$$

In (8), K denotes the mean curvature at $y \in \Gamma$. Equation (8) holds only for a closed smooth surface, while in equation (7) the unit normal n exterior to Ω^+ is used in the definition (6) of θ. Generalization of above formulas to piecewise smooth surfaces is available [8].

The adjoint problem approach.

This approach for deriving an analytical expression of the first variation of $J(p_\Gamma)$ with respect to Γ is known e.g. in structural shape optimization [6]. Applications in BIE context can be found in [4], [7].

The problem (5) may be viewed as a constrained optimization problem, where $J(p)$ is to be minimized under the constraint $p = p_\Gamma$ (1). The latter can be expressed in weak form as:

$$\forall w \in \mathcal{V} \qquad \mathcal{A}(p, w; \Gamma) = 0 \tag{11}$$

where

$$\begin{aligned} \mathcal{V} &= \{w \in H^1_{loc}(\Omega^+),\ [\![w]\!](y) = 0\ (y \in \Gamma)\} \\ \mathcal{A}(p, w; \Gamma) &\equiv \int_{\Omega^+} \left(\frac{1}{\rho^+\omega^2}\nabla p \cdot \nabla \overline{w} - \kappa^+ p\overline{w}\right) dV_y \\ &+ \int_{\Omega^-} \left(\frac{1}{\rho^-\omega^2}\nabla p \cdot \nabla \overline{w} - \kappa^- p\overline{w}\right) dV_y \\ &- \int_\Gamma [\![\frac{1}{\rho\omega^2}p^I_{,n}]\!]\overline{w}\, dS_y \end{aligned} \tag{12}$$

and $\kappa^\pm = (\rho^\pm c_\pm^2)^{-1}$ is the acoustic compressibility. A lagrangian functional $\mathcal{L}$ is then introduced:

$$\mathcal{L}(p, w, \Gamma) = J(p) + \mathcal{A}(p, w; \Gamma) \tag{13}$$

where $w \in \mathcal{V}$, a trial function, is the Lagrange multiplier. Application of formulas (7), (8) to (13) gives:

$$\begin{aligned} \frac{d}{d\tau}\mathcal{L} &= \int_C j'(p - \hat{p})p_{,\tau}\, dC_y \\ &+ \int_{\Omega^+} \left(\frac{1}{\rho^+\omega^2}\nabla p_{,\tau} \cdot \nabla \overline{w} - \kappa^+ p_{,\tau}\overline{w}\right) dV_y \\ &+ \int_{\Omega^-} \left(\frac{1}{\rho^-\omega^2}\nabla p_{,\tau} \cdot \nabla \overline{w} - \kappa^- p_{,\tau}\overline{w}\right) dV_y \\ &+ \int_\Gamma \theta[\![\frac{1}{\rho\omega^2}\nabla p \cdot \nabla \overline{w} - \kappa p_{,\tau}\overline{w}]\!]\, dS_y \\ &+ \int_\Gamma \left(\overline{w}[\![\frac{1}{\rho\omega^2}p^I_{,n}]\!]_{,\tau} + \overline{w}_{,n}[\![\frac{1}{\rho\omega^2}p^I_{,n}]\!]\theta\right) dS_y \\ &+ \int_\Gamma \left([\![\frac{1}{\rho\omega^2}p^I_{,nn}]\!] - 2K[\![\frac{1}{\rho\omega^2}p^I_{,n}]\!]\right) \overline{w}\theta\, dS_y \\ &+ \mathcal{A}(p, w_{,\tau}; \Gamma) \end{aligned} \tag{14}$$

in which derivatives with respect to τ are taken for $\tau = 0$. Some simplifications can be made in (14):

- As $w_{,\tau} \in \mathcal{V}$, one has from (12):

$$\mathcal{A}(p, w_{,\tau}; \Gamma) = 0 \tag{15}$$

- Splitting the gradients into tangential and normal parts (see Appendix) gives:

$$\nabla p \cdot \nabla \overline{w} = \nabla_S p \cdot \nabla_S \overline{w} + p_{,n}\overline{w}_{,n} \tag{16}$$

Moreover, as shown in the Appendix, one has:

$$(p^I_{,n})_{,\tau} + \theta p^I_{,nn} = \left(2Kp^I_{,n} - k^2p^I\right)\theta - \mathrm{div}_S\,(\theta\nabla_S p^I) \tag{17}$$

so that one gets:

$$\begin{aligned} &\int_\Gamma \theta[\![\frac{1}{\rho\omega^2}\nabla_S p \cdot \nabla_S \overline{w} - \kappa p_{,\tau}\overline{w}]\!]\, dS_y \\ &+ \int_\Gamma \overline{w}[\![\frac{1}{\rho\omega^2}\left((p^I_{,n})_{,\tau} + [p^I_{,nn} - 2Kp^I_{,n}]\theta\right)]\!]\, dS_y \\ &= \int_\Gamma [\![\frac{1}{\rho\omega^2}\left(\theta\nabla_S p\nabla_S \overline{w} - \overline{w}\mathrm{div}_S\,(\theta\nabla_S p^I)\right)]\!]\, dS_y \\ &\quad - \int_\Gamma \theta\left\{[\![\kappa p\overline{w}]\!] + \frac{1}{c_+^2}[\![\frac{1}{\rho}p^I\overline{w}]\!]\right\} dS_y \\ &= \int_\Gamma [\![\frac{1}{\rho\omega^2}\theta\nabla_S(p + p^I)\nabla_S\overline{w}]\!]\, dS_y \\ &\quad - \int_\Gamma \theta\left\{[\![\kappa p\overline{w}]\!] + \frac{1}{c_+^2}[\![\frac{1}{\rho}p^I\overline{w}]\!]\right\} dS_y \end{aligned} \tag{18}$$

where identity (42) for integration by parts has been applied to $f \equiv \theta\overline{w}\nabla_S p^I$.

Finally, eqn. (14) is rewritten using (15) and (18) and taking into account the constraint $p = p_\Gamma$. The first variation of $\mathcal{L}$ is then expressed as

$$\frac{d}{d\tau}\mathcal{L} = \frac{\partial\mathcal{L}}{\partial p}p_{,\tau} + \frac{\partial\mathcal{L}}{\partial\Gamma}\theta \tag{19}$$

$$\begin{aligned} \frac{\partial\mathcal{L}}{\partial p}p_{,\tau} &= \int_C j'(p - \hat{p})p_{,\tau}\, dC_y \\ &+ \int_{\Omega^+} \left(\frac{1}{\rho^+\omega^2}\nabla p_{,\tau} \cdot \nabla \overline{w} - \kappa^+ p_{,\tau}\overline{w}\right) dV_y \\ &+ \int_{\Omega^-} \left(\frac{1}{\rho^-\omega^2}\nabla p_{,\tau} \cdot \nabla \overline{w} - \kappa^- p_{,\tau}\overline{w}\right) dV_y \end{aligned} \tag{20}$$

$$\begin{aligned} \frac{\partial\mathcal{L}}{\partial\Gamma}\theta &= \int_\Gamma [\![\frac{1}{\rho\omega^2}\theta(p_{,n} + p^I_{,n})\overline{w}_{,n}]\!]\, dS_y \\ &+ \int_\Gamma [\![\frac{1}{\rho\omega^2}\theta\nabla_S(p + p^I) \cdot \nabla_S\overline{w}]\!]\, dS_y \\ &- \int_\Gamma \theta\left\{[\![\kappa p\overline{w}]\!] + \frac{1}{c_+^2}[\![\frac{1}{\rho}p^I\overline{w}]\!]\right\} dS_y \end{aligned} \tag{21}$$

Now the choice of the Lagrange multiplier w is restricted in such a way that $\frac{d}{d\tau}\mathcal{L} = 0$ for $\theta \equiv 0$, that is, we put:

$$\frac{\partial\mathcal{L}}{\partial p}p_{,\tau} = 0, \qquad \forall p_{,\tau} \in \mathcal{V} \tag{22}$$

In view of eqn. (20), the variational problem defined by (22) has a unique solution $w \equiv w_\Gamma$, called the *adjoint field*. Equation (22) is the weak formulation of the *adjoint problem*, which strong formulation reads:

$$\left\{\begin{array}{ll} \Delta w + k_+^2 w = -\rho^+\omega^2 j'(p_\Gamma - \hat{p})\delta_C & \text{in } \Omega^+ \\ \Delta w + k_-^2 w = 0 & \text{in } \Omega^- \\ [\![w]\!] = 0 & \text{on } \Gamma \\ [\![\frac{1}{\rho\omega^2} w_{,n}]\!] = 0 & \text{on } \Gamma \\ \text{(radiation condition)} & \end{array}\right. \quad (23)$$

In view of (20) or (23), the state and adjoint problems are governed by the same operator.

As a result, the *shape derivative* of J is given in terms of $p_\Gamma, w_\Gamma, \theta$ by:

$$\frac{dJ}{d\tau} = \frac{d}{d\tau}\mathcal{L}(p_\Gamma, w_\Gamma, \Gamma) \quad (24)$$

Taking into account (22), (21) and the jump conditions which appear in (1), (23), one obtains as a result the following expression:

$$\begin{aligned} \frac{dJ}{d\tau} &= \frac{1}{\rho_+\omega^2}\frac{1-a}{a}\int_\Gamma \theta \nabla_S(p_\Gamma + p^I)\cdot\nabla_S\overline{w}_\Gamma\, dS_y \\ &- \kappa^+\int_\Gamma \theta\left\{\frac{1-ab^2}{ab^2}p_\Gamma\overline{w}_\Gamma + \frac{1-a}{a}p^I\overline{w}_\Gamma\right\} dS_y \\ &+ \int_\Gamma \frac{1}{\rho_+\omega^2}\theta[(p_\Gamma^+)_{,n} + p^I_{,n}][\![(\overline{w}_\Gamma)_{,n}]\!]\, dS_y \end{aligned} \quad (25)$$

where $a = \rho^-/\rho^+$, $b = c^-/c^+$. The following comments can be made:

1. Expression (25) of $dJ/d\tau$ is clearly a linear form over θ.
2. The adjoint field w does not depend on θ. Therefore, the adjoint problem approach needs the solution of two distinct boundary-value problems. More generally, the computation of the gradient of N distinct functionals needs the solution of $N+1$ boundary-value problems over the same geometry. This is in contrast with the direct differentiation approach [6], where $D+1$ boundary-value problems are to be solved, D being the number of design parameters, whatever the number of functionals to be minimized.
3. Moreover, the direct and adjoint problems share the same governing operator. Therefore, each computation of the complete gradient (25) needs the building and factorization of only one matrix operator.

In the case where Ω^- is illuminated by N incident waves p_i^I ($i = 1,\ldots,N$) in succession, generating pressure fields p_Γ^i in Ω^+ and N sets of measurements $\hat{p}_i$ on C, the functional J in (5) becomes:

$$J(p) = \sum_{i=1}^{N}\int_C j(p^i - \hat{p}^i)(y)\, dC_y = \sum_{i=1}^{N} J_i \quad (26)$$

The shape derivative of $J(p_\Gamma)$ is then given by:

$$\frac{dJ}{d\tau} = \sum_{i=1}^{N}\frac{dJ_i}{d\tau} \quad (27)$$

where $dJ_i/d\tau$ is given by (25) with $p_\Gamma = p_\Gamma^i, w_\Gamma = w_i^\Gamma$ and the w_i^Γ solve the N adjoint problems obtained by putting $p_\Gamma = p_\Gamma^i, \hat{p} = \hat{p}^i$ for $i =,\ldots,N$ in turn in the right-hand side of (23).

BIE formulation for the adjoint problem approach.

From (2), (3), the adjoint problem (23) may now be formulated in terms of two coupled BIE:

$$\begin{aligned} w(x) &+ \int_\Gamma w(y)[G^+_{,n}(x,y) - G^0_{,n}(x,y)]\, dS_y \\ &+ \int_\Gamma [w(y) - w(x)]G^0_{,n}(x,y)\, dS_y \\ &- \int_\Gamma w^+_{,n}(y)G^+(x,y)\, dS_y \\ &= \rho^+\omega^2\int_C \overline{j'(p_\Gamma - \hat{p})}G(x,y)\, dC_y \end{aligned} \quad (28)$$

$$\begin{aligned} &\int_\Gamma w(y)[G^-_{,n}(x,y) - G^0_{,n}(x,y)]\, dS_y \\ &+ \int_\Gamma [w(y) - w(x)]G^0_{,n}(x,y)\, dS_y \\ &+ \int_\Gamma a w^+_{,n}(y)G^-(x,y)\, dS_y = 0 \end{aligned} \quad (29)$$

Then, in order to evaluate $dJ/d\tau$, one has to

1. Solve the state coupled BIE (2), (3) for $p_\Gamma, (p_\Gamma)_{,n}$.
2. Solve the *adjoint BIE* (28), (29) for $w_\Gamma, (w_\Gamma)_{,n}$.
3. Use the obtained values in expression (25) of $dJ/d\tau$.

Special case of a hard obstacle

The result (25) includes the special case of a hard obstacle [2]. The latter is obtained by putting $\kappa^- = 0$, $\rho^- = \infty$ (i.e. $a = \infty, b = infty$) in (25), and reads:

$$\frac{dJ}{d\tau} = \int_\Gamma \theta\left[\nabla_S\overline{w}_\Gamma\cdot\nabla_S(p_\Gamma + p^I) - k_+^2\overline{w}_\Gamma(p_\Gamma + p^I)\right] dS \quad (30)$$

where the adjoint field w_Γ solves:

$$\left\{\begin{array}{ll} (\Delta + k^2)w = -(\overline{j'(p_\Gamma - \hat{p})})\delta_C & \text{in } \Omega^+ \\ w_{,n} = 0 & \text{on } \Gamma \\ \text{(radiation condition)} & \end{array}\right. \quad (31)$$

The system of coupled BIEs (28), (29) reduces to a single adjoint BIE as follows [2]:

$$\begin{aligned} w(x) &+ \int_\Gamma w(y)[G_{,n}(x,y) - G^0_{,n}(x,y)]\, dS_y \\ &+ \int_\Gamma [w(y) - w(x)]G^0_{,n}(x,y)\, dS_y \\ &= \int_C \overline{j'(p_\Gamma - \hat{p})}G(x,y)\, dC_y \end{aligned} \quad (32)$$

Numerical implementation.

Description of the unknown geometry. In the examples presented here, the unknown surface is a 'n-ellipsoid' defined by means of 10 parameters $\hat{d}_1, \ldots, \hat{d}_{10} = y_1^G, y_2^G, y_3^G$ (coordinates of the gravity center y^G), a, b, c (principal axes), ϕ, θ, ψ (Euler angles associated to the principal directions), n (exponent). Denote by $\mathcal{S}$ the unit "n-sphere" of equation $x_1^n + x_2^n + x_3^n = 1$ (hence $n = 2$ and $n = \infty$ yield respectively the ordinary unit sphere and the cube of vertices $(\pm 1, \pm 1, \pm 1)$). The unknown surface is then mapped on $\mathcal{S}$ as follows:

$$Y \in \mathcal{S} \rightarrow y = y^G + R(AY) \in \Gamma \tag{33}$$

where $AY = (aY_1, bY_2, cY_3)$ and $R = R(\phi, \theta, \psi)$ denotes the rotation which transforms the coordinate axes onto the principal axes of the n-ellipsoid (several (ϕ, θ, ψ) triplets may define the same rotation). Thus, the unit sphere and the unit cube are respectively transformed into ellipsoids and boxes, with arbitrary center, size and orientation.

BEM implementation The present study has been done using our BEM research code ASTRID. A BE mesh $\mathcal{M}_0$ of the unit n-sphere $\mathcal{S}$ (using 8-noded isoparametric elements here) is created. Then the BE mesh $\mathcal{M}$ of the current surface Γ is obtained either by application of the mapping (33) to the nodes of $\mathcal{M}_0$ (NI = 10) or by direct updating of the nodal coordinates of $\mathcal{M}$ (NI = $3M$). Similarly, the definition (34) has been applied to the BE nodes of $\mathcal{M}$, the field $\theta_k(y)$ on Γ being interpolated using these nodal values and the shape functions associated to the boundary elements. The numerical search of the n-ellipsoid has been done by taking either the 10 parameters $y_1^G, y_2^G, y_3^G, a, b, c, \phi, \theta, \psi, n$ or the $3M$ nodal coordinates of the $3M$ BE mesh nodes coordinates as unknowns. Moreover, taking $\tau \equiv d_k$ ($1 \leq k \leq$ NI) in turn defines $\theta(y) = \theta_k(y)$ as follows:

$$\theta_k(y) = n.y_{,d_k} \tag{34}$$

Equations (25) or (30) with $\theta(y) = \theta_k(y)$ thus yield the value of the partial derivative $\partial J / \partial d_k$.

Minimization algorithm. After preliminary tests, the best results, in terms of both accuracy and computational efficiency, have been found to be obtained using the BFGS Quasi-Newton unconstrained minimization algorithm. The computational efficiency of the algorithm, and sometimes the level of convergence, is found to be strongly dependent on the line search algorithm, and especially on the initialization of its 'bracketing' [5], [9] step. The latter has been implemented as described in [5], chap.2 and features a user-set parameter $\sigma \in]0, 1[$, which allows one to choose from high accuracy ($\sigma \sim 0$) to low accuracy ($\sigma \sim 1$) line searches. Low-accuracy line search ($\sigma = .9$) has been used for the results displayed in Tables 2 to 5.

Numerical examples.

The numerical examples presented below illustrate the case of a hard obstacle.

They have been run on HP-Apollo 400 workstations. The surfaces Γ are meshed using 24 boundary elements, totalizing 74 nodes. The measurement surface C is a sphere of radius 10m, on which the (numerically simulated) pressure values are given at 290 points and interpolated using boundary elements for functional and gradient evaluations.

Example 1: 10 unknowns $d_1, \ldots, d_{10}$

We consider the case where $\hat{\Gamma}$ is defined by the values given in table 1 of parameters $\hat{d}_1, \ldots, \hat{d}_9$ ($\omega = 0.5, c = 1$). Four cases are considered:

- Exponent n excluded (cases 1,2 – NI = 9) or included (cases 3,4 – NI = 10) from the search.
- The 'true' obstacle is an ellipsoid ($n = 2$, cas 1,3) or a rectangular box ($n = \infty$, cases 2,4).

Thus, in case 2, the 'true' obstacle cannot be reached exactly by the minimization process.

In some cases, measurement noise has been artificially simulated my multiplication of the data values $\hat{p}$ by $1 + r$, r being random numbers uniformly distributed in $[-\epsilon, \epsilon]$, with $\epsilon = 0, 10^{-3}, 10^{-2}, 10^{-1}$ in the results displayed below.

As the same n-ellipsoid can result from many combinations of Euler angles and permutations of principal axes in (33), it is difficult to measure the accuracy of the identification of Γ by merely comparing the identified parameters d_k with thode used to create the simulated 'true' Γ. Instead, the relative errors e_V, e_A, e_I for the volume, boundary area and geometrical inertia tensor (with respect to the *fixed* coordinate system $Ox_1x_2x_3$) of the hard obstacle have been computed. The indicator e_I accounts for the quality of the recovery of the spatial location and orientation of $\hat{\Gamma}$.

These results (numerical values of J_{final}/J_0, e_V, e_A, e_I) are displayed in tables 2, 3, 4, 5, together with the number of functional (and gradient) evaluations spent. The data $\hat{p}^i$ come from three incident plane pressure waves p_i^I, propagating along the coordinate directions e_i ($i = 1, 2, 3$). Table 5 also indicates the value of n reached (the 'true' value in this case being $n = \infty$).

- Cases 1,3,4 exhibit very good convergence and accuracy, especially for non-perturbed data, see tables 2, 4, 5. This is a clear indication of the good performance of the APA for the computation of the gradient of J. At least in the range $\epsilon = 10^{-3}$ to 10^{-1} considered here, the error indicators e_V, e_S, e_I are often found to vary linearly with ϵ in the results presented here, and J_{final}/J_0 to vary quadratically. The numerical solution of the inverse problem hence behaves well with respect to measurement noise.

- The convergence and accuracy remains good for case 2 (table 3), where the 'true' cavity is a rectangular box and exact convergence is hence impossible. Moreover, the results appear to be less sensitive to data noise than in cases 1, 3, 4.

- Convergence is much slower when $d_{10} \equiv n$ plays an active role and is included in the search, see Case 4. Moreover, the recovery of n has been found to be sensitive to implementation details like how the BFGS updating formula is written or the initialization of the line-search. This suggests that the recovery of n is a more ill-posed problem than the recovery of Euler angles, principal axes and center coordinates.

  Upon examination of the convergence process for Cases 2 and 4 (i.e. search of a box with n respectively excluded and included), it has been noticed that they are almost identical until the Case 2 termination point. In view of the respective function/gradient evaluation counts for Cases 2 and 4, one concludes that most (typically two-thirds of) computing effort in Case 4 is spent to recover n alone.

- Each evaluation of J and $\partial J/\partial d_k$, takes about 20 seconds, while the evaluation of J alone takes about 15 seconds. The overall computer time spent for solving the inverse problem varies from about 15 to 75 minutes. The recovery of the exponent n is time-consuming.

- The efficiency of the line search is an important issue since here most computer time is spent on evaluations of J and $\partial J/\partial d_k$. In our case, most BFGS iterations use only one evaluation of J and $\partial J/\partial d_k$.

Example 2: Nodal coordinates $d_1, \ldots, d_{3M}$ as unknowns

In this example an ellipsoidal hard obstacle ($n = 2$), defined by table 6 is searched using the coordinates of the BE mesh nodes as unknowns, so that there is no a priori information on the unknown shape.

Figure ?? depicts the values of J_{final}/J_0, e_V, e_S, e_I taken at each step of the minimization process, using $3M = 222$ unknowns (nodal coordinates) and (for comparison purposes) 9 unknowns $(y_1^G, y_2^G, y_3^G, a, b, c, \phi, \theta, \psi)$.

These curves show that the final values taken by the indicators e_V, e_I in the 222 unknowns case, although satisfactory, are not as good as in the 9 unknowns case. They show that the location, size and orientation of the obstacle are correctly reconstructed. On the contrary, the final value taken by e_A ($\sim 34\%$) indicates that the reconstructed surface oscillates strongly.

Concluding comments and perspectives.

The approach presented here on a restricted class of shape identification problems can be extended to many more physical contexts: elastodynamics, heat conduction,... It is essentially a numerical tool, in that it allows an optimal use of classical unconstrained minimization methods using gradient evaluations, applied to the physical model and data at hand. On the other side, it provides no insight on the fundamental characteristics of the identification problem, such as existence or uniqueness of the solution for the available data.

The numerical results presented here show the efficiency of our approach for the computation of functional gradients. The very good results obtained on cases where a moderate number of parameters is used for the description of the unknown surface show that the basic components of the inversion strategy perform well. This can be viewed as a qualification step.

When applied to more complex descriptions of the unknown surface, allowing in principle the recovery of more general shapes, ill-posedness appears, in the form of highly oscillating reconstructed surfaces, as is shown in our second exemple. Hence the next step of our numerical approach should involve a *regularization* [10] of the inverse problem, by means of a stabilizing positive functional $P(\Gamma)$, so that the unknown surface Γ is searched as a minimizer of $R(\Gamma, \alpha) = J(p_\Gamma) + \alpha P(\Gamma)$ $(0 < \alpha \ll 1)$ instead of $J(\Gamma)$ alone. A suggestion for the functional $P(\Gamma)$ is:

$$P(\Gamma) = \frac{1}{2}\int_\Gamma (\text{div}_S\, \boldsymbol{n})^2\, dS + \beta \int_L (1 - \boldsymbol{n}^+ . \boldsymbol{n}^-)\, ds \quad (35)$$

where L denotes the edges present on the current Γ; $\boldsymbol{n}^+, \boldsymbol{n}^-$ are the unit normals adjacent to an edge and β is an adjustable coefficient which ensures that the two terms of $P(\Gamma)$ are dimensionally consistent. The first integral term allows the penalization of high curvatures, which may appear even to the continuous reconstructed shape as a result of data uncertainties, while the second is more specifically to damp numerical oscillations of the BE-discretized surface: the jump of unit normals between elements contribute notably to the oscillations one wants to avoid. L is then taken as the set of all element boundaries.

Finally, let us mention that technical problems arise due to the discontinuity of $\boldsymbol{n}$ accross element boundaries:

- The modelling of moving surfaces should ideally, in a BE context, use only *one* unknown per node (θ), instead of three as in our example 2. But, owing to representation (6) of the moving surface, this amounts to introduce vector nodal values $\theta\boldsymbol{n}$, which is problematic at all boundary nodes. Other representations of the BE-discretized moving surface should be investigated in order to use only one unknown per node while still allowing for general shapes.

- The oscillations of numerically reconstructed surfaces may primarily come from the normal jumps at element boundaries, hence the need of an additional term in a regularizing functional like (35) in order to cater for this specific issue.

These aspects, as well as extension of the present approach to elastodynamics, are currently under examination.

References

[1] Bonnet M. - Méthode des équations intégrales régularisées en élastodynamique tridimensionnelle. PhD thesis (Ecole Nationale des Ponts et Chaussées, Paris, France), Bulletin EDF/DER série C, n^0 1/2, 1987.

[2] Bonnet M. - A numerical approach for shape identification problems using BIE and shape differentiation. In CA Brebbia, J. Dominguez, and F. Paris, editors, *Boundary Elements XIV, vol.2: stress analysis and computational aspects*, pp 541–553. Computational Mechanics Publications, 1992.

[3] Bonnet M. and Bui H.D. - Regularization of the Displacement and Traction BIE for 3D Elastodynamics using indirect methods, in *Advances in Boundary Element Techniques*, J.H. Kane, G. Maier, N. Tosaka and S.N. Atluri, eds., Springer-Verlag, 1992.

[4] Choi J.O., Kwak B.M. - Boundary Integral Equation Method for Shape Optimization of Elastic Structures. *Int. J. Num. Meth. in Eng.* **26**, *pp. 1579-1595, 1988.*

[5] Fletcher R. *Practical Methods of Optimization*, J. Wiley & Sons.

[6] Haug, Choi, Komkov - *Design Sensitivity Analysis of Structural Systems*, Academic Press, 1986.

[7] Meric R.A. - Shape optimization of thermoelastic solids. J. Therm. Stresses. **11**, pp. 187-206, 1988.

[8] Petryk H., Mroz Z. - Time derivatives of integrals and functionals defined on varying volume and surface domains. Arch. Mech. **38**(5-6), pp 697-724, 1986.

[9] Press W.H., Flannery B.P., Teukolsky S.A., Vetterling W.T. - *Numerical recipes: the art of scientific computing.* Cambridge press, 1986.

[10] Tikhonov A.N., Arsenin V.Y. - *Solutions to ill-posed problems*, Winston-Wiley, New York, 1977.

Tangential differential operators and integration by parts

Let S be a twice continuously differentiable *closed* (C^2) surface, of unit normal $\boldsymbol{n}$ (open surfaces can be considered as well, see e.g. [3]). Consider a scalar field $u(\boldsymbol{y})$, $\boldsymbol{y} \in S$, which may be undefined outside S (e.g. $u = n_i(\boldsymbol{y})$, $u = \theta(\boldsymbol{y})$). In this case, the cartesian derivatives $u_{,i}$ are generally meaningless, and one has to introduce tangential differential operators. The domain of definition of u is extended in a neighbourhood V of S by introducing a continuation $\hat{u}$ of u outside S defined as: $\forall(\boldsymbol{y} \in V)$, $\hat{u}(\boldsymbol{y}) = u(P(\boldsymbol{y}))$, where $P(\boldsymbol{y})$ is the orthogonal projection of $\boldsymbol{y}$ onto S. Clearly the restriction of $\hat{u}$ to S is equal to u. Moreover the normal derivative of $\hat{u}$ is equal to zero, i.e. the vector $\nabla\hat{u}$ is tangent to S; therefore it may be used to define the tangential gradient $\nabla_S u$ of the function u;

$$\nabla_S u = \nabla_S \hat{u} = \nabla \hat{u} \tag{36}$$

If u is an arbitrary scalar function defined in V, one has, consistently with (36):

$$\boldsymbol{\nabla}_S u = \boldsymbol{\nabla} u - \boldsymbol{n} u_{,n} \tag{37}$$

The symbol ($\hat{}$) is now omitted, keeping in mind if necessary the extension.

One can also introduce the *surface divergence* div_S of a vector field $\boldsymbol{u} = u_s \boldsymbol{e}_s$:

$$\text{div}_S = \text{div} - (\cdot)_{,n}.\boldsymbol{n} = \boldsymbol{\nabla}_S u_s \cdot \boldsymbol{e}_s \tag{38}$$

An interesting consequence of (37) is the following identity for the Laplace operator:

$$\Delta u = u_{,nn} - 2K u_{,n} + \text{div}_S(\boldsymbol{\nabla}_S u) \tag{39}$$

$$2K = -\text{div}_S\, \boldsymbol{n} \tag{40}$$

The classical Stokes' identity for a vector field $\mathbf{U}$ defined over V reads:

$$\int_S \boldsymbol{n}.\text{rot}\boldsymbol{U}\, \mathrm{d}S = 0 \tag{41}$$

Application of identity (41) to $\boldsymbol{U} = (\boldsymbol{n} \wedge e_j) f$ yields integration by parts identities associated to the tangential gradient (37):

$$\int_S (-\boldsymbol{n} K f + \boldsymbol{\nabla}_S f)\, \mathrm{d}S = 0 \tag{42}$$

An auxiliary formula for shape differentiation.

The following formula can be proved using elementary differential geometry:

$$\boldsymbol{n}_{,\tau} = -\boldsymbol{\nabla}_S \theta \tag{43}$$

A combined application of (10) and (43) leads to:

$$(u_{,n})_{,\tau} = (u_{,\tau})_{,n} - \boldsymbol{\nabla}_S u.\boldsymbol{\nabla}_S \theta \tag{44}$$

In the particular case $u = p^I$, one has:

$$p^I_{,\tau} = 0 \tag{45}$$

since the definition of the incident field does not depend on the actual location of Γ. Moreover, since $u = p^I$ solves the Helmholtz equation, identity (39) gives:

$$u_{,nn} = \Delta u + 2K u_{,n} - D_s D_s u \tag{46}$$

Eqn. (45), together with (39), gives eqn. (17).

	y_1^G	y_2^G	y_3^G	ϕ	θ	ψ	a	b	c
'true'	1.0	0.0	−2.0	0.4	0.9	0.6	1.0	3.0	1.0
initial	0.0	0.0	0.0	0.0	0.0	0.0	1.5	1.5	1.5

Table 1: Example 2: 'True' and initial values $d_1, \ldots, d_9$

Case 1	$\epsilon = 0.$	$\epsilon = 10^{-3}$	$\epsilon = 10^{-2}$	$\epsilon = 10^{-1}$
J_{final}/J_0	$1.59\,10^{-7}$	$4.19\,10^{-7}$	$2.41\,10^{-5}$	$2.42\,10^{-3}$
e_V	$8.50\,10^{-7}$	$1.56\,10^{-4}$	$1.56\,10^{-3}$	$1.52\,10^{-2}$
e_S	$5.98\,10^{-6}$	$1.17\,10^{-4}$	$1.11\,10^{-3}$	$1.06\,10^{-2}$
e_I	$7.46\,10^{-5}$	$1.67\,10^{-4}$	$1.01\,10^{-3}$	$8.80\,10^{-3}$
nb. eval.	37	37	35	40

Table 2: Results for Case 1

Case 2	$\epsilon = 0.$	$\epsilon = 10^{-3}$	$\epsilon = 10^{-2}$	$\epsilon = 10^{-1}$
J_{final}/J_0	$2.76\,10^{-3}$	$2.77\,10^{-3}$	$2.82\,10^{-3}$	$6.36\,10^{-3}$
e_V	$1.00\,10^{-2}$	$1.02\,10^{-2}$	$1.16\,10^{-2}$	$2.54\,10^{-2}$
e_S	$1.34\,10^{-1}$	$1.34\,10^{-1}$	$1.35\,10^{-1}$	$1.42\,10^{-1}$
e_I	$2.60\,10^{-2}$	$2.61\,10^{-2}$	$2.69\,10^{-2}$	$3.39\,10^{-2}$
nb. eval.	41	30	42	60

Table 3: Results for Case 2.

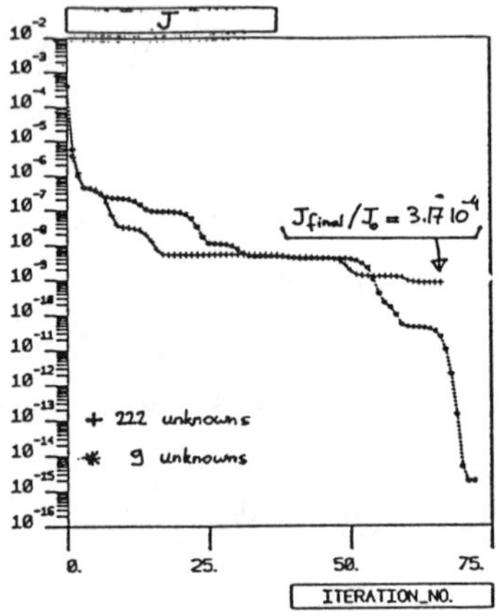

Figure 1: Example 2: J_{final}/J_0 vs. iteration no

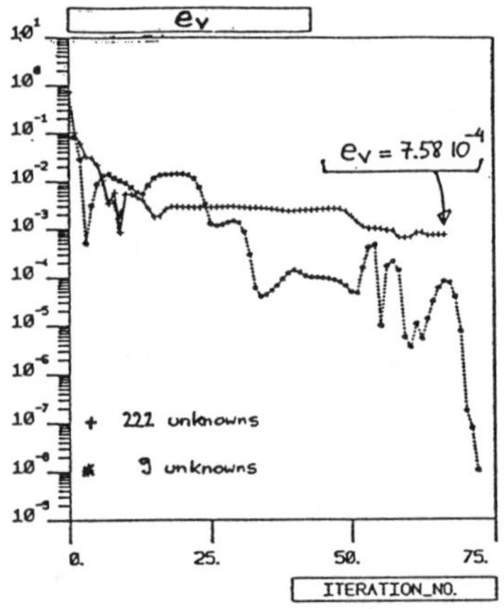

Figure 2: Example 2: e_V vs. iteration no

Case 3	$\epsilon = 0.$	$\epsilon = 10^{-3}$	$\epsilon = 10^{-2}$	$\epsilon = 10^{-1}$
J_{final}/J_0	$1.59\,10^{-7}$	$4.18\,10^{-6}$	$2.41\,10^{-5}$	$2.35\,10^{-3}$
e_V	$8.87\,10^{-6}$	$1.69\,10^{-4}$	$1.63\,10^{-3}$	$1.40\,10^{-2}$
e_S	$3.95\,10^{-5}$	$4.30\,10^{-5}$	$8.13\,10^{-4}$	$1.39\,10^{-2}$
e_I	$5.57\,10^{-5}$	$1.35\,10^{-4}$	$8.91\,10^{-4}$	$1.07\,10^{-2}$
nb. eval.	71	63	76	60

Table 4: Results for Case 3.

Case 4	$\epsilon = 0.$	$\epsilon = 10^{-3}$	$\epsilon = 10^{-2}$	$\epsilon = 10^{-1}$
J_{final}/J_0	$7.58\,10^{-8}$	$4.46\,10^{-7}$	$3.35\,10^{-5}$	$3.35\,10^{-3}$
e_V	$2.06\,10^{-5}$	$1.72\,10^{-4}$	$1.32\,10^{-3}$	$1.50\,10^{-2}$
e_S	$9.39\,10^{-4}$	$1.03\,10^{-3}$	$1.88\,10^{-3}$	$1.03\,10^{-2}$
e_I	$1.91\,10^{-4}$	$2.39\,10^{-4}$	$9.32\,10^{-4}$	$8.71\,10^{-2}$
n (final)	1035.	1057.	1062.	1045.
nb. eval.	126	121	123	98

Table 5: Results for Case 4.

	y_1^G	y_2^G	y_3^G	ϕ	θ	ψ	a	b	c
'true'	1.0	1.0	0.5	0.4	0.9	0.6	1.0	2.0	1.0
initial	1.0	1.0	0.5	0.4	0.9	0.6	1.2	2.4	1.2

Table 6: Example 2: 'True' and initial values $d_1, \ldots, d_9$

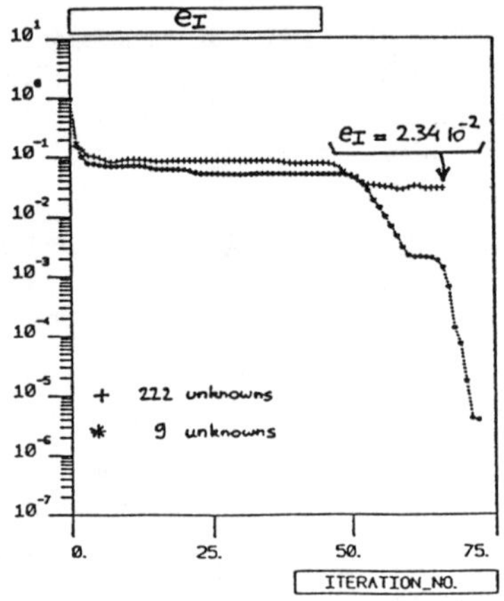

Figure 3: Example 2: e_I vs. iteration no.

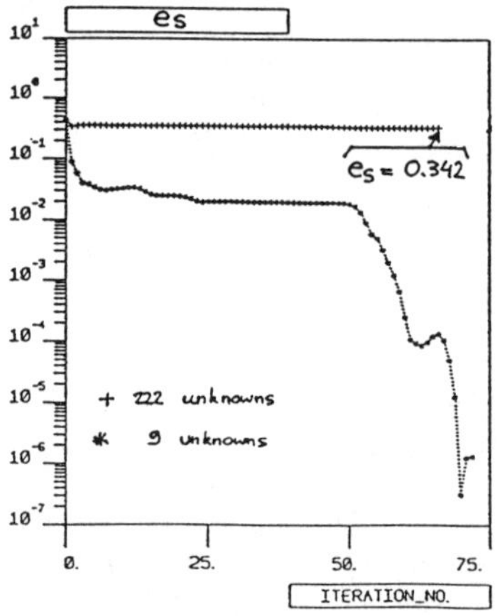

Figure 4: Example 2: e_S vs. iteration no.

Inverse Problems in Engineering: Theory and Practice
ASME 1993

STABLE INVERSION AND ITS APPLICATION TO OUTPUT TRACKING FOR FLEXIBLE MULTI-LINK MANIPULATORS

Degang Chen and Hong Chao Zhao
Department of Electrical and Computer Engineering
Iowa State University
Ames, Iowa

ABSTRACT

This paper studies the stable inversion of nonlinear nonminimum phase systems and its application to output tracking from the perspective of nonlinear geometric control. The inverse problem here is to find an appropriate control input trajectory which when applied to the dynamic system will reproduce exactly a prespecified time trajectory of the output. Using the notions of zero dynamics and stable/unstable manifolds, a new approach is presented for constructing stable inverses for nonlinear nonminimum phase systems. Local existence and uniqueness of such stable inverses is established. A numerical procedure is presented based on iterative linearization and solutions of Riccati like equations. These results are further applied to the tip trajectory tracking for flexible manipulators. The forward dynamics and the inverse dynamics are worked out in detail. The iterative procedure is carried out to obtain the nominal input which is then applied to the forward system to achieve fast slewing tip trajectory. Simulation results demonstrate that the stable inverse is very effective in reproducing the desired output trajectories.

1. INTRODUCTION

Research interest in the control of articulated flexible structures has considerably increased in the past few years, as evidenced by the large number of such papers at recent CDC's and ACC's. This is motivated by the need for space-based manipulators which are necessarily lightweight and therefore flexible, due to the high transportation cost. The expense of large motors and amplifiers required to drive massive earth-bound industrial manipulators is additional motivation for the design and control of lightweight manipulators. Furthermore, even for robot manipulators normally considered rigid, link flexibility cannot be neglected during fast speed motion control.

The study of flexible manipulator control was pioneered by Cannon and Schmitz [1] where a linear quadratic optimal control approach was successfully applied to the end-effector tracking control of a one link flexible robot arm. In this paper, the nonminimum phase effect was first demonstrated and the difficulty in controlling the output of such a system was raised. After that, many researchers have considered different approaches to the control of flexible one link arm which is a linear system for small deflections. For example, Siciliano and Book [2] used a singular perturbation approach to deal with the flexible modes. These results achieved stable closed-loop system, but performances were poor. Bayo [3] presented a new inverse dynamics approach where Fourier transform was used to obtain stable but noncausal control input to be used in an open-loop fashion. Experimental results demonstrated the effectiveness of such an approach and very fast and almost exact tip tracking was obtained without structural vibration. In [4], this approach was applied to multilink flexible manipulators in an approximate linearization fashion, and passive joint feedback control was used to achieve exponentially stable closed-loop systems.

Despite successful experimental results [4] on the output tracking control of multi-link flexible manipulators, all the above mentioned techniques apply only to linear systems, ie, the one link case, and dose not apply to multi-link manipulators which are necessarily nonlinear. Nonlinear control of flexible manipulators is recent and limited. Lucibello Di Benedetto [5] applied the recently developed nonlinear regulation theory [6] to the control of nonlinear flexible arms and achieved asymptotic tracking of periodic output trajectories. In a similar approach by De Luca, et al [7], simulation results demonstrated asymptotic tracking of a finite trajectory with transient errors existing at the beginning and at the end of the maneuver. This transient error phenomenon is a fundamental limitation of the regulation approach.

Another approach to output tracking is based on inversion which was first studied by Brockett and Mesarovic [8]. Later on, Silverman [9] developed an easy-to-follow step-by-step procedure for the inversion of multivariable linear systems. The linear inversion results were extended by Hirschorn [10] to real analytic nonlinear systems. All these inversion algorithms produce causal inverses for a given desired output $y_d(t)$ and a fixed initial condition $x(t_0)$, and for nonminimum phase systems produce unbounded $u(t)$ and $x(t)$. This fundamental difficulty has been noted for a long time. Singh and Schy [11] and De Luca and Siciliano [12] have applied these inversion techniques to the control of flexible manipulators. Simulation and experimental results verify that, although exact output tracking can be achieved for some time interval, internal vibration builds up.

Motivated by the success of the noncausal inverse dynamics approach and the difficulties in both classical inversion and recent nonlinear regulation, the second author has recently developed the notion of stable inversion[13 14] and solved the problem for a class of nonlinear nonminimum phase systems with well defined relative degree whose zero dynamics have hyperbolic fixed point. A numerical procedure is also developed[15] for constructing stable inverses based iterative linearization and decomposition of the stable/unstable subspaces. This approach to output tracking avoids difficulties in both regulation and classical inversion while preserves the advantages of both. Thus exact output tracking can be obtained without transient errors nor internal oscillation build-up. In this paper, we apply the stable inversion approach to achieve exact tip trajectory tracking for multi-link flexible manipulators.

The remainder of the paper is organized as follows. In the next section the problem of stable inversion is presented and is shown to reduce to a two-point boundary value problem of reduced-order ordinary differential equations together with conditions on local existence and uniqueness of solution. In section 3, a numerical procedure is presented which involves iterative linearization and, in each step, integration of Ricatti-like equations. Section 4 develops a mathematical model for a two-flexible-link robot manipulator using the assumed mode technique. Then an inverse model of this manipulator is derived in section 5 together with an appropriate two-point boundary value problem which ensures the stability of the inverse. A numerical procedure for iteratively constructing the stable inverses is given at the beginning of section 6. This procedure is used to obtain a nominal control input torque for a given tip position trajectory. This nominal input is applied to the forward system for output tracking. Simulation results are presented together with comparison to those by computed torque method. Finally, some conclusion remark are given in section 7.

2. Framework and Problem Statement

We consider a nonlinear system of the form

$$\dot{x} = f(x) + g(x)u \tag{1}$$

$$y = h(x), \tag{2}$$

where

$$y = (y_1, y_2, \cdots, y_m)^T,$$

$$u = (u_1, u_2, \cdots, u_m)^T,$$

$$h(x) = [h_1(x), h_2(x), \cdots, h_m(x)]^T,$$

$$g(x) = [g_1(x), g_2(x), \cdots, g_m(x)].$$

defined on a neighborhood X of the origin of $\mathbf{R}^n$, with input $u \in \mathbf{R}^m$ and output $y \in \mathbf{R}^m$. $f(x)$, $g_i(x)$, $i = 1, 2, \cdots, m$ are smooth vector fields and $h_i(x)$ for $i = 1, 2, \cdots, p$ are smooth functions on X, with $f(0) = 0$ and $h(0) = 0$. For such a system, the stable inversion problem is first developed in [13] and stated as follows.

Stable Inversion Problem: Given a smooth reference output trajectory $y_d(t)$ with compact support, find a control input $u_d(t)$ and a state trajectory $x_d(t)$ such that

1) u_d and x_d satisfy the differential equation

$$\dot{x}_d(t) = f(x_d(t)) + g(x_d(t))u_d(t)$$

2) exact output tracking is achieved:

$$h(x_d(t)) = y_d(t),$$

3) u_d and x_d are bounded and

$$u_d(t) \to 0, \quad x_d(t) \to 0 \quad \text{as} \quad t \to \pm\infty.$$

Note that here we require $y_d(t)$ to have compact support, that is, there exist t_0 and t_f such that $y_d(t) = 0$ for all $t \le t_0$ and all $t \ge t_f$. However, the development in this paper can be extended with little effort to cover desired trajectories whose first derivatives have compact support. The extension covers a large class of realistic trajectories. We call x_d the desired state trajectory and u_d the nominal control input. These can be incorporated into a dead-beat controller by using the nominal control input as a feed-forward signal and $x - x_d$ as an error signal for feedback.

Assume that the system has well-defined relative degree $r = (r_1, r_2, \cdots, r_m)^T \in \mathbf{N}^m$ at the equilibrium point 0, that is, in an open neighborhood of 0,
(i) for all $1 \le j \le m$, for all $1 \le i \le m$, for all $k < r_i - 1$ and $k \ge 0$, and for all x,

$$L_{g_j} L_f^k h_i(x) = 0, \tag{3}$$

(ii) the $m \times m$ matrix $\beta(x) \triangleq L_g^{(1)} L_f^{(r-1)} h(x)$ is nonsingular.

Under this assumption, the system can be partially linearized. To do this, we differentiate y_i until at least one u_j appears explicitly. This will happen at exactly the r_ith derivative of y_i due to (3). Define $\xi_k^i = y_i^{(k-1)}$ for $i = 1, \cdots, m$ and $k = 1, \cdots, r_i$, and denote

$$\begin{aligned}\xi &= (\xi_1^1, \xi_2^1, \cdots, \xi_{r_1}^1, \xi_1^2, \cdots, \xi_{r_2}^2, \cdots, \xi_{r_m}^m)^T \\ &= (y_1, \dot{y}_1, \cdots, y_1^{(r_1-1)}, y_2, \cdots, y_2^{(r_2-1)}, \cdots, y_m^{(r_m-1)})^T.\end{aligned}$$

Choose η, an $n - |r|$ dimensional function on $\mathbf{R}^n$ such that $(\xi^T, \eta^T)^T = \psi(x)$ forms a change of coordinates with $\psi(0) = 0$ [16]. In this new coordinate system, the system dynamics of equation (1) becomes

$$\begin{cases} \xi_1^i = \xi_2^i \\ \cdots \\ \xi_{r_i-1}^i = \xi_{r_i}^i \\ \xi_{r_i}^i = \alpha_i(\xi,\eta) + \beta_i(\xi,\eta)u \end{cases} \quad \text{for } i = 1,\cdots,m,$$

$$\dot\eta = q_1(\xi,\eta) + q_2(\xi,\eta)u,$$

which, in a more compact form, is equivalent to

$$y^{(r)} = \alpha(\xi,\eta) + \beta(\xi,\eta)u,$$

$$\dot\eta = q_1(\xi,\eta) + q_2(\xi,\eta)u, \quad (4)$$

where

$$\alpha(\xi,\eta) = L_f^r h(\psi^{-1}(\xi,\eta)),$$

$$\beta(\xi,\eta) = L_g^1 L_f^{r-1} h(\psi^{-1}(\xi,\eta)),$$

$\alpha(0,0) = 0$ since $f(0) = 0$, and α_i and β_i are the ith row of α and β respectively. Since by the relative degree assumption, $\beta(\xi,\eta)$ is nonsingular, the following feedback control law

$$u \triangleq \beta^{-1}(\xi,\eta)[y_d^{(r)} - \alpha(\xi,\eta)] \quad (5)$$

is well defined and leads to

$$y^{(r)} = y_d^{(r)},$$

and therefore

$$\xi = \xi_d \triangleq (y_{d1}, \dot y_{d1}, \cdots, y_{d1}^{(r_1-1)}, y_{d2}, \cdots, y_{d2}^{(r_2-1)}, \cdots, y_{dm}^{(r_m-1)}).$$

Then equation (4), which we call the *reference dynamics*, or the zero dynamics driven by the reference output trajectory, becomes

$$\dot\eta = p(y_d^{(r)}, \xi_d, \eta), \quad (6)$$

where

$$p(y_d^{(r)}, \xi_d, \eta) \triangleq q_1(\xi_d, \eta)$$
$$+ q_2(\xi_d,\eta)\beta^{-1}(\psi^{-1}(\xi_d,\eta))[y_d^{(r)} - \alpha(\psi^{-1}(\xi_d,\eta))].$$

It is clear now that an integration of the reference dynamics gives rise to a trajectory of the original state through the inverse coordinate transformation $x = \Psi^{-1}(\xi,\eta)$ and an input trajectory by equation (5). Now the question is how to integrate the reference dynamics to generate a bounded input solving the stable inversion problem, since the reference dynamics may be unstable in both positive and negative time directions in general.

For reference trajectories with compact support, the reference dynamics become autonomous zero dynamics for t outside the compact interval $[t_0, t_f]$. Assuming that $\eta = 0$ is a hyperbolic equilibrium point of the autonomous zero dynamics, then there exists stable and unstable manifolds W^s and W^u for $t \le t_0$ and for $t \ge t_f$. Locally W^u can be defined by an equation $B^s(\eta) = 0$ and, similarly, W^s can be defined by $B^u(\eta) = 0$. The following theorems can be established [15].

Theorem 1: Let the system described by (1) and (2) with $p = m$ have well defined relative degree and its zero dynamics have a hyperbolic equilibrium at 0. Then the stable inversion problem has a solution if and only if the following two-point boundary value problem has a solution:

$$\dot\eta = p(y_d^{(r)}, \xi_d, \eta), \quad (6)$$

subject to

$$B^s(\eta(t_0)) = 0,$$
$$B^u(\eta(t_f)) = 0. \quad (7)$$

Theorem 2: The two point boundary value problem of (6) and (7) has a unique solution provided $\|\bar\xi_d\|_\infty \triangleq \sup_{t\in[t_0,t_f]} \|\bar\xi_d(t)\|_2$ is sufficiently small, where $\bar\xi_d \triangleq ((y^{(r)})^T, \xi_d^T)^T$.

3. An Iterative Solution to the TPBV Problem

Now let us consider an iterative linearization approach to the solution of the TPBV problem. In each iteration, we will linearize the equations in (6) and (7) along the solution obtained in the previous step. This gives rise to a new linear time varying two point boundary value problem. This linear problem will be solved involving integration in two directions and the solution is taken to be this step's new approximation. This process continues until some convergence criterion is met.

First we make precise the linearization. To initialize, we take $\eta^0 = 0$ for all t. Let η^{k-1} be the solution obtained in the (k-1)th step. Let η^k denote the new corrected solution to be solved in the k-th step, the current step. Linearizing the right hand sides of equation (6) and equation (7) along η^{k-1} and setting η to η^k, we have

$$\dot\eta^k = \frac{\partial p}{\partial\eta}(\bar\xi_d, \eta^{k-1})(\eta^k - \eta^{k-1}) + p(\bar\xi_d, \eta^{k-1}), \quad (8)$$

subject to

$$\frac{\partial B^s}{\partial\eta}(\eta^{k-1}(t_0))(\eta^k(t_0) - \eta^{k-1}(t_0)) + B^s(\eta^{k-1}(t_0)) = 0, \quad (9a)$$

$$\frac{\partial B^u}{\partial\eta}(\eta^{k-1}(t_f))(\eta^k(t_f) - \eta^{k-1}(t_f)) + B^u(\eta^{k-1}(t_f)) = 0. \quad (9b)$$

By defining the following symbols,

$$A^k(t) \triangleq \frac{\partial p}{\partial\eta}(\bar\xi_d(t), \eta^{k-1}(t))$$

$$b^k(t) \triangleq -\frac{\partial p}{\partial\eta}(\bar\xi_d(t), \eta^{k-1}(t))\eta^{k-1}(t) + p(\bar\xi_d(t), \eta^{k-1}(t)),$$

$$C_s^k \triangleq \frac{\partial B^s}{\partial\eta}(\eta^{k-1}(t_0)), \quad C_u^k \triangleq \frac{\partial B^u}{\partial\eta}(\eta^{k-1}(t_f)),$$

$$\alpha^k \triangleq -\frac{\partial B^s}{\partial\eta}(\eta^{k-1}(t_0))\eta^{k-1}(t_0) + B^s(\eta^{k-1}(t_0)),$$

$$\beta^k \triangleq -\frac{\partial B^u}{\partial\eta}(\eta^{k-1}(t_f))\eta^{k-1}(t_f) + B^u(\eta^{k-1}(t_f)),$$

we can rewrite equations (8) and (9) in the following format,

$$\dot\eta^k = A^k(t)\eta^k + b^k(t),$$

$$C_s^k \eta^k(t_0) = \alpha^k, \quad C_u^k \eta^k(t_f) = \beta^k,$$

which is a linear time varying two point boundary value problem we have to solve in each step.

Next, we need to solve the linear time varying two point boundary value problem. The idea is to try to separate the stable and unstable dynamics, and integrate the stable part forward in time and the unstable part backward in time. If $A^k(t)$ is time invariant, this separation is easily done through eigenspace decomposition. In the time varying case, the stable and unstable subspaces are coupled through time variation and the problem is a little more complicated. Here we use a technique from linear quadratic optimal control. For convenience, we will drop the superscript k in the following.

Assume that the matrix $\begin{pmatrix} C_s \\ C_u \end{pmatrix}$ is invertible. We apply a change of coordinates:

$$z = \begin{pmatrix} z_1 \\ z_2 \end{pmatrix} \triangleq \begin{pmatrix} C_s \\ C_u \end{pmatrix} \eta.$$

Since $B^s(\eta) = 0$ is the condition for η to be on the unstable manifold, therefore $B^s(\eta)$ can be viewed as the stable part of η. Hence, in the linear approximation, z_1 is, roughly speaking, picking up the stable part of η, and similarly, z_2 is picking up the unstable part.

Let $T = (T_s \ T_u)$ be the inverse transformation matrix, then,

$$\eta = (T_s \ T_u) \begin{pmatrix} z_1 \\ z_2 \end{pmatrix}$$

Then in the new coordinates, the system becomes:

$$\dot{z} = \begin{pmatrix} \dot{z}_1 \\ \dot{z}_2 \end{pmatrix} = \begin{pmatrix} C_s \\ C_u \end{pmatrix} \dot{\eta} = \begin{pmatrix} C_s \\ C_u \end{pmatrix} [A(t)\eta + b(t)].$$

Substituting the inverse transformation into the above, we can rearrange the above equation into

$$\dot{z}_1 = A_{11}(t) z_1 + A_{12}(t) z_2 + b_1(t), \tag{10a}$$

$$\dot{z}_2 = A_{21}(t) z_1 + A_{22}(t) z_2 + b_2(t), \tag{10b}$$

with initial and final conditions specified, respectively, as

$$z_1(t_0) = \alpha \quad z_2(t_f) = \beta,$$

and

$$A_{11}(t) \triangleq C_s A(t) T_s \quad A_{12} \triangleq C_s A(t) T_u$$

$$A_{21}(t) \triangleq C_u A(t) T_s \quad A_{22} \triangleq C_u A(t) T_u$$

$$b_1(t) \triangleq C_s b(t) \quad b_2(t) \triangleq C_u b(t).$$

Since z_1 and z_2 satisfy a pair of linear coupled differential equations, it is easy to see that the solutions are also linearly related. Therefore, there exist a vector function $g(t)$ and a matrix function $S(t)$ of suitable dimensions, such that

$$z_2(t) = S(t)\, z_1(t) + g(t), \tag{11}$$

with a suitable final value condition

$$S(t_f) = 0 \quad g(t_f) = \beta. \tag{12}$$

Differentiating both side of equation (11) yields

$$\dot{z}_2(t) = \dot{S}(t)\, z_1(t) + S(t)\, \dot{z}_1(t) + \dot{g}(t).$$

Substituting the value of $\dot{z}_1$ and $\dot{z}_2$ from (10) results in

$$A_{21} z_1 + A_{22} z_2 + b_2 = \dot{S} z_1 + \dot{g} + S\{A_{11} z_1 + A_{12} z_2 + b_1\}$$

$$A_{21} z_1 + A_{22}[S\, z_1 + g] + b_2$$

$$= \dot{S}\, z_1 + \dot{g} + S\,\{A_{11} z_1 + A_{12}[S\, z_1 + g] + b_1\}.$$

Since this equation holds for all values of $z_1(t)$, comparing coefficients leads to

$$\dot{S} = A_{21} + A_{22} S - S\, A_{11} - S\, A_{12} S \tag{13a}$$

$$\dot{g} = [\, A_{22} - S\, A_{12}\,]\, g + [\, b_2 - S\, b_1\,] \tag{13b}$$

with final value conditions specified in equation (12). Since equation (13a) contains only known function except S, it can be integrated backward in time to get $S(t)$. Once this is done, equation (13b) can also be integrated backward in time to solve for $g(t)$. With S and g as known functions, equation (10a) can be rewritten as

$$\dot{z}_1 = [\, A_{11}(t) + A_{12}(t) S(t)\,]\, z_1 + b_1(t) + A_{12}(t) g(t), \tag{14}$$

which can be integrated forward in time to obtain $z_1(t)$. Finally, the algebraic equation (11) is used to obtain $z_2(t)$.

4. Forward Dynamics of a Two-Flexible-Link Robot Arm

A robot can be considered as the assembly of many flexible links. For simplicity, we consider a robot arm with two elastic links. Both joints are revolute and input torques are applied at these points. A model of a two-link flexible arm is shown in Figure 1. Each link i has total length l_i, mass per unit length ρ_i, area moment of inertia I_i, Young's modulus E_i. Attached at one end of link i is a tip mass m_{e_i}, and at the other end a hub of inertia I_{b_i}. We assume that the links are maneuvered in the horizontal plane and that the out-of-plane deflections are negligible.

By the assumed modes method we may approximate the continuous deflection of a flexible link by a set of assumed shape functions and their time-dependent generalized coordinates. Let the flexible displacements of link 1 and link 2 be $y_1(z_1, t)$ and $y_2(z_2, t)$ respectively. Also, let $\phi_{1i}(z_1)$ and $\phi_{2i}(z_2)$ be the ith necessary admissible shape functions, $q_{1i}(t)$ and $q_{2i}(t)$ be the corresponding generalized coordinates. Then the distributed deflections of the two links are approximated by

$$y_1(z_1, t) = \sum_{i=1}^{n_1} \phi_{1i}(z_1) q_{1i}(t) = \phi_1^T q_1 \tag{15a}$$

$$y_2(z_2, t) = \sum_{i=1}^{n_2} \phi_{2i}(z_2) q_{2i}(t) = \phi_2^T q_2. \tag{15b}$$

The dynamics of the arm is given by Lagrange equation:

$$\frac{d}{dt}\frac{\partial L}{\partial \dot{\psi}} - \frac{\partial L}{\partial \psi} = \tau \tag{16}$$

where $\psi = (\theta_1, \theta_2, q_{11}, q_{12}, q_{21}, q_{22})^T$ is a set of generalized coordinates for the system; τ is the generalized forces acting the generalized coordinates; and L, the Lagrangian, is the difference, $K - P$, between the kinetic energy K and the potential energy P, which are given respectively by

$$K = \frac{1}{2}\dot{\psi}^T M(\psi)\dot{\psi}, \quad P = \frac{1}{2}\psi^T K\psi, \tag{17}$$

where $M(\psi)$ is a positive definite symmetric inertia matrix and is a nonlinear function of ψ, and K is the stiffness matrix composed of the stiffness matrices of the flexible links. For the two-link case, it has the form shown as

$$K = \begin{bmatrix} K_{11 2\times2} & K_{12 2\times4} \\ K_{21 4\times2} & K_{22 4\times4} \end{bmatrix} = \begin{bmatrix} 0_{2\times2} & 0_{2\times2} & 0_{2\times2} \\ 0_{2\times2} & K^1{}_{2\times2} & 0_{2\times2} \\ 0_{2\times2} & 0_{2\times2} & K^2{}_{2\times2} \end{bmatrix}.$$

Substituting (17) into (16) we get

$$M\ddot{\psi} + \dot{M}\dot{\psi} - \frac{1}{2}\frac{\partial(\dot{\psi}^T M\dot{\psi})}{\partial\psi} + K\psi = Bu - d \tag{18}$$

where $u = (u_1, u_2)^T$ is the input torque vector applied at the joints, $B = (I_{2\times2}\ 0_{2\times4})^T$ determines how joint torques affect the generalized coordinates, d is the Rayleigh dissipation force due to structural damping of the flexible links which has the form: $d = C\dot{\psi}$ where C is the damping matrix taken to be a proportion of the stiffness matrix as a common practice which is of the form:

$$C = \begin{bmatrix} C_{11 2\times2} & C_{12 2\times4} \\ C_{21 4\times2} & C_{22 4\times4} \end{bmatrix} = \begin{bmatrix} 0_{2\times2} & 0_{2\times2} & 0_{2\times2} \\ 0_{2\times2} & C^1{}_{2\times2} & 0_{2\times2} \\ 0_{2\times2} & 0_{2\times2} & C^2{}_{2\times2} \end{bmatrix}$$

with $C^1{}_{ij} = \alpha_1 K^1{}_{ij}$, $C^2{}_{ij} = \alpha_2 K^2{}_{ij}$. Hence, equation (18) can be written as

$$M\ddot{\psi} + H(\psi,\dot{\psi}) + C\dot{\psi} + K\psi = Bu. \tag{19}$$

There are many ways to choose the system output. Depending on which point along the links the output point is located, the whole system can be either minimum or nonminimum phase. If the output is selected to be the joint angles, i.e, the sensors and actuators collocated, the system is known to be minimum phase. A more meaningful choice of output, as is our choice, is the tip position. But this choice renders the system nonminimum phase which will be demonstrated later on by the eigenvalues of inverse system. Both the Cartesian coordinates and angular coordinates can be used for tip positions. Here for simplicity, we choose the tip angular positions as the system output, which is given by

$$y = \theta + \left[\tan^{-1}\left(\frac{y_1(l_1,t)}{l_1}\right), \tan^{-1}\left(\frac{y_2(l_2,t)}{l_2}\right)\right]^T$$

where $\theta = (\theta_1, \theta_2)^T$, $y = (y_1, y_2)^T$. For small elastic deformations, the output simplifies to

$$y = \theta + \left[\frac{y_1(l_1,t)}{l_1}, \frac{y_2(l_2,t)}{l_2}\right]^T. \tag{20}$$

Substituting (15) into equation (20), we obtain

$$y = D\psi \tag{20}$$

where $D = \begin{bmatrix} D_1 & D_2 \end{bmatrix}$, $D_1 = I_{2\times2}$, and

$$D_2 = \begin{bmatrix} \frac{\phi_{11}(l_1)}{l_1} & \frac{\phi_{12}(l_1)}{l_1} & 0 & 0 \\ 0 & 0 & \frac{\phi_{21}(l_2)}{l_2} & \frac{\phi_{22}(l_2)}{l_2} \end{bmatrix}.$$

5. Inverse Dynamics and Its Equivalent TPBV Problem

Based on the above forward dynamic equations, we here derive the linearized inverse dynamic equations. It can be easily verified that the following two approaches will end up with the same results. One way is to linearize the forward dynamic equations first, then to combine equations (19) and (20) to get the result. Another way is to get the inverse dynamic equations first, then to do the linearization. We here choose the second approach.

Equation (19) can be written in two parts:

$$M_{11}(\psi)\ddot{\theta} + M_{12}(\psi)\ddot{q} + H_1(\psi,\dot{\psi}) = B_1 u \tag{21a}$$

$$M_{21}(\psi)\ddot{\theta} + M_{22}(\psi)\ddot{q} + H_2(\psi,\dot{\psi}) + M_2\dot{q} + M_3 q = 0 \tag{21b}$$

where $M_2 = C_{22}$ and $M_3 = K_{22}$. Given a reference output trajectory $y_d(t)$, equation (20) can be written as

$$\theta = D_1^{-1}y_d - D_1^{-1}D_2 q.$$

Substituting this equation into (21b), we obtain the nonlinear inverse dynamic equations:

$$M_{21}(y_d,q)(D_1^{-1}\ddot{y}_d - D_1^{-1}D_2\ddot{q}) + M_{22}(y_d,\dot{y}_d,q,\dot{q}) + M_2\dot{q} + M_3 q = 0.$$

Combining terms, we get

$$M_1\ddot{q} + M_2\dot{q} + M_3 q + H_2 = M_4\ddot{y}_d \tag{22}$$

where

$$M_1 = M_{22}(y_d(t),q(t)) - M_{21}(y_d(t),q(t))D_1^{-1}D_2$$

$$M_4 = -M_{21}(y_d(t),q(t))D_1^{-1}$$

In order to carry out the iterative algorithm to do the inversion, we need to linearize the inverse dynamic equation (22). Let us do it term by term at point defined as $(q_0^T, \dot{q}_0^T)^T$. For convenience, we write terms only as functions of q's instead of both q's and y_d's.

Let $M(x)$ be a $k\times l$ matrix function of $x \in \mathbf{R}^n$ and $y \in \mathbf{R}^n$ be a column vector. The derivative of M at a point x_0 in the direction of y is defined as

$$D_x^{x_0}My \triangleq \sum_{i=1}^{n}\frac{\partial M}{\partial x_i}\bigg|_{x=x_0} y_i.$$

Using this notation and neglecting higher order terms, the first term $M_1(q)\ddot{q}$ in equation (22) can be linearized as:

$$M_1(q)\ddot{q} = \left[M_1^0 + D_q^0 M_1\left[q - q_0\right]\right]\left[\ddot{q}_0 + \left[\ddot{q} - \ddot{q}_0\right]\right]$$

$$= M_1^0\ddot{q} + (D_q^0 M_1 q)\ddot{q}_0 - (D_q^0 M_1 q_0)\ddot{q}_0,$$

where the superscript 0 stands for evaluation along q_0 or $\dot{q}_0$ whichever is applicable. Since it can be easily verified that

$$\left[D_x My\right]z = \left[D_x(Mz)\right]y$$

where z is an appropriately dimensioned vector or matrix, we obtain

$$M_1(q)\ddot{q} = M_1^0\ddot{q} + D_q^0(M_1\ddot{q}_0)q - D_q^0(M_1\ddot{q}_0)q_0. \quad (23)$$

Since M_2 and M_3 both are constant matrices, they are already of the required linear form. For the term $H_2(q,\dot{q})$, we have

$$H_2(q,\dot{q}) = H_2^0 + D_q^0 H_2\left[q - q_0\right] + D_{\dot{q}}^0 H_2\left[\dot{q} - \dot{q}_0\right]$$
$$= H_2^0 - D_q^0 H_2 q_0 - D_{\dot{q}}^0 H_2 \dot{q}_0 + D_q^0 H_2 q + D_{\dot{q}}^0 H_2 \dot{q} \quad (24)$$

Similar to the derivation for the first term, we can get the linearized form of $M_4\ddot{y}_d$ as :

$$M_4\ddot{y}_d = M_4^0\ddot{y}_d - D_q^0(M_4\ddot{y}_d)q + D_q^0(M_4\ddot{y}_d)q \quad (25)$$

Thus combining equations (23), (24) and (25) together, the linearized inverse dynamics can be expressed as:

$$A_1\ddot{q} + A_2\dot{q} + A_3 q = A_4 \quad (26)$$

for some suitably defined matrices A_i, $i = 1, 2, 3, 4$. The state space form of the equation (26) can be written as

$$\dot{x}(t) = A(t)x(t) + b(t) \quad (27)$$

where $x = (q^T, \dot{q}^T)^T$, and

$$A(t) = \begin{bmatrix} 0 & I \\ -A_1^{-1}A_3 & -A_1^{-1}A_2 \end{bmatrix},$$

$$b(t) = \begin{bmatrix} 0 \\ A_1^{-1}A_4 \end{bmatrix}.$$

It is a generally accepted fact that a flexible link manipulator with tip position as output is a nonminimum phase system. Thus its inverse system has eigenvalues in both left and right half planes. In addition, since sustained oscillation without any tip movement is physically impossible, the inverse system has no eigenvalues on the $j\omega$-axis and thus must have a hyperbolic fixed point. These properties are verified in the next section when the eigenvalues of the inverse system are actually computed.

Let us form the matrix X_s by taking as columns the eigenvectors and the generalized eigenvectors of $A(t)$ at some fixed time t corresponding to eigenvalues having negative real parts, and X_u, the eigenvectors and generalized eigenvectors corresponding to eigenvalues having positive real parts. Then, we have

$$A(t)\begin{bmatrix} X_s & X_u \end{bmatrix} = \begin{bmatrix} X_s & X_u \end{bmatrix}\begin{bmatrix} \Lambda_s & 0 \\ 0 & \Lambda_u \end{bmatrix}$$

where Λ_s and Λ_u are the corresponding Jordan forms. Denoting

$$\begin{bmatrix} Y_s \\ Y_u \end{bmatrix} = \begin{bmatrix} X_s & X_u \end{bmatrix}^{-1},$$

we obtain

$$\begin{bmatrix} \Lambda_s & 0 \\ 0 & \Lambda_u \end{bmatrix} = \begin{bmatrix} Y_s \\ Y_u \end{bmatrix} A(t)\begin{bmatrix} X_s & X_u \end{bmatrix}$$

$$Y_s A(t) X_u = 0 \quad (28a)$$

$$Y_u A(t) X_s = 0 \quad (28b)$$

Since we know that $x(t)$ belongs to W^u, for all t less than or equal to t_0 and $x(t)$ belongs to W^s, for all t greater than or equal to t_f, where W^u and W^s denote the unstable and stable manifolds of the origin respectively Therefore, at time t_0, $x(t_0)$ belongs to W^u which means $x(t_0)$ can be written as the linear combination of the columns of X_u. Thus, we have

$$x(t_0) = X_u Z_u \quad (29a)$$

where Z_u is a column vector of coefficients. Combining equations (28a) and (29a), we have

$$Y_s A(t_0)x(t_0) = Y_s A(t_0) X_u Z_u = 0 \quad (30a)$$

Similarly while at time t_f, we have

$$x(t_f) = X_s Z_s \quad (29b)$$

and thus by (28b) and (29b),

$$Y_u A(t_f)x(t_f) = Y_u A(t_f) X_s Z_s = 0 \quad (30b)$$

Denoting $C_s = Y_s A(t_0)$ and $C_u = Y_u A(t_f)$ and combining equations (30) with equations (27), we obtain the linear time varying TPBV problem as follows:

$$\dot{x}(t) = A(t)x(t) + b(t) \quad (31a)$$

subject to

$$C_s x(t_0) = 0 \quad (31b)$$

$$C_u x(t_f) = 0 \quad (31c)$$

where all matrices are defined as above.

6. Simulation Analysis

In this section, we present the digital simulation results to illustrate the performance of the inverse dynamics method. To find out the required torques corresponding to the given desired tip trajectories, our simulation goes through the following steps:

1: Set $q_0(t) = 0$ for all t.

2: Linearize (22) along $q_0(t)$ to get (31) and (10-15).

3: Integrate equation (13a) backward in time to get $S(t)$.

4: Integrate equation (13b) backward in time to get $g(t)$.

5: Integrate equation (14) forward in time to get $z_1(t)$ and get $z_2(t)$ by (11).

6: Compute $q(t) = \begin{pmatrix} C_s \\ C_u \end{pmatrix}^{-1} \begin{pmatrix} z_1 \\ z_2 \end{pmatrix}$

7: If $\|q - q_0\| >$ threshold, then $q_0 = q$ and go to step 2, else continue.

8: Compute nominal input $u_d(t)$ from equation (21a).

The two-link flexible arm under consideration has parameters as listed in Table 1. The desired tip trajectories of the flexible arm, the angular displacements together with their corresponding acceleration profiles, are used as the input of the inverse dynamic equations. The trajectories are shown in Figure 2. The above procedure is implemented to obtain the nominal torque input. Figure 3 shows the joint torques needed to produce the desired tip trajectories. As expected, the torques needs to be applied to preshape the links some time before the tip starts moving. These calculated torques are applied to the forward dynamic equations together with some small joint feedback. The results are shown in Figure 4 indicating that the tip follows the desired trajectories exactly without any undershoot or overshoot.

To illustrate the effectiveness of our stable inverse dynamics approach, let us compare it with the well known computed torque method. In this scheme, the input torque is calculated as

$$\tau^* = M(\theta_d)\ddot{\theta}_d + H_1(\theta_d, \dot{\theta}_d) + K_d(\dot{\theta}_d - \dot{\theta}) + K_p(\theta_d - \theta).$$

The tracking performance is shown in Figure 5, which exhibit significant errors including undershoot and overshoot.

7. Conclusion

The iterative approach to stable inversion of nonlinear nonminimum phase systems is successfully applied to the tip trajectory tracking for a two-link flexible robot manipulator. The key assumptions on well defined relative degree and hyperbolicity of the fixed point of the zero dynamics are satisfied. Simulation results demonstrate that the stable inversion approach is very effective for obtaining exact output tracking for flexible manipulators. This approach is expected to perform equivalently well for other realistic nonminimum phase systems. Future work will be on efficient numerical algorithms for constructing stable inverses and on new applications of stable inversion.

References

1. R.H. Cannon and E. Schmitz, "Initial Experiments on the End Point Control of a Flexible One-link Robot," *Int. J. Robotics Research*, vol. 3, pp. 62-75, 1984.
2. B. Siciliano and W.J. Book, "A Singular Perturbation Approach to Control Lightweight Flexible Manipulators," *Int. J. Robotics Research*, vol. 7, pp. 79-90, 1988.
3. E. Bayo, "A Finite-Element Approach to Control the End-Point Motion of a Single-Link Flexible Robot," *J. Robotic Systems*, vol. 4(1), pp. 63-75, 1987.
4. B. Paden, D. Chen, R. Ledesma, and E. Bayo, "Exponentially Stable Tracking Control for Multi-Link Flexible Manipulators," *to appear in ASME J Dynamics, Measurement and Control.*
5. P. Lucibello and M.D. Di Benedetto, "Output Tracking for a Nonlinear Flexible Arm," *submitted to J. of ASME.*
6. A. Isidori and C.I. Byrnes, "Output Regulation of Nonlinear Systems," *IEEE Trans. on Automatic Control*, vol. 35, no. 2, pp. 131-140, 1990.
7. A. De Luca, L. Lanari, and G. Ulivi, "Output Regulation of a Flexible Robot Arm," *9th Int. Conference on Analysis & Optimization of Systems INRIA*, 1990.
8. R.W. Brockett and M.D. Mesarovic, "The Reproducibility of Multivariable Systems," *J. Mathematical Analysis and Applications*, vol. 11, pp. 548-563, 1965.
9. L.M. Silverman, "Inversion of Multivariable Linear Systems," *IEEE Trans. on Automatic Control*, vol. 14, no. 3, pp. 270-276, 1969.
10. R.M. Hirschorn, "Invertibility of Multivariable Nonlinear Control Systems," *IEEE Trans. on Automatic Control*, vol. 24, no. 6, pp. 855-865, 1979.
11. S.N. Singh and A.A. Schy, "Control of Elastic Robotic Systems by Nonlinear Inversion and Modal Damping," *J. Dynamic Systems Measurement and Control*, vol. 108, 1986.
12. B. Siciliano, "Trajectory Control of a Nonlinear One-link Flexible Arm," *Int. J. Control*, vol. 50(5), pp. 1699-1715, 1989.
13. D. Chen, *Nonlinear Adaptive Control of Electro-Mechanical Systems*, PhD Dissertation, University of California, Santa Barbara, 1992.
14. D. Chen and B. Paden, "Stable Inversion of Nonlinear Nonminimum Phase Systems," *Proceedings of Japan/USA Symposium on Flexible Automation*, pp. 791-797, 1992.
15. D. Chen, "An Iterative Solution to Stable Inversion of Nonlinear Nonminimum Phase Systems," *Proceedings of American Control Conference*, San Francisco, 1993.
16. A. Isidori, *Nonlinear Control Systems: An Introduction*, Springer-Verlag, New York, 1989.

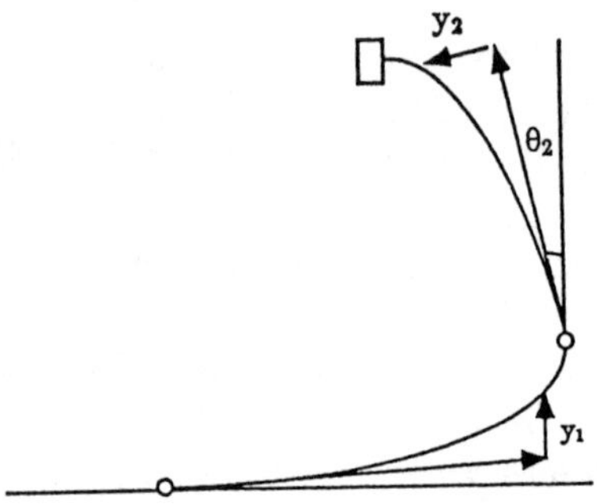

Figure 1. Two-Link Flexible Arm

	Link One	Link Two
l (m)	1	1
rho (kg/m)	0.3	0.1
EI (N*m*m)	22.5	2.5
Me (kg)	0.15	0.10
Ib (kg*m*m)	0.2	0.067

Table 1.

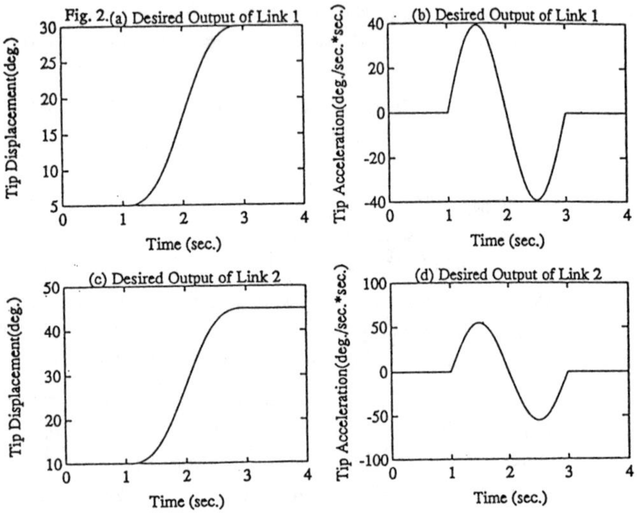

Figure 3. Input Torques by Inverse Dynamics

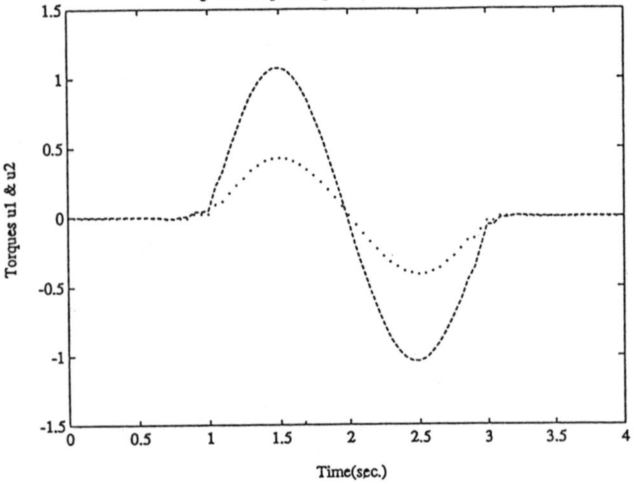

Figure 5. (a) Desired and Simulated Outputs by Computed Torque Technique

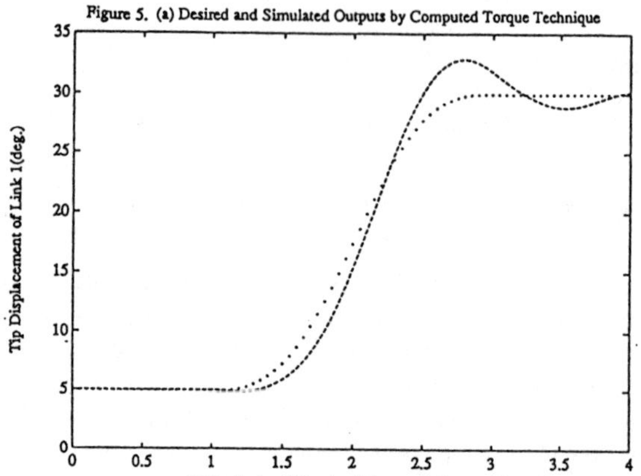

(b) Desired and Simulated Outputs by Inverse Dynamics

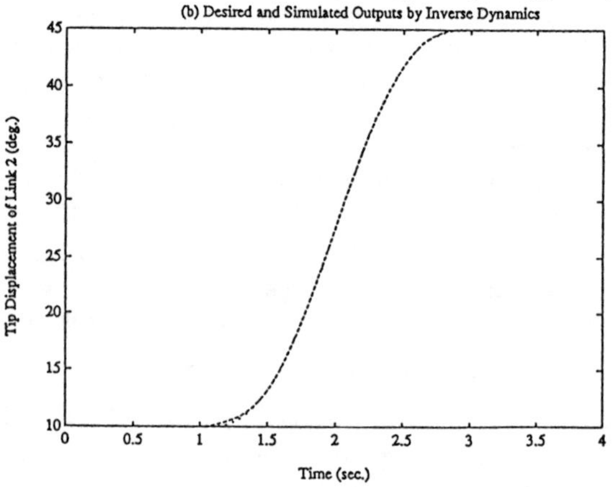

Figure 4. (a) Desired and Simulated Outputs by Inverse Dynamics

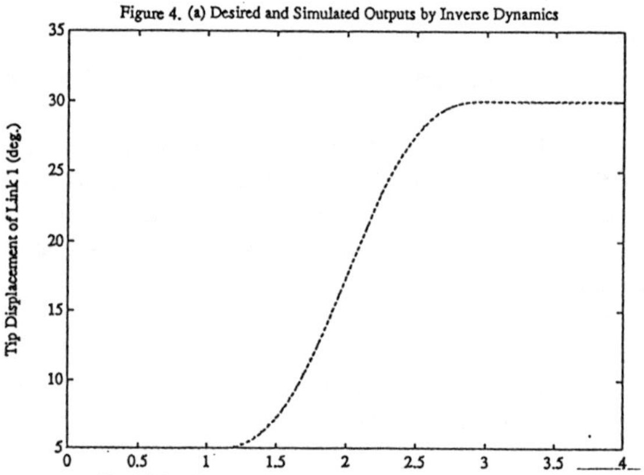

Fig. 5.6(b) Desired and Simulated Outputs by Computed Torque Technique

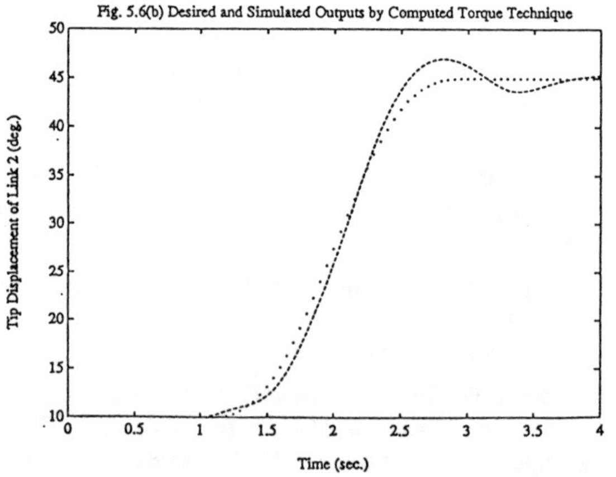

ITERATIVE ALGORITHMS FOR ESTIMATING TEMPERATURE-DEPENDENT THERMOPHYSICAL CHARACTERISTICS

Eugene A. Artyukhin*
Laboratoire de Thermocinetique
ISITEM
Université de Nantes
Nantes, France

ABSTRACT

Algorithms based on the iterative regularization method are analyzed in application to the inverse problem solution of estimating thermophysical characteristics of solids as arbitrary enough and smooth temperature functions from transient temperature measurements. The parametrization of unknown functions is introduced and the inverse problem is reduced to the estimation of unknown parameters. A priori information on positiveness and the smoothness of desired functions is taken into account.

INTRODUCTION

Inverse heat transfer problems are usually ill-posed and special regularizing algorithms are needed to solve such problems [1]. The comparative analysis has shown [2, 3] that the effective enough method for building computational algorithms to solve ill-posed inverse problems is the method of iterative regularization [4]. Algorithms based on this method are usually refered to as iterative algorithms. High efficiency of iterative algorithms has been demonstrated by solving numerically inverse heat transfer problems of different types [2, 4].

The quality of ill-posed problem solutions depends sufficiently on a priori information being taken into account in the computational algorithms [2, 4, 5]. Importance of that for parameter estimation techniques has been discussed in [6] but there are only a few works on the subject at present [6, 7, 8]. One of the sufficient advantages of the iterative algorithms is the possibility of taking into consideration different a priori information about desired characteristics in the frame of general, universal computational scheme [4]. The main goal of this work is to develop such techniques for estimating temperature-dependent thermophysical characteristics of solids in the form of arbitrary enough smooth temperature functions.

* Permanent address:
College of Cosmonautics, Moscow Aviation Institute,
Moscow, Russia

INVERSE PROBLEM FORMULATION

Consider one-dimensional heat conduction process in a specimen which has the form of a slab and made of a material with temperature-dependent thermophysical characteristics. The mathematical model of the process under analysis is written as follows :

$$C(T)\frac{\partial T}{\partial t} = \frac{\partial}{\partial x}\left(k(T)\frac{\partial T}{\partial x}\right), \qquad 0 < x < L,\ 0 < t < t_f \quad (1)$$

$$T(x,0) = T_0(x), \qquad 0 \le x \le L \quad (2)$$

$$\alpha_1 k\left(T(0,t)\right)\frac{\partial T(0,t)}{\partial x} + \mu_1 T(0,t) = v_1(t) \quad (3)$$

$$\alpha_2 k\left(T(L,t)\right)\frac{\partial T(L,t)}{\partial x} + \mu_2 T(L,t) = v_2(t) \quad (4)$$

where α_1 , μ_1 , α_2 , and μ_2 are parameters, by fixing which boundary conditions of the first, second or third kind can be analyzed. The mathematical model (1) to (4) is usually refered to as the direct heat conduction problem. If all characteristics of the model are given, temperature field in the slab as a function of space and time can be computed.

We consider another situation when temperature-dependent thermophysical characteristics C(T) or/and k(T) are unknown but all other characteristics of the model are given. In this case, it is impossible to compute the temperature field without a certain additional information on temperature in the slab. To get the additional information, temperature histories at some number of space points are usually measured. Suppose that temperature histories are measured at N internal points with coordinates $x = d_n$, $n = 1,..., N$ as well as at the boundaries $x = d_0 = 0$ and $x = d_{N+1} = L$. Measurement results can be written as :

$$T_{meas}(d_n,t) = f_n(t)\,, \qquad n = 0, \dots, N+1 \tag{5}$$

The problem of estimating functions C(T) and k(T) from conditions (1) to (5) is usually refered to as the coefficient inverse heat conduction problem [4] or parameter estimation problem [9].

Different problem statements are possible for coefficient inverse heat conduction problems according to a priori information available on the characteristics C(T) and k(T). It is possible to estimate any one temperature function only or two functions simultaneously. The problem of estimating two functions is the most interesting because the maximal information is received in this case by using measurements in only one experiment. The inverse problem of estimating any one characteristic is a particular problem.

It is extremely important to know conditions under which inverse problems (1) to (5) have unique solutions. The generalized review of basic results on that can be found in [4]. Main results can be formulated as follows. The inverse problem of estimating any one thermophysical characteristic, as an arbitrary enough temperature function, has unique solution when the boundary conditions of any kind are given and temperature history is measured at least at one point of the [0 , L] interval. If the boundary condition of the first kind is given at any boundary, temperature sensor can not be located at this boundary. If two temperature functions are estimated, the boundary condition for heat flux entering the slab must be known at least at one boundary and this heat flux must not be equal to zero. Temperature histories must be measured at least at two points which can be located at the boundaries with prescribed heat flux.

Very important a priori information is available on desired characteristics. As it follows from physical viewpoint, they must be strongly positive. It means that the desired functions must satisfy the following conditions :

$$C(T) > 0\,, \qquad T \in [a\,,b] \tag{6}$$

$$k\,(T) > 0\,, \qquad T \in [a\,,b] \tag{7}$$

where a and b are minimal and maximal temperatures respectively being realized in the specimen. Equations (6) and (7) are constraints inequalities on the desired functions. In addition to the positiveness of unknown functions, the values of these functions can be sometimes known for given temperature, for example, for room temperature. There could be more then one such given temperatures in general case. This type of a priori information on desired functions is written as constraints-equalities

$$C(T_i) = C_i^*\,, \qquad i = 1, \dots, I_1 \tag{8}$$

$$k\,(T_i) = k_i^*\,, \qquad i = 1, \dots, I_2 \tag{9}$$

The inverse problem of interest is to estimate functions C(T) or/and k(T) satisfying, in general case, constraints (6) to (9).

It should be noted that heat flux history at the boundary must be measured to satisfy the uniqueness requirements when estimating two temperature functions. It can be a difficult enough technical problem in some cases. As an example, heating of materials by high-enthalpy gas flows can be pointed out [10]. Temperature at the boundaries can be measured much easier.

There are two cases when boundary conditions of the first kind can be used to estimate two temperature functions. In the first case, values of unknown functions must be known at least for one temperature [11]. It leads to the inverse problem formulation with constraints in the form of equalities (8) and (9). The second case is to use two- or multilayer specimen in which the material of one layer has unknown thermophysical characteristics. The direct problem is then written in the form of heat conduction problem for multilayer slab. Both cases can be used in practice. In this work, we consider one-layer specimen and analyse the use of different types of a priori information to estimate temperature-dependent thermophysical characteristics.

ITERATIVE ALGORITHMS

Variational formulation of the inverse problem

We use the variational formulation of the inverse problem to build computational algorithms. The problem is to find such unknown temperature functions, for which temperature histories computed from the mathematical model (1) to (4) at the sensor locations are close to measured histories. Computed and measured temperatures are time functions if one sensor is used and vector functions for a number of sensors. In this formulation, a measure defining the discrepancy between two functions or vector functions is needed. In other words, a metric functional space must be introduced. When solving ill-posed inverse problems, the space L_2 of square integrable functions is usually used[4]..

Consider unknown functions as a vector function :

$$u(T) = [u_1(T), u_2(T)]^T =$$

$$\{u_1(T) = C(T), u_2(T) = k(T), T \in [a, b]\} \qquad (10)$$

In such notations, the discrepancy or the residual functional is written as :

$$J(u(T)) = \frac{1}{2} \sum_{n=0}^{N+1} \int_0^{t_f} [T(d_n, t, u(T)) - f_n(t)]^2 dt \qquad (11)$$

where $T(d_n, t, u(T))$, $n = 0, ..., N+1$ are temperature histories computed at the sensor locations by solving the direct problem (1) to (4) with given values of desired characteristics. In variational formulation, the inverse problem is to minimize the residual functional with respect to desired functions. In order to complete the problem formulation, the space of solutions is needed to be pointed out. Denote this space by U and different choice of U will be discussed further.

Parametrization of desired functions

When solving coefficient inverse heat conduction problems with temperature-dependent desired characteristics, a parametrization of unknown functions is needed (see, for example, [4]). The universal enough form for a parametrization can be writen as :

$$C(T) = \sum_{m=1}^{M_1} C_m B_{1,m}(T), \qquad T \in [a, b] \qquad (12)$$

$$k(T) = \sum_{m=1}^{M_2} k_m B_{2,m}(T), \qquad T \in [a, b] \qquad (13)$$

where $C_1, ..., C_{M_1}$ and $k_1, ..., k_{M_2}$ are unknown parameters, $B_{1,1}(T), ..., B_{1,M_1}(T)$ and $B_{2,1}(T), ..., B_{2,M_2}(T)$ are given basic functions. Note that unknown functions belong to the U space of inverse problem solutions and the basic functions must also belong to this space. When introducing the parameterization, a subspace $U_M \subset U$ of parameterized functions is isolated in the U space.

Desired characteristics and, consequently, basic functions $B_{1,m}(T)$, $m = 1, ..., M_1$ and $B_{2,m}(T)$, $m=1, ..., M_2$ should satisfy certain smoothness requirements. These requirements are defined by the conditions of the residual functional differentiability. In particular, the thermal conductivity as a temperature function must be twice differentiable, and the volumetric heat capacity must be one time differentiable [4]. Therefore, basic functions should be chosen by taking into account that requirements.

By using parametrizations of unknown functions (11) and (12), the inverse problem is reduced to the estimation of a vector of parameters

$$p = [p_1, p_2, ..., p_M]^T, \qquad M = M_1 + M_2, \quad p \in R_M \qquad (14)$$

where $p_m = C_m$, $m = 1, ..., M_1$, $p_m = k_m$, $m = M_1 + 1, ..., M$, R_M is M-dimensional Eucledian space. The residual functional is a function of M variables in this case.

Minimization procedure

The residual functional minimization with respect to desired parameters is the most important procedure in algorithms for solving inverse problems. To do that, we use constrained gradient-type optimization methods, in particular, the conjugate gradient projection method [12]. The successive improvements of desired parameters are built via formula

$$p^{s+1} = P_w(p^s + \gamma^s g(J'^{(s)})), \qquad s = 0, 1, ... \qquad (15)$$

where s is an iteration number, P_w is the operator of projecting on a space of admissible solutions denoted by W, which is built taking into account constraints (6) to (9), γ is the descent parameter, $g(J')$ is the descent direction, J' is the residual functional gradient, p^0 is an initial guess for unknown parameters given a priori. The descent direction is calculated via formula

$$g(J'^{(s)}) = J'^{(s)} + \beta^s g^{s-1} \qquad (16)$$

where parameter β^s is defined as

$$\beta^0 = 0, \qquad \beta^s = ||J'^{(s)}|| / ||J'^{(s-1)}|| \qquad (17)$$

$||\cdot||$ denotes the norm. The descent parameter in calculated from the condition

$$\gamma^s = \text{Arg}\min J(P_w[p^s + \gamma g(J'^{(s)})]),\ \ \gamma > 0 \qquad (18)$$

The procedure of successive improvements (16) can be implemented in different spaces. The R_M space is usually used in practice [4,9]. In this case, the residual functional gradient J'_p is calculated in the R_M space. But desired characteristics are functions. To take into consideration that, it is necessary to develop the minimization procedure in the U_M space. We consider such procedure for the volumetric heat capacity u(T) = C(T) as an example.

In U_M space, the residual functional gradient has the form

$$J'_u = \sum_{m=1}^{M_1} g_m B_{1,m}(T) \qquad (19)$$

where $g = [g_1, ..., g_{M1}]$ is the descent direction in the U_M space. Formulas for successive improvements of the desired function are built for unknown parameters, but the descent direction must be determined in the U_M space. If the residual functional gradient J'_p in the R_M space is known, the descent direction in the U_M space can be calculated by the following manner [4]. Let parameters p_m, $m = 1, ..., M_1$, receive variations Δp, $m = 1,..., M_1$. The function u(T) has then a variation

$$\Delta u(T) = \sum_{m=1}^{M_1} \Delta p_m B_{1,m}(T) \qquad (20)$$

The following relationship takes place for the residual functional variation

$$\delta J = (\Delta u(T), J'_u)_{U_M}$$

$$= \left(\sum_{m=1}^{M_1} \Delta p_m B_{1,m}(T), \sum_{m=1}^{M_1} g_m B_{1,m}(T) \right)$$

$$= (\Delta p, Gg) = \langle \Delta p, J'_p \rangle_{RM} \qquad (21)$$

where $G = \{G_{j,m} = (B_{1,j}, B_{1,m})_{U_M},\ j, m = 1, ..., M_1\}$ is the Gram's matrix for basic functions. It follows from (23), that :

$$Gg = J'_p \qquad (22)$$

Different spaces can be used as the U_M space. It is possible, for example, to consider the function C(T) belonging to the L_2 space. In this case, elements of the Gram's matrix are calculated as follows :

$$G_{j,m} = \int_a^b B_{1,j}(T), B_{1,m}(T)\, dt \qquad (23)$$

The Gram's matrix is symmetric and positively defined. The system of linear algebraic equations can be solved effectively by the square root method [13]. The descent direction for the thermal conductivity is calculated by the same manner.

Iterative regularization

The iterative procedure of minimizing the residual functional with respect to desired parameters is highly effective if temperature histories $f_n(t)$, n = 0, ..., N+1, are smooth functions. In the inverse problem, that functions are measurement results and always corrupted by random noise. The noise is a random fluctuating process and the fluctuations have no physical sence. When the residual functional is going to be as minimal as possible, temperature histories computed at sensor locations can follow the noised measurements. That is possible only if desired functions have enough flexibility to permit the direct problem solution to follow noised measurements. In that case, desired functions also begin having fluctuations which have no physical sence. It means that the inverse problem solution becomes unstable. This is the reason why regularizing methods and algorithms are needed to solve ill-posed problems [1].

In the iterative regularization method, the process of successive improvements of the inverse problem solution is stopped when the residual functional value is in agreement with measurement errors computed in the L_2 space as a space of measured temperature histories. The regularizing stop condition is

$$s^* : J(u^{s^*}) \cong \delta^2 \qquad (24)$$

where s^* is the number of the last iteration, δ^2 is the residual functional level defined as

$$\delta^2 = \frac{1}{2} \sum_{n=0}^{N+1} \int_0^{t_f} \sigma_n^2(t)\, dt$$

$\sigma_n^2(t)$ is the estimate of the time-dependent standard deviation for temperature history measurement by nth sensor.

For linear ill-posed problems, it has been proved mathematically for the unconstrained residual functional minimization by gradient-type methods of the first kind that when iterations are stopped according to the residual principle (24), the approximate solution of the ill-posed inverse problem is close enough to the exact solution [4]. For nonlinear coefficient inverse heat conduction problems, the high capacity of work has been demonstrated by extensive computational experiments [2,4].

Computation of the residual functional gradient

The most effective method for calculating the gradient is based on the theory of extreme problems [4]. Following this approach, an adjoint problem is introduced and analytical formulas are then derived for the gradient.

It can be shown that the adjoint problem for the inverse problem of interest can be written as follows [4] :

$$- C(T)\frac{\partial\psi_n}{\partial t} = \frac{\partial}{\partial x}\left(k(T)\frac{\partial\psi_n}{\partial x} - \frac{\partial k}{\partial T}\frac{\partial T_n}{\partial x}\frac{\partial\psi_n}{\partial x}\right)$$

$$d_{n-1} < x < d_n, \quad 0 \le t < t_f, \quad n = 1, ..., N+1 \qquad (25)$$

$$\psi_n(x,t_f) = 0 \qquad d_{n-1} \le x \le d_n \qquad (26)$$

$$\alpha_1 k(T_1(0,t))\frac{\partial\psi_1(0,t)}{\partial x} + \mu_1\psi_1(0,t) = T_1(0,t) - f_0(t) \qquad (27)$$

$$k(T_n(d_n,t))\left[\frac{\partial\psi_n(d_n,t)}{\partial x} - \frac{\partial\psi_{n+1}(d_n,t)}{\partial x}\right] = T_n(d_n,t) - f_n(t)$$

$$n = 1, ..., N \qquad (28)$$

$$\psi_n(d_n,t) = \psi_{n+1}(d_n,t)\, dt, \quad n = 1, ..., N \qquad (29)$$

$$\alpha_2 k(T_{N+1}(L,t))\frac{\partial\psi_{N+1}(L,t)}{\partial x} + \mu_2\psi_{N+1}(L,t)$$

$$= T_{N+1}(L,t) - f_{N+1}(t) \qquad (30)$$

Let functions C(T) and k(T) receive variations $\Delta C(T)$ and $\Delta k(T)$ respectively. Following the technique of work [4], it is possible to show that the variation of the residual functional can be written in the form

$$\delta J = \sum_{i=0}^{N+1}\int_0^{t_f}\int_{d_{n-1}}^{d_n}\psi_n\left(\Delta k(T)\frac{\partial T_n}{\partial x^2} + \frac{\partial\Delta k}{\partial T}\left(\frac{\partial T_n}{\partial x}\right)^2\right)dx\,dt$$

$$+ \sum_{i=0}^{N+1}\int_0^{t_f}\int_{d_{n-1}}^{d_n}\psi_n\left(-\frac{\partial T_n}{\partial x}\Delta C(T)\right)dx\,dt$$

$$+ \alpha_1\int_0^{t_f}\psi_1(0,t)\frac{\partial T_1(0,t)}{\partial x}\Delta k(T_1(0,t))\,dt$$

$$- \alpha_2\int_0^{t_f}\psi_{N+1}(L,t)\frac{\partial T_{N+1}(L,t)}{\partial x}\Delta k(T_{N+1}(L,t))\,dt \qquad (31)$$

Taking into account parametrizations (12) and (13), the following expressions for the residual functional gradient can be derived:

$$\frac{\partial J}{\partial C_m} = -\sum_{n=1}^{N+1}\int_0^{t_f}\int_{d_{n-1}}^{d_n}\psi_N(L,t)\frac{\partial T_N}{\partial t}B_{1,m}(T)\,dx\,dt$$

$$m = 1 ..., M_1 \qquad (32)$$

$$\frac{\partial J}{\partial k_m} =$$

$$\sum_{i=0}^{N+1}\int_0^{t_f}\int_{d_{n-1}}^{d_n}\psi_n\left(B_{2,m}(T)\frac{\partial^2 T_n}{\partial x^2} + \frac{\partial B_{2,m}}{\partial T}\left(\frac{\partial T_n}{\partial x}\right)^2\right)dx\,dt$$

$$+ \alpha_1\int_0^{t_f}\psi_1(0,t)\frac{\partial T_1(0,t)}{\partial x}B_{2,m}(T(0,t))\,dt$$

$$- \alpha_2\int_0^{t_f}\psi_{N+1}(1,t)\frac{\partial T_{N+1}(1,t)}{\partial x}B_{2,m}(T_{N+1}(L,t))\,dt$$

$$m = 1, ..., M_2 \qquad (33)$$

Formulas (33) and (34) define the residual functional gradient in the R_M space.

Projection on a subspace of admissible solutions

Constraints (6) to (9) define the subspace $W \subset U_M$ of admissible inverse problem solutions as

$$W = \left\{\begin{array}{l} C(T) > 0,\ k(T) >,\ \ 0\ T \in [a, b] \\ C(T_i) = C_i^*\ i = 1 ..., I_1 \\ k(T_i) = k_i^*\ i = 1 ..., I_2 \end{array}\right\} \qquad (34)$$

The projection procedure is to transform an arbitrary enough element $u(T) \in U_M$ into another element $w(T) \in W$.

Taking into account constraints-equalities (6) and (7) is the first step of the projection. A part of unknown parameters in approximations (12) and (13) should be excluded. As a result, the dimension of the vector of desired parameters is reduced, and a set of basic functions is changed. Transformed approximations can be written as :

$$C(T) = \sum_{m=1}^{M_1} C_m \varphi_{1,m}(T), \quad T\in[a,b] \tag{35}$$

$$k(T) = \sum_{m=1}^{M_2} k_m \varphi_{2,m}(T), \quad T\in[a,b] \tag{36}$$

where I_1 parameters in C(T) and I_2 parameters in k(T) are known, $\varphi_{1,m}(T)$, m = 1, ..., M_1 and $\varphi_{2,m}(T)$, m = 1, ..., M_2 are transformed basic functions.

The second step is to project the solution on the subspace of positive functions. We use the following algorithm for the projection. The volumetric heat capacity is considered here as an example

The projection $w^* = \sum_{m=1}^{M_1} w_m \varphi_{1,m}(T)$ of a given element $u(T) \in U_M$, on the subspace $W \subset U_M$ is defined as follows [14] :

$$||u - w^*||^2 = \inf \Psi(w), \quad \Psi(w) = ||u - w||^2, \quad w \in W \tag{37}$$

where the norm is determined in the U_M space. The following representation of the space W is used [7] :

$$W = \left\{ u(T) = \sum_{m=1}^{M_1} p_m \varphi_{1,m}(T) \ge C_{min}, \quad T\in[a,b] \right\} \tag{38}$$

where C_{min} is a given value. The additional grid is then built in the [a,b] interval as

$$\omega = \{T_j = a + (j-1)h, \; h = (b-a)/(n-1) \quad n > M_1\} \tag{39}$$

where n is a given number. Conditions (38) are required to be satisfied for each node of the grid (39). As a result, a set of linear constraints is formulated as

$$\sum_{m=1}^{M_1} w_m \varphi_{1,m}(T_j) = C_{min}, \qquad j = 1,\dots,n \tag{40}$$

or in matrix notations

$$Dw \ge \alpha \tag{41}$$

where D is M_1*n matrix with elements $D_{m,j} = \varphi_{1,m}(T_j)$, α is a vector with dimension n, $n > M_1$, and elements $\alpha_j = C_{min}$ j = 1,..., n. As a result, the extremal problem (37) can be writen as

$$\Psi(w^*) = \inf ||u - w||^2_{U_M}, \quad Dw \ge \alpha \tag{42}$$

The expression for $\Psi(w)$ is then transformed as

$$\Psi(w) = ||u - w||^2_{U_M}$$

$$= \int_a^b \left[\sum_{m=1}^{M_1} (p_m - w_m)\varphi_m\right]\left[\sum_{k=1}^{M_1} (p_k - w_k)\varphi_k\right] dt$$

$$= \sum_{m=1}^{M_1} (p_m - w_m)\left[\sum_{k=1}^{M_1} (p_k - w_k)\int_a^b \varphi_m\varphi_k \, dt\right]$$

$$= \langle p-w, G(p-w) \rangle = w^T Gw - 2p^T Gw + p^T Gp \tag{43}$$

The Gram's matrix G can be transformed by using the Holessky decomposition as

$$G = H^T H \tag{44}$$

where H is an upper thridiagonal matrix. By using ξ = Hp notation, the expression (43) can be written as

$$\Psi(w) = (Hw - \xi, Hw - \xi)_{R_M} = ||Hw - \xi||^2_{R_M} \tag{45}$$

The extremal projection problem with constraints (40) is reduced, as a result, to the following form:

$$\Psi(w^*) = \inf ||Hw - \xi||^2_{R_M} \tag{46}$$

This is the well-known problem of the least square method [15]. The same procedure is used for the thermal conductivity.

Calculation of the descent parameter

In the constrained minimization procedure, the descent parameter is calculated for each iteration from the condititon (19). We use the following algorithm to solve the problem. The descent parameter γ is represented as a vector

$$\gamma = [\gamma_1, \gamma_2] \tag{47}$$

where γ_1 is a descent parameter component for C(T) and γ_2 is that for k(T). Linear estimates for components γ_1 and γ_2 are then calculated [4]. The residual functional in sth iteration can be written as

$$J^{s+1} = \sum_{n=0}^{N+1} \int_0^{t_f} [A_1 + A_2 + A_3]^2 \, dt \tag{48}$$

where

$$A_1 = T(d_n, t, u^s) - f_n(t)$$

$$A_2 = \gamma_1 \vartheta_1(d_n, t, g_1(J'^{s}_{u1}))$$

$$A_3 = \gamma_2 \vartheta_2(d_n, t, g_2(J'^{s}_{u2}))$$

where $\vartheta_1(x,t,g(J'))$ and $\vartheta_2(x,t,g(J'))$ are temperature variations, which are determined when giving a variation to only one desired characteristic, $\Delta C(T)$ or $\Delta k(T)$ respectively [4]. By minimizing J^{s+1} with respect to γ_1 and γ_2, the following system of linear algebraic equations can be derived :

$$\sum_{j=1}^{2} \gamma_j \left[\sum_{n=1}^{N+1} \int_0^{t_f} \vartheta_j(d_n,t)\, \vartheta_k(d_n,t)\, dt \right]$$

$$= \sum_{n=1}^{N+1} \int_0^{t_f} \left[T_j(d_n,t) - f_n(t) \right] \vartheta_k(d_n,t)\, dt \,, \quad k = 1, 2 \tag{49}$$

The system (49) has the symmetric and positively defined matrix. It can be solved effectively by using, for example, the square root method [13].

After calculating the descent parameter, the new approximation for desired characteristics is determined as

$$u_1^{s+1} = u_1^s + \gamma_1 g_1 (J'^{(s)}_{u1}) \tag{50}$$

$$u_2^{s+1} = u_2^s + \gamma_2 g_2 (J'^{(s)}_{u2}) \tag{51}$$

Then, the operation of projecting on the space of positive functions is conducted.

SOME RESULTS

The application of developed algorithms is illustrated here by solving the inverse problem of estimating temperature-dependent thermal conductivity. Some results of computational experiments are presented.

Initial data for the test case are similar to [16] and they are the following. The slab thickness is L=0.0172 m, the experiment duration is 220 s and initial temperature distribution is $T_0(x) =$ 300 K. The volumetric heat capacity is known $C(T) = 2.4\ 10^6$ J/(m.K). Given boundary conditions of the first kind are shown in Fig. 1. One temperature sensor is used to estimate unknown thermal conductivity k(T). The sensor coordinate is d = 0.002 m. The function to be estimated is a priori given and shown in Fig. 2. Exact temperature history computed at the sensor location is shown in Fig.1. Random number generator is used to simulate measurement errors. Noised temperature "measurements" are shown in Fig.1 too. The thermal conductivity is supposed to be an element of the L_2 space. Cubic B-splines with $M_2 = 4$ are used to approximate unknown function. Results of the estimation for noised "measurements" are given in Fig. 2.

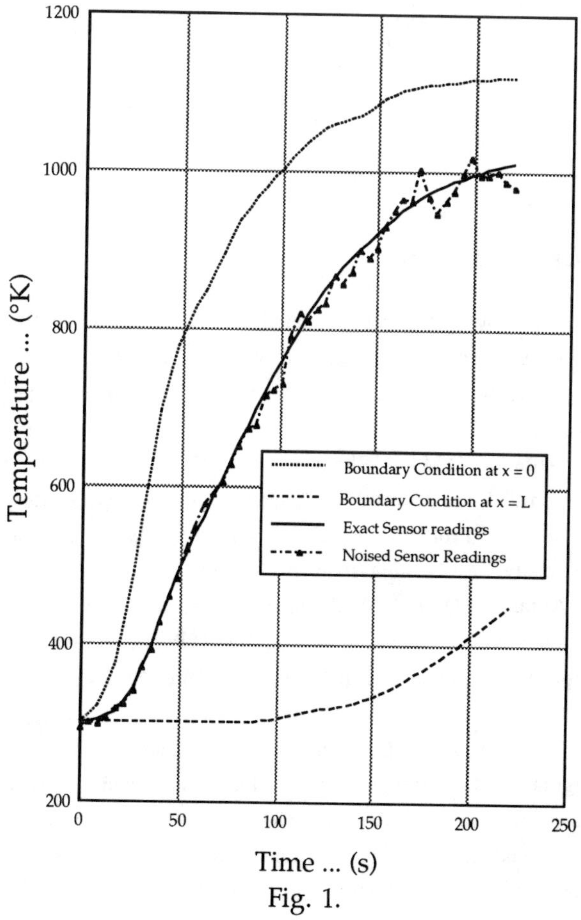

Fig. 1.

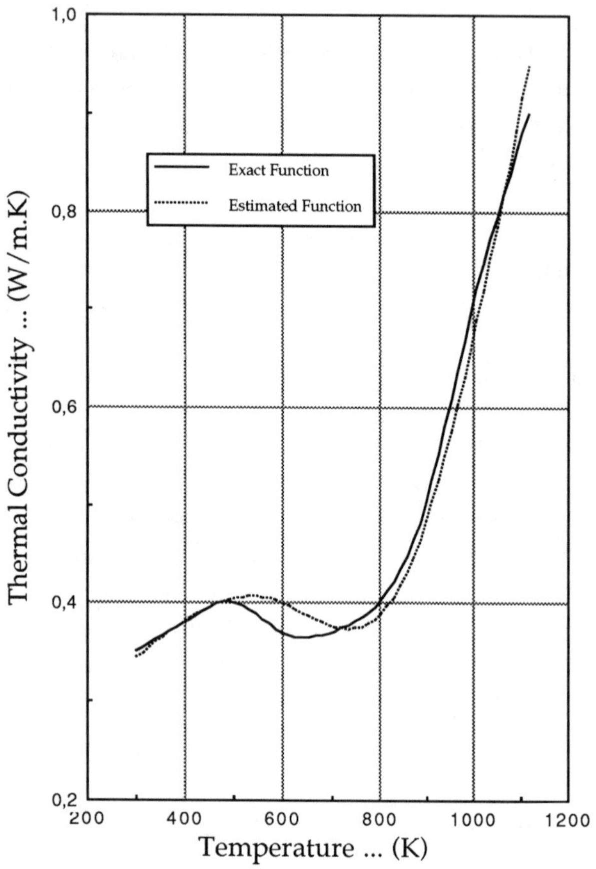

Fig. 2.

REFERENCES

[1] Tikhonov, A.N., and Arsenin, V.Ya., Solutions of Ill Posed Problems, V.H. Winston and Sons, Washington, D.C., 1977.

[2] Alifanov, O.M., Inverse Heat Transfer Problems, Mashinostroenie, Moscow, 1988.

[3]. Alifanov, O.M., "Methods of Solving Inverse Heat Transfer Problems", Final Report: Joint American-Russian NSF Workshop on Inverse Problems in Heat Transfer, Michigan State University, East Lansing, June 13-15, 1992.

[4] Alifanov, O.M., Artyukhin, E.A., and Rumyantsev, S.V, Extreme Methods of Solving Ill-Posed Problems and Their Application to Inverse Heat Transfer Problems, Nauka, Moscow, 1988.

[5] Tikhonov, A.N., Goncharskiy, A.V., Stepanov, V.V., and Yagola, A.G., Regularizing Algorithms and A priory Information, Nauka, Moscow, 1983.

[6] Artyukhin, E.A., "Directions and Problems of Research in Heat Transfer Parameter Estimation",Final Report: Joint American-Russian NSF Workshop on Inverse Problems in Heat Transfer, Michigan State University, East Lansing, June 13-15, 1992.

[7] Artyukhin,E.A.,Ivanov,A.G., and Nenarokomov, A.V., "Solution of Coefficient Inverse Heat Conduction Problems Taking into account A Priori Information on Desired Function Values", Journal of Engineering Physics (to be published).

[8] Artyukhin, E.A., "Inverse Problems and Optimal Experiment Design in Unsteady Heat Transfer' Processes Identification", Proceedings of the Third International Conference on Inverse Design Concepts and Optimization in Engineering Sciences, October, 23-25, 1991, Washington D.C., pp. 513-527, 1991.

[9] Beck, J.V., and Arnold, K.I., Parameter Estimation in Engineering and Science, John Wiley and Sons, 1977.

[10] Polezhaev, Yu.V., and Yurevitch F.B., Thermal Protection, Energiya, Moscow, 1976.

[11] Muzilev, N.V., "On a Uniqueness of Simultaneous Determination of the Thermal Conductivity and Heat Capacity per Unit Volume", USSR Journal of Computational Mathematics and Mathematical Physics, Vol. 32, No. 1, pp. 102-108, 1983.

[12] Vasilyev, F.P., Numerical Methods for Solving Extreme Problems, Nauka, Moscow, 1980.

[13] Faddeev, D.K., and Faddeeva, V.N., Computational Methods ofLinear Algebra, Fizmatgiz, Moscow - Leningrad, 1963.

[14] Kantorovich, L.V., and Akilov, G.P., The Functional Analysis, Nauka, Moscow, 1963.

[15] Lawson,C.L., and Hanson, R.L., Solving Least Square Problems, Prentice Hall, Englewood Cliffs, New York, 1974.

[16] Artyukhin, E.A., Guseva, L.I., Tryanin,A.P., and Shibin, A.G., "Data Processing and Planning of Nonstationary Thermophysical Exsperiments", Journal of Engineering Physics, Vol. 56, No. 3, pp. 286-290, 1989.

Inverse Problems in Engineering: Theory and Practice
ASME 1993

THERMOPHYSICAL PROPERTIES ESTIMATION FROM TRANSIENT DATA: KALMAN VERSUS GAUSS APPROACH

F. Scarpa and G. Milano
Dipartimento di Ingegneria Energetica

D. Pescetti
Istituto di Fisica di Ingegneria

Università di Genova
Genova, Italy

ABSTRACT

After a short historical review of the estimation theory from Gauss to Kalman, the MAP sequential estimator, a modern implementation of the deterministic Gauss approach, is compared with the discrete Kalman filter. The last one, owing to its statistical formulation, provides in general more reliable estimates because it takes into account the noise affecting the measured input variables (control functions) and the measured initial conditions. Both estimators are developed with reference to the problem of identification of temperature dependent thermophysical properties of materials from transient heat conduction data. Thermal conductivity and volumetric heat capacity are simultaneously reconstructed as a function of temperature using data of a single test. Several results of simulated experiments are analysed by means of two distinct methodologies, the *Monte Carlo* technique and the *covariance analysis*. The main characteristics of the algorithms are compared and the results show the better performance of the Kalman filter.

NOMENCLATURE

B% - percentage parameter bias
C - volumetric heat capacity
C_j - jth volumetric heat capacity coefficient
I_e - efficiency index
I_r - reliability index
N_t - total number of time steps
N_z - number of discretized spatial intervals
T - temperature
b - slab thickness
d - spatial discretization interval
h - discretized time interval
k - thermal conductivity
k_j - jth thermal conductivity coefficient
q - specific heat flux
t - time
t_m - total test duration
t_k - discretized time
u - control function vector
v - measurement noise vector
w - control function noise vector
$\hat{x}$ - state estimate vector
x - real state vector
z - spatial coordinate

Greek symbols

β - parameter vector
δ - initial temperature error
σ_c - ensemble standard deviation
σ_{cov} - standard deviation from the *covariance analysis*
σ_f - standard deviation predicted by the filter
σ_{id} - standard deviation predicted by the ideal filter
σ_v - measurement noise standard deviation
σ_w - control measurement noise standard deviation
τ - comparison ratio

Subscripts

0 - initial value
cov - covariance analysis
f - evolutive process
g - measurement process
i - iteration index or current index
id - ideal filter
m - effective measurement
max - maximum
ref - reference
k - time index

Superscripts

t - transposition operator
* - reference trajectory
^ - estimate

INTRODUCTION

A large class of estimation problems is concerned with finding an optimal estimate of some quantity, e.g. a random variable or an unknown parameter, when a function of this quantity, corrupted by an additive noise, is available from a measurement process.

One of the first studies about this class of problems was performed by Gauss who developed in 1794, at the age of seventeen, the method of least squares applied to the determination of the orbit of celestial bodies. The astronomical studies that prompted the invention of this technique were successively described in (Gauss, 1808).

He recognized the possibility of a dual approach to regression analysis, giving rise respectively to a deterministic optimization problem (minimization of the sum of the squared errors) or to a stochastic estimation problem (evaluation of the most probable parameter estimates, which implies calculation about probability density functions).

In the early 1940's Wiener and Kolmogorov dealt with the class of problems related to the estimation of random signals. The result from this work is an integral equation (the Wiener-Hopf equation) that, unfortunately, has only limited application because it can be solved explicitly in a few special cases.

In the 1950's the idea of generating least-squares estimates recursively was introduced and stimulated by the increased usage of digital computers. Kalman (1960) realized that digital computers are more effective in solving differential rather than integral equations and transformed the W-H equation into an equivalent differential form. Rather than try to solve it, he recognized that it was better to put the computational burden on the computer.

In the "classical control theory" by Wiener, the emphasis was on the analysis and synthesis of systems in terms of their input-output characteristic by means of Laplace and Fourier transforms, the usual electric engineering approach. Kalman, with a different perspective, stressed the "state-space" description of a system in which one deals with the basic models that give rise to the observed output. An interesting discussion, directed to least square estimation theory, from its inception by Gauss and Legendre to its modern form as developed by Kalman, can be found in (Sorenson, 1970).

Until now the practicality of the Kalman Filter has made it immensely popular in aerospace applications, such as navigation and tracking. Relatively few Authors addressed to thermal problems this technique which, in this particular field, appears more popular in eastern countries. In (Simbirskii, 1976) the Kalman technique was used to identify the temperature dependent thermal conductivity. A local iterative modification of the Kalman algorithm was proposed in (Matsevityi and Multanovskii, 1987) while in (Scarpa et al., 1991) the Kalman approach was used to identify the temperature dependent thermal diffusivity for the class of transient experiments in which temperature is measured with sensors embedded inside the sample. In that paper, a method to include in the inverse solution the identification of the true location of the internal sensors was also given.

In the present paper, the discrete Kalman filter is compared with the MAP sequential estimator, a recursive way to implement the original Gauss' idea. The comparison is done both by means of the *Monte Carlo* technique and with the use of the *covariance analysis* method. The test problem is the simultaneous identification of temperature dependent thermal conductivity and volumetric heat capacity from a single simulated experiment of one-dimensional transient heat conduction. Temperature and heat flux are measured at the two boundaries of the sample and no information coming from the interior is utilized. This kind of experimental procedure, already used in (Garnier et al., 1991) in the assumption of constant thermophysical properties, appears very promising because it reduces the total test time and does not require any intrusion of sensors inside the specimen.

ESTIMATION ALGORITHMS

Consider a discretized heat conduction model, with unknown parameters, whose deterministic evolution is described by the following nonlinear vector difference equations:

$$\boldsymbol{x}(t_k)=\boldsymbol{f}\,[\boldsymbol{x}(t_{k-1}),\boldsymbol{u}_f(t_{k-1})] \tag{1}$$

were $\boldsymbol{x}(t_0)$ is initial conditions and $\boldsymbol{u}_f(t_{k-1})$ represents the process control. Here $\boldsymbol{x}$ is the true (unknowable) state vector defined as

$$\boldsymbol{x}\equiv[\boldsymbol{T}^t,\boldsymbol{\beta}^t\,]^t \tag{2}$$

where $\boldsymbol{T}$ is the temperature vector and $\boldsymbol{\beta}$ the parameter vector. System (1) is observed through measurement data $\boldsymbol{y}_m$ obtained at discrete instants of time t_k. These data are assumed to be related to the state according to

$$\boldsymbol{y}_m(t_k)=\boldsymbol{g}[\boldsymbol{x}(t_k),\boldsymbol{u}_g(t_k)] + \boldsymbol{v}(t_k) \tag{3}$$

$$\boldsymbol{v} \longrightarrow N(\boldsymbol{0},\mathbf{R}_v)$$

where $\boldsymbol{v}$ is the noise affecting the measurement process and the standard notation $q \longrightarrow N(m_q,R_q)$ refers to a random variable q characterized by a white normal distribution with mean m_q and variance R_q.

To perform a simulation of the process, the control functions $\boldsymbol{u}_f$ and $\boldsymbol{u}_g$ have to be measured . Furthermore, some knowledge about the initial condition $\boldsymbol{x}(t_0)$ is required. Thus we have at our disposal

$$\hat{\boldsymbol{x}}(t_0) = \boldsymbol{x}(t_0) + \boldsymbol{\delta}\,, \qquad \boldsymbol{\delta} \longrightarrow N(\boldsymbol{0},\mathbf{P}_0) \tag{4}$$

$$\boldsymbol{u}_{fm}(t_k) = \boldsymbol{u}_f(t_k) + \boldsymbol{w}_f(t_k)\,, \quad \boldsymbol{w}_f \longrightarrow N(\boldsymbol{0},\mathbf{R}_{wf}) \tag{5}$$

$$\boldsymbol{u}_{gm}(t_k) = \boldsymbol{u}_g(t_k) + \boldsymbol{w}_g(t_k)\,, \quad \boldsymbol{w}_g \longrightarrow N(\boldsymbol{0},\mathbf{R}_{wg}) \tag{6}$$

where $\boldsymbol{\delta}$ represents the uncertainty relative to the initial condition while $\boldsymbol{w}_f$ and $\boldsymbol{w}_g$ are white gaussian zero-mean noises superposed onto the measurement of the control functions u_f and u_g; $\mathbf{P}_0$, $\mathbf{R}_{wf}$ and $\mathbf{R}_{wg}$ are known variance/covariance (var/covar) matrices; the vectors w_f and w_g are assumed uncorrelated with v and $\hat{x}(t_0)$.

Under the above statements, an approximation $\hat{x}$ of the unbiased minimum variance estimate of x, and thus of the parameter vector $\boldsymbol{\beta}$, is given by the *linearized* Kalman-Bucy Filter algorithm (LKF). If some of the above conditions are dropped then suboptimal techniques can be formulated which require less computational effort. Among these we consider the MAP (Maximum a posteriori) sequential estimator, in which the calculated temperature is considered deterministic. The MAP algorithm can be directly derived from the following LKF equations:

Initial conditions iteration loop on **i**

$$x^*(0) \equiv [T^{*t}(0), \beta_i^{*t}]^t,\ \hat{x}(0 \mid 0) \text{ and } \mathbf{P}(0 \mid 0) \tag{7}$$

.... time loop on k

Reference evolution

$$x^*(k) = f[x^*(k\text{-}1), u_{fm}(k\text{-}1)] \tag{8}$$

State prediction

$$\hat{x}(k \mid k\text{-}1) = x^*(k) + \mathbf{A}(k\text{-}1)\cdot[\hat{x}(k\text{-}1 \mid k\text{-}1) - x^*(k\text{-}1)] \tag{9}$$

$$\mathbf{P}(k \mid k\text{-}1) = \mathbf{A}(k\text{-}1)\cdot\mathbf{P}(k\text{-}1 \mid k\text{-}1)\cdot\mathbf{A}^t(k\text{-}1) + \mathbf{Q}(k\text{-}1)\cdot\mathbf{R}_{wf}(k\text{-}1)\cdot\mathbf{Q}^t(k\text{-}1) \tag{10}$$

Measurement prediction

$$y(k) = g[x^*(k), u_g(k)] + \mathbf{C}(k)\cdot[\hat{x}(k \mid k\text{-}1) - x^*(k)] \tag{11}$$

$$\mathbf{R}_y(k) = \mathbf{C}(k)\cdot\mathbf{P}(k \mid k\text{-}1)\cdot\mathbf{C}^t(k) + \mathbf{D}(k)\cdot\mathbf{R}_{wg}(k)\cdot\mathbf{D}^t(k) \tag{12}$$

Gain

$$\mathbf{K}(k) = \mathbf{P}(k \mid k\text{-}1)\cdot\mathbf{C}^t(k)\cdot[\mathbf{R}_y(k) + \mathbf{R}_v(k)]^{-1} \tag{13}$$

Correction

$$\hat{x}(k \mid k) = \hat{x}(k \mid k\text{-}1) + \mathbf{K}(k)\cdot[y_m(k) - y(k)] \tag{14}$$

$$\mathbf{P}(k \mid k) = [\,\mathbf{I} - \mathbf{K}(k)\cdot\mathbf{C}(k)\,]\cdot\mathbf{P}(k \mid k\text{-}1) \tag{15}$$

.... evolution on $k=1,..,N_t$

Reference update

$$\beta_{i+1}^* = \hat{\beta}(N_t) \quad i \longleftarrow i+1 \tag{16}$$

- $\hat{x}$ - state estimate vector
- x^* - state reference vector
- u_{fm} - process/control vector (flux)
- u_{gm} - measurement/control vector (flux)
- y - measurement vector, estimate (temperature)
- y_m - measurement vector, measure (temperature)
- e - innovation vector
- $\mathbf{A}$ - state transition jacobian matrix
- $\mathbf{Q}$ - process/control transmission matrix
- $\mathbf{C}$ - measurement jacobian matrix
- $\mathbf{D}$ - measur./control transmission matrix
- $\mathbf{K}$ - *Kalman gain* matrix
- $\mathbf{P}$ - state var/covar matrix
- $\mathbf{R}_y$ - measurement var/covar matrix
- $\mathbf{R}_e$ - innovation var/covar matrix
- $\mathbf{R}_{wf}$ - process/control noise var/covar matrix
- $\mathbf{R}_{wg}$ - measur./control noise var/covar matrix
- $\mathbf{R}_v$ - measurement noise var/covar matrix

where:
the matrices **A**, **Q**, **C** and **D** are all computed at $x=x^*(\cdot)$; the notation k stands for t_k ; $\hat{x}(k \mid k\text{-}1)$ stands for "estimate of $x(k)$ by means of the information available at $t=t_{k-1}$"; $\mathbf{R}_{wf}$, $\mathbf{R}_{wg}$ and $\mathbf{R}_v$ are diagonal matrices; N_t is the number of time steps;

Because of the general nonlinearity of the models, the algorithm has been obtained by linearizing equations (1) and (3) by means of a first order Taylor expansion with respect to a reference "trajectory"

$$x^*(t_k) \equiv [T^{*t}(t_k), \beta^{*t}]^t \tag{17}$$

obtained by driving the process equation with a constant parameter vector β^*, see Eq.(8).

In the MAP algorithm the initial temperature distribution is assumed known and both the calculated temperature and the process control u_{fm} are considered deterministic. Thus, if we set to zero the initial elements of **P** relative to the temperature in Eq. (7) and the elements of $\mathbf{R}_{wf}$ in Eq. (10), then the Kalman filter is reduced to the sequential MAP estimator. The algorithm (7÷16), especially in the MAP reduced form, does not represent a numerically efficient method to employ because of the large computational effort required by the error covariance equation (10). However, in this way, we can utilize exactly the same computer procedure for both LKF and MAP estimators.

One can note that the vector β^* does not change during the iterations over the time index k, in fact it represents the driving parameter vector for the reference trajectory x^*. On the contrary, the vector $\hat{\beta}$ is a function of the time and its initial value $\hat{\beta}(0)$, following the Ordinary Least Squares (OLS) perspective, is set equal to β^* and it is continuously updated after each iteration loop. At the end of an iteration ($k=N_t$) the value of β^* is updated with the use of the final value $\hat{\beta}(N_t)$ and the run over the time index restarts as described in (Beck and Arnold, 1977) for the MAP sequential estimator.

A different kind of linearization if often adopted in literature. As already stated, in the algorithm (7÷16) the linearization is made with respect to the reference trajectory, that is

$$f(\hat{x}) = f(x^*) + \mathbf{A} \cdot (\hat{x} - x^*) \qquad (18)$$

If, on the other hand, one assumes as reference each new state estimate as soon as it becomes available, that is $x^* = \hat{x}$, we obtain the *extended* version of the algorithm. This choice assures, on average, the use of the best reference available, so that the effects of fails in linearity assumptions, due to possible large errors in the initial value of x^*, are minimized. This modified version can be simply accomplished by putting the equation $x^*(k\text{-}1) = \hat{x}(k\text{-}1 \mid k\text{-}1)$ between Eq.(8) and Eq.(9).

In this work the MAP and the Kalman algorithms are implemented with the use of the two above linearization techniques and the results of the estimation process are compared.

STATEMENT OF THE PROBLEM AND IMPLEMENTATION

We consider a homogeneous, isotropic slab subjected to a monodimensional transient conduction and we assume that the nonlinear governing heat conduction equation can be written in the form:

$$C(T)\frac{\partial T}{\partial t} = \frac{\partial}{\partial z}\left[k(T)\frac{\partial T}{\partial z}\right], \quad T \equiv T(z,t) \qquad (19)$$

$$0<z<b \qquad 0<t\leq t_m$$

where $k(T)$ and $C(T)$ are the temperature dependent thermal conductivity and volumetric heat capacity, z is the spatial coordinate, t the time, T the temperature, b the slab thickness and t_m the total test duration.

Equation (19) is coupled with the following initial and boundary conditions (2nd kind):

$$T(z,0) = \Phi(z) \qquad (20)$$

$$u_{f1}(t) = -k(T)\frac{\partial T}{\partial z}\bigg|_{x=0}, \quad u_{f2}(t) = -k(T)\frac{\partial T}{\partial z}\bigg|_{x=b} \qquad (21)$$

$$0<z<b \qquad 0<t\leq t_m$$

As we want to identify the unknown temperature dependent thermophysical properties from a set of noisy temperature and heat flux measurements on the boundaries of the slab, thermal conductivity and volumetric heat capacity have been parameterized in the polynomial approximation as:

$$k(T) = \sum_{j=0}^{n_1} k_j\,[T - T_{ref}]^j, \qquad C(T) = \sum_{j=0}^{n_2} C_j\,[T - T_{ref}]^j \qquad (22)$$

where T_{ref} is a reference temperature, n_1 and n_2 are integer numbers. Thus the vector $\boldsymbol{\beta}$ of unknown parameters is defined as:

$$\boldsymbol{\beta} = \{ k_0,.., k_{n1}, C_0,.., C_{n2} \} \qquad (23)$$

Equations (19÷21) have been discretized with an explicit two-level finite difference scheme, because it provides a simple way to implement the evolutive Eqs. (8÷10) of the Kalman algorithm. The numerical scheme is locally stable provided that

$$h < \frac{d^2 C}{2k} \qquad (24)$$

where h is the time step and d is the spatial discretization interval.

The modeling of the measurement process starts from the Fourier law applied at $z=0$ and $z=b$. With the use of a spatial grid of N_z+1 points $(0,..,N_z)$ and a three points differentiation formula, the following recursive equations for the temperatures T_0 and T_{Nz} can be obtained:

$$T_0 = \frac{1}{3}\left[4T_1 - T_2 + \frac{2d}{k(T_0)}u_{g1}\right] \qquad (25)$$

$$T_{Nz} = \frac{1}{3}\left[4T_{Nz\text{-}1} - T_{Nz\text{-}2} - \frac{2d}{k(T_{Nz})}u_{g2}\right] \qquad (26)$$

$$0<t\leq t_m; \qquad d = b/N_z$$

where $u_{g1} = q(0,t)$, $u_{g2} = q(b,t)$ and q is the specific heat flux.

ANALYSIS METHODOLOGY

To evaluate and compare the performance of the two algorithms under investigation, we follow both the brute-force *Monte Carlo* method and the less expensive *covariance analysis* method as reported for example in (Candy, 1986).

The first method consists in the generation of an ensemble of estimation errors $e_\beta = \hat{\boldsymbol{\beta}} - \boldsymbol{\beta}$, where $\boldsymbol{\beta}$ is the exact (simulated) parameter vector and $\hat{\boldsymbol{\beta}}$ the filter estimate. The required noisy measurements are produced by means of a simulator (the truth model) and a random number generator. Then, by computation of sample statistics of the mean and the variance of the vector e_β, we are able to evaluate if the filter under investigation produces biased estimates and/or if their sample variances are too different from those given by the filter itself.

The *covariance analysis* requires the development of an ideal, optimal but practically unrealizable filter, in order to estimate the (minimum) error covariance $\mathbf{P_{id}}$ (ideal). The value of $\mathbf{P_{id}}$ will provide the basis to compare the performance of the investigated filters. As in our case the truth model is nonlinear, then a good approximation of $\mathbf{P_{id}}$ can be obtained by linearizing the LKF model about the trajectory generated by the truth model. The suboptimal gain sequence (**K**) produced by the filter under investigation is then used to evolve $\mathbf{P}_{cov}$ (covariance analysis) by means of the ideal filter covariance propagation model. In the ideal filter we use the truth model to obtain $\mathbf{P}_{id}$ as follows:

Prediction

$$\mathbf{P}_{id}(k \mid k\text{-}1) = \mathbf{A}_{id}(k\text{-}1)\cdot\mathbf{P}_{id}(k\text{-}1 \mid k\text{-}1)\cdot\mathbf{A}^t_{id}(k\text{-}1) +$$

$$+ \mathbf{Q}_{id}(k\text{-}1)\cdot\mathbf{R}_{wf}(k\text{-}1)\cdot\mathbf{Q}^t_{id}(k\text{-}1) \qquad (27)$$

Correction

$$\mathbf{P}_{id}(k \mid k) = [\mathbf{I} - \mathbf{K}_{id}(k)\cdot\mathbf{C}_{id}(k)\ ]\cdot\mathbf{P}_{id}(k \mid k\text{-}1) \qquad (28)$$

In the filter under test, the actual error covariance is calculated using the truth model and the suboptimal gain **K**, that is

$$\mathbf{P}_{cov}(k \mid k\text{-}1) = \mathbf{A}_{id}(k\text{-}1)\cdot\mathbf{P}_{cov}(k\text{-}1 \mid k\text{-}1)\cdot\mathbf{A}^t_{id}(k\text{-}1) +$$

$$+ \mathbf{Q}_{id}(k\text{-}1)\cdot\mathbf{R}_{wf}(k\text{-}1)\cdot\mathbf{Q}^t_{id}(k\text{-}1) \qquad (29)$$

and

$$\mathbf{P}_{cov}(k \mid k) =$$

$$= [\mathbf{I} - \mathbf{K}(k)\cdot\mathbf{C}_{id}(k)]\cdot\mathbf{P}_{cov}(k \mid k\text{-}1)\cdot[\mathbf{I} - \mathbf{K}(k)\cdot\mathbf{C}_{id}(k)]^t +$$

$$+ \mathbf{K}(k)\cdot[\ \mathbf{R}_v(k) + \mathbf{D}_{id}(k)\cdot\mathbf{R}_{wg}(k)\cdot\mathbf{D}^t_{id}(k)\ ]\cdot\mathbf{K}^t(k) \qquad (30)$$

The cumbersome form of Eq.(30) is due to the non optimality of **K**.

For this problem, the *Cramer-Rao* bound is given by the matrix $\mathbf{P}_{id}$. As known, the Cramer-Rao lower bound indicates the "best" (minimum error covariance) that any estimator can achieve. Since **K** is suboptimal we have

$$\mathbf{P}_{cov}(k \mid k) > \mathbf{P}_{id}(k \mid k)$$

So by comparing square roots of the diagonal entries of the above matrices, the degradation of performances of the filter can be evaluated.

PERFORMANCE COMPARISON

The simulated experiments have been generated with the use of the same heat transfer model and numerical scheme adopted in the estimation algorithms. The test conditions reported here are typical for insulating materials: sample thickness b= 0.08 m; time measuring interval h= 8 s; number of simulated measurements for each sensor N_t= 400; total test duration t_m= 3200 s; uniform initial temperature equal to reference temperature $T_0=T_{ref}$=20 °C; maximum temperature reached during thermal transient T_{max} = 250 °C; white gaussian noise superposed onto temperature and heat flux signals having a standard deviation respectively σ_v= 0.05 °C, σ_w ranging from 0.0 to .8 Wm^{-2}. Sensors: one thermocouple and one fluxmeter at z=0,the same at z=b.

As boundary conditions the following excitation functions have been tested:

test T: heat flux variations giving rise to a step temperature variation from T_0 to T_{max} in z=0 and constant temperature T_0 in z=b, that is:

$$q(0,t) \quad \text{so that} \quad T(0,t)=T_{max}$$

$$q(b,t) \quad \text{so that} \quad T(0,t)=T_0, \qquad 0<t\le t_m$$

test F: step variation of heat flux in z=0 with insulated back face in z=b,that is:

$$q(0,t)=q_{max}$$

$$q(b,t)=0, \qquad 0<t\le t_m$$

A second order polynomial approximation for thermal conductivity and heat capacity has been assumed:

$$k(T) = k_0 + k_1(\ T - T_{ref}\ ) + k_2\ (\ T - T_{ref})^2 \qquad (31)$$

$$C(T) = C_0 + C_1(\ T - T_{ref}) + C_2\ (\ T - T_{ref})^2 \qquad (32)$$

The six exact values of the parameters are:
k_0= 3.0·10^{-2} W·m^{-1}·K^{-1}, k_1= 1.0·10^{-4} W·m^{-1}·K^{-2} and k_2= 5.0·10^{-7} W·m^{-1}·K^{-3} ; C_0= 50000 J·m^{-3}·K^{-1}, C_1= 50 J·m^{-3}·K^{-2} and C_2= 0.25 J·m^{-3}·K^{-3}.

The *Monte Carlo* results are reported in Figs.1÷4, where noise of increasing variance is superimposed onto the process/control functions. Tab. 1 shows the performance of the extended Kalman filter, while in Tab.2 the effectiveness of the *covariance analysis* can be observed. Finally, in Figs. 5-6, the effect of a little error in the initial temperature distribution is presented.

In all the cases here considered, the ideal covariance matrix $\mathbf{P}_{id}$, has been assumed as reference value.

Monte Carlo (200 runs)

To quantify the results of the *Monte Carlo* analysis we arbitrarily introduce two performance indexes.
The *efficiency index* I_e, defined, for a given parameter, as the ratio between the ideal standard deviation σ_{id} given by the ideal filter and its sample counterpart σ_c resulting from the *Monte Carlo* analysis; the *reliability index* I_r, defined as the ratio between the standard deviation σ_f estimated by the filter under investigation and σ_c, that is:

$$I_e = \sigma_{id} / \sigma_c; \qquad I_r = \sigma_f / \sigma_c \qquad (33)$$

Since σ_{id} represents the square root of the Cramer-Rao bound, I_e gives a measure of the optimality of the filter. On the other side, as σ_f is the starting line for the construction of the confidence bounds of an estimated parameter, the index I_r gives the degree of reliability of the confidence bounds predicted by the filter.

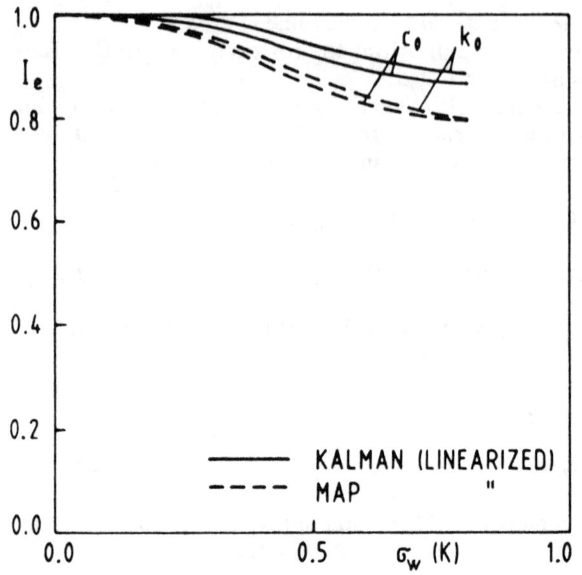

Fig. 1 Comparison of the efficiency of Kalman and MAP filters as a function of the control noise σ_w ; test T (temperature step variation)

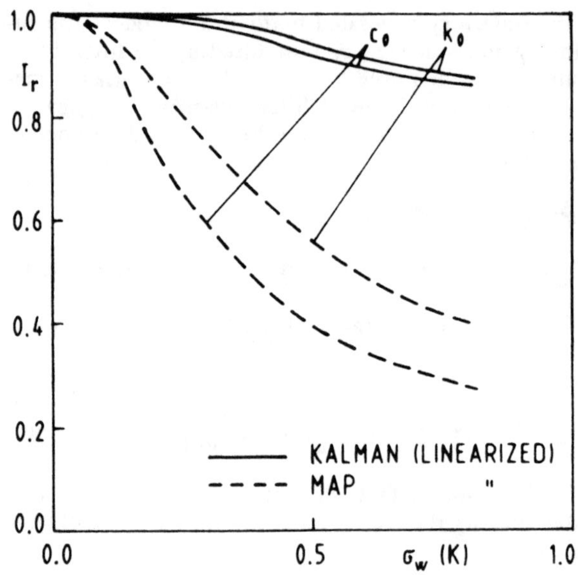

Fig. 2 Comparison of the reliability of Kalman and MAP filters as a function of the control noise σ_w ; test T (temperature step variation)

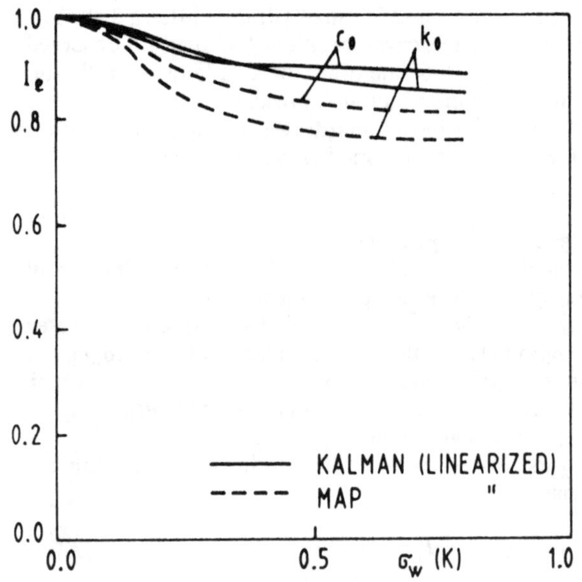

Fig. 3 Comparison of the efficiency of Kalman and MAP filters as a function of the control noise σ_w ; test F (heat flux step variation)

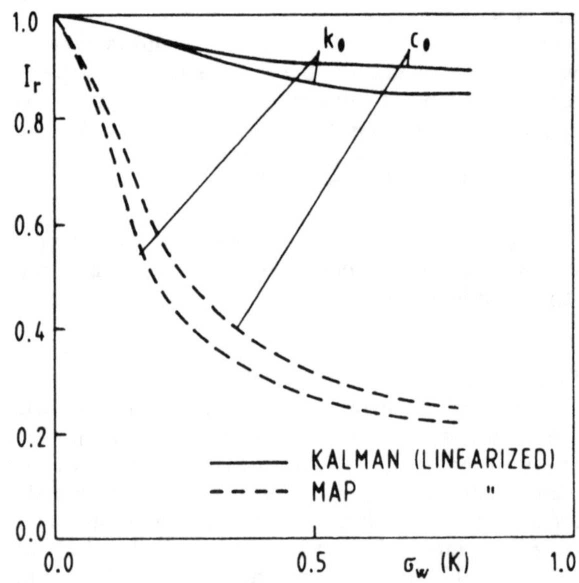

Fig. 4 Comparison of the reliability of Kalman and MAP filters as a function of the control noise σ_w ; test F (heat flux step variation)

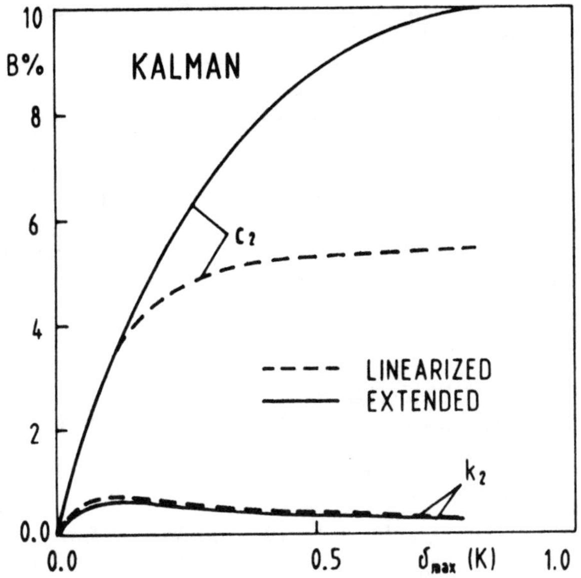

Fig. 5 Influence of the error in the initial temperature distribution; *linearized* and *extended* Kalman filter; test F (heat flux step variation)

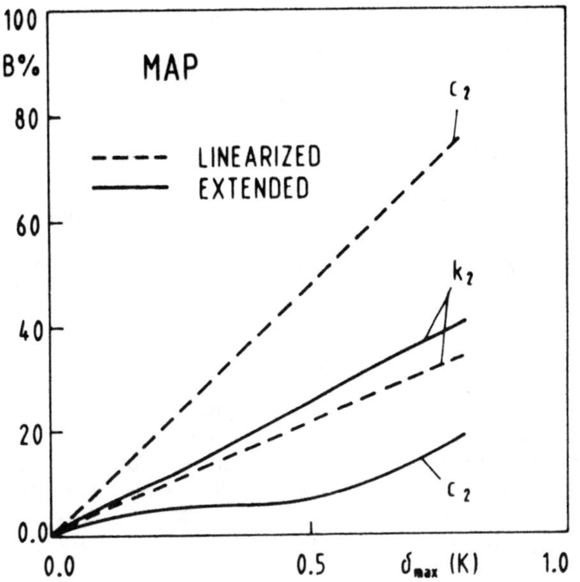

Fig. 6 Influence of the error in the initial temperature distribution; *linearized* and *extended* MAP estimator; test F (heat flux step variation)

In Figs. 1-4 the indexes I_e and I_r are compared for the linearized Kalman and MAP algorithms. The results are reported as a function of the standard deviation σ_w of the noise affecting the boundary conditions and refer to the parameters k_0 and C_0.

From Figs. 1-2 (temperature step variation) the superior performance of the LKF can be drawn. Despite the presence of noisy boundary conditions both filters show good values of I_e ($I_e > .87$ for LKF, $I_e > .80$ for MAP) but the MAP algorithm fails in the prediction of the parameter variances, which appear underestimated (Fig. 2, dashed lines); in the worst case $I_r = .29$, i.e. the effective standard deviation of the reconstructed parameter C_0 results more then three time the value estimated by the MAP algorithm. All the other parameters (k_1, k_2, C_1, C_2) have a similar trend.

Figs. 3-4 show the results obtained with test F (step variation of heat flux). The behavior of the two estimators is qualitatively similar to that described for test T (step temperature variation). The results of Figs. 1-2 are slightly better than those reported in Figs. 3-4; this can be probably attributed to the higher quality of the information available in a step temperature test (Scarpa, 1993).

Table 1 shows the behavior of the extended version of the algorithms for test F without noise superimposed onto the control (note that for $\sigma_w = 0$ the MAP and the Kalman filter coincide). The standard deviations σ_f predicted by the filters are greatly underestimated in respect to the sample ones σ_c. In general the extended version of both estimators has shown a behavior much less reliable than that given by the linearized version. In the authors' opinion, the use of the algorithms in extended version is advisable only when a real time application is needed.

Covariance analysis

The results of the *covariance analysis* are not depicted because they are practically indiscernible from those of Figs. 1÷4. In particular, the variance values obtained from $\mathbf{P_{cov}}$ are more accurate than those given by the *Monte Carlo* method. This is an important statement because a few runs of the *covariance analysis* (one in case of linear system) provide the same information of hundreds of tests.

In the *Monte Carlo* method the variance σ_c^2 of a parameter β can be computed as follows:

$$\sigma_c^2 = \frac{1}{N} \Sigma_j{}_1^N \, (\hat{\beta}_j - \beta)^2 \tag{34}$$

while in the *covariance analysis* we have:

$$\sigma_{cov}^2 = \frac{1}{N} \Sigma_j{}_1^N \, \sigma_{cov\,j}^2 \tag{35}$$

Table 2 shows the values of the above quantities and the ratio τ between the sample variances of $\sigma_{cov\,j}^2$ and $(\hat{\beta}_j - \beta)^2$. Table 2 refers to the test F with $\sigma_w = 0.2 \ \mathrm{W \cdot m^{-2}}$ and the algorithm considered is the linearized version of the MAP estimator.

TABLE 1 Performance of the *extended* Kalman filter; Monte Carlo results (200 runs); Test F; σ_w = 0.0

param.	σ_c	σ_f	$\hat{\beta}$
k_0	$4.99 \cdot 10^{-5}$	$2.12 \cdot 10^{-5}$	$2.996 \cdot 10^{-2}$
k_1	$7.12 \cdot 10^{-7}$	$5.83 \cdot 10^{-7}$	$1.002 \cdot 10^{-4}$
k_2	$3.41 \cdot 10^{-9}$	$3.38 \cdot 10^{-9}$	$5.003 \cdot 10^{-7}$
C_0	75.9	18.1	$4.993 \cdot 10^{+4}$
C_1	1.65	0.68	$5.131 \cdot 10^{+1}$
C_2	$8.76 \cdot 10^{-3}$	$5.03 \cdot 10^{-3}$	$2.439 \cdot 10^{-1}$

TABLE 2 Comparison between *Monte Carlo* and the *covariance analysis*; Test F; σ_w = 0.2

param.	σ_c	σ_{cov}	τ
k0	$4.48 \cdot 10^{-5}$	$4.37 \cdot 10^{-5}$	$1.74 \cdot 10^{-5}$
k1	$1.24 \cdot 10^{-6}$	$1.26 \cdot 10^{-6}$	$2.54 \cdot 10^{-6}$
k2	$7.12 \cdot 10^{-9}$	$6.67 \cdot 10^{-9}$	$1.49 \cdot 10^{-6}$
C0	32.4	31.3	$4.20 \cdot 10^{-4}$
C1	1.24	1.29	$1.42 \cdot 10^{-4}$
C2	$9.62 \cdot 10^{-3}$	$9.32 \cdot 10^{-3}$	$4.84 \cdot 10^{-5}$

One can observe the very small values of τ for all the identified parameters; one run of the *covariance analysis* provides, in this case, a more accurate result than 200 runs of *Monte Carlo*. On average, the relative 95% confidence associated to σ_c is about $\pm$ 10%, while for σ_{cov} the confidence results $\pm$ 1.0%.

Error in the initial temperature distribution

In this test, the effect of a little parabolic-shaped error in the initial temperature distribution is investigated. This kind of error occurs, for example, when the specimen has not completely reached the thermal equilibrium. Here we consider, as boundary condition, a step variation of heat flux without noise superposed onto the control functions ($\sigma_w = 0$). Figs. 5-6 show the percentage bias B% of the parameters C_2 and k_2 as a function of the maximum error δ_{max} (at $z = b/2$) affecting the initial temperature.

Fig. 5 (note the change of scale) shows the superior performance of LKF. This is due to the presence, in the matrix **P**, of the elements relative to the initial temperature values while in the MAP algorithm the same elements are set to zero.

CONCLUSION

The discrete Kalman filter and the MAP sequential estimator, in the *linearized* and *extended* version, are compared as parameter identificators. The test case is the identification of the temperature dependent thermal conductivity and volumetric heat capacity of materials by means of a single transient heat conduction experiment.

The *linearized* Kalman filter shows the best performance and it closely follows the ideal filter. The sequential MAP estimator performs well in comparison to the LKF in presence of low rate of noise ($\sigma_w \leq 0.1$ W $\cdot$ m^{-2}), otherwise it provides good parameter estimates but unsafe confidence bounds. For both algorithms the linearized version appears to be the most reliable.

The *covariance analysis* provides a powerful tool for investigating the performance of MAP and Kalman filters, even in the nonlinear case here considered. It represents a less expensive alternative to the crude *Monte Carlo* technique; in fact, a single run of the *covariance analysis* is statistically equivalent to hundreds of runs required by the *Monte Carlo* simulation.

ACKNOWLEDGEMENTS

This work was supported by Italian MURST (40%).

REFERENCES

Beck, J.V., and Arnold, K.J., 1977, *Parameter Estimation in Engineering and Science*, John Wiley & Sons, New York.

Candy, J.V., 1986, *Signal Processing: The Model based Approach*, McGraw-Hill, New York.

Garnier, B., D.Delaunnay, D., and Beck, J.V., 1991, "Measurement of Surface Temperature of Composite Materials for the Optimal Estimation of Their Thermal Properties", *Proceedings of the Fourth Annual Inverse Problems In Engineering Seminar*, J.V. Beck editor, Michigan State University, Lansing.

Gauss, K.G., 1808, *Theoria Motus Corporum Coelestium in Sectionibus Conicis Solem Ambientium*, Società delle Scienze di Gottinga, Gottinga.

Kalman, R.E., 1960, "A New Approach to Linear Filtering and Prediction Problems", *Trans. ASME*, Vol. 82D, pp. 35-45.

Matsevityi, Y.M., and Multanovkii, A.V., 1987, "Simulation of Thermal Process and Identification of Heat Transfer Parameters", *Syst. Anal. Model. Simul.*, Vol. 4, N° 5, pp. 371-385.

Scarpa, F., Bartolini, R., and Milano, G., 1991, "State-Space (Kalman) Filter in the Reconstruction of Thermal Diffusivity from Noisy Temperature Measurements", *High Temperatures-High Pressures*, Pion Limited, London, Vol. 23, pp. 633-642.

Scarpa, F., and Milano, G., 1993, "Identification of Temperature Dependent Thermophysical Properties from Transient Data: a Kalman Filtering Approach", Internal report, Dipartimento di Ingegneria Energetica, Università di Genova, Genova, pp. 1-16.

Simbirskii, D.F., 1976, "Solution of the Inverse Thermal Conductivity Problem with application to Optimum Filtration", *High Temperature* (translated from Teplofizika Vysokikh Temperatur), Vol. 14 , N° 5, pp. 925-931.

Sorenson, H.W., 1970, "Least-squares estimation: From Gauss to Kalman", *IEEE Spectrum*, Vol. 7, pp. 63-68.

Inverse Problems in Engineering: Theory and Practice
ASME 1993

DETERMINATION OF THE REACTION FUNCTION IN A REACTION-DIFFUSION PARABOLIC PROBLEM

Helcio R. B. Orlande and M. N. Özişik
Department of Mechanical and Aerospace Engineering
North Carolina State University
Raleigh, North Carolina

ABSTRACT

In this paper we use the *Conjugate Gradient Method* with *Adjoint Equation* in order to estimate the reaction function in a reaction-diffusion parabolic problem. The accuracy of this method of inverse analysis is verified by using simulated measurements as the input data for the inverse problem. Functional forms containing sharp corners and discontinuities are generally very difficult to recover by an inverse analysis. The present approach is capable of handling such situations quite readily and accurately. A comparison of the present function estimation approach with the parameter estimation technique, using B-Splines to approximate the reaction function, revealed that the use of function estimation reduces the computer time requirements.

NOMENCLATURE

C	volumetric heat capacity
d	direction of descent defined by equation (16)
g	reaction function
J	functional defined by equation (4)
J'	gradient of the functional J, given by equation (14)
k	thermal conductivity
L	thickness of the region
M	number of sensors in the region
q	heat flux
T	estimated temperature
t	time
t_f	final time
x	space variable
Y	measured temperature

Greek

β	search step size determined from equation (19)
ΔT	sensitivity function satisfying problem (8)
ε	real number
Φ	dimensionless heat flux defined in equation (26.e)
ϕ_L	heat flux at x=L
Γ	dimensionless reaction function defined in equation (26.b)
γ	conjugation coefficient defined in equation (17)
η	dimensionless space variable defined in equation (26.d)
λ	Lagrange multiplier satisfying the adjoint problem given by equations (12)
θ	dimensionless temperature defined in equation (26.a)
σ	standard deviation of the measurement errors
τ	dimensionless time defined in equation (26.c)

Subscript

ε	perturbed quantities

Superscript

p	number of iterations

INTRODUCTION

The study of reaction-diffusion problems has several applications, including, among others, nonlinear heat conduction (Joseph,1965 and Joseph and Sparrow,1970), chemical reactor analysis (Aris, 1975 and Kamenetskii, 1969), combustion (Gel'fand, 1963 and Kamenetskii, 1969), enzyme kinetics(Kernevez, 1980) and population dynamics (Cosner,

1990). The existence, uniqueness and stability of the solution of direct reaction-diffusion problems have been addressed by different investigators (Cohen, 1971 and Pao, 1978, 1985) .

A vast amount of literature exists on the analysis and solution of linear inverse diffusion problems. In the case of nonlinear inverse diffusion, the available works are mostly concerned with the estimation of temperature dependent properties, such as thermal conductivity and heat capacity(Artyukhin, 1975, 1982, 1987, Goryachev and Yudin, 1981, and Jarny et al, 1986). In such cases, the dependence of the unknown quantity on temperature was approximated by a polynomial(Artyukhin, 1975), B-splines(Artyukhin, 1982, 1987) or piecewise linear continuous functions(Jarny et al, 1986 and Goryachev and Yudin, 1981). With such approaches, the inverse analysis is reduced to the determination of the constant coefficients of the functional form assumed for the unknown (i.e., a finite dimensional minimization problem). The steepest descent (Artyukhin, 1975 and 1987, and Jarny et al, 1986) or the conjugate gradient method (Artyukhin, 1982, and Goryachev and Yudin, 1981) have been used for the solution of such parameter estimation problems.

If no information is available on the functional form of the unknown quantity, the minimization has to be performed on an infinite dimensional space of functions; but to the author's knowledge, such an approach has not been used for the determination of temperature dependent properties.

In this work we apply a function estimation approach based on the *conjugate gradient method* of inverse analysis with *adjoint equation* to estimate the unknown reaction function in a reaction-diffusion parabolic problem. It is assumed that no prior information is available on the functional form of the unknown quantity. The accuracy of the method is examined under strict conditions, by using transient simulated measured data in the inverse analysis.

INVERSE ANALYSIS FOR ESTIMATING REACTION FUNCTION

The inverse analysis of function estimation approach, utilizing the conjugate gradient method with adjoint equation considered here, consists of the following basic steps(Jarny et al, 1991):

1. The direct problem;
2. The inverse problem;
3. The sensitivity problem;
4. The adjoint problem and the gradient equation;
5. The conjugate gradient method of minimization;
6. The stopping criterion; and
7. The computational algorithm.

We present below the salient features of each of these steps, as applied to the estimation of the unknown reaction function.

The Direct Problem

For the present study, the direct problem is taken as the one dimensional reaction-diffusion system given by:

$$C(T)\frac{\partial T}{\partial t} - \frac{\partial}{\partial x}\left[k(T)\frac{\partial T}{\partial x}\right] - g(T) = 0 \qquad \text{in } 0 < x < L \text{ ; for } t > 0 \qquad (1.a)$$

$$\frac{\partial T}{\partial x} = 0 \qquad \text{at } x = 0 \text{ ; for } t > 0 \qquad (1.b)$$

$$k(T)\frac{\partial T}{\partial x} = \phi_L(t) \qquad \text{at } x = L \text{ ; for } t > 0 \qquad (1.c)$$

$$T(x,0) = F(x) \qquad \text{for } t = 0 \text{ ; in } 0 < x < L \qquad (1.d)$$

Such a problem is encountered in the mathematical modelling of several physical processes, involving different forms of the reaction function g(T). Consider, for example, the passage of electrical current through an one dimensional solid conductor with temperature-dependent electrical resistivity. For such a case, g(T) represents a temperature-dependent heat source term. In combustion, as in other chemical reactions, the dependence of the reaction rate on temperature is usually given by the Arrhenius expression (Aris, 1975 and Kamenetskii, 1969) in the form

$$g(T) = Q\,Z\,e^{-E/RT} \qquad (2)$$

where Q is the heat of reaction, Z is the frequency factor, E is the activation energy and R is the universal gas constant.

The *direct problem* defined above by equations (1) is concerned with the determination of the temperature distribution T(x,t) in the medium, when the physical properties C(T) and k(T), the boundary and initial conditions, and the reaction function g(T) are known. In order to solve this direct problem, we used the combined method of finite differences with θ=2/3 and the resultant nonlinear system of algebraic equations was linearized by the expansions:

$$k^{n+1} = k^n + \left(\frac{dk}{dT}\right)^n (T^n - T^{n-1}) \qquad (3.a)$$

$$C^{n+1} = C^n + \left(\frac{dC}{dT}\right)^n (T^n - T^{n-1}) \qquad (3.b)$$

$$g^{n+1} = g^n + \left(\frac{dg}{dT}\right)^n (T^n - T^{n-1}) \qquad (3.c)$$

where the superscript "n" denotes the time step.

The Inverse Problem

For the *inverse problem*, the reaction function g(T) is regarded unknown but everything else in equations (1) is known. In addition, temperature data are considered available at some appropriate locations within the medium at various time steps.

The inverse analysis utilizing the conjugate gradient method requires the solution of the direct, sensitivity and adjoint problems, together with the gradient equation. The development of sensitivity and adjoint problems are discussed next.

The Sensitivity Problem

The solution of the direct problem (1) with reaction function g(T) unknown, can be recast as a problem of optimum control, that is, choose the control function g(T) such that the following functional is minimized:

$$J[g(T)] \equiv \frac{1}{2}\int_{t=0}^{t_f}\sum_{m=1}^{M}\{T[x_m,t\,;\,g(T)]-Y_m(t)\}^2\,dt \qquad (4)$$

where $Y_m(t)$ and $T[x_m,t;g(T)]$ are the measured and estimated temperatures, respectively, at a location x_m in the medium. If an estimate is available for g(T), the temperature $T[x_m,t\,;\,g(T)]$ can be computed from the solution of the direct problem given by equations (1).

In order to develop the sensitivity problem, we assume that the reaction function g(T) is perturbed by an amount $\varepsilon\Delta g(T)$. Then, the temperature T(x,t) undergoes a variation $\varepsilon\Delta T(x,t)$, that is,

$$T_\varepsilon(x,t) = T(x,t) + \varepsilon\Delta T(x,t) \qquad (5.a)$$

where ε is a real number.

Due to the nonlinear character of the problem, the perturbation of temperature causes variations on the temperature-dependent physical properties, as well as on the reaction function. The resulting perturbed quantities are linearized as

$$k_\varepsilon(T_\varepsilon) = k(T+\varepsilon\Delta T) \approx k(T) + \left(\frac{dk}{dT}\right)\varepsilon\Delta T \qquad (5.b)$$

$$C_\varepsilon(T_\varepsilon) = C(T+\varepsilon\Delta T) \approx C(T) + \left(\frac{dC}{dT}\right)\varepsilon\Delta T \qquad (5.c)$$

$$g_\varepsilon(T_\varepsilon)=g(T+\varepsilon\Delta T) + \varepsilon\Delta g(T) \approx g(T) + \left(\frac{dg}{dT}\right)\varepsilon\Delta T + \varepsilon\Delta g(T) \qquad (5.d)$$

For convenience in the subsequent analysis, the differential equation (1.a) of the direct problem is written in operator form as

$$\mathcal{L}(T) \equiv C(T)\frac{\partial T}{\partial t} - \frac{\partial}{\partial x}\left[k(T)\frac{\partial T}{\partial x}\right] - g(T) = 0 \qquad (6.a)$$

and the perturbed form of this equation becomes

$$\mathcal{L}_\varepsilon(T_\varepsilon) \equiv C_\varepsilon(T_\varepsilon)\frac{\partial T_\varepsilon}{\partial t} - \frac{\partial}{\partial x}\left[k_\varepsilon(T_\varepsilon)\frac{\partial T_\varepsilon}{\partial x}\right] - g_\varepsilon(T_\varepsilon) = 0 \qquad (6.b)$$

To develop the sensitivity problem, we apply a limiting process for the differential equations (6.a,b) in the form

$$\lim_{\varepsilon\to 0}\frac{\mathcal{L}_\varepsilon(T_\varepsilon) - \mathcal{L}(T)}{\varepsilon} = 0 \qquad (7)$$

and similar limiting processes are applied for the boundary and initial conditions (1.b-d) of the direct problem. After some manipulations, the following sensitivity problem results for the determination of the sensitivity function $\Delta T(x,t)$:

$$\frac{\partial(C\,\Delta T)}{\partial t} - \frac{\partial^2(k\,\Delta T)}{\partial x^2} - \frac{dg}{dT}\Delta T - \Delta g = 0 \qquad \text{in } 0<x<L\,;\ \text{for } t>0 \qquad (8.a)$$

$$\frac{\partial(k\Delta T)}{\partial x} = 0 \qquad \text{at } x=0\,;\ \text{for } t>0 \qquad (8.b)$$

$$\frac{\partial(k\Delta T)}{\partial x} = 0 \qquad \text{at } x=L\,;\ \text{for } t>0 \qquad (8.c)$$

$$\Delta T(x,0) = 0 \qquad \text{for } t=0\,;\ \text{in } 0<x<L \qquad (8.d)$$

where $C\equiv C(T)$, $k\equiv k(T)$, $\Delta T\equiv\Delta T(x,t)$, $g\equiv g(T)$ and $\Delta g\equiv\Delta g(T)$.

The Adjoint Problem and the Gradient Equation

To derive the adjoint problem and the gradient equation, we multiply equation (1.a) by the Lagrange Multiplier $\lambda(x,t)$ and integrate over the time and space domains. The resulting expression is then added to the functional given by equation (4) to obtain

$$J[g(T)] = \frac{1}{2}\int_{x=0}^{L}\int_{t=0}^{t_f}\sum_{m=1}^{M}(T-Y)^2\,\delta(x-x_m)\,dt\,dx\ +$$

$$+\int_{x=0}^{L}\int_{t=0}^{t_f}\left\{C(T)\frac{\partial T}{\partial t} - \frac{\partial}{\partial x}\left[k(T)\frac{\partial T}{\partial x}\right] - g(T)\right\}\lambda(x,t)\,dt\,dx \qquad (9)$$

where $\delta(\bullet)$ is the Dirac delta function.

The directional derivative of the functional J[g(T)] in the direction of the perturbation $\Delta g(T)$, is defined as (Jarny et al, 1991):

$$D_{\Delta g}\,J[g(T)] = \lim_{\varepsilon\to 0}\frac{J[g_\varepsilon(T_\varepsilon)] - J[g(T)]}{\varepsilon} \qquad (10)$$

where the term $J[g_\varepsilon(T_\varepsilon)]$ is obtained by writing equation (9) for the perturbed quantities given by equations (5). The following expression results:

$$D_{\Delta g}J[g(T)] = \int_{x=0}^{L}\int_{t=0}^{t_f}\sum_{m=1}^{M}\Delta T\,(T-Y)\,\delta(x-x_m)\,dt\,dx\ +$$

$$+\int_{x=0}^{L}\int_{t=0}^{t_f}\frac{\partial (C\,\Delta T)}{\partial t}\lambda(x,t)\,dt\,dx-\int_{t=0}^{t_f}\int_{x=0}^{L}\frac{\partial^2 (k\,\Delta T)}{\partial x^2}\lambda(x,t)\,dx\,dt-$$

$$-\int_{x=0}^{L}\int_{t=0}^{t_f}\lambda(x,t)\,\Delta g\,dt\,dx-\int_{x=0}^{L}\int_{t=0}^{t_f}\frac{dg}{dT}\Delta T\,\lambda(x,t)\,dt\,dx \qquad (11)$$

The inner integrals in the 2nd and 3rd terms in equation (11) are integrated by parts and the resulting expression for $D_{\Delta g}J[g(T)]$ is allowed to go to zero. After some manipulations, the following adjoint problem is obtained for the determination of the Lagrange multiplier $\lambda(x,t)$:

$$-C\frac{\partial \lambda}{\partial t}-k\frac{\partial^2 \lambda}{\partial x^2}-\frac{dg}{dT}\lambda+\sum_{m=1}^{M}(T-Y)\,\delta(x-x_m)=0 \quad \text{in } 0<x<L\,;\ \text{for } t>0 \qquad (12.a)$$

$$\frac{\partial \lambda}{\partial x}=0 \quad \text{at } x=0\,;\ \text{for } t>0 \qquad (12.b)$$

$$\frac{\partial \lambda}{\partial x}=0 \quad \text{at } x=L\,;\ \text{for } t>0 \qquad (12.c)$$

$$\lambda(x,t_f)=0 \quad \text{for } t=t_f\,;\ \text{in } 0<x<L \qquad (12.d)$$

and equation (11) reduces to

$$D_{\Delta g}\,J[g(T)]=-\int_{x=0}^{L}\int_{t=0}^{t_f}\lambda(x,t)\,\Delta g(T)\,dt\,dx\rightarrow 0 \qquad (13.a)$$

The directional derivative of a functional $J[g(x,t)]$ in the direction $\Delta g(x,t)$, for a reaction function $g(x,t)$ belonging to the space of square integrable functions in the domain $(0,t_f)\times(0,L)$, is given by(Jarny et al, 1991):

$$D_{\Delta g}\,J[g(x,t)]=\int_{x=0}^{L}\int_{t=0}^{t_f}J'[x,t;g(x,t)]\,\Delta g(x,t)\,dt\,dx \qquad (13.b)$$

If it is assumed that there exists one-to-one correspondence between the temperature T and the pair (x,t), that is, $g(T)\equiv g(x,t)$ and $\Delta g(T)\equiv\Delta g(x,t)$, a comparison of equations (13.a) and (13.b) yields the gradient of the functional $J'[T,g(T)]$ as

$$J'[T,g(T)]=-\lambda(x,t) \qquad (14)$$

The sensitivity function $\Delta T(x,t)$ obtained from the solution of problem (8) and the gradient of the functional given by equation (14) are used in the conjugate gradient method of minimization as discussed next.

The Conjugate Gradient Method of Minimization

The iterative procedure for the determination of the reaction function is taken as(Jarny et al, 1991):

$$g^{p+1}(T)=g^{p}(T)-\beta^{p}\,d^{p}(T) \qquad (15)$$

and the direction of descent $d^p(T)$ is given by:

$$d^{p}(T)=J'[T,g^{p}(T)]+\gamma^{p}\,d^{p-1}(T) \qquad (16)$$

where the superscript "p" denotes the number of iterations and the conjugation coefficient γ^p is determined from (Lasdon et al, 1967):

$$\gamma^{p}=\frac{\displaystyle\int_{x=0}^{L}\int_{t=0}^{t_f}\{J'[T;g^{p}(T)]\}^2\,dt\,dx}{\displaystyle\int_{x=0}^{L}\int_{t=0}^{t_f}\{J'[T;g^{p-1}(T)]\}^2\,dt\,dx} \quad \text{for } p=1,2,\ldots$$

$$\text{with} \quad \gamma^{0}=0 \qquad (17)$$

Equation (16) shows that the direction of descent $d^p(T)$ is a conjugation of the gradient direction $J'[T,g^p(T)]$ and of the previous direction of descent $d^{p-1}(T)$, except for p=0 when $\gamma^0=0$. We note that if the term involving γ^p is omitted, equation (16) reduces to the steepest descent method. Although simpler, the steepest descent method does not converge as fast as the conjugate gradient method (Lasdon et al, 1967).

The coefficient β^p, which determines the step size in going from iteration p to p+1 in equation (15), is obtained by minimizing $J[g^{p+1}(T)]$ given by equation (4) with respect to β^p, that is,

$$\min_{\beta} J[g^{p+1}] = \min_{\beta} \frac{1}{2} \int_{t=0}^{t_f} \sum_{m=1}^{M} [\, T(x_m,t\,;\, g^p+\beta^p d^p) - Y_m(t)\,]^2 \, dt \tag{18}$$

The estimated temperature $T(x_m,t\,;\,g^p+\beta^p d^p)$ in equation (18) is linearized using a Taylor series expansion. The resulting expression is minimized and then solved for β^p to yield

$$\beta^p = \frac{\int_{t=0}^{t_f} \sum_{m=1}^{M} [\, T(x_m,t\,;\,g^p) - Y_m(t)\,]\, \Delta T(x_m,t\,;\,d^p)\, dt}{\int_{t=0}^{t_f} \sum_{m=1}^{M} [\, \Delta T(x_m,t\,;\,d^p)\,]^2\, dt} \tag{19}$$

where $\Delta T(x_m,t\,;\,d^p)$ is the solution of the sensitivity problem at position x_m and time t, which is obtained from equations (8) by setting $\Delta g(T)=d^p(T)$.

Once $d^p(T)$ is computed from equation (16) and β^p from equation (19), the iterative process defined by equation (15) can be applied to determine $g^{p+1}(T)$, until a specified stopping criterion based on the discrepancy principle described below is satisfied.

The Stopping Criterion

If the problem involves no measurement errors, the traditional check condition specified as

$$J[g^{p+1}(T)] < \varepsilon_J \tag{20}$$

where ε_J is a small specified number, can be used. However, the observed temperature data contains measurement errors; as a result, the inverse solution will tend to approach the perturbed input data and the solution will exhibit oscillatory behavior as the number of iterations is increased. The computational experience shows that it is advisable to use the discrepancy principle to stop the iteration process,that is, by assuming

$$T[x_m,t\,;\,g(T)] - Y_m(t) \approx \sigma = \text{constant} \tag{21}$$

ε_J is obtained from equation (4) as:

$$\varepsilon_J = \frac{M}{2}\sigma^2 t_f \tag{22}$$

where σ is the standard deviation of the measurement errors.

Hence, equation (20) with ε_J determined from equation (22) is used to stop the iterations.

The Computational Algorithm

The algorithm for the iterative scheme given by the conjugate gradient method is summarized below.

Suppose $g^p(T)$ is available at iteration p, then:

Step 1. Solve the direct problem given by equations (1) and compute $T(x,t)$;

Step 2. Check the stopping criterion given by equation (20) with ε_J determined from equation (22). Continue if not satisfied;

Step 3. Knowing $T(x_m,t;g^p)$ and $Y_m(t)$, solve the adjoint problem given by equations (12) to obtain $\lambda(x,t)$;

Step 4. Knowing $\lambda(x,t)$, compute the gradient of the functional from equation (14);

Step 5. Knowing $J'[T,g^p(T)]$, compute first the conjugation coefficient from equation (17) and then the direction of descent from equation (16);

Step 6. Solve the sensitivity problem given by equations (8) by setting $\Delta g(T)=d^p(T)$, to determine $\Delta T(x,t)$;

Step 7. Knowing $\Delta T(x_m,t;d^p)$, compute the search step size β^p from equation (19);

Step 8. Knowing β^p and $d^p(T)$, compute the new estimate $g^{p+1}(T)$ from equation (15) and go to step 1.

RESULTS AND DISCUSSION

In order to examine the accuracy of the function estimation approach using the conjugate gradient method as applied to the analysis of the inverse problem previously described, we studied test cases by using simulated measured temperatures as the input data for the inverse analysis. The simulated temperature data were generated by solving the direct problem for a specified reaction function. The temperatures calculated in this manner are considered exact measurements, T_{ex}, and the simulated measured temperature data, Y, containing measurement errors, are determined as

$$Y = T_{ex} + \alpha\,\sigma \tag{23}$$

where $\alpha\sigma$ is the error term and σ is the standard deviation of the measurements. For normally distributed errors, with zero mean and a 99% confidence level, α lies within the range

$$-2.576 < \alpha < 2.576 \tag{24}$$

The values of α were randomly determined with the subroutine DRNNOR from the IMSL (1987).

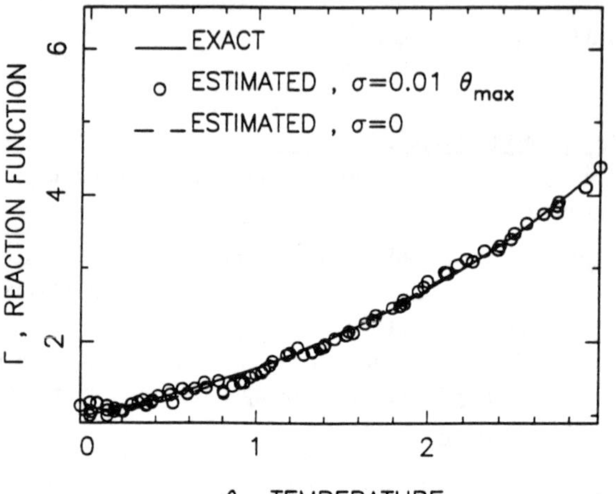

FIGURE 1. Inverse solution with exponential variation for the reaction function

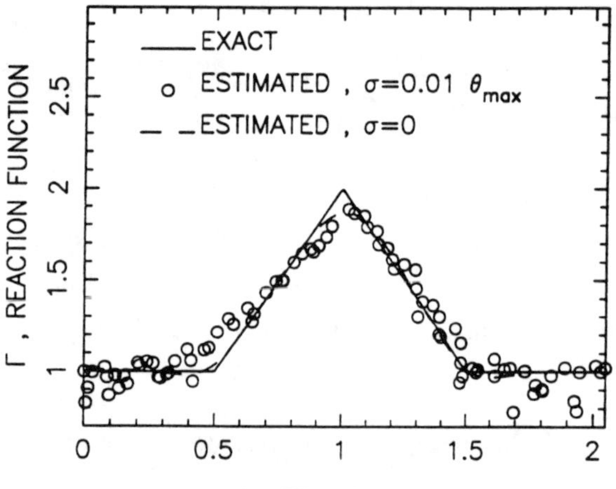

FIGURE 2. Inverse solution with triangular variation for the reaction function

To generate the simulated measurements, the direct problem given by equations (1) was expressed in dimensionless form by taking the coefficients k(T) and C(T) as

$$k(T) = k_0 (1 + k_1 T) \quad ; \quad C(T) = C_0 (1 + C_1 T) \qquad (25.a,b)$$

and by introducing the following dimensionless variables:

$$\theta = \frac{T - T_0}{\left(\frac{\phi_L}{k_0}\right) L} \; ; \; \Gamma = \frac{g(T)\, L}{\phi_L} \; ; \; \tau = \frac{k_0\, t}{L^2\, C_0} \; ; \; \eta = \frac{x}{L} \; ; \; \Phi = \frac{q}{\phi_L} \qquad (26.a\text{-}e)$$

where k_0 , k_1 , C_0 and C_1 are constants, T_0 is the initial temperature in the medium which is assumed to be uniform, and ϕ_L is the heat flux applied at the boundary x=L, which is assumed to be constant.

The accuracy of the present method of inverse analysis was verified under strict conditions by using a single sensor in the region and by considering reaction functions with exponential behavior, sharp corners and discontinuities. The sensor should be located in a place where maximum temperature variations occur for a fixed variation of the sought quantity. In linear problems, such a location may be estimated by examining the sensitivity functions (Mikhailov, 1989). However, for the nonlinear problem considered here, this kind of analysis cannot be performed because the sensitivity problem depends on the functional form of g(T), which is not known a priori. Therefore, numerical experiments were made here in order to estimate the optimum sensor position. Generally, the optimum sensor position was in the left half of the medium, i.e., $0 < \eta < 0.5$. This fact is probably due to the unsymmetrical boundary conditions considered for the direct problem.

For all test cases analyzed here, we considered $\sigma=0$ (errorless measurements) and $\sigma=0.01\ \theta_{max}$, where θ_{max} is the maximum temperature measured by the sensor.

Figure 1 shows the results for an exponential variation of the dimensionless reaction function in the form

$$\Gamma(\theta) = e^{0.5\,\theta} \qquad (27)$$

Clearly, the agreement between the estimated and the exact reaction functions is excellent, for both situations of errorless and inexact measurements, and such was the case for other smooth functional forms tested. The optimum sensor position for this case was at $\eta=0.2$.

Figure 2 shows similar results for a reaction function with a triangular variation. A comparison of the exact reaction function with the one estimated by using errorless measurements ($\sigma=0$) indicates that the present function estimation approach can resolve sharp corners, although some smoothing is noticed at the corner at $\theta=1$. The agreement between the exact solution and the results obtained by using measurements with random errors is good. The optimum sensor position for this case was at $\eta=0.2$.

Figure 3 presents the results obtained for a step variation of the reaction function. The curve obtained with errorless measurements ($\sigma=0$) is in very good agreement with the exact solution, although some oscillations are observed near the discontinuities. The results obtained by using measurements with random errors is also in good agreement with the exact functional form of the reaction function. The optimum sensor position for this case was at $\eta=0.3$.

Table 1 presents the number of iterations, the RMS error and the CPU times for the function estimation approach based on the

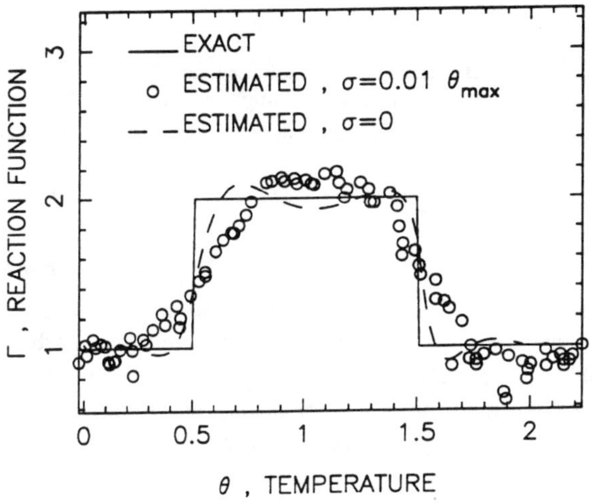

FIGURE 3. Inverse solution with step variation for the reaction function

conjugate gradient method, as applied to the cases shown in figures 1 to 3, respectively, for measurements containing random errors. The calculations were performed on a Cray Y-MP supercomputer and the RMS error is defined as:

$$e_{RMS} = \sqrt{\frac{1}{P}\sum_{i=1}^{P} [\, g_{ex}(T_i) - g_{est}(T_i)\,]^2} \qquad (28)$$

where the subscripts "ex" and "est" denote "exact" and "estimated" quantities, respectively, and P denotes the number of temperature measurements used to compute the RMS error.

In order to compare the results obtained by using the function estimation approach with those of parameter estimation, the inverse analysis was also performed by using cubic B-splines to approximate the reaction function. We present in appendix A the finite dimensional minimization problem for such a case. Table 2 presents the results obtained with the parameter estimation approach for the same functional forms considered above.

A comparison of tables 1 and 2 reveals that for the same order of magnitude of RMS errors, the CPU times for the parameter estimation are larger than those for the function estimation. It appears that the evaluation of the B-splines during each iteration of the conjugate gradient method, causes the increase in CPU time for the parameter estimation approach. Tables 1 and 2 show that the number of iterations is very similar for the function and parameter estimation approaches, except for the case of exponential variation of the reaction function. The initial guess used for both approaches was the exact value of the reaction function at the final temperature measured by the sensor, so that the instabilities inherent of the conjugate gradient method at the final temperature value could be avoided.

TABLE 1 - Results obtained with function estimation approach

Function (figure)	Number of Iterations	RMS error	CPU Time (sec)
Exponential (1)	5.	0.1047	2.13
Triangular (2)	6	0.0885	2.14
Step (3)	8	0.2097	2.85

TABLE 2 - Results obtained with parameter estimation approach using cubic B-splines to approximate the reaction function

Function (figure)	Number of B-Splines	Number of Iterations	RMS error	CPU Time (sec)
Exponential (1)	4	50	0.0989	31.07
Triangular (2)	15	7	0.0923	4.52
Step (3)	20	7	0.2080	4.60

CONCLUSIONS

The inverse analysis utilizing the conjugate gradient method of minimization with adjoint equation, provides an efficient approach for the estimation of the reaction function, with no prior information on the functional form of the unknown quantity.

Results obtained with test cases using simulated measurements with random errors indicate that the present approach is accurate, even for reaction functions involving sharp corners and discontinuities.

A comparison of the present function estimation approach with the parameter estimation using cubic B-Splines to approximate the reaction function, reveals that the CPU times are larger for the parameter estimation than for the function estimation approach.

ACKNOWLEDGEMENT

The CPU time for this work has been provided by the North Carolina Supercomputing Center. One of the authors (H.R.B.O.) would like to acknowledge the support provided by CNPQ, an agency of the Brazilian government.

REFERENCES

Aris, R., 1975, *The Mathematical Theory of Diffusion and Reaction in Permeable Catalysts, Vol.1* , Clarendon Press, Oxford.

Artyukhin,E.A., 1975, "Determination of Thermal Diffusivity from Experimental Data," *J. Engr. Phys.*, Vol. 29, pp. 878-881.

Artyukhin,E.A., 1982, "Recovery of the Temperature Dependence of the Thermal Conductivity Coefficient from the

Solution of the Inverse Prolem," *High Temp.*, Vol. 19, pp. 698-702.

Artyukhin,E.A. and Nenarokomov,A.V., 1987, "Coefficient Inverse Heat-Conduction Problem," *J. Engr. Phys.*, Vol.53, pp. 1085-1090.

Cohen,D.S., 1971, "Multiple Stable Solutions of Nonlinear Boundary Value Problems Arising in Chemical Reactor Theory," *SIAM J. Appl. Math,* Vol. 20, pp. 1-13.

Cosner,C., 1990, "Eigenvalue Problems with Indefinite Weights and Reaction-Diffusion Models in Population Dynamics," *Reaction-Diffusion Equations*, K.J. Brown and A.A. Lacey, ed. Clarendon Press, Oxford.

Gel'fand,I.M., 1963, "Some Problems in the Theory of Quasilinear Equations," *Am. Math. Soc. Transl.,* Vol. 29, pp. 295-381.

Goryachev,A.A. and Yudin,V.M., 1981, "Solution of the Inverse Coefficient Problem of Heat Conduction, "*J. Engr. Phys.* Vol. 43, pp.1148-1154 .

IMSL Library Edition 10.0, 1987, *User's Manual Math Library*, Houston,Texas.

Jarny,Y., Delaunay,D. and Bransier,J., 1986, "Identification of Nonlinear Thermal Properties by an Output Least Square Method," *Proccedings, 8th International Heat Transfer Conference*, pp. 1811-1816.

Jarny,Y., Özişik,M.N. and Bardon,J.P., 1991, "A general Optimization Method using Adjoint Equation for Solving Multidimensional Inverse Heat Conduction," *Int. J. Heat Mass Transfer,* Vol.34, pp.2911-2929.

Joseph,D.D.,1965, "Nonlinear Heat Generation and Stability of the Temperature Distribution in Conducting Solids," *Int. J. Heat Mass Transfer* , Vol. 8, pp. 281-288.

Joseph,D.D. and Sparrow,E.M., 1970, "Nonlinear Diffusion Induced by Nonlinear Sources," *Quart. Appl. Math,* pp.327-342.

Kamenetskii, D.A.F., 1969, *Diffusion and Heat Transfer in Chemical Kinetics* ,Plenum Press, New York.

Kernevez,J.P., 1980, *Enzyme Mathematics* , North-Holland, Amsterdam.

Lasdon,L.S., Mitter,S.K. and Warren,A.D., 1967, "The Conjugate Gradient Method for Optimal Control Problem," *IEEE Trans. Automatic Control* , Vol.12, pp.132-138.

Mikhailov,V.V.,1989, "Arrangement of the Temperature Measurement Points and Conditionality of Inverse Thermal Conductivity Problem," *J. Engr. Phys.,* Vol.57, pp.1369-1373.

Pao, C.V., 1978, "Asymptotic Behavior and Nonexistence of Global Solutions for a Class of Nonlinear Boundary Value Problems of Parabolic Type," *J. Math. Anal. and Appl.,* Vol. 65, pp. 616-637.

Pao,C.V., 1985, "Monotone Iterative Methods for Finite Difference System of Reaction-Diffusion Equations," *Numer. Math.,* Vol. 46, pp. 571-586.

APPENDIX A

Finite Dimensional Minimization Problem

For the parameter estimation approach considered here, we approximate the reaction function using B-splines in the form

$$g(T) = \sum_{j=1}^{N} g_j B_j(T) \tag{A.1}$$

where the vector $\mathbf{g}=(g_1, g_2, ..., g_N)$ of the unknown B-splines coefficients will be determined with an inverse analysis based on the conjugate gradient method of minimization (Artyukhin, 1982, and Goryachev and Yudin, 1981). It can be shown that the direct, sensitivity and adjoint problems remain the same and are given by equations (1), (8) and (12), respectively; but from equation (A.1) we have

$$\Delta g(T) = \sum_{j=1}^{N} \Delta g_j B_j(T) \tag{A.2.a}$$

and

$$\frac{d\,g(T)}{d\,T} = \sum_{j=1}^{N} g_j \frac{d\,B_j(T)}{d\,T} \tag{A.2.b}$$

The iterative procedure of the Conjugate Gradient Method for estimating the vector **g** is given by

$$\mathbf{g}^{p+1} = \mathbf{g}^p - \beta^p \mathbf{d}^p \tag{A.3}$$

The vector with the direction of descent, $\mathbf{d}^p$, is determined as:

$$\mathbf{d}^p = \mathbf{J}'^p + \gamma^p \mathbf{d}^{p-1} \tag{A.4}$$

and the conjugation coefficient γ^p is obtained from:

$$\gamma^p = \frac{\mathbf{J}'^p \cdot \mathbf{J}'^p}{\mathbf{J}'^{p-1} \cdot \mathbf{J}'^{p-1}} \quad \text{for } p=1,2,... \quad \text{with } \gamma^0 = 0 \tag{A.5}$$

where "·" denotes the scalar product.

It can be shown that the search step size β^p remains the same as that for the function estimation and is obtained from equation (19). The gradient vector components, J'_j, are given by

$$J'_j = -\int_{x=0}^{L} \int_{t=0}^{t_f} \lambda(x,t)\, B_j[T(x,t)]\, dt\, dx \quad \text{for } j=1,...,N \tag{A.6}$$

Similarly to the function estimation approach, the iterative procedure given by equation (A.3) to estimate **g** is stopped when the discrepancy principle given by equation (20), with ε_J determined from equation (22), is satisfied.

EVALUATION OF AN INVERSE HEAT CONDUCTION PROCEDURE FOR DETERMINING LOCAL CONVECTIVE HEAT TRANSFER RATES

D. Naylor and P. H. Oosthuizen
Department of Mechanical Engineering
Heat Transfer Laboratory
Queen's University
Kingston, Ontario, Canada

ABSTRACT

A transient method for measuring the distribution of local convective heat transfer rates on the surface of a body has been evaluated numerically and experimentally. In this method, temperature-time variations are measured at several subsurface points as a model cools in a test flow. Using a simple Inverse Heat Conduction (IHC) technique, these temperature-time data are then used to obtain the local convective heat transfer rate distribution. As a test case, the method has been applied to measure two-dimensional forced convective heat transfer from a square prism in cross flow. In addition, the sensitivity of the method to experimental uncertainties has been evaluated. The data obtained using the present IHC method were found to compare well with published experimental data measured using more conventional techniques.

NOMENCLATURE

A_i	coefficients used to define the local Biot number distribution
Bi	local Biot number based on w', hw'/k_m
E	sum squared temperature error
h	local convective heat transfer coefficient
h_r	average radiation coefficient (defined by eq. (12))
k_a	thermal conductivity of air
k_m	thermal conductivity of the model material
M	total number of measuring points
N	total number of A_i coefficients
Nu	local Nusselt number, hw'/k_a
n'	unit normal vector
n	n'/w'
Re	Reynolds number based on w'
T'_∞	ambient temperature
T'_s	average model surface temperature
T	dimensionless temperature
T_c	calculated dimensionless model temperature
T_o	initial dimensionless model temperature
T_m	measured dimensionless model temperature
t	dimensionless time, $\alpha t'/w'^2$
t_{max}	maximum dimensionless time
U'_∞	free stream velocity
x,y	dimensionless cartesian coordinates
w'	side length of the model

Greek

α	thermal diffusivity
β	angle between the model and air stream
ϵ	emissivity of the model
σ	Stefan-Boltzmann constant

Superscripts

'	dimensional quantity
j	measuring point

INTRODUCTION

There are several well established experimental methods for measuring local convective heat transfer coefficients. Interferometry, naphthalene sublimation and surface temperature measurements on constant heat flux generating metal foils are the most common techniques. However, these methods either require expensive experimental equipment, or have limited application to three dimensional geometries.

The purpose of the present study is to evaluate an inverse heat conduction (IHC) method for measuring the distribution of convective local heat transfer rates on the surface of a body. In this method, a body constructed of a low conductivity material is heated to a uniform temperature and then exposed to a test flow. During the transient cooling process, the temperature-time variations are measured at several subsurface points. The local heat transfer coefficients are then calculated using a simple inverse heat conduction (IHC) procedure. One advantage of this method over more conventional techniques is that measurements can be obtained with relatively simple, low cost experimental models and equipment. Also, this method can, in general, be applied to three dimensional geometries. However, it should be noted that considerable amounts of computational power may be required to solve the IHC problem.

The adequacy of the present IHC procedure has been examined numerically in a previous study (Oosthuizen and Naylor (1992)) for different numbers of measuring points and different forms of the assumed local heat transfer distribution. This study indicates that the procedure may be a viable measurement technique. In the present study, this method has been applied to the test case of two-dimensional forced convective heat transfer from a square prism in cross flow. For this geometry, the procedure has been evaluated both numerically and experimentally. In addition, the sensitivity of the method to experimental uncertainties has been evaluated.

There are several previous studies that have used measured temperature-time variations, either at the surface or within the body to determine the local surface heat transfer rate. However, in many of these studies, the conduction within the model is assumed to be locally one-dimensional (for example, see Camci et al. (1991), Ireland and Jones (1985)).

THE IHC METHOD

The IHC method used in the present study is similar to the method used by Dorri and Chandra (1991) to measure contact resistance variations with temperature. An essential part of the present IHC method is the numerical solution of the transient conduction problem for an arbitrary convective coefficient distribution on the model surface. With the assumption that the temperature field is two dimensional and assuming that the material properties are constant, the governing equation in cartesian coordinates is:

$$\frac{\partial^2 T'}{\partial x'^2} + \frac{\partial^2 T'}{\partial y'^2} = \frac{1}{\alpha}\frac{\partial T'}{\partial t'} \tag{1}$$

where α is the thermal diffusivity and the primes (′) denote dimensional quantities.

The boundary condition at any point on the surface is:

$$-k\frac{\partial T'}{\partial n'} = h(T' - T'_\infty) \tag{2}$$

where n′ is the unit normal to the surface. In the present study, it is assumed that the model is initially at an isothermal temperature T'_o above the ambient temperature.

The following dimensionless variables have been introduced:

$$x = x'/w \qquad y = y'/w$$
$$t = t'\alpha/w^2 \qquad T = (T' - T'_\infty)/(T'_o - T'_\infty) \tag{3}$$

In terms of these variables, the eq.(1) becomes:

$$\frac{\partial^2 T}{\partial x^2} + \frac{\partial^2 T}{\partial y^2} = \frac{\partial T}{\partial t} \tag{4}$$

The surface boundary condition is:

$$-\frac{\partial T}{\partial n} = Bi\, T \tag{5}$$

where Bi is the Biot number defined as:

$$Bi = hw'/k_m \tag{6}$$

The distribution of Biot number on the model surface is assumed to be defined in terms of a series of coefficients, A_1, A_2,...A_N. Of course, there are many possible ways to approximate the Biot number distribution. In the present study, the local Biot number distribution is approximated by a piece-wise function consisting of N constant values. So, each coefficient represents the locally averaged Biot number over a sub-region of the surface.

If the values of the coefficients A_1, A_2,...A_N are guessed, the corresponding variation of temperature (T_c^j) at each of the measurement locations with time can be calculated. Note that the experimental temperature variation (T_m^j) at each of the measurement points is known. So, the square of the difference between the calculated and the measured temperatures, summed over all thermocouples and all time steps, is then calculated as follows:

$$E = \sum_{t=0}^{t_{max}} \sum_{j=1}^{M} (T_c^j - T_m^j)^2 \tag{7}$$

where t_{max} is the maximum time to which the calculations are taken and M is the number of measurement points. The superscript j refers to the measurement point.

The optimum local heat transfer rate distribution is then obtained by minimizing the sum square error E with respect to each unknown coefficient:

$$\frac{\partial E}{\partial A_i} = 2\sum_{t=0}^{t_{max}} \sum_{j=1}^{M} (T_c^j - T_m^j)\frac{\partial T_c^j}{\partial A_i} \tag{8}$$

Let T_{c0}^j be the calculated temperature variations for an initial set of coefficients. The term ($T_c^j - T_m^j$) is then linearized using a Taylor series as follows:

$$(T_c^j - T_m^j) = (T_{c0}^j - T_m^j) + \frac{\partial T_c^j}{\partial A_1}\Delta A_1 + \frac{\partial T_c^j}{\partial A_2}\Delta A_2 + \ldots + \frac{\partial T_c^j}{\partial A_N}\Delta A_N \tag{9}$$

Substituting eq.(9) into eq.(8) gives an equation for the corrections to the coefficients ΔA_i required to minimize E:

$$\sum_{t=0}^{t_{max}} \sum_{j=1}^{M} \left((T_{c0}^j - T_m^j)\frac{\partial T_c^j}{\partial A_i} + \frac{\partial T_c^j}{\partial A_1}\frac{\partial T_c^j}{\partial A_i}\Delta A_1 + \frac{\partial T_c^j}{\partial A_2}\frac{\partial T_c^j}{\partial A_i}\Delta A_2 + \ldots + \frac{\partial T_c^j}{\partial A_N}\frac{\partial T_c^j}{\partial A_i}\Delta A_N \right) = 0 \tag{10}$$

Applying eq.(10) for each of the coefficients A_1, A_2,...A_N gives a set of N linear equations in ΔA_1, ΔA_2,...ΔA_N. The values of $\partial T_c^j/\partial A_i$ in eq.(10) are obtained by calculating the temperature variations (T_{c0}^j) from the initial guessed values of A_i and then, one by one, adding a small amount δA_i (e.g. $\delta A_i = 0.01A_i$) to each of the A_i values and recalculating the temperature values T_{c1}^j. The derivatives at each time step are approximated by:

$$\frac{\partial T_c^j}{\partial A_i} = \frac{T_{c1}^j - T_{c0}^j}{\delta A_i} \tag{11}$$

Because of approximations in the above analysis, the following iterative solution has been used:

1. Guess the values of all the A_i coefficients, i=1,2,..,N.
2. Calculate the variations of dimensionless temperature up to the maximum dimensionless time at all the measurement points, j=1,2,...,M.
3. Increase A_1 by a small amount, δA_1, and recalculate the temperature variations.
4. Repeat step (3) for coefficients $A_2, A_3, .., A_N$.
5. Calculate the derivatives using eq.(11). Using these derivatives, apply eq.(10) for each coefficient to obtain N equations in the N unknown ΔA_i values. Solve these equations.
6. Use the ΔA_i values to get improved values for the coefficients, i.e., $A_1+\Delta A_1$, $A_2+\Delta A_2$,..., $A_N+\Delta A_N$.
7. Repeat steps (2) to (6) until the coefficients cease to change (within a selected tolerance) from one iteration to the next.

NUMERICAL EVALUATION

The experimentally measured local Nusselt number distributions around a square prism obtained by Igarashi (1985) have been used to evaluate the procedure numerically. These known local convective heat transfer rate distributions have been applied as the boundary condition in the numerical solution of transient cooling of a solid square prism. Then, using the calculated temperature-time variations at selected points as measured variations (i.e. simulated experimental data), the IHC procedure has been applied to predict the input local Nusselt number distribution.

As previously mentioned, a piece-wise function has been used to describe the local Biot number distribution for the IHC procedure. Note that only one half of the square prism is considered because of symmetry. So, for example, if a four coefficient distribution is used, then:

A_1 = Bi on the upper half front face
A_2 = Bi on the upstream half of the side face
A_3 = Bi on the downstream half of the side face
A_4 = Bi on the upper half rear face

Calculations have been made for a model conductivity to air conductivity ratio of $k_m/k_a=10$. Also, the maximum dimensionless time for the calculations was selected as $t_{max}=0.02$. (This corresponds to a maximum time of about 27 minutes for the plexiglass model to be used in the experiments: w'=9.52cm, $\alpha=1.12\times10^{-7}m^2/s$). Although this value of t_{max} is quite arbitrary, the results were found to be very insensitive to the value chosen.

The results shown in Fig. 2 have been obtained for the measuring point distribution shown in Fig. 1. As shown in Fig. 1, thirteen evenly spaced measurement points at a depth of w'/8 were used on one side of the line of symmetry. Figure 2 shows the input distribution (from Igarashi (1985)) and the derived data using an eight coefficient piece-wise function for the local Biot number.

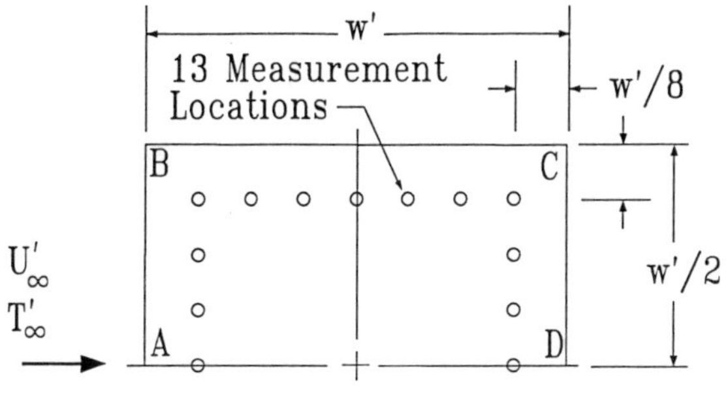

Fig. 1: Distribution of the temperature measuring points for the numerical evaluation.

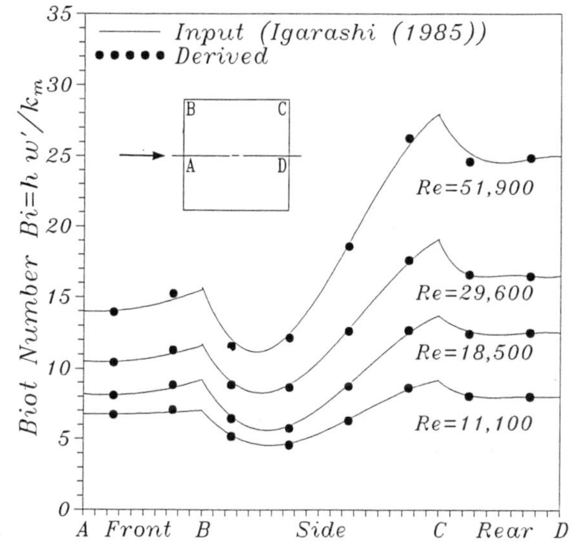

Fig. 2: Comparison of the input (experimental data) and derived Biot number distributions around a square prism, using thirteen measurement points and an eight coefficient function.

(Note that the Biot number in Fig. 2 is the Nusselt number multiplied by the conductivity ratio k_a/k_m). It can be seen that, in the absence of experimental uncertainties in the "measured" temperature-time data, the present IHC procedure gives results very close to the input distributions. Also, using this approximate distribution, the average Nusselt numbers were recovered at each Reynolds number within two percent.

Of course, in an actual experiment the measured data will contain some degree of experimental uncertainty. For this reason, the above numerical approach was also used to determine the sensitivity of the IHC procedure to experimental uncertainties. The main sources of error stem from (i) uncertainties in the thermal properties of the model (k_m, α), (ii) temperature measurement errors, and (iii) temperature measurement location uncertainties.

The transient cooling of a body by convection is a function of the Biot number (hw'/k_m) and the Fourier number (dimensionless time $\alpha t'/w'^2$). Via these dimensionless numbers, uncertainties in the thermal conductivity (k_m) and thermal diffusivity (α) of the model material introduce errors into the predicted local transfer

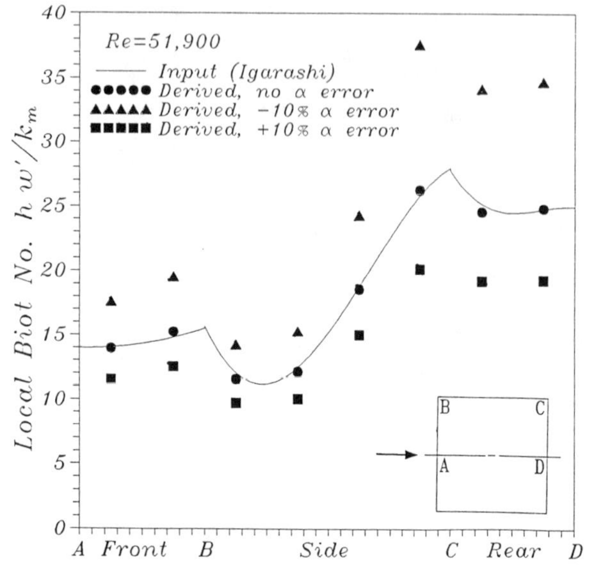

Fig. 3: The effect of uncertainty in the model thermal diffusivity on the predicted local Biot number distribution for Re=51,900 (using thirteen measurement points and an eight coefficient function).

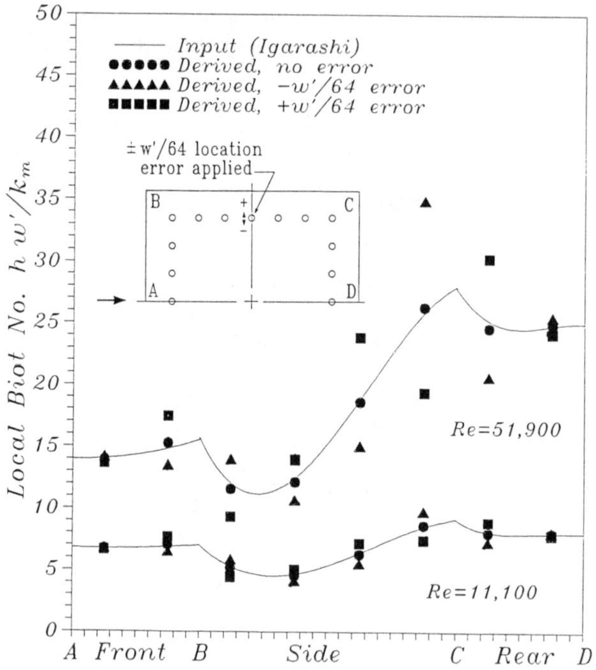

Fig. 4: The effect of uncertainty in the measurement locations on the predicted local Biot number distribution for Re=11,100 and Re=51,900 (using thirteen measurement points and an eight coefficient function).

rate distribution.

The effect of the uncertainty in the conductivity of the model can be seen easily. When the IHC procedure is applied to experimental temperature-time data, the results are obtained in terms of the local Biot number distribution. Of course, in order to obtain the local convective coefficients, the local Biot numbers must be multiplied by k_m/w'. So, if the model conductivity is, for example, ten percent higher than the true value, this will cause a ten percent bias error in the local heat transfer rate distribution.

The effect of the thermal diffusivity (α) error is not linear and was determined numerically. This was done by recalculating the dimensionless time scale of the simulated experimental data to account for an error of ± 10 percent in α. Then, the IHC procedure was applied to the rescaled data. A positive error in thermal diffusivity (i.e., assumed value greater than actual value) lengthens the amount of dimensionless time that the model takes to cool. As shown in Fig. 3, this causes the IHC procedure to under-predict the local Biot number distribution. Similarly, a negative error in α causes the predicted Biot number distribution to be high. Note that an error in thermal diffusivity causes primarily a shift in the distributions. Unfortunately, this bias error in the average Biot number is substantially larger than the corresponding error in α. An error of -10 percent in α, causes a +35 percent error in the average Biot number at Re=51,900. At lower Reynolds numbers (lower model average Biot numbers), the results were found to be slightly less sensitive to thermal diffusivity errors; an error of -10 percent in α, causes about a +20 percent error in the average Biot number at Re=11,100.

The effect of temperature measurement uncertainties caused by thermocouple calibration errors has also been evaluated numerically. Again, this was done by modifying the simulated experimental data to include an error of ± 0.3C° in the temperature measurements. The IHC procedure was then applied to the modified data. Compared to the other error sources, the effect of this error was found to be small. For example, even with all the thermocouples reading high by +0.3C°, the average Biot number was only 1.3 percent low at Re=11,100 and 2.6 percent low at Re=51,900.

Figure 4 shows the results of numerically simulating the effect of uncertainties in the measurement locations, i.e. thermocouple tip position errors. In this case, the simulated experimental data from only one of the thirteen measurement locations were modified to have a location error of $\pm w'/64$. As shown in Fig. 4, the data from the measurement point located at the mid-point of the side were modified. These data were modified by taking the temperature-time variations (from the transient conduction solution) from one node closer to the surface and one node farther away from the surface. It can be seen in Fig. 4 that although only one measurement location is in error, the entire predicted distribution is affected. Near the measurement error location, the effect on the Bi distribution is easily understood. For example, if the temperature measurement device is positioned closer to the surface than assumed ($+w'/64$ error), the temperature decreases more rapidly with time, so locally the Biot numbers are too high. Farther away from the location of the position error, oscillations in the Biot number are produced; the oscillations occur because of the nature of the least squares method. At Re=51,900, the scatter in the results was as much as ± 30 percent. At Re=11,100, the scatter was reduced to about less than ± 14 percent; at lower average Biot numbers, the temperature gradients are less severe, so the results are less sensitive to small measurement location errors. Nevertheless, these results indicate that relatively large experimental models will be required unless substantial care is taken to position the thermocouple tips accurately. It should also be noted that the average Biot numbers were only weakly affected

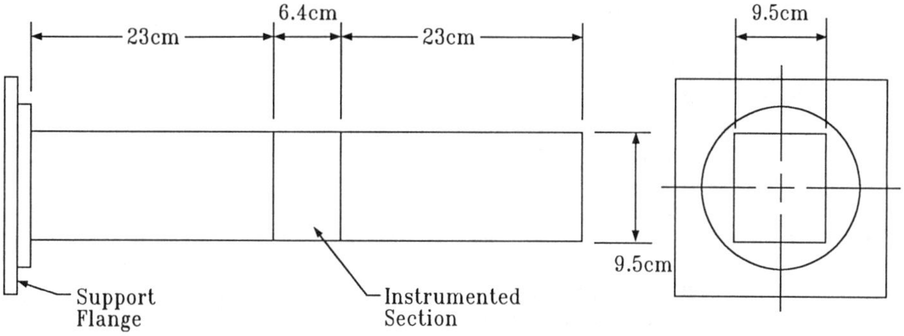

Fig. 5: Drawing showing the dimensions of the plexiglass model.

by the measurement position error. The average Biot numbers were affected by less than three percent.

EXPERIMENTAL APPARATUS AND PROCEDURE

To evaluate the IHC procedure experimentally, a plexiglass model was constructed (k=0.195 W/mK, $\alpha=1.12\text{x}10^{-7}\text{m}^2/\text{s}$). The dimensions of the model are shown in Fig. 5. The model was machined in three sections and had a 9.52cmx9.52cm square cross section. The three sections were held together by four nylon rods that threaded partway into the middle section. Copper-constantan thermocouples (30 gauge) were installed in a short instrumented section located half way along the square prism's length. The thermocouples were installed from both ends, routed through forty-eight evenly spaced 2.4mm dia. holes (3/32") drilled at a depth of w'/8 from the model surface, as shown in Fig. 6. In order to position the thermocouple tips more accurately, a 1mm dia. (#60 drill) countersunk hole was drilled at the bottom of the 2.4mm dia. hole. In total, the drilling operations removed only about one percent of the material in the instrumented section. The thermocouple tip locations were estimated to have an uncertainty of less than ±0.5mm (i.e., less than ±w'/190).

The model was tested in a low speed wind tunnel with a 61cmx76cm cross section. Hence, the model blockage was about eleven percent of the wind tunnel cross section. The air velocity ranged from 2m/s to 10m/s and was measured by a pitot static tube connected to a Barocel electronic pressure transducer. The Reynolds number, based on w', ranged from 11,300 to 55,200.

Prior to being inserted in the test flow, the model was uniformly heated to about 70°C in an oven. One additional thermocouple (which was not used in the IHC analysis) was installed in the center of the model so that uniform temperature conditions could be confirmed. In all cases, prior to starting the test, the model was isothermal to within 1.4C°, which is about three percent of the overall temperature difference. As shown in Fig. 5, the model was made with a flange at one end so that it could be suspended from the top of the wind tunnel. This allowed rapid positioning of the model in the wind tunnel; the transfer time from the oven to the wind tunnel was less than five seconds. Since the total cooling time for the experiment was more than forty minutes (dimensionless time of t=0.03, corresponds to 42 minutes), the error associated with the model transfer time was negligible.

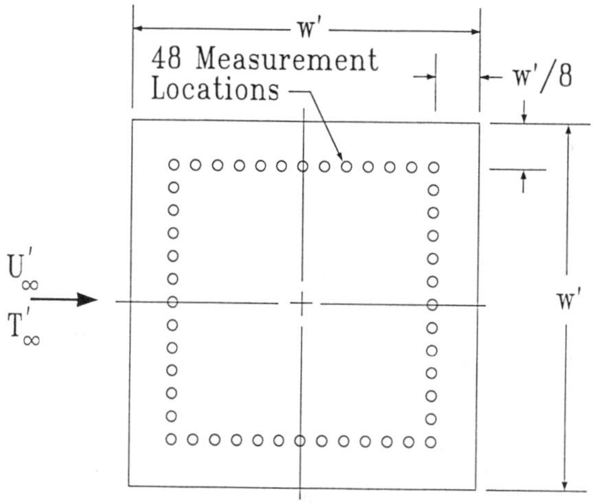

Fig. 6: Cross sectional view of the instrumented section showing the temperature measurement locations.

The temperature decay at each of the measurement points was recorded using a data acquisition system with an accuracy of about ±0.3C°. The data acquisition system had a scanning frequency of about 1.5 thermocouples per second. Plots of the temperature-time data showed that this scanning frequency was more than sufficient to resolve the cooling process.

Of course, when the IHC procedure is applied to actual experiment data, the predicted local heat transfer rate distribution is a combination of both the convection and radiation heat losses. In order to obtain the local convective heat transfer rate, the radiation component must be subtracted from the total heat transfer rate. This was done by defining an average radiation coefficient as follows:

$$h_r = \frac{\epsilon\sigma(T_s'^4 - T_\infty'^4)}{T'_s - T'_\infty} \tag{12}$$

where T'_s is the average model surface temperature obtained by taking the arithmetic average of the numerically predicted model surface temperature at t'=0 and $t'=t'_{max}$. The emissivity was assumed to be $\epsilon = 0.8$. This average radiation coefficient was then subtracted from each of the local heat transfer coefficients predicted by the IHC procedure. The radiation correction was about twenty-five percent of the total heat transfer at the lowest Reynolds number (Re=11,300) and about ten percent of the total heat transfer at the highest Reynolds number (Re=55,200).

The errors associated with the above radiation correction will primarily cause a shift in the results i.e. an error in the average Nusselt numbers. Although the correction is quite crude, the general form of the local convective heat transfer distribution will be affected only slightly; and it is the ability of the present IHC method to predict the distribution that is the primary concern of this study. Nevertheless, pending the success of the present investigation, a future refinement to the procedure should be the inclusion of radiation losses directly in the boundary condition (eq.(5)) of the conduction solution.

EXPERIMENTAL RESULTS

The IHC calculations were performed on the experimental temperature-time data for three values of a maximum dimensionless time, $t_{max}=0.03$, 0.04, 0.05. The results from these calculations confirmed that the Bi distributions are insensitive to the value of t_{max} used. For these three values of t_{max}, the local Biot numbers were the same within ± 1.5 percent and the average Biot numbers were within ± 0.2 percent.

Figure 7 shows a comparison of the local Nusselt number distributions from the present experiment with the experimental results of Igarashi (1985) and Goldstein et al. (1990). In Fig. 7, the local Nusselt number distribution were obtained using an eight coefficient piece-wise function and twenty-five measurement locations. Prior to applying the IHC procedure, the temperature-time data from thermocouples in symmetrical locations across the line of symmetry have been averaged. To give an indication of the

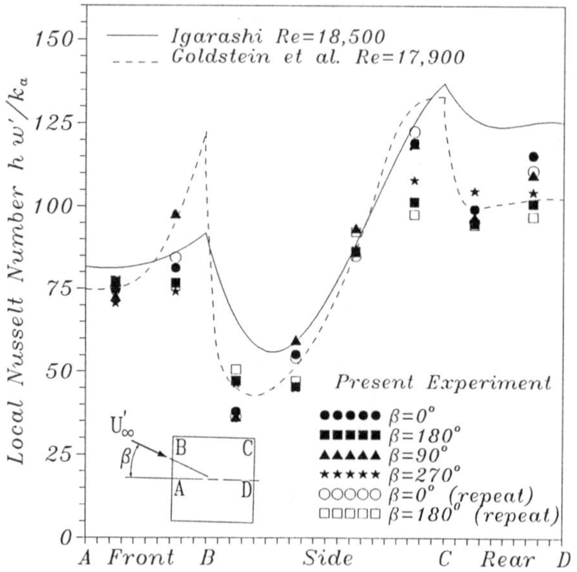

Fig. 7: Comparison of the experimental local Nusselt number distributions obtained using 25 measurement points and an 8 coefficient distribution with the experimental data of Igarashi and Goldstein et al..

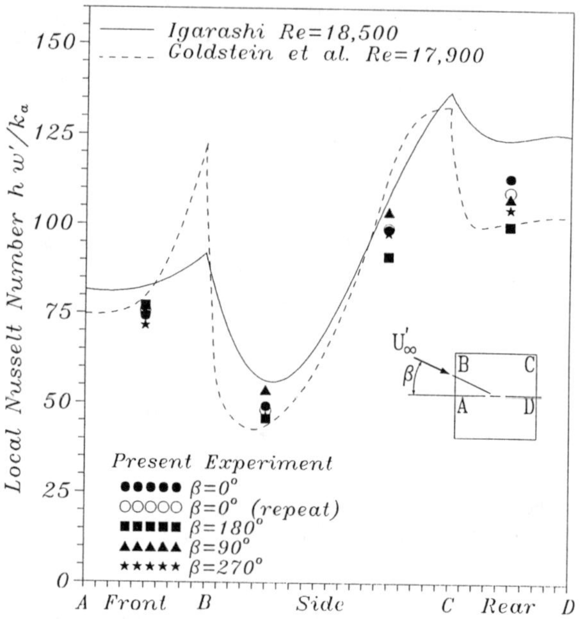

Fig. 8: Comparison of the experimental local Nusselt number distributions obtained using 25 measurement points and a 4 coefficient distribution with the experimental data of Igarashi and Goldstein et al..

accuracy of the method, experimental data are shown for the model rotated through ß=0°, 90°, 280°, 270°. It can be seen in Fig. 7 that the general form of local heat transfer distribution is in fair agreement with the previously published data. However, the scatter is as high as ± 15 percent. As indicated by the results of the numerical evaluation, this scatter is caused primarily by uncertainties in the thermocouple tip locations. In several repeated experiments for a fixed angle (ß), the local Nusselt number data were reproducible to within about ± 3 percent.

In Fig. 8, the results are shown using a four coefficient piece-wise function and twenty-five measurement locations. In comparison to the results shown in Fig. 7, it can be seen that scatter in the local Nusselt number distributions decreases when the surface is divided into fewer (four) subsections. In this case, the scatter is reduced because more thermocouples contribute to each individual local heat transfer coefficient. Hence, the random errors in the temperature measurements (caused primarily by location errors) are averaged to a greater extent. For this same reason, reducing the number of measurement points was also found to increase the scatter of the data.

The variation of the average Nusselt number with Reynolds number is shown in Fig. 9. For comparison, correlations by Igarashi (1985) and Hilpert (1933) are also shown. The data from the present experiment given in Fig. 9 have been corrected for wind tunnel blockage effects. The increase in the average Nusselt number caused by blockage was calculated to be about fifteen percent, using the correction equation of Morgan (1975) for circular cylinders. The error bars shown on the data were estimated using the method of Kline and McClintock (1953). It can be seen that the present method gives reasonable values for the average Nusselt numbers over the full Reynolds number range of the present experiment. Also, the average Nusselt numbers obtained using an eight coefficient distribution are almost identical to those obtained using a four coefficient distribution.

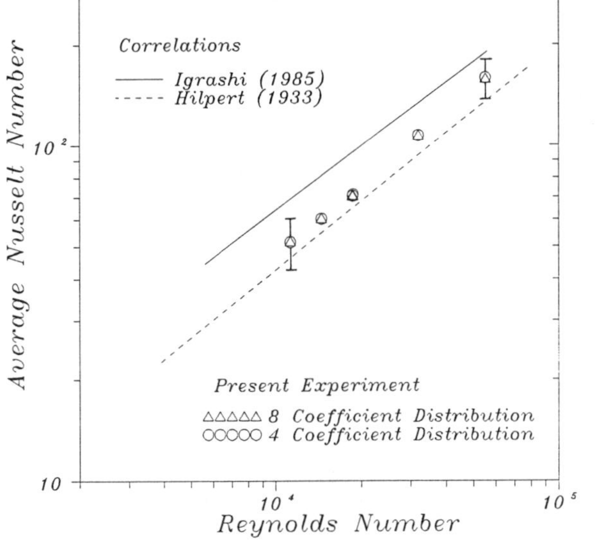

Fig. 9: Comparison of the average Nusselt number data from the present experiment with published correlation equations for forced convection from a square prism in cross flow.

SUMMARY

The results of this study indicate that transient temperature measurements combined with a simple IHC procedure can be used to measure local convective heat transfer rate distributions. However, it was found that in practise this technique is sensitive to small errors in the temperature measurement locations. In the present experimental work, thermocouple tip location uncertainty was the primary source of scatter in the local Nusselt number distributions. Also, it has been shown that uncertainties in the thermal properties of the test model (particularly thermal diffusivity) can lead to substantial bias errors in the results. Nevertheless, it was found that local convective heat transfer rate distributions can be obtained which are sufficiently accurate for many practical purposes.

ACKNOWLEDGEMENTS

This work was supported by the Natural Sciences and Engineering Research Council of Canada.

REFERENCES

Camci, C., Kim, K., Hippensteele, S.A., Poinsate, P.E., 1991, "Convection Heat Transfer at the Curved Bottom Surface of a Square to Rectangular Transition Duct Using a New Hue Capturing Based Liquid Crystal Technique", Fundamental Measurements of Heat Transfer, D.E. Beasley ed., ASME HTD-vol. 179, pp.7-22.

Dorri, B. and Chandra, U., 1991, "Determination of Thermal Contact Resistances using Inverse Heat Conduction Procedure", Numerical Methods in Thermal Problems: Proceedings of the Seventh International Conference, R.W. Lewis ed., vol. VII, part 1, pp.213-223.

Goldstein, R.J., Yoo, S.Y. and Chung, M.K., 1990, "Convective Mass Transfer from a Square Cylinder and its Base Plate", International Journal of Heat and Mass Transfer, vol.33, no. 1, pp.9-18, 1990.

Hilpert, R., 1933, "Wärmeabgabe von geheizten drähten und rohrem im luftstrom", Gebiete Ingenieurw, vol. 4-5, pp. 215-224.

Igarashi, T., 1985, "Heat Transfer from a Square Prism to an Air Stream", International Journal of Heat and Mass Transfer, vol.28, no.1, pp.175-181.

Ireland, P.T., Jones, T.V., 1985, "The Measurement of Local Heat Transfer Coefficients in Blade Cooling Geometries", AGARD Conference Proceedings on Heat Transfer and Cooling, CP390, Paper 28, Bergen.

Kline, S.J., McClintock, F.A., 1953, "Describing Uncertainties in Single-Sample Experiments", Mechanical Engineering, vol. 75, pp. 3-8.

Morgan, V.T., 1975, "The Overall Convective Heat Transfer from Smooth Circular Cylinders", Advances in Heat Transfer, T.F. Irvine and J.P. Hartnett ed., Academic Press, New York, vol. 11, pp. 199-264.

Oosthuizen, P.H., Naylor, D., 1992, "A Numerical Investigation of a Procedure for Determining Local Heat Transfer Rates From Transient Temperature Measurements", The 28th National Heat Transfer Conference, San Diego, California, General Papers in Heat Transfer, ASME HTD-vol. 204, pp. 57-64.

Inverse Problems in Engineering: Theory and Practice
ASME 1993

DEVELOPMENT OF AN ANALYTICAL MODEL FOR THE THERMAL DIFFUSIVITIES OF SPHERICAL OBJECTS SUBJECTED TO COOLING

Ibrahim Dincer
Department of Energy Systems
TUBITAK-Marmara Research Center
Gebze, Kocaeli, Turkey

ABSTRACT

Transient heat transfer taking place during cooling of the spherical products was defined, and a simple equation for determining the variation of the thermal diffusivity of an individual spherical product was obtained. The experiments were performed to carry out temperature distributions at the centers of the individual spherical products. Using the present approach, the thermal diffusivity values were determined easily. The results of the present model were compared with the data obtained by the Riedel's correlation.

NOMENCLATURE

a = thermal diffusivity of the product, m^2/s
a_w = thermal diffusivity of water at the product temperature ($=0.148 \cdot 10^{-6}$ m^2/s)
A = constant in Eq.(6)
Bi = Biot number
D = diameter, m
f = function
Fo = Fourier number
h = heat transfer coefficient, W/m^2K
k = thermal conductivity, W/mK
N = root of transcendental equation
Pr = Prandtl number
r = radial coordinate
R = radius, m
Re = Reynolds number
t = time, s
T = temperature, °C or K
U = average flow velocity, m/s
W = water content, in decimal unit

Greek Symbols

ν = kinematic viscosity, m^2/s
ρ = density, kg/m^3
ϕ = temperature difference, °C or K
θ = dimensionless temperature
Γ = dimensionless radial distance
μ = dynamic viscosity, kg/ms

Subscripts

f = fluid medium
i = initial
n = refers to nth number characteristic value
s = surface
w = water
1 = refers to 1st number characteristic value

INTRODUCTION

Many fruits and vegetables are perishable, and their storage life can be extended by refrigeration.

These fruits and vegetables are stored at temperatures just slightly above freezing to prolong their life.

In order to design an efficient and effective cooling system for the products, engineers and researches must make an exact analysis of heat transfer during food-cooling and must determine the thermal properties in terms of thermal diffusivity, thermal conductivity, specific heat and heat transfer coefficient. The optimum processing conditions will give a minimum cost for certain specified conditions.

Several investigations on the determination of the thermal conductivities of the food products have been undertaken by Gaffney et al.(1980), Ansari et al.(1984, 1985), Ansari and Afaq (1986) and Bhowmik and Hayakawa (1979). Gaffney et al.(1980) reported a set of the thermal diffusivities for several food commodities which were obtained by several researchers. These reported values by different researchers vary considerably for a given product and as a result, a great range of thermal diffusivity values have been found for the same product by different researchers. Ansari et al. (1985) investigated the calculations of bulk thermal diffusivity both in the frozen and in unfrozen states on the basis of transient heat transfer analysis. Ansari and Afaq (1986) developed a model for estimating the thermal diffusivity during cool down of spherical products, such as apples, oranges and potatoes. On the other hand, numerous researchers also developed several techniques for the thermal diffusivities of some products (Ansari et al.,1984; Bhowmik and Hayakawa, 1979).

The recommended method for determination of the thermal diffusivity at the present time is its calculation it from the experimentally measured values of thermal conductivity, specific heat and mass density. In many cases, the specific heat can probably be estimated with a sufficient accuracy from the product composition, so the basic data needed from the experimental measurements are the thermal conductivity and the mass density. However, the present model, using temperature values measured in the product, is an easier method for estimating the thermal diffusivity.

The aim of the present work is to analyze the heat transfer during cooling of the spherical products and to develop a simple model for determining the thermal diffusivities of these products.

ANALYSIS

Consider the spherical product maintained at some initial temperature (T_i). The product is immersed into the cold water flow at temperature (T_f) and the product surface is exposed to the cooling. The temperature distribution within the product is a function of time and radius.

The differential heat transfer equation for the temperature distribution in one dimensional spherical coordinate is

$$(\partial^2 T/\partial r^2) + (2/r)(\partial T/\partial r) = (1/a)(\partial T/\partial t) \quad \text{and}$$

$$(\partial^2 \phi/\partial r^2) + (2/r)(\partial \phi/\partial r) = (1/a)(\partial \phi/\partial t) \qquad (1)$$

where $\phi = (T - T_f)$.

The boundary and initial conditions are

$$\phi(r,0) = \phi_i = (T_i - T_f),$$

$$\phi(0,t) = \text{finite},$$

$$-k[\partial \phi(R,t)/\partial r] = h\phi(R,t).$$

This is a problem which may be solved by the separation of the variables method. The detailed solution may be found in Arpaci (1966).

The expression for the dimensionless temperature distribution becomes

$$\theta = \sum_{n=1}^{\infty} A_n[(\mathrm{Sin}N_n \cdot \Gamma)/(N_n \cdot \Gamma)]\exp(-N_n^2 \cdot Fo) \qquad (2)$$

At the center, $\Gamma = r/R = 0$ and therefore, Eq.(2) is rewritten as follows.

$$\theta = \sum_{n=1}^{\infty} A_n \cdot \exp(-N_n^2 Fo) \qquad (3)$$

By neglecting the Fourier number values smaller than 0.2, we have n=1. Thus, the dimensionless temperature distribution becomes

$$\theta = A_1 \cdot \exp(-N_1^2 Fo) \qquad (4)$$

After some algebraic manipulation, the present model for determining the thermal diffusivities of the spherical products is found as

$$a = [(R^2 \cdot \ln(A_1/\theta))/(N_1^2 \cdot t)] \quad (5)$$

where

$$A_1 = [2\cdot(\mathrm{Sin}N_1 - N_1\cdot\mathrm{Cos}N_1)/(N_1 - \mathrm{Sin}N_1\cdot\mathrm{Cos}N_1)] \quad (6)$$

$$\theta = (T - T_f)/(T_i - T_f) \quad (7)$$

$$N_1 = (1 - Bi)\cdot \tan N_1 \quad \text{or}$$

$$N_1 \approx [(10.3\cdot Bi)/(3.2 + Bi)]^{0.5} \quad (8)$$

$$Bi = h\cdot R/k \quad (9)$$

$$Fo = a\cdot t/R^2 \quad (10)$$

It can be seen in Eq.(5) that the thermal diffusivity of the product is a function of its temperature as given below.

$$a = f(\theta(\Gamma,t)) \quad (11)$$

The heat transfer coefficient is estimated using the correlation of Vliet and Leppert for the spherical bodies (Holman, 1976).

$$h = (k_w/D)\cdot(1.2 + 0.53\cdot Re^{0.54})\cdot(\mu/\mu_s)^{0.25}\cdot Pr^{0.3} \quad (12)$$

$$\text{where } Re = (U\cdot D/\nu) \quad (13)$$

The thermal conductivity of the food products strongly depends on the water content and this value is estimated by means of the Sweat's correlation (Sweat, 1985).

$$k = 0.148 + 0.493\cdot W \quad (14)$$

In addition to the present model, the well-known Riedel's correlation is used to estimate the thermal diffusivity of fruits and vegetables (ASHRAE Handbook of Fundamentals, 1981). This correlation is

$$a = 0.088\cdot 10^{-6} + (a_w - 0.088\cdot 10^{-6})\cdot W \quad (15)$$

EXPERIMENTAL

In order to obtain temperature measurements during hydrocooling experiments, the products were cooled in the immersion type hydrocooling unit, shown schematically in Fig.1. Main components of the hydrocooling unit included a mechanical vapor-compression refrigerating machine, and cold water pool. The cold water pool was used as the test section. Cooling of the chilled water, involved a refrigerating equipment large enough to recool the water very fast, is obtained by circulating the water in the test section. This chilled water provides the product cooling load. Immersion cooling was accomplished by dipping the case containing the products in a vessel which was insulated with a 0.1 m thick glass wool. Apricots, plums and peaches were selected as perfectly spherical products. The batch of 5 kg from each commodity were placed into the polyethylene cases. Each of 12 product samples in a batch had a copper-constantan thermocouple probe embedded at the near the core of apricots, plums and peaches (just outside the seed of the product). The other three probes were positioned at the inlet, middle and outlet portions in the cold water pool to measure the water temperatures. Subsequently, the experimental procedure as described, i.e., measuring the dimensions and weighting the products, placing the products into the case, turning on the experimental system, dipping the case into cold water pool, recording the data, and measuring the temperatures of the products and water, was repeated for each food commodity and ended when center temperatures of the products reached to their storage temperatures. The temperature data were taken and recorded using an ELLAB CMC 821 series microprocessor device coupled to a Seikosha Printer. Thermocouple measurement resolution using this system was approximately 0.1°C. The flow velocity of the water was measured by a digital flowmeter (Hontzsch GmbH, Germany). The water contents of the products were determined using dry matter method in the laboratory.

The experimental conditions and the physical properties of the products are summarized as follows:
$T_f=1\pm0.1$, $U_f=0.05$ m/s for water,
$T_i=22\pm0.5$°C, $T_e=2$°C, W=85%, $D=0.047\pm2\cdot10^{-3}$ m for apricots,
$T_i=22\pm0.5$°C, $T_e=2$°C, W=86%, $D=0.037\pm2\cdot10^{-3}$ m for plums, and

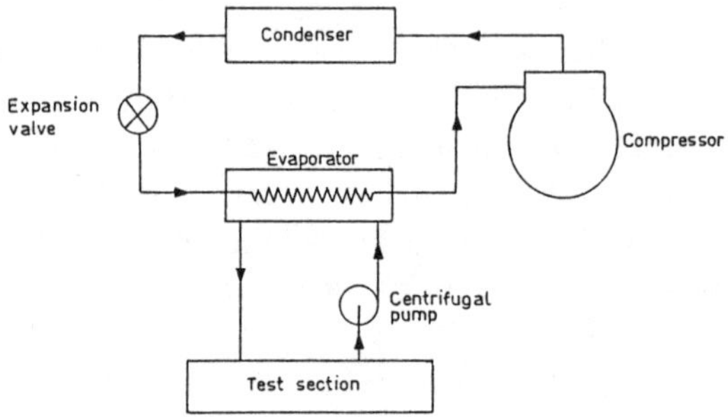

Fig.1 Schematic drawing of experimental apparatus

T_i=22±0.5°C, T_e=4°C, W=89%, D=0.056±3·10^{-3} m for peaches.

The experimental apparatus used in this investigation, along with the details of the facility construction, operation and accuracy, has been reported previously (Dincer, 1991; 1992; Dincer et al., 1992).

RESULTS AND DISCUSSION

The estimated thermal and physical properties of the spherical products (peaches, apricots and plums) are given in Table 1. Also, the thermal diffusivity values obtained by Riedel's correlation are given in this table. The thermophysical properties of the water used in the calculations of the convective heat transfer coefficients of the spherical products were taken from Holman (1976).

Using Eq.(15), the thermal diffusivity can be calculated from experimentally determined water content of the food product. The values of thermal diffusivity given in Table 1 were estimated as a constant value from Eq.(15) (Riedel's correlation), depending on their water contents. This approach requires considerable time and elaborate instrumentation. However, it provides low accurate value, because it only depends on the product water content, except changes in the product temperature. Actually, the thermal diffusivity of the product is effected by both its

Table 1. Thermal and physical properties of the products being cooled in a water pool

	Peaches	Apricots	Plums
Re	1610.86	1351.97	1064.32
h (W/m^2K)	749.00	815.20	915.48
k (W/mK)	0.5868	0.5670	0.5720
Bi	35.70	33.78	29.60
N_1	3.0521	3.0474	3.0339
A_1	1.9925	1.9916	1.9890
a (mm^2/s)	0.1414	0.1390	0.1396

temperature and water content, as well as food's composition and porosity. In the hydrocooling experiments, the water contents of the product remained constant due to non-existence of the moisture evaporation.

Figures 2-4 present the variations of the thermal diffusivity for the spherically shaped food products in batches of 5 kg cooled with water having temperature of 1°C and flow velocity of 0.05 m/s. It is clear from Figs.2-4 that the thermal diffusivity decreases with decreasing the product temperature. For each product, two distinct regimes are observed. These two regimes are referred to as the "initial stage" and "later stage". The initial stage exists for a very short period of time, and at the start there is a large temperature difference between the product and water medium. During this stage, the thermal diffusivity profile decreases rapidly. The later stage is the period of time following this initial stage . This stage involves a large period, and the results of the model developed at this stage approaches the constant profile of Riedel's correlation.

The values of the thermal diffusivity obtained by the present model ranged between 0.17-0.98 mm^2/s for peaches, 0.17-0.73 mm^2/s for apricots and 0.14-0.47 mm^2/s for plums.

The results presented here indicate that the thermal diffusivity remained constant for the temperature difference of 5°C and for the lower differences than 5°C, between the product and cooling medium. Therefore, if the difference between the product and cooling medium is small, the effect of the product temperature on the thermal diffusivity can be neglected, and in this case, Riedel's correlation may be

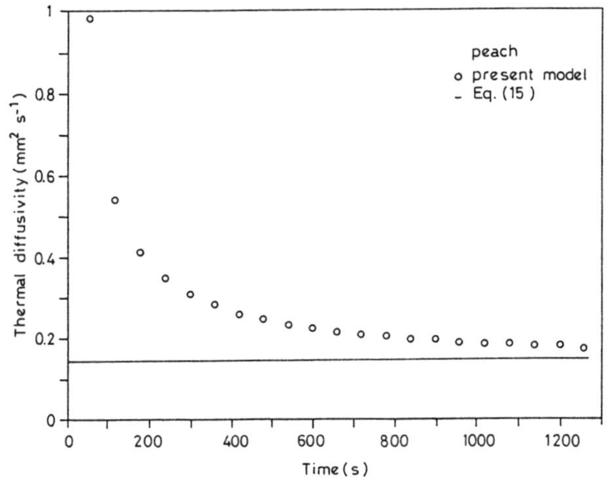

Fig.2 Variation in the thermal diffusivity of an individual peach

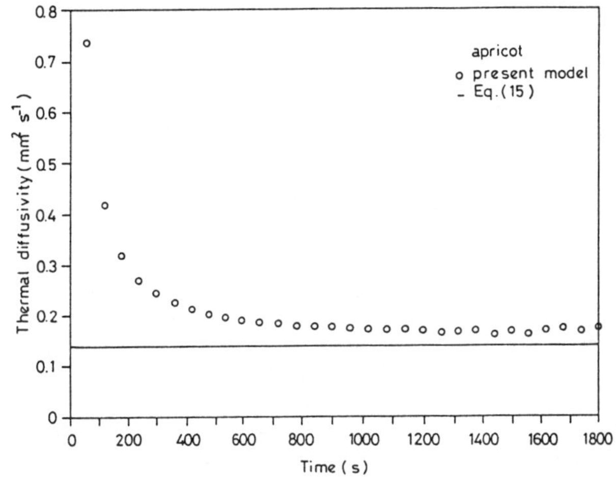

Fig.3 Variation in the thermal diffusivity of an individual apricot

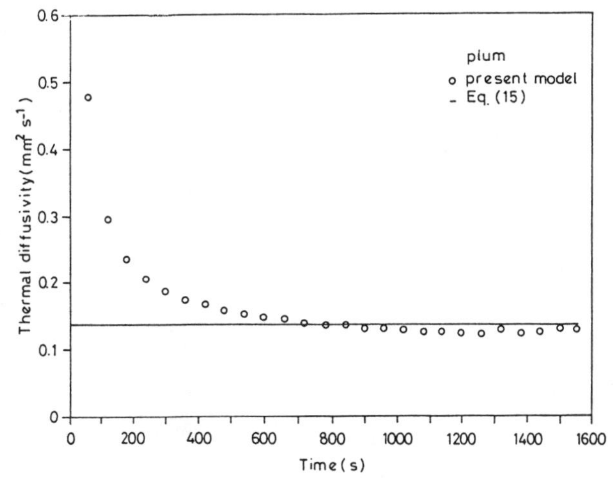

Fig.4 Variation in the thermal diffusivity of an individual plum

used to calculate the thermal diffusivity. When the temperature differences between the product and cooling medium is more than 5° C, the present model which gives the variation of the thermal diffusivity is more accurate and applicable than the well-known Riedel's correlation, and other correlations depending on the water content.

CONCLUSIONS

Temperature measurements of the spherical products, namely peaches, apricots and plums for the immersion cooling experiments were taken, and used to determine the thermal diffusivities of the products by means of the present model. The changes in the thermal diffusivities of the spherical products were studied and plotted against the time. The obtained thermal diffusivity values showed a clearly noticeable variation with the temperature. The model developed was in a good agreement with the Riedel's correlation while the variation of the thermal diffusivity remains constant with time. The results of the present study showed that the thermal diffusivity depending on the product temperature decreased with time under the experimental conditions. The thermal diffusivities determined by present model ranged between 0.17-0.98 mm^2/s for peaches, 0.17-0.73 mm^2/s for apricots and 0.14-0.47 mm^2/s for plums. The results also demonstrated that the model developed is capable of determining the thermal diffusivities of spherical products, and it can easily be applied to spherical products on a large scale.

ACKNOWLEDGMENT

The author wishes to express his gratitude to the Department of Food and Refrigeration Technology, TUBITAK-Marmara Research Centre for the experimental facilities.

REFERENCES

Ansari, F.A. and Afaq, A., 1986, "New Method of Measuring Thermal Diffusivity of Spherical Produce", *International Journal of Refrigeration*, Vol.9, pp.158-160.

Ansari, F.A.; Charan, V. and Varma, H.K., 1984, "Heat and Mass Transfer in Fruits and Vegetables and Measurements of Thermal Diffusivity",*International*

Communications in Heat and Mass Transfer, Vol.11, pp.583-590.

Ansari, F.A.; Charan, V. and Varma, H.K., 1985, "Measurements of Thermophysical Properties and Analysis of Heat Transfer during Freezing of Slab-Shaped Food Commodities", *International Journal of Refrigeration*, Vol.8, pp.85-90.

Arpaci, V.S., 1966, *Conduction Heat Transfer*, Addison-Wesley, Reading, Mass.

ASHRAE Handbook of Fundamentals, 1981, American Society of Heating, Refrigerating and Air-conditioning, Inc., Atlanta, GA.

Bhowmik, S.R. and Hayakawa, K., 1979, "A New Method for Determining the Apparent Thermal Diffusivity of Thermally Conductivity Food", *Journal of Food Science*, Vol.44, pp.469-474.

Dincer, I., 1991, "A Simple Model for Estimation of the Film Coefficients During Cooling of Certain Spherical Foodstuffs with Water", *International Communications in Heat and Mass Transfer*, Vol.18, pp.431-443.

Dincer, I., 1992, "Methodology to Determine Temperature Distributions in Cylindrical Products Exposed to Hydrocooling", *International Communications in Heat and Mass Transfer*, Vol.19, pp.359-371.

Dincer, I.; Yildiz, M.; Loker, M. and Gun, H., 1992, "Process Parameters for Hydrocooling Apricots, Plums and Peaches", *International Journal of Food Science and Technology*, Vol.27, pp.347-352.

Gaffney, J.J.; Baird, C.D. and Eshleman, W.D., 1980, "Review and Analysis of the Transient Method for Determining Thermal Diffusivity of Fruits and Vegetables", *ASHRAE Transactions*, Vol.86, pp.261-280.

Holman, J.P., 1976, *Heat Transfer*, McGraw-Hill, New York.

Sweat, V.E., 1985, "Thermal Conductivity of Food: Present State of the Data", *ASHRAE Transactions*, Vol.91, pp.299-311.

Inverse Problems in Engineering: Theory and Practice
ASME 1993

DISTRIBUTION OF A THERMAL CONTACT RESISTANCE: INVERSION USING EXPERIMENTAL LAPLACE AND FOURIER TRANSFORMATIONS AND AN ASYMPTOTIC EXPANSION

Jean Christophe Batsale, Abdelhakim Bendada,
Denis Maillet, and Alain Degiovanni
Laboratoire d'Energétique et de Mécanique et Appliquée
Institut National Polytechnique de Lorraine
Vandoeuvre-lès-Nancy, France

ABSTRACT

The depth and the thermal contact resistance of a defect within a material can be identified using methods based on time Laplace transformation for cases where heat transfer produced by a thermal stimulation can be considered as one-dimensional (for appropriate geometries).

It is proposed here to extend the preceding methods to cases where heat transfer cannot be considered as one-dimensional anymore. In that purpose, a space Fourier transform is applied to the time Laplace transform of signal produced by an infrared scanning analyser. The lateral dimensions of the defect constitute then additional parameters. It is assumed that the defect thermal resistance can be considered as small with respect to the sane slab resistance. An asymptotic expansion (method of perturbations) allows the construction of an approximate solution that is very convenient for inversion.

Both validity and limits of this method have been studied starting from experimental results. These have been obtained using a calibrated sample of known characteristics.

NOMENCLATURE

a = defect dimension in x direction
a_x, a_y, a_z = diffusivities in the 3 directions
b = defect dimension in y direction
A, B, C, D = quadripole coefficients
e = slab thickness
e_1, e_2 = thickness of each layer on each side of the defect
f, F = functions in Laplace space
I = integral
K = coefficient
L, ℓ = lateral dimensions of the slab
M, M' = quadripole matrices
M_k = Laplace contrast
p = Laplace variable
Q = density of energy absorbed by the slab
s = coefficient
t = time
T = temperature
u = coefficient ($= \sqrt{p+\alpha^2+\beta^2}$)
v, w = coefficients
x, x_1, x_2 = direction x and limits of defect in this direction
y, y_1, y_2 = direction y and limits of defect in this direction
z = direction normal to the slab plane
z_d = defect depth ($= e_1$)
z_s = reduced face coordinate (= 0 or 1)
α, β = eigenvalues in x and y directions
Δ_q = difference $q_2 - q_1$ ($q = x$ or y)
Δ = contrast
ε = reduced contrast resistance ($= R_c$)
φ = heat flux density
ϕ = Laplace and double Fourier transform (LFF) of φ
γ = square root of p
$\lambda_x, \lambda_y, \lambda_z$ = conductivities in the 3 directions
ρc = volumetric heat capacity
Σ_q = sum ($q_1 + q_2$)
τ = L transform of T
θ = LFF transform of T
ψ = L transform of φ

Subscripts

c = relative to the defect center

i = relative to each layer (i = 1 or 2) on each side of the defect

j, k = relative to α_j and β_k

m = relative to defect number m

n = relative to the order of the perturbation method

Superscripts

sup, inf = relative to each side of the defect

* = dimensionless quantity

— = space average

INTRODUCTION

For a few decades composite materials are extensively used in various industrial fields (aerospace, nuclear, microelectronics...). Control of their quality requires the use of non destructive evaluation techniques.This control can be implemented at different times of the material's life: during manufacturing and assembling, on the finished product and later on the ground or in factory for maintenance. Infrared thermography is one of the possible methods that can be developed for this purpose.

A technique of non destructive thermal evaluation of delaminations in stratified composites has already been developed and tested on carbon fiber reinforced polymers -see Maillet et al. (1991,1992): the composite slab is excited by a uniform heat pulse on one of its sides (front) while the rear or front side transient temperature field is recorded by an IR camera. A direct analytical one-dimensionnal (1D) model has been constructed. Estimation of the defect parameters (delamination depth and thermal resistance) is done directly in Laplace space, the signal being the thermal contrast, that is a locally normalized temperature difference between the studied point of the checked side of the slab and a reference area considered as sane.

In order to make the inversion technique more efficient, especially for cases where the in-plane dimensions of the delamination become small compared to the slab thickness, it is interesting to extend it to cases where heat transfer, in the same experiment as before, can no longer be considered as 1D, even on a local basis. A direct model will be developed here that gives the exact analytical solution of heat transfer in a slab containing a two-dimensional delamination (in the defect plane). The Laplace time transform previously used for data compression of the different thermal frames produced by the IR camera, will be followed by a bi-dimensional space Fourier transform.

The air delamination is characterized here by a thermal contact resistance without any capacitive property. In the case where this resistance is small compared with the slab resistance, a perturbation method, based on an asmyptotic expansion of the front or rear side temperature, can provide an approximate solution that is very convenient for further development of an inversion technique. In the present article this new multiple transformation (1 Laplace + 2 Fourier) associated with the perturbation method used at its first order, has been experimentally tested in order to estimate the delamination area.

THREE DIMENSIONAL HEAT TRANSFER MODEL

FORMULATION IN LAPLACE SPACE

The case of a rectangular (L x ℓ) planar slab of thickness e, that contains a resistive defect of finite width a (= x_2 - x_1) and of finite length b (= y_2 - y_1), with a uniform contact resistance R_c on its whole area, is typical of a delamination in a composite stratified material - see figure 1. One assumes that the thermal excitation of the slab if a Dirac heat pulse characterized by a uniform absorbed energy by unit area Q (at time t = 0) on the front face (z = 0), that the slab is insulated from the outside and the material temperature is equal to zero before excitation.

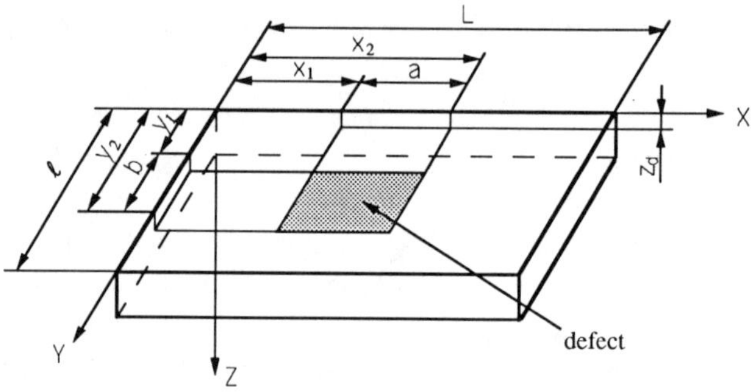

Fig. 1 Geometric presentation of the finite medium containing a defect

The Laplace transform $\tau(x, y, z, p)$ (Laplace variable p) of temperature $T(x, y, z, t)$ in the material is the solution of the following system:

$$\frac{\partial^2\tau}{\partial z^2} + \frac{\lambda_x}{\lambda_z}\frac{\partial^2\tau}{\partial x^2} + \frac{\lambda_y}{\lambda_z}\frac{\partial^2\tau}{\partial y^2} - \frac{p}{a_z}\tau = 0 \quad (1a)$$

$$\text{for} \quad x = 0, L \qquad \frac{\partial\tau}{\partial x} = 0 \quad (1b)$$

$$\text{for} \quad y = 0, \ell \qquad \frac{\partial\tau}{\partial y} = 0 \quad (1c)$$

$$\text{for} \quad z = 0 \qquad -\lambda_z\frac{\partial\tau}{\partial z} = Q \quad (1d)$$

for $z = z_d$ $$\frac{\partial \tau^{sup}}{\partial z} = \frac{\partial \tau^{inf}}{\partial z} \tag{1e}$$

$$\tau^{sup} - \tau^{inf} = R_c\, s(x,y) \left[- \lambda_z \frac{\partial \tau}{\partial z} \right] \tag{1f}$$

where $s(x,y) = 1$ if $(x,y) \in [x_1, x_2] \times [y_1, y_2]$

$s(x,y) = 0$ elsewhere

for $z = e$ $$\frac{\partial \tau}{\partial z} = 0 \tag{1g}$$

The upper indexes sup and inf relate to the upper and lower side of the defect. In order to reduce the number of parameters in the problem, that can be fully anisotropic (with the assumption that the slab faces are parallel to the principal direction of anisotropy of the composite), it is very convenient to write it using dimensionless variables.

DIMENSIONLESS FORMULATION

The different variables of the problem are made dimensionless the following way:

$\tau^* = \tau / (Qe/\lambda_z)$ $\quad a^* = \frac{a}{e} (\lambda_z/\lambda_x)^{1/2}$

$\psi^* = \psi / Q$ $\quad b^* = \frac{b}{e} (\lambda_z/\lambda_y)^{1/2}$

$x^* = \frac{x}{e} (\lambda_z/\lambda_x)^{1/2}$ $\quad p^* = e^2 p/a_z$

$y^* = \frac{y}{e} (\lambda_z/\lambda_y)^{1/2}$ $\quad R_c^* = R_c / (e/\lambda_z)$

$z^* = z/e$ $\quad z_d^* = z_d/e$

ψ being the Laplace transform of the z component of the heat flux density $\varphi(= -\lambda_z \partial T/\partial z)$. In the remaining part of this article the asterisk superscript will be omitted fot simplicity reasons. Partial differential equation (1a) then becomes:

$$\frac{\partial^2 \tau}{\partial^2 z} + \frac{\partial^2 \tau}{\partial^2 x} + \frac{\partial^2 \tau}{\partial^2 y} - p\,\tau = 0 \tag{2}$$

DOUBLE FOURIER TRANSFORM AND 3D QUADRIPOLES

The double Fourier cosinus space transform of function $\tau(x,y,z,p)$ is the following integral transform:

$$\theta(\alpha,\beta,z,p) = \int_0^L \int_0^\ell \tau(x,y,z,p) \cos(\alpha x) \cos(\beta y)\, dx\, dy \tag{3}$$

where dimensionless space pulsations take discrete values (the eigenvalues of the temperature field that are met in the solution by the method of separation of variables):

$$\alpha_j = j\,\pi / L \quad \text{and} \quad \beta_k = k\,\pi/\ell$$

where j and k are non-negative integers.

This form applied to equation (2) gives:

$$\frac{\partial^2 \theta}{\partial z^2} - (p + \alpha^2 + \beta^2) = 0 \tag{4}$$

whose general solution that respect homogeneous equations in x and y of system (1) has the following form:

$$\theta = F \cosh(uz) + G \sinh(uz) \tag{5}$$

with: $u = \sqrt{p + \alpha^2 + \beta^2}$

and F and G being two constant that can be determined by the boundary and interface conditions in z. If θ is known, return to Laplace space is given by:

$$\tau(x,y,z,p) = \frac{1}{L\,\ell} \Big[\theta_{00} + 2 \sum_{j=1}^{\infty} \theta_{j0}\, v_j + 2 \sum_{k=1}^{\infty} \theta_{0k}\, w_k + 4 \sum_{j=1}^{\infty} \sum_{k=1}^{\infty} \theta_{jk} v_j\, w_k \Big] \tag{6}$$

with: $\theta_{jk} = \theta(\alpha_j, \beta_k, z, p)$

$v_j = \cos(\alpha_j x)$ $\quad w_k = \cos(\beta_k y)$

If ϕ is the double Fourier cosinus transform of Laplace heat flux density ψ, wich is equal to unity in $z = 0$ and to zero in $z = 1$ - see equations (1d) and (1g), one has:

for $z = 0$ $$\phi(\alpha,\beta,0,p) = \frac{\sin(\alpha L)}{\alpha} \frac{\sin(\beta\,\ell)}{\beta} \tag{7}$$

for $z = 1$ $$\phi(\alpha,\beta,0,p) = 0 \tag{8}$$

If the arguments different from $z(\alpha,\beta,0,p)$ are omitted in the notation of the Laplace and double Fourier transforms (LFF) θ and ϕ of temperature and heat flux density, equation (5) can lead to a linear relationship between the two quantities on the front ($z = 0$) and rear ($z = 1$) side of the slab:

$$\begin{aligned} \theta(0) &= A\,\theta(1) + B\,\phi(1) \\ \phi(0) &= C\,\theta(1) + D\,\phi(1) \end{aligned} \qquad (9)$$

Let us note that we are confronted here to an extension of the quadripole method - see Carslaw and Jaeger (1959) and Degiovanni (1988) to three dimensional unsteady problems.

Boundary conditions (1d to g) can be put under the following matrix forms:

$$\begin{bmatrix} \theta(0) \\ \frac{\sin(\alpha L)}{\alpha}\ \frac{\sin(\beta \ell)}{\beta} \end{bmatrix} = \begin{bmatrix} A_1 & B_1 \\ C_1 & D_1 \end{bmatrix} \begin{bmatrix} \theta^{sup} \\ \phi(z_d) \end{bmatrix} \qquad (10a)$$

$$\begin{bmatrix} \theta^{sup} \\ \phi(z_d) \end{bmatrix} = \begin{bmatrix} \theta^{inf} + R_c\, I \\ \phi(z_d) \end{bmatrix} \qquad (10b)$$

$$\begin{bmatrix} \theta^{inf} \\ \phi(z_d) \end{bmatrix} = \begin{bmatrix} A_2 & B_2 \\ C_2 & D_2 \end{bmatrix} \begin{bmatrix} \theta(1) \\ 0 \end{bmatrix} \qquad (10c)$$

with:

$$I = \int_{x_1}^{x_2} \int_{y_1}^{y_2} \psi(x,y,z_d,p)\cos(\alpha x)\cos(\beta y)\,dx\,dy \qquad (10d)$$

$$A_i = D_i = \cosh(u\,e_i)$$
$$B_i = \frac{1}{u}\sinh(u\,e_i) \qquad \text{for } i = 1, 2$$
$$C_i = u\sinh(u\,e_i)$$

with: $u = \sqrt{p+\alpha^2+\beta^2}$

$e_1 = z_d \qquad e_2 = 1 - z_d$

Solution of system (10) in terms of the LFF transforms θ of temperature on the two sides of the slab can be found exactly using the double inverse Fourier transform (6) written for the pair (ψ,ϕ) instead of (τ,θ): the intermediate unknowns are LFF fluxes $\phi(z_d)$ written for pulsations α_j and β_k; They are the solution of a linear system based on the series of this inverse transform and on the definition of the I integral of equation (10d) - see Batsale (1992) for exact solution in a two dimensional case.

FORMULATION USING A PERTURBATION METHOD

A more rapid method can be used to solve system (10). It is only valid for small values of the reduced contact resistance R_c. The method of perturbations consists in writing asymptotic series expansion of the variables of the model with respect to the small parameter R_c that will be noted ε from now on:

$$\begin{aligned} \theta(\alpha,\beta,z,p) &= \sum_{n=0}^{\infty} \theta_n(\alpha,\beta,z,p)\,\varepsilon^n \\ \phi(\alpha,\beta,z,p) &= \sum_{n=0}^{\infty} \phi_n(\alpha,\beta,z,p)\,\varepsilon^n \end{aligned} \qquad (11)$$

These series expansions replace θ and ϕ in system (10) and term by term indentification of the coefficients of ε^n lead to a nested series of linear problems that allow calculation of the different θ_n and ϕ_n's.

Identification at Order ε^0

System (11) writes out the following way for order zero:

$$\begin{bmatrix} \theta(0) \\ \frac{\sin(\alpha L)}{\alpha}\ \frac{\sin(\beta \ell)}{\beta} \end{bmatrix} = M_1 M_2 \begin{bmatrix} \theta_0(1) \\ 0 \end{bmatrix} \qquad (12)$$

with:

$$M_i = \begin{bmatrix} A_i & B_i \\ C_i & D_i \end{bmatrix} \qquad \text{for } i = 1, 2$$

This zero order system obviously represents 1D heat transfer inside a sane slab ($\varepsilon = 0$). Return to simple Laplace space can be done by setting $\alpha = \beta = 0$ in the coefficients of the M_i matrices that are now noted M_i' in that case:

$$\begin{bmatrix} \tau_0(0) \\ 1 \end{bmatrix} = M_1' \begin{bmatrix} \tau_0(z_d) \\ \psi_0(z_d) \end{bmatrix} \qquad \begin{bmatrix} \tau_0(z_d) \\ \psi_0(z_d) \end{bmatrix} = M_2' \begin{bmatrix} \tau_0(1) \\ 0 \end{bmatrix} \qquad (13)$$

The zero order Laplace flux at the defect level is therefore:

$$\psi_0(z_d) = \sinh[\sqrt{p}\,(1-z_d)] \,/\, \sinh(\sqrt{p}) \qquad (14)$$

Identification at Order ε^1

For order ε^1, system (10) becomes:

$$\begin{bmatrix} \theta_1(0) \\ 0 \end{bmatrix} = M_1 \begin{bmatrix} \theta_1^{sup} \\ \phi_1(z_d) \end{bmatrix} \qquad (15a)$$

$$\begin{bmatrix} \theta_1^{sup} \\ \phi_1(z_d) \end{bmatrix} = \begin{bmatrix} \theta_1^{inf} + I_0 \\ \phi_1(z_d) \end{bmatrix} \quad (15b)$$

$$\begin{bmatrix} \theta_1^{inf} \\ \phi_1(z_d) \end{bmatrix} = M_2 \begin{bmatrix} \theta_1(1) \\ 0 \end{bmatrix} \quad (15c)$$

I_0 having the same definition as I in equation (10d), just replacing ψ by ψ_0. Substitution of ψ_0 defined by equation (14) into this definition allows the calculation of integral I_0 and therefore the solution of system (15) in terms of first order LFF temperatures on the two sides of the slab:

$$\theta_1(1) = -\frac{4}{\alpha\beta} K \sinh(u\, e_1) \sinh(\sqrt{p}\, e_2) / [\sinh(\sqrt{p}) \sinh(u)] \quad (16a)$$

$$\theta_1(0) = \frac{4}{\alpha\beta} K \sinh(u\, e_2) \sinh(\sqrt{p}\, e_2) / [\sinh(\sqrt{p}) \sinh(u)] \quad (16b)$$

with:

$$K = \sin(\alpha\, \Delta_x/2) \cos(\alpha\, \Sigma_x/2) \sin(\beta\, \Delta_y/2) \cos(\beta\, \Sigma_y/2)$$

and:

$$\Delta_q = q_2 - q_1 \qquad \Sigma_q = q_1 + q_2 \quad \text{for} \quad q = x, y$$

Products of equations (16) by ε represent the LFF transforms of the thermal contrast ΔT on rear and front side. Return to the real (t,x,y) space can be implemented using FFT and Stehfest (1970) algorithms.

Interest of First Order Perturbations

If expansion (11) is stopped at the first order ($n = 1$), one immediately notices that:

- the LFF transform $\varepsilon\theta_1$ of thermal contrast ΔT - see Maillet (1992) - has a very nice simple analytical expression (16) for rear or front side measurement even in this 3D transient heat transfer situation.

- this expression is linear in ε, the reduced thermal resistance of the delamination, which means that, for small ε's, contrasts produced by two delaminations, of resistances ε_1 and ε_2 located at different depths with different lateral extent, simply add up (which is not the case for the temperatures fields):

$$\tau(x,y,z_s,p) = \frac{(1-z_s)\cosh(\sqrt{p}) + z_s}{\sqrt{p}\,\sinh(\sqrt{p})} \left\{ 1+\varepsilon_1 f_1+\varepsilon_2 f_2 \right\} + o(\varepsilon) \quad (17)$$

where $f_m(x,y,z_s,p)$ is the FF original of function $\theta(z_s)$ - equation (16) - multiplied by $\sqrt{p}\,\sinh(\sqrt{p})$ ($z_s = 1$, rear side) or by $\sqrt{p}\,\tanh(\sqrt{p})$ ($z_s = 0$, front side), this function being written with the location parameters of defect number m ($m = 1$ or 2). Let us note that expression (17) can also been written by replacing in its right member the term between braces by $\{ 1- \varepsilon_1 f_1 - \varepsilon_2 f_2 \}^{-1}$. It has the advantage of " sticking " to the exact solution for larger ε_k's while being equivalent to equation (17) for small values of these parameters.

Expression of the Space Averaged Laplace Contrast

The space average of the Laplace contrast is defined as:

$$\overline{\Delta\tau}\,(z_s,p) = \frac{1}{L\,\ell} \int_0^L \int_0^\ell \Delta\tau(x,y,z_s,p)\, dx\, dy \quad (18)$$

for $\quad z_s = 0$ or 1

where $\Delta\tau$ is the difference between the Laplace transform of temperature at one point of the face of the slab and its value in the absence of any defect. With the preceding first order perturbation method, the integral in the second member of equation (18) is equal to $\varepsilon\, \theta_1(0,0,z_s,p)$. Application of equations (16) allow the calculation of $\overline{\Delta\tau}$ for each face:

$$\overline{\Delta\tau}(1, p) = -\frac{\varepsilon\, ab}{L\,\ell} \sinh(\sqrt{p} z_d) \sinh[\sqrt{p}\,(1-z_d)] / \sinh^2(\sqrt{p}) \quad (19a)$$

$$\overline{\Delta\tau}(0, p) = \frac{\varepsilon\, ab}{L\,\ell} \sinh^2[\sqrt{p}\,(1-z_d)] / \sinh^2(\sqrt{p}) \quad (19b)$$

This shows that the average Laplace contrast (and therefore the instantaneous averaged contrast $\overline{\Delta T}$) is proportional to the area ab of the defect.

The locally 1D technique mentioned in the introduction allows the calculation of the value of the reduced thermal resistance of the defect at the level ($x_c = \Sigma_x/2$, $y_c = \Sigma_y/2$) of its center in a rear side experiment for wide enough delaminations ($a, b > 2$ in the case $\lambda_x = \lambda_y$):

$$\varepsilon = \frac{M_1^2[1+\gamma_2 \sinh(\gamma_2)\, M_2] \sinh(\gamma_1) \tanh(\gamma_1)}{[1+M_1\gamma_1 \sinh(\gamma_1)][-M_1+M_2\cosh(\gamma_1)-1.5\, M_1 M_2 \gamma_1 \sinh(\gamma_2)]} \quad (20)$$

with:

$$\gamma_i = \sqrt{p_i} \qquad M_i = \Delta\tau(x_c,y_c,1,p_i) \quad i = 1,2 \qquad \text{and } p_2 = 4\, p_1$$

In the case of a defect of small reduced resistance ε, application of equations (19a) and (20) allow the determination of the defect area ab if its depth z_d is known, using a rear side experiment. The determination of its depth can be done through a front side experiment and use of the 1D technique at the defect center level. Let us note that, for the same reasons as before (better approximation for more resistive defects) equation (19a) can be put under the following form:

$$\overline{\Delta\tau}\,(1,p) = \frac{1}{\sqrt{p}\,\sinh(\sqrt{p})}\left[\frac{1}{1+\varepsilon F(p)} - 1\right] \qquad (21)$$

A similar form can be found for front face.

EXPERIMENTAL VALIDATION

MEASUREMENT TECHNIQUE

A test slab (60 x 60 mm, 2 mm thick) made out of T300 carbon epoxy laminate (14 plies) having the following thermal properties:

$$\lambda_z = 0.67 \text{ W m}^{-1}\text{ K}^{-1} \qquad \lambda_x = \lambda_y = 2.40 \text{ W m}^{-1}\text{ K}^{-1}$$

$$\rho c = 1.62 \times 10^{-6} \text{ J m}^{-3}$$

has been used to validate the preceding inversion technique. It contained a 10 x 10 mm insert made of two 25 µm thick teflon films located at midslab depth (z_d = 0.5). This sample has been manufactured by Dassaut Aviation. Heat pulse excitation was produced by an assembly of four flash tubes located on the sides of a 10 mm vertical square . The temperature field on the rear side of this test slab was recorded by a 782 SW AGEMA IR camera and acquisition and data storage was done on a DATAMIN board located in a 386 PC computer. The software allowed acquisition of frames of 64 x 128 pixels at a rate of 25 frames per second.

RESULTS

An instantaneous frame of the rear side experiment is shown in figure 2 and an average of the frames over a period of 2 seconds allows a better detection of the defect in figure 3. Application of equation (20) - with p_1 = 1 - at the level of the defect center, with a reference point (ε = 0) equidistent to both defect center and slab edges, produced the following value for the reduced defect resistance:

$$\varepsilon = 0.185$$

This value is relatively high for application of the preceding first order perturbation method. It is about three times higher than the nominal value corresponding to teflon and carbon epoxy. But previous experiments have already shown that air layers of a few micrometers trapped between the teflon films or between them and the matrix are sufficient to justify this departure. Application of equation (19a) with the preceding value of ε, and z_d equal to 0.5, can provide a value for the ab area once a value chosen for the reduced Laplace variable p. The optimal value for p was found to be equal to 6.25 (for the closest estimate to the real 10 x 10 mm area). Since our model does not take into account the lateral heat losses, the integration domain has been reduced to reduce the edge effects ($L \times \ell$ = 48 x 42 mm instead of 60 x 60 mm).

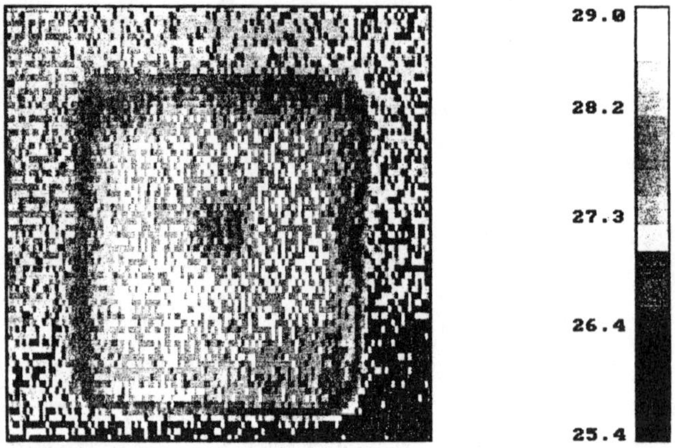

Fig. 2 Instantaneous frame n° 37, time = 1.44 s T(°c)

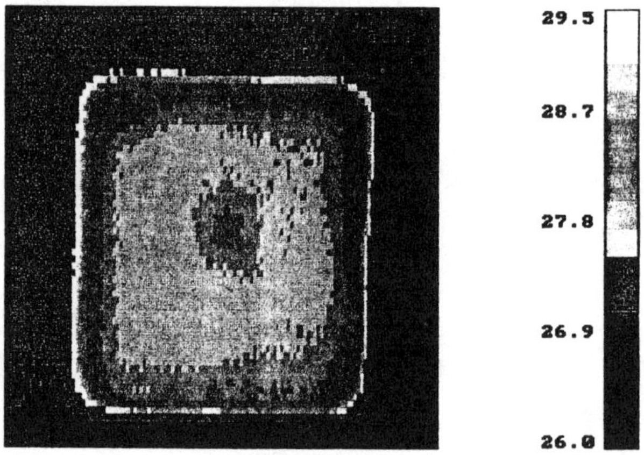

Fig. 3 Time average of frames 34 to 53 T(°c)

The experimental 2D Laplace contrast field (for p = 6.25) is shown in figure 4. A line space profile on this contrast, passing through the level of the defect center, is shown in figure 5; It is affected by quite an important noise. The corresponding average value of the Laplace contrast is:

$$\overline{\Delta\tau} = -18 \times 10^{-4}$$

Assuming $a = b$, the estimated value of a given by equation (19a) is:

$$a = 8.9 \text{ mm}$$

that is a departure of -11 % from the real value.

Application of equation (21) gives

$$a = 10.9 \text{ mm}$$

which means a +9 % departure.

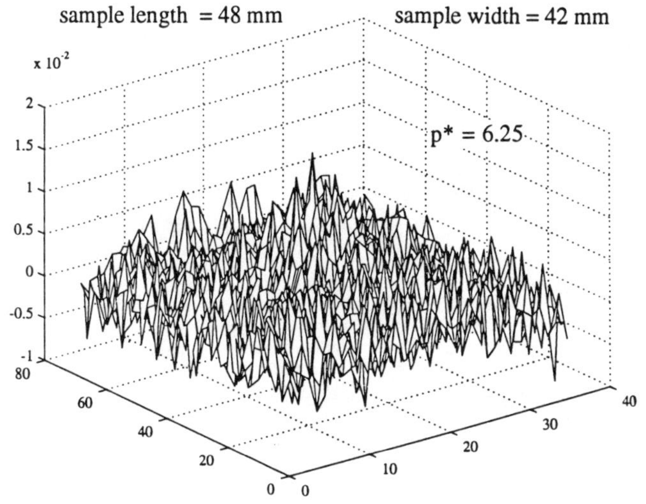

Fig. 4 - Reduced Laplace experimental contrast

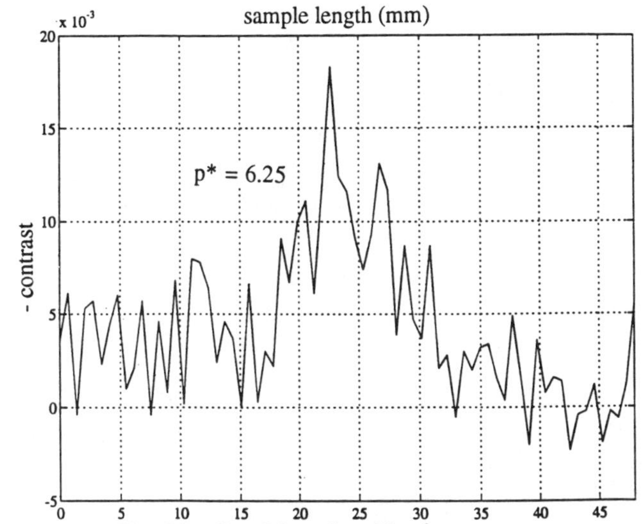

Fig. 5 Profile of the reduced Laplace contrast

A recalculated Laplace contrast field for $p = 6.25$, $z_d = 0.5$, $a = b =$ 10 mm and $\varepsilon = 0.185$ has been obtained using equations (16a) and inverse FFT -equation (6). It is plotted in figure 6 and has an extremum of 14 x 10^{-3} instead of the experimental maximum at 18 x 10^{-3} in figure 5. When compared, these two figures show that the area affected by the Laplace contrast for $p = 6.25$ should be about 20 x 20 mm, which is not the case for the experiment where noise spreads outside this zone. This indicates that lower values of p should give larger estimated area since the noise on ΔT is less reduced. This is in fact the case: a p value of 1 gives an estimated side a of 27.6 mm by the same technique. On the opposite, a too high value for p, that would clear the Laplace contrast of its noise outside the area affected by the defect presence should damper it too much at the location of the defect center, to have a good signal over noise ratio at this level.

The difficulty stems here from the delicate choise of the integration area (and of the related optimum value for p) in order to estimate $\overline{\Delta\tau}$ wich represents the $\alpha = \beta = 0$ component of the FF transform of $\Delta\tau$ (its constant term): if L and ℓ goes to infinity, $\overline{\Delta\tau}$ goes to zero, but it is not the case for the defect area! Choise of an estimation based on non zero eigenvalues α and β in the experimental calculation of the LFF contrast $\Delta\theta$ should therefore be tested in the future.

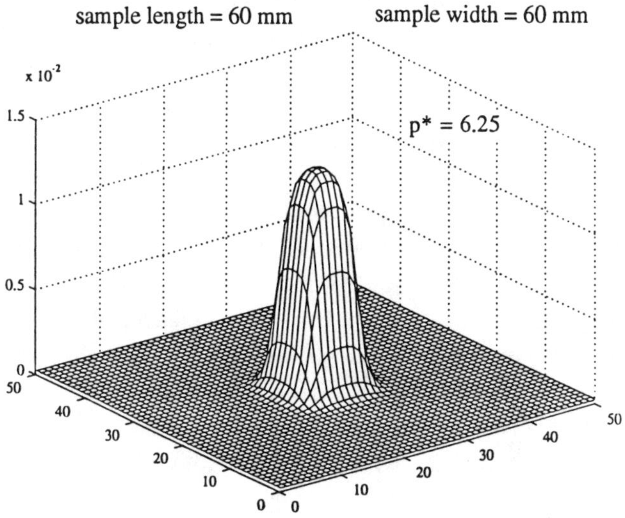

Fig. 6 Rear side reduced Laplace contrast

CONCLUSION

The direct problem of transient three-dimensional diffusion in an anisotropic composite slab, that contains a resistance of limited extent located in a plane parallel to the slab faces, has been solved using Laplace and Fourier cosinus integral transforms and two different methods. The first one consists in solving a linear system and provides an exact solution. The second one, that uses a perturbation method written at its first order, produces a very simple expression for the transform of the thermal contrast produced by the defect presence, under the condition that the small parameter, the reduced contact resistance, has low values.

A parameter estimation method based on this expression, written for the zero space frequencies, has been implemented, starting from a non destructive flash experiment, on a test slab containing a defect, with rear side temperature measurement using an infrared camera. It has provided an estimate of the defect area in this slab. Discussion of the results, that depend on the integration area and on the chosen values for the Laplace variable, has been done.

As a continuation of this work, the property of additivity of the contrasts produced by low resistance defects will be used in the future to develop a more general analytical procedure of inversion of thermographic measurements on a multidelaminated slab, using experimental Laplace and Fourier transforms.

REFERENCES

Batsale, J.C., Maillet, D., Degiovanni, A., 1992, "Extension de la Méthode des Quadripôles à l'Aide des Transformations Integrales - Application au Defaut Plan Bidimensionnel," Article submitted to the *International Journal of Heat and Mass Transfer.*

Carslaw, H.S. and Jaeger, J.C., 1959, "Conduction of Heat in Solids, " Clarendon Press, Oxford, 2nd Edition, p.326.

Degiovanni,A., 1988, "Conduction dans un Mur Multicouche avec Sources: Extension de la Notion de Quadripôles," *International Journal of Heat and Mass Transfer,* vol 31, 3 pp.553-557.

Maillet, D., Houlbert A.S, Didierjean S., Degiovanni, A., 1991, "Identification of a Contact Resistance Located at any Depth Inside a Laminated Composite Material Using Experimental Laplace Transforms and Thermography," *4th Annual Inverse Problems in Engineering Seminar, East Lansing,* June 10-11.

Maillet, D., Didierjean S., Houlbert A.S., Degiovanni, A., 1992, "Non Destructive Transient Thermal Evaluation of Delaminations Inside a Laminate: a Thermal Processing Technique of Thermal Images," *Proceedings of Quantitative Infrared Thermography, Eurotherm Seminar 27,* D.Balageas et al. , ed. , Editions Européennes Thermique et Industrie, Paris, pp.212-217.

Stehfest, H., 1970, "Remarks on Algorithm 368, Numerical Inversion of Laplace Transforms, " *Com. A.C.M.* , p.624.

Inverse Problems in Engineering: Theory and Practice
ASME 1993

A STEPWISE TECHNIQUE FOR INVERSE PROBLEM IN OPTIMAL BOUNDARY CONTROL OF THERMAL SYSTEMS

Boris Rohál'-Il'kiv, Zuzana Országhová, and Tomáš Hrúz
Department of Automation and Measurement
Faculty of Mechanical Engineering
Slovak Technical University
Bratislava, Slovak Republic

Keywords: Distributed parameter systems; spline approximation; predictive control; inverse modelling; regularization.

Abstract

In the paper a procedure for stepwise solution of an inverse heat conduction problem is proposed. Based on the procedure a control system for optimal boundary control of one class of thermal systems is submitted. The controlled thermal system is considered to be decomposed to two subsystems - a subsystem which is easy to control by a feedback using measurable outputs of the system and to a subsystem which distributed state - usually a spatial temperature distribution inside a heated material - is inacessible to direct measurement. The dynamics of the state is supposed to be described by a suitable distributed parameter model with a boundary excitation performed via the measurable system output. The control task then consists of optimal varying the measurable system output, that governs the boundary excitation of the distributed parameter subsystem, until it is calculated that the required shape of the distributed state has been reached. In the paper an optimal reference values for the system output, which should be tracked by a controller, are generated using stepwise technique for inversion of the distributed parameter model. The models of both subsystems are considered in continuous time, nonparametric convolutional integral forms. Using a spline approximation of the convolutional integral describing the first subsystem a predictive controller with receding horizon strategy is designed. The own inverse problem is solved by an iterative regularization method. The resulting stepwise procedure is illustrated on a one-dimensional problem of the boundary heating of a thin metal bar.

1 Introduction

A heating of solid materials is one of typical technological operations in industry. In many of this operations the aim is to remove the solid once a *centre* temperature of the solid has reached a specified value, or once the temperature within the solid has reached a specified spatial distribution. Moreover the heating process should be as quick as possible and optimal from technological, economical and ecological point of view. Modelling of this processes naturally leads to the problems of distributed parameter systems where state variables depend on spatial positions. More often than not only some of these spatially distributed variables are accesible to direct measurement. In this case the point is how to manipulate the unmeasured distributed state variables by utilizing other data measured on the thermal system only and the knowledge of the physical laws governing the process at hand.

The thermal state inside a solid object during the heating operation in a furnace is a typical example. Since it is difficult to monitor this state by routine measurement techniques, the control aim can be realized only by means of the surface temperature control. The thermal system can be easily decomposed to a pair of subsystems - a subsystem with measurable outputs (surface - boundary temperature of the heated object) which is easy to control by a feedback and a subsequent subsystem which is driven by the preceding subsystem and whose distributed state is inaccessible to direct measurement (inside temperature of heated object). Control of this inside temperature is achieved not by a feedback but by maintaining a pre-calculated temperature time profile at the boundary of the object.

The aim of this paper is to submit a predictive control system working with the measurable output of the thermal system and tracking a specified, optimally pre-calculated reference signal for boundary control of the system in order to obtain a required spatial temperature profile in the heated object at a selected time instants t_v. For given spatial temperature profile the reference signal for the system boundary control is obtained by inverting a distributed parameter model, which describes the dynamics of the unmeasurable temperature distribution in the second subsystem of the given thermal system. In the paper the inverse problem is converted to some *regularization* problem and is solved by a *stepwise* technique. This technique seems to be suitable for on-line control of thermal systems under a condition of stochastic disturbances acting on the controlled systems. The dynamics of the first subsystem is modelled by continuous-time convolutional integrals with finite-support kernels. The input and output signals of the subsystem are considered to be a polynomial splines. The B-splines are taken as base functions of these splines. The control synthesis is based on minimization of an integral continuous-time quadratic loss function, which after spline approximation is transformed to simple matrix quadratic form. To minimize the form the quadratic programming is employed. The allowed control input signal is then defined by a set of a suitable selected linear equality and inequality constraints which act on the vector of the polynomial coefficients of this signal.

In the following parts of this paper we will briefly discuss only the main ideas of the proposed boundary control system and for simplicity we will concentrate on simple one-dimensional heating problem: boundary heating of a thin metal bar. The heating apparatus is considered to be the controlled subsystem of the thermal system with the input signal u and the system output signal y - the measured boundary temperature of the metal bar. This temperature is the manipulated input to the second subsystem - heated metal bar - where the *unmeasured* spatial distribution of temperature in the direction of the bar length is modelled by known equation of the heat conduction.

2 Spline-based predictive controller

The controlled subsystem is assumed to be represented by the following time-invariant linear continuous convolution model:

$$\int_0^t a(t-\tau)\, y(\tau)\, d\tau + \\ + \int_0^t b(t-\tau)\, u(\tau)\, d\tau + o = \epsilon(t) \qquad (1)$$

where the finite-support kernels $a(t-\tau)$, $b(t-\tau)$ and the output signal $y(t)$ are considered to be approximated by suitable chosen spline functions, $\epsilon(t)$ is a noise of the model, o is an offset term. The input signal $u(t)$ is a polynomial spline of defined order which is generated by submitted control synthesis. The model (1) was originally developed in [3] mainly to improve discrete-time control with high sampling rate. The finite supports of the kernels a, b determine the finite lengths T_a, T_b of the past history of the signals y and u over which it is reasonable to perform integration in model (1) for any t.

Let's approximate the kernels and the signals in equation (1) by spline function. Then we can obtain the following discrete form of the original model:

$$c_a^T\, Q_a(t)\, c_y + c_b^T\, Q_b(t)\, c_u + o = e(t) \qquad (2)$$

where
c_a, c_b are vectors of model parameters (vectors of spline coefficients of the unknown kernels a, b)
c_u, c_y are vectors of spline coefficients of input and output signals u, y
$Q_a(t)$, $Q_b(t)$ are matrices of integrals of spline base functions products:

$$Q_a(t) = \int_{t-T_a}^t m_a(t-\tau)\, m_y^T(\tau)\, d\tau \qquad (3)$$

$$Q_b(t) = \int_{t-T_b}^t m_b(t-\tau)\, m_u^T(\tau)\, d\tau$$

$e(t)$ is noise term of the model modified by the approximation errors, the vectors m contain the spline base functions used for the approximation tasks. It is easy to show that the matrices $Q_a(t)$, $Q_b(t)$ do not depend on time and can be calculated in advance, before the regulation starts. Using this matrices we can form useful filters for the measured variables which enable us to keep low order models for identification and simultaneously high sampling rate for measurement and control.

To determine the optimal control input signal we will start with minimization of the following integral continuous-time quadratic loss function:

$$J(u) = \frac{1}{T_h} \int_{t_k}^{t_k+T_h} [(y(t) - y_r(t))^2 \ w_y(t) +$$
$$+(u(t) - u_r(t))^2 \ w_u(t)] \ dt, \ u(t) \ \in \ U_{ad}(t) \qquad (4)$$

where:
$y_r(t)$, $u_r(t)$ are reference signals; $w_y(t)$, $w_u(t)$ are weighting functions; T_h is a control horizon and $U_{ad}(t)$ is a set of allowed control input signals.

Let's consider the input and output signal in (4) (together with reference values) in the form of spline function. The projected control input signal will be wanted in its piecewise polynomial representation. Then after substitution to (4) we can find that:

$$J(p_h) = (y_h - y_h^r)^T \ Q_y \ (y_h - y_h^r) +$$
$$+(p_h - p_h^r)^T \ Q_u \ (p_h - p_h^r), p_u \ \in \ C_{ad}(T_h) \qquad (5)$$

where the penalty matrices Q_y, Q_u fully depend on the used spline base functions and can be calculated in advance; the vector p_h contains the coefficients of all polynomial pieces which form the projected control input signal on the time interval $[t_k, t_k + T_h]$ and the vector y_h includes the sampled values of the continuous-time output signal $y(t)$ to be predicted over the above time interval. To minimize the matrix form (5) the quadratic programming technique is well suited. Two classes of constraints are simultaneously used:
- the constraints which are inevitable to formulate the input signal as the spline function of given order
- the constraints which are due to physical limitations on the actuator or process (more frequently amplitude and rate limitations).

The quadratic programming technique can easily cover other types of constraints which are interesting for practice. Remarcable *tuning knob* in the spline based synthesis is a distance between the spline knots of the projected input signal in the time interval $[t_k, t_k + T_h]$:
- the polynomial pieces of the input signal can be projected on the same time interval as they will be really applied
- the polynomial pieces can be projected on time intervals which are a multiple of the interval as they will be really applied; it means that nonrealised (fake) control periods have been inserted into control horizon $[t_k, t_k + T_h]$.

The positive features of the technique consist in significant reduction of computation burden of quadratic programming and in zero control weighting for the control of non-minimum phase systems. For details see [6].

3 Formulation of the inverse problem

Let the behaviour of the unmeasured temperature field of the metal bar $s(x,t)$ at the time instant t and the position x of the bar is described by the parabolic partial differential equation:

$$\frac{\partial}{\partial t} s(x,t) - a^2 \frac{\partial^2}{\partial x^2} s(x,t) + bs(x,t) = 0$$
$$s(x,t_0) = s_0(x), s(0,t) = y(t), \frac{\partial s}{\partial x}(L,t) = 0$$
$$0 \le x \le L, \quad t \ge t_0, \quad a \ne 0$$
$$a^2 = \frac{\lambda}{c.\rho}, \quad b = \frac{h}{c.\rho} \qquad (6)$$

with known Green's function $G(x, \xi, t)$:

$$G(x,\xi,t) = \frac{2}{L} \sum_{n=0}^{\infty} \sin r_n x \sin r_n \xi \exp\left(-bt - a^2 {r_n}^2 t\right)$$
$$r_n = (2n+1)\frac{\pi}{2L}$$

where: L is the length of the bar; λ is thermal conductivity coefficient; c is specific heat; ρ is specific mass of the bar and h is heat-transfer coefficient.
Then the solution of the equation can be given in the following integral form:

$$s(x,t) = \int_{t_0}^{t} \int_{0}^{L} G(x,\xi,t-\tau) \ w(\xi,\tau) \ d\tau \qquad (7)$$

where $w(x,t)$ is a *standardizing* function (see [1]):

$$w(x,t) = s_0(x)\delta(t) - a^2 \delta'(x) y(t) \qquad (8)$$

which includes an exiting function, boundary and initial conditions and $\delta(.)$ is Dirac function. The heating of the bar is controlled through the boundary temperature $y(t) = s(0,t)$ and the task is to find such function $y(t)$ - boundary heating of the bar - which ensures us attainment of the required spatial distribution of the bar temperature $s(x,t)$ at specified time instant t_v. In this situation the relation (7) simplifies to the following form:

$$s(x,t_v) = \int_{t_0}^{t_v} \frac{\partial}{\partial \xi} \ G(x,\xi,t_v - \tau) \mid_{\xi=0} \ y(\tau) \ d\tau +$$
$$+ \int_0^L \ G(x,\xi,t_v - t_0) \ s_0(\xi) \ d\xi \qquad (9)$$

where

$$s_0(x) = s(x,0)$$

is given initial condition. The second part of equation (9) is known in the case of known initial condition. Let's denote it as $s_c(x, t_v)$ and define modified state $s_m(x, t_v)$ as:

$$s_m(x, t_v) = s(x, t_v) - s_c(x, t_v)$$

then

$$s_m(x, t_v) = \tag{10}$$

$$= -a^2 \int_{t_0}^{t_v} \frac{\partial}{\partial \xi} G(x, \xi, t_v - \tau) \mid_{\xi=0} y(\tau) d\tau$$

The last equation can be written in an operator form:

$$s_m = Ay \qquad s_m \in S \quad , y \in Y \subseteq Z \tag{11}$$

where A is the linear integral operator of relation (10), Z and S are Hilbert spaces, Y is closed convex set, build by a priori limitations of this task . The relation (11) represents an integral equation of the first type and solution of this equation fulfils the definition of the *ill-posed problems* in the Hadamard's sense. Therefore it is necessary to use some *regularization* method, which will give satisfactory results. In this paper we employ a method of Tikhonov [7], where the task of solving the equation (11) is replaced by the task of minimization of following *smoothing functional* $M_\alpha[y]$:

$$M_\alpha[y] = \parallel A_h y - s_{m\delta} \parallel^2 + \alpha \parallel y \parallel^2 \tag{12}$$

where $\alpha > 0$ is the regularization parameter.

A_h is an operator which approximates the operator A with defined error h, that means

$$\parallel A_h - A \parallel \leq h, \tag{13}$$

$s_{m\delta}$ is the left side of (11) which is specified by error δ:

$$\parallel s_m - s_{m\delta} \parallel \leq \delta, \tag{14}$$

Let's define so-called *generalized deviation* as :

$$\rho(\alpha) = \parallel A_h\ y_\alpha - s_{m\delta} \parallel^2 - \\ -(\delta + h \parallel y_\alpha \parallel^2) - (\mu\ (s_{m\delta}, A_h))^2 \tag{15}$$

where

$$\mu\ (s_{m\delta}, A_h) = \inf_{y \in Y} \parallel A_h y - s_{m\delta} \parallel$$

is the *degree of inconsistency.*

The regularization parameter α of the smoothing functional is chosen by *generalized principle of deviation*, which is following. If the condition:

$$\parallel s_{m\delta} \parallel^2 > \delta^2 + (\mu(s_{m\delta}, A_h))^2$$

is not fulfilled,the approximate solution of the equation (11) is $y = 0$. If the condition (3) is fulfilled, so the generalized deviation (3) has a positive root α^* and solution of the equation (11) is *minimum* y_{α^*} of the smoothing functional (12).

The solution of the equation (11) can be find by following iterative procedure :

1. Choose of arbitrary (sufficiently large) value of the parameter α
2. Minimization of the functional $M_\alpha[y]$ on the bounded region $Y \subseteq Z$
3. Evaluation of the generalized deviation $\rho(\alpha)$
4. Search for the root α^* of the equation $\rho(\alpha) = 0$ with accuracy ε, it means check the condition:

   $$\mid \rho(\alpha) \mid \leq \varepsilon \tag{16}$$

   where $\varepsilon = C * \delta$ and $C < 1$ is a constant which depends on the desired accuracy of the root α^*
5. If the condition (16) is not fulfilled the procedure is repeated from the point 2. Otherwise the *minimum* y_{α^*} is the solution of the equation (11).

As regards to the accuracy of the proposed procedure the fulfillment of the inequality (14) for found solution $y(t)$ is always guaranteed for suitably chosen operator error h and the inaccuracy δ.

4 Stepwise technique for the inverse problem

In preceding section the inverse task for the distributed model (6) was transformed to the problem of solving the operator equation (11) for a given left side. The solution $y_{\alpha^*} = y_r(t)$ of this equation forms the reference signal for optimal boundary control of the heated bar. The iterative regularization method used for solving equation (11) is valid only for in advance given and constant integral bounds t_0, t_v. This fact is necessary to take into account in designing the generator of the reference signal $y_r(t)$ for on-line boundary control of the thermal system. One way is to base the generator structure on *stepwise triggering* the inversion task in ekvidistantly located discrete time instants t_v. The distance between the time instants determines the integral bounds in the numerical solution of the equation (11) and in next explanation we will call the distance *inversion horizon* T_v. From practical point of view the length of the horizon T_v depends on several factors:

-technological needs for the heating process and the goal of the heating

-dynamical properties of the thermal system

-time behaviour of disturbances acting on the measurable system output.

In the process of the stepwise triggering of the inversion task with the time period T_v it is necessary to know at the particular starting time instant a *true* profile of the unmeasurable temperature distribution $s(x,t)$ in the heated bar. The true profile $s(x,t)$, which is really reached at the end of a preceding period, creates the initial condition for the inversion in a subsequent period. Because the temperature profile $s(x,t)$ is not measured its true time developement can be only *simulated* using a responce of the model (6) due to the really measured system output signal $y(t)$. For numerical calculation of the true responce $s(x,t)$ it is advantageous to utilize again the operator form (11) of the model. The length of the time interval during which the integration in (11) with the real signal $y(t)$ is performed we will call a *simulation horizon* T_m. For numerical reasons it is suitable to choose $T_m = T_v/ni$, where ni is given integer.

The starting point for the numerical solution of the simulation and the inversion tasks consists in suitable discretization of the basic relation (11) and its transformation to a matrix form. The resulting matrix form oriented to the simulation we will call a *simulation* model and the matrix form aimed at the inversion we will call an *inversion* model. Based on the above models the generator of the reference signal $y_r(t)$ is constructed. The required profiles $s_z(x)$ enter the generator with time period T_v and the real measured signal $y(t)$ enters the generator with period T_m.

5 Discretization of the simulation model

Consider the time instants t_j, $j = 1,2,\dots$, in which the simulation tasks have to performed and let $\Delta t_j = t_j - t_{j-1} = T_m$.Then based on equation (10) for the time $\langle t_{j-1}, t_j\rangle$ it holds:

$$s(x,t_j) = s_m(x,t_j) + s_c(x,t_j) \tag{17}$$

$$s_m(x,t_j) = = -a^2 \int_{t_{j-1}}^{t_j} \frac{\partial}{\partial\xi} G(x,\xi,t_j-\tau)\,|_{\xi=0}\; y(\tau)\; d\tau \tag{18}$$

$$s_c(x,t_j) = \int_0^L G(x,\xi,t_j - t_{j-1})\; s(\xi,t_{j-1})\; d\xi \tag{19}$$

In further treatment let us replace the continuous functions $s(x,t), s_m(x,t), s_c(x,t)$ by following vectors containig the values of the functions at spatial points x_i at time instants t_j:

$$\begin{aligned} \mathbf{sm}_j^T &= [s_m(x_1,t_j),\dots,s_m(x_i,t_j),\dots,s_m(x_{nx},t_j)] \\ \mathbf{sc}_j^T &= [s_c(x_1,t_j),\dots,s_c(x_i,t_j),\dots,s_c(x_{nx},t_j)] \\ \mathbf{s}_j^T &= [s(x_1,t_j),\dots,s(x_i,t_j),\dots,s(x_{nx},t_j)] \end{aligned} \tag{20}$$

where

$$x_1 = 0,\; x_i = x_{i-1} + \Delta x,\; i = 2,\dots,nx,$$

$$\Delta x = L/(nx-1)$$

In the case of the simulation model the function $s(x,t)$ represents the *true* state which was reached by the system during the time period T_m and in the case of the inversion task the function $s(x,t)$ represents the desired profile the achievement of which during the time period T_v is the goal of control.

5.1 Determination of vector $\mathbf{sc}_j$

The spatial discretization in the vector (20) is done for points $x = x_i$, i=1,2,...,nx, :

$$s_c(x_i,t_j) = \int_0^L G(x_i,\xi,T_m)\; s(\xi,t_{j-1})\; d\xi \tag{21}$$

By taking into account spline approximations of the profiles $s(x_i,t_{j-1})$:

$$s(x_i,t_{j-1}) = \sum_{k=1}^{nx} cs_k\; M_k(x_i) \tag{22}$$

where: cs_k are spline coefficients and $M_k(x_i)$ are B-splines, then it is possible to arrive to the following discrete form:

$$\begin{aligned} \mathbf{sc}_j &= \mathbf{G}_{m0}\; \mathbf{s}_{j-1} \\ \mathbf{G}_{m0} &= \mathbf{G}_{mj}\; \mathbf{M}_{n\xi}\; \mathbf{M}_{nx}^{-1} \end{aligned} \tag{23}$$

The matrix $\mathbf{G}_{m0}$ with dimensions $[nx.nx]$ can be set in advance before the control starts. For the entries of the above matrices one gets:

$$\begin{aligned} \mathbf{G}_{mj} &= \{G(x_i,\xi_k,T_m)\Delta\xi\}_{i=1,nx;\; k=1,n\xi} \\ \mathbf{M}_{nx} &= \{M_i(x_k)\}_{i=1,nx;\; k=1,nx} \\ \mathbf{M}_{n\xi} &= \{M_i(\xi_k)\}_{i=1,nx;\; k=1,n\xi} \end{aligned}$$

5.2 Determination of vector $\mathbf{sm}_j$

The elements of the vector $\mathbf{sm}_j$ are evaluated using the equation (18) written for the points $x = x_i$, i=1,2,...,nx, :

$$s_m(x_i,t_j) = = -a^2 \int_{t_{j-1}}^{t_j} \frac{\partial}{\partial\xi} G(x_i,\xi,t_j-\tau)\,|_{\xi=0}\; y(\tau)\; d\tau \tag{24}$$

Let us approximate the real measured output signal $y(t)$ (controlled boundary temperature of the bar) by spline function $v_{ym}(t)$:

$$v_{ym}(t) = \sum_{k=1}^{mt} cym_k\; M_k(t) \tag{25}$$

$$v_{ym}(t) = \mathbf{m}_{mt}^T(t)\ \mathbf{c}_{ym} \tag{26}$$
$$\mathbf{c}_{ym}^T = [cym_1, \ldots, cym_{mt}]$$
$$\mathbf{m}_{mt}^T(t) = [M_1(t), \ldots, M_{mt}(t)]$$

where $M_k(t)$ are suitable chosen B-splines. After substitution (26) to (24) and minor derivation the following relation is valid:

$$\begin{aligned} sm_j &= \mathbf{G}_{mb}\ \mathbf{M}_{mt}\ \mathbf{c}_{ym} = \\ &= \mathbf{G}_{md}\ \mathbf{c}_{ym} \\ \mathbf{G}_{md} &= \mathbf{G}_{mb}\ \mathbf{M}_{mt} \end{aligned} \tag{27}$$

For matrix $\mathbf{G}_{mb}$ - $[nx.n\tau]$ it holds:

$$\mathbf{G}_{mb} = \{gmb(i,k)\}_{i=1,nx;\ k=1,n\tau} \tag{28}$$
$$gmb(i,k) =$$
$$= -a^2 \frac{\partial}{\partial \xi} G(x_i, \xi, T_m - (k-1)\Delta\tau)\ |_{\xi=0}\ \ \Delta\tau$$

the matrix $\mathbf{M}_{mt}$ - $[n\tau.mt]$ contains the values of the mention B-splines:

$$\mathbf{M}_{mt} = \{M_k(\tau_i)\}_{i=1,n\tau;\ k=1,mt}$$

The matrix $\mathbf{G}_{md}$ is also possible to evaluate in advance. Using the relations obtained for the vectors $\mathbf{s}c_j$ and $\mathbf{s}m_j$ we can find resulting discrete form for vector $\mathbf{s}_j$, see (17):

$$\begin{aligned} \mathbf{s}_j &= \mathbf{s}c_j + \mathbf{s}m_j = \\ &= \mathbf{G}_{m0}\ \mathbf{s}_{j-1} + \mathbf{G}_{md}\ \mathbf{c}_{ym} \end{aligned} \tag{29}$$

6 Discretization of the inversion model

The philosophy of discretization is similar with the simulation model. Now we are interested to find an optimal boundary condition - reference signal for $y(t)$ in the time interval $T_v = t_v - t_0$ with the aim to bring the state $s(x,t)$ at time instant $t_v = t_0 + T_v$ to the required state $s_z(x)$ as close as possible. Let us approximate the signal $y(t)$ by linear combination of nt B-splines with spline coefficient vector:

$$\mathbf{c}_{yv}^T = [cyv_1, \ldots, cyv_{nt}] \tag{30}$$

and set the required state $s_z(x)$ through a vector of its values at points x_i, i=1,2,...,nx, :

$$\mathbf{s}z = [s_z(x_1), \ldots, s_z(x_{nx})] \tag{31}$$

The dimensions of other vectors $(nx, n\xi, n\tau)$ are the same as in previous subsection. Similar to (29) it is possible to write:

$$\mathbf{s}m_{ni} = \mathbf{s}z - \mathbf{s}c_{ni} \tag{32}$$

where the vector $\mathbf{s}c_{ni}$ is calculated according to (23) for given initial state vector $\mathbf{s}_0$ using of ni steps of the simulation:

$$\begin{aligned} \mathbf{s}c_{ni} &= \mathbf{G}_{v0}\ \mathbf{s}_0 \\ \mathbf{G}_{v0} &= \mathbf{G}_v\ \mathbf{M}_{n\xi}\ \mathbf{M}_{nx}^{-1} \end{aligned} \tag{33}$$

The components of matrix $\mathbf{G}_v$ - $[nx.n\xi]$ are:

$$\begin{aligned} \mathbf{G}_v &= \{gv(i,j)\}_{i=1,nx;\ j=1,n\xi} \\ gv(i,j) &= G(x_i, \xi_j, T_v)\ \Delta\xi \end{aligned} \tag{34}$$

Based on the obtained discrete relations it is possible to approximate the smoothing functional (12) by following resulting form [4]:

$$\begin{aligned} \hat{M}_\alpha[y] &= \mathbf{c}_{yv}^T\ \mathbf{F}\ \mathbf{c}_{yv} - 2\ \mathbf{h}^T\ \mathbf{c}_{yv} + \\ &+ \mathbf{s}m_{ni}^T\ \mathbf{s}m_{ni}\ \Delta x + \alpha\ \mathbf{c}_{yv}^T\ (\mathbf{M}_y + \mathbf{M}_{yd})\ \mathbf{c}_{yv} = \\ &= \mathbf{c}_{yv}^T\ \mathbf{R}\ \mathbf{c}_{yv} - 2\ \mathbf{h}^T\ \mathbf{c}_{yv} + \mathbf{s}m_{ni}^T\ \mathbf{s}m_{ni}\ \Delta x \end{aligned} \tag{35}$$

with

$$\mathbf{R} = \mathbf{F} + \alpha\ (\mathbf{M}_y + \mathbf{M}_{yd}) \tag{36}$$

where

$$\begin{aligned} \mathbf{F} &= \mathbf{G}_{vd}^T\ \mathbf{G}_{vd} \Delta x \quad \mathbf{G}_{vd} = \mathbf{G}_{vb}\ \mathbf{M}_{nt} \\ \mathbf{G}_{vb} &= \{gvb(i,k)\}_{i=1,nx;\ k=1,n\tau} \\ & gvb(i,k) = \\ & = -a^2 \frac{\partial}{\partial \xi} G(x_i, \xi, T_v - (k-1)\Delta\tau)\ |_{\xi=0}\ \Delta\tau \\ \mathbf{M}_{nt} &= \{M_k(\tau_i)\}_{i=1,n\tau;\ k=1,nt} \\ \mathbf{M}_y &= \Delta t \sum_{i=1}^{nt} \mathbf{m}_{\tau_i}\ \mathbf{m}_{\tau_i}^T \\ \mathbf{M}_{yd} &= \Delta t \sum_{i=1}^{nt} \mathbf{md}_{\tau_i}\ \mathbf{md}_{\tau_i}^T \\ \mathbf{m}_{\tau_i} &= [M_1(\tau i), \ldots, M_{nt}(\tau i)] \\ \mathbf{md}_{\tau_i} &= [M'_1(\tau i), \ldots, M'_{nt}(\tau i)] \end{aligned}$$

and

$$\begin{aligned} \mathbf{s}m_{ni} &= \mathbf{G}_{vd}\ \mathbf{c}_{yv} \\ \mathbf{h}^T &= \mathbf{s}m_{ni}^T\ \mathbf{G}_{vd}\ \Delta x \end{aligned} \tag{37}$$

The procedure of minimization of the smoothing functional (35) can be numerically solved by various quadratic programming methods, we utilized the efficient algorithm of Powell [5]. The algorithm enables us to set various technologically inspired constraints on the vector $\mathbf{c}_{yv}$ and to limit the optimal reference signal for $y(t)$.

7 Illustrative example

The complex software for the controller and the inversion tasks was build under PC-KOS Conversational Monitor using SIC (Simulation Identification Control) library (for detail information see [2]) for the model (6) of boundary heated bar.

The heater was simulated by a second order system with time delay. The inversion tasks were solved for
- realizable profiles, it means that the required profiles $s_z(x)$ are exact solutions of equation (6)
- nonrealizable profiles, the required profiles $s_z(x)$ do not belong to the class of possible solutions of (6).

The simulations with the first type of the profiles are illustrated on Fig.1. The inversion task was solved for time instants T_v, $2T_v$ and $3T_v$ (indicated by vertical dash lines) with $T_v = 350s$ and $T_m = 25s$. The required temperature profiles $s_z(x)$ chosen for the above time instants are marked on Fig.1c,d,e by asterisks. The corresponding optimal reference signal $y_r(t)$ obtained through stepwise solving the inversion task is drawn on Fig.1.a by a full line, the real controlled boundary temperature $y(t)$ is marked by the asterisks. The control input signal $u(t)$ to the heater is on Fig.1.b. The white noise disturbances acting on the controlled signal $y(t)$ are considered. The slight deviation between the required and reached profile in Fig.1.e is due to inconsistency of the inversion task for the time instants $2T_v$. The required profile for the time instants is fully reachable only from zero initial state.

The simulations with the nonrealizable profile are introduced on Fig.2. The results show good numerical stability of the proposed inversion algorithm for the case of inconsistent - physically not real tasks.

8 Conclusion

The procedure for regularization of the inversion task (11) is numerically tested on the problem of boundary heated bar. The optimal solution of the inversion was taken as the reference trajectory for designed predictive controller. It was found that the regularization method seek for minimum power consumption solution. The advantage of the described method was also possibility to specify the operator inaccuracy h and the zone δ for the selection of the modified state $s_m(x,t)$.

This paper describes a relatively early stage of the research in this area and gives only the main ideas which must be further elaborated for real-time implementation.

References

[1] BUTKOVSKIY, A.G., Structural theory of distributed parameter systems, *Elis Horwood Limited*, UK, (1983).

[2] KÁRNÝ,M. - HALOUSKOVÁ,A. - BÖHM,J. - KULHAVÝ,R. - NEDOMA,P., Design of Linear Quadratic Control:Theory and Algorithms for Practice, *Kybernetika*, Vol. 21, Academia, Prague, (A supplement), (1989).

[3] KÁRNÝ, M. - NAGY, I. - BÖHM, J. - HALOUSKOVÁ, A., Design of spline-based selftuners, *Kybernetika*, Vol. 26, pp. 17-30, (1990).

[4] ORSZÁGHOVÁ, Z., Control of a class of distributed parameter systems based on model inversion. PhD thesis (in slovak), Slovak Technical University, Bratislava, (1993).

[5] POWELL, M.J.D., ZQPCVT a Fortran subroutine for convex quadratic programming. Report DAMTP/1983/NA17, Department of Applied Mathematics and Theoretical Physics, University of Cambridge, (1983)

[6] ROHÁĽ - IĽKIV, B. - ORSZÁGHOVÁ, Z. - RICHTER, R. - ZELINKA, P., Predictive control with constrained spline input signal, *IFAC Symposium MICC*, Prague, (1992).

[7] TIKHONOV, A.N. - GONCARSKIY, A.V. - STEPANOV, V.V - JAGODA, A.G., Regularization algorithms and apriory information. (In russian). Moskva. Nauka, (1983).

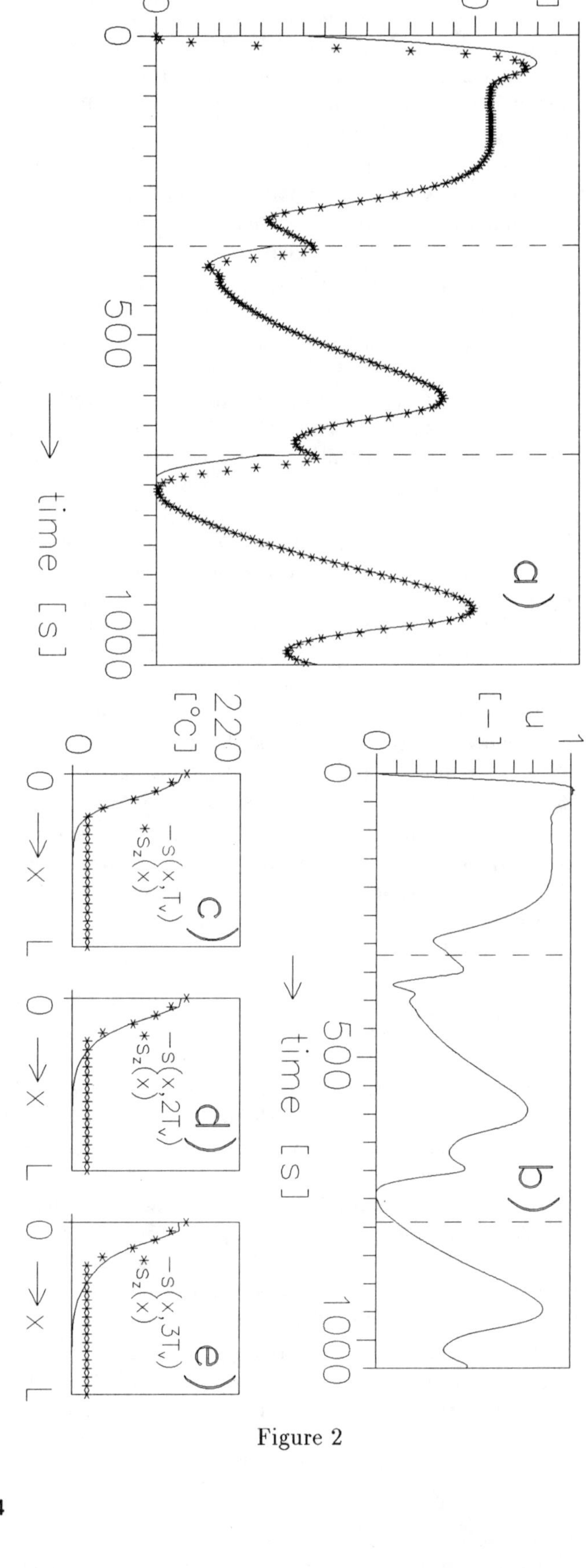

Figure 2

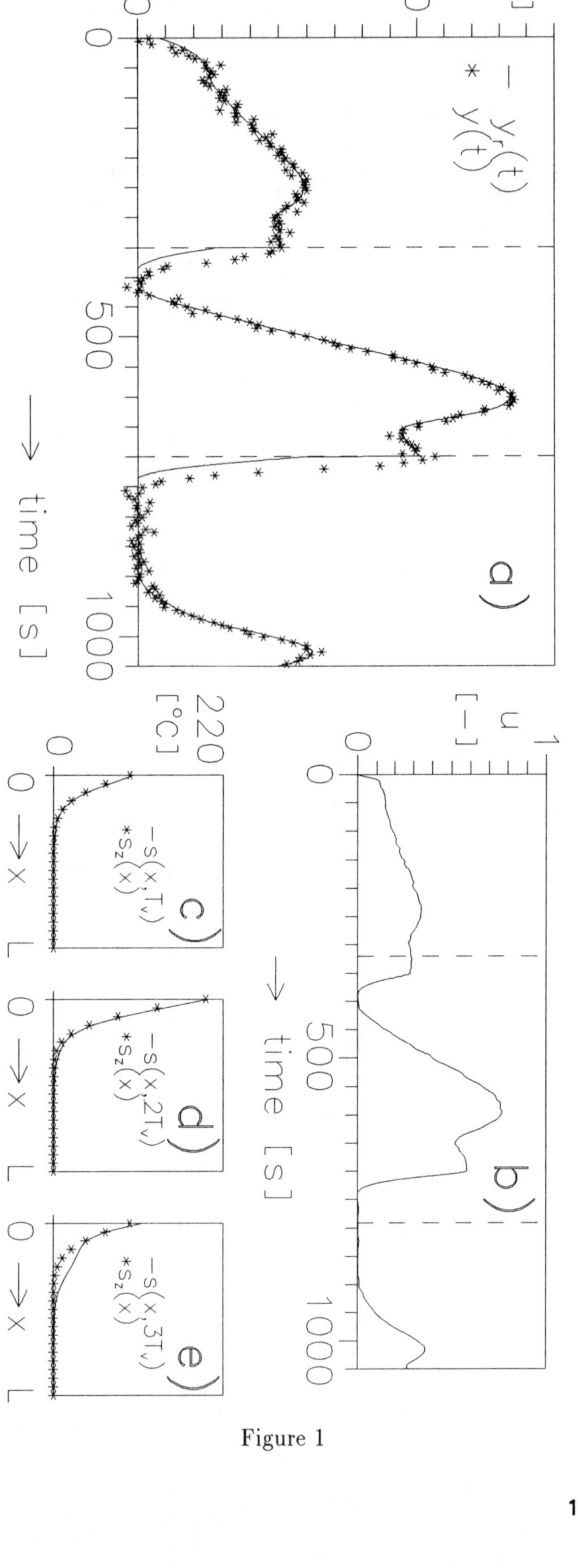

Figure 1

Inverse Problems in Engineering: Theory and Practice
ASME 1993

AN INVERSE PROBLEM FOR A PLATE UNDER PULSE LOADING

Henry R. Busby
Department of Mechanical Engineering
Ohio State University
Columbus, Ohio

David M. Trujillo
TRUCOMP
Fountain Valley, California

ABSTRACT

Most structural dynamics problems are concerned with the determination of the response of the structure due to a known force. The inverse dynamics problem in structural dynamics is to estimate the unknown forcing function using a minimum number of response measurements. A method is presented here to determine an unknown forcing function acting on a simply-supported elastic plate using a minimum number of velocity measurements at discrete points. Different model approximations were used to represent the system matrix and to investigate the effects on their results. Generalized cross validation was used to determine the optimal smoothing parameter.

INTRODUCTION

Most structural dynamic problems are concerned with the determination of the response of the structure due to some known input. These problems are termed 'direct' problems. On the other hand the inverse structural dynamics problem is concerned with the estimation of unknown applied forces based on the measured transient data. This is a difficult problem and falls into a class of problems called ill-conditioned because the solution is extremely sensitive to the noise that is always present in the measurements. One very successful approach to these problems is to combine several rather abstract mathematical concepts, which together produce practical and excellent solutions to the problem.

These are:

1. Least squares minimization with regularization (sometimes referred to as Tikhonov's method).
2. Dynamic programming, which produces a very efficient method for solving the regularized least squares problem.
3. Generalized cross-validation to select the optimal regularization parameters.

Trujillo (1978) and Simonian (1981) used dynamic programming to handle the inverse dynamics problem. Dynamic programming is the most straightforward method for inverse problems and offers a great deal of flexibility in the type of model, the number and location of measurements and the number and location of unknown input forces. The concept of modal analysis was used by Busby and Trujillo (1987) and Hollandsworth and Busby (1989) to reduce the order of the system.

Lim and Pilkey (1992) investigated a large truss structure subjected to an unknown dynamic force. They showed how to select the important modes and how to determine the measurement locations. Their method was restrictive in that it requires the simultaneous measurement of displacement, velocity, and acceleration at each measurement location. Michaels and Pao (1985) investigated the inverse source problem for an infinite plate. They presented an iterative method of deconvolution which was used to determine the orientation and time-dependent amplitude of the force from the transient response of the plate surface at a minimum of two locations given the source location. Lin and Datseris (1988) employed the theory of plates to solve an unknown inverse plate problem with application to position and force sensing. Manoach, Karagiozova, and Hadjikov (1990) solved an inverse problem of an irregularly heated thin circular plate subjected to a pulse loading. They considered two cases: (i) when a restriction on the maximum center deflection is given, and (ii) when the displacement of the plate center as a known function in time is required. D'Cruz, Crisp and Ryall (1991) present a method for determining the location and magnitude of a static point force acting on a simply-supported elastic rectangular plate from a number of displacement readings at discrete points on the plate. D'Cruz, Crisp, and Ryall (1992) also solved a slightly different inverse problem in that the applied force was known to be a harmonic but its location was unknown. In our investigation the location of the force is known but the time history is unknown.

MATHEMATICAL MODEL

Although the mathematical model can be very general and represent any linear dynamic system, a simply-supported square orthotropic plate with first order shear deformation will be used to investigate the performance of the methods. The simply-supported plate will be converted to a vector-matrix differential equation using a finite element formulation (see Reddy (1984)).

The differential equation for a plate with first order shear deformation is given by

$$\begin{aligned} L_1(w,\phi_x,\phi_y)+q &= m_1\,\partial^2 w/\partial t^2 \\ L_2(w,\phi_x,\phi_y) &= m_2\,\partial^2\phi_x/\partial t^2 \\ L_3(w,\phi_x,\phi_y) &= m_2\,\partial^2\phi_y/\partial t^2 \end{aligned} \tag{1}$$

where L_1, L_2, and L_3 are linear differential operators defined by

$$L_1 = A_{55}\frac{\partial}{\partial x}\left(\phi_x+\frac{\partial w}{\partial x}\right)+A_{44}\frac{\partial}{\partial y}\left(\phi_y+\frac{\partial w}{\partial y}\right) \tag{2}$$

$$\begin{aligned} L_2 = {}&\frac{\partial}{\partial x}\left(D_{11}\frac{\partial\phi_x}{\partial x}+D_{12}\frac{\partial\phi_y}{\partial y}\right)+D_{66}\frac{\partial}{\partial y}\left(\frac{\partial\phi_x}{\partial y}+\frac{\partial\phi_y}{\partial x}\right) \\ &-A_{55}\left(\phi_x+\frac{\partial w}{\partial x}\right) \end{aligned} \tag{3}$$

$$\begin{aligned} L_3 = {}&D_{66}\frac{\partial}{\partial x}\left(\frac{\partial\phi_x}{\partial y}+\frac{\partial\phi_y}{\partial x}\right)+\frac{\partial}{\partial y}\left(D_{12}\frac{\partial\phi_x}{\partial x}+D_{22}\frac{\partial\phi_y}{\partial y}\right) \\ &-A_{45}\left(\phi_y+\frac{\partial w}{\partial y}\right) \end{aligned} \tag{4}$$

and $m_1=\rho h$ and $m_2=\rho h^3/12$, ρ being the density. ϕ_x and ϕ_y denote the rotation of the transverse normal about the y and x-axis, and D_{11}, D_{12}, D_{22}, D_{66}, A_{44}, and A_{55} are the orthotropic plate stiffnesses

$$D_{11}=\frac{E_1h^3}{12(1-\nu_{12}\nu_{21})},\; D_{22}=\frac{E_2}{E_1}D_{11} \tag{5}$$

$$D_{12}=\nu_{12}D_{22}=\nu_{21}D_{11} \tag{6}$$

$$A_{44}=G_{23}hk,\; A_{55}=G_{13}hk \tag{7}$$

where k is the shear correction coefficient taken to be 5/6.

Let the two-dimensional region Ω be subdivided into a number of finite elements $\Omega_e(e=1,2,\ldots N)$. Over each element the generalized displacements (w, ϕ_x, ϕ_y) are interpolated according to the form

$$w_e(x,y,t)=\sum_{i=1}^{n} w^e_i(t)\psi^e_i(x,y) \tag{8}$$

$$\phi_{x_e}(x,y,t)=\sum_{i=1}^{n} \phi_{x_i}{}^e(t)\psi^e_i(x,y) \tag{9}$$

$$\phi_{y_e}(x,y,t)=\sum_{i=1}^{n} \phi_{y_i}{}^e(t)\psi^e_i(x,y) \tag{10}$$

where ψ_i is the interpolation function corresponding to the ith node in the element. The interpolation function can be linear, quadratic, or cubic.

Substituting Eq. (8-10) into the Galerkin integrals associated with the operator Eq. (1), which must hold in each element Ω_e, gives

$$\int_{\Omega_e}\left([\mathbf{L}]\{\delta\}-\{\mathbf{f}\}\right)\{\psi_i\}dxdy=0 \tag{11}$$

where $\{\delta\}=\{w, \phi_x, \phi_y\}^T$. Integrating by parts once, yields an equation of the form

$$\overline{\mathbf{M}}\ddot{\mathbf{U}}+\overline{\mathbf{K}}\mathbf{U}=\overline{\mathbf{F}} \tag{12}$$

A practical method to reduce the order of the system is static condensation. This condensation relates the dependent or secondary degrees of freedom (rotational) to the primary degrees of freedom (transverse). The relation between the secondary and primary degrees of freedom is found by establishing the static relation between them. This method is simple to apply, but is only an approximation that may produce large errors if care is not taken. The final reduced dynamic system is given by

$$\mathbf{M}\ddot{\mathbf{u}}+\mathbf{K}\mathbf{u}=\mathbf{F} \tag{13}$$

where **u** is a vector containing all the transverse displacements of the model, **M** is the assembled mass matrix, **K** is the assembled stiffness matrix, and **F** represents the applied forces. Eq. (13) as a first order system is written as

$$\begin{Bmatrix}\dot{\mathbf{x}}_1\\ \dot{\mathbf{x}}_2\end{Bmatrix}=\begin{bmatrix}\mathbf{0} & \mathbf{I}\\ -\mathbf{M}^{-1}\mathbf{K} & \mathbf{0}\end{bmatrix}\begin{Bmatrix}\mathbf{x}_1\\ \mathbf{x}_2\end{Bmatrix}+\begin{Bmatrix}\mathbf{0}\\ \mathbf{M}^{-1}\mathbf{F}\end{Bmatrix} \tag{14}$$

or

$$\dot{\mathbf{x}}=\mathbf{K}^*\mathbf{x}+\mathbf{f} \tag{15}$$

These differential equations are converted to discrete equations using the exponential matrix. The final discrete model is

$$\mathbf{x}_{j+1} = \mathbf{T}\mathbf{x}_j + \mathbf{P}\mathbf{f}_j \tag{16}$$

where $\mathbf{x}$ represents a vector of length 2n containing the displacements and velocities of the nodes. In our example, due to biaxial symmetry, only one quarter of the square simply supported plate was modeled with 4 quadratic elements. Thus, n=16 in our example, but in practical problems n can easily reach 1000. **T** is a matrix which represents the dynamics of the model, **f** is a vector of length n_g representing the unknown applied forces, and **P** is a matrix (2n x n_g) relating the forces to the system. Values for **T** and **P** are given by

$$\mathbf{T} = e^{\mathbf{K}^* h}, \mathbf{P} = \mathbf{K}^{*-1}(\mathbf{T} - \mathbf{I}) \tag{17}$$

Typically n_g is much less than n. In our example, there is only one unknown force so that n_g is equal to one. A timestep h represents the difference between the state variable states $\mathbf{x}_j$ and $\mathbf{x}_{j+1}$.

For the backward implicit method the discrete equation is given by Eq. (16) where

$$\mathbf{T} = \begin{bmatrix} \left(\mathbf{I} - \overline{\mathbf{K}}^{-1}\mathbf{K}h^2\right) & \overline{\mathbf{K}}^{-1}\mathbf{M}h \\ -\overline{\mathbf{K}}^{-1}\mathbf{K}h & \overline{\mathbf{K}}^{-1}\mathbf{M} \end{bmatrix}, \mathbf{P} = \begin{Bmatrix} \overline{\mathbf{K}}^{-1}h^2 \\ \overline{\mathbf{K}}^{-1}h \end{Bmatrix} \tag{18}$$

and $\overline{\mathbf{K}} = \mathbf{M} + \mathbf{K}h^2$. In the same manner the Newmark generalized acceleration method with δ=1/2 and α=1/4 yields

$$\mathbf{T} = \begin{bmatrix} \left(\mathbf{I} - \overline{\mathbf{K}}^{-1}\mathbf{K}\frac{h^2}{2}\right) & \left\{\mathbf{I} + \overline{\mathbf{K}}^{-1}\left(\mathbf{M} - \mathbf{K}\frac{h^2}{4}\right)\right\}\frac{h}{2} \\ -\overline{\mathbf{K}}^{-1}\mathbf{K}h & \overline{\mathbf{K}}^{-1}\left(\mathbf{M} - \mathbf{K}\frac{h^2}{4}\right) \end{bmatrix} \tag{19}$$

$$\mathbf{P} = \begin{Bmatrix} \overline{\mathbf{K}}^{-1}\frac{h^2}{2} \\ \overline{\mathbf{K}}^{-1}h \end{Bmatrix}, \overline{\mathbf{K}} = \mathbf{M} + \mathbf{K}\frac{h^2}{4} \tag{20}$$

PROBLEM STATEMENT

Now, suppose that a series of measurements have been taken and are replaced by the vector $\mathbf{d}_j$, where the length of dj is m. The number of measurements m is usually much less than the number of variables n but greater than n_g. These measurements are related to the state variables $\mathbf{x}_j$ by

$$\mathbf{d}_j \sim \mathbf{Q}\mathbf{x}_j \tag{21}$$

where Q is an (m x n) matrix which associates the measurements to the state variables. The measurements could represent displacements, velocities, accelerations. In our example, the velocity will be used.

The problem is to find the unknown forces $\mathbf{f}_j$ that when they are used in Eq. (16) will force the model to best match the measurements represented by Eq. (21). It is obvious that an exact match with the measured data will not work. This is due to the fact that all measurements have some degree of noise while for the models, smooth derivatives of all orders have been assumed. One of the most common methods of adjoining the data to the model is with the use of least squares. In vector form, this is represented with a vector inner product (**u**,**v**) and would be represented by an error sum over all the data points N.

$$E = \sum_{j=1}^{N}\left(\mathbf{d}_j - \mathbf{Q}\mathbf{x}_j, \mathbf{d}_j - \mathbf{Q}\mathbf{x}_j\right) \tag{22}$$

Even this least squares criteria is not sufficient because a mathematical solution that will minimize E will usually end up with the model exactly matching the data. This situation should be avoided. This is where the regularization method is used. By adding a term to the above least squares error

$$E = \sum_{j=1}^{N}\left(\mathbf{d}_j - \mathbf{Q}\mathbf{x}_j, \mathbf{d}_j - \mathbf{Q}\mathbf{x}_j\right) + b\left(\mathbf{f}_j, \mathbf{f}_j\right) \tag{23}$$

one can control the amount of smoothness that occurs in the solution by varying the parameter b. This method is sometimes referred to as Tikhonov's method. What is now required of the solution is to best match the data (the first term of Eq.(23)), but to have some degree of smoothness (the second term of Eq.(23)). This immediately bring up the question of what should be the value of the smoothing parameter b. In this paper we will use the method of Generalized Cross Validation to determine the optimal value of b.

One of the techniques used to find the unknown functions $\mathbf{f}_j$ is dynamic programming which was developed by Richard Bellman (see Bellman (1957) and Bellman and Dreyfus(1962)). Dynamic programming is a mathematical technique used to solve certain optimization problems of which the above is just one kind. It is closely related to the optimal control problem, with $\mathbf{f}_j$ being the control forces. For this class of problem it is the most natural technique to use, because it possesses computational stability and efficiency.

Thus, for a fixed value of the smoothing parameter b, the optimal forces $\mathbf{f}_j$ are determined. Then, on another level of optimization, the optimal value of b is determined.

GENERALIZED CROSS VALIDATION

The method of generalized cross-validation (see Golub, Heath, and Waba (1979), and Craven and Waba (1979)) is used to determine the optimal value of b. The method is as follows: Let the true signal before it was corrupted with noise be represented by s_j. A measurement d_j is the equal to $s_j + \varepsilon_j$, where ε_j is a random noise component. Solving the above model for various values of the parameter b generates estimates y_j. The least squares error between the true signal s_j and the estimate y_j is given by

$$S = \sum_{j=1}^{N} \left(y_j - s_j\right)^2 \tag{24}$$

The optimal value of b is the one that minimizes S. However, the problem is that s_j is not known. If s_j were known then it would be a simple matter to determine the optimal parameter b. Therefore, the concept of generalized cross validation allows one to construct another function V, without knowing s_j, such that minimizing V gives the same b as if one could minimize S (see Trujillo and Busby (1989)). The method works on the idea of deleting one of the data points and then using the rest of the data to predict the deleted point. This gives a measure of a predicted value against an independent measurement. This method can be repeated for all the data points and, in this way, an entire predicted error analysis can be developed.. The value of b that minimizes the predicted error is the optimal one. Hence, the technique uses the data itself to find the optimal smooth curve. Dohrmann, Busby, and Trujillo (1988) have shown how the generalized cross-validation function is computed as an extension of the dynamic programming formulas.

NUMERICAL EXAMPLE

In order to demonstrate the proposed procedure the simply-supported square orthotropic plate, with first order shear deformation, shown in Fig. 1 is employed. The plate contains 4, nine noded elements and has 75 DOF (three DOF per node).

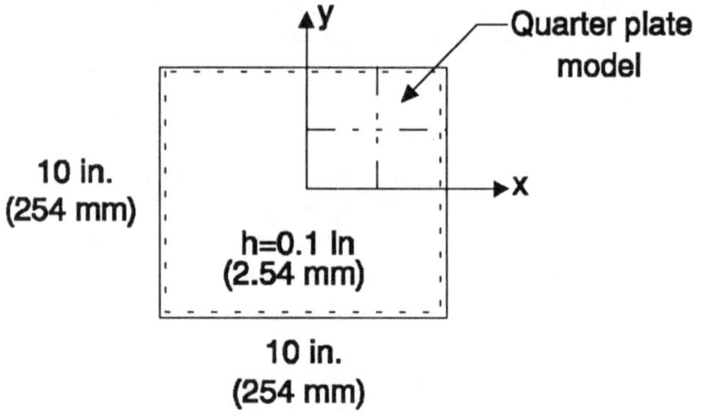

FIG. 1 SQUARE PLATE MODEL

However, the rotational degrees of freedom have been condensed out to reduce the order of the system. Thus, the final finite element model has 16 transverse displacements as shown in Fig. 2.

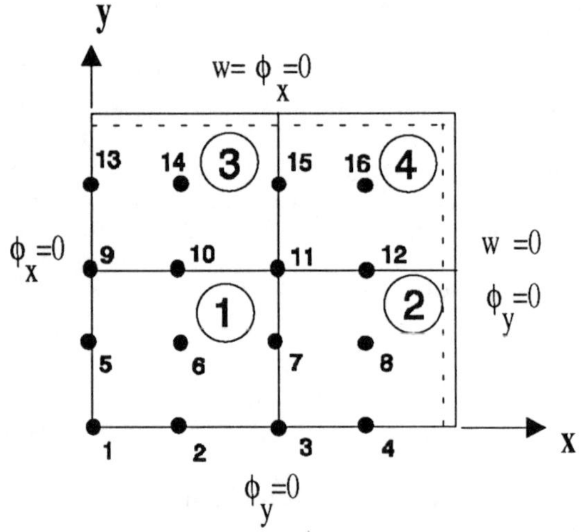

Fig. 2 FINITE ELEMENT MODEL

The orthotropic plate was assumed to be isotropic and the following properties were used:

$$E = E_1 = E_2 = 10.4 \times 10^6 \text{ psi. (72 MPa)}$$

$$G = G_{12} = G_{13} = G_{23} = 3.94 \times 10^6 \text{ psi. (27 MPa)}$$

$$\rho = 0.1 \text{ lb/in}^3 \text{ (2800 kg/m}^3\text{)}, \nu = 0.32$$

The load was applied at the center of the plate, node 1. The load applied was a double rectangular pulse given by the following:

$$F = 50.0 \text{ lb. (222.4 N)}; \quad 0.005 < t < 0.002; \quad 0.0035 < t < 0.005$$

$$F = 0.0; \quad 0 < t < 0.005; \quad 0.002 < t < 0.0035; \quad t > 0.005$$

A direct solution of the model was use to solve for the displacement and velocity time histories. A timestep of 5×10^{-5} seconds was used in the analysis. Using velocity as the measurement, a noise level using a standard deviation of 5 was added to the simulated measurement. This represents approximately one percent of the peak value.

The zeroth order regularization problem was solved using the exponential technique. Only the velocity measurement at node 11 (this corresponds to state variable 27) was used in determining the unknown forcing function. The results using the exponential matrix technique are shown in Fig. 3, where the b value was determined using GCV to be 5.8×10^{-2}. Although the results are noisy, they still reflect the true nature of the unknown impact force.

In the general formulation of the inverse problem, it is possible to replace the regularization term with a derivative of unknown forces in place of the forces themselves. This is

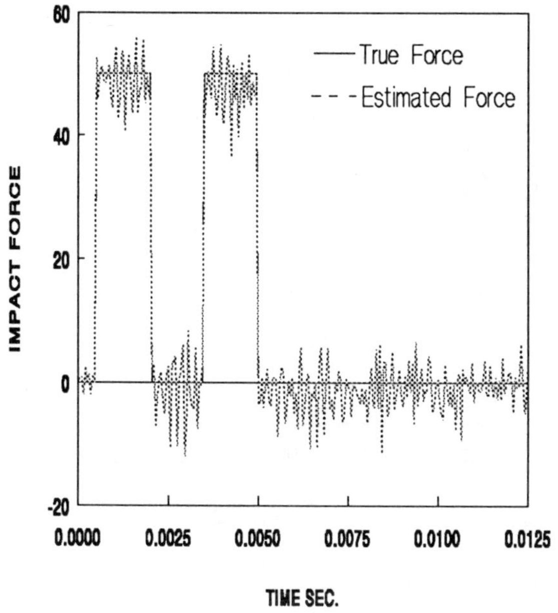

FIG. 3 IMPACT FORCE, ZERO ORDER REGULARIZATION

easily accomplished by adjoining the forces to the state variable and solving for the derivatives of the forces, r_j. Thus let the forces be represented by

$$\mathbf{f}_{j+1} = \mathbf{f}_j + \mathbf{r}_j \tag{25}$$

The new system is now

$$\begin{Bmatrix} \mathbf{x}_{j+1} \\ \mathbf{f}_{j+1} \end{Bmatrix} = \begin{bmatrix} \mathbf{T} & \mathbf{P} \\ \mathbf{0} & \mathbf{I} \end{bmatrix} \begin{Bmatrix} \mathbf{x}_j \\ \mathbf{f}_j \end{Bmatrix} + \begin{Bmatrix} \mathbf{0} \\ \mathbf{r}_j \end{Bmatrix} \tag{26}$$

and the error equation becomes

$$E = \sum_{j=1}^{N} \left(\mathbf{d}_j - \mathbf{Q}\mathbf{x}_j, \mathbf{d}_j - \mathbf{Q}\mathbf{x}_j\right) + b\left(\mathbf{r}_j, \mathbf{r}_j\right) \tag{27}$$

This is known as the first order regularization. The results using first-order regularization are shown in Fig. 4. It is noted that the resulting impact force is much smoother and gives a very good representation of the true impact force. The optimal b value for this case was found to be 1.35.

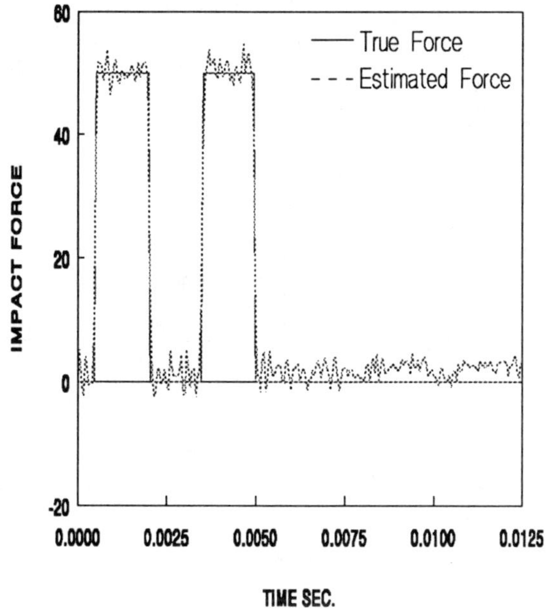

FIG. 4 IMPACT FORCE WITH FIRST ORDER REGULARIZATION

Newmark Method

The inverse problem for the plate under pulse loading was also investigated using the Newmark method described above. The results are shown in Fig. 5. using first order regularization. It is noted that this method tends to smooth out some of the high frequency content of the forcing function. This is due to numerical damping. The regularization parameter b determined for this case by GCV was found to be 18.5.

Backward Implicit Method

In addition to the exponential method and the Newmark method the backward implicit method was also used to evaluate the inverse pulse problem. The results for this case are shown in Fig. 6. The backward implicit method tends to smooth out the high frequency content even more than that obtained from the Newmark method. The results, were obtained using a first order regularization. The value of b for this case was found to be 250.

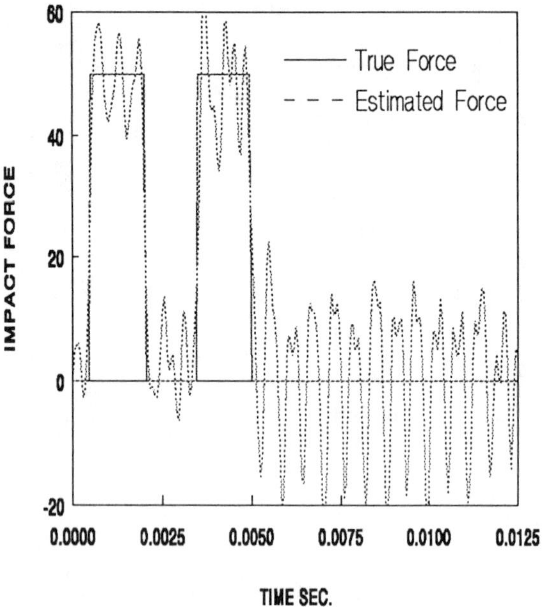

FIG. 5 IMPACT FORCE USING NEWMARK METHOD

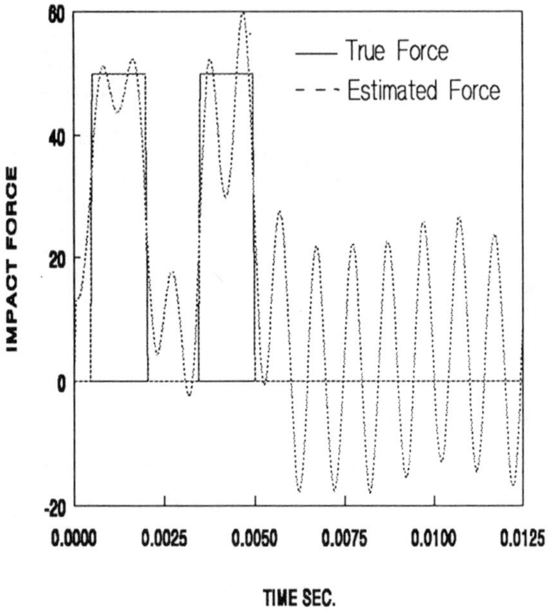

FIG. 6 IMPACT FORCE USING BACKWARD IMPLICIT METHOD

CONCLUSIONS

The results of the above example give one confidence that the exponential matrix along with generalized cross-validation and dynamic programming is an excellent method for solving the inverse structural dynamics problem. The method will work well for problems that have a few hundred degrees of freedom. The Newmark and backward implicit methods can also be used, but are not as accurate as the exponential method. In addition these methods have numerical damping which also leads to inaccuracy of results.

REFERENCES

Bellman, R., 1957, *Dynamic Programming*, Princeton University Press, Princeton, New Jersey.

Bellman, R., and Dreyfus, S., 1962, *Applied Dynamic Programming*, Princeton University Press, Princeton, New Jersey.

Busby, H. R., and Trujillo, D. M., 1987, "Solution of an Inverse Dynamics Problem Using an Eigenvalue Reduction Technique," *Computer and Structures*, Vol. 25, pp. 109-117.

Craven, P., and Wahba, G., 1979, "Smoothing Noisy Data With Spline Functions," *Numerische Mathematik*, Vol. 31, pp. 377-403.

Dohrmann, C. R., and Busby, H. R., and Trujillo, D. M., 1988, "Smoothing Noisy Data Using Dynamic Programming and Generalized Cross-Validation," ASME *Journal of Biomechanical Engineering*, Vol. 110, pp. 37-41.

Golub, G. H., Heath, M., and Wahba, G., 1979, "Generalized Cross-Validation as a Method for Choosing a Good Ridge Parameter," *Technometrics*, Vol. 21, pp. 215-223.

Hollandsworth, P. E., and Busby, H. R., 1989, "Impact Force Identification Using the General Inverse Technique," *International Journal Impact Engineering*, Vol. 8, pp. 315-322.

Lim, T. W., and Pilkey, W. D., 1992, "A Solution to the Inverse Dynamics Problem for Lightly Damped Flexible Structures Using a Modal Approach," *Computer and Structures*, Vol. 43, pp. 53-59.

Lin, P. P., and Dateris, P., 1988, "Inverse Problem Via Plate Theory With Application to Position and Force Sensing," ASME *Journal of Applied Mechanics*, Vol. 55, pp. 489-491.

Manoach, E., Karagiozova, D., and Hadjikov, L., 1991, "An Inverse Problem for an Initially Heated Circular Plate Under a Pulse Loading," *ZAMM,* Vol. 71, pp. 413-416.

Michaels, J. E., and Pao, Yih-Hsing, 1985, "The Inverse Source Problem for an Oblique Force on an Elastic Plate," *Journal Acoustical Society of America*, Vol. 77, pp. 2005-2011.

Reddy, J. N., 1984, *An Introduction to the Finite Element Method*," McGraw-Hill Book Company, New York.

Simonian, S. S., 1981, "Inverse Problems in Structural Dynamics - I. Theory," *International Journal Numerical Methods in Engineering*, Vol. 17, pp. 357-365.

Simonian, S. S., 1981, "Inverse Problems in Structural Dynamics - II. Applications," *International Journal Numerical Methods in Engineering*, Vol. 17, pp. 367-386.

Trujillo, D. M., 1978, "Application of Dynamic Programming to the General Inverse Problem, "*International Journal Numerical Methods in Engineering*, Vol. 12, pp. 613-624.

Trujillo, D. M., and Busby, H. R., 1989, "Optimal Regularization of the Inverse Heat-Conduction Problem," *Journal of Thermophysics and Heat Transfer*, Vol. 3, pp. 423-427.

Inverse Problems in Engineering: Theory and Practice
ASME 1993

AN APPLICATION OF OPTIMIZATION TECHNIQUES TO MEASUREMENT OF TWO- AND THREE-DIMENSIONAL CRACKS BY THE ELECTRIC POTENTIAL CT METHOD

Shiro Kubo and Kiyotsugu Ohji
Department of Mechanical Engineering
Faculty of Engineering
Osaka University
Osaka, Japan

Kenji Nakatsuka
West Japan Railway Company
Osaka, Japan

Hiroshi Fujito
Toyota Motor Corporation
Aichi, Japan

ABSTRACT

The present authors proposed the electric potential CT (computed tomography) method for identifying location, shape and size of cracks embedded in an electric conductive body from electric potential distributions observed on the surfaces of the body. As inverse analysis schemes for the crack identification, the inverse boundary integral equation method and the least residual method were proposed. In this paper the optimization techniques were introduced in the crack identification based on the least residual method by taking the residual as the objective function of optimization.

The least residual method incorporating optimization techniques was applied to the identification of a two-dimensional inclined crack in a plate specimen from potential readings on the side faces of the plate. To ensure the uniqueness of the identification, the multiple current application technique was employed. It was found that the modified Powell optimization technique was useful for an automatic identification of the inclined crack. Selection of the variables of the optimization was discussed for obtaining good estimates.

The least residual method incorporating optimization techniques was also applied to the identification of a surface crack from back surface potential readings. The conjugate gradient method and the modified Powell method were utilized as optimization techniques. A hierarchical optimization procedure was proposed, in which the variables of optimization were varied by referring to the results of previous analyses. It was found that the optimization techniques were effective for the identification of the surface crack.

INTRODUCTION

The electric potential method has been used for monitoring crack growth as well as for measuring crack length (Wilkowski and Maxey, 1983, Murai and Kagawa, 1986). Crack identification from electric potential distribution can be one of the domain/boundary inverse problems, according to the present author's classification of inverse problems related to analyses of distribution of field quantities (Kubo and Ohji, 1987, Kubo, 1988a, 1992). Many research work has been done on the application and extension of the electric potential method for crack identification (Michael, et al., 1982, Miyoshi and Nakano, 1986, Coffin, 1988, Abe, et al., 1988).

The present authors proposed the electric potential CT (computed tomography) method for identifying location, shape and size of cracks embedded in an electric conductive body from D.C. electric potential distributions observed on the surfaces of the body (Ohji, et al., 1985, Kubo, et al., 1986, 1991b). For the crack identification by the electric potential CT method, two inverse analysis schemes were proposed: the inverse boundary integral equation method and the least residual method. In the crack identification based on the least residual method, the plausible crack was defined as that giving the smallest value of residual R defined as a square sum of difference between the calculated and measured potential values. Hierarchical schemes were incorporated in the analyses to find effectively the plausible crack (Kubo, et al., 1988b, 1988c, Sakagami, et al., 1990a, 1990b). The applicability of the method was demonstrated by numerical simulations and experiments. Those procedures, however, made an automatic identification difficult.

In this paper the optimization techniques are introduced in the crack identification based on the least residual method, in order to find automatically the most plausible crack. Residual R and crack configuration parameters are taken as the objective function and variables of the optimization, respectively.

As an example, the optimization techniques are applied to the identification of an inclined internal crack in a two-dimensional plate from potential readings on side faces. An identification of a surface crack from back surface potential readings is also made. The applicability of the conjugate gradient method and the modified Powell method in the identification is examined.

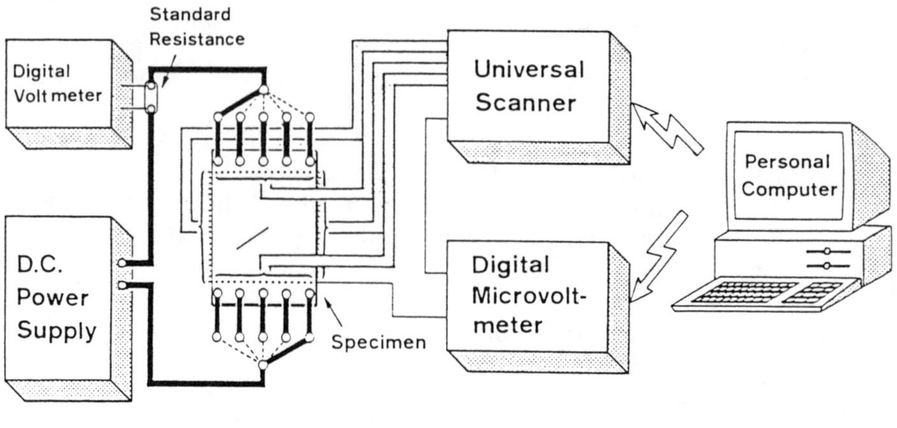

Fig. 2 The system used for measuring potential distribution for a two-dimensional internal crack

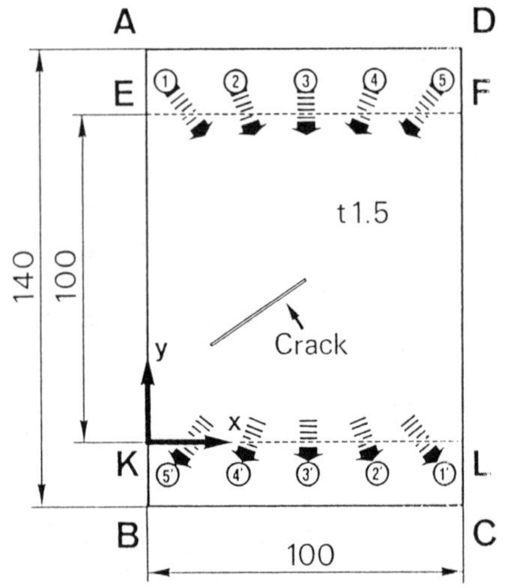

Fig. 1 A two-dimensional inclined internal crack and location of electrodes used for multiple current applications

LEAST RESIDUAL METHOD INCORPORATING OPTIMIZATION TECHNIQUES

Both of the inverse boundary integral equation method and the least residual method are proposed for the inverse analysis of crack identification based on boundary element formulation. The least residual method is used in the present study. In the least residual method a quasi-solution is sought as in the followings. Based on subsidiary information, such as a priori information, physical limits, or experiences, type of crack and range of crack parameters are prescribed. Cracks satisfying this subsidiary information is called admissible. When an admissible crack is assumed, the potential distribution with the existence of the crack is obtained by solving the governing equation of D.C. potential, i.e. the Laplace equation. The boundary element method is effective in calculating the potential distribution. A quasi-solution is defined as the crack giving the smallest residual between the measured and calculated potential distribution among admissible cracks. Residual R is given by the following surface integral over observation surface Γ_{ob}.

$$R = \int_{\Gamma_{ob}} w\,[\phi^{(m)} - \phi^{(c)}]^2\, d\Gamma \tag{1}$$

where w is a weighting factor, $\phi^{(m)}$ and $\phi^{(c)}$ are measured and calculated potential distributions.

When the crack is expressed by parameters q_1, q_2, $\cdots$, residual R can be expressed in terms of these parameters. The quasi-solution corresponding to plausible crack is found by minimizing R:

$$R = R(q_1, q_2, \cdots) \rightarrow \min \tag{2}$$

Combination of the parameters giving minimum R value can be sought by using optimization techniques (Sakagami, et al., 1990c). Optimization process is truncated when R value goes below a threshold, which is determined based on the level of error in measurement.

When the solution is not stable, regularization may be introduced using smoothing functional $\Lambda(q_1, q_2, \cdots)$ and regularization parameter α: crack giving the smallest value of the following function Ω is sought.

$$\Omega = R(q_1, q_2, \cdots) + \alpha \cdot \Lambda(q_1, q_2, \cdots) \rightarrow \min \tag{3}$$

In this case the selection of the value of α is important.

IDENTIFICATION OF AN INCLINED INTERNAL CRACK IN A TWO-DIMENSIONAL PLATE

Method of Identification

An identification of a two-dimensional inclined crack embedded in a stainless steel strip was made using the least residual method incorporating hierarchical schemes (Sakagami, et al., 1988). In the present study the optimization techniques are applied to this problem.

The crack in a strip is shown in Fig. 1. It was assumed that there was only one straight crack in the plate. The crack location, angle and size were unknown in advance and to be determined from the potential distributions measured on flux free side faces EK and LF.

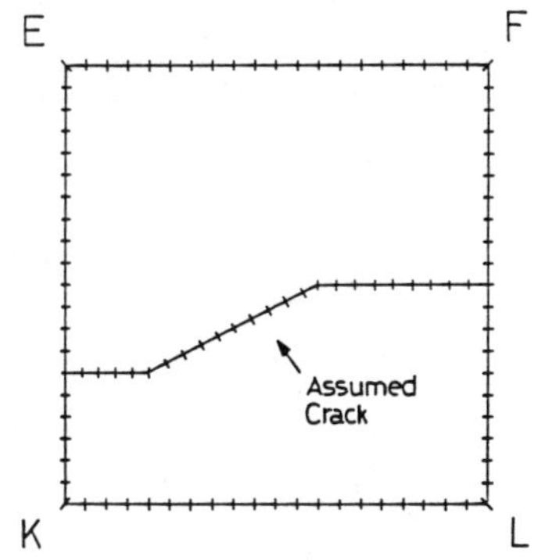

Fig. 3 Boundary element discretization of a specimen with a inclined crack

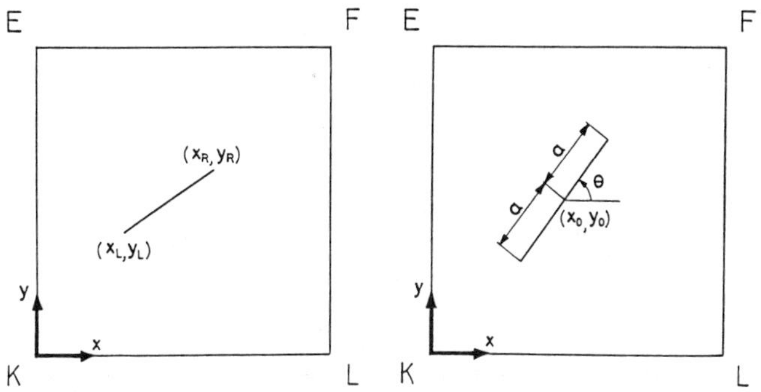

(a) Variable set (1) (b) Variable set (2)

Fig. 4 Selection of variables in the optimization

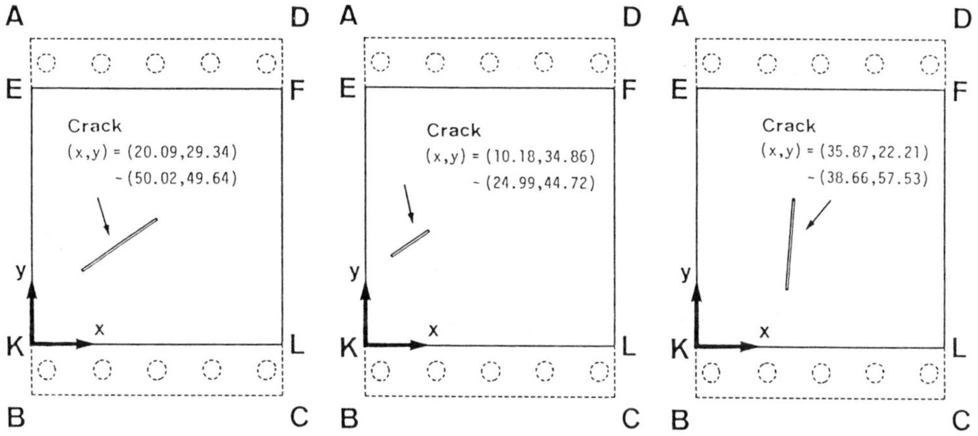

(a) Long inclined crack (b) Short inclined crack (c) Long inclined crack with large angle

Fig. 5 Inclined internal cracks identified

It was shown that the inclined crack cannot be uniquely identified from the potential distribution under only one current application condition (Kubo, et al., 1989, Kubo, 1991). When the plane containing the crack is unknown electric potential distributions under two current application conditions are necessary to identify a single two-dimensional crack. To identify a single three-dimensional crack in an unknown plane electric potential distributions under three current application conditions are necessary. In the identification of the inclined crack, the multiple current application technique (Sakagami, et al., 1988) was applied to ensure the uniqueness of the crack identification. A system for measuring potential distributions with multiple current applications is shown in Fig. 2. A D.C. current was supplied between one of the five pairs of electrodes (1-1'), (2-2'), ···, and (5-5') consecutively placed on the top and bottom ends of the specimen as shown in Fig. 1. The electric potential readings at 40 measuring points on EK and LF were obtained for each electrode locations. To use the potential distributions under five current application conditions simultaneously in the identification, the sum of residuals R_S defined by the following equation was introduced.

$$R_S = \sum_k \int_{EK+LF} w\, [\phi^{(m)}{}_k - \phi^{(c)}{}_k]^2 \, d\Gamma \qquad (4)$$

where $\phi^{(m)}{}_k$ and $\phi^{(c)}{}_k$ denote the measured and calculated potential distributions for the k-th current application condition, and w is a weighting factor. Distribution $\phi^{(c)}{}_k$ was calculated by the boundary element method for assumed cracks. An example of boundary element discretization is shown in Fig. 3. As the weighting factor the Dirac delta function was employed, whose center coincided with the location of potential probes. The plausible crack was identified as the crack giving the smallest R_S value. R_S was then taken as an objective function of the optimization. The modified Powell method, which did not require the calculation of derivatives, was applied as an optimization technique.

Two variable sets, shown in Fig. 4, were used in the optimization. Variable set (1) simply used the coordinates of the two crack tips of the two-

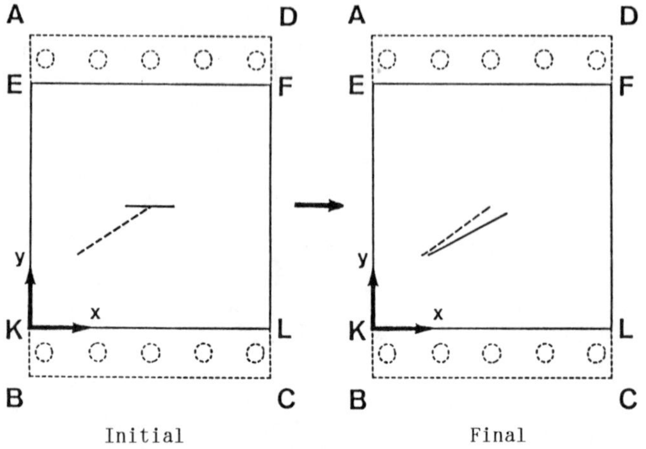

(a) Long inclined crack

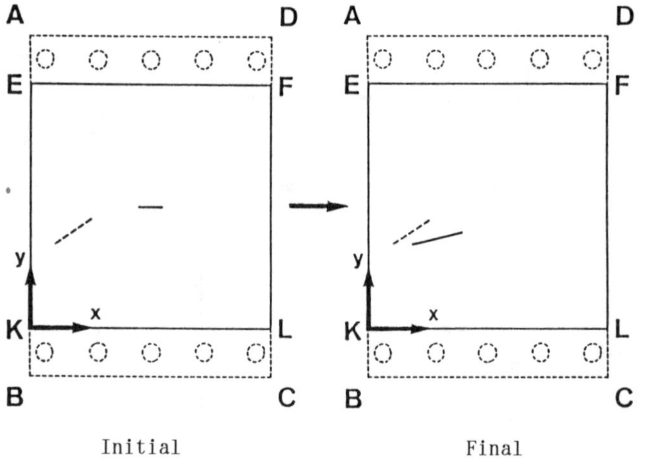

(b) Short inclined crack

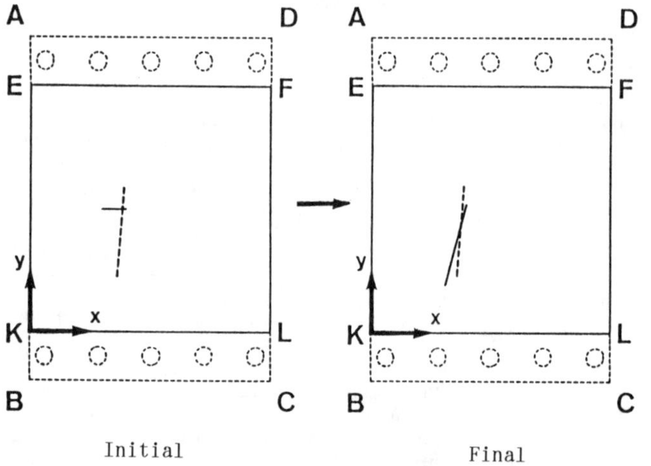

(c) Long inclined crack with large angle

Fig. 6 Identification of inclined cracks using variable set (1)

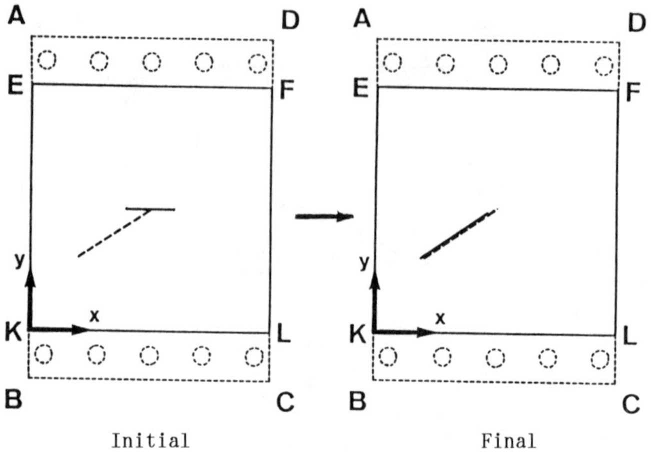

(a) Long inclined crack

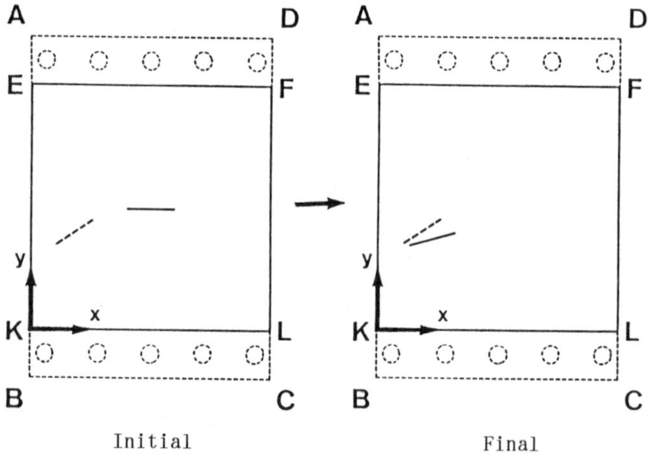

(b) Short inclined crack

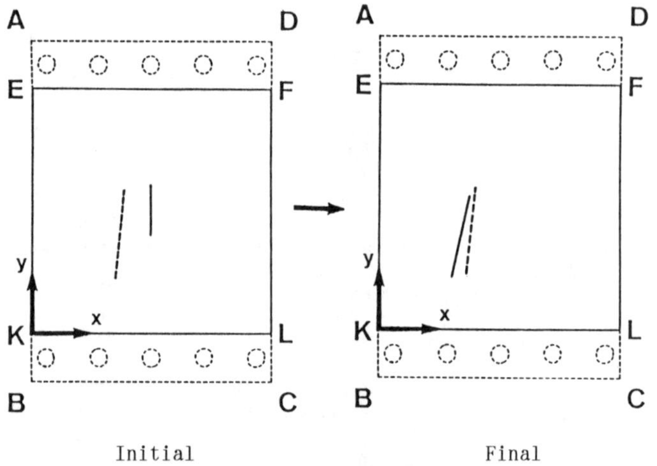

(c) Long inclined crack with large angle

Fig. 7 Identification of inclined cracks using variable set (2) dimensional crack. Variable set (2) employed the crack length, which has a major influence on the potential distribution, as well as the location of crack center and the angle of the crack.

Results of Identification

The identification was made for (a) a long inclined crack, (b) a short inclined crack and (c) a long inclined crack with large angle with the x-axis, shown in Fig. 5.

Identification Using Variable Set (1). The cracks identified by the proposed method using variable set (1) are shown in Figs. 6 (a), (b) and (c), for crack configuration (a), (b) and (c), respectively. The actual cracks are depicted by the dashed lines. The estimated cracks as well as the initially assumed cracks are shown by the solid lines.

As can be seen in Fig. 6 (a), the long inclined crack (a) is identified reasonably with variable set (1) without difficulties. In the estimation of (b) short inclined crack and (c) long inclined crack with large angle, however, the estimated cracks were sensitive to the initially assumed crack. The cracks which gave the minimum R_S values among several initial cracks employed in the identification are shown in Figs. 6 (b) and (c). It can be seen from these figures that the cracks are identified reasonably, when small R_S values are attained.

Identification Using Variable Set (2). The cracks identified using variable set (2) are shown in Figs. 6 (a), (b) and (c). The actual cracks are depicted by the dashed lines, the estimated cracks as well as the initially assumed cracks being shown by the solid lines. The standard values of crack parameters of initially assumed crack are a =10, x_0 = 50, y_0 = 50, θ = 0. Estimated cracks using the standard values of initial crack parameters for (a) a long inclined crack and (b) a short inclined crack are shown in Figs. 7 (a) and (b), respectively. In these cases reasonable estimates of cracks are obtained irrespective of the initial values of crack parameters. From the comparison of Fig. 7 (a) with Fig. 6 (a), the long inclined crack (a) is identified better with variable set (2) than with variable set (1). The short inclined crack (b) is also identified better and faster with variable set (2). For the identification of (c) a long inclined crack with large angle, the vertical crack shown by the solid line on left of Fig. 7 (c) was used as the initial crack. It is seen that a reasonably estimate is obtained for the crack with large angle also.

Thus the optimization technique was successfully applied to the crack identification.

IDENTIFICATION OF A THREE-DIMENSIONAL SURFACE CRACK

Method of Identification

Numerical simulations were conducted of the identification of a three-dimensional semi-elliptical surface crack in a rectangular block shown in Fig. 8 from back surface potential data. Assuming that there existed only one semi-elliptical crack whose surface was perpendicular to the z-axis, the location and size of the crack were identified. The electric potential readings for ϕ = 1 on top surface ABCD and ϕ = -1 on bottom surface EFGH were used in the identification. The location of 15 measuring points placed on back surface CGHD is shown with (•) in Fig. 9. Boundary element direct analysis was made to calculate the potential distribution for the crack configuration shown in Fig. 8. Potential values at 15 measuring points were used as measured potential readings in this numerical simulation. To take account of error in electric potential measurement, electric potential readings were rounded off to four digits of data. Residual defined by Eq. (1) was evaluated between this measured potential readings and the calculated potential readings for an assumed crack. An example of the boundary element discretization used for potential calculation is shown in Fig. 10. As the weighting factor w in Eq. (1), the Dirac delta function whose center coincided with the location of potential measuring points was employed. The plausible crack was identified as the crack giving the smallest R value.

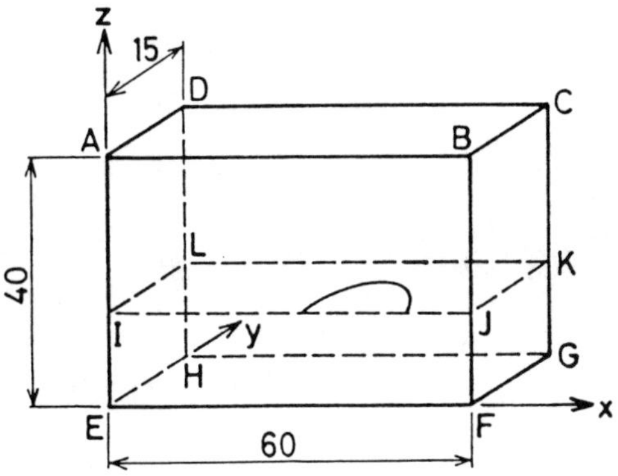

(a) Configuration of a cracked specimen

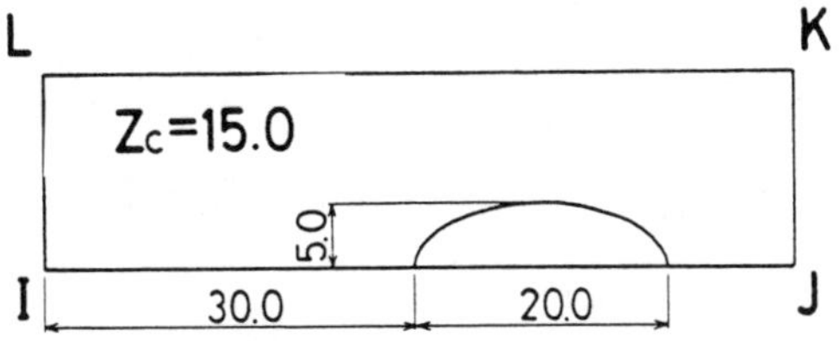

(b) Crack shape

Fig. 8 A three-dimensional surface crack

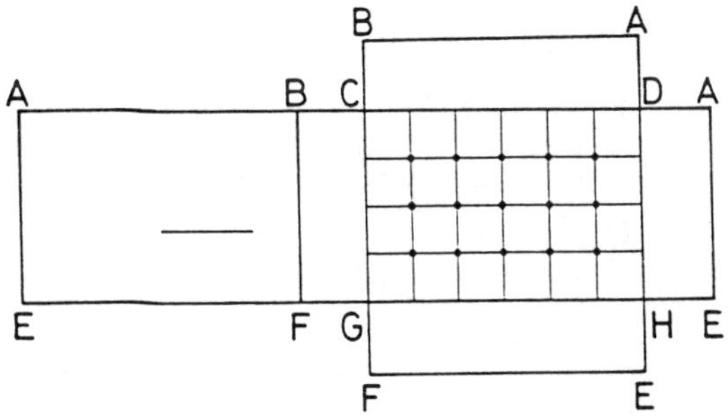

Fig. 9 Location of measuring points

As crack parameters, z_c and x_c representing the crack location, a and c representing crack depth and crack length were used. Combination of crack parameters z_c, x_c, a and c giving the minimum value of R was sought automatically by applying optimization schemes. As an initial combination of crack parameters z_c = 20, x_c = 30, a = 7.5, c = 7.5 were used. The step of modification of parameters in

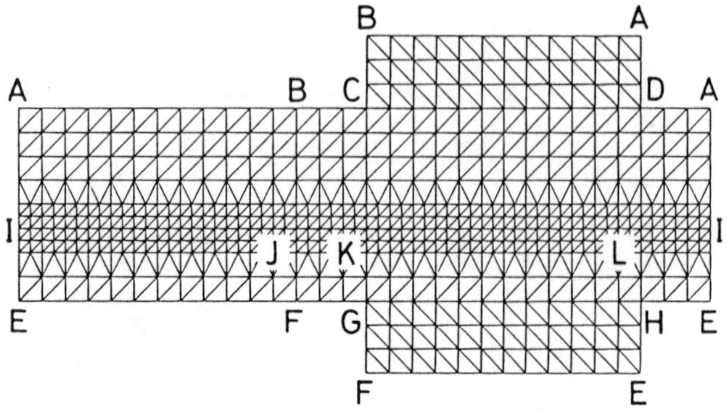

(a) Surface of cracked body

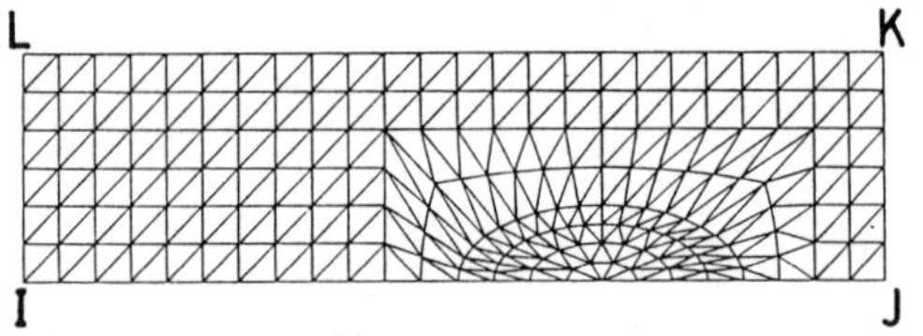

(b) Plane containing crack

Fig. 10 Boundary element discretization of a specimen with a surface crack

optimizing procedure was decreased gradually. A crack which gave R value smaller than the threshold value was employed as the solution.

As optimization procedures the conjugate gradient method and the modified Powell method were applied. In some cases R value changed slightly after certain iterations of optimization and did not go below the threshold value. A hierarchical optimization procedure consisting of three stages was then introduced in the modified Powell method to overcome the difficulties and to reduce the computation time. This procedure was based on the present authors' results of numerical simulations and experiments that parameters representing crack location rather than those representing the crack shape was easily estimated in the early stage of crack identification. In stage I all of the crack parameters were used as the variables of the optimization. In stage II following stage I, a and c representing crack shapes were used as the variables of optimization, assuming that a rough estimate of crack location was already obtained. In stage III all the crack parameters were used as the variables of optimization again.

Results of Identification

The conjugate gradient method and the modified Powell method were used to identify the crack shown in Fig. 8. Parameters x_c, a and c were used as variables of optimization, assuming that height z_c of the crack was known in advance. R was evaluated from 4 digits of potential values at 15 measuring points on back surface CGHD.

Figures 11 and 12 show the results obtained by applying the conjugate gradient method and the modified Powell method with the hierarchical optimization procedure, respectively. The solid lines in the figures designate the estimated crack shapes or initial crack shapes, dashed lines being real ones.

As can be seen from these figures, the estimated crack shapes agree very well with the real ones. For

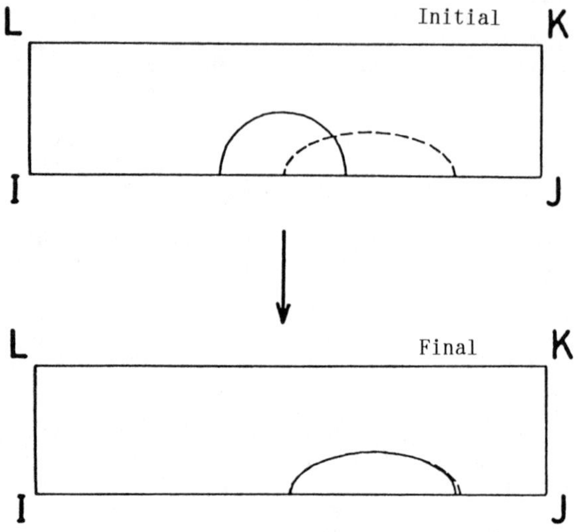

Fig. 11 Identification of a surface crack by using the conjugate gradient method

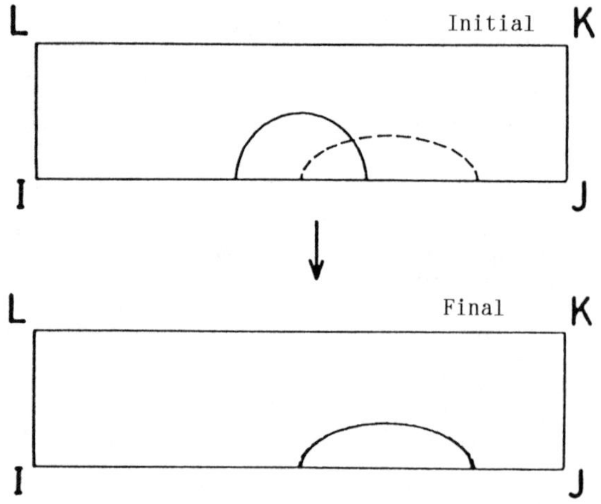

Fig. 12 Identification of a surface crack by using the modified Powell method

some crack configurations the conjugate gradient method attained small R values faster than the modified Powell method. For other crack configurations, however, the conjugate gradient method failed to attain small R values. The modified Powell method was found to be applicable with ease even to the identification of cracks, for which convergence could not be attained by the conjugate gradient method.

CONCLUSIONS

The optimization techniques were introduced in the crack identification by the electric potential CT method based on the least residual method, by taking residual R as the objective function of the optimization, to find automatically the plausible crack. As an example of the application of the method using the optimization techniques, a two-dimensional inclined crack in a plate specimen was identified from potential readings on the side faces of the plate. To ensure the uniqueness of the identification of the inclined internal crack, the multiple current application technique was employed. The modified Powell method was utilized as an optimization scheme. The conjugate gradient method and the modified Powell method were applied to the identification of a surface crack from back surface potential readings. A hierarchical optimization procedure was proposed, in which the variables to be optimized were varied by referring to the results of previous analyses. It was found that the optimization techniques were useful for the automatic identification of the cracks. Experiences are found to be useful to find good combination of variables of the optimization and to construct good procedures for accomplishing good estimations.

ACKNOWLEDGEMENTS

The work is partly supported by the Ministry of Education, Culture and Science under Grant-in-Aid for Scientific Research.

REFERENCES

Abe, H., Saka, M., Wachi, T., and Kanoh, Y., 1988, "An Inverse Problem in Non-Destructive Inspection of a Crack in a Hollow Cylinder by Means of Electrical Potential Method", in "Computational Mechanics '88", Vol. 1 (Ed. Atluri, S.N. and Yagawa, G.), Proc. Int. Conf. on Computational Engineering Science, Springer, Berlin, pp. 12.ii.1-12.ii.4.

Coffin, L.F., 1988, "Role of Damage Tolerance and Fatigue Crack Growth in the Power Generation Industry", ASTM STP 969, Amer. Soc. Testing Mater., pp. 235-259.

Kubo, S., Sakagami, T., and Ohji, K., 1986, "Electric Potential CT Method Based on BEM Inverse Analyses for Measurement of Three-Dimensional Cracks", in "Computational Mechanics '86", (Ed. Yagawa, G. and Atluri, S.N.), Proc. Int. Conf. on Computational Mechanics, Springer-Verlag, Berlin, pp. V-339 to V-344.

Kubo, S., and Ohji, K., 1987, "Applications of the Boundary Element Method to Inverse Problems", Chapter 10, "Applications of the Boundary Element Method", (Ed. Japan Soc. for Computational Methods in Eng.), Corona Publ., Tokyo, pp.181-198.

Kubo, S., 1988a, "Inverse Problems Related to the Mechanics and Fracture of Solids and Structures", JSME Int. J., Ser.I, Japan Soc. Mech. Engrs., Vol.31, pp. 157-166.

Kubo, S., Sakagami, T., and Ohji, K., 1988b, "Reconstruction of a Surface Crack by Electric Potential CT Method", in "Computational Mechanics '88", Vol. 1 (Ed. Atluri, S.N. and Yagawa, G.), Proc. Int. Conf. on Computational Engineering Science, Springer-Verlag, Berlin, pp. 12.i.1-12.ii.5.

Kubo, S., Sakagami, T., Ohji, K., Hashimoto, T., and Matsumuro, Y., 1988c, "Quantitative Measurement of Three-Dimensional Surface Cracks by the Electric Potential CT Method, Trans. Japan Soc. Mech. Eng., Ser.A, Vol.54, pp. 218-225.

Kubo, S., Sakagami, T., and Ohji, K., 1989, "On the Uniqueness of the Inverse Solution in Crack Determination by the Electric Potential CT Method", Trans. Japan Soc. Mech. Eng., Ser.A, Vol.55, pp. 2316-2319.

Kubo, S., 1991a, "Requirements for Uniqueness of Crack Identification from Electric Potential Distributions", in "Inverse Problems in Engineering Science" (Ed. Yamaguti, M., Hayakawa, K., Iso, Y., Mori, M., Nishida, T., Tomoeda, K. and Yamamoto, M.), ICM-90 Satellite Conf. Proc., Springer-Verlag, Tokyo, pp. 52-58.

Kubo, S., Sakagami, T., and Ohji, K., 1991b, "Electric Potential CT Method for Measuring Two- and Three-Dimensional Cracks", in "Fracture Mechanics" (Current Japanese Materials Research, Vol. 8) (Ed. Okamura, H. and Ogura, K.), Elsevier, London, pp. 235-254.

Kubo, S., 1992, "Inverse Problems", Baifukan, Tokyo.

Michael, D.H., Waechter, R.T., and Collins, R., 1982, "The Measurement of Surface Cracks in Metals by Using A.C. Electric Fields", Proc. Roy. Soc. Lond. A, Vol.381, pp. 139-157.

Miyoshi, T., and S. Nakano, 1986, "A Study of the Determination of Surface Crack Shape by the Electric Potential Method", Trans. Japan Soc. Mech. Eng., Ser.A, Vol.52, pp. 1097-1102.

Murai, T., and Kagawa, Y., 1986, "Boundary Element Iterative Techniques for Determining the Interface Boundary Between Two Laplace Domain - A Basic Study of Impedance Plethysmography as an Inverse Problem", Int. J. Numerical Methods in Eng., Vol. 23, pp. 35-47.

Ohji, K., Kubo, S., and Sakagami, T., 1985, Electric "Potential CT Method for Measuring Location and Size of Two- and Three-Dimensional Cracks (Development of Boundary Element Inverse Analysis Method for Electric Potential Problems and Its Application to Non-Destructive Crack Measurement)", Trans. Japan Soc. Mech. Eng., Ser.A, Vol.51, pp. 1818-1826.

Sakagami, T., Kubo, S., Hashimoto, T., Yamawaki, H., and Ohji, K., 1988, "Quantitative Measurement of Two-Dimensional Inclined Cracks by the Electric Potential CT Method with Multiple Current Applications", JSME Int. J., Ser.I, Vol.31, pp. 76-86.

Sakagami, T., Kubo, S. and Ohji, K., 1990a, "A Hierarchical Inversion Scheme for Reconstructing a Three-Dimensional Internal Crack from an Electric Potential Distribution", Int. J. Pressure Vessels & Piping, Vol.44, pp. 35-47.

Sakagami, T., Kubo, S., Ohji, K., Yamamoto, K. and Nakatsuka, K., 1990b, "Identification of a Three-Dimensional Internal Crack by the Electric Potential CT Method", Trans. Japan Soc. Mech. Eng., Ser.A, Vol.56, pp. 27-32.

Sakagami, T., Kubo, S., and Ohji, K., 1990c, "Crack Identification by the Electric Potential CT Inverse Analyses Incorporating Optimization Techniques", Engineering Analysis, Vol.7, pp. 59-65.

Wilkowski, G.M., and Maxey, W.A., 1983, "Review and Applications of the Electric Potential Method for Measuring Crack Growth in Specimens, Flawed Pipes, and Pressure Vessels", ASTM STP 791, Amer. Soc. Testing Mater., pp. II-266 to II-294.

Inverse Problems in Engineering: Theory and Practice
ASME 1993

SHAPE OPTIMIZATION IN ELASTICITY AND ELASTO-VISCOPLASTICITY

Xin Wei
Department of Aerospace
and Mechanical Engineering
University of Arizona
Tucson, Arizona

Liang-Jenq Leu
Department of Theoretical
and Applied Mechanics
Cornell University
Ithaca, New York

Abhijit Chandra
Department of Aerosapce
and Mechanical Engineering
University of Arizona
Tucson, Arizona

Subrata Mukherjee
Department of Theoretical
and Applied Mechanics
Cornell University
Ithaca, New York

ABSTRACT

The objective of this paper is to present shape optimization results in nonlinear solid mechanics involving material nonlinearities. Optimal shape designs are carried out by coupling the standard and sensitivity analyses with an optimizer. The optimization algorithm adopted here is based on a sequential quadratic programming method with a line search. Shape sensitivities are obtained by direct differentiation of the derivative boundary element method formulation of the problem. Numerical solutions are presented for optimal shapes of cutouts in thin plates. Optimal elastic design is also carried out. The difference between optimal shapes of solids undergoing elastic and elasto-viscoplastic deformation is shown clearly by these examples

INTRODUCTION

In the current literature, many efforts have been made on optimal design. However, most researchers focus on linear structural mechanics, such as designs of truss structures with minimum-weight, beams with maximum limit loads, columns with maximum buckling loads, plates with maximum fundamental frequencies etc. For example, Banichuk (1990) and Haftka and Gürdal (1992) give many optimal designs for problems of this class and many optimal algorithms are also discussed there. The interest here, however, is in optimal design of problems of nonlinear continuum mechanics.

Optimal shape design of solids undergoing small-strain elasto-viscoplastic deformation is investigated in this paper. The goal here is to maximize or minimize an objective function without violating constraints. The objective functions and constraints are typically functions of stresses and/or displacements, which, in turn, depend on the initial shape of a body. The stresses and displacements are time dependent and also dependent on the loading history . In this study, shape parameters that define the initial shape of part or all of a body are taken to be the design variables, and the objective function and constraint values are evaluated at a fixed time T from the start of the simulation.

An optimization process starts with a preliminary design. Next, the standard and sensitivity analyses are carried out up to a fixed time T. The values of the objective function, constraints and their sensitivities (gradients) at this time are calculated and input to an optimizer, which then provides a new design. If the new design is acceptable (this is typically decided by such criteria as satisfaction of the Kuhn-Tucker conditions), the process stops. Other convergence criteria such as convergence of the design variables, convergence of the objective functions and/or constraints, or achievement of small values of the gradients of the objective functions and/or constraints have been used as well. (see, for example, Papalambros and Wilde, 1988). Otherwise, i.e., if the new design is not acceptable, the iterative process is continued, producing a succession of designs, until an optimal design is achieved. This iterative algorithm in optimal design is illustrated in Figure 1.

To achieve an optimal design, accurate determination of design sensitivity coefficients (DSCs) is required. This fact has been recognized by Wei et al.(1993). DSCs are the rates of change of response quantities, such as stress or displacement in a loaded body, with respect to design variables. The calculation of DSCs for linear problems, such as linear elasticity, is well established. For more details, the reader is referred to Haug et al. (1986). Basically, three different approaches have been used: the finite difference approach (FDA), the adjoint structure approach (ASA), and the direct differentiation approach (DDA). These three approaches have been used in conjunction with numerical methods such as the finite element method (FEM) and the boundary element method (BEM) to obtain the DSCs.

In this paper, the DDA is used together with the BEM to

obtain DSCs. However, instead of linear problems, nonlinear problems are attacked here. The determination of DSCs for nonlinear problems has attracted much attention recently, and many efforts have been made on this topic. For example, Mukherjee and his co-workers have published a series of papers aiming at accurately determining DSCs in nonlinear continuum mechanics, involving material and/or geometric nonlinearities. (see, Mukherjee and Chandra 1989; Mukherjee and Chandra 1991; Zhang et al. 1992a; Zhang et al. 1992b; Leu and Mukherjee 1993). For a more detailed discussion and recent development on this topic, the reader is referred to those papers and the many references cited therein. In general, for problems of this class, DSCs are time dependent and stress driven. Therefore, accurate determination of the stress history is of crucial importance. For this reason, a derivative boundary element method (DBEM) is used in the above work of Mukherjee and his co-workers.

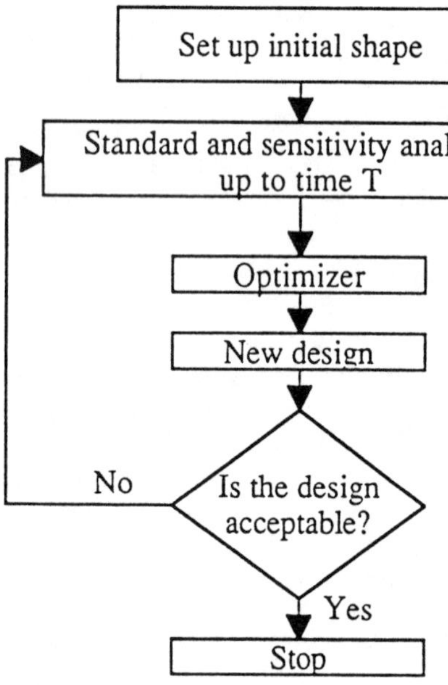

Fig. 1 Optimal shape design algorithm

This paper first reviews the calculation of DSCs for plane strain small-deformation elasto-viscoplastic problems. The equations for the DSCs are obtained by the DDA of the relevant BEM equations of the problem. The formulation given by Zhang et al.(1992a) is followed, but some modifications have been made to some of the equations presented there. The rest of this paper is concerned with shape optimization. First, numerical implementation and optimal algorithms are discussed. Next, parameterization of the shape of a body and calculation of geometric sensitivities are given. Finally, numerical examples are provided for a thin plate with an elliptical cutout.

BEM FORMULATION FOR PLANE-STRAIN SMALL-DEFORMATION PROBLEMS

In this section, a derivative boundary element (DBEM) formulation given by Zhang et al. (1992a), is reviewed. Here, only the formulation for plane strain is given; the plane stress version is quite analogous to that described by Zhang et al. (1992a) and is not repeated here in the interest of brevity.

Boundary Equations

The rate form of the DBEM formulation for two-dimensional elasto-viscoplasticity in a simply connected domain is $(i,\ j,\ k = 1,\ 2)$

$$\begin{aligned} 0 = &\int_{\partial B} [U_{ij}(\mathbf{b},P,Q)\dot{\tau}_i(\mathbf{b},Q) - W_{ij}(\mathbf{b},P,Q)\dot{\Delta}_i(\mathbf{b},Q)]\, dS(\mathbf{b},Q) \\ &+ 2G\dot{\epsilon}_{ik}^{(n)}(\mathbf{b},P)\int_{\partial B} U_{ij}(\mathbf{b},P,Q)n_k(\mathbf{b},Q)\, dS(\mathbf{b},Q) \\ &+ 2G\int_B U_{ij,k}(\mathbf{b},P,q)[\dot{\epsilon}_{ik}^{(n)}(\mathbf{b},q) - \dot{\epsilon}_{ik}^{(n)}(\mathbf{b},P)]\, dA(\mathbf{b},q) \qquad (1) \end{aligned}$$

This is a modified form of equation (1) in Zhang et al. (1992a). The domain integral in this equation is completely regular. A two-dimensional body B, has the boundary ∂B in the x_1, x_2 plane, and $\dot{\Delta}_i = \frac{\partial v_i}{\partial s}$, $\dot{\tau}_i$ are the components of the tangential derivative of the velocity (s is the curvilinear length coordinate measured around ∂B in the anticlockwise sense), and the traction rate, respectively. Here, $\mathbf{b}$ denotes a design variable vector; P (or p) and Q (or q) are source and field points, with capital letters denoting points on ∂B and lower case letters denoting points inside B; n_k are components of the unit outward normal to ∂B at a point Q on it. A comma denotes a derivative with respect to the coordinates of a field point. The kernel U_{ij} is given in many references (for example, Mukherjee 1982), and the explicit form of W_{ij} is given by Ghosh et al. (1986). They are both $ln\, r$ singular. In this DBEM formulation, the rates of traction and the displacement derivative vectors are the primary unknowns on ∂B. One advantage of this formulation is that stress rates at a boundary node can be determined accurately and directly from those primary unknowns by the equation

$$\dot{\sigma}_{ij} = A_{ijk}\dot{\tau}_k + B_{ijk}\dot{\Delta}_k + C_{ijkl}\dot{\epsilon}_{kl}^{(n)} + D_{ij}\dot{\epsilon}_{kk}^{(n)} \qquad (2)$$

where A_{ijk} etc., are functions of the components of the unit outward normal ($\mathbf{n}$) and unit (anticlockwise) tangent ($\mathbf{t}$) vectors to ∂B at P, and the shear modulus and Poisson's ratio G and ν, respectively. Explicit expressions for A_{ijk}, etc., are given by Zhang et al. (1992a).

Since conforming boundary elements are used in the formulation, two more boundary equations are required for each corner. For a discussion of modeling of corners, the reader is referred to Zhang et al. (1992a). Another two useful equations, representing continuity of the velocity, are

$$\int_{\partial B} \dot{\Delta}_i\, dS = 0 \qquad (3)$$

These, together with the corner equations and boundary integral equations , constitute the system equations. Therefore, the system is overdetermined. However, as long as the system equations are consistent, the solution is unique.

Internal Equations

Strain and stress rates are also required at internal points for the nonelastic problem. To this end, the version of equation (1) at an internal point is first differentiated and then regularized by the "addition-subtraction" method. The final form, with a $O(1/r)$ domain singularity, is $[i, j, k, \bar{l} = 1, 2; \bar{l} = \frac{\partial}{\partial x_l(p)}]$

$$
\begin{aligned}
\dot{u}_{j,\bar{l}}(\mathbf{b},p) = {} & \int_{\partial B} [U_{ij,\bar{l}}(\mathbf{b},p,Q)\dot{\tau}_i(\mathbf{b},Q) - W_{ij,\bar{l}}\dot{\Delta}_i(\mathbf{b},Q)]\, dS(\mathbf{b},Q) \\
& - 2G\dot{\epsilon}_{ik}^{(n)}(\mathbf{b},p)\int_{\partial B} U_{ij,k}(\mathbf{b},p,Q)n_l(\mathbf{b},Q)\, dS(\mathbf{b},Q) \\
& - 2G\int_B U_{ij,kl}(\mathbf{b},p,q)[\dot{\epsilon}_{ik}^{(n)}(\mathbf{b},q) - \dot{\epsilon}_{ik}^{(n)}(\mathbf{b},p)]dA(\mathbf{b},q) \qquad (4)
\end{aligned}
$$

The stress rates for an internal point can be obtained from Hooke's law as $(i, j, k = 1, 2)$

$$
\dot{\sigma}_{ij} = \lambda \dot{u}_{k,k}\delta_{ij} + G(\dot{u}_{i,j} + \dot{u}_{j,i}) - 2G\dot{\epsilon}_{ij}^{(n)} \qquad (5)
$$

where $\lambda = 2G\frac{\nu}{(1-\nu)}$ is the first Lamé constant. It is noted that the inelastic strain rate is assumed to be incompressible. This assumption is also used in deriving the boundary and the internal integral equations above.

SENSITIVITY FORMULATION FOR PLANE STRAIN

Sensitivity equations, based on the direct analytical differentiation of the relevant DBEM equations, are given in this section. The plane stress sensitivity equations are analogous to these equations. Unless otherwise indicated, the ranges of indices in this section is 1, 2.

Boundary Equations

The first step here is the differentiation of equation (1) with respect to a design variable b (which is any component of the design variable vector $\mathbf{b}$) of a variable of interest. Let a superscribed (*) denote the design derivative of a variable and a superscribed (°) denote the design derivative of its rate, respectively. Now, one obtains, by the chain rule of differentiation, the equation

$$
\begin{aligned}
0 = {} & \int_{\partial B} [U_{ij}(\mathbf{b},P,Q)\overset{\circ}{\tau}_i(\mathbf{b},Q) - W_{ij}(\mathbf{b},P,Q)\overset{\circ}{\Delta}_i(\mathbf{b},Q)]\, dS(\mathbf{b},Q) \\
& + \int_{\partial B} [\overset{*}{U}_{ij}(\mathbf{b},P,Q)\dot{\tau}_i(\mathbf{b},Q) - \overset{*}{W}_{ij}(\mathbf{b},P,Q)\dot{\Delta}_i(\mathbf{b},Q)]\, dS(\mathbf{b},Q) \\
& + \int_{\partial B} [U_{ij}(\mathbf{b},P,Q)\dot{\tau}_i(\mathbf{b},Q) - W_{ij}(\mathbf{b},P,Q)\dot{\Delta}_i(\mathbf{b},Q)]\, d\overset{*}{S}(\mathbf{b},Q) \\
& + 2G\overset{\circ}{\epsilon}{}_{ik}^{(n)}(\mathbf{b},P)\int_{\partial B} U_{ij}(\mathbf{b},P,Q)n_k(\mathbf{b},Q)\, dS(\mathbf{b},Q) \\
& + 2G\dot{\epsilon}_{ik}^{(n)}(\mathbf{b},P)\int_{\partial B} \overset{*}{U}_{ij}(\mathbf{b},P,Q)n_k(\mathbf{b},Q)\, dS(\mathbf{b},Q) \\
& + 2G\dot{\epsilon}_{ik}^{(n)}(\mathbf{b},P)\int_{\partial B} U_{ij}(\mathbf{b},P,Q)\overset{*}{n}_k(\mathbf{b},Q)\, dS(\mathbf{b},Q) \\
& + 2G\dot{\epsilon}_{ik}^{(n)}(\mathbf{b},P)\int_{\partial B} U_{ij}(\mathbf{b},P,Q)n_k(\mathbf{b},Q)\, d\overset{*}{S}(\mathbf{b},Q) \\
& + 2G\int_B \overset{*}{U}_{ij,k}(\mathbf{b},P,q)[\dot{\epsilon}_{ik}^{(n)}(\mathbf{b},q) - \dot{\epsilon}_{ik}^{(n)}(\mathbf{b},P)]dA(\mathbf{b},q) \\
& + 2G\int_B U_{ij,k}(\mathbf{b},P,q)[\overset{\circ}{\epsilon}{}_{ik}^{(n)}(\mathbf{b},q) - \overset{\circ}{\epsilon}{}_{ik}^{(n)}(\mathbf{b},P)]dA(\mathbf{b},q) \\
& + 2G\int_B U_{ij,k}(\mathbf{b},P,q)[\dot{\epsilon}_{ik}^{(n)}(\mathbf{b},q) - \dot{\epsilon}_{ik}^{(n)}(\mathbf{b},P)]d\overset{*}{A}(\mathbf{b},q) \qquad (6)
\end{aligned}
$$

Formulas for $\overset{*}{U}_{ij}, \overset{*}{W}_{ij}, d\overset{*}{S}, d\overset{*}{A}$, and $\overset{*}{n}_i$, are given by Zhang et al. (1992a).

The sensitivity of stress rates can be obtained by direct differentiation of equation (2). This is of the form

$$
\begin{aligned}
\overset{\circ}{\sigma}_{ij} = {} & A_{ijk}\overset{\circ}{\tau}_k + B_{ijk}\overset{\circ}{\Delta}_k + C_{ijkl}\overset{\circ}{\epsilon}{}_{kl}^{(n)} + D_{ij}\overset{\circ}{\epsilon}{}_{kk}^{(n)} \\
& + \overset{*}{A}_{ijk}\dot{\tau}_k + \overset{*}{B}_{ijk}\dot{\Delta}_k + \overset{*}{C}_{ijkl}\dot{\epsilon}_{kl}^{(n)} + \overset{*}{D}_{ij}\dot{\epsilon}_{kk}^{(n)} \qquad (7)
\end{aligned}
$$

Also, the sensitivity version of equation (4) can be written as

$$
\int_{\partial B} \overset{\circ}{\Delta}_i\, dS + \int_{\partial B} \dot{\Delta}_i\, d\overset{*}{S} = 0 \qquad (8)
$$

Sensitivity Equations at an Internal Point

By following an analogous approach to that used in deriving the boundary equations, the sensitivity equations at an internal point can be obtained by the direct differentiation of the corresponding equations for mechanics problems. These are not repeated here.

NUMERICAL IMPLEMENTATION AND OPTIMIZATION ALGORITHM

Numerical Implementation and Solution Strategy

Numerical implementation of the standard as well as the sensitivity problems for elasto-viscoplastic materials follows usual practice (see Zhang et al. 1992a). The boundary ∂B of the body is subdivided into piecewise quadratic, conforming boundary elements. The variables $\dot{\tau}_i, \dot{\Delta}_i$, their sensitivities, $\overset{\circ}{\tau}_i, \overset{\circ}{\Delta}_i$, and the scalar $d\overset{*}{S}/dS$ are assumed to be piecewise quadratic on the boundary elements. The domain of the body is divided into $Q4$ internal cells. The nonelastic strain rate components $\dot{\epsilon}_{ik}^{(n)}$, and their sensitivities $\overset{\circ}{\epsilon}{}_{ik}^{(n)}$, as well as the quantity $d\overset{*}{A}/dA$, are interpolated on the $Q4$ internal cells. It is noted that the nonelastic strain rate is determined by a so called unified elasto-viscoplastic model developed by Anand (1982) (see, also Zhang et al. 1992a and 1992b).

Logarithmically singular integrands are integrated with log-weighted Gaussian integration formulae. The $O(1/r)$ singular domain integrals are regularized by polar coordinate mapping (Mukherjee 1982, pp. 91-92) and then evaluated by Gaussian quadrature on a square. Typical numbers of Gauss points are 20 and 16 for regular and log-singular boundary integrals and 3×3 for regular and regularized domain integrals.

The solution strategy for the standard and the sensitivity problems, which involves solutions of appropriate equations at the start of each time step and then marching forward in time, is described in detail by Zhang et al. (1992a). Time integration has been carried out with fixed time steps in an explicit manner.

Optimization Algorithm

Optimal shape design is carried out by coupling the above standard and sensitivity analyses with an optimizer, as illustrated in Figure 1. For each iteration, values of the objective function, constraints, and their sensitivities, at a fixed time T from the start of the simulation, are taken as input to the optimizer.

The optimizer adopted here–subroutine DN0ONF, available from the IMSL/MATH library, is based on subroutine NLPQL (Schittkowski 1986). It uses a sequential quadratic programming method with a line search to solve the general nonlinear programming problems. Details of the algorithm follow.

A typical nonlinear optimization problem is stated as:

$$\min \ \phi(\mathbf{b}), \qquad \mathbf{b} \in R^n \tag{9a}$$

$$\text{subject to } h_j(\mathbf{b}) = 0, \qquad \text{for } j = 1, \dots m \tag{9b}$$

$$g_k(\mathbf{b}) \geq 0, \qquad \text{for } k = 1, \dots p \tag{9c}$$

$$b_{il} \leq b_i \leq b_{iu}, \quad \text{for } i = 1, \dots n \tag{9d}$$

where $\phi(\mathbf{b})$ is the objective function, $\mathbf{b}$ is the design variable vector with n components, b_{il} and b_{iu} are the lower and upper bounds of the ith component of $\mathbf{b}$, and $h_j(\mathbf{b})$ and $g_k(\mathbf{b})$ are equality and inequality constraints, respectively.

The sequential quadratic programming algorithm NLPQL uses a quadratic approximation of the Lagrangian and linearization of the constraints to define a sequence of subproblems. This requires the evaluation of a positive definite approximation of the Hessian of ϕ.

Let $\mathbf{d}_k$ be the solution of a subproblem at the kth iterative step. A line search is used to find a new design $\mathbf{d}_{k+1}$, which is defined as

$$\mathbf{b}_{k+1} = \mathbf{b}_k + \lambda_k \mathbf{d}_k, \quad 0 \leq \lambda_k \leq 1 \quad \text{(no sum on } k) \tag{10}$$

such that the augmented Lagrangian function has a lower function value at the new design. Here, λ_k is the line search or step length parameter.

The iterative process stops when the Kuhn-Tucker conditions are satisfied within an acceptable tolerance. Schittkowski (1986) shows that, under some mild assumptions, the algorithm converges globally, i.e., starting from an arbitrary initial point, at least one accumulation point of the iterates will satisfy the Kuhn-Tucker optimality conditions.

Coupling of the optimizer with the mechanics and sensitivity calculations is straightforward. One must input values of the functions

$$\phi, \ \frac{\partial \phi}{\partial b_i}, \ h_j, \ g_k, \ \frac{\partial h_j}{\partial b_i} \text{ and } \frac{\partial g_k}{\partial b_i} \tag{11}$$

at each iteration. Typically, ϕ, h_j and g_k depend on quantities such as stress or displacement, and are, therefore, implicit as well as explicit functions of $\mathbf{b}$. The gradients of these functions with respect to b_i, are obtained from the sensitivities such as $\frac{d\sigma}{d\mathbf{b}}$, by the chain rule of differentiation. It is important to note that, for elasto-viscoplastic problems, the above functions and their gradients are evaluated at a preset time T from the start of the simulation.

The experience of the authors of this paper is that this optimizer performs very well for the problems considered in this work, provided that the sensitivities are obtained with sufficient accuracy.

OPTIMIZATION OF PLATES WITH CUTOUTS

This section is concerned with the determination of optimal shapes of elliptical cutouts in thin plates subjected to constant ratios of far field stresses (see Figure 2). By symmetry, only one quarter of the plate is analyzed and the corresponding boundary conditions are also shown in this figure. The outline of this section is as follows. First, parameterization of the cutout is discussed. Next, the objective function and constraints employed are addressed. Finally, numerical results, including both optimal elastic and elasto-viscoplastic designs, are presented.

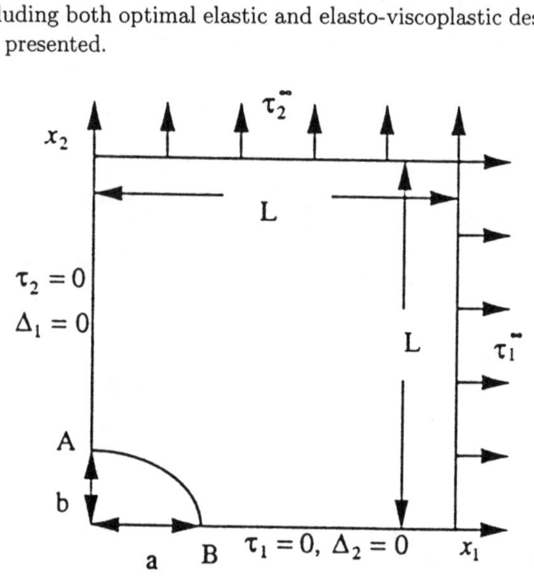

Fig. 2 Geometry and boundary conditions

Parameterization of Cutout and Geometric Sensitivities

Parameterization of geometry (shape) is required in the calculation of geometric sensitivities such as $\overset{*}{x}, d\overset{*}{A}/dA$, and $d\overset{*}{S}/dS$. Here, the elliptical cutout boundary, as shown in Figure 2, is modeled as

$$x_1 = a \cos\theta \tag{12a}$$

$$x_2 = b \sin\theta \tag{12b}$$

Parameterization of the geometries of cutouts of general shapes can be found in Lekhnitski (1968) or Wei et al. (1993).

In this work, a is fixed and $\beta = b/a$ is taken as the design variable. The corresponding sensitivities of geometrical quantities, referred to Figure 3, are as follows:

(a) Boundary Points

on BC,CD,and DE

$$\overset{*}{x}_1 = \overset{*}{x}_2 = 0, \quad \frac{d\overset{*}{S}}{dS} = 0, \quad \frac{d\overset{*}{A}}{dA} = 0 \tag{13}$$

on EA

$$\overset{*}{x}_1 = 0, \quad \overset{*}{x}_2 = \frac{L - x_2}{L - b} \quad \text{(linear assumption)}$$
$$\frac{d\overset{*}{S}}{dS} = \frac{-1}{L-b}, \quad \frac{d\overset{*}{A}}{dA} = \frac{-1}{L-b} \tag{14}$$

on AB

$$\overset{*}{x}_1 = 0, \quad \overset{*}{x}_2 = \sin\theta = \frac{x_2}{b}$$
$$\frac{d\overset{*}{S}}{dS} = \frac{b^3 x_1^2}{a^4 x_2^2 + b^4 x_1^2}, \quad \frac{d\overset{*}{A}}{dA} = \frac{1}{b} - \frac{x_2^2}{b^3} \tag{15}$$

(b) Internal Points

inside the rectangle BCDF

$$\overset{*}{x}_1 = \overset{*}{x}_2 = 0, \quad \frac{d\overset{*}{S}}{dS} = 0, \quad \frac{d\overset{*}{A}}{dA} = 0 \tag{16}$$

inside the region BFEA, using a linear assumption for the design velocity $\overset{*}{x}_2$

$$\overset{*}{x}_{1P} = 0, \quad \overset{*}{x}_{2P} = \frac{L - x_{2P}}{L - x_{2Q}} \overset{*}{x}_{2Q}$$
$$\frac{d\overset{*}{A}}{dA} = -\frac{\sqrt{1 - x_{1P}/a^2}}{L - b\sqrt{1 - x_{1P}^2/a^2}} \tag{17}$$

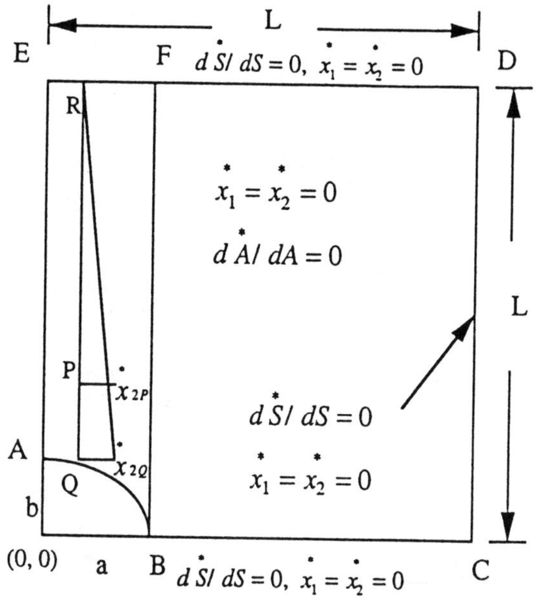

Fig. 3 Sensitivities of geometrical quantities

Objective Function and Constraints

Suppose that the far-field stresses along the x_1 and x_2 axes are given by the equations

$$\tau_1^\infty = S(t) = 8 + 4t \quad \text{and} \quad \tau_2^\infty = 0.75\, S(t) \quad \text{(MPa)} \tag{18}$$

The loading time is $T = 4$ seconds, and the time increments adopted in the explicit integration are: $\Delta t = 0.2\,s$, for $0 \le t \le 2\,s; 0.05\,s$, for $2 \le t \le 4\,s$.

The objective function is the value of ϕ at time T, where

$$\phi = \frac{1}{L_c} \int_{\partial B_c} (\sigma_{tt}(s) - \bar{\sigma}_{tt})^2 \, dS \tag{19}$$

with ∂B_c the cutout boundary, σ_{tt} the tangential stress on the cutout boundary, $\bar{\sigma}_{tt}$ the mean value of σ_{tt}, and L_c the total length of the cutout boundary. This objective function expresses the requirement of minimizing the variance of the tangential stress on the cutout, thereby requiring the tangential stress on the cutout to be as uniform as possible. *It is very important to note that, for the time-dependent elasto-viscoplastic problems discussed here, an objective function is defined as the value of ϕ at a fixed time T from the start of the deformation process.*

The sensitivity of ϕ can be obtained by the direct differentiation of equation (19). This is of the form

$$\dot{\phi} = \frac{1}{L_c} \int_A^B 2(\sigma_{tt} - \bar{\sigma}_{tt})(\dot{\sigma}_{tt} - \dot{\bar{\sigma}}_{tt}) \, dS$$

$$- \frac{\dot{L}_c}{L_c} \phi + \frac{1}{L_c} \int_A^B (\sigma_{tt} - \bar{\sigma}_{tt})^2 \, d\dot{S} \qquad (20)$$

where

$$\dot{\bar{\sigma}}_{tt} = - \frac{\dot{L}_c}{L_c} \bar{\sigma}_{tt} + \frac{1}{L_c} \int_A^B \dot{\sigma}_{tt} \, dS + \frac{1}{L_c} \int_A^B \sigma_{tt} \, d\dot{S}$$

$$L_c = \int_A^B dS \quad \text{and} \quad \dot{L}_c = \int_A^B d\dot{S}$$

The constraint is that $0.5 \leq \beta \leq 1$.

Numerical Results

As shown in Figure 2, the length of each side of the plate is assumed to be $5\,m$ and the axis of the elliptical hole along the x_1 axis is $a = 1\,m$. The mesh used in the numerical examples here is given in Wei et al. (1993). The boundary is modeled by 34 quadratic elements and the domain by 192 $Q4$ internal cells.

The results for successive iterations are shown in Table 1, where the starting value of β is 1. The optimal elasto-viscoplastic design for such a loading history is $\beta = 0.69$. It is noted that the optimizer gives a convergent solution at the 5th iterative step. The corresponding tangential stress concentration distributions along the cutout boundary, for those successive designs, at the final time T, are shown in Figure 4. Also depicted in Figure 4 is the result for the well-known optimal elastic design $\beta = 0.75$ (Banichuk 1983). Clearly, this best elastic design starts with a uniform tangential stress along the cutout. However, as inelastic deformation develops and stress redistribution occurs, the uniform pattern does not remain. Therefore, it is no longer the optimal. Instead, the best design should start with a non-uniform tangential stress distribution on the cutout and finally end up with a uniform one. For a detailed discussion of these numerical results, the reader is referred to Wei et al. (1993).

CONCLUSIONS

The first results for shape optimization of solids involving material nonlinearities, presented in this paper, are extremely encouraging. The optimal elasto-viscoplastic design is quite different from the optimal elastic design. The reasons for that are stress redistribution and path (or time) dependence of such materials undergoing inelastic deformation.

Work on optimal design of large deformation problems, involving both material and geometric nonlinearities, with applications in the design of manufacturing processes, is currently in progress.

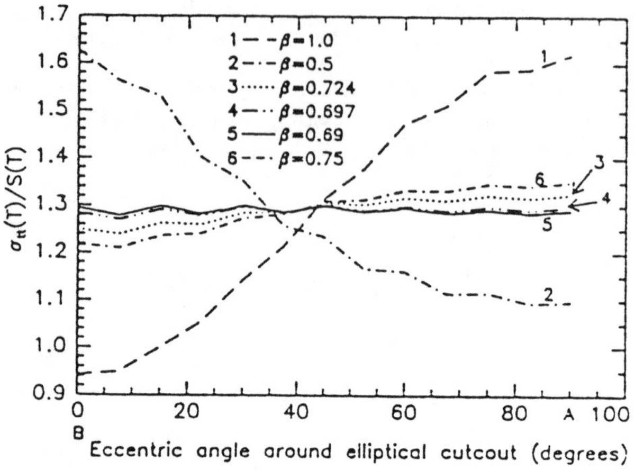

Fig. 4 Tangential stress concentration around elliptical cutout, at final time T, for different values of β

Number of Iterations	$\beta = b/a$	ϕ (MPa^2)
1	1.000	31.56600
2	0.500	14.13300
3	0.724	0.41076
4	0.697	0.03175
5	0.690	0.01052

Table 1 Values of β and ϕ at different iterations for the elasto-viscoplastic shape optimization problem.
CPU time=1.278 hours on an IBM 3090 supercomputer.

ACKNOWLEDGMENTS

The authors gratefully acknowledge the financial support provided by the U.S. National Science Foundation through Grant MSS 8922185 to Cornell University and The University of Arizona. All computing for this research was performed at the Cornell National Supercomputer Facility.

REFERENCES

Anand, L. 1982, "Constitutive Equations for the Rate-Dependent Deformation of Metals at Elevated Temperatures," *Journal of Engineering Materials and Technology*, ASME, Vol. 104, pp. 12-17.

Banichuk, N. V., 1983, *Problems and Methods of Optimal Structural Design*, Plenum, New York.

Banichuk, N. V., 1990, *Introduction to Optimization of Structures*, Springer-Verlag, New York.

Ghosh, N., Rajiyah, H., Ghosh, S. and Mukherjee, S., 1986, "A New Boundary Element Method Formulation for Linear

Elasticity," *Journal of Applied Mechanics*, ASME, Vol. 53, pp. 69-76.

Haftka, R. T. and Gürdal, Z., 1992, *Elements of Structural Optimization*, Third Revised and Expanded Edition, Kluwer Academic Publishers, Dordrecht, The Netherlands.

Haug, E. J., Choi, K. K. and Komkov, V., 1986, *Design Sensitivity Analysis of Structural Systems*, Academic Press, Baltimore and London.

Lekhnitski, S. G., 1968, *Anisotropic Plates*, 2nd Ed. (translated by Tsai, S. W. and Cheron, T.), Gordon and Breach Science Publisher, New York.

Leu, L. J. and Mukherjee, S., 1993, "Implicit Objective Integration for Sensitivity Analysis in Nonlinear Solid Mechanics," Submitted for Publication.

Mukherjee, S., 1982, *Boundary Element Methods in Creep and Fracture*, Elsevier Applied Science Publishers, New York.

Mukherjee, S. and Chandra, A., 1989, "A Boundary Element Formulation for Design Sensitivities in Materially Nonlinear Problems," *Acta Mech.*, Vol. 78, pp. 243-253.

Mukherjee, S. and Chandra, A., 1991, "A Boundary Element Formulation for Design Sensitivities in Problems Involving both Geometric and Material Nonlinearities," *Math. Comp. Modelling*, Vol. 15, pp. 245-255.

Papalambros, P. Y. and Wilde, D. J., 1988, *Principles of Optimal Design: Modeling and Computation*, Cambridge University Press.

Schittkowski, K., 1986, "NLPQL : a FORTRAN Subroutine Solving Constrained Nonlinear Programming Problems," *Annals of Operations Research*, Vol. 5, pp. 485-500.

Wei, X., Leu, L. J., Chandra, A. and Mukherjee, S., 1993, "Shape Optimization in Elasticity and Elasto-Viscoplasticity," *International Journal of Solids and Structures*, Submitted for Publication.

Zhang Q., Mukherjee, S. and Chandra, A., 1992a, "Design Sensitivity Coefficients for Elasto-Viscoplastic Problems by Boundary Element Methods," *International Journal for Numerical Methods in Engineering*, Vol. 34, pp. 947-966.

Zhang Q., Mukherjee, S. and Chandra, A., 1992b, "Shape Design Sensitivity Analysis for Geometrically and Materially Nonlinear Problems by the Boundary Element Method," *International Journal of Solids and Structures*, Vol. 20, pp. 2503-2525.

Inverse Problems in Engineering: Theory and Practice
ASME 1993

A NUMERICAL SOLUTION WITH ERROR ESTIMATING FOR INVERSE ELASTICITY PROBLEMS

Antoinette M. Maniatty
Department of Mechanical Engineering,
Aeronautical Engineering, and Mechanics
Rensselaer Polytechnic Institute
Troy, New York

ABSTRACT

This paper is concerned with the solution of a class of inverse problems in linear elasticity which involve the determination of unknown traction boundary conditions on an elastic body using measured displacements from another region of the body. A statistical method is used to determine the unknown traction boundary condition and to estimate bounds on the error in the solution.

INTRODUCTION

In many applications in solid mechanics, the problem arises where the boundary conditions on the body of interest are not sufficiently known in order to give a direct solution. For example, consider a contact problem where it may be difficult to measure accurately the conditions on the boundary in the contact region. On the other hand, additional information regarding parts of the solution or over-specified boundary conditions on another part of the boundary may be more easily measured. For the applications considered herein, that could be in the form of measured internal displacements or measured tractions and displacements on part of the boundary near the region with unknown boundary conditions. This results in an inverse problem where the goal is to use this additional information to determine the unknown boundary condition. Once the unknown boundary condition is known, the direct problem can then be solved for the displacement, stress, and strain fields. In general, inverse problems are unstable and may not have unique solutions.

Problems of this classification involving linear elastic isotropic materials were studied previously by Maniatty et al. (1989) who used simple diagonal regularization in a finite element framework to determine the unknown traction boundary conditions. Spatial regularization was introduced in conjunction with the boundary element method by Zabaras et al. (1989) and with the finite element method in Schnur and Zabaras (1990). A polynomial approximation technique referred to as the keynode method was also presented in Schnur and Zabaras (1990). Linear elastic static problems will be the focus of the investigation herein as well. A statistical approach following the method presented in Tarantola (1987) will be presented which has the added feature of providing an easy method for estimating bounds on the error in the solution. Both isotropic and anisotropic cases will be investigated in a finite element framework. A more detailed analysis with a comparison of this method to the spatial regularization method is presented in Maniatty and Zabaras (1993).

INVERSE ELASTICITY PROBLEM DEFINITION

The definition of the inverse elasticity problem discussed herein follows that of the usual two-dimensional direct elasticity problem with the exception that the boundary conditions are unspecified on part of the boundary. Instead, additional displacements are specified approximately at discrete locations either internal to the domain or on another part of the boundary where tractions are already specified. Summarizing the problem in equation form

$$div\ \mathbf{T} = \mathbf{0} \qquad \text{on B} \qquad (1)$$

$$\mathbf{E} = \frac{1}{2}[\nabla \mathbf{u} + (\nabla \mathbf{u})^T] \tag{2}$$

$$\mathbf{T} = \boldsymbol{\mathcal{L}}[\mathbf{E}] \tag{3}$$

$$\mathbf{e}_i \cdot \mathbf{u} = \hat{u}_i \qquad \text{on } \partial B_{1i} \tag{4}$$

$$\mathbf{e}_i \cdot (\mathbf{Tn}) = \hat{t}_i \qquad \text{on } \partial B_{2i} \tag{5}$$

$$\mathbf{e}_i \cdot \mathbf{u}(\mathbf{x}_\beta) \approx \hat{u}_i^*(\mathbf{x}_\beta), \qquad \beta = 1, N_s \tag{6a}$$

$$\mathbf{x}_\beta \in (B \cup \partial B_{2i}), \quad \mathbf{x}_\beta \notin (\partial B_{1i} \cup \partial B_{3i}). \tag{6b}$$

The first 3 equations are the usual field equations prescribed on the body B for linear elasticity where **T** is the stress tensor, **E** is the strain tensor, **u** is the displacement field, and $\boldsymbol{\mathcal{L}}$ is the fourth order elasticity tensor which operates on the strain tensor. Equations (4) and (5) are the usual displacement $\hat{u}_i$ and traction $\hat{t}_i$ boundary conditions specified on the parts of the boundary ∂B_{1i} and ∂B_{2i}, respectively, where $\partial B_{1i} \cap \partial B_{2i} = \varnothing$ for i=1,2. In addition, there remains a third part of the boundary $\partial B_{3i} = \partial B_i - (\partial B_{1i} \cup \partial B_{2i})$ with no prescribed boundary condition, where ∂B_i is the boundary of B. Equations (6a) and (6b) define the additional displacements $\hat{u}_i^*$ prescribed approximately at discrete locations, $\mathbf{x}_\beta$, $\beta = 1, N_s$, either internal to the domain or on ∂B_{2i} where tractions have already been prescribed. It is assumed that these additional displacements are known from measurements which have some random error.

FINITE ELEMENT DISCRETIZATION

A discretized form of the problem resulting from applying the finite element method has the following form

$$\{\mathbf{f}(\{\hat{\mathbf{t}}\})\} + \{\mathbf{f}(\{\tilde{\mathbf{t}}\})\} = [\mathbf{K}]\{\mathbf{u}\} \tag{7}$$

where $\{\mathbf{f}(\{\hat{\mathbf{t}}\})\}$ and $\{\mathbf{f}(\{\tilde{\mathbf{t}}\})\}$ are the parts of the force vector due to the known and unknown tractions, respectively, and where $[\mathbf{K}]$ and $\{\mathbf{u}\}$ are the usual linear elastic stiffness matrix and vector of nodal displacements. Writing a Taylor expansion for $\{\mathbf{f}(\{\tilde{\mathbf{t}}\})\}$ around $\{\tilde{\mathbf{t}}\} = \{\mathbf{0}\}$ and using equation (7) above gives

$$\{\mathbf{f}(\{\tilde{\mathbf{t}}\})\} = [\mathbf{K}]\left[\frac{\partial \mathbf{u}}{\partial \tilde{\mathbf{t}}}\right]\{\tilde{\mathbf{t}}\} = [\mathbf{K}]\{\mathbf{u}\} - \{\mathbf{f}(\{\hat{\mathbf{t}}\})\}$$

or

$$\left[\frac{\partial \mathbf{u}}{\partial \tilde{\mathbf{t}}}\right]\{\tilde{\mathbf{t}}\} = \{\mathbf{u}\} - \{\bar{\mathbf{u}}\} \tag{8}$$

where

$$\{\bar{\mathbf{u}}\} = [\mathbf{K}]^{-1}\{\mathbf{f}(\{\hat{\mathbf{t}}\})\}$$

is the vector of nodal displacements due to the tractions on ∂B_{2i} with zero tractions on ∂B_{3i}. Equation (8) is exact since the system is linear in $\{\tilde{\mathbf{t}}\}$. The matrix $[\partial \mathbf{u}/\partial \tilde{\mathbf{t}}]$ is easily computed from

$$\left[\frac{\partial \mathbf{u}}{\partial \tilde{\mathbf{t}}}\right] = [\mathbf{K}]^{-1}\left[\frac{\partial \mathbf{f}}{\partial \tilde{\mathbf{t}}}\right].$$

Since only part of $\{\mathbf{u}\}$ is prescribed, i.e. $\{\mathbf{u}^*\} = [\mathbf{Q}]\{\mathbf{u}\}$ where $\{\mathbf{u}^*\}$ is the vector of modeled displacements corresponding to the measured displacements $\{\hat{\mathbf{u}}^*\}$, equation (8) reduces to

$$[\mathbf{S}^*]\{\tilde{\mathbf{t}}\} = \{\mathbf{u}^*\} - [\mathbf{Q}]\{\bar{\mathbf{u}}\} \tag{9}$$

where

$$[\mathbf{S}^*] = [\mathbf{Q}]\left[\frac{\partial \mathbf{u}}{\partial \tilde{\mathbf{t}}}\right]$$

is the sensitivity matrix. Now the problem is to minimize the following error measure

$$\begin{aligned} E &= \frac{1}{2}(\{\mathbf{u}^*\} - \{\hat{\mathbf{u}}^*\})^T(\{\mathbf{u}^*\} - \{\hat{\mathbf{u}}^*\}) \\ &= \frac{1}{2}([\mathbf{S}^*]\{\tilde{\mathbf{t}}\} + [\mathbf{Q}]\{\bar{\mathbf{u}}\} - \{\hat{\mathbf{u}}^*\})^T \\ &\quad ([\mathbf{S}^*]\{\tilde{\mathbf{t}}\} + [\mathbf{Q}]\{\bar{\mathbf{u}}\} - \{\hat{\mathbf{u}}^*\}) \end{aligned} \tag{10}$$

with respect to $\{\tilde{\mathbf{t}}\}$. The solution to this minimization for $\{\tilde{\mathbf{t}}\}$ is generally unstable and may not be unique. Therefore, a special technique for solving inverse problems is required. A statistical method is employed here so that error estimates in the solution can be obtained.

STATISTICAL APPROACH

Since inverse problems, such as the one described herein, are usually unstable, additional a priori information regarding the solution is often used to impose stability on the solution. In this case, instead of solving the minimization given in equation (10), another minimization close to it will be solved which incorporates such a priori information. This new minimization can be derived using a statistical interpretation of the problem. Following the method outlined in Tarantola (1987), if the errors in the data and the model are assumed to follow Gaussian distributions, and the a priori information on the unknown tractions is also Gaussian, then the probability density function $p_t(\{\tilde{\mathbf{t}}\})$ representing the information on the solution $\{\tilde{\mathbf{t}}\}$ is

$$p_t(\{\tilde{t}\}) = C\exp[-m(\{\tilde{t}\})] \tag{11a}$$

where

$$m(\{\tilde{t}\}) = \frac{1}{2}[([S^*]\{\tilde{t}\} - \{u'\})^T [C_D]^{-1} ([S^*]\{\tilde{t}\} - \{u'\}) + \{\tilde{t}-\tilde{t}_o\}^T [C_\Omega]^{-1}\{\tilde{t}-\tilde{t}_o\}] \tag{11b}$$

and

$$\{u'\} \equiv \{\hat{u}^*\} - [Q]\{\bar{u}\} .$$

The matrix $[C_D]$ is a covariance operator combining uncertainties in both the data and the model, $[C_\Omega]$ is a covariance operator expressing uncertainties in the a priori information on the solution, $\{\tilde{t}_o\}$ is an initial guess for $\{\tilde{t}\}$, and C is a constant. It can be shown that the Gaussian assumption has the simplifying result that the errors in the data $\{u'\}$ and the modelization errors simply combine by addition of the respective covariance operators, i.e.

$$[C_D] = [C_{u'}] + [C_T]$$

where $[C_{u'}]$ and $[C_T]$ represent covariance operators expressing uncertainties in the data and the model, respectively. The expected solution for $\{\tilde{t}\}$ maximizes $p_t(\{\tilde{t}\})$ and therefore minimizes $m(\{\tilde{t}\})$. Solving this new minimization problem gives the following

$$\langle\{\tilde{t}\}\rangle = ([S^*]^T[C_D]^{-1}[S^*] + [C_\Omega]^{-1})^{-1} ([S^*]^T[C_D]^{-1}\{u'\} + [C_\Omega]^{-1}\{\tilde{t}_o\}) \tag{12}$$

where $\langle\{\tilde{t}\}\rangle$ indicates the mean value of $\{\tilde{t}\}$ which is taken to be the estimated solution. Furthermore, the covariance operator for the solution can be shown to be (Tarantola, 1987)

$$[C_t] = ([S^*]^T[C_D]^{-1}[S^*] + [C_\Omega]^{-1})^{-1} . \tag{13}$$

The standard deviation in the solution $\{\tilde{t}\}$ is then just the square root of the diagonal terms of $[C_t]$. This information can be used to estimate error bounds.

In reality, it is unlikely that either the errors in the data, $\{u'\}$, or in the model, which is a finite element approximation, are actually Gaussian. Furthermore, note that the "data", $\{u'\}$, includes not only the measured displacements $\{\hat{u}^*\}$, but also $\{\bar{u}\}$ which is a function of the tractions $\{\hat{t}\}$ on ∂B_{2i} and is defined using a finite element approximation. It is postulated that the Gaussian assumption can be used here because, despite the inaccuracy in describing the overall distributions, it still provides the desired information and has the added advantage of significantly simplifying the problem. Observe that the only characteristics of interest here regarding the density function of $\{\tilde{t}\}$ are the position of the "center" and the "error bounds". The position of the "center" (maximum likelihood point) will not be affected as long as the actual density functions are symmetric. Furthermore, the behavior of the density functions far from the center is only crucial if stray points far from the center exist. If the error bounds on the data and the model are known fairly precisely and no outliers exist, then a normal distribution will give a conservative estimate of the error. This is often the case because the error bounds on the measured data should be known from the sensor characteristics, and the error in the finite element method is readily estimated using established techniques, such as that described in Johnson (1987).

The results in equations (12) and (13) can be further simplified by neglecting any correlation in the errors in both the data and model and letting the standard deviations in the data $\sigma_{u'}$ and the model σ_T be constant, i.e. not vary from point to point in the structure. Then $[C_D]$ becomes

$$[C_D] = (\sigma_{u'}^2 + \sigma_T^2)[I] = \sigma_D^2[I]$$

where σ_D is the standard deviation which combines the uncertainty of the data and the model. This simplifies equations (12) and (13) to

$$\langle\{\tilde{t}\}\rangle = ([S^*]^T[S^*] + \sigma_D^2[C_\Omega]^{-1})^{-1} ([S^*]^T\{u'\} + \sigma_D^2[C_\Omega]^{-1}\{\tilde{t}_o\}) \tag{14}$$

$$[C_t] = \sigma_D^2([S^*]^T[S^*] + \sigma_D^2[C_\Omega]^{-1})^{-1} . \tag{15}$$

The error in the finite element model is generally correlated. Assuming uncorrelated error gives a conservative estimate, so the assumption is taken to be acceptable, although more work needs to be done in this area to obtain better estimates for $[C_T]$ and $[C_{u'}]$.

It is still necessary to define $[C_\Omega]$, which contains the a priori information on the solution. This can be done by assuming some smoothness in the difference between the solution and the initial guess as is done in the regularization method described in Beck et al. (1985) and Tikhonov and Arsenin (1977). Comparing the function m to be minimized with respect to $\{\tilde{t}\}$ given in equation (11b) to that arising in the regularization method, it can be seen that $[C_D]^{-1}$ acts like a regularization matrix. Assuming spatial regularization like that used in Schnur and Zabaras [3], the last term in equation (11b), $\{\tilde{t}-\tilde{t}_o\}^T[C_\Omega]^{-1}\{\tilde{t}-\tilde{t}_o\}$, can then be defined as the discretized form of the following smoothing functional

$$\Omega = \sum_{i=1}^{2}\left[\alpha_0^i\int_{\partial B_{3i}}(\tilde{t}_i-\tilde{t}_{oi})^2ds+\alpha_1^i\int_{\partial B_{3i}}\left(\frac{\partial\tilde{t}_i}{\partial s}-\frac{\partial\tilde{t}_{oi}}{\partial s}\right)^2ds+\alpha_2^i\int_{\partial B_{3i}}\left(\frac{\partial^2\tilde{t}_i}{\partial s^2}-\frac{\partial^2\tilde{t}_{oi}}{\partial s^2}\right)^2ds\right] \quad (16)$$

where α_0^i, α_1^i, and α_2^i $(i = 1,2)$ are the zeroth, first, and second order regularization parameters, respectively for the ith degree of freedom. So $[C_\Omega]^{-1}$ can be expressed as

$$[C_\Omega]^{-1} = \sum_{i=1}^{2}(\alpha_0^i[R_0]^i+\alpha_1^i[R_1]^i+\alpha_2^i[R_2]^i)$$

where $[R_j]^i$ $(j = 0,1,2)$ can be evaluated by discretizing equation (16) and using the finite element interpolation functions. Therefore, the regularization parameters are inversely related to the degree of uncertainty in the a priori information regarding the smoothness. In other words, as the degree of uncertainty in the jth order smoothness increases, the values of α_j^i should decrease.

In general, the regularization parameters must be selected using a trial and error method. The regularization parameters must be small enough so as not to affect (dominate) the solution, but large enough to stabilize it. In this work, the largest regularization parameters that do not start to significantly affect the solution are sought because they give the smallest error bars. The solution should be fairly insensitive to regularization parameters over one or more orders of magnitude. If the solution is always sensitive to the choice of regularization parameters, then a stable solution cannot be found using this method.

NUMERICAL EXAMPLES

Three cases involving the same problem geometry but different material symmetries are presented to demonstrate the algorithm and to investigate the effect of anisotropy on the results. In order to test the algorithm in each case, the "experiment" will be performed numerically by prescribing the "unknown" traction conditions and then solving the direct problem with the finite element method to determine the displacements at the "sensor" locations. Uniform random error will be superimposed on these computed displacements to simulate errors in the measurements. Then these approximate displacements will be used as input to the inverse algorithm to solve for the "unknown" tractions. These predicted tractions can then be compared to the original prescribed tractions to determine the accuracy of the solution.

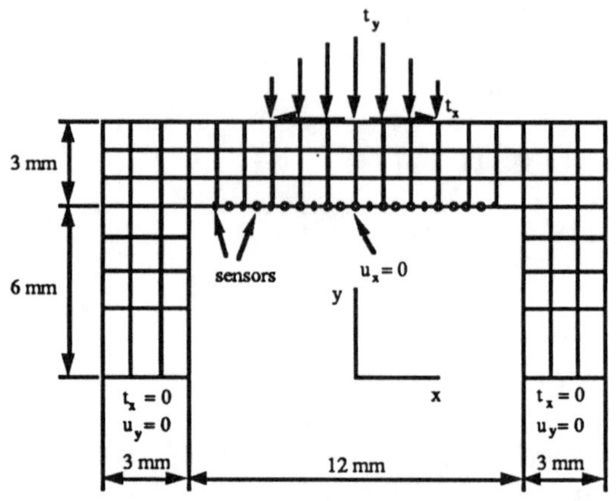

FIGURE 1. PROBLEM GEOMETRY.

The problem to be analyzed involves a channel section on a relatively frictionless support with the unknown traction condition on part of the outside of the channel. The geometry of the problem with the finite element mesh is shown in Figure 1. Eight node isoparametric elements are used. The problem is assumed to be in plane strain. The boundary conditions are traction free everywhere except in the regions indicated. Specifically, the bottom edges at $y = 0$ are fixed in the y-direction and free in the x-direction. To prevent rigid body motion, one node at $x = 0$ on the inside of the channel (with the arrow indicating $u_x = 0$) is fixed in the x-direction, but free in the y-direction. Shear and normal traction distributions are applied on the top of the channel on the region $-3\ mm < x < 3\ mm$. The equations describing these traction distributions are

$$t_x = 0.2x^3 + 3x$$

$$t_y = \frac{5}{9}x^2 - 10$$

where x is in *mm* and t_x and t_y are in MPa. The displacements resulting from these applied tractions are determined at the 21 "sensor" locations shown in Figure 1. Then these displacements with superimposed random error are used to inversely determine the applied traction condition on the top of the channel.

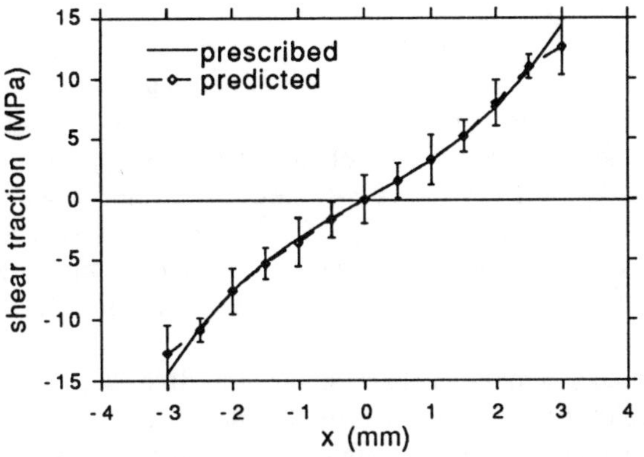

FIGURE 2. PREDICTED SHEAR TRACTION DISTRIBU-TIO FOR ISOTROPIC CASE.

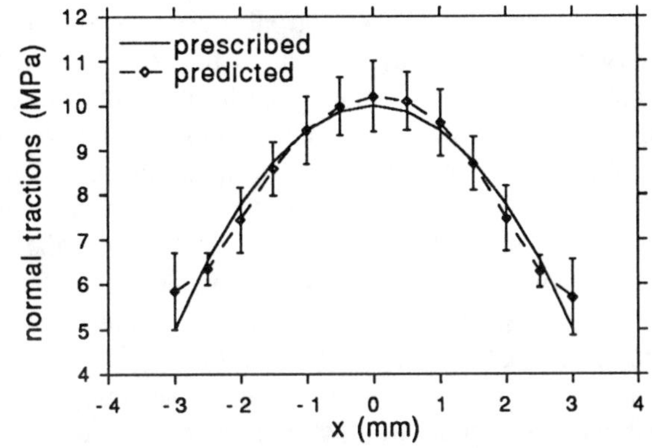

FIGURE 3. PREDICTED NORMAL TRACTION DISTRIBU-TION FOR ISOTROPIC CASE.

In the first case, the material is assumed to be isotropic, so

$$\mathbf{T} = 2\mu\mathbf{E} + \lambda(tr\mathbf{E})\,\mathbf{I}$$

where μ and λ are the Lamé parameters, *tr* is the trace operator and **I** is the identity tensor. The Lamé parameters are taken to be μ =75.4 GPa and λ =121.4 GPa. Superimposed uniform random error is added to the displacements where the random error ε is such that $-a < \varepsilon < a$, and $a = 10^{-7}$ *mm* which is approximately 0.1% of the "measured" displacements in the y-direction. Since no known tractions are applied $\{\mathbf{u}'\} = \{\hat{\mathbf{u}}^*\}$ so $\sigma_{u'} = \sigma_*$ is just the standard deviation in the displacements at the "sensors". Furthermore, since the random error in the displacement data is a uniform distribution, the standard deviation is $\sigma_* = 0.577a$. In this case modelization error is neglected since the same model (finite element model) is used in both the experiment and in the inverse problem formulation, so

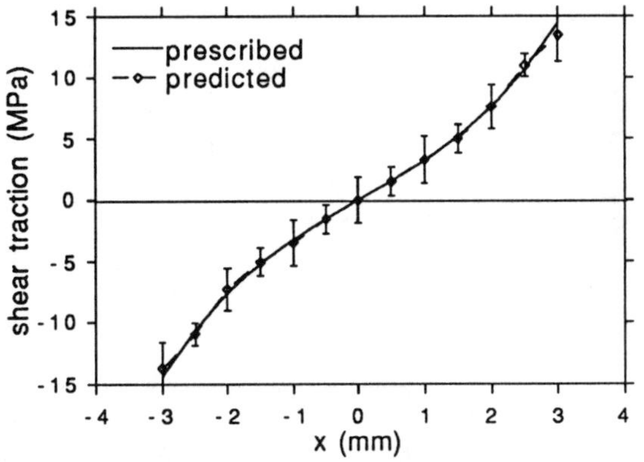

FIGURE 4. PREDICTED SHEAR TRACTION DISTRIBU-TION FOR ANISOTROPIC CASE WITH THE ORIENTA-TION OF THE LATTICE AT 0°.

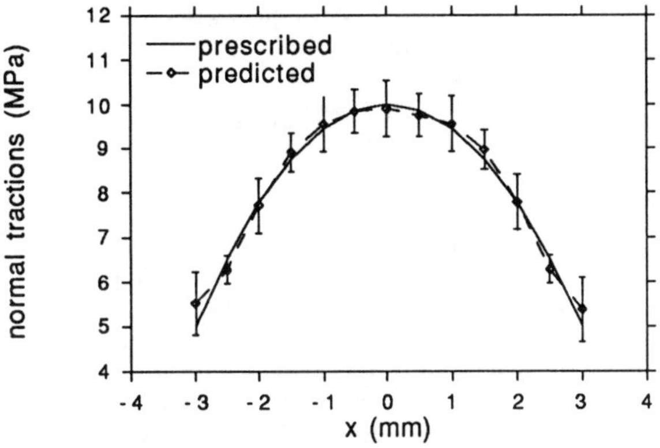

FIGURE 5. PREDICTED NORMAL TRACTION DISTRIBU-TION FOR ANISOTROPIC CASE WITH THE ORIENTA-TION OF THE LATTICE AT 45°.

$\sigma_D = \sigma_{u'}$ in this case. Only first order regularization was used with $\alpha_1^1 = 0.3\ mm/MPa^2$ and $\alpha_1^2 = 3 mm/MPa^2$. Previous work by Schnur and Zabaras (1990) and Maniatty and Zabaras (1993) shows that first order regularization is often the most effective in these problems. Finally, the initial guess for the unknown tractions is taken to be $\{\mathbf{t}_o\} = \{\mathbf{0}\}$ since it is assumed no other information is known about the tractions. The resulting predicted shear and normal tractions are plotted in Figures 2 and 3 with the original prescribed tractions and the predicted error bar. The error bars are three standard deviations in the tractions computed from the covariance matrix given in equation (15). Three standard deviations for a Gaussian distribution gives a probability of 99.7% that the solution falls within the error bars. Since the distributions here may not actually be Gaussian, three standard deviations does not have the same meaning,

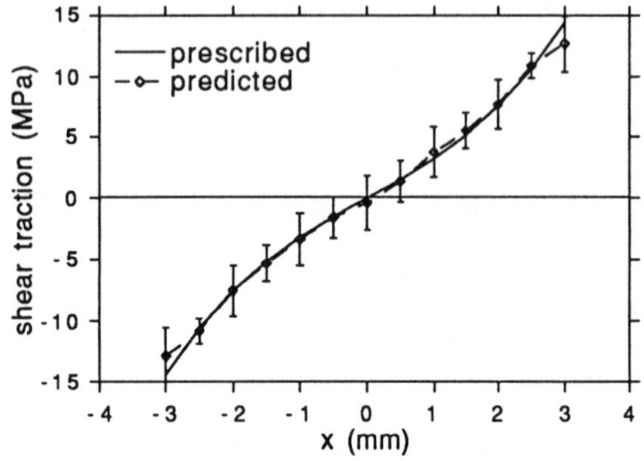

FIGURE 6. PREDICTED SHEAR TRACTION DISTRIBUTION FOR ANISOTROPIC CASE WITH THE ORIENTATION OF THE LATTICE AT 45°.

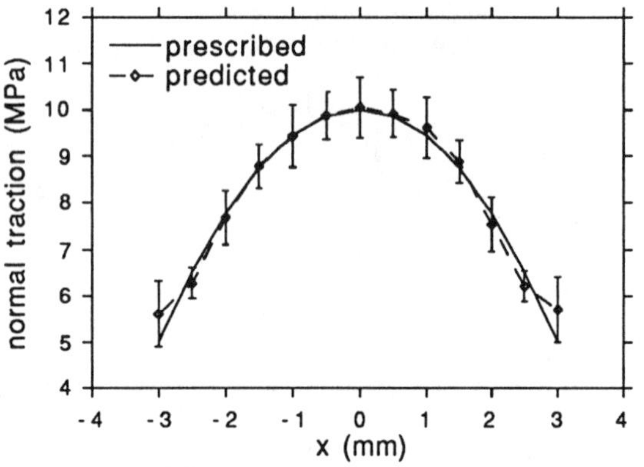

FIGURE 7. PREDICTED NORMAL TRACTION DISTRIBUTION FOR ANISOTROPIC CASE WITH THE LATTICE ORIENTATION AT 45°.

although it may be expected that three standard deviations is conservative since the actual error distribution was a uniform distribution which cannot have any outlier points. In both the shear and normal tractions, the error bars seem to be conservative except towards the ends of the distributions. This may be due to the way the finite element method handles traction boundary conditions at discontinuities such as those that occur at the ends of the distributions. Additional investigation of this problem is needed. This behavior was also observed in Schnur and Zabaras (1990) who found that they needed to specify the end conditions.

In the second and third example, the channel is assumed to be made of copper which is formed as a single crystal. Copper crystals have a cubic symmetry and have a high degree of anisotropy. For a material with cubic symmetry, equation (3) can be expressed as

$$\mathbf{T} = 2\mu\mathbf{E} + \lambda(tr\mathbf{E})\mathbf{I} - 2\beta(\mathcal{C}:\mathbf{E})$$

where μ, λ, and β are the material constants and $\mathcal{C}$ is a fourth order tensor which is determined by the symmetry of the cube. In the coordinates of the lattice the components of $\mathcal{C}$ are $\mathcal{C}_{ijlk} = 1$ if $i = j = k = l$ and $\mathcal{C}_{ijlk} = 0$ otherwise. The parameters for copper are $\mu = 75.4$ GPa, $\lambda = 121.4$ GPa, and $\beta = 51.9$ GPa (Hertzberg, 1976). In both examples, the same random error is applied to the displacement data, and the same regularization parameters are chosen as in the first example assuming an isotropic material. In example 2, the lattice coordinate coincides with the global coordinates, and in example 3, the lattice is rotated 45 degrees in the plane of the problem. The results for example 2 are shown in Figures 4 and 5, and the results from example 3 are shown in Figures 6 and 7. There is little difference between each of the three examples. The anisotropic results appear to be slightly better. The displacements at the sensors for both anisotropic cases were higher than those for the isotropic case. For the case where the lattice and global coordinates coincided, the displacements were approximately double those in the isotropic case, and for the case where the lattice was oriented at 45 degrees from the global axis, the displacements were approximately 20% higher than in the isotropic case. Since the same error was added to the displacement data in each case, the error was less as a percentage of the total displacement in the anisotropic cases which may explain the slight difference in the results.

CONCLUSION AND FUTURE WORK

An algorithm based on statistical theory in conjunction with the finite element method for solving inverse problems in linear elasticity has been presented. The problems investigated involve the solution of unknown traction boundary conditions on elastic bodies given approximately "measured" displacements either on another part of the boundary where the tractions are already known or internal to the body. Numerical examples involving both isotropic and cubic materials were presented. The algorithm was shown to handle each problem fairly well, and there was little difference between the results for the different material symmetries. The algorithm only seemed to fail in predicting the tractions at the endpoints of the distributions well. This may be due to the way the finite element method approximates the discontinuity at the endpoints in the traction distribution, in which case, a modification to the sensitivity matrix may be required.

Additional work is needed to examine the effect of using Gaussian distributions and neglecting correlations in the errors in both the model and the data. Monte Carlo simulations should be carried out to investigate these effects. In this work, the results were compared against a direct finite element model of the actual problem. To better understand the modelization errors, it will be necessary to investigate classical problems with analytic solutions.

ACKNOWLEDGEMENTS

The author wishes to thank J. Papa for his assistance in programming and running some of the test cases. This work has been supported by the Henry Luce Foundation.

REFERENCES

Beck, J. V., Blackwell, B., and St. Clair, C. R., 1985, *Inverse Heat Conduction: Ill-Posed Problems*, Wiley-Interscience, New York.

Hertzberg, R. W., 1976, *Deformation and Fracture Mechanics of Engineering Materials*, Wiley, New York.

Johnson, C., 1987, *Numerical Solution of Partial Differential Equations by the Finite Element Method*, Cambridge University Press, Cambridge.

Maniatty, A. M., Zabaras, N. J., and Stelson, K., 1989, "Finite Element Analysis of Some Inverse Elasticity Problems," *ASCE Journal of Engineering Mechanics*, Vol. 115, pp. 1302-1316.

Maniatty, A. M. and Zabaras, N. J., 1993, "Investigation of Regularization Parameters and Error Estimating in Inverse Elasticity Problems," *International Journal for Numerical Methods in Engineering*, accepted for publication.

Schnur, D. S. and Zabaras, N. J., 1990, "Finite Element Solution of Two-Dimensional Inverse Elasticity Problems Using Spatial Smoothing," *International Journal for Numerical Methods in Engineering*, Vol. 30, pp. 57-75.

Tarantola, A., 1987, *Inverse Problem Theory: Methods for Data Fitting and Model Parameter Estimation*, Elsevier, New York.

Tikhonov, A. N. and Arsenin, V. Y., 1977, *Solution of Ill-Posed Problems*, V. H. Winston, Washington D. C.

Zabaras, N. J., Morellas, V., and Schnur, D. S., 1989, "A Spatially Regularized Solution of Inverse Elasticity Problems Using the Boundary Element Method," *Communications in Applied Numerical Methods*, Vol. 5, pp. 547-553.

Inverse Problems in Engineering: Theory and Practice
ASME 1993

CONE-BEAM TOMOGRAPHY: AN INTRODUCTION AND ITS ENGINEERING CHALLENGES

Bruce D. Smith
Department of Mathematical Sciences
University of Cincinnati
Cincinnati, Ohio

Abstract

Tomographic scanners have been developed to the point that today they are capable of producing quality two-dimensional images of the internal structure of the human body despite the fact that forming these images are ill-posed. To make three-dimensional images accurately and economically future scanners may make use of a three-dimensional data collection scheme known as "cone-beam" tomography. However, to produce images from cone-beam data new computer algorithms need to be developed. Of course, these algorithms such be computationally efficient and accurate. Computationally efficiency is largely determined by the type of algorithm used. Accuracy is greatly influenced by the trajectory the scanner uses when collecting the data. Unfortunately, the trajectories that result in the more accurate images are more difficult to implement mechanically. Consequentially, accuracy is connected to the mechanical complexity of the scanner.

1 INTRODUCTION

1.1 What Is Tomography

The familiar medical procedure referred to as a "CAT scan" involves taking a series of x-rays of a patient and combining them with the help of a computer to form a composite image of the patient. CAT scans have been a success story in medical technology. First introduced in the early 1970's, the x-ray CAT scanners have evolved to the point that today they are routinely used to produce high quality images of the interior structures of the human body.

Tomography has been successfully applied to fields other than medicine. Tomography is routinely used in nondestructive evaluation, geology, and electron microscopy. Other more exotic applications involve determining the distribution of water temperatures in the ocean, the distribution of the air-born pollutants in the atmosphere, the distribution of the temperatures of the gases inside a flame, the distribution of radioactivity in reactor rods inside commercial nuclear power plants, and the distribution of liquid gas interface in two phase flows in pipes. As witnessed by the large number of successful applications that have followed, Godley Hounsfield and Allen M. Cormack were rightfully awarded a Nobel prize in medicine and physics in 1979 for their pioneering work in two-dimensional tomography.

1.2 Why consider cone-beam tomography

There are many instances where the whole three--dimensional reconstruction of an object is needed rather than simply one two-dimensional cross-section of the object. The most common way that three-dimensional images are produced today is by producing a series of two-dimensional cross-sectional images and then "stacking" them one on top of the other. However, stacking has several drawbacks. One such limitation stems from the fact that the x-ray beam used in collecting the data is of finite width. This results in cross-sectional reconstructions that actually have finite width rather than, as many times assumed, being infinitesimally thin. The only way to increase the spatial resolution in the axial direction (that is, the direction normal to the cross-sectional plane) beyond the width of the x-ray beam is to overlap the cross-sectional reconstructions (rather than simply stacking one on top of the other) and to perform a deconvolution of the overlapping cross-sections. Unfortunately, this overlapping of measurements increases the dose to the patient.

Other drawbacks of stacking result from the substantial amount of time that is required to produce the cross-sectional images needed. The ability to obtain an accurate *in vivo* measurement of the cardiovascular blood volume and blood flow, and the ability to see the motion of the heart as it beats would significantly aid in the research, diagnosis, and treatment of heart disease. To achieve this would require measuring all the data needed to reconstruct the whole three-dimensional heart in about 50 milliseconds, which is not feasible by stacking.

Making three-dimensional images using cone-beam tomography, which is a relatively new area in tomography, has several advantages over the more conventional "stacking" of cross-sections. First, it eliminates the need for stacking by making it possible to collect a sufficient amount of information so that a complete three-dimensional image can be produced from the data collected in one "scan" of the object. In another medical application – Single Proton Emission Computed Tomography (SPECT) – tomography can improve the SNR.

The relatively poor SNR associated with SPECT is due to the small number of photons that are counted. It has been demonstrated (Jaszczak, 1986) that equipping large-field-of-view scintillation cameras with cone-beam collimators substantially increases the number of photons that are counted, and hence, increases the SNR. The use of cone-beam collimator requires the use of cone-beam tomography.

1.3 What is cone-beam tomography.

Cone-beam tomography is best explained by comparing it with other tomographic procedures that have been used. The earliest CT scanners used the so called "parallel" geometry. As seen in the top section of Figure 1, the x-rays are emitted on one side of the object and are detected on the other. From the measurements of the emitted and detected x-rays intensities, an estimate can be made of the line integral along the line containing the x-ray source and detector. Note that in this geometry the x-rays are parallel to each other and are restricted to a two-dimensional plane. After one "projection" of the object is obtained, the x-ray source and detector are rotated and a second " projection" of the object is obtained. The process is repeated until the x-ray source and detector have been rotated by 180 degrees. This gives a complete data set for obtaining a two-dimensional cross-section of the object.

To speed up the data collection process the "fan-beam" geometry was developed. As illustrated in the middle section of Figure 1 the x-rays in this geometry diverge two-dimensional from a point x-ray source forming a "fan" of x-rays. Once again, the x-rays and detector are rotated 360 degrees around the object to obtain a complete data set.

The cone-beam geometry is illustrated in the bottom portion of Figure 1. The x-rays diverge three-dimensionally from the source forming a "cone" of rays. To detect the x-rays a two-dimensional detector system is placed on the opposite side of the object. The data set that is used to reconstruct the whole three-dimensional object is obtained by moving the source and detector about the object along some specified trajectory.

2 DEVELOPING FAST AND ACCURATE ALGORITHMS

In order to achieve the potential benefits associated with cone-beam tomography, algorithms that form accurate images from cone-beam data sets need to be developed. Developing algorithms that have acceptable execution times as well as being accurate are exceptionally challenging to develop. The additional "dimensionality" posed by being a three-dimensional algorithm rather than the conventional two-dimensional algorithm substantially increases the number of computer operations that need to be performed. Tomographic algorithms can be broadly classified as being an iterative or an non-iterative method. After reviewing both methods the advantages and disadvantages of each method will be discussed. Then an aspect that is unique to cone-beam tomography which greatly effects the speed and accuracy obtainable will be discussed.

2.1 Non-iterative methods

The key to many of the non-iterative methods is the central slice theorem. The central slice theorem relates the data measured to the two-dimensional Fourier transform of the object. Consider, as shown in Figure 2, the case where parallel x-rays penetrating the object at an angle θ with respect to the x-axis. The data measured for a fixed θ provides a one-dimensional profile or "projection" of the object. This projection of the object is denoted by the function $P(\ell, \theta)$ in Figure 2 where the scalar ℓ is the distance along the face of the detector. The central slice theorem states that the one-dimensional Fourier transform of such a projection is equal to the portion or "slice" of the two-dimensional Fourier transform of the object. This slice intersects the origin in the frequency domain. (Note the work "central" in the name of the theorem.) In particular, the one-dimensional Fourier transform of a projection is equal to the two-dimensional Fourier transform of the object along the line that intersects the origin which makes an angle θ with the x-axis in the frequency domain (which is denoted by ω_x in Figure 2).

To see how the central slice theorem can be used to reconstruct an object, consider the following. If what was described in the last paragraph is repeated for each θ between 0 and 180 degrees, then one would produce the two-dimensional Fourier transform of the object along every line that intersects the origin. Hence, the two-dimensional Fourier transform of the object would be known at each point in its domain. Knowing the two-dimensional Fourier transform at each point in its domain enables one to perform an inverse two-dimensional Fourier transform to obtain the object. Hence, the central slice theorem used in the manner just described can be used to reconstruct the object from its projections.

The methods just described for reconstructing the object is referred to as the "Direct Fourier Method." Its numerical implementation has been discussed in (Mersereau, 1976) and (Sezan and Stark, 1984). The central slice theorem can be used to develop reconstruction methods other than the "Direct Fourier Method." The "ρ-filter" method developed in (P. Smith,1973) is an example of such a method. Another is the "convolution backprojection" method was proposed in (Bracewell, 1960) and (Ramachandran and Laksminarayanan, 1971). Historically speaking, the convolution backprojection method has been the most widely used method.

2.2 Iterative methods

Historical speaking, perhaps the most widely used iterative method is the method known as ART. In ART the problem of forming images from their projections is formulated as a matrix inversion problem. In this formulation the projection data, is placed in a column vector and is equal to the matrix product of a matrix, which is large and sparse, with a second column vector whose entries are the values of the pixels that represent the image. ART is an iterative technique specially developed to solve the matrix equation that results in this formulation. ART is similar to the Gauss–Seidel method, which is a well-known technique for solving matrix equations in general.

Another iterative method that has been proposed recently for tomographic imaging is Projections Onto Convex Sets (POCS) (Sezan and Stark, 1982). In POCS the problem of forming an image from the projection data is formulated in a Hilbert space setting. It is assumed that there are m constraints that are known a priori about the object. An example of such an constraint is assuming that the object has compact support. Furthermore, it is assumed that each constraint implies the object lies within a certain closed convex set. The object is considered to be a function that lies in the intersection of all the convex sets. POCS involves iteratively projecting an initial guess onto each set until a function is found in (or at least "close" to) the intersection of all the sets.

Another iterative method that has found favor recently is the so called EM – Expectation Maximization method (Shepp and Vardi, 1982) (Lange and Carlson, 1984). In EM the problem of forming an image from projections is formulated as an estimation problem. In this method a probability density function for the projection data is formulated using a model of the physical process involved in the collection of the data. This probability density function is a function of the (unknown) image being reconstructed. The objective in the EM method is to find an image that maximizes the probability density function for the observed data. An image that maximizes this density function is referred to as the maximum likelihood estimation of the image. Numerically finding such an image is difficult. The iterative procedure used to find such an image results from two steps: forming a conditional expectation (the "E" step) and then maximizing the conditional expectation (the "M" step). The EM method has been found to be useful with the poor SNR data associated with SPECT.

2.3 Advantage and disadvantages of the methods

The iterative and the non–iterative methods each have their own advantages and disadvantages. First consider speed. It has been demonstrated that non–iterative methods can produce accurate images using fewer computer operations than the iterative methods. Hence, when it comes to speed non–iterative methods have an advantage.

Next, consider accuracy. A factor that greatly influences accuracy is the amount of information contained in the data set measured. To see how the amount of information can vary, consider the two–dimensional parallel case illustrated in the top portion of Figure 1. If, as previously mentioned, the source and detector are moved 180 degrees about the object, it can be seen that every possible line integral has been measured. It is well known that both the iterative and non–iterative methods can produce accurate images with this data set.

However, there are applications in which it is not possible or undesirable to move the source and detector the whole 180 degrees. With such incomplete data sets the non–iterative methods generally result in images with substantial artifacts. On the other hand, iterative methods can be used to produce images with less severe artifacts (Sezan and Stack, 1984). Hence, with regards to accuracy, when incomplete data sets are used the iterative methods, generally speaking, have an advantage.

2.4 An important aspect of cone–beam tomography

Because of their speed and accuracy, it seems desirable to develop algorithms for cone–beam tomography that are non–iterative. Recall, however, that producing accurate images using non–iterative methods generally requires a complete data set. But what is a complete data set in the case of cone–beam data? Answering this question for the two–dimensional parallel case is easy to do. Observe from the top portion of Figure 1 that once the source and detector has been moved 180 degrees about the object that every possible data point is measured, and hence, the data set is complete. Unfortunately, the analogue argument for cone–beam data doesn't hold for cone–beam data. It is not obvious how the source and detector should be moved about the object so that sufficient information is measured to make it possible to obtain accurate images using non–iterative methods.

The following condition, which was first developed in (B. Smith, 1985), specifies whether a given trajectory of the source and detector will result in a complete cone–beam data set.

> *If on every plane that intersects the object there lies a source, then one has complete information about the object.*

A circle centered about the object is an example of a trajectory that does not provide a complete data set. Note that the planes with normals perpendicular and nearly perpendicular to the plane containing the circle do not intersect the circle. For an example of a curve that does provide a complete data set consider the "sine–on–the–cylinder" curve. One can imagine this three–dimensional curve being made by drawing two periods of a sinusoid on a piece of paper and then connecting the two end points of the sinusoid together by wrapping the paper into a cylindrical shape. Note that if the amplitude of the sinusoid is sufficiently large any plane that intersects an object placed inside the cylinder will intersect the curve as well. Hence, this curve satisfies the completeness condition.

3 ALGORITHMS VERSUS OTHER DESIGN CONSTRAINTS

If the source and detector are moved along a trajectory that provides a complete data set, then at least in theory, it is possible to obtain images that are accurate using non–iterative methods. On the other hand, if a trajectory which doesn't provide a complete data is used, then to obtain accurate images it seems likely that iterative methods will be needed. Unfortunately, iterative methods are usually slow. Consequentially, if developing fast and accurate algorithms were the only consideration in designing cone–beam tomographic imaging systems then only trajectories that provide complete data sets would be considered.

3.1 Other design constraints

Whatever trajectory is used some mechanical device will be needed to move the source and detector about the object. Minimizing the mechanical complexities of such a device certainly needs to be considered in designing a cone–beam imaging system in addition to the need to developing algorithms that are fast and accurate. Unfortunately, the trajectories that provide complete data sets tend to be more difficult to mechanically implement than a circle trajectory which, as mentioned previously, doesn't provide a complete data set. For example consider the sine–on–the–cylinder trajectory mentioned earlier. In this trajectory, not only does the source and detector need to be moved about the object just as in the circle trajectory, but in addition the source and detector need to be moved up and down as well. It is interesting to note in passing that one of the reasons why that the parallel data collection scheme was replaced by the fan–beam collection scheme years ago was the mechanical simplicity provided by the fan scheme.

Recently, it has been demonstrated in SPECT imaging of the head that the SNR (signal to noise ratio) of the data depends on the trajectory of the detector (Manglos and Smith, 1992). A factor that determines the SNR of the data is the focal length of the collimator. For some trajectory it is necessary to increase the focal length in order to keep the head within the field of view of the camera. The longer the focal length is the poorer the SNR. Unfortunately, the trajectories that provide complete data (e.g., the sine–on–the–cylinder) require a focal length longer than those trajectories that do not provide complete data (e.g., the circle)

3.2 More work is needed

Hence, several competing interests arise when the selection of a data collection trajectory is being contemplated. To minimize the mechanical complexities that are involved it is desirable to use a circular trajectory. To maximize the SNR in SPECT imaging of the head it is again desirable to considered using a circle. However, to produce algorithms that are fast and accurate it is desirable to consider non–circular trajectories that provide a complete data set. Hence, there are several trade–offs that need to be consider when selecting the trajectory of the source and detector.

A number of questions need to be addressed before an informed decision can be made with regards to the design of the source and detector trajectory. For example, can algorithms be developed for a circle that have acceptable speed and accuracy? Another closely related question is: Does the improved speed and accuracy associated with a complete trajectory justify the associated increased in mechanical complexities? To make an informed discussion with regards to the selection of a trajectory these questions need to be addressed in the future.

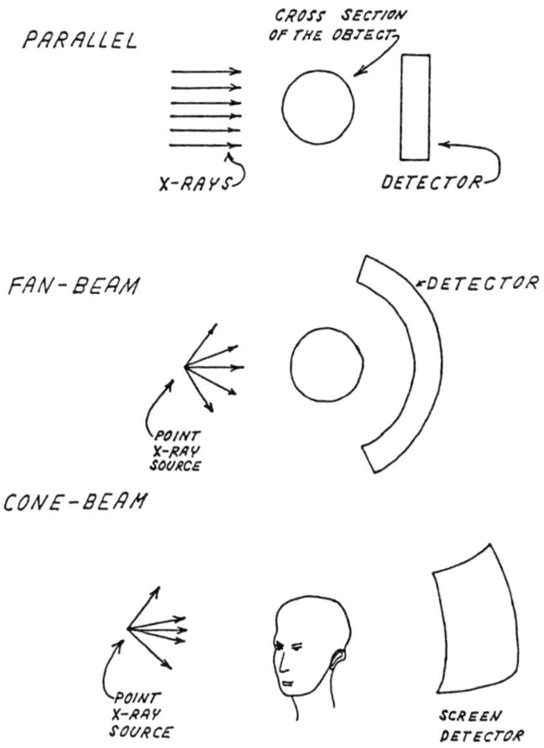

Figure 1. Three geometries that have been used in CT scanners.

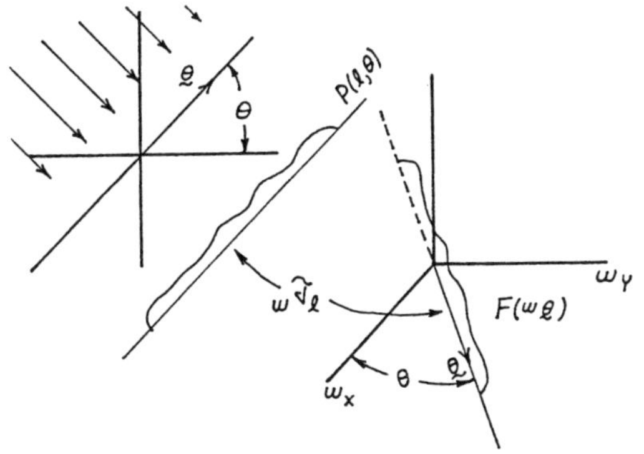

Figure 2. An illustration of the central section theorem.

Acknowledgment. This paper was supported in part by grant number 1 R01 CA1077-01A1. Its contents are solely the responsibility of the author and do not necessarily represent the official views of the National Cancer Institute.

References

Bracewell, R.N., "Strip integration in radio astronomy," *Aust. J. Phys.,* Vol. 9, pp. 198, 1956.

Jaszczak, R.J., Floyd, Jr., Manglos, K.L., Greer, K.L., and Coleman, R.E., "Cone beam collimation for single photon emission computed tomography: Analysis, simulation, and image reconstruction using filter backprojection," *Medical Physics,* Vol. 13, pp. 484, 1986.

Lange, K., and Carson, R., "EM Reconstruction algorithms for emission and transmission tomography," *Journal of Computer Assisted Tomography,* Vol. 8, No. 2, pp. 306, 1984.

Manglos, S.H., and Smith, B.D., "Practical Evaluation of Several Cone Beam Orbits for SPECT," *IEEE 1992 Nuclear Science Symposium, Medical Imaging Conference 92,* Orlando, Florida, October, 1992.

Mersereau, R.M., " Direct Fourier transform techniques in 3-D image reconstruction," Vol. 6, pp. 247, 1976.

Ramachandran, G.N., and Lakshiminarayanan, "Three-dimensional reconstruction from radiographs and electron micrographs and electron micrographs: Applications of convolutions instead of Fourier transform," *Proc. Natl., Acad. Sci. USA,* vol. 68, pp. 2236, 1971.

Sezan, M.I., and Stark, H., "Tomographic image reconstruction from incomplete view data by convex projections and direct Fourier inversion," Vol. 3,pp. 91, 1984.

Shepp, L.A., and Vardi, Y., "Maximum Likelihood Reconstruction for Emission Tomography," Vol. MI–1, No. 2, pp. 113, 1982.

Smith, B.D.," Image reconstruction from cone–beam projections: Necessary and sufficient conditions and reconstruction methods," *IEEE Med. Img.,* Vol. MI-4, pp. 75, 1985.

Smith, P.R., Peters, T.M., and Bates, R.H.T., "Image reconstruction from finite numbers of projections," *J. Phys. A: Math. Nucl. Gen.,* vol. 6, pp. 361, 1973.

Inverse Problems in Engineering: Theory and Practice
ASME 1993

DIFFERENTIAL CEPSTRAL FILTER FOR INVERSE SCATTERING

R. V. McGahan, J. B. Morris, F. C. Lin, and M. A. Fiddy
Department of Electrical Engineering
University of Massachusetts, Lowell
Lowell, Massachusetts

Abstract

This paper presents a new approach to determine permittivity distribution functions of strongly scattering objects having inhomogeneity scales either comparable with or larger than the illuminating wavelength. The measured scattered field data, for each illuminating direction, are backpropagated into the object domain to obtain a noisy image which is the product of the permittivity distribution function and the total field. The differential cepstral filtering algorithm is then used to reconstruct the permittivity distribution function.

1. Introduction

In diffraction tomography, many inversion methods usually make the assumption that the object is a weak scatterer having a physical size either very small or very large compared with the illumination wavelength. Under this assumption, the methods based on adopting the first-order Born approximation (BA) or the Rytov approximation (RA) become Fourier-based inversion algorithms (see, e.g., Fiddy, 1986). The measured scattered field data can then be mapped onto the Ewald sphere in k-space and inverse transformed to reconstruct the permittivity profile of the object.

Recent developments in the inverse scattering problem have been elaborated by extending the domain of validity of the BA and RA. These improved methods either assume that scattering is still sufficiently weak so that the Born or Rytov series converge or they assume that some *a priori* information about the scatterer is available which can provide a (strongly scattering) background, against which small fluctuations in permittivity can be imaged. The latter approach, which we have been pursuing via the application of the distorted-wave Born and Rytov approximations (Wombell and Fiddy, 1988, Lin and Fiddy, 1990) had been incorporated with the spectral estimation technique (Byrne and Fiddy, 1987 and 1988) to improve the image resolution. These distorted-wave methods based on Fourier inversion scheme are also computationally fast and simple. However, their use is still limited to cases in which much is known a priori about the more strongly scattering component.

The purpose of this paper is to describe a new inversion approach which invokes a procedure in differential cepstrum theory to recover the permittivity distribution function $V(\mathbf{r})$ from the backpropagated image. This image is the product of $V(\mathbf{r})$ and the total field $\Psi(\mathbf{r})$ in the object domain. This inversion method, based on the differential cepstral filtering, is equally simple computationally and can further extend the range of validity of the existing Fourier-based techniques. In addition, this method avoids the phase wrapping problem usually associated with the homomorphic filtering algorithm (McGahan, et al., 1991 and 1992).

2. Theoretical Background

Consider the scatterer illuminated by a time-harmonic plane wave with a time factor $e^{-i\omega t}$ and the incident field $\Psi_o(\mathbf{r},k\hat{\mathbf{r}}_o) = e^{ik\hat{\mathbf{r}}_o\cdot\mathbf{r}}$ where the free-space wavenumber k is equal to $\omega\sqrt{\mu_0\varepsilon_0}$. The free-space permeability and permittivity are μ_0 and ε_0, respectively. The plane wave propagates in the direction $\hat{\mathbf{r}}_o$. From the scalar Helmholtz equation, the total field $\Psi(\mathbf{r},k\hat{\mathbf{r}}_o)$ in free-space can be expressed by the inhomogeneous Fredholm integral equation of the first kind (see, e.g., Lin and Fiddy, 1990),

$$\Psi(\mathbf{r},k\hat{\mathbf{r}}_o) = \Psi_o(\mathbf{r},k\hat{\mathbf{r}}_o) - k^2\int_D d\mathbf{r}' G_o(\mathbf{r},\mathbf{r}')V(\mathbf{r}')\Psi(\mathbf{r}',k\hat{\mathbf{r}}_o)$$

$$= \Psi_o(\mathbf{r},k\hat{\mathbf{r}}_o) + \Psi_s(\mathbf{r},k\hat{\mathbf{r}}_o) \qquad (1)$$

where $\Psi_s(\mathbf{r},k\hat{\mathbf{r}}_o)$ is the scattered field. The dependence of fields on the illumination direction is explicitly indicated in order to reveal the fact that each of reconstructed images from the scattered field data corresponds to one specific illumination direction. The Born approximation (BA) linearizes this integral equation by substituting $\Psi(\mathbf{r}',k\hat{\mathbf{r}}_o)$ for the known incident field $\Psi_o(\mathbf{r}',k\hat{\mathbf{r}}_o)$ in the integral above. This approximation is valid when $ka\,|V(\mathbf{r})|_{max} \ll 1$ where a is the linear dimension of the scatterer. Since the scattered field data were collected in the far-field $|k\mathbf{r}| \to \infty$, the free-space Green's function $G_o(\mathbf{r},\mathbf{r}')$ can be approximated as

$$G_o(\mathbf{r},\mathbf{r}') \approx -\frac{e^{ikr}}{4\pi r}e^{-ik\hat{\mathbf{r}}\cdot\mathbf{r}'} \qquad (2)$$

Hence, if we define a scattering amplitude in the scattering direction $\hat{\mathbf{r}}$ as $f(k\hat{\mathbf{r}},k\hat{\mathbf{r}}_o)$ so that

$$\Psi_s(\mathbf{r},k\hat{\mathbf{r}}_o) \equiv \frac{e^{ikr}}{4\pi r}f(k\hat{\mathbf{r}},k\hat{\mathbf{r}}_o), \qquad (3)$$

under the BA the scattering amplitude then, has a Fourier relationship with $V(\mathbf{r})$, namely,

$$f^{BA}(k\hat{\mathbf{r}},k\hat{\mathbf{r}}_o) = k^2\int_D d\mathbf{r}' e^{-ik\hat{\mathbf{r}}\cdot\mathbf{r}'}V(\mathbf{r}')e^{ik\hat{\mathbf{r}}_o\cdot\mathbf{r}'}. \qquad (4)$$

For most cases the internal field, that is the total field $\Psi(\mathbf{r},k\hat{\mathbf{r}}_o)$ within the object domain D, cannot be assumed as the incident field. However, when this Fourier relationship is naively used to compute $V(\mathbf{r})$ from the scattered field, it is clear that the reconstructed function is the effective permittivity distribution function given by

$$V_B(\mathbf{r},k\hat{\mathbf{r}}_o) = \left[V(\mathbf{r})\Psi(\mathbf{r},k\hat{\mathbf{r}}_o)/\Psi_o(\mathbf{r},k\hat{\mathbf{r}}_o)\right]. \qquad (5)$$

Since the scattered field can always be written as

$$\Psi_s(\mathbf{r},k\hat{\mathbf{r}}_o) = -k^2\int_D d\mathbf{r}' G_o(\mathbf{r},\mathbf{r}')\frac{V(\mathbf{r}')\Psi(\mathbf{r}',k\hat{\mathbf{r}}_o)}{\Psi_o(\mathbf{r}',k\hat{\mathbf{r}}_o)}\Psi_o(\mathbf{r}',k\hat{\mathbf{r}}_o) \qquad (6)$$

one can always formulate the Fourier relationship

$$f(k\hat{\beta},k\hat{\alpha}) = \int_D d\mathbf{r}' e^{-ik(\hat{\beta}-\hat{\alpha})\cdot\mathbf{r}'}V_B(\mathbf{r}',k\hat{\alpha}) \qquad (7)$$

between the scattering amplitude and the effective permittivity distribution function. Eq. (7) can be used to reconstruct a set of single view, back-propagated images: each corresponds to the product of $V(\mathbf{r})\Psi(\mathbf{r},k\hat{\mathbf{r}}_o)$ for each illumination direction $\hat{\mathbf{r}}_o$. In the following section, we apply the differential cepstral filtering to extract $V(\mathbf{r})$ from the backpropagated images.

3. Differential Cepstral Filtering Technique

We can regard the internal field $\Psi(\mathbf{r},k\hat{\mathbf{r}}_o)$, which is the total field in the object domain, as a noise term, containing a certain range of spatial frequencies. With respect to the spatial frequency content of the permittivity distribution function $V(\mathbf{r})$, this *multiplicative noise* component can be removed by homomorphic filtering techniques (Oppenheim and Schafer, 1975, Gonzalez and Wintz, 1977, and Pratt, 1978). For a weak scatterer the internal field is approximately equal to the incident field and, as such, will have a characteristic spatial frequency in the direction of propagation but be essentially a plane wave normal to this. As the degree of scattering increases, the internal field will become increasingly oscillatory in all directions, but it necessarily must retain a characteristic correlation length (i.e., periodicity) determined by the spatial frequency of the incident field in the scatterer. One can therefore expect the spatial frequency content of $\Psi(\mathbf{r},k\hat{\mathbf{r}}_o)$ to be concentrated around that spatial frequency.

In the straightforward homomorphic filtering technique, by taking the logarithm of $V\Psi$, which is the abbreviation for $V(\mathbf{r})\Psi(\mathbf{r},k\hat{\mathbf{r}}_o)$, we can convert this multiplicative relation into an additive one. Then, by filtering the cepstrum of $\log(V) + \log(\Psi)$, we can suppress the randomly varying "noise-like" component Ψ. However, the phase wrapping problem arises in this procedure. This is attributed to the fact the imaginary part of $\log(V\Psi)$, which is the phase of $V\Psi$, can be highly discontinuous because the phase will normally be wrapped into $[-\pi, \pi]$. Hence, abrupt but regular discontinuities in phase will generate unwanted harmonics in the cepstrum. As a result, it demands tremendous effort to identify and remove these spurious spectral components. On the other hand, to develop a phase unwrapping technique in order to solve this problem has been shown to be extremely difficult (Scivier, et al., 1984 and 1986, Scivier and Fiddy, 1985).

A solution to avoid this phase wrapping problem is to use the differential cepstral filtering technique (Raghuramireddy and Unbehauen, 1985, Reddy and Rao, 1987, Rossmanith and Unbehauen, 1989). After taking partial derivatives of $\log(V\Psi)$ with respect to each Cartesian coordinate x, y, and z and summing up these derivatives, we obtain a quantity S given by

$$S \equiv \frac{\partial(V\Psi)/\partial x + \partial(V\Psi)/\partial y + \partial(V\Psi)/\partial z}{V\Psi} \quad . \tag{8}$$

Numerically, we can use Fourier transformation to calculate derivatives of $V\Psi$. When the product $V\Psi$ contains small values or zeros, it requires that a regularization procedure be used to avoid ill-conditioning problem. This can be accomplished by multiplying both numerator and denominator of S with the complex conjugate, $(V\Psi)^*$, of $V\Psi$ and then, adding a small and positive regularization parameter δ to the new denominator. Thus, we can rewrite S as

$$S \approx \frac{(V\Psi)^*\left[\partial(V\Psi)/\partial x + \partial(V\Psi)/\partial y + \partial(V\Psi)/\partial z\right]}{|V\Psi|^2 + \delta} \quad . \tag{9}$$

The differential cepstrum of S will retain the spatial frequencies of the internal field Ψ added to those from V; this allows filtering, integrating (also carried out by Fourier transformation), and taking the exponential of the integrated outcome to recover V.

It was mentioned by Raghuramireddy and Unbehauen (1985) that to just avoid the phase wrapping problem, it is sufficient, in the differential cepstral filtering technique, to consider the partial derivative of $\log(V\Psi)$ with respect to only one of the coordinates. In practice, we suggest that it is advisable to use the symmetric form of Eq. (8) in order to be able to select notch filters with spherical symmetry for a wider class of permittivity distribution functions.

4. Illustration of Differential Cepstral Filtering

In this section, the merit of the differential cepstral filtering technique is illustrated for image reconstructions from synthesized data [Figures 1 and 2], simulated scattered field data [Figures 3 and 4], and measured scattered field data [Figure 5]. All reconstructed images are one-dimensional (1-D) slices of different V confined in the object domains which are discretized to 256 points on the x-axis. A notch filter $L(x)$ of the form

$$L(x) = \frac{1}{d^4}\left[d - e^{-\left(\frac{x - x_l}{w}\right)^2}\right]^2 \left[d - e^{-\left(\frac{x - x_r}{w}\right)^2}\right]^2 \qquad (10)$$

is chosen for both real and imaginary parts of the differential cepstrum of log(VΨ). The quantities d, w, x_l, and x_r are the depth, half-width of the Gaussian function at its e^{-1} value, location of the left minimum, and location of the right minimum, respectively. Locations of minima of the notch filter can be specified according to the number of cycles Ψ plotted within a 256-point range. We also introduce an apodizer $A(x)$ of the form

$$A(x) = \frac{1}{d^4}\left[1 - e^{-\left(\frac{x - 1}{w'}\right)^2}\right]^2 \left[1 - e^{-\left(\frac{x - 256}{w'}\right)^2}\right]^2 \qquad (11)$$

where w′ is the half-width of the Gaussian function at its e^{-1} value.

In the following image reconstructions, the regularization parameter δ is chosen to be 0.001, the depth d is 1.1, and the half-widths w and w′ are fixed at 9 and 256 points, respectively.

For generating synthesized data, we create two real permittivity profiles, a Gaussian function having a height of $0.06\varepsilon_0$ and a half-width of 32 points in a 256-point object domain [Figure 1a] and a three-step function (having heights $0.02\varepsilon_0$, $0.06\varepsilon_0$, and $0.04\varepsilon_0$ which occupy 45, 76, and 40 points, respectively [Figure 2a]. Then, we multiply these profiles with a real cosine function cos(kx) having the spatial frequency of 20 cycles within 256 points. These two products are plotted in figures 1b and 2b, respectively. In figures 1c and 2c, we show the reconstructions after eliminating the fundamental spatial frequency, which becomes 40 cycles after taking the modulus of the cosine field cos(kx) in Eq. (9) increases the spatial frequency to 2k. In figures 1d and 2d, we demonstrate the removal of some ripples in the reconstructions after the fundamental and the second harmonics are removed. In figures 1e and 2e, we show the further improved reconstructions after the fundamental, the second, and the third harmonics are removed. Note that in the above reconstructions, the apodizer is used once only. We find that if we apply the apodizer twice, the reconstruction is improved even more for the step-wise permittivity profile with sharp edges [Figure 2f]. This improvement due to the "nonlinear" filtering process of the apodizer for the step-wise permittivity profiles is an on-going research topic.

For simulated cases, the scattered fields in the far-field for two cylinders, A and B, are calculated on the azimuthal plane for the 3cm-wavelength microwave radiation. Each cylinder consists of the inner cylinder of radius 3.6cm with a permittivity ε_{in} and the outer annular cylinder of thickness 6.3cm with a permittivity ε_{out}. In figures 3a and 3b, we show the real and imaginary parts of the 1-D slice of the backpropagated VΨ, for the cylinder A with $\varepsilon_{in} = 1.03\varepsilon_0$ ($V_{in} = 0.03$) and $\varepsilon_{out} = 1.1\varepsilon_0$ ($V_{out} = 0.1$), in the 256-point domain which corresponds to a range of 64cm. After removing the fundamental, the second, and the third harmonics at 18, 36, and 72 cycles in the 256-point cepstrum domain, we plot the reconstructed V in figure 3c for the cylinder A. Similarly, for the cylinder B with $\varepsilon_{in} = 1.1\varepsilon_0$ ($V_{in} = 0.1$) and $\varepsilon_{out} = 1.03\varepsilon_0$ ($V_{out} = 0.03$), figures 4a and 4b are the plots for the real and imaginary parts of the 1-D slice of the backpropagated VΨ and figure 4c is the reconstruction after the removal of the fundamental, the second, and the third harmonics at 18, 36, and 72 cycles and the application of the apodizer once in the 256-point cepstrum domain.

We have also applied the method to experimental scattered field data collected in the far-field from a hollow cylindrical cardboard tube of permittivity $\varepsilon = 2.7\varepsilon_0$ (V = 1.7) having an outermost radius of 15cm and illuminated by 3cm-wavelength microwave radiation. In this case, the real and imaginary parts of the 1-D slice of the backpropagated $V\Psi$ for one illumination direction are plotted in figures 5a and 5b. It shows that after removing two sets of harmonics (20, 40, and 80 cycles) and (30, 60, 90 cycles) and apodizing once in the cepstrum domain, we obtain an excellent reconstruction of V in figure 5c on the 64cm range (discretized to 256 points). Note that in figure 5c, the contribution to the reconstructed V beyond its expected edge does not vanish. We believe that if there are several 1-D slices of the backpropagated $V\Psi$ corresponding to several illumination directions, the averaging process will reduce this spurious contribution beyond the expected edge of the reconstructed V.

5. Conclusions

The differential cepstral filtering technique should be particularly effective for strongly scattering objects. It is expected that many sets of data for more complicated $V\Psi$ will have to be processed in order to properly characterize the internal fields for different illumination directions. These internal fields will become less periodic and distinctive as the scattering strength, i.e., the degree of multiple scattering increases. Our assumption is that values of the internal fields for different illumination directions can be well approximated by a small bandwidth of higher-order harmonics for very strongly scattering objects. Conversely, for a weakly scattering object, the internal field remains periodic in the illumination direction and well represented by a plane wave in the orthogonal direction through the cross section of the scatterer.

We have demonstrated the feasibility of the differential cepstral filtering technique and presented five reconstructions of V from 1-D slices of backpropagated $V\Psi$. Obviously, in practice, the objective is to recover an unknown V from $V\Psi$ obtained by backpropagation or Fourier inversion techniques. For these cases, the problem of limited availability of data to execute this inversion will arise. The recovered $V\Psi$ will necessarily be convolved with the Fourier transform of the data window used. This introduces a second level of difficulty in that a low pass filtered version of V has to be estimated. This suggests using techniques of bicepstral analysis in order to recover V from scattered field data obtained from many different illumination directions. This work is in progress and will be reported at a later date. Also, since a nonlinear operation is performed on the backpropagated field data, as it is on the scattered field with the Rytov approximation, a comparison between the reconstructions obtained in this way and those obtained using the Rytov approximation is being studied.

Acknowledgements

M. A. Fiddy and F. C. Lin acknowledge the support of Office of Naval Research grant N00014-89-J-1158.

References

Fiddy, M. A., 1986, "Inversion of optical scattered field data," J. Phys., D19, pp. 301-317.

Wombell, R. J. and Fiddy, M. A., 1988, "Inverse scattering within the distorted-wave Born approximation," Inverse Problems, 4, pp. L23-L27.

Lin, F. C., and Fiddy, M. A., 1990, "Image estimation from scattered field data," Int. J. Imaging Systems and Technology, 2, pp. 76-95.

Byrne, C. L., and Fiddy, M. A., 1987, "Estimation of continuous object distributions from limited Fourier magnitude measurements," J. Opt. Soc. Am., A4, pp. 112-117.

Byrne, C. L., and Fiddy, M. A., 1988, "Images as power spectra; reconstruction as a

Wiener filter approximation," Inverse Problems, 4, pp. 399-409.

McGahan, R., Lin, F. C., and Fiddy, M. A., 1991, "New methods for inverting scattered field data from strongly scattering targets," Inverse Problems Conference on Computational Algorithms, Texas A&M University, College Station, TX, March 10-14.

McGahan, R., Lin, F. C., and Fiddy, M. A., 1992, "Cepstral filtering for recovery of object from scattered field data," Proceedings, SPIE'92 Conference on Inverse Problems in Scattering and Imaging, Fiddy, M. A., ed., San Diego, CA, July 19-24.

Oppenheim, A. V., and Schafer, R. W., 1975, Digital Signal Processing, Prentice Hall, Englewood Cliffs, New Jersey.

Gonzalez, R. C., and Wintz, P., 1977, Digital Image Processing, Addison-Wesley Publishing Co., Reading, Massachusetts.

Pratt, W. K., 1978, Digital Image Processing, John Wiley & Sons, New York.

Scivier, M. S., Hall, T. J., and Fiddy, M. A., 1984, "Phase unwrapping using the complex zeros of a band-limited function and the presence of ambiguities in two dimensions," Optica Acta, 31, pp. 619-623.

Scivier, M. S., Fiddy, M. A., and Burge, R. E., 1986, "Estimating SAR phase for complex SAR imagery," J. Phys., D19, pp. 357-362.

Scivier, M. S., and Fiddy, M. A., 1985, "Phase ambiguities and the zeros of multidimensional band-limited functions," J. Opt. Soc. Am., A2, pp. 693-697.

Raghuramireddy, D., and Unbehauen, R., 1985, "A simple formula for computation of the two-dimensional differential cepstrum," Frequenz, 39, pp. 310-312.

Reddy, G. R., and Rao, V. V., 1987, "Signal delay and waveform estimatiion through differential cepstrum averaging," IEEE Trans. Acoustics, Speech, and Signal Processing, ASSP-35, pp. 1487-1489.

Rossmanith, H., and Unbehauen, R., 1989, "Formulas for computation of 2-D logarithmic and 2-D differential cepstrum," Signal Processing, 16, pp. 209-217.

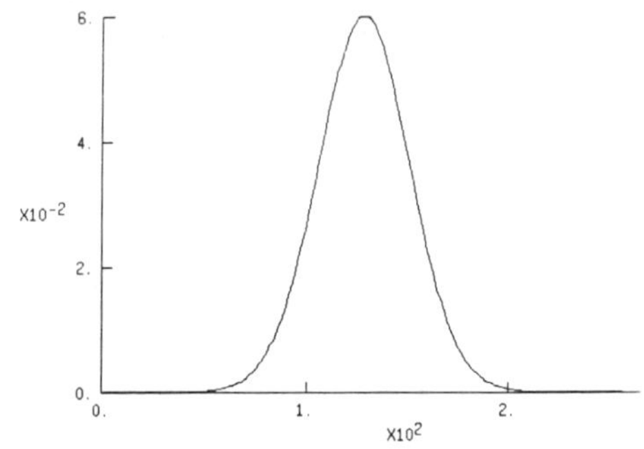

Figure 1a

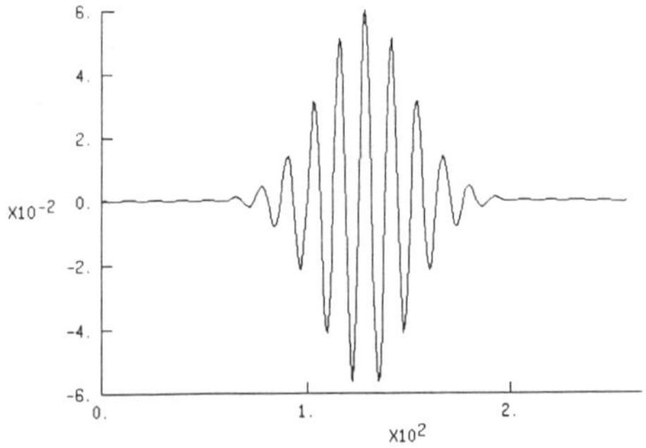

Figure 1b

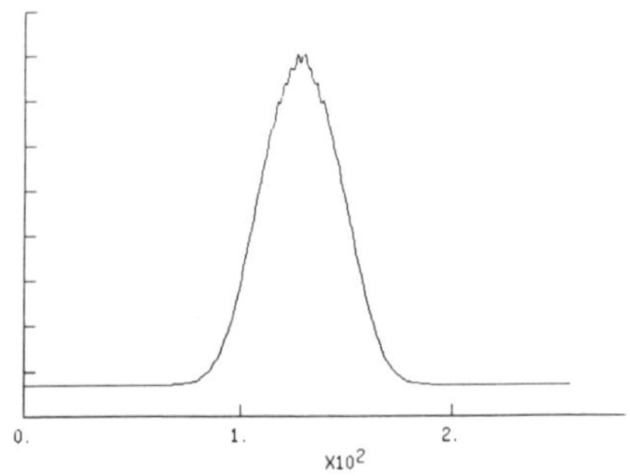

Figure 1c

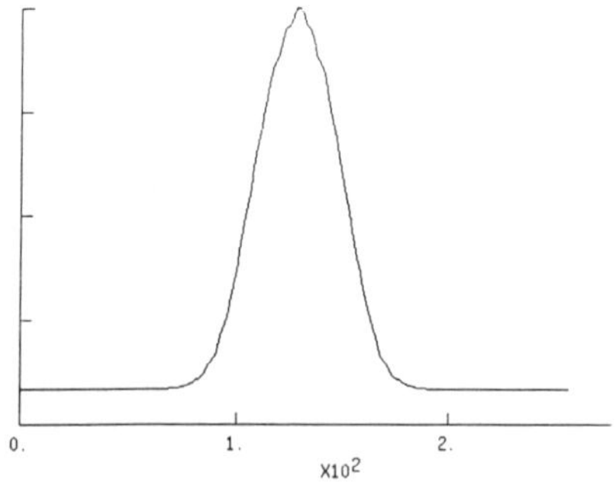

Figure 1d

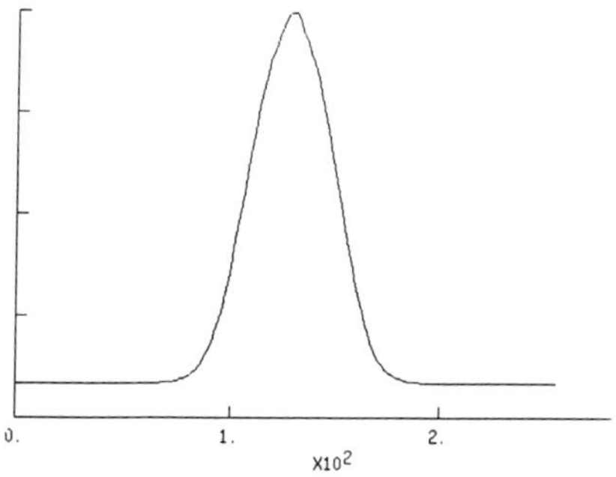

Figure 1e

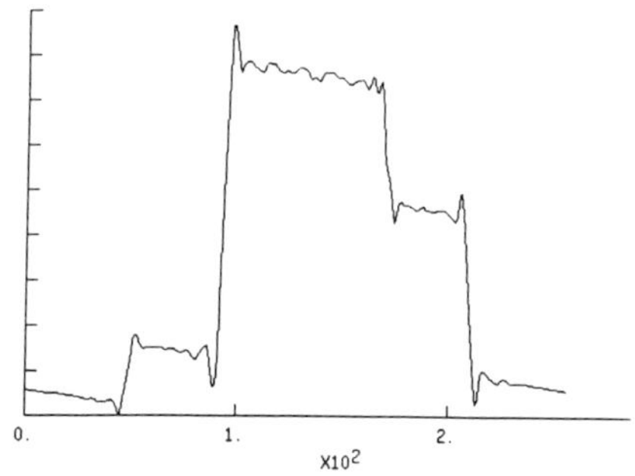

Figure 2d

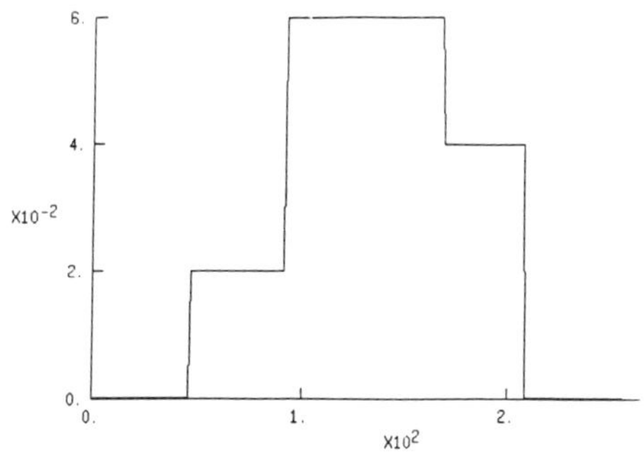

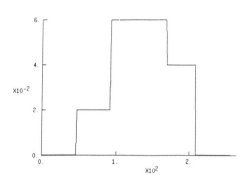

Figure 2a

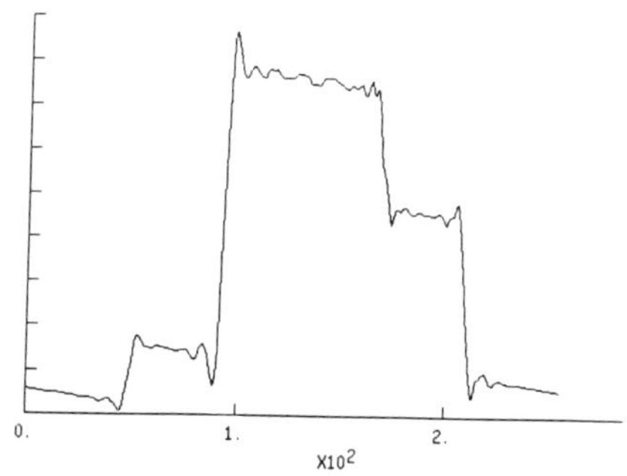

Figure 2e

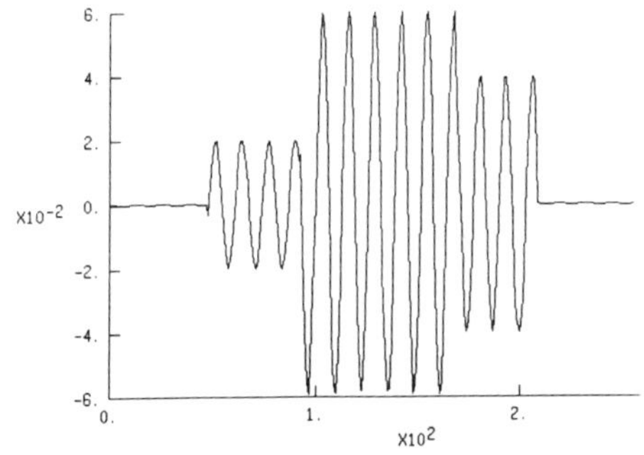

Figure 2b

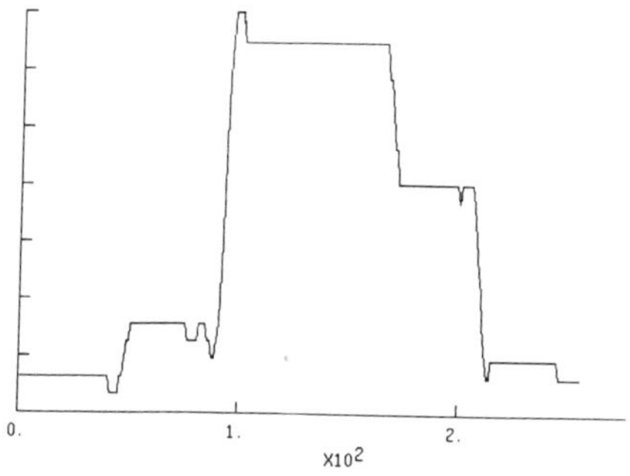

Figure 2f

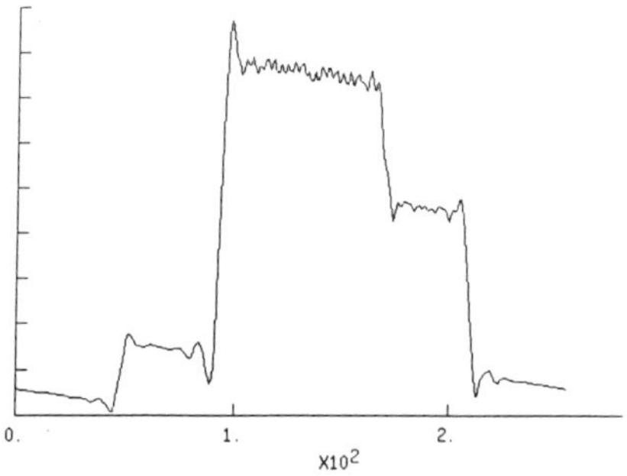

Figure 2c

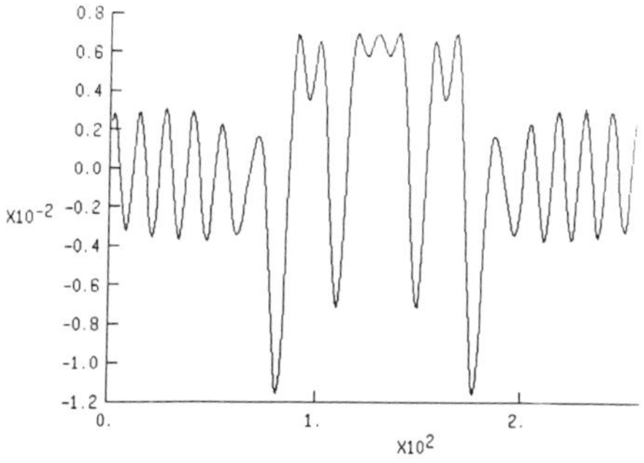

Figure 3a

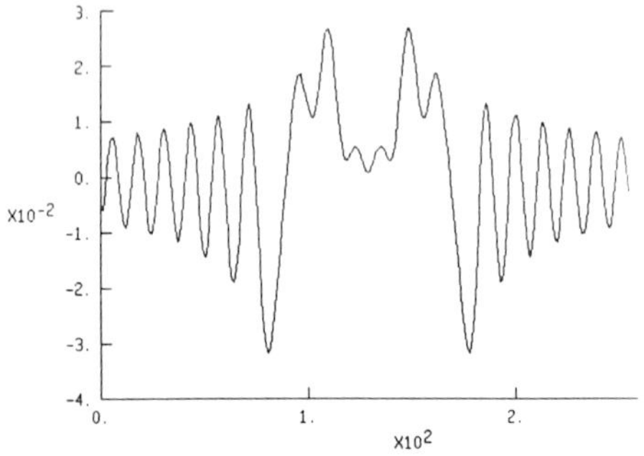

Figure 3b

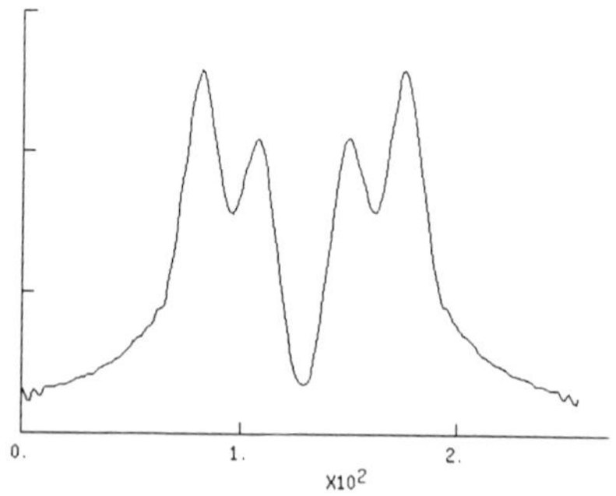

Figure 3c

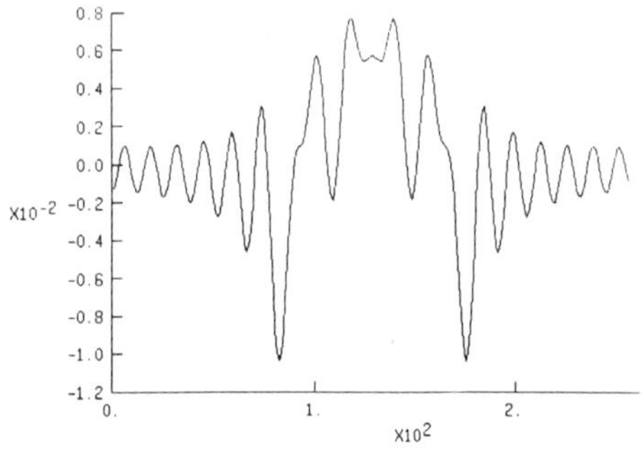

Figure 4a

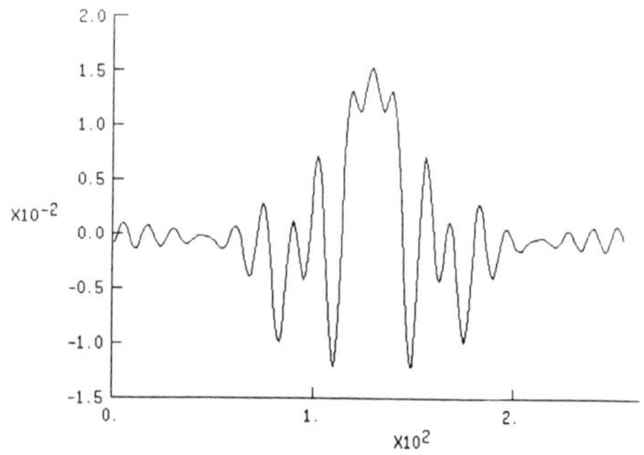

Figure 4b

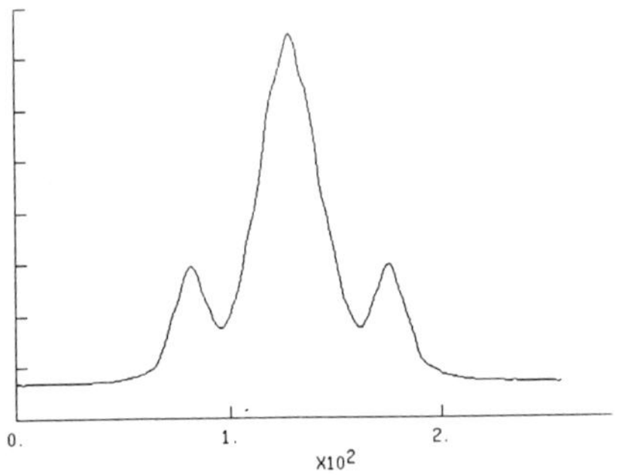

Figure 4c

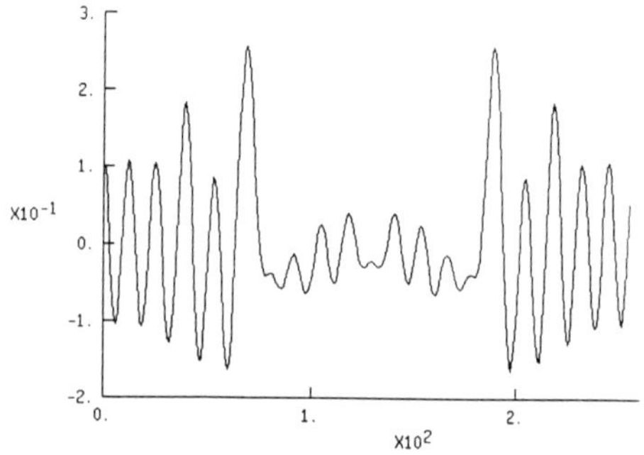

Figure 5a

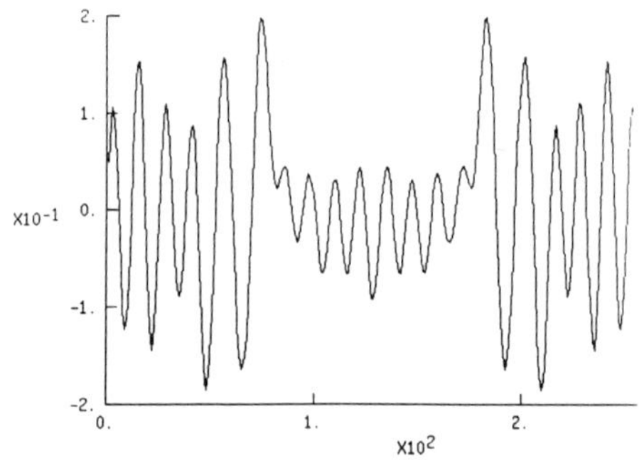

Figure 5b

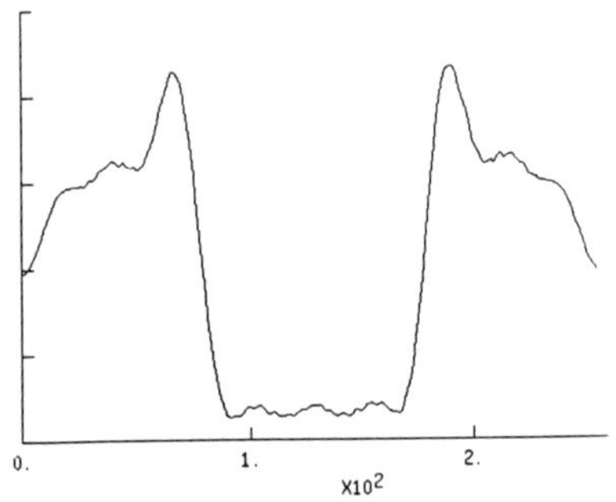

Figure 5c

PROGRESS IN IMPEDANCE IMAGING FOR GAS-LIQUID FLOWS: PART 1 – ANALYTICAL AND NUMERICAL DEVELOPMENTS

Jen-Tai Lin and Owen C. Jones
Department of Nuclear Engineering
and Engineering Physics
Center for Multiphase Research
Rensselaer Polytechnic Institute
Troy, New York

ABSTRACT

This paper describes developments undertaken in the framework of a finite element model for inverse calculation of electrical field properties from known response to a boundary excitation. Specifically, a new "exponential block decomposition," method is described in conjunction with improved handling of the Hessian matrix for fast impedance imaging of multi-region systems. The method yields more than three-orders improvement in computational efficiency over results previously reported. In addition, this paper outlines a method whereby 2-D scanning can be used coupled with a voltage conversion preconditioning correction for 3-D distortional effects which has the potential to allow true 3-D images to be reconstructed using several 2-D scanned image planes.

INTRODUCTION

Work has been in progress at Rensselaer for the past several years to develop a fast, nonintrusive imaging system based on electrical impedance methods.[1-6] The concept of impedance imaging is relatively straightforward. A volume of a liquid-vapor flow is excited electrically through means of boundary electrodes (Fig. 1).[7] The response is measured and the internal distribution of electrical properties is iteratively determined which "best" matches the measured response. Methods for solution to this nonlinear problem are iterative in nature. An internal property distribution is assumed, response to the applied excitation is calculated and compared with the measured response, and variations in the original distribution are undertaken which push the calculations in the direction of the measurements.

This problem is termed an "inverse problem" since it is directly inverse to the "forward problem" where the field properties and boundary conditions are known and the electrical field is computed. Since both electrical resistivity and permittivity of water and vapor are significantly different in each fluid, development of this method having sufficient rapidity would allow interfacial structure and evolution to be determined for two-phase flows.

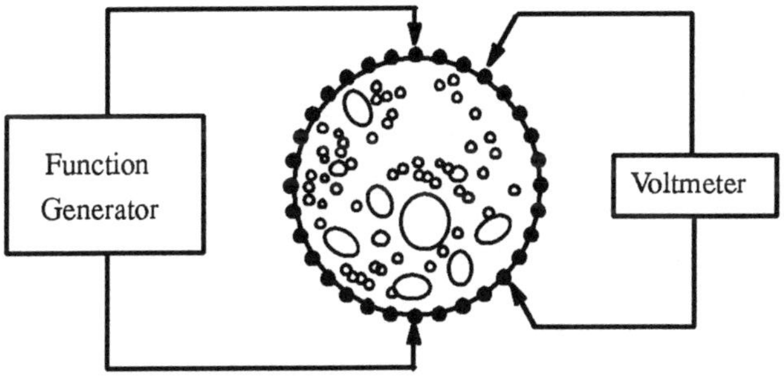

Figure 1. Schematic of excitation and measurement system.

HISTORICAL REVIEW

Introduction

Impedance imaging was begun for geological assay c.f. Dynes and Lytle[8], and recently developed for biomedical uses (c.f. Seagar, Barber, and Brown[9]). To date, the best of applications applied to real systems produce a very fuzzy planar "picture" of resistivity or permittivity variations but the results are encouraging.

Maxwell's equations for the behavior of the electrical field are utilized to determine the internal distribution of electrical properties which minimizes in the least squares sense, the difference between the computed boundary response given the excitation, and the measured response. If there are N-electrodes, and all possible independent measurement combinations of excitation and response are utilized, there are N(N-1)/2 independent excitation combinations, each having N separate measurements of voltage and/or current. Least squares minimization allows less than the full compliment of excitations to be utilized with commensurate time saved.

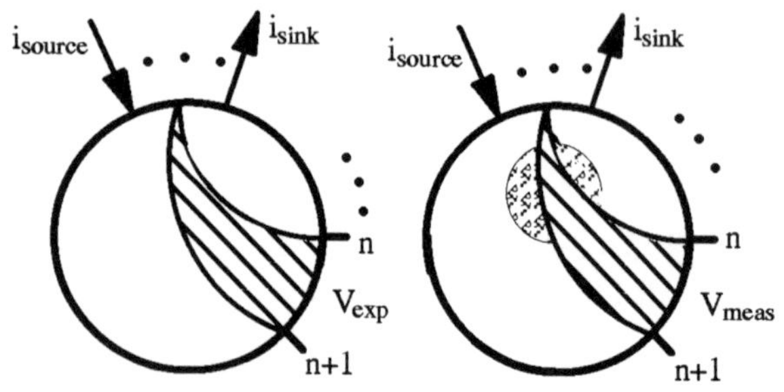

Figure 2. If the measured voltage is higher than the calculated voltage, multiply the resistivity in the shaded area by that factor.

Increasing the number of electrodes can improve the image only up to a point, after which better imaging comes only by improving the accuracy of measurement.[10,11] Fuks et al.[12] has provided methods of estimating the degree of accuracy obtained with digital conversion of data.

Swanson[13] proposed the first frontal plane impedance imaging technique where current was directed front–to–back through a body. Henderson et al.[14] and Henderson and Webster designed the impedance camera. They applied a constant 100kHz voltage to a large electrode on the chest, which resulted in varying currents collected through a 100–electrode array on the back. All electrodes except for the measuring electrode were maintained at ground potential.

Recent work has been reported on application of tomographic methods in the multiphase area. Xia, Plaskowski and Beck discuss a capacitance method of imaging where electrodes were placed on the outside of a test apparatus and measurements taken of charge due to an applied potential.[15,16] The method was based on a stray–immune circuit developed by Huang.[17] However, the imaging was undertaken by back projection and therefore approximate. Nor does this method produce unique results. Halow and Fashing also have reported the use of capacitance methods for visualizing fluidized bed systems, again using the method of back projection for imaging.[18]

Computational Algorithms

All methods are iterative if they attempt to converge to a suitable image. On the other hand, the NOSER method utilizes an exact solution to the first Taylor iteration.[19] The iterative methods fall into the following categories: finite difference; finite element; back projection. Most iterative methods use some variational or least squares principal. By far the most popular and extensively used are back projection methods for two reasons. First, finite difference methods used in iterative sequences produce such extreme amounts of artificial diffusion that it is difficult to achieve reasonable images. Secondly, finite element methods, until recently as a result of the work in this project, have been extremely resource intensive, requiring hours of computer time on state–of–the–art machines.

Back Projection. The method of Barber and Brown[20,21,22] involved back–projection based on linearization around a constant conductivity. 50 kHz current was injected into various combinations of neighboring electrodes. If the measured voltage on a real object was high by some factor, they multiplied the model resistivity by that factor everywhere in the shaded region Figure 2. This fastest of all methods to date was subsequently improved upon by Santosa and Vogelius[23] but with mixed results. Beck and his co–workers Huang, et al.[24], Beck and Williams[25], have also developed back–plane projection methods for analysis of gas–liquid pipe flows of gas and oil.

Lytel and Dines[26] developed a square computer model of a two–dimensional transverse plane and iterated using Laplace's equation in ten iterations. Tasto and Schomberg[27] also simulated a square computer model of a two–dimensional transverse plane using up to 432 projection angles. Kim et al.[28] simulated a two–dimensional transverse plane and obtained the sensitivity matrix of current changes due to resistivity changes for all electrodes.

Very slow transient results were obtained by Brown, Barber, and Seagar[29] when a dish of heated saline solution was reconstructed showing the thermal patterns of convection. From comparison of their results with Price's estimates of resistivity Price[30], it seems that changes of the order of 1.5–10 Ω–cm were easily resolved.

To date, while methods of back projection are only approximate and results possibly not unique, the rapidity of computation coupled with the reasonableness of the results for the intended purposes has made this method by far the method of choice in the biomedical field. However, for purposes of definitively locating the boundaries in heterogeneous media, this method appears to hold only little promise.

Variational Methods. Wexler et al.[31] simulated a geophysical cubic model of a three–dimensional solid containing four layers of 7x7 volume elements pixels. Measurements were made at the top surface only. Using 1700 iterations, they demonstrated fair reconstruction of an object in layers 2 and 3 whose conductivity was five times that of the host medium.

A variational method developed by Kohn and Vogelius[32] is similar to that of Wexler et al.[33] but guaranteed to converge. It was shown by Kohn and McKenney,[34] however, to produce results no better than those of Wexler[33]. Murai and Kagawa[35] used a "matrix regularization" method based on Akaike's information criterion to eliminate the problem of ill–conditioning in their methods.

Finite Element Methods. Starting with the suggestions of Kim, Tompkins, and Webster,[36] this work has been the basis for the most successful inversions reported to that date (Dynes and Lytle[8], Murai and Kagawa[35], and Yorkey[37-44]). Systems were very ill posed due to the extreme lack of sensitivity of boundary elements to changes in the central regions of the field, resulting in many orders of magnitude difference between eigenvalues of the sensitivity matrix. Furthermore, the best results in this area reported in the literature were criticized as being artificially enhanced by making nodal element boundaries coincide with the discontinuities in conductivity of the medium being imaged. The problem becomes much more difficult when the two do not correspond and in the latter case, results tend to be worse than other methods. Both difficulties have been the subject of recent research at Rensselaer.

Yorkey, Webster, and Tompkins YWT, used Marquardt's conditioning method which they stated to be better than Akaike's method to improve the problem of ill–posedness. Their results appear quite successful in the inversion of two carefully–chosen numerical experiments (Yorkey[37], Yorkey and Webster[38], Yorkey, Webster, and Tompkins,[39-44]). Finite element methods were used to obtain accurate reconstructions in four iterations. No reconstruction of real situations has yet been reported and Kohn and McKenney[34] indicate the YWT tests were "biased by the nature of the synthetic data."

Yorkey et al.[44] examined several other methods including the perturbation method used by Kim et al.[36], the equipotential lines method used by Barber et al.[45] and by Barber and Brown[22], the iterative equipotential lines method the original one proposed did not iterate, and the method used by Wexler et al., and similarly by Kohn and Vogelius[32] referenced by YWT. Of the five methods tried, only the YWT method converged to zero error in overall resistivity, and seemed to obtain the correct result locally, in spite of the fact that they only utilized adjacent electrodes for excitation--a pattern guaranteed to produce the most difficult problems with sensitivity. Other methods either did not converge or converged with some error.

NOSER. The NOSER method Newton One Step Error Reconstruction, is a noniterative reconstruction which uses exact solution of the uniform field problem and exact computation of the first Taylor-series corrections in the iterative process (Cheney[19]). The method is characterized by multiple simultaneous excitations and utilizes specific nodalization geometry termed the *Joshua tree*: 496 cylindrical sectors of approximately equal area, carefully comprised to yield reasonable results.

Distinguishability. The discreteness of changes within the body being imaged has an effect on the resultant image reconstruction. Certainly, internal changes smaller than the smallest local element will not be imaged correctly, if at all. Furthermore, variations in boundary response can decrease with decreasing disturbance until the instrumentation can no longer respond adequately to the changes. The question arises, then, as to the relationship between sensitivity and accuracy of measurement.

Isaacson and coworkers Isaacson[10], Gisser, Isaacson, and Newell[46], Isaacson and Cheney[47], described a method to estimate the conditions necessary to distinguish a homogeneous cylindrical body of one size, centered in another cylindrical region of homogeneous structure.

As part of this effort, Fuks et al.[48] showed that number of electrodes, measurement accuracy, and digital conversion resolution all play important roles in image accuracy, 16 bits being adequate for resolving 496 elements. Newell[11] compared multiple current excitation, opposite pair current, and adjacent pair current pattern showing the superiority of using multiple simultaneous excitations in trigonometric patterns is demonstrated (Fig. 3).

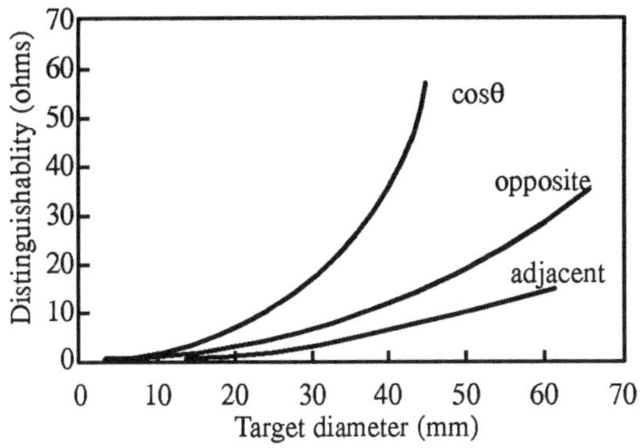

Figure 3. Distinguishability of conductive targets having different diameters placed at the center of the phantom.[11]

Three-Dimensional Modeling

Liu et al.[49] developed a quasi-two-dimensional back projection, cubic model of a three-dimensional solid. The model contained three layers of 4x4 voxels image elements). They obtained better results when electrodes were placed on a diagonal between two planes than when they were in only one plane. They were able to successfully resolve up to 18 objects at random locations in three iterations. Goble and Isaacson[50] constructed a three-dimensional saline tank phantom with 64 electrodes. The phantom had four layers of electrodes and each layer had 16 equidistantly placed stainless steel electrodes. Using stacked-cosine excitation patterns and two-dimensional NOSER algorithms, they were able to obtain quasi-three-dimensional images.

Summary of EIT Methods

To date, the best images have come from two-dimensional algorithms using the finite element method or NOSER methods. The only three-dimensional methods utilized effectively have been based on the method of back projection, and these have been quasi-three-dimensional using extended two-dimensional algorithms. No true three dimensional methods have been reported.

Back projection methods are heuristic, approximate, and not single-valued. Results appear fair at best, sufficient for preliminary diagnostic purposes, but appear to have no potential for the clarity necessary for interfacial imaging in gas-liquid systems. On the other hand, they are vary fast and may possibly be used to provide a preliminary indication of gross structure valuable for initializing the more complete, finite element model.

The NOSER method is quick, has been optimized for high sensitivity, and produces results equivalent to the best images obtained to date. However, this method appears to be at the peak of its capability. Again, its potential for further development as required for interfacial imaging seems limited.

Finite element methods seem to have been remarkably useful in reconstruction tomography of electrical fields. Starting with the suggestions of Kim, Tompkins, and Webster[36], this work has been the basis for the most successful inversions reported on to date (Dynes and Lytle[8], Murai and Kagawa[35], and Yorkey[37-44]).

Only the finite element methods reported in the literature, despite their obvious limitations, appeared to have potential for significant development and precise imaging. The methods developed by Webster, Yorkey, and their co-workers had only scratched the surface of development and hence were chosen as the basis of our analytical and numerical development.

ANALYSIS

Newton-Raphson Method (Yorkey's Method)

Yorkey's method of the Newton-Raphson solution is adopted starting with Poisson's equation for the field potential Φ as

$$\nabla \cdot (\sigma + j\omega\varepsilon)\nabla\Phi = 0. \qquad (1)$$

The equivalent network of field equations obtained using Kirchoff's law is

$$Y_{NxN} V_{NxP} = C_{NxP} \qquad (2)$$

where Y is the admittance matrix, V is the voltage matrix, and C is the current matrix. Also, N is the number of total nodes, M is the number of total elements, P is the number of total current excitations, and E

is the number of total measured electrodes. Following Yorkey's algorithm, we define and error function ϕ

$$\phi(\sigma) = \frac{1}{2}[f(\sigma)-V_o]^T[f(\sigma)-V_o] \tag{3}$$

with V_o and $f(\sigma)$, the measured and the estimated computed, electrode voltages respectively for a resistivity distribution σ, and T denotes transpose of a matrix. To find a σ–distribution which minimizes $\phi(\sigma)$, we let the derivative of the error function vanish. Thus,

$$\phi'(\sigma) = [f'(\sigma)]^T[f(\sigma) - V_o] = 0 \tag{4}$$

where

$$\sigma_{1xM} = [\sigma_1, \sigma_2, \sigma_3, ...\sigma_m, ...\sigma_M]. \tag{5}$$

The differential of the computed boundary electrode voltages is defined as

$$f'(\sigma) = \frac{df(\sigma)}{d\sigma} \tag{6}$$

where f' is called the Jacobian matrix (J). The second derivative

$$\phi''(\sigma) = [f'(\sigma)]^T f'(\sigma). \tag{7}$$

We take the Taylor series expansion of $\phi'(\sigma)$, around a point σ^k and keep the linear terms

$$\phi'(\sigma) \cong \phi'(\sigma^k) + \phi''(\sigma^k)\nabla(\sigma^k) = 0 \tag{8}$$

where ϕ'' is called the Hessian matrix, and find the updating equation for σ

$$\Delta\sigma^k = -[f'(\sigma^k)^T f'(\sigma^k)]^{-1} f'(\sigma^k)^T[f(\sigma^k) - V_o]. \tag{9}$$

We have now derived an iterative procedure which estimates σ^k. At iteration k we have an estimate of σ^k and updated σ^k so that

$$\sigma^{k+1} = \sigma^k + \Delta\sigma^k. \tag{10}$$

The positive definiteness of the approximate Hessian matrix guarantees that σ is minimized.

A New Algorithm to Calculate the Hessian Matrix $(J^T \cdot J)$

Since the majority of computational effort is involved with the Jacobian and subsequently the Hessian, significant reductions can be made if part of this process can be eliminated. Defining matrices **A** as the square of the Jacobian

$$A \equiv [J^T_{MxEP} * J_{MxEP}]_{MxM} \tag{11}$$

and **B** as

$$B \equiv -J^T_{MxEP}[f(\sigma^k)_{EPx1} - V_{o_{EPx1}}] \tag{12}$$

then Eq. (9) will become

$$A_{MxM}d\sigma^k_{Mx1} = B_{Mx1}. \tag{13}$$

The correction term for the conductivity/permittivity of each element for kth iteration can be solved by

$$d\sigma^k_{Mx1} = A_{MxM}^{-1} \cdot B_{Mx1} \tag{14}$$

if the matrices A and B can be calculated. Equation (14) shows that the Jacobian itself is not needed for the inverse calculation. Recall that the Jacobian matrix is defined by

$$J^k \equiv \frac{\partial f^k}{\partial\sigma} \quad \text{with} \quad [J_{EPxM}]_m = \frac{\partial f_{EPx1}}{\partial\sigma_m}. \tag{15}$$

Since Eq. (15) is a derivative of one matrix with respect to another, the Jacobian matrix consists of M vectors given by

$$J_{EPxM} = [J_{1_{EPx1}}, J_{2_{EPx1}}, ..., J_{m_{EPx1}}, ..., J_{M_{EPx1}}] = \left\{ Tran_{NP\to EP} * \left[vec\left(-Y^{-1}_{NxN} * \frac{\partial Y_{NxN}}{\partial\sigma_m} V_{NxP} \right)_{NxP} \right]_{NPx1} \right\}_{EPx1} \tag{16}$$

where $Tran_{NP\to EP}$ is the transition matrix which transforms from the matrix of size NxP size of all nodes, N, for all current excitations, P, into the matrix of size ExP having all measured electrodes for all current excitations. Then,

$$\left[Y^{-1}_{NxN} \frac{\partial Y_{NxN}}{\partial\sigma_m} V_o \right] = \begin{bmatrix} \sum_{\alpha=1}^{N} y^{-1}_{1,\alpha} \sum_{\beta=1}^{N} y'_{m_{\alpha,\beta}} V_{\beta,j} \\ \sum_{\alpha=1}^{N} y^{-1}_{2,\alpha} \sum_{\beta=1}^{N} y'_{m_{\alpha,\beta}} V_{\beta,j} \\ \vdots \\ \sum_{\alpha=1}^{N} y^{-1}_{N,\alpha} \sum_{\beta=1}^{N} y'_{m_{\alpha,\beta}} V_{\beta,j} \end{bmatrix}. \tag{17}$$

Each element of the Jacobian matrix **J** is to pick up node electrodes from the matrix of $[Y^{-1}(dY/d\sigma_m)V)_{NPx1}]_{NPxM}$ so that

$$A \equiv [J^T_{MxEP} * J_{MxEP}]_{MxM} = J^2. \tag{18}$$

Therefore, element-i,j of the square of the Jacobian is

$$(J^2)_{(i,j)} = \sum_{e=1}^{EP} J^*_{e,i}\ J_{e,j} \tag{19}$$

where e is the index for excitation and electrodes given by $(e_a - 1) * P + e_b$ where e_a is the electrode index, from 1 to E and e_b is the excitation index, from 1 to P. Thus, (19) may be written as

$$(J^2)_{(i,j)} = \sum_{e=1}^{EP} \left(\sum_{\alpha=1}^{N} y^{-1}_{nl(e_i),\alpha} \sum_{\beta=1}^{N} y'_{i_{\alpha,\beta}} V_{\beta,e_j} \right)_e \cdots \left(\sum_{\alpha=1}^{N} y^{-1}_{nl(e_i),\alpha} \sum_{\beta=1}^{N} y'_{j_{\alpha,\beta}} V_{\beta,e_j} \right)_e. \tag{20}$$

Here, $nl(e_a)$ is the node number of electrode index e_a, $ni(\alpha)$ and $ni(\beta)$ are node numbers of the neighborhood of element i, where α and β are dummy variables from 1 to 4, $nj(\alpha)$ and $nj(\beta)$ are node numbers of the neighborhood of element j, where α and β are also dummy variables from 1 to 4. Note too that $V_{ni(\beta),e_b}$ is the voltage on the node of $ni(\beta)$ for excitation e_b, $V_{nj(\beta),e_b}$ is the voltage on the node of $nj(\beta)$ for the excitation e_b, $y'_{i_{ni(\alpha),ni(\beta)}}$ is the derivative of the admittance from node $ni(\alpha$, to node $ni(\beta$, with respect to the conductivity of element i, and $y'_{j_{nj(\alpha),nj(\beta)}}$ is the derivative of the admittance from node $nj(\alpha$, to node $nj(\beta$, with respect to the conductivity of element j. Also, $y'_{j_{nl(e_a),ni(\alpha)}}$ is the element having the coordinates node of the electrode e_a, node on $ni(\alpha)$, in the matrix Y^{-1}.

Combining and rearranging results in

$$(J^2)_{(i,j)} = \sum_{e=1}^{EP}\left(\sum_{\alpha=1}^{4} y^{-1}_{nl(e_a),ni(\alpha)} \sum_{\beta=1}^{4} y'_{i_{ni(\alpha),ni(\beta)}} V_{ni(\beta),e_b}\right)_e \cdots$$
$$\cdots\left(\sum_{\alpha=1}^{4} y^{-1}_{nl(e_a),nj(\alpha)} \sum_{\beta=1}^{4} y_{j_{nj(\alpha),nj(\beta)}} V_{nj(\beta),e_b}\right)_e \qquad (21)$$

where

$$B_{Mx1} \equiv -J^T_{MxEP}\left[f(\sigma^k)_{EPx1} - V_{o_{EPx1}}\right] = (B_j)_{Mx1} \qquad (22)$$

and finally

$$B_j = \sum_{e=1}^{EP}(f_e - V_e)\left(\sum_{\alpha=1}^{4} y^{-1}_{nl(e_a),nj(\alpha)} \sum_{\beta=1}^{4} y'_{j_{nj(\alpha),nj(\beta)}} V_{nj(\beta),e_b}\right)_e . \qquad (23)$$

Note too that

$$y'_{m_{(i,j)}} \equiv element\ \ of\ \left(\left(\frac{\partial Y_{NxN}}{\partial \sigma_m}\right)_{NxN}\right)_{(i,j)} \qquad (24)$$

with

$$J_{e,i} \equiv \left(Y^{-1}\frac{\partial Y}{\partial \sigma_i}V\right)_e . \qquad (25)$$

The savings in computer storage is approximately one order of magnitude.

Block Decomposition Method

The exponential or exact numerical scheme developed by Baliga and Patankar[51] used for forward solutions is adapted for the solution of this inverse problem in finite element form. Patankar and his coworkers [51, 52, 53] present an exponential or exact numerical scheme for both finite element and finite difference methods when strong property variations are present.

A streamline upwind scheme was used by Hughes[54] in the Galerkin FE method for first-derivative dominant problems. A streamline exponential scheme was also proposed by Raithby[55] and subsequently extended by Ramadhyani and Patankar[52] and by Hookey et al.[56]

The concept of "block decomposition" BD, is to allow the conductivity within each element to vary in a manner that the results can be continuous from element-to-element within the mesh rather than constant within a given element. Then once the large blocks have been computed, each element is subdivided decomposed, into smaller elements.

The potential $\phi(x,y$, inside the element "e" can be represented by the voltages V_1, V_2, V_3, V_4 on nodes at the corner of the rectangular element by

$$\phi(x,y) = \sum_{m=1}^{m=4} f_m V_m \qquad (26)$$

where f's are the interpolation functions. The element K_{lm} of stiff matrix can be defined as

$$K_{lm} \equiv \int_{V(\Omega_e)}\left[\sigma\frac{\partial f_l}{\partial x}\left(\frac{\partial f_m}{\partial x}\right) + \sigma\frac{\partial f_l}{\partial y}\left(\frac{\partial f_m}{\partial y}\right)\right]dxdy \qquad (27)$$

so the result is

$$K_{lm}\ V_m = I_l . \qquad (28)$$

Note that the net current on the node is

$$I_l \equiv -\int_{A_e}(\ W_l\ q^{\ e}\ ) \cdot d\vec{A} \qquad (29)$$

satisfying Kirchhoff's law.

Consider the one dimensional governing equation

$$\frac{d^2\phi}{dx^2} + \frac{\sigma_x}{\sigma}\frac{d\phi}{dx} = 0. \qquad (30)$$

The solution of Eq. (30) is

$$\phi(x) = D_0 + D_1\int e^{-\int\frac{\sigma_x}{\sigma}dx}\,dx \qquad (31)$$

If conductivity is exponential in the element, σ_x/σ is a constant the approximation solution for two-dimensions is [51, 52, 53]

$$\phi(x,y) = C_0 + C_1 e^{g_r r} + C_2 n + C_3 e^{(g_x \cdot x + g_y \cdot y)} \qquad (32)$$

where r is the direction of the maximum directional derivative σ_r, and n is the direction which is normal to the direction r, and both x and y are Cartesian coordinates.

The four unknowns C_1, C_2, C_3, C_4 are solved to match four given positions at nodes 1–4 of the element for given voltages V_1–V_4 so

$$\phi(x,y) = \sum_{m=1}^{m=4} f_m V_m \qquad (33)$$

where f_1, f_2, f_3, f_4 can be found in terms of x, y, r, n.

The solutions in the Cartesian coordinates obtained are:

$$f_1 = \left(1 - \frac{\xi'}{a'}\right)\left(1 - \frac{\eta'}{b'}\right) \quad \text{and} \quad f_2 = \left(\frac{\xi'}{a'}\right)\left(1 - \frac{\eta'}{b'}\right) \qquad (34)$$

and also

$$f_3 = \left(\frac{\xi'}{a'}\right)\left(\frac{\eta'}{b'}\right) \quad \text{and} \quad f_4 = \left(1 - \frac{\xi'}{a'}\right)\left(\frac{\eta'}{b'}\right) \qquad (35)$$

where

$$x' \equiv e^{g_x x}, \quad \xi' \equiv x' - x'_1 \quad \text{and} \quad a' \equiv x'_2 - x'_1 \qquad (36)$$

and also

$$y' \equiv e^{g_y y}, \quad \eta' \equiv y' - y'_1 \quad \text{and} \quad b' \equiv y'_2 - y'_1 \qquad (37)$$

where

$$x \in [x_1, x_2] \quad \text{and} \quad y \in [y_1, y_2]. \tag{38}$$

From Eq. (27) we can calculate admittance matrix, Y_{lm}, which is identical to the stiff matrix K_{lm}. An algebraic set can be obtained for

$$Y\,V = I \tag{39}$$

Preconditioned Voltage Conversion

The method of incomplete preconditioned voltage conversion PVC, can be used to correct both geometrical distortion and errors from the numerical modelling. For this example, a first order correction was made by using the incomplete set of eigenvalues and preconditioning the measured voltages on the periphery.

The straightforward solution of Eqs. (1) and (39) is Kirchhoff's Law of electrical currents. As presented in Ref. 1–5, Kirchhoff's Law can be described in finite element form as

$$Yf = C \tag{40}$$

where Y is the FEM admittance matrix, f is the FEM–predicted voltage vector, and C is the measured excitation current matrix. The experimental case, on the other hand, is described by

$$\Gamma V = C \tag{41}$$

where Γ is the real admittance matrix and V is the measured voltage vector. Then, since the current matrices are identical in each case

$$Yf = \Gamma V. \tag{42}$$

Now at this point, we have Y, the computed FE admittance matrix, we have f, the computed FE voltage vector, but we do not have V, the voltage vector which, like f, includes the internal nodes as well as the boundary values. Therefore, we can not utilize (42) directly for determination of Γ, the real admittance matrix. For this, we need to perform additional steps and make one major assumption.

Premultiplication of (42) by Y^{-1} yields

$$f = (Y^{-1}\Gamma)V = AV. \tag{43}$$

If the computation of Y is reasonably accurate, the matrix A should be close to a diagonal matrix so that $A \approx \{a_{ii}\}$ where $\{a_{ii}\}$ are the eigenvalues of A. If Y were exact it would be identical to the actual admittance matrix and the matrix A would be an identity matrix.

Now, we know neither Γ nor V since the latter includes both boundary and internal nodal voltages. However, we can say that the results of carrying out the products in (43) is

$$f_i \approx a_{ii}V_i \tag{44}$$

which provides a mapping between the measured and computed voltage domain. In other words, the method has converted the vector space of measured voltages from the experimental domain to the computational domain which is possible for the numerical modelling. Equivalently the measured voltages are preconditioned before the computation.

Since A should be close to an identity matrix, the off–diagonal terms should be orders of magnitude smaller than unity––i.e., A is expected to be a very stiff matrix. The same should be true in either the homogeneous case or the target case. It can thus be assumed that $A_{homog} = A_{target}$. This is a first–order approximation, a Jacobi–like preconditioning, in that it allows us to use the computed values of A as determined from the homogeneous case when computing target conductivity distributions. Similar developments can also be found in the theory of space decomposition and subspace correction by Xu.[57]

Consider now the balance of the process. Also, keep in mind that only the boundary voltages are known; i.e., V is an incomplete vector. We now write

$$\Gamma V = (YA)V = C. \tag{45}$$

However, we can simply combine the A with the V and resolve the problem. Thus,

$$Y(AV) = C \tag{46}$$

or

$$YV' = C \tag{47}$$

where V' is the corrected voltage vector based on the matrix A. Note that while the corrections apply to the entire vector, boundary and internal values, only the boundary values are known. This, now, presents a new problem to be solved

$$Yf' = C \tag{48}$$

where Y and f' are a new admittance matrix and a voltage vector which solve the minimization problem with respect to V'. Thus, (48) is solved for the admittance and voltage distributions such that

$$\phi^2 = (f' - V_o')^T(f' - V_o') \tag{49}$$

is minimized, where V'_o is the "preconditioned" boundary voltage based on Eq. (46); i.e.,

$$V_o' = AV_o. \tag{50}$$

For the present, the matrix A is determined from the homogeneous case, and then applied to the target cases without change. However, it seems reasonable that the matrix A can be computed from many different target cases and then a method found to apply the variations of this matrix with a conductivity distribution in a way which will let us apply the method directly to the unknown distribution problem.

CONCLUSIONS

This paper described the development of improved finite element methods including an improvement in the method of treating the Jacobian and the sensitivity matrix, an exponential block decomposition method which increased computational speed by more than three orders, and a preconditioned voltage conversion method which utilizes comparisons between 2–D and 3–D homogeneous results or between numerical and experimental homogeneous results to form a correction matrix. These new methods have the potential capability of decreasing certain errors including, but not limited to: geometrical measurement, numerical modelling, dimensional distortion and electrical measurement. Application to experimental data is presented in the companion to this paper.[58]

ACKNOWLEDGEMENTS

The authors are grateful to both Hitachi, Ltd., and to the U.S. Department of Energy for their generous support of this research.

REFERENCES

1. Lin, J. T., Ovacik, L, and Jones, O.C., 1992, "Investigation of Electrical Impedance Imaging for Two-Phase, Gas-Liquid Flows." **Chemical Engineering Communications, 118**, pp. 299-325.

2. Lin, J.T., Ovacik, L.,Jones, O.C., Newell, J.C.,Cheney, M, and Suzuki, H., 1991, "Use of Electrical Impedance Imaging in Two-Phase, Gas-Liquid flows." Presented at the 1991 National Heat Transfer Conference, Minneapolis, July 28-31.

3. Lin, J. T., Suzuki, H., Ovacik, L., Jones, O. C., Newell, J. C., and Cheney, M., 1991, "Use of Electrical Impedance Imaging in Two-phase, Gas-Liquid flows." Proc. **7th Ann. Rev. of Prog. in Appl. Computational Electromagnetics**, U. S. Naval Postgraduate School, Monteray, March 18-22.

4. Jones, O.C., 1991, "Developments in Impedance Imaging of Two-Phase Flows." **Proc. Int. Conf. on Multiphase Flows**. Tsukuba, Japan, Sept. 24-27,.

5. Jones, O.C., Lin, J. T. and Ovacik, L.,1991, "Electrical Impedance Imaging in Two-Phase, Gas-Liquid Flows:1. Initial Investigation." **Proc. Conf. on Inverse Design Concepts and Optimization in Engineering Science-III**, Washington, Oct.23-25.

6. Jones, O. C., Lin, J. T.,Ovacik, L. and Shu, H., 1992, "Advances in Impedance Imaging of Gas-Liquid Systems." **Proc. U.S. Japan Seminar on Two-Phase Flow Dynamics**, Berkeley, CA, July 5-11, pp. 205-218.

7. Jones. O.C. 1992, "Cooperative Research for the Development of Electrical Impedance Computed Tomography in Two-Phase Flows," progress report, March.

8. Dynes, K.A., and Lytle, R.J., 1981, "Analysis of Electrical Conductivity Imaging," **Geophysics, 46**, pg. 1025-1036.

9. Seagar, A.D., Barber, D.C., and Brown, B.H., 1987, "Electrical Impedance Imaging," **IEE Proc. 134, Pt. A**, No. 2, pg. 201-210.

10. Isaacson, D.G., 1986, "Distinguishability of conductivities by Electric Current Computed Tomography," **IEEE Med. Imaging MI-5**, 91-95

11. Newell, J.C. Gisser, David G. Isaacson, David., 1988, "An Electrical Current Tomograph," **IEEE trans. on biomedical eng. Vol. 35**, No. 10, Oct.

12. Fuks, L.F., 1989, "Reactive Effects in Impedance Imaging," PhD Thesis, Rensselaer Polytechnic Institute, May.

13. Swanson, D.K. 1976, "Measurement errors and the origin of electrical impedance changes in the limb," PhD Thesis.

14. Henderson, R.P. et al. "An impedance Camera for Spatially Specific Measurements of the Thorax," **IEEE Trans. Biomed. Eng. BME-25** 250-4

15. Xie, C.G., Plaskowski, A., and Beck, M.S., 1989, "8-electrode capacitance system for two-component flow identification. Part 1: Tomographic flow imaging," **Science, 136A**(4), pg. 173-183.

16. Xie, C.G., Plaskowski, A., and Beck, M.S., 1989, "8-electrode capacitance system for two-component flow identification. Part 2: Flow regime identification," **Science, 136A**(4), pg. 173-183.

17. Huang, S., 1986, **Capacitance transducers for concentration measurement in multi-component flow processes**, Ph.D. Thesis, Department of Instrumentaiton and Analytical Science, UMIST, Manchester, UK.

18. Halow, J.S., and Fashing, G.E., 1988, "Preliminary capacitance imaging experiments of a fluidized bed," **AIChE Sym. Ser. 86**(276), pg. 41-50.

19. Cheney, M., Isaacson, D., Newell, J.C., Simske, S., and Goble, J., 1990, "NOSER: An algorithm for solving the inverse conductivity problem," **Int. J. Imaging. Systems and Tech., 2**, pg. 66-75.

20. Barber, D..C., Brown, B.H. 1983, "Imaging spatial distributions of resistivity using applied potential tomography," **Electronics letters. Vol.19** pp.933-935. 27 Oct.

21. Barber, D.C., and Brown, B.H., 1984, "Applied Potential Tomography," **J. Phys. E: Sci. Instrum., 17**, pg. 723-733.

22. Barber, D.C., and Brown, B.H., 1985, "Recent developments in applied potential tomography--APT," in **Proc. 9th Int. Conf. Info. Proc. Med. Imaging**, Washington, D.C.

23. Santosa, F., and Vogelius, M., 1988, "A Backprojection Algorithm for Electrical Impedance Imaging," Institute for Physical Science & Technology, Univ. Maryland, Tech. Note BN-1081, July.

24. Huang, S.M., Plaskowski, A.B., Xie, C.G. and Beck, M.S., 1989, "Tomographic Imaging of Two-component Flow Using Capacitance Sensors," **J. Phys. E: Sci. Instrum. 22**, pp. 173-177.

25. Beck, M., and Williams, R., 1990, "Looking into Process Plant," **The Chem. Engr., 26** July, pg. 14-15.

26. Lytle, R.J. Dines, K.A. 1970, "A system for Determining the Spatial Variation of Electrical Conductivity' Lawrence Livermore Lab.

27. Tasto, M. 1978, "Object Reconstruction from Projections and Some Nonlinear Extensions," **Pattern Recognition and Signal Processing**, C.H. Chen, Ed., Alphen aan den Rijn, Sijthott & Noordhott.

28. Kim, Y., Webster, J.G. and Tompkins, W.J., 1983, "Electric impedance Imaging of Thorax," **Journal of Microwave Power, 18**(3, pp. 245-257.

29. Brown, B.H., Barber, D.C., and Seagar, A.D., 1985, "Applied potential tomography - clinical applications," **IEEE Conf. Publ. Inst. Electr. Eng.**, No. 257, pg. 74-78.

30. Price, L.R., 1979, "Electrical impedance computed tomography (ICT): a new CT imaging technique," **IEEE Trans. Nucl. Sci. USA, NS-26, 2**, pg. 2736-2739.

31. Wexler, A. 1988, "Electrical impedance imaging in two and three dimensions," **Clin Phys. Physiol. Meas. 9** Supply. A 29-33.

32. Kohn, R.V., and Vogelius, M., 1987, "Relaxation of a Variational Method for Impedance Computed Tomography," **Comm. Pure Appl. Math., 40**, pg. 745-777.

33. Wexler, A., Fry, B., and Neiman, M.R., 1985, "Impedance-computed tomography algorithm and system," **Appl. Opt., 24**, 23, pg. 3985-3992.

34. Kohn, R.V., and McKenney, A., 1989, "Numerical Implementation of a Variational Method for Electrical Impedance Tomography," Courant Institute of Mathematical Sciences, private communication.

35. Murai, T., and Kagawa, Y., 1985, "Electrical impedance computed tomography based on a finite element model," **IEEE Trans. Biomed. Eng., BME-32, 3**, pg. 177-184.

36. Kim, Y., Webster, J.G., and Tompkins, W.J., 1983, "Electrical impedance imaging of the thorax," **J. Microwave Power, 18**, 3, pg. 245-257.

37. Yorkey, T.J., 1986, **Comparing Reconstruction Methods for Electrical Impedance Tomography**," Ph.D. Thesis, Dep. Elec. Comput. Eng., Univ. Wisc., Madison, WI 53706, August.

38. Yorkey, T.J., and Webster, J.G., 1987, "A comparison of impedance tomographic reconstruction algorithms," **Clin. Phys. Physiol. Meas., 8**, suppl. A, pg. 55-62.

39. Yorkey, T.J., Webster, J.B., and Tompkins, W.J., 1985, "Errors caused by contact impedance in impedance imaging," **Proc. Ann. Conf. IEEE Eng. Med. Biol. Soc., 7**, 1, pg. 632-637.

40. Yorkey, T.J., Webster, J.G., and Tompkins, W.J., 1986, "An optimal impedance tomographic reconstruction algorithm," **Proc. Ann. Conf. IEEE Eng. Med. Biol. Soc., 8**, 1, pg. 339-342.

41. Yorkey, T.J., Webster, J.G. and Tompkins, W.J., 1986, "An Optimal Impedance Tomographic Reconstruction Algorithm," IEEE Eight Annual Conf. of the Eng. in Medicine and Biology Society, pp. 339-342.

42. Yorkey, J.T., 1986, "Comparing Reconstruction Methods for Electrical Impedance Tomography," PhD Thesis, University of Wisconsin-Madison, August.

43. Yorkey, T.J., Webster, J.G., and Tompkins, W.J., 1987a, "An improved perturbation technique for electrical impedance imaging with some criticisms," **IEEE Trans. Biomed. Eng., 34**, 11, pg. 898-901.

44. Yorkey, T.J., Webster, J.G., and Tompkins, W.J., 1987b, "Comparing reconstruction algorithms for electrical impedance tomography," **IEEE Trans Biomed. Eng., 34**, pg. 843-852.

45. Barber, D.C., Brown, B.H., and Freeston, I.L., 1983, "Imaging spatial distributions of resistivity using applied potential tomography," **Elec. Lett., 19**, pg. 933-935.

46. Gisser, D.G., Isaacson, D., and Newell, J.C., 1987, "Current Topics in Impedance Imaging," **Clin. Phys. Physiol. Meas., 8**, Suppl. A, pg. 39-46.

47. Isaacson, D., and Cheney, M., 1990, "Current Problems in Impedance Imaging," in **Inverse Problems in Partial Differential Equations**, D. Coliton, R. Ewing, and W. Rundell, Eds., SIAM, Philadelphia.

48. Fuks, L.F., Isaacson, D., Gisser, D.G., and Newell, J.C., 1989, "Tomographic Images of Dielectric Tissue Properties," **IEEE-Trans. Biomed. Eng.**

49. Liu, W.P. 1988, "Three dimension reconstruction in electrical impedance tomography," **Clin Phys. Physiol. Meas. 9** Supply. A, pg. 131-135.

50. Goble, J., and Isaacson, D., 1989, "Optimal current patterns for three-dimensional electrical current computed tomography," **Proc. Annu. Int. Conf. IEEE Eng. in Med. and Biology Soc. 11**, pg. 463-464.

51 .B. R. Baliga, and S. V. Patakar, 1980, "A New Finite-Element Formulation for Convection-Diffusion Problems." **Numerical Heat Transfer, 3**, pg. 393-409.

52. S. Ramadhyani and S. V. Patankar, 1985, "Solution of the Convection-Diffusion Equation by a Finite-Element Method Using Quadrilateral Elements." **Numerical Heat Transfer, 8**, pg. 595-612.

53. S. V. Patankar, 1980, **Numerical Heat Transfer and Fluid Flows**. Hemisphere, Washington.

54. T. J. R. Hughes, W. K. Liu, and Brooks, 1979, "Finite Element Analysis of Incompressible Viscous Flows by the Penalty Function Formulation." **J. Comput. Phys., 30**, pg. 1-60.

55. G. D. Raithby, 1976, "Skew Upstream Differencing Schemes for Problems Involving Fluid Flow." **Comput. Methods Appl. Mech. Eng., 19**, pg. 59-98.

56. N. A. Hookey, B. R. Baliga and C. Parash, 1988, "Evaluation and Enhancements of Some Control Volume Finite-Element Methods-Part 1. Convection-Diffusion Problems." **Numerical Heat Transfer, 14**, pg. 255-272.

57. Jinchao Xu, 1992, "Iterative Methods by Subspace Decomposition and Subspace Correction." **SIAM Review, 34**, No. 4, pg.581-613.

58. H. Shu, L. Ovacik, J-T Lin, and O.C. Jones, 1993, "Progress in Impedance Imaging for Gas-Liquid Flows: 2. Experimental Developments. Companion to this paper, this conference.

PROGRESS IN IMPEDANCE IMAGING FOR GAS-LIQUID FLOWS: PART 2 – EXPERIMENTAL RESULTS

Hongjun Shu, Levent Ovacik, Jen-Tai Lin, and Owen C. Jones
Department of Nuclear Engineering
and Engineering Physics
Center for Multiphase Research
Rensselaer Polytechnic Institute
Troy, New York

ABSTRACT

This paper describes experimental results using a new "exponential block decomposition," finite element method developed for fast impedance imaging of multi-region systems using nonintrusive boundary excitation and measurement. Both two-dimensional and three-dimensional situations are shown. Application of algebraic conditioning methods show significant improvement in overall image quality especially regarding noise near the boundary. Additional image conditioning methods are discussed and results shown.

INTRODUCTION

The overall introduction and background for impedance imaging was provided in Ref. 1. This paper continues the description of work undertaken in impedance imaging relative to gas–liquid flows with particular reference to application and evolution of the finite element method. Emphasis is placed on using true two-dimensional methods for algorithm development.

The concept of impedance imaging is relatively straightforward. A volume of a liquid–vapor flow is excited electrically through means of boundary electrodes (Fig. 1).[2] The response is measured and the internal distribution of electrical properties is determined which "best" matches the measured response. This iterative problem is termed an "inverse problem" since it is directly inverse to the "forward problem" where the field properties and boundary conditions are known and the electrical field is computed. Since both electrical resistivity and permittivity of water and vapor are significantly different in each fluid, development of this method having sufficient rapidity would allow interfacial structure and evolution to be determined for two-phase flows.

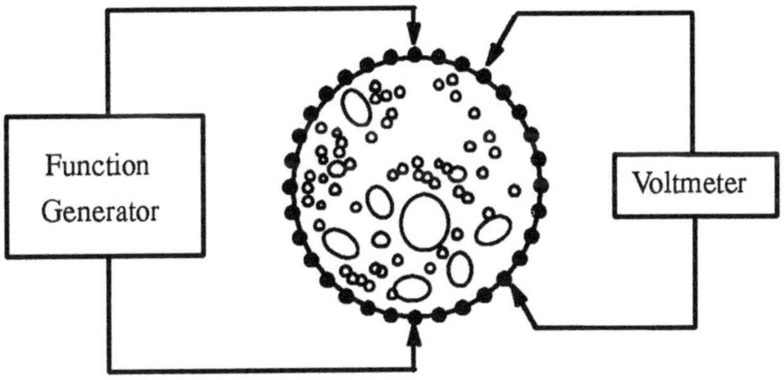

Figure 1. Schematic of excitation and measurement system.

HISTORICAL REVIEW OF ELECTRICAL IMPEDANCE IMAGING

Introduction

A detailed review of impedance tomography was provided in Ref. 1 and shall not be repeated herein. The body of unknown internal electrical field properties (Fig. 1) is surrounded by electrodes on the bounding surface[2] and excited electrically in various combinations. From the measured boundary response solutions to the electrical field equation are iterated while perturbing the conductivity distribution until the error between computed and measured boundary response is minimized.

Recent work has been reported on application of tomographic methods in the multiphase area by Xia, Plaskowski and Beck,[3] Huang,[4] and by Halow and Fashing.[5] All have used capacitance methods and a rapid but nonunique imaging method termed "back projection" for their imaging. It is considered, however, that this method is insufficiently accurate to allow interfacial structure to be determined.

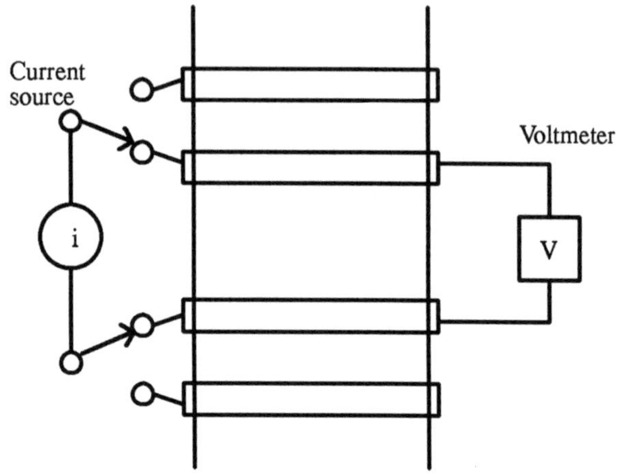

Figure 2. Two–electrodes system vs. four–electrodes system.

Measurements Required

In general, measurements of impedance are performed by connecting low–impedance electrodes to a phantom, and driving a current between them. With reference to Fig. 1, it is seen that both input and output may be multiplexed and imaging measurements undertaken automatically. The resulting voltage and phase may then be measured. Concern has been expressed by many investigators relative to errors induced by contact impedance between electrodes and surrounding material. As a result, numerous different methods have been devised for excitation and measurement. Furthermore, there is controversy regarding the relative advantages of applied current vs. applied voltage methods.

Simple Excitation Methods

Two Electrodes vs. Four Electrodes. With the switches in the inner position Fig. 2 shows a two–electrode (two–pole) system typical of some of the early work. With the switches in the outer position, Fig. 2 shows a four–electrode system. The contact impedance of the voltage–sensing electrodes do not cause error because ideally no current flows through them. Most EICT methods can, therefore, be classified by the number of poles used to make a single measurement, and the method of excitation. Some feel that the four–pole method eliminates errors due to contact resistance at excitation electrodes, but this is not clearly a benefit [Newell et al.[6,7]]. Usually alternating current is utilized since this eliminates polarization effects in electrolytic media.

Two–Pole Methods. Contact impedance was minimized by Barber et al.[8], using a two–pole method and high–impedance measurement methods, but results were quite blurred. Two–pole methods were also used with little success by Dynes and Lytle[9] and by Starzyk and Dai[10].

Three–Pole Methods. Price[11], although unsuccessful in obtaining images from his work, appears to have been the first in the biomedical field to attempt obtaining impedance tomographs utilizing the three–pole method. His suggestion of the use of "guarding" methods was followed by others, all of whom were unsuccessful [Bates et al.[12], Schneider[13], Seagar et al.[14]]. Furthermore, in the three–pole method, only the voltages on individual electrodes are measured. As a result, small voltage differences are obtained by subtracting the measured voltages leading to substantial errors [Smith[15]].

Four–Pole Methods. Reasonable results were obtained by Wexler[16] using a four–pole applied potential method with real domain reconstructions even where there were widely varying conductivities in an overall conducting medium––i.e., metal and plastic shapes in a conducting water field. His results were some of the earliest reported where recognizable images were obtained, even though they were still quite blurred.

Frequency Considerations

At high frequencies there will be interaction with the target, and stray capacitors become increasingly important. Furthermore, over a few MHz, space charge effects become important and microwave transmission modes begin to dominate. Thus, frequencies used in impedance imaging are generally in the range of a few dozen Hz up to several hundred KHz, or even a MHz.

Electrodes and the Phantom

An electrode can be considered to be a transducer that converts the electronic current in a wire to an ionic current in an electrolyte. Contact impedance between the electrode and the medium must be small and known in order to design instruments and reconstruct impedance images. Electrodes must be utilized which are constructed of reasonable materials and sized to interface properly between the metal of the instrument and the ionic conductors in the phantom. They must also be designed to correspond with the geometries capable of being properly modeled numerically. In addition, because in some cases the voltage differences due to the presence of a body in the field are only a few millivolts relative to the homogeneous case, the geometric accuracy of electrode size and placement as well as the phantom geometry itself relative to the model being used in the imaging process is quite important.

There are a number of engineering tradeoffs which must be considered when selecting optimal electrodes. Measuring voltage on an electrode different from the current electrode minimizes the effect of contact impedance. Another design consideration relates to the size of electrodes. With point–electrode models, the larger an electrode, the larger an error is introduced into a point–electrode model, this error increasingly preventing adequate images from being obtained. However, as electrode size decreases, the current for a given potential will decrease resulting in loss in sensitivity. Furthermore, as the electrodes increase in size and approach each other, they may cause an unmodeled current–shunting effect between electrodes.

Excitation Patterns

Many different excitation patterns can be used depending on the application and desire for sensitivity in imaging. Systems which force current through the central regions of the phantom increase sensitivity. Systems which increase current flow also increase sensitivity in general. Some systems attempt to do both. Lytle and Dines[17] applied to all electrodes stepped voltages that would produce a uniform current distribution in a homogeneous media. Kim and Woo[18] presented a very flexible data collection system, where each of 125 electrodes was programmed separately and various pairs of voltage–injecting

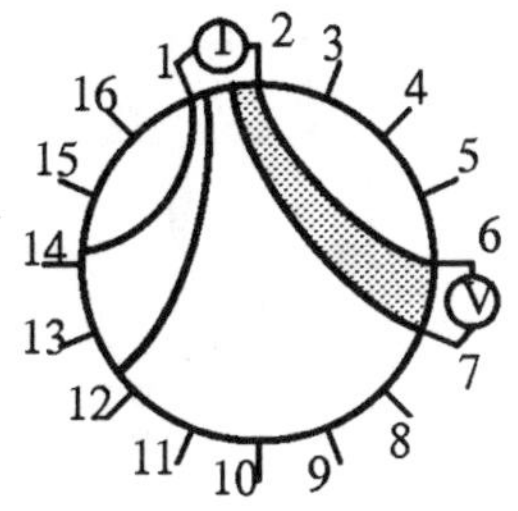

Figure 3. Neighboring method

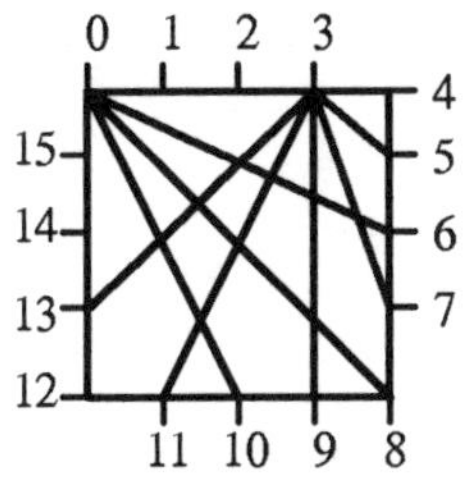

Figure 4. Cross method

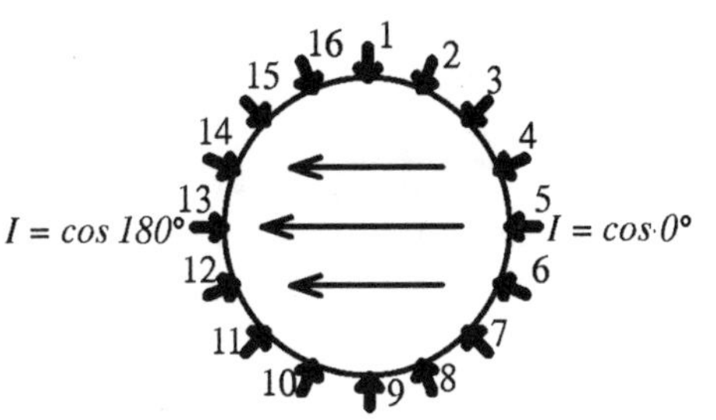

Figure 7. Adaptive method

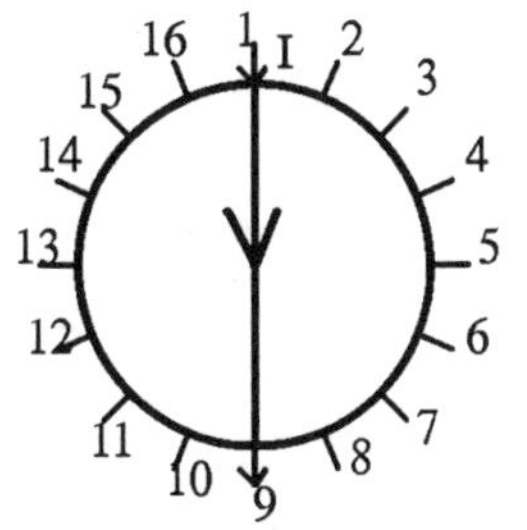

Figure 5. Opposite method

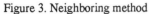

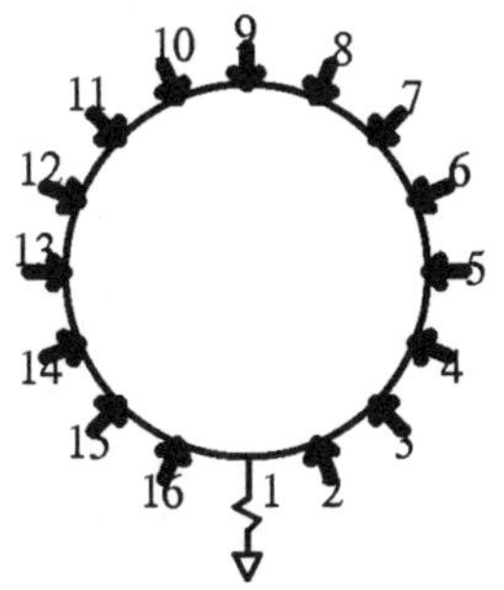

Figure 6. Multireference methc

and current-measuring electrodes could be selected. Gisser et al.[19] applied current to all electrodes in *sin(k*Θ) and *cos(k*Θ) pattern where *k* was half the number of electrodes and measured voltages for all electrodes. Tasto and Schomberg[20] obtained the voltage-current data points by placing electrodes of different dimensions, insulated from each other, on opposite sides of a tank filled with saline.

Neighboring Method. Brown and Seagar[21] used the neighboring method of data collection (Fig. 3). They applied the current through two neighboring electrodes and measured the voltage from all other successive pairs of adjacent electrodes in this four-electrode method. All 208 combinations of neighboring electrodes were excited, although because of Kirchoff's law only 104 are independent.

It is seen that the neighboring method of data collection has a very nonuniform current distribution. Most of the current travels near the peripheral electrodes and hence good sensitivity is obtained at the periphery. This method is least sensitive to central regions in the field.

Cross Method. Hua et al.[22] used the cross method of data collection, where the currents are injected between a pair of electrodes separated by large dimensions (Fig. 4). Compared to the neighboring method, the current is more uniformly distributed. For a 16-electrodes system, there are 182 combinations of which 91 are independent. The cross method does not have as good a sensitivity in the periphery as the neighboring method, but has better matrix conditioning and sensitivity for computation over the entire region.

Opposite Method. Hua[22] also used the opposite or 180° method (Fig. 5). The voltage reference electrode was adjacent to the current-injecting electrodes, the voltages were measured with respect to the reference at all the electrodes, except the current-injecting electrodes. For 16 electrodes, this method has more uniform current density and hence good sensitivity.

Multi-Reference Method. Hua[22] also tried the multireference method injecting from 0.0 to 5mA simultaneously into all electrodes (Fig. 6). The reference electrode was chosen for all combinations to obtain a total of 240 data points for a 16-electrode system, out of which 120 were independent.

To reduce the data collection time and obtain only the independent measurement, approximately eight reference electrodes were used. The current flowed through all the electrodes simultaneously, so the error due to the skin contact impedance is reduced, if it is assumed to be constant throughout the periphery. In practice, it is not constant, hence the multireference method of data collection introduces error due to skin contact impedance. For a nonhomogeneous medium of circular cross section, a somewhat uniform current distribution can be obtained by suitable section of the amplitude of current through the current-injecting electrodes. Due to the somewhat uniform current distribution, this method gives good sensitivity.

Adaptive Method. In an attempt to improve the difficulties associated with ill conditioning, Newell, Gisser, and Isaacson and their coworkers at Rensselaer have been developing the multi-pole current distribution (MPCD) method which increases the sensitivity of the entire field and especially the central regions of the field furthest from the bounding electrodes. This method resulted from a mathematical analysis showing the "best" application of electrical current in a radially-symmetric system to be sin(kθ) and cos(kθ), k=1...K where K is half the number of circumferential electrodes (Gisser et al.[23], Newell et al.[6,7], Fuks et al.[24], Isaacson and Cheney[25], Cheng et al.[26]). This distribution is optimum in effect for an axisymmetric conductivity distribution because at any instant all electrodes are simultaneously excited and the total input current is the sum of individual electrode-pair currents thereby increasing the sensitivity and decreasing the effects of noise in the system.

Gisser et al.[27,19] proposed the adaptive method of data collection, in which almost any desired current distribution can be obtained by injecting current of appropriate magnitude through all the electrodes simultaneously to obtain optimum sensitivity. The sensitivity in the desire zone is measured, and an algorithm keeps changing the current until the optimal current distribution is obtained. For a 16-electrode system, 16 current sources, which yield a total of 120 independent data points, are required.

Figure 7 shows that *for a homogeneous resistivity distribution*, the amplitude of the current in the electrodes follows a cosine curve to give a perfectly uniform current distribution. This method gives the best distinguishability *of an axisymmetric target* and is the most versatile method of data collection. In maximizing distinguishability, an

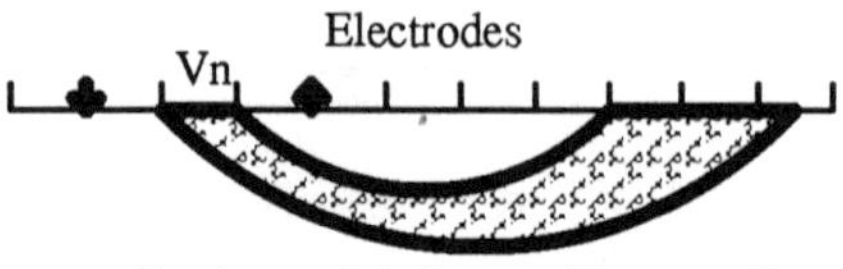

Equipotentials for a uniform medium

Figure 8. Linear array

interactive experimental process must be pursued to approximate a 'best' current j to distinguish between σ_o and σ_1. Therefore, this method is not suitable for fast imaging methods required in rapidly changing systems.

Linear Array. It is not always possible to surround the region of interest with the electrodes, especially in geological applications. Powell et al.[28] used a linear array of electrodes on the top of the surface with constant current excitation. Figure 8 shows the equipotential lines for the current injected through a pair of electrodes. The voltage was measured at intermediate electrodes producing 104 independent measurements for a 16–electrode system. Note that the resolution and sensitivity are very poor away from the array.

Comparison of Data Collection Methods

The data collection methods discussed are shown in Table 1. It is seen that of all the methods, those that utilize multiple sources simultaneously have the best characteristics. Ultimately, this is anticipated to be the method of choice for impedance imaging in general, and for imaging of gas–liquid flows in particular. It should be mentioned that a point model does not do a good job in reconstructing data obtained with plate electrodes and that a plate model may be difficult to implement

Automated Processing

Scan vs. Parallel Processing. Sequential measurements at different positions are often not as fast as those made simultaneously, nor do they allow as accurate comparison during overall changes. It is clearly faster, all else being equal, to have many electrodes and recorders acting in parallel for simultaneous recording from separated regions which are changing rapidly in time. In the case of sequential measurement processes, if we are to image transient processes, the sum total of all measurement time for each image must be short compared with time periods of interesting events. For instance, if elements are moving in space, say at 1 m/s, across the imaging plane, an imaging rate of 1000 frames per second will still allow for a minimum of 1 mm spatial blur in each image simply due to the time it takes to obtain data.

Table 1. Comparison of different data collection methods in terms of their performance.[29]

Data collection method	Current Distribution in the Cross Section	Spatial Resolution and Distinguishability
Neighboring method	Very poor at the center	Poor
Cross method	Uniform	Good
Opposite method	Uniform	Good
Multireference method	Very uniform	Very good
Adaptive method	Very uniform	Very good
Linear array	High only near the surface	Good for depth up to 2 electrode spacing

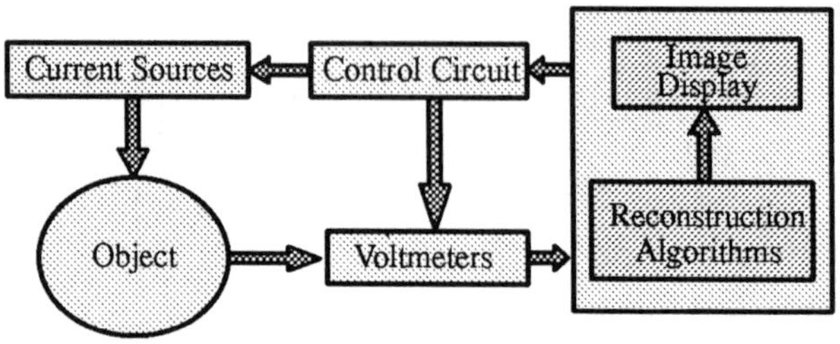

Figure 9. An electrical impedance tomography system

Digital Control and Data Collection Speed. Figure 9 shows how the various stages of a typical, applied current impedance imaging system are controlled by a computer. This includes stages from the current generation through to the display. The computer also has a matrix decoder, which collects data and later does the computing. The oscillator generates an AC signal. The magnitude of the AC in the current generators is controlled by a multiplying digital–to–analog converter (DAC). The digital I/O port of a computer sequentially selects a DAC through the address demultiplexer and sets the amplitude of the current generator. Voltage is measured at the voltage electrodes by connecting them to a precision voltmeter. The measured voltages are digitized by an analog–to–digital converter(ADC). Thus the computer controls the current and voltage data collection by programming the appropriate I/O ports.

Sensitivity, Ill Conditioning, and Nodalization Effects

To date, none of the images obtained as reported in the literature appear to be more than a fuzzy representation of the true situation, and represent only a distant relationship with the quality of imaging by other means. At first, this difference was associated with errors in measurement accuracy, lack of guarding, or other difficulties in experimental technique; however, there can be orders of magnitude differences between the sensitivity of a given boundary measurement to variations in field conductivity depending on the location of the variation. Similar order differences can thus occur in the eigenvalues of the solution matrix thereby making the inversion problem severely ill–posed and difficult to solve [Tarassenko and Rolph[30], Murai and Kagawa[31,32]]. This ill–posedness has also been blamed for much of the fuzziness in images to date.

Additionally, Seagar et al.[33] contend that the blurring of two–dimensional results in a continuously variable conservative field is due to nonzero effective wave number (infinite wave length) of the applied signal. They show, however, that successful reconstructions can be made for certain classes of piecewise constant media (similar to two–phase systems), and that the process is relatively simple when the discrete zones are circular in shape. Certainly, the use of algorithms which assume constant conductivity within finite elemental

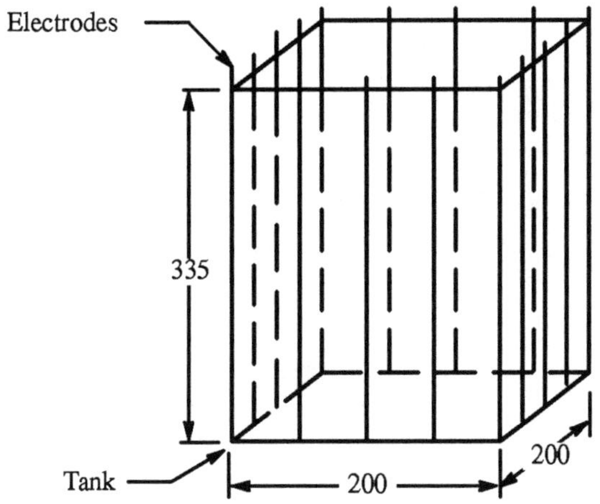

Figure 10. Point-electrode two-dimensional phantom used in CMR.

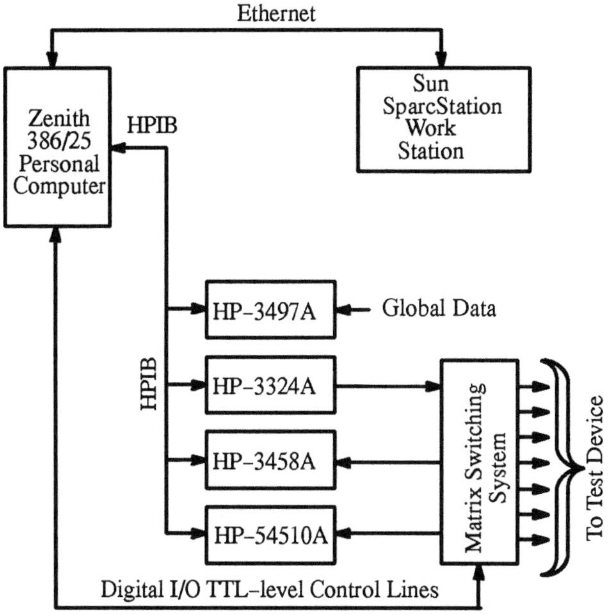

Figure 11. Schematic of the high accuracy excitation and measurement system.

regions cause a wave number truncation. Higher spatial frequencies are lost and this loss can result in numerical instabilities, spatial oscillations in converged results, and other difficulties. This is a problem which has yet to be resolved in the literature.

EXPERIMENT

A two-dimensional experiment was devised which consists of a glass tank 200 x 200 x 335-mm deep (Fig. 10). Stainless steel, vertical, 1.5-mm diameter electrodes were placed around the periphery of this tank, held in place by Plexiglass templates top and bottom. Distilled water was utilized as the homogeneous field fluid. Experimental data were obtained with an accurate, slow speed excitation/acquisition system shown schematically in Fig. 11 Negligible field distortion was found in the axial dimension. Coaxial cables connected the excitation/measurement system through a fully-guarded matrix switching system to the end of each electrode. Excitation was provided by a programmed wave form generator. A 10-nV-resolution voltmeter recorded each sample amplitude. Sequential sample averaging was used until variations due to single measurements were reduced to $1{:}10^4$. The system could also measure phase with less than 1 ns resolution using zero-crossover comparisons on a 1 GHz digital storage oscilloscope. The entire process was under programmed computer control as data collection for a single image required several hours to obtain for a 32 electrode test. Three-dimensional data were obtained similarly but for the electrodes which were insulated except for a 50-mm length in the center of each.

RESULTS AND DISCUSSIONS

Two-Dimensional Results

Uncorrected Results. Figure 12 shows the results of imaging four 2-D targets: two conductors, one only a little over 3-mm diameter; two insulators. Figure 12a shows the actual geometry while Fig. 12b shows the reconstructed results based using the BC method with dashed lines showing the actual locations of the targets. No image enhancement procedures were used but the results show that all four elements are discernable. Location and shape, however, need improvement.

In most of the images obtained to date, both voltages and currents have been measured. However, due to the time required to obtain a complete set of accurate data, Kirchoff's law was used to determine some of the current sink data. That is, the current was measured on the source electrodes, and was not measured on the ground electrodes.

The number of excitations required for accurate imaging using least squares minimization of effort depends on the error in the measurements. Figure 13 indicates that 64 excitations are sufficient to accurately image nonhomogeneous targets when the measurement level is accurate within one part in 10^4. In this figure, it is seen that the relative error per electrode in converged images decreases to a level consistent with the measurement error.

Heuristic Image Correction. Ref 1 described analytically-based correction methods. However, there are a number of common sense methods which might be used including: subtraction of the homogeneous image from the target image to produce a difference image; division of the target image by the homogeneous image to produce a reference image; trigger-level filtering which should be advantageous for binary fields such as considered herein.

Figure 14, 15, 16, and 17 show the results of combinations of these methods. the locations of the targets are shown as the dashed lines in each image. In all cases the contrast is improved over noncorrected images. Furthermore, multiple excitations using Walsh functions seems to provide marginally better results than pair excitations. It is obvious that such manipulations provide some improvement, especially with regard to background noise and noise near the boundaries, although with the actual shape of the targets eluding the final result.

Correction with the A-Matrix. Tests of the PVC correction method using the A-matrix were undertaken on a 2-D geometry

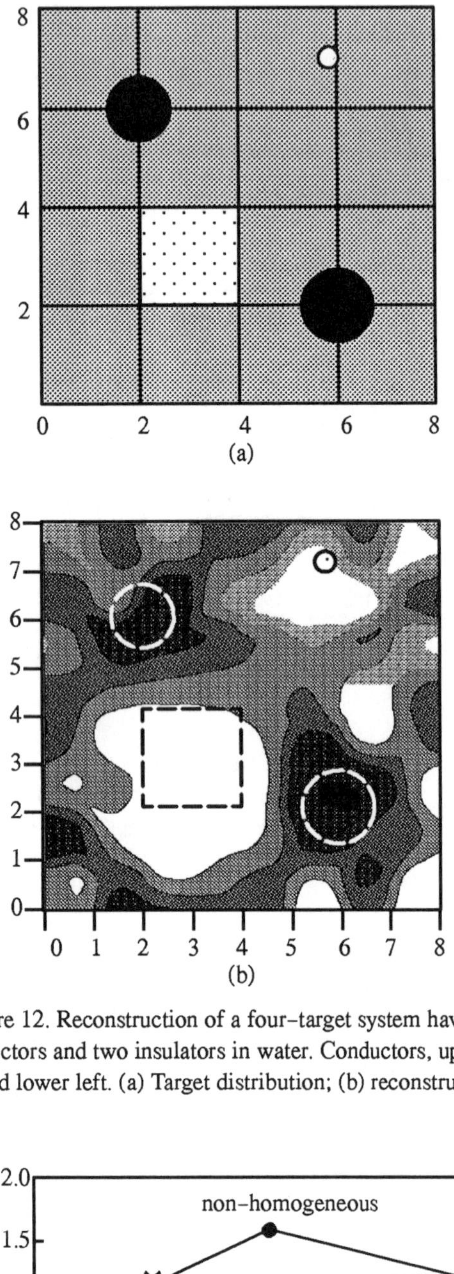

Figure 12. Reconstruction of a four-target system having two conductors and two insulators in water. Conductors, upper right and lower left. (a) Target distribution; (b) reconstruction.

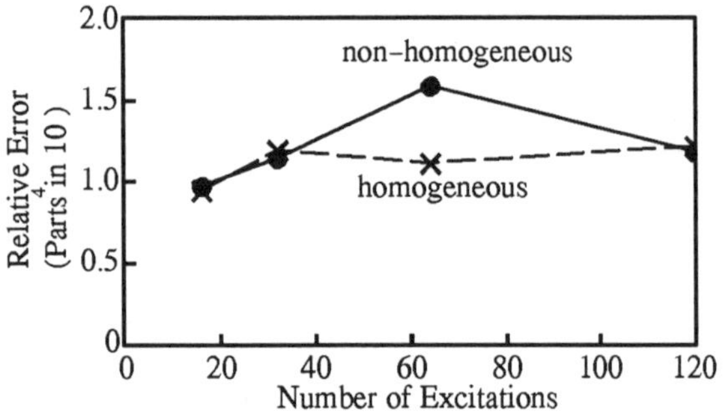

Figure 13. RMS error in boundary reconstruction as a function of number of excitations.

shown in Fig. 18 This consisted of two 25-mm square insulating rods and a single 50-mm square insulator. Results are shown in Fig. 19.

One thing is clear. The larger shapes appear to be in the correct location and with approximately the correct shape. However, the smaller shapes either blend in with the larger, or disappear altogether. Considerable work needs to be done in this area but the results appear promising.

Three-Dimensional Results

Figure 20 diagrams the progression of a three-dimensional conductive shape. This shape was moved progressively from one location to another in a 3-D phantom having 50-mm long electrodes. Images for this case are shown in Fig. 21 indicating that the approximate shape and size of the resulting scanned image were representative of the actual situation, although noise is still present in the images. Note that scanning is the method whereby we would image multiphase flows.

Discussion

The results, while promising, indicate that considerable improvement will be required before cost and effort of impedance imaging will be worth the investment. Heuristic methods offer some improvement over virgin imaging but are lacking in the ability to reproduce accurate shapes. Many other methods not shown have similar characteristics. In comparison with previously reported results, however, the PVC method offers substantial improvement in imaging especially in the immediate vicinity of the electrodes. At least three major improvements in experimental technique appear to be needed: 1. use plate electrodes to increase the boundary coverage; 2. find the optimum conductivity of liquid relative to contact impedance which will maximize imaging accuracy; 3. increase the voltage and current to improve measurement accuracy and sensitivity.

In addition, there are other analytical improvements which may improve imaging accuracy. The one which comes first to mind for the current situation of binary fields with sharp, discontinuous boundaries between conductivities is to allow decomposed elements to vary in size and shape to accommodate boundaries while providing the correct link between the decomposed elements and the gross elements. Furthermore, there may be methods whereby the binary nature of the field can be fed into the analytical/numerical stream; however, note that simply placing binary trigger levels into the computational stream is demobilizing and prevents convergence.

Construction of a plate-electrode phantom is relatively straightforward. Modeling of the boundary in the FEM is not. Furthermore, increasing the voltage is unattractive after a given point due to safety concerns. Conductivity of the liquid can be increased by additives but this also increases the contact resistance of the electrodes. It appears that there may be an optimum in this respect with initial increases in conductivity increasing the overall sensitivity and accuracy of imaging followed by increased boundary noise and uncertainty due to contact effects as conductivity is increased further.

Generally, one sure way to get higher contrast ratio between discrete regions and better spatial definition is to increase the number of elements which can be used to describe the steep change of the conductivity at boundaries, but this increases the computational effort and is not a desirable direction. Application of adaptive grid methods together with better methods to handle discontinuous fields appears to be the direction of most promise.

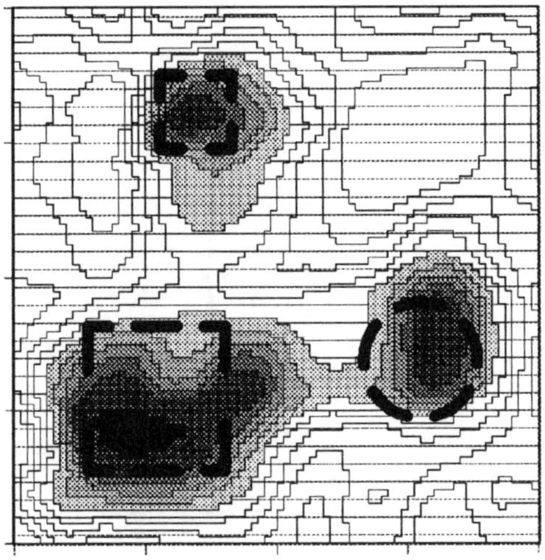

Figure 14. 16 electrodes, 15 Walsh function excitations, division processing.

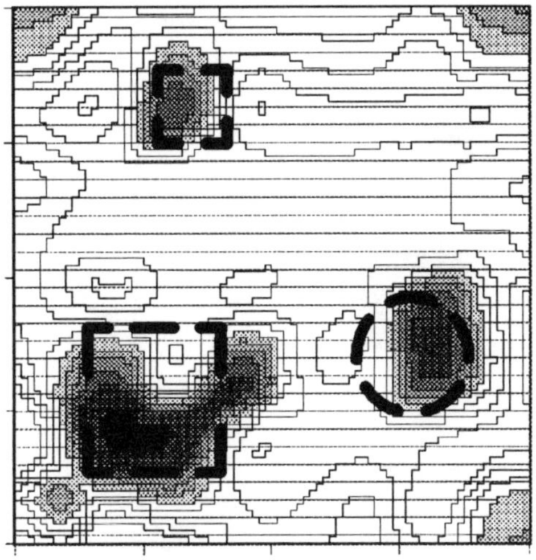

Figure 16. 32 Pair excitations, division processing.

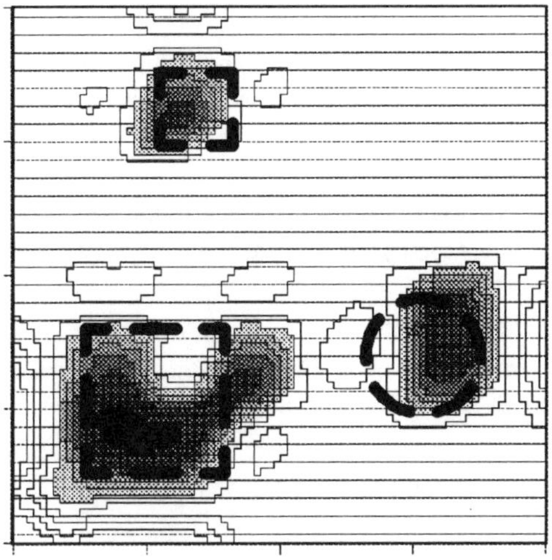

Figure 15. 16 electrodes 15 Walsh function, division & filtering (Level=0.5).

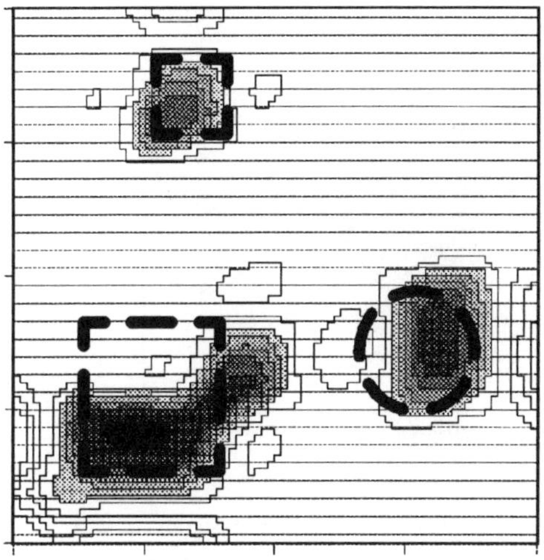

Figure 17. 32 pair excitations, division and filtering (Level=0.5)

CONCLUSIONS

This paper described experimental results using newly improved finite element methods including an exponential block decomposition method which increased computational speed by more than three orders, and a preconditioned voltage conversion method which utilizes comparisons between 2-D and 3-D homogeneous results or between numerical and experimental homogeneous results to form a correction matrix. In addition, application of heuristic image enhancing methods were described with promising results. These new methods have the potential capability of decreasing certain errors including, but not limited to: geometrical measurement, numerical modelling, dimensional distortion and electrical measurement.

Reconstruction of impedance images without using the correction algorithm retains noise, especially in the peripheral regions of high voltage gradients near electrodes. The methods described either with admittance matrix or heuristic corrections are shown to eliminate the error especially is these regions. However, the methods can not yet adequately recognize a small target close to a larger target since a contrast of less than 20% exists between the two. It is felt that the contrast ratio between the large and small targets can be improved by using more local information inside the element and through improvement in experimental procedures.

Finally, there is yet considerable room for improvement in the quality of the image based only on the processing of existing data, within the accuracy of the system. While improvements are continually being made in methods of excitation and measurement, the biggest advances have been made because of the existence of a high accuracy system able to obtain well qualified data of known characteristics (i.e., known static targets in a truly two-dimensional experiment). Comparison with equivalent three dimensional targets in a geometrically similar three-dimensional system has also proven invaluable in our numerical methods development.

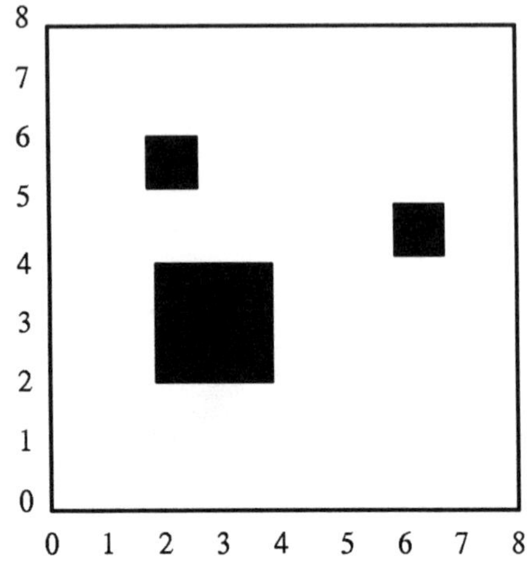

Figure 18. Original size and location of insulator targets in 20x20 cm tank filled with distilled water.

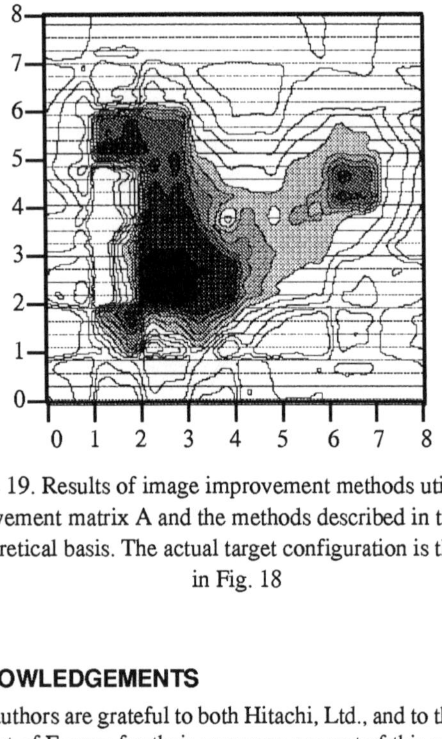

Figure 19. Results of image improvement methods utilizing the improvement matrix A and the methods described in the section of theoretical basis. The actual target configuration is that shown in Fig. 18

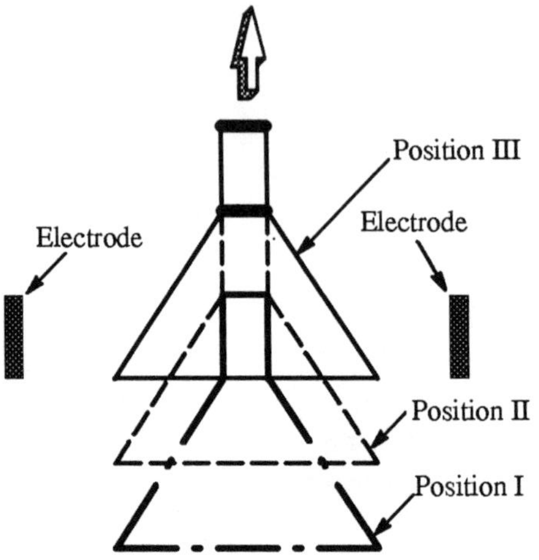

Figure 20. Positions of cone target moving upward between the electrodes. The bottom diameter is 6" and upper diameter 1".

ACKNOWLEDGEMENTS

The authors are grateful to both Hitachi, Ltd., and to the U.S. Department of Energy for their generous support of this research.

REFERENCES

1. J-T Lin, and O.C. Jones, "Progress in Impedance Imaging for Gas-Liquid Flows: 1. Analytical and Numerical Results." This conference.

2. O.C. Jones. 1992, "Cooperative research for the development of electrical impedance computed tomography in two-phase flows," progress report, March.

3. C.G. Xie, A. Plaskowski, and M.S. Beck, 1989, "8-electrode capacitance system for two-component flow identification. Part 1: Tomographic flow imaging," **Science, 136A**(4), pg. 173-183.

4. S. Huang, 1986, **Capacitance transducers for concentration measurement in multi-component flow processes**, Ph.D. Thesis, Department of Instrumentaiton and Analytical Science, UMIST, Manchester, UK.

5. J.S. Halow, and G.E. Fashing, 1988, "Preliminary capacitance imaging experiments of a fluidized bed," **AIChE Sym. Ser. 86**(276), pg. 41-50.

6. J.C. Newell, D. Isaacson, and D.G. Gisser, 1989, "Rapid Assessment of Electrode Characteristics for Impedance Imaging," **IEEE-Trans., Biomed. Eng.**, in press.

7. J.C. Newell, D.G. Gisser, and D. Isaacson, 1988, "An Electric Current Tomograph," **IEEE-Trans. Biomed. Eng., 35** (10), pg. 828-833.

8. D.C. Barber, B.H. Brown and I.L. Freeston, 1983, "Imaging spatial distributions of resistivity using applied potential tomography," **Elec. Lett., 19**, pg. 933-935.

9. K.A. Dynes and R.J. Lytle, 1981, "Analysis of electrical conductivity imaging," **Geophysics, 46**, pg. 1025-1036.

10. J.A. Starzyk, and H. Dai, 1985, "Element evaluation in the resistive networks," **Midwest Sym. Circuits Syst., 28**, pg. 178-181

11. L.R. Price, 1979, "Electrical impedance computed tomography (ICT): a new CT imaging technique," **IEEE Trans. Nucl. Sci. (USA), NS-26, 2**, pg. 2736-2739.

12. R.H.T. Bates, G.C. Mckinnon, and A.D. Seager, 1980, "A limitation on systems for imaging electrical conductivity distributions," **IEEE Trans. Biomed. Eng., BME-27, 7**, pg. 418-420.

13. H. Shomberg and M. Tasto, 1981, "Reconstruction of spatial resistivity distribution," Phillips GMDH, Hamburg, Germany (FRG), MS-H 2715/81.

14. A.D. Seagar, D.C. Barber and B.H. Brown, 1987, "Electrical Impedance Imaging," **IEE Proc. 134, Pt. A**, No. 2, pg. 201-210.

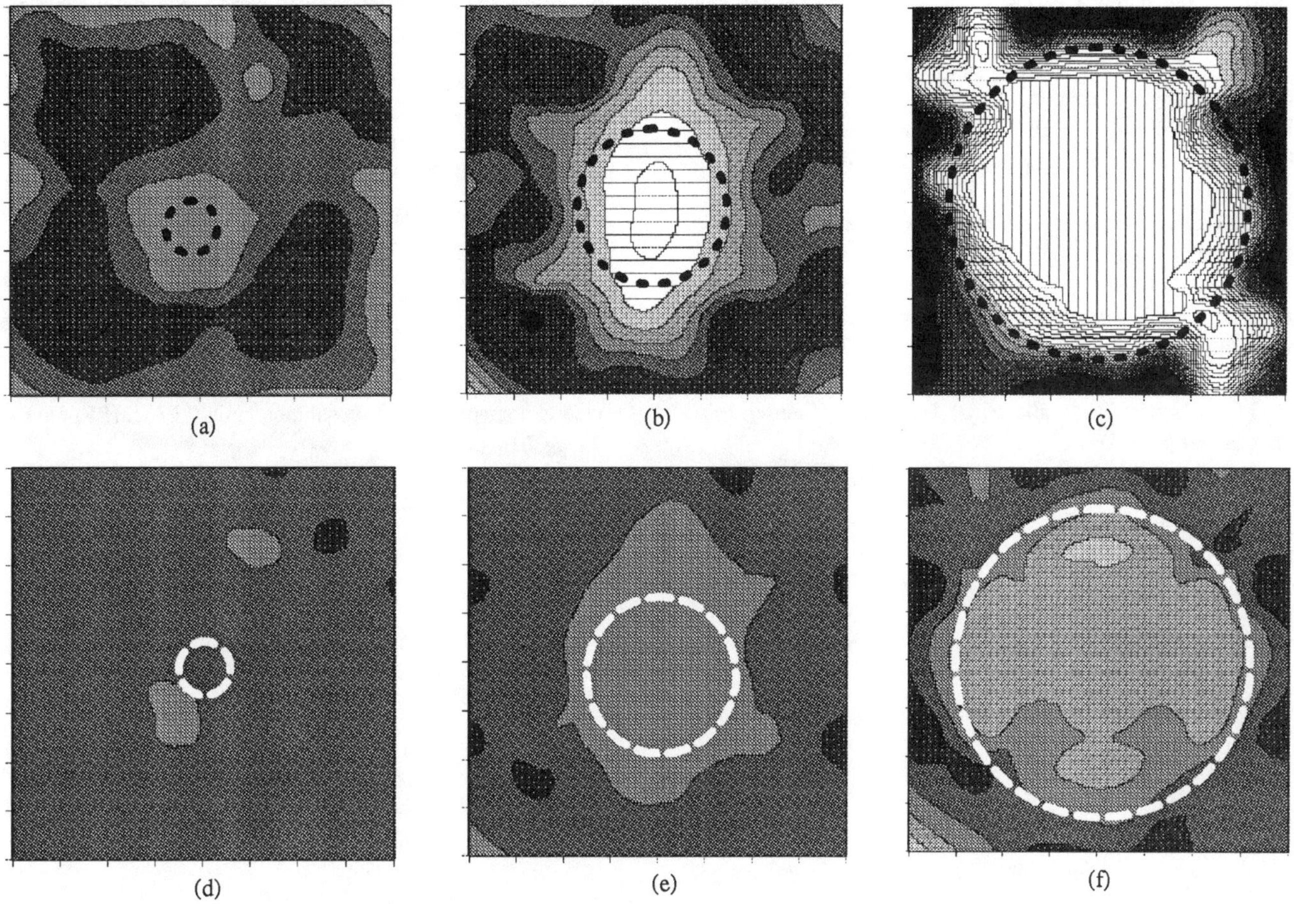

Figure 21. Reconstructed images of 3-D conductor cone target. In this case, the voltage / current ratio is corrected. (a)-(c) are the computed images for Position I, II, and III respectively. The voltage/current data is corrected using the experimental method. (d)-(f) correspond to the same target but the voltage/current data is corrected numerically. Dashed lines indicate the actual size and location of the target in the plane.

15. D.N. Smith, 1985, "Determination of impedance using numerous simultaneous currents (DINSC) – system design and practical applications," **IEEE Conf. Publ. (Inst. Electr. Eng.)**, No. 257, pg. 69–73

16. A. Wexler, B. Fryand M.R. Neiman, 1985, "Impedance-computed tomography algorithm and system," **Appl. Opt., 24**, 23, pg. 3985-3992.

17. R.J. Lytle and K.A. Dines, 1977, "A system for determining the spatial variation of electrical conductivity' Lawrence Livermore Lab.

18. Y. Kim, J.G. Webster and W.J. Tompkins, 1983, "Electric impedance Imaging of Thorax", **Journal of Microwave Power, 18**(3), pp. 245–257.

19. Cook, R.D., 1992, **ACT-3: High Speed, High Precision Electrical Impedance Tomography**, PhD thesis, Rensselaer Polytechnic Institute.

20. M. Tasto, 'Object reconstruction from projections and some nonlinear extensions," Pattern recognition and signal processing, C.H. Chen, Ed.

21. B.H. Brown, and A.D. Segar, 1987, "The sheffield data collection system," **Clin phys. Med. Boil. 5** 431–47,1987

22. P. Hua, P., 1987, "Electrical impedance tomography in medicine'. Ph.D Preliminary proposal.

23. D.G. Gisser, D. Isaacson, and J.C. Newell, 1987, "Current Topics in Impedance Imaging," **Clin. Phys. Physiol. Meas., 8**, Suppl. A, pg. 39–46.

24. L.F. Fuks, D. Isaacson, D.G. Gisser and J.C. Newell, 1989, "Tomographic Images of Dielectric Tissue Properties," **IEEE-Trans. Biomed. Eng.**, in review.

25. D. Isaacson and M. Cheney, 1990, "Current Problems in Impedance Imaging," in **Inverse Problems in Partial Differential**

Equations, D. Coliton, R. Ewing, and W. Rundell, Eds., SIAM, Philadelphia.

26. K–S Cheng, D. Isaacson, J.C. Newell, and D.G. Gisser, 1989, "Electrode Models for Electric Current Computed Tomography," **IEEE–Trans. Biomed. Eng., 36**(9), pg. 918-924.

27. D.G. Gisser, "Current topics in impedance imaging," **Clin. phys. physiol. Meas. 8** Supply. A, pg. 39–46

28. H.M. Powell, 1987, "Impedance imaging using a linear electrode array',Clin phys. physiol. meas. 8 Supply.,A, pg. 109–18.

29. J.G. Webster, 1990, **Electrical Impedance Tomography**, Adam Hilger, New York.

30. L. Tarassenko and P. Rolfe, 1984, "Imaging spatial distributions of resistivity – an alternative approach," **Electron. Lett., 20,** 14, pg. 574–576.

31. T. Murai and Y. Kagawa, 1985, "Electrical impedance computed tomography based on a finite element model," **IEEE Trans. Biomed. Eng., BME–32, 3**, pg. 177–184.

32. T. Murai and Y. Kagawa, Y., 1986, "Boundary element iterative techniques for determining the interface boundary between two Laplace domains–a basic study of impedance plethysmography as an inverse problem," **Int. J. Numer. Methods Eng. (GB), 23**, 1, pg. 35–47

33. A.D. Seagar, T.S. Yeo and R.H.T. Bates, 1984, "Full wave computed tomography, part 2: Resolution limits," **Proc. IEE, part A, 131**, pg. 616–622.

Inverse Problems in Engineering: Theory and Practice
ASME 1993

A TWO DIMENSIONAL INVERSE HEAT CONDUCTION ALGORITHM

Keith A. Woodbury and Sunil K. Thakur
Department of Mechanical Engineering
University of Alabama
Tuscaloosa, Alabama

ABSTRACT

Some industrial cooling applications give rise to multi-dimensional heat removal. Examples include quenching of continuously cast ingot and spray cooling on continuous heat treatment lines.

This report details an inverse method for determining surface cooling rates which are time and space dependent. The method presented is for a two-dimensional transient conduction process in Cartesian coordinates, but the technique can be generalized to the third spatial dimension. In the present study, all boundaries other than the cooled one are assumed adiabatic.

The technique used is Beck's function specification method, and is applicable to conduction problems which are linear (i.e., with temperature independent coefficients). The unknown heat flux is assumed to be a function of the surface coordinate and time. A piece-wise constant variation is assumed for both space and time. The method is tested by a numerical experiment.

NOMENCLATURE

b	thickness of the body (dimensional)
G_{XIJ}	x Green's Function
G_{YMN}	y Green's Function
k	sensor-index
k	thermal conductivity in Eq. 5
L	width of the body (dimensional)
M	time-index
N_p	number of pulses
N_s	number of sensors
q_o	initial heat flux
q	heat flux
r	number of future time steps
t	time
t^+	$\alpha t/L^2$, time (non-dimensional)
T	temperature (dimensional)
T_0	initial temperature
T^+	$(T - T_0)/(q_0 L/k)$ dimensionless temperature
x, y	distance coordinates
x^+	x/L (non-dimensional)
y^+	y/L (non-dimensional)
x_1^{*+}	pulse starting point(non-dimensional)
x_2^{*+}	pulse end point(non-dimensional)

Greek Symbols

α	material thermal diffusivity
γ	L/b, ratio of width to thickness
Δ	$x_2^{*+} - x_1^{*+}$,width of the pulse
Δ^+	Δ/L (non-dimensional)
ϕ	sensitivity coefficients

INTRODUCTION

The IHCP

In the 'direct' heat conduction problems, the heat flux or the temperature histories at the surface of the object are known as a function of time and the interior temperature distribution can easily be found. However, in many dynamic heat transfer problems, the surface heat flux and temperature histories must be determined from transient temperature measurements at one or more interior locations. Such a problem belongs to the class of inverse problems, and is known as an Inverse Heat Conduction Problem (IHCP).

Due to the ill-posedness of the IHCP, it is more difficult to solve than the direct problem. A mathematical problem is considered ill-posed if the result does not depend continuously on the input data. The IHCP is ill-posed because inherent errors in the measurement of the required input temperature data cause instability in the solution.

IHCP's are classified according to the object of their analysis. A *boundary* IHCP determines information missing at one or more boundaries during the process. A *coefficient* IHCP determines one or more parameters appearing in the solution of the heat conduction equation, e.g., the thermal diffusivity α. Examples of applications of the IHCP to both classes of problems are numerous; they have been employed in situations involving the determination of heat-transfer coefficients (Osman and Beck, 1987), internal thermal resistances (Beck, 1988), internal energy sources (Silva-Neto and Özisik, 1992), and even computation of phase boundary locations in solidification or melting problems (Zabaras et al., 1988). The focus of the present effort is on the boundary IHCP.

Multi-dimensional Boundary IHCP's

In the past two decades, considerable attention has been paid to the one-dimensional IHCP, but the multi-dimensional IHCP has come under scrutiny only recently. Hensel and Hills (1989) approached the steady-state two-dimensional IHCP by employing an adjoint formulation to approximate a set of sensitivity-coefficients that relate temperature and heat flux observations to unknown surface conditions. Larsen (1985) approached the two-dimensional IHCP as a parameter-estimation problem while Imber (1974) prescribed an analytical solution called 'Temperature Extrapolation Mechanism' that is applicable to two-dimensional conduction systems for geometries of arbitrary shape. Other work on the two-dimensional problem includes a finite element analysis by Bass et al. (1980) and several articles published by Soviet researchers, mainly by O. M. Alifanov (1981). A detailed review of multi-dimensional IHCP can be found in Hensel (1986).

ALGORITHM DEVELOPMENT

Forward Solver

In the present study, an inverse algorithm is developed by following Beck's function specification method. To construct the inverse solver, a solution to the corresponding forward problem must first be found. Figure

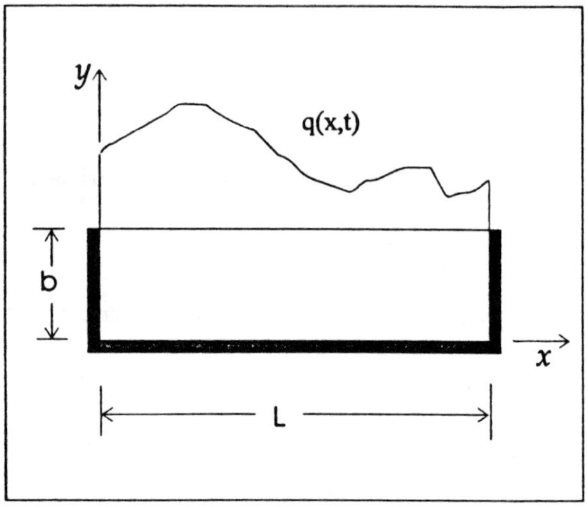

FIGURE 1 - GEOMETRY OF THE PROBLEM

1, depicting a rectangular slab insulated on all sides except one, illustrates the geometry and boundary conditions. The forward problem may be stated in dimensionless variables mathematically as follows. Given:

$$\frac{\partial T^+}{\partial t^+} = \left[\frac{\partial^2 T^+}{\partial x^{+2}} + \frac{\partial^2 T^+}{\partial y^{+2}}\right]$$

Subject to the boundary conditions:

$$\left.\frac{\partial T^+}{\partial x^+}\right|_{x^+=0} = \left.\frac{\partial T^+}{\partial x^+}\right|_{x^+=1} = \left.\frac{\partial T^+}{\partial y^+}\right|_{y^+=0} = 0$$

and

$$q^+(x^+, 1/\gamma, t^+) = \left.\frac{\partial T^+}{\partial y^+}\right|_{y^+=1/\gamma} = f(x^+, t^+)$$

along with an initial condition

$$T^+(x^+, y^+, 0) = T_0^+$$

Find:

$$T^+(x^+, y^+, t^+), \qquad t^+ > 0$$

The solution to the forward problem uses a Green's function approach and exploits the linearity of the problem. The continuous surface heat flux is replaced by a series of independent time-varying heat pulses of uniform width (see Figure 2). The total solution is found as the linear superposition of the solution for each heat flux taken independently. That is,

$$\begin{aligned} T^+(x^+, y^+, t^+) = \\ T^+_{pulse_1}(x^+, y^+, t^+) + T^+_{pulse_2}(x^+, y^+, t^+) + \\ \dots + T^+_{pulse_{N_p}}(x^+, y^+, t^+) \end{aligned} \qquad (1)$$

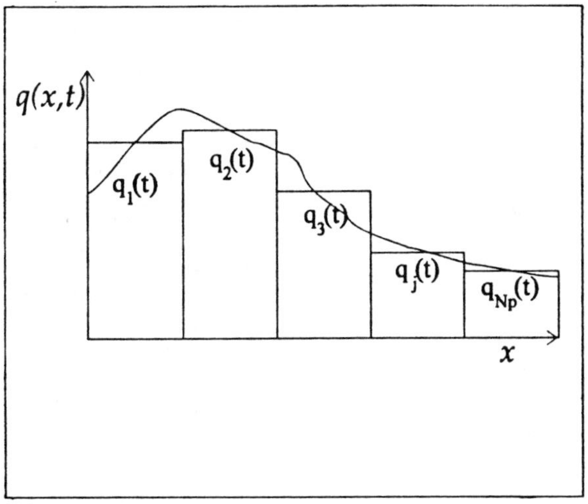

FIGURE 2 - HEAT FLUX APPROXIMATION

The solution for each pulse is found in a generic way by applying the two-dimensional Green's Function to a single pulse. The solution for an arbitrary variation in surface heat flux is given in dimensional form by the integral (Beck et al., 1992):

$$T(x,y,t) = \int_{x'=0}^{L}\int_{y'=0}^{b} G_{XIJ}(x,t|x',0)G_{YMN}(y,t|y',0) \times g(x',y')dx'dy' + \frac{\alpha}{k}\int_{\tau=0}^{t}\int_{x'=0}^{L} G_{XIJ}(x,t|x',\tau) \times G_{YMN}(y,t|b,\tau)q(x',b,t')dx'd\tau. \quad (2)$$

Here $g(x',y')$ represents the initial condition, and $q(x',b,t')$ the heat flux imposed on the upper surface of the domain. The required Green's Functions (GFs) depend on the geometry of the problem. The notation G_{XIJ} refers to the GF specific to the rectangular coordinate type of the boundary condition on the boundaries $x' = 0$ and $x' = L$. Similarly, G_{YMN} refers to the GF for the type of the boundary condition at $y' = 0$ and $y' = b$. Green's functions for our present boundary conditions are given as:

$$G_{XIJ}(x,t|x',\tau) = \frac{1}{L}\{1 + 2\sum_{m=1}^{\infty} e^{-m^2\pi^2\alpha(t-\tau)/L^2} \times \cos(m\pi x/L)\cos(m\pi x'/L)\} \quad (3)$$

and,

$$G_{YMN}(y,t|y',\tau) = \frac{1}{b}\{1 + 2\sum_{n=1}^{\infty} e^{-n^2\pi^2\alpha(t-\tau)/b^2} \times \cos(n\pi y/b)\cos(n\pi y'/b)\} \quad (4)$$

In the present case, zero initial conditions are assumed, and the first integration vanishes from Eq. (2). A pulse of width $\Delta^+ = (x_2^+ - x_1^+)$ centered at location $x_*^+ = (x_2^+ + x_1^+)/2$ is used to arrive at the general form of the solution for a single arbitrary heat pulse. The second integral in Eq. (2) is integrated using $q(x^+,b,t^+)$ as:

$$q(x^+,b,t^+) = \begin{cases} 1 & x_1^+ \le x^+ \le x_2^+ \\ 0 & \text{otherwise} \end{cases}$$

Thus, the solution for a step change in an arbitrary one of the surface heat pulses is found. A discrete Duhamel's sum, which is described below, is used to account for the time variation of the heat pulse.

The above Green's functions and boundary condition multiplied and integrated in Eq. (2) yield the response at any point (x^+,y^+) due to the step change in surface heat flux. The result, in a non-dimensional form is:

$$\begin{aligned} T^+(x^+,y^+,t^+) &= t^+\Delta^+\gamma \\ &+ \frac{2\Delta^+}{\pi^2\gamma}\sum_{n=1}^{\infty}\frac{-1^n}{n^2}(1-\exp^{-n^2\pi^2\gamma^2t^+})\cos(n\pi\gamma y^+) \\ &+ \frac{2\gamma}{\pi^3}\sum_{m=1}^{\infty}\frac{1}{m^3}(1-\exp^{-m^2\pi^2t^+}) \times \\ &\quad \cos(m\pi x^+)\{\sin(m\pi x_2^{*+}) - \sin(m\pi x_1^{*+})\} \\ &+ \frac{4\gamma}{\pi^3}\sum_{m=1}^{\infty}\sum_{n=1}^{\infty}\frac{-1^n}{m(m^2+n^2\gamma^2)} \times \\ &\quad (1-\exp^{-(m^2+n^2\gamma^2)\pi^2t^+})\cos(m\pi x^+)\cos(n\pi\gamma y^+) \\ &\quad \{\sin(m\pi x_2^{*+}) - \sin(m\pi x_1^{*+})\} \quad (5) \end{aligned}$$

Discrete Duhamel's Theorem

The sensitivity coefficients relates the response of a sensor at a particular location (x_k^+, y_k^+) to the specific heat flux disturbance q_i. In the current context, there exists a sensitivity coefficient for each sensor corresponding to each of the assumed surface heat pulses. Denoting the sensor located at (x_k^+, y_k^+) as sensor k, the sensitivity coefficient for sensor k to heat flux pulse i is

$$\frac{\partial T_k^+}{\partial q_i^+} \equiv \phi_{k,i}(t^+) = T^+(x_k^+, y_k^+, t^+) \quad (6)$$

That is, the sensitivity coefficient is the same as the dimensionless temperature response given by Eq. (5) evaluated at the sensor location.

The total response of a sensor as a function of time can be found for a linear problem by superposition. That is, the summation of the effect of all the surface pulses shown in Fig. 2 must be made in order to find the total response at any point in the domain. This results in a finite sum, in a form similar to a Duhamel integral. If time is considered discrete according to $t^+ = M\Delta t^+$, the discrete form of Duhamel's theorem is used in calculating the total response for sensor k (that is, $T^+(x_k^+, y_k^+, M\Delta t^+)$ is

$$T^+_{M,k} = T^+_o + \sum_{i=1}^{N_p}\sum_{n=1}^{M} q^+_{n,i}[\Delta\phi_{M-n}]_{k,i} \tag{7}$$

This is an extension of the result for a single pulse given by Beck (Beck et al., 1985). Note that, due to the superposition, the terms $[\Delta\phi_{M-n}]_{k,i}$ occur. This quantity is the simple time difference in the sensitivity of the kth sensor to the ith pulse at time $M - n$. The $\Delta\phi_m$ notation is the same as that of Beck and indicates this temporal difference. Thus,

$$[\Delta\phi_m]_{k,i} = [\phi_{m+1}]_{k,i} - [\phi_m]_{k,i} \tag{8}$$

Matrix Formulation

The presentation follows that of Beck (Beck et al., 1985) By expressing Eq. (7) for a number of sensors N_s and for times t_M through t_{M+r-1}, the temperature in a one-,two-, or three dimensional body with temperature-independent thermal properties can be given in the standard form of

$$\mathbf{T} = \mathbf{T}|_{\mathbf{q}=0} + \mathbf{X}\mathbf{q} \tag{9}$$

Here $\mathbf{X}$ is a matrix of the sensitivity time differences $\Delta\phi$, and $\mathbf{q}$ is a vector of heat fluxes. The quantity $\mathbf{T}|_{\mathbf{q}=0}$ is the vector of nodal temperature computed from all previously estimated values of the heat flux vector; these represent the "initial conditions" on the current estimation step. $\mathbf{T}$ is a vector of discrete temperatures corresponding to N_s temperature sensors from the present time t_M through r future time steps to t_{M+r-1}:

$$\mathbf{T} = \begin{Bmatrix} \mathbf{T}(M) \\ \mathbf{T}(M+1) \\ \mathbf{T}(M+2) \\ \vdots \\ \mathbf{T}(M+r-1) \end{Bmatrix} \qquad \mathbf{T}(m) = \begin{Bmatrix} T_1(m) \\ T_2(m) \\ \vdots \\ T_{N_s}(m) \end{Bmatrix}$$

Thus $\mathbf{T}$ is an $(N_s \times r) \times 1$ matrix; that is an $N_s \times r$ vector. The vector $\mathbf{q}$ contains the N_p values of heat flux from the present time (M) through r future time steps:

$$\mathbf{q} = \begin{Bmatrix} \mathbf{q}(M) \\ \mathbf{q}(M+1) \\ \cdot \\ \cdot \\ \mathbf{q}(M+r-1) \end{Bmatrix} \qquad \mathbf{q}(m) = \begin{Bmatrix} q_1(m) \\ q_2(m) \\ \cdot \\ \cdot \\ q_{N_p}(m) \end{Bmatrix}$$

Thus, $\mathbf{q}$ is an $(r \times N_p) \times 1$ matrix, that is a vector of length $r \times N_p$. The matrix $\mathbf{X}$ contains the appropriate values of the sensitivity differences $\Delta\phi$:

$$\mathbf{X} = \begin{bmatrix} \mathbf{a}(1) & 0 & 0 & 0 & 0 \\ \mathbf{a}(2) & \mathbf{a}(1) & 0 & 0 & 0 \\ \mathbf{a}(3) & \mathbf{a}(2) & \mathbf{a}(1) & 0 & 0 \\ \vdots & \vdots & \vdots & \ddots & 0 \\ \mathbf{a}(r) & \mathbf{a}(r-1) & \cdots & \mathbf{a}(2) & \mathbf{a}(1) \end{bmatrix}$$

Here $\mathbf{X}$ is an $(N_s \times r) \times (N_p \times r)$ matrix and $\mathbf{a}(m)$ is defined by

$$\mathbf{a}(m) = \begin{bmatrix} a_{11}(m) & a_{12}(m) & .. & a_{1N_p}(m) \\ a_{21}(m) & a_{22}(m) & .. & a_{2N_p}(m) \\ \cdot & \cdot & .. & \cdot \\ \cdot & \cdot & .. & \cdot \\ a_{N_s1}(m) & a_{N_s2}(m) & .. & a_{N_sN_p}(m) \end{bmatrix}$$

More specifically, for the present case, where the solution for the temperature response at the sensor location is given by the discrete Duhamel's theorem (Eq. 7), the components of $\mathbf{a}(m)$ are given by:

$$\mathbf{a}(m) = \begin{bmatrix} \Delta\phi_{1,1}(m) & \Delta\phi_{1,2}(m) & \cdots & \Delta\phi_{1,N_p}(m) \\ \Delta\phi_{2,1}(m) & \Delta\phi_{2,2}(m) & \cdots & \Delta\phi_{2,N_p}(m) \\ \vdots & \vdots & \ddots & \vdots \\ \Delta\phi_{N_s,1}(m) & \Delta\phi_{N_s,2}(m) & \cdots & \Delta\phi_{N_s,N_p}(m) \end{bmatrix}$$

Here the generic entry $\Delta\phi_{k,i}(m)$ corresponds to the temporal difference in the sensitivity coefficient for sensor k to pulse i at time m and is identical to the expression in Eq. (8).

The Eq. (9) is an equation for $N_p \times r$ values of q with $N_s \times r$ temperature data values available from an experiment. As long as $N_s \geq N_p$ an overdetermined problem for the q's results and the following minimization problem makes sense. However, a customary procedure is to introduce future time information (Beck, 1968) to counter the ill-posedness of the problem by temporarily assuming $\mathbf{q}(M+j) = \mathbf{q}(M)$. This is only a temporary assumption that is used to obtain $\mathbf{q}(M)$. To facilitate this, $\mathbf{q}$ is set equal to $\mathbf{q} = \mathbf{AB}$; the dimension of the $\mathbf{A}$ matrix is $(N_p \times r) \times N_p$. $\mathbf{A}$ and $\mathbf{B}$ matrices are defined by

$$\mathbf{A} = \begin{bmatrix} \mathbf{A}(1) \\ \mathbf{A}(2) \\ \vdots \\ \mathbf{A}(r) \end{bmatrix} \qquad \mathbf{B} = \begin{bmatrix} q_1(M) \\ q_2(M) \\ \vdots \\ q_{N_p}(M) \end{bmatrix}$$

$\mathbf{A}(m)$ is an $N_p \times N_p$ matrix given by the identity matrix:

$$\mathbf{A}(m) = \begin{bmatrix} 1 & 0 & \cdots & \cdots & \cdots & 0 \\ 0 & 1 & 0 & & \cdots & \cdots & 0 \\ 0 & 0 & 1 & 0 & \cdots & 0 \\ 0 & 0 & 0 & 1 & \cdots & 0 \\ \vdots & \vdots & \vdots & \vdots & \ddots & \vdots \\ 0 & 0 & 0 & 0 & \cdots & 1 \end{bmatrix}$$

With this temporary assumption, the equation (9) becomes

$$\begin{aligned} \mathbf{T} &= \mathbf{T}|_{\mathbf{q}=0} + \mathbf{X}\mathbf{q} \\ &= \mathbf{T}|_{\mathbf{q}=0} + \mathbf{XAB} \\ &= \mathbf{T}|_{\mathbf{q}=0} + \mathbf{ZB} \end{aligned} \qquad (10)$$

Where the matrix $\mathbf{Z} = \mathbf{XA}$. This Eq. (10) represents an equation for N_p components of $\mathbf{q}(M)$ with $N_s \times r$ temperature data values. Now, so long as $N_s \times r \geq N_p$, an overdetermined problem results and the following minimization problem makes sense.

The idea now is to minimize the discrepancy between the model-predicted values given by Eq. (10) and data acquired from an experiment, $\mathbf{Y}$. This discrepancy is quantified by the square of the difference, which in matrix form becomes:

$$S = (\mathbf{Y} - \mathbf{T}|_{\mathbf{B}=0} - \mathbf{ZB})^T(\mathbf{Y} - \mathbf{T}|_{\mathbf{B}=0} - \mathbf{ZB})$$

To minimize this function, the matrix derivative with respect to $\mathbf{B}$ is set equal to zero. The result gives the estimator for $\mathbf{B}$:

$$\hat{\mathbf{B}} = (\mathbf{Z}^T\mathbf{Z})^{-1}\mathbf{Z}^T(\mathbf{Y} - \mathbf{T}|_{\mathbf{B}=0})$$

Sequential application of this equation yields the $\hat{\mathbf{q}}(M)$ vector for successive time intervals.

ALGORITHM TESTING

Numerical Experiment

The 2DIHCP algorithm from the previous section was verified by means of a numerical experiment. The idea of a numerical experiment is to impose a known heat flux on the domain (Figure 1) and compute the response of the domain by a suitably accurate numerical solver.

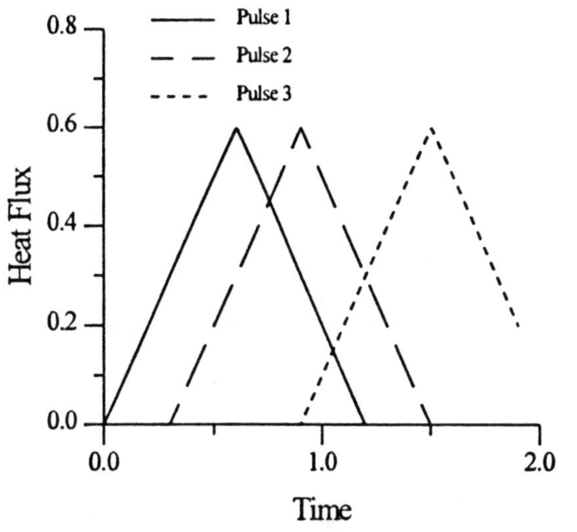

FIGURE 3 – TIIME HISTORY OF THE THREE PULSES

The domain response at a fixed number of points is recorded, and these individual temperature responses form the basis for sensor data for the inverse algorithm. These responses are "poisoned" slightly by adding random noise to simulate measurement error. This synthetic data is then input to the IHCP algorithm, and its output compared to the original assumed heat flux.

An assumed heat flux profile is needed to initiate the process. A simple, yet realistic, heat flux is desired. The choice of three pulses of fixed width, with varying magnitudes, is chosen as it mimics a plate which is cooled on its surface from one end. Figure 4 shows the assumed arrangement of the discrete pulses. A triangular time history for each of the heat flux pulses is assumed (similar to Beck's assumption for the one-dimensional case (1985, page193)); see Figure 3.

The geometric model is composed of three sensors and three pulses equally spaced over the body. Figure 4 shows the sensors in their relation of the pulses. The thermocouples are assumed to be located immediately below the centerline of each individual pulse, on the back plane of the plate. The plate has an assumed aspect ratio $\gamma = 7.10$.

Forward Solver

The discrete form of Duhamel's Theorem (Eq. 7) is used to compute the temperature. Through trials, it was learned that to compute the temperature response to a good accuracy, a large number of m and n terms in Eq. 5 must be evaluated and a small time step must be utilized. The accuracy was judged by computing for a single heat flux into a one-dimensional domain using the standard triangle flux of Beck and comparing to his

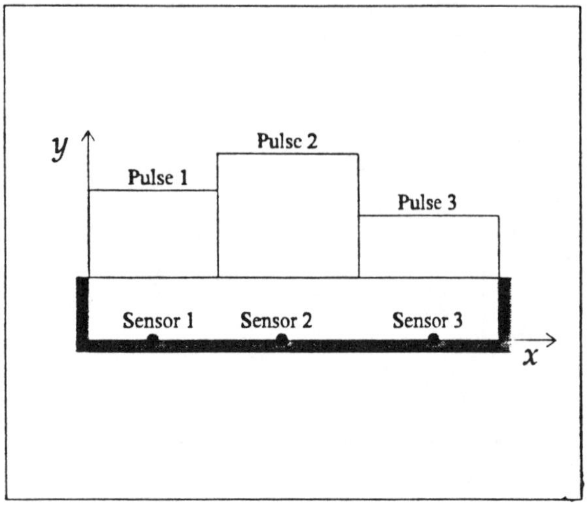

FIGURE 4 – PULSES AND LOCATION OF SENSORS FOR THE NUMERICAL EXPERIMENT

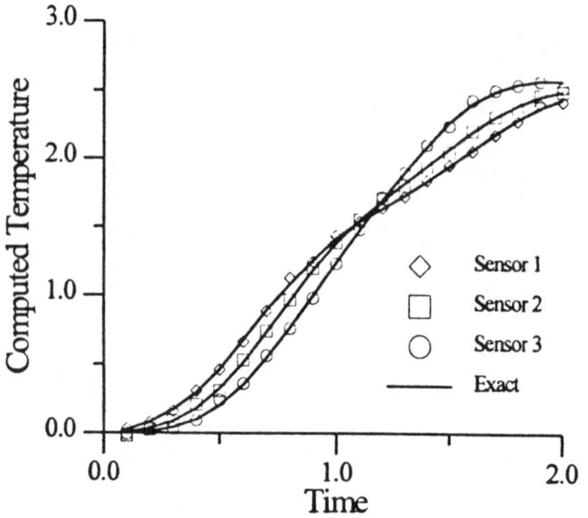

FIGURE 5 – ARTIFICIAL DATA FOR TEST CASE

published tabular values (Beck et al., 1985). The finding was that $m = n = 500$ terms and $\Delta t^+ = 0.001$ was required to match the values published by Beck. Note that this requires 250,000 terms in the double sum of Eq. 5! This was alleviated by modifying the summation to periodically compute the contribution for the next, say, 20 terms of each of the three series in Eq. 5. When this contribution became "small" ($< 10^{-10}$), no further terms were evaluated for that series.

Artificial Data

The assumed flux profiles are now input into the forward solver to obtain the exact solution for the thermocouples. The resulting temperatures have been verified by comparing with the results obtained using a finite element analysis. The pure data for the three thermocouples are seen in Fig.5 as the solid lines. This pure data is perturbed by adding a random disturbance with zero mean and a known standard deviation. A random number generator was used to add noise to the pure data according to

$$T_{data} = T_{pure} + \sigma \times \text{rand}(-1, 1)$$

where rand$(-1, 1)$ is a "random" number in the range of -1 to +1.

Two sets of data were generated; one with $\sigma_Y^+ = 0.0017$ and one with $\sigma_Y^+ = 0.017$. The first corresponds to a 3σ scatter of ± 0.005 and is the same as that used by Beck (1985, page 172) The second data set, with $\sigma_Y^+ = 0.017$, corresponds to a 3σ scatter of ± 0.05, and results in a discernible deviation from the true (pure) values (see Fig. 5). These values are shown as the data points in the figure, while the solid lines represent the pure data.

RESULTS

Heat Flux

The 2DIHCP program has been used to analyze both the pure and corrupted data. Figure 6 shows the comparison with the exact (known) heat flux history for the pure data. These results were computed with $r = 1$ future time step; that is, an exact matching algorithm. For these computations, $\Delta t^+ = 0.10$ was used. Good agreement, as expected, is seen in the figure.

Figure 7 shows the results for the corrupted data having $\sigma_Y^+ = 0.0017$ along with the known heat flux. The comparison is again quite favorable. Again, these computations were obtained with $r = 1$ future time step and $\Delta t^+ = 0.10$.

As a stringent test of the algorithm's ability to resolve heat fluxes in the face of noise, the data having $\sigma_Y^+ = 0.017$ was processed. The result are shown in Fig. 8 for r=2 future time steps. While the predictions are not beautiful, they do a good job of recovering the heat flux history, in spite of large errors in the input data.

Sensitivity Coefficients

The sensitivity coefficients are shown in Table 1 for the first two sensors to each of the three heat pulses. The sensitivity coefficients for the last sensor are not shown, as they are the reflexive image of those for Sensor 1 due to symmetry. Note that all of the sensitivity values are non-negligible; in other words, even for the relatively high aspect ratio γ, every sensor has strong

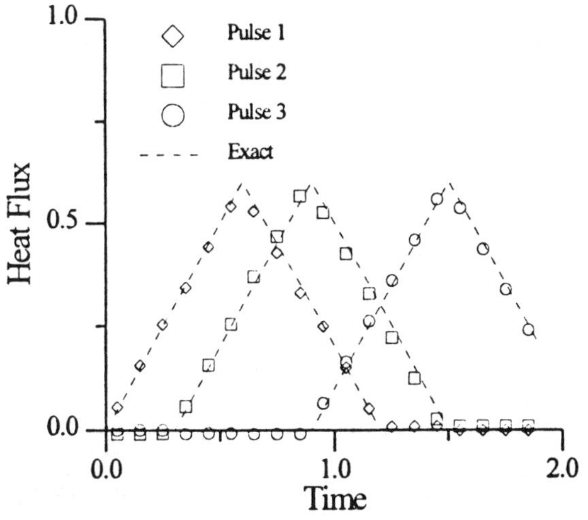

FIGURE 6 - CALCULATED SURFACE HEAT FLUX WITH ERRORLESS DATA

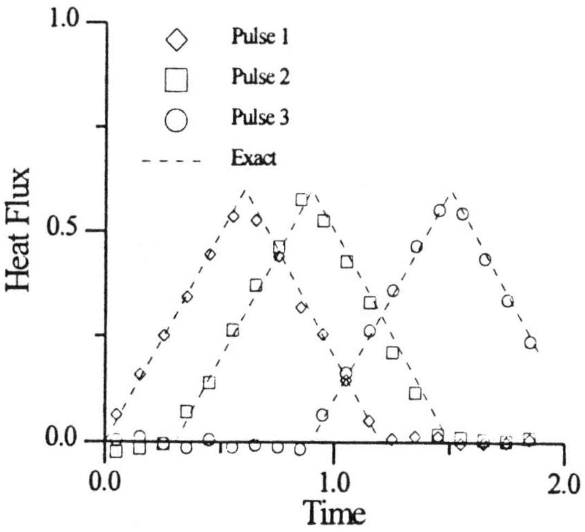

FIGURE 7 - CALCULATED SURFACE HEAT FLUX WITH RANDOM ERRORS: $\sigma_Y = 0.0017$

sensitivity to each pulse. This suggests that a locally one-dimensional heat flow assumption might not be appropriate.

For these computations, $m = n = 100$ was used to compute the sensitivity coefficients, with the time steps $\Delta t^+ = 0.10$. Based on the previous investigations with the forward solver, it would seem that this is not enough terms nor a small enough time step to accurately compute the sensitivity coefficients. This suggests that precise knowledge of the sensitivity coefficients is not necessary to obtain reasonable estimates.

CONCLUSIONS

The following conclusions can be summarized from this work:

- For linear problems, a Green's Function approach, coupled with a Duhamel's Summation, provides a general framework for development of multi-dimensional IHCP algorithms.

- For the test case presented, the sensitivity coefficients suggest that local one-dimensionality is NOT a good assumption.

- It seems that precise knowledge of the sensitivity coefficients is not necessary for the test case presented.

- For the test case presented, m=n=100 terms with $\Delta t^+ = 0.10$ and $r = 2$ gives reasonable results, even in the face of relatively large amounts of noise in the input data.

REFERENCES

Alifanov, O. M. and Kerov, N. V. (1981). "Determination of External Thermal Load Parameters by Solving the Two-dimensional Inverse Heat Conduction Problem,". *Journal of Engineering Physics, 41*, 1049–1053.

Bass, B. R., Drake, J. B., and Ott, L. J. (1980). "ORDMIN: A Finite Element Program for Two-Dimensional Nonlinear Inverse Heat Conduction Analysis,". Technical Report NUREG/CR-1709, Oak Ridge National Laboratories, Oak Ridge, Tennessee.

Beck, J. V. (1968). "Surface Heat Flux Determination Using an Integral Method,". *Nuclear Engineering and Design, 7*, 170–178.

Beck, J. V. (1988). "Combined Parameter and Function Estimation in Heat Transfer with Application to Contact Conductance,". *Transactions of the ASME Journal of Heat Transfer, 110*, 1046–1058.

Beck, J. V., Blackwell, B., and St. Clair, C. (1985). *Inverse Heat Conduction: Ill-posed Problems.* New York: Wiley-Interscience.

Beck, J. V., Cole, K. D., Haji-Sheikh, A., and Litkouhi, B. (1992). *Heat Conduction using Green's Functions.* New York: Hemishpere Publishing Company.

Hensel, E. (1986). *Multi-dimensional Inverse Heat Conduction.* PhD thesis, New Mexico State University, Mechanical Engineering Deparment.

Hensel, E. and Hills, R. (1989). "Steady-State Two-Dimensional Inverse Heat Conduction,". *Numerical Heat Transfer, Part B, Applications, 15*, 227–240.

Imber, M. (1974). "Temperature Extrapolation Mechanism for Two-Dimensional Heat flow,". *AIAA*

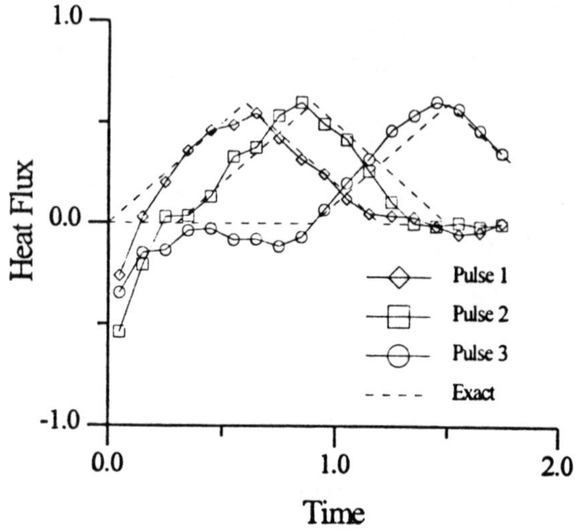

FIGURE 8 – CALCULATED SURFACE HEAT FLUX WITH RANDOM ERRORS: $\sigma_Y = 0.017$

Journal, *12*(8), 1089–1093.

Larsen, M. E. (1985). "An Inverse Problem: Heat Flux and Temperature Prediction for a High Heat Flux Experiment,". Technical Report SAND-85-2671, Sandia National Laboratory, Alburquerque, New Mexico.

Osman, A. M. and Beck, J. V. (1987). "Nonlinear Inverse Problem for the Estimation of Time-and-Space dependent Heat Transfer Coefficients,". Reno, Nevada. AIAA 25th Aerospace Sciences Meeting. AIAA-87-0150.

Silva-Neto, A. J. and Özisik, M. N. (1992). "Simultaneous Estimation of Location and Strength of a Plane Heat Source,". Presented at Fifth Seminar on Inverse Problems in Engineering.

Zabaras, N., Mukerjee, S., and Richmond, O. (1988). "An Analysis of Inverse Heat Transfer Problems with Phase Changes Using an Integral Method,". *Transactions of the ASME Journal of Heat Transfer*, *110*, 554–561.

Sensor $k = 1$			Sensor $k = 2$		
$i = 1$	$i = 2$	$i = 3$	$i = 1$	$i = 2$	$i = 3$
0.4576	0.1825	0.0465	0.1825	0.3215	0.1825
0.7751	0.4183	0.2033	0.4183	0.5601	0.4183
1.0417	0.6549	0.4101	0.6549	0.7968	0.6549
1.2896	0.8916	0.6356	0.8916	1.0335	0.8916
1.5304	1.1283	0.8681	1.1283	1.2702	1.1283
1.7686	1.3650	1.1033	1.3650	1.5069	1.3650
2.0059	1.6017	1.3394	1.6017	1.7436	1.6017
2.2428	1.8384	1.5759	1.8384	1.9803	1.8384
2.4796	2.0751	1.8125	2.0751	2.2169	2.0751
2.7163	2.3117	2.0491	2.3117	2.4536	2.3117
2.9530	2.5484	2.2858	2.5484	2.6903	2.5484
3.1897	2.7851	2.5225	2.7851	2.9270	2.7851
3.4264	3.0218	2.7592	3.0218	3.1637	3.0218
3.6630	3.2585	2.9958	3.2585	3.4004	3.2585
3.8997	3.4952	3.2325	3.4952	3.6371	3.4952
4.1364	3.7319	3.4692	3.7319	3.8738	3.7319
4.3731	3.9685	3.7059	3.9685	4.1104	3.9685
4.6098	4.2052	3.9426	4.2052	4.3471	4.2052
4.8465	4.4419	4.1793	4.4419	4.5838	4.4419
5.0832	4.6786	4.4160	4.6786	4.8205	4.6786

TABLE 1: SENSITIVITY COEFFICIENTS FOR THE TEST PROBLEM

Inverse Problems in Engineering: Theory and Practice
ASME 1993

IDENTIFICATION OF DIFFUSIVITY COEFFICIENT AND INITIAL CONDITION BY DISCRETE MOLLIFICATION

Carlos E. Mejía and Diego A. Murio
Department of Mathematical Sciences
University of Cincinnati
Cincinnati, Ohio

Abstract
We discuss the problem of simultaneously identifying the diffusivity coefficient and the initial temperature distribution in a one dimensional parabolic equation. A suitable space marching implementation of the Mollification Method is introduced that provides accurate and stable numerical solutions whenever the boundary conditions at the active surface are not known exactly. Several numerical examples are presented to illustrate the accuracy of the new algorithm.

1 NOMENCLATURE

1.1 Capital Letters

A : Computed diffusivity coefficient
B : Set $[0,1] \times [0,T]$
$C, C_1, C_2, ..., C_7$: Generic constants
C^0 : Space of continuous functions
C^2 : Space of two times differentiable functions whose second partial derivatives are continuous.
D : Set $(0,1) \times (0,1)$
E : Error function
G : Generic discrete function
I : Generic interval
J_δ : Mollification operator
L, M, N : Grid size parameters
L^2 : Space of square integrable functions
T : Time interval right endpoint
U : Computed temperature function
V : Computed heat flux
W : Computed time derivative of the temperature

1.2 Lowercase Letters

a : Exact diffusivity coefficient
d_1, d_2, d_3 : Constants
f : Forcing term
h, k : Grid step sizes
l^2 : Space of discrete functions whose 2-norm is finite
t : Time variable
u : Exact temperature function
$u_t \equiv w$: Exact time derivative of the temperature
$u_x \equiv v$: Exact heat flux
u_{xx} : Second partial derivative of the temperature with respect to x twice
x : Space variable

1.3 Greek Letters

α, β : Generic constants
δ : Mollification parameter
ϵ : Maximum noise in data
σ : Exact boundary coefficient
ϕ : Exact boundary heat flux
ψ : Exact boundary temperature
ρ_δ : Gaussian kernel

1.4 Superscripts and Subscripts

j : Space index
m : Identifies measured data functions
n : Time index

1.5 Other Symbols

ΔA : Error in computed coefficient
ΔU : Error in computed temperature
ΔV : Error in computed heat flux
ΔW : Error in computed time derivative of the temperature
$|\ |$: Infinity norm of discrete functions of one variable
$\|\ \|_{\infty}$: Infinity norm of discrete functions of two variables
$\|\ \|_2$:l^2−norm
$|\ |_2$:l^2−Relative error

2 INTRODUCTION

The identification of diffusivity coefficients in parabolic equations is receiving considerable attention from researchers in a variety of fields, namely, Heat Conduction, Oil Recovery, Groundwater Flow, etc. A detailed treatment of some problems in these areas can be found in Chavent and Jaffré (1986), Murio (1993) and Wheeler (1988).

The use of a space marching scheme along with some kind of regularization, has proved to be an effective way of solving coefficient identification problems. We only mention the identification of a coefficient dependent only on the space variable through the implementation of a Hyperbolic Regularization and a space marching scheme that relies on the knowledge of the exact initial condition (Ewing and Lin, 1989), the combination of the Hyperbolic Regularization and the Mollification Method that allows noisy initial and boundary data and carries out the same identification problem of Ewing and Lin (Mejía and Murio, 1993), and the space marching implementation of the Mollification Method that, based on noisy temperature data available throughout the domain, is able to accurately identify a heat transfer coefficient dependent on space and time (Hinestroza and Murio, 1993).

In this paper, we develop a stable space marching implementation of the Mollification Method that, based only on measured noisy boundary data, is able to approximate a diffusivity coefficient dependent on space and time and the initial condition. The paper is organized as follows:

The identification problem and the regularization properties of the Mollification Method are introduced in Section 2.

Section 3 contains a careful derivation of the marching scheme and a description of its implementation in algorithmic form.

The analysis of the method is carried out in Sections 4 and 5. The main results are the stability analysis of Theorem 3 and the error estimate of Theorem 5.

Section 6 contains some illustrative numerical experiments.

3 PRELIMINARIES

3.1 The Identification Problem

Let $D = (0,1) \times (0,1)$. The coefficient identification problem is the following:
Identify $a(x,t)$, $(x,t) \in D$, and $u(x,0)$, $x \in (0,1)$, in

$$\begin{array}{ll} u_t = au_{xx} + f(x,t), & (x,t) \in D \\ a(0,t) = \sigma(t), & 0 < t, \\ u(0,t) = \psi(t), & 0 < t, \\ u_x(0,t) = \phi(t), & 0 < t. \end{array} \quad (1)$$

Since it is not realistic to have exact boundary data, we assume that the boundary data for problem (1) are obtained from measurements. They are given by $C^0(0,1)$ functions $\sigma_m(t)$, $\psi_m(t)$ and $\phi_m(t)$ satisfying $\|\sigma - \sigma_m\|_{\infty,(0,1)} \leq \epsilon$, $\|\psi - \psi_m\|_{\infty,(0,1)} \leq \epsilon$ and $\|\phi - \phi_m\|_{\infty,(0,1)} \leq \epsilon$, where ϵ is a positive tolerance. The ill-posedness of parameter estimation problems and the absence of an initial temperature distribution, configure problem (1) as a challenging and interesting inverse problem.

In order to restore some kind of continuity with respect to the boundary data, we develop a space marching implementation of the Mollification Method that is able to accurately estimate the heat transfer coefficient $a(x,t)$ and the temperature $u(x,t)$ throughout the domain D. Furthermore the combination of the Mollification and a linear extrapolation procedure that follows the ideas in Murio and Hinestroza (1993), enables us to estimate the initial data at $t = 0$.

3.2 Mollification

The Mollification Method is a filtering procedure that is appropriate for the regularization of a variety of ill-posed problems; its description and several of its applications can be found in Murio (1993). We use the Gaussian kernel

$$\rho_\delta(t) = \frac{1}{\delta\pi^{1/2}} \exp(\frac{-t^2}{\delta^2})$$

as mollifier and define the δ-mollification of a square integrable function $f(t)$ by

$$J_\delta f(t) = (\rho_\delta * f)(t) = \int_{-\infty}^{\infty} \rho_\delta(t-s) f(s) ds.$$

$J_\delta f(t)$, the one-dimensional convolution of ρ_δ and f, is a C^∞ function. The mollifier $\rho_\delta(t)$ is positive, falls to nearly zero outside $[-3\delta, 3\delta]$ and its total integral is 1. The success of the Mollification Method as a regularization procedure is due, in part, to the following properties:

Proposition 1 *(Consistency) If $f(t) \in C^2(I)$, $I \subset R$, then there exists a constant C independent of δ such that*

$$\|J_\delta f - f\|_{\infty,I} \leq C\delta \text{ and } \|J_\delta f' - f'\|_{\infty,I} \leq C\delta.$$

Proof. (Murio, 1993).□

If we only know a measured approximation $f_m(t)$ of $f(t)$ and a tolerance $\epsilon > 0$ such that $\|f - f_m\|_{\infty,I} \leq \epsilon$, then we have the following stability estimates:

Proposition 2 *(Stability) If $f_m(t) \in C^0(I)$ and $\|f - f_m\|_{\infty,I} \leq \epsilon$, then*

$$\|J_\delta f - J_\delta f_m\|_{\infty,I} \leq \epsilon$$

and

$$\|(J_\delta f)' - (J_\delta f_m)'\|_{\infty,I} \leq \left(\frac{2}{\sqrt{\pi}}\right)\frac{\epsilon}{\delta}.$$

Proof. (Murio, 1993).□

If I is a bounded interval with end points a and b, the convolution $\rho_\delta \star f$ requires either the extension of f to a slightly bigger set $I' \supset I$ or the consideration of f restricted to a suitable compact set $K \subset I$. Both sets, I' and K depend on δ. Our approach is the first one. Assume $\rho_\delta(t) = 0$ if $|t| > 3\delta$. We seek extensions $\tilde{f}$ of f to the interval $[a, b + 3\delta]$, minimizing the L^2 norm

$$\left\| J_\delta \tilde{f} - f \right\|_{L^2[b-3\delta,b]}. \tag{2}$$

If $\tilde{f}$ is assumed to be a polynomial on $[b, b+3\delta]$, it can be proved that there is a unique least squares solution for problem (2). Similar results hold for extensions to the interval $[a - 3\delta, b]$. Complete details on this approach can be found in Mejía (1993).

The implementation of the Mollification Method for Problem (1) with measured data, leads us to the consideration of the following problem:

Identify $a(x,t)$, $(x,t) \in D$, and $u(x,0), x \in (0,1)$, in

$$\begin{array}{ll} u_t = a u_{xx} + f(x,t), & (x,t) \in D, \\ u(0,t) = J_\delta \psi_m(t), & 0 < t, \\ u_x(0,t) = J_\delta \phi_m(t), & 0 < t, \\ a(0,t) = J_\delta \sigma_m(t), & 0 < t, \end{array} \tag{3}$$

where δ is the radius of mollification, and $J_\delta \psi_m(t)$, $J_\delta \phi_m(t)$ and $J_\delta \sigma_m(t)$ are the mollifications in t of $\psi_m(t)$, $\phi_m(t)$ and $\sigma_m(t)$ respectively.

4 THE MARCHING SCHEME

Let M and N be positive integers, $h = \frac{1}{M}$, $k = \frac{1}{N}$, $x_j = jh$, $j = 0, 1, ..., M$, $t_n = nk$, $n = 0, 1, ...$ We denote $v(x,t) = u_x(x,t)$ and $w(x,t) = u_t(x,t)$. For $n \geq 0$, let

$$\begin{array}{rcll} a_0^n &=& \sigma(nk) & \\ u_0^n &=& \psi(nk) & \\ v_0^n &=& \phi(nk) & \\ w_0^n &=& \psi'(nk) & \\ f_j^n &=& f(jh, nk), & j \geq 0 \\ u_j^n &=& u(jh, nk), & j \geq 1 \\ v_j^n &=& v(jh, nk), & j \geq 1 \\ w_j^n &=& w(jh, nk), & j \geq 1 \\ a_j^n &=& a(jh, nk), & j \geq 1. \end{array} \tag{4}$$

For the solution u, the coefficient a and the forcing term f of the PDE (1), we make the following assumptions:

Assumption 1 *Let $B = [0,1] \times [0,T]$, where T depends on h and k in a way to be specified later.*

a. (Regularity) $u(x,t) \in C^2(B)$ and $a, f \in C^0(B)$.

b. (Parabolicity) There exists a constant α such that $0 < \alpha \leq a(x,t)$, $(x,t) \in B$.

c. (Identifiability) There exists a constant β such that $0 < \beta \leq |u_{xx}(x,t)|$, $(x,t) \in B$.

Let the variables of the numerical method be U_j^n, V_j^n, W_j^n and A_j^n. They are discrete functions defined on the grid with discretization steps h and k. Their starting values are given for all n in order to proceed with a space marching scheme. They are:

$$\begin{array}{rcl} U_0^n &=& J_\delta \psi_m(nk) \\ V_0^n &=& J_\delta \phi_m(nk) \\ W_0^n &=& (J_\delta \psi_m)'(nk) \\ A_0^n &=& J_\delta \sigma_m(nk). \end{array}$$

The space marching numerical scheme is defined by the equations

$$U_{j+1}^n = U_j^n + h V_j^n \tag{5}$$

$$W_{j+1}^n = W_j^n + \frac{h}{k}\left(V_j^{n+1} - V_j^n\right) \tag{6}$$

$$V_{j+1}^n = V_j^n + \frac{h}{A_j^n}\left(W_j^n - f_j^n\right) \tag{7}$$

and

$$A_{j+1}^n = A_j^n \frac{\left(W_{j+1}^{n-1} - f_{j+1}^n\right)}{\left(W_j^n - f_j^n\right)}. \tag{8}$$

The calculations are performed in a triangular region in the (x,t)-plane. A sufficient amount $L+1$ of point values of the boundary data at $x = 0$ should be read in order to be able to recover, not only the values of the coefficient a but also the temperature u and the heat flux u_x for $(x,t) \in D$. We define T by setting $(L+1)k = T$.

The space marching scheme can be described as follows:

For $j = 0, 1, ..., M-1$,

1. Compute U_{j+1}^n by (5) for $n = 1, 2, ..., L+1-j$.
2. Compute U_{j+1}^0 by linear extrapolation.
3. Compute W_{j+1}^n by (6) for $n = 1, 2, ..., L-j$.
4. Compute V_{j+1}^n by (7) for $n = 1, 2, ..., L-j$.
5. Compute A_{j+1}^n by (8) for $n = 2, 3, ..., L-j$.
6. Mollify A_{j+1}^n for $n = 2, 3, ..., L-j$.
7. Compute A_{j+1}^1.

The computation of A_{j+1}^1 is carried out by the extension procedure discussed in the last Section. In this case, we use constant extensions. For more details, consult Mejía, (1993).

Some equations satisfied by the exact discrete functions defined in (4) are essential now. They are at the same time, a motivation for the definition of our numerical scheme and an important step toward the proof of error estimates. The first one, similar to (5), is a first order Taylor expansion of the temperature solution $u(x,t)$, i.e.,

$$u_{j+1}^n = u_j^n + h v_j^n + O(h^2). \tag{9}$$

The second one, corresponding to (6), is a first order discretization of the equation $u_{tx} = u_{xt}$,

$$w_{j+1}^n = w_j^n + \frac{h}{k}\left(v_j^{n+1} - v_j^n\right) + O(h^2) + O(k), \tag{10}$$

and the third one, corresponding to (7), is an approximation of the differential equation in problem (1):

$$v_{j+1}^n = v_j^n + \frac{h}{a_j^n}\left(w_j^n - f_j^n\right) + O(h^2). \tag{11}$$

A motivation for equation (8) requires several steps:

1. First order discretization of the time derivative:

$$w_j^n = \frac{1}{k}\left(u_j^{n+1} - u_j^n\right) + O(k). \tag{12}$$

2. Centered difference approximation of the differential equation (1):

$$\begin{aligned} \tfrac{1}{2k}\left(u_j^{n+1} - u_j^n\right) - f_j^n &= a_j^n\left(\tfrac{1}{2h}\left(v_{j+1}^n - v_j^n\right)\right) \\ &\quad + O(h^2) + O(k^2). \end{aligned} \tag{13}$$

3. Forward difference approximation of the differential equation (1):

$$\begin{aligned} \tfrac{1}{k}\left(u_j^{n+1} - u_j^n\right) - f_j^n &= a_j^n\left(\tfrac{1}{h}\left(v_{j+1}^n - v_j^n\right)\right) \\ &\quad + O(h) + O(k). \end{aligned} \tag{14}$$

From equations (13) and (14) we obtain

$$\begin{aligned} \tfrac{1}{k}\left(u_{j+1}^n - u_{j+1}^{n-1}\right) - f_{j+1}^n &= a_{j+1}^n\left(\tfrac{1}{h}\left(v_{j+1}^n - v_j^n\right)\right) \\ &\quad + O(h) + O(k), \end{aligned}$$

and combining this equation with (12) and (14), we get

$$a_{j+1}^n = a_j^n \frac{\left(w_{j+1}^{n-1} - f_{j+1}^n\right)}{\left(w_j^n - f_j^n\right)} + O(h) + O(k). \tag{15}$$

5 STABILITY

In this section, we prove a stability estimate for the numerical variables U_j^n, V_j^n, W_j^n and A_j^n. In the next section we apply this result to the proof of an error estimate that takes into consideration the presence of noise in the boundary data. We start with the definition of maximum norms for discrete functions. Let G_j^n be a discrete function. We denote

$$|G_j| = \max_n \left|G_j^n\right| \tag{16}$$

and

$$\|G\|_\infty = \max_j |G_j|. \tag{17}$$

We also need some assumptions on the numerical solution in order to develop our estimates. They are the discrete counterparts of Assumptions 1 b. and 1 c. respectively.

Assumption 2 *For all j, n,*
a. $0 < \alpha \le A_j^n$.
b. $W_j^n - f_j^n$ does not change sign and $0 < \alpha\beta \le \left|W_j^n - f_j^n\right|$.

Theorem 3 *If Assumptions 1 and 2 hold, and there are constants d_1 and d_2 such that $\max\{h, \frac{h}{\alpha}\} \le d_1$ and $\frac{2h}{k} \le d_2$, then $\|V\|_\infty$, $\|U\|_\infty$, $\|W\|_\infty$ and $\|A\|_\infty$ are finite.*

Proof. The proof is an induction over j.

$j = 0$: The convolution that defines the mollification of a square integrable function yields a C^∞ function. Thus, the numerical boundary conditions are well defined bounded discrete functions of n.

Suppose $|V_j|$, $|U_j|$, $|W_j|$ and $|A_j|$ are bounded. We prove that so are $|V_{j+1}|$, $|U_{j+1}|$, $|W_{j+1}|$ and $|A_{j+1}|$.

$$\begin{aligned} \left|V_{j+1}^n\right| &= \left|V_j^n + \tfrac{h}{A_j^n}\left(W_j^n - f_j^n\right)\right| \\ &\le \left|V_j^n\right| + \tfrac{h}{A_j^n}\left(\left|W_j^n\right| + \left|f_j^n\right|\right) \\ &\le |V_j| + \tfrac{h}{\alpha}\left(|W_j| + \|f\|_\infty\right). \end{aligned}$$

Hence,

$$|V_{j+1}| \le |V_j| + d_1\left(|W_j| + \|f\|_\infty\right).$$

$$
\begin{aligned}
|W_{j+1}^n| &= \left|W_j^n + \tfrac{h}{k}\left(V_j^{n+1} - V_j^n\right)\right| \\
&\le \left|W_j^n\right| + \tfrac{h}{k}\left(\left|V_j^{n+1}\right| + \left|V_j^n\right|\right) \\
&\le |W_j| + d_2\,|V_j|.
\end{aligned}
$$

Thus,

$$|W_{j+1}| \le |W_j| + d_2\,|V_j|.$$

For the diffusivity coefficient, before mollification, we have the following nonlinear estimate:

$$
\begin{aligned}
|A_{j+1}^n| &= A_j^n \frac{|W_{j+1}^{n-1} - f_{j+1}^n|}{|W_j^n - f_j^n|} \\
&\le \tfrac{A_j^n}{\alpha\beta}\left(\left|W_{j+1}^{n-1}\right| + \left|f_{j+1}^n\right|\right) \\
&\le \tfrac{|A_j|}{\alpha\beta}\left(|W_{j+1}| + \|f\|_\infty\right) \\
&\le \tfrac{|A_j|}{\alpha\beta}\left(|W_j| + d_2\,|V_j| + \|f\|_\infty\right).
\end{aligned}
$$

The mollified coefficient, denoted $A_{\delta,j+1}^n$, satisfies, according to Proposition 1, the slightly different estimate

$$\left|A_{\delta,j+1}^n\right| \le \frac{|A_j|}{\alpha\beta}\left(|W_j| + d_2\,|V_j| + \|f\|_\infty\right) + C\delta,$$

which implies

$$|A_{\delta,j+1}| \le \frac{|A_j|}{\alpha\beta}\left(|W_j| + d_2\,|V_j| + \|f\|_\infty\right) + C\delta.$$

Finally, it is clear that

$$|U_{j+1}| \le |U_j| + d_1\,|V_j|,$$

and this concludes the induction. Recalling definition (17), we conclude that $\|V\|_\infty$, $\|U\|_\infty$, $\|W\|_\infty$ and $\|A\|_\infty$ are finite.□

6 ERROR ANALYSIS

We start with the definition of the discrete error functions:

$$
\begin{aligned}
\Delta U_j^n &= u_j^n - U_j^n, \\
\Delta V_j^n &= v_j^n - V_j^n, \\
\Delta W_j^n &= w_j^n - W_j^n, \\
\Delta A_j^n &= a_j^n - A_j^n,
\end{aligned}
$$

and (recall definition (16)),

$$E_j = |\Delta U_j| + |\Delta V_j| + |\Delta W_j| + |\Delta A_j|. \tag{18}$$

A word of caution is in order now. The error in $A_{\delta,j}^n$, according to Proposition 1, satisfies

$$
\begin{aligned}
\left|a_j^n - A_{\delta,j}^n\right| &= \left|a_j^n - A_j^n + A_j^n - A_{\delta,j}^n\right| \\
&\le \left|a_j^n - A_j^n\right| + \left|A_j^n - A_{\delta,j}^n\right| \\
&\le \left|a_j^n - A_j^n\right| + C\delta,
\end{aligned}
$$

which indicates that the part of this error that requires our attention is $|\Delta A_j| = \left|a_j^n - A_j^n\right|$. As the next Lemma shows, an $O(\delta)$ term is already present in E_0 and we have no need to add it again to the errors E_j for $j > 0$.

The error at the boundary, due to the use of measured approximations of the values at the active boundary, deserves special consideration.

Lemma 4 *There exists a constant C_1 independent of δ, ϵ, h and k such that*

$$E_0 \le C_1\left(\epsilon + \delta + \frac{\epsilon}{\delta}\right). \tag{19}$$

Proof. The proof is an application of Propositions 1 and 2.

$$
\begin{aligned}
|\Delta U_0^n| &= |u_0^n - U_0^n| \\
&= |\psi(0, nk) - J_\delta \psi_m(nk)| \\
&\le C\delta + \epsilon,
\end{aligned}
$$

then,

$$|\Delta U_0| \le C\delta + \epsilon. \tag{20}$$

Similarly,

$$|\Delta V_0| \le C\delta + \epsilon, \tag{21}$$

and

$$|\Delta A_0| \le C\delta + \epsilon. \tag{22}$$

For the approximation of the boundary time derivative we have

$$
\begin{aligned}
|\Delta W_0^n| &= |w_0^n - W_0^n| \\
&= |\psi'(0, nk) - (J_\delta \psi_m)'(nk)| \\
&\le C\delta + \left(\tfrac{2}{\sqrt{\pi}}\right)\tfrac{\epsilon}{\delta},
\end{aligned}
$$

and this implies

$$|\Delta W_0| \le C\delta + \left(\frac{2}{\sqrt{\pi}}\right)\frac{\epsilon}{\delta}. \tag{23}$$

We choose $C_1 = \max\{4C, 3\}$. By adding inequalities (20)-(23) we obtain the desired result.□

The main result of this paper is the following error estimate:

Theorem 5 *If Assumptions 1 and 2 hold, and there are constants d_2 and d_3 such that $\frac{2h}{k} \le d_2$ and $\max\left\{h, \frac{h}{\alpha^2}\right\} \le d_3$, then there exists a constant C_2 independent of δ, ϵ, h and k such that*

$$\|E\|_\infty \le C_2 E_0 + O(h) + O(k). \tag{24}$$

Proof. Subtracting (7) from (11) we obtain

$$\begin{aligned}\triangle V_{j+1}^n &= \triangle V_j^n + \tfrac{h}{a_j^n}\left(w_j^n - f_j^n\right) \\ &\quad -\tfrac{h}{A_j^n}\left(W_j^n - f_j^n\right) \\ &\quad +O(h^2).\end{aligned}$$

Since

$$\begin{aligned}&\tfrac{h}{a_j^n}\left(w_j^n - f_j^n\right) - \tfrac{h}{A_j^n}\left(W_j^n - f_j^n\right) \\ &= \tfrac{h}{a_j^n A_j^n}\left(A_j^n \triangle W_j^n - W_j^n \triangle A_j^n\right. \\ &\qquad \left. + f_j^n \triangle A_j^n\right),\end{aligned}$$

then

$$\begin{aligned}\left|\triangle V_{j+1}^n\right| &\leq \left|\triangle V_j^n\right| + \tfrac{h}{\alpha^2}\left[A_j^n\left|\triangle W_j^n\right|\right. \\ &\quad \left. + \left|W_j^n\right|\left|\triangle A_j^n\right| + \left|f_j^n\right|\left|\triangle A_j^n\right|\right] \\ &\quad +O(h^2) \\ &\leq \left|\triangle V_j\right| + d_3\left[\|A\|_\infty\left|\triangle W_j\right|\right. \\ &\quad \left. + \left(\|W\|_\infty + \|f\|_\infty\right)\left|\triangle A_j\right|\right] \\ &\quad +O(h^2),\end{aligned}$$

and this implies

$$\left|\triangle V_{j+1}\right| \leq C_3 E_j + O(h^2), \tag{25}$$

where $C_3 = \max\left\{1, d_3\|A\|_\infty, d_3\left(\|W\|_\infty + \|f\|_\infty\right)\right\}$.

Now, from (8) and (15) we get

$$\begin{aligned}&a_{j+1}^n\left(w_j^n - f_j^n\right) - A_{j+1}^n\left(W_j^n - f_j^n\right) \\ &= a_j^n\left(w_{j+1}^{n-1} - f_{j+1}^n\right) \\ &\quad -A_j^n\left(W_{j+1}^{n-1} - f_{j+1}^n\right) \\ &\quad +O(h) + O(k),\end{aligned}$$

and then

$$\begin{aligned}\triangle A_{j+1}^n\left(w_j^n - f_j^n\right) &= -A_{j+1}^n \triangle W_j^n \\ &\quad +\triangle A_j^n\left(w_{j+1}^{n-1} - f_{j+1}^n\right) \\ &\quad +A_j^n \triangle W_{j+1}^{n-1} \\ &\quad +O(h) + O(k).\end{aligned}$$

By Assumption 1,

$$\alpha\beta \leq a_j^n\left|u_{xx}(jh, nk)\right| = \left|w_j^n - f_j^n\right|,$$

and we conclude

$$\begin{aligned}\alpha\beta\left|\triangle A_{j+1}^n\right| &\leq A_{j+1}^n\left|\triangle W_j^n\right| + A_j^n\left|\triangle W_{j+1}^{n-1}\right| \\ &\quad +\left|\triangle A_j^n\right|\left(\left|w_{j+1}^{n-1}\right| + \left|f_{j+1}^n\right|\right) \\ &\quad +O(h) + O(k) \\ &\leq \|A\|_\infty\left|\triangle W_j^n\right| + \|A\|_\infty\left|\triangle W_{j+1}^{n-1}\right| \\ &\quad +\left(\|w\|_\infty + \|f\|_\infty\right)\left|\triangle A_j^n\right| \\ &\quad +O(h) + O(k), \\ &\leq 2\|A\|_\infty\left|\triangle W_j\right| \\ &\quad +\left(\|w\|_\infty + \|f\|_\infty\right)\left|\triangle A_j\right| \\ &\quad +O(h) + O(k),\end{aligned}$$

which indicates that we can find a constant C_4 such that

$$\left|\triangle A_{j+1}\right| \leq C_4 E_j + O(h) + O(k). \tag{26}$$

Subtracting (6) from (10) we find

$$\begin{aligned}\triangle W_{j+1}^n &= \triangle W_j^n + \tfrac{h}{k}\left(\triangle V_j^{n+1} - \triangle V_j^n\right) \\ &\quad +O(h^2) + O(k).\end{aligned}$$

This implies,

$$\begin{aligned}\left|\triangle W_{j+1}^n\right| &\leq \left|\triangle W_j\right| + \tfrac{2h}{k}\left|\triangle V_j\right| \\ &\quad +O(h^2) + O(k),\end{aligned}$$

which provides the estimate

$$\begin{aligned}\left|\triangle W_{j+1}\right| &\leq \left|\triangle W_j\right| + d_2\left|\triangle V_j\right| \\ &\quad +O(h^2) + O(k).\end{aligned}$$

As before, we can easily choose a constant C_5 so that

$$\left|\triangle W_{j+1}\right| \leq C_5 E_j + O(h^2) + O(k). \tag{27}$$

Finally, subtracting (5) from (9) we readily obtain

$$\left|\triangle U_{j+1}\right| \leq C_6 E_j + O(h^2), \tag{28}$$

where $C_6 = \max\{1, d_3\}$.

Now, we add inequalities (25)-(28) and get the estimate

$$E_{j+1} \leq C_7 E_j + O(h) + O(k),$$

where $C_7 = C_3 + C_4 + C_5 + C_6$, and we iterate this inequality to obtain

$$E_{j+1} \leq \left(C_7\right)^{j+1} E_0 + O(h) + O(k).$$

Since $1 \leq C_6 < C_7$, we set $C_2 = \left(C_7\right)^M$ and this ends the proof.□

7 NUMERICAL RESULTS

In this section we discuss the implementation of the numerical scheme developed in the last section and present the numerical results obtained from two examples. In all cases, the discretization parameters are the following: $M = N$, $h = k = \frac{1}{M}$, $x_j = jh$, $j = 0, 1, ..., M$, $L + 1 = 3N$, $t_n = nk$, $n = 0, 1, ..., L + 1$.

The discretized measured approximations of the boundary data $\sigma_m(t_n)$, $\psi_m(t_n)$ and $\phi_m(t_n)$ are simulated by adding random errors to $\psi(t_n)$, $\sigma(t_n)$ and $\phi(t_n)$ respectively. Specifically,

$$\begin{aligned} \sigma_m(t_n) &= \sigma(t_n) + \epsilon_n^1, \\ \psi_m(t_n) &= \psi(t_n) + \epsilon_n^2, \\ \phi_m(t_n) &= \phi(t_n) + \epsilon_n^3, \end{aligned}$$

where the ϵ_n^i 's, $i = 1, 2, 3$, are Gaussian random variables with variance ϵ^2.

The measure of errors is based on the weighted l^2-norms defined as follows:

- If G_j is a discrete function of j, $j = 1, 2, ..., M$, its weighted l^2- norm is given by

$$||G||_2 = \left[\frac{1}{M} \sum_{j=1}^{M} |G_j|^2 \right]^{\frac{1}{2}}.$$

- If G_j^n is a discrete function of j and n, $j = 1, 2, ..., M$ and $n = 1, 2, ..., N$, then its weighted l^2- norm is defined by

$$||G||_2 = \left[\frac{1}{MN} \sum_{j=1}^{M} \sum_{n=1}^{N} |G_j^n|^2 \right]^{\frac{1}{2}}.$$

To test the stability and accuracy of the numerical scheme, we use different average perturbations ϵ and appropriate values for the regularization parameter δ. In the Tables and Figures, we use the relative errors in A_j^n and U_j^0 which are defined by

$$|\Delta A|_2 = \frac{||\Delta A||_2}{||a||_2}$$

and

$$|\Delta U^0|_2 = \frac{||\Delta U^0||_2}{||u^0||_2}$$

respectively.

Example 1. Identify $a(x, t)$ and $u(x, 0)$ in

$$\begin{aligned} &u_t = au_{xx} + f(x, t), && 0 < x < 1, 0 < t, \\ &a(0, t) = 1 + 0.01t, && 0 < t, \\ &u(0, t) = \exp(t), && 0 < t, \\ &u_x(0, t) = \exp(t), && 0 < t, \end{aligned}$$

where $f(x, t) = -0.01(x + t)\exp(x + t)$.

The exact solutions are

$$a(x, t) = 1 + 0.01(x + t) \quad \text{and} \quad u(x, 0) = \exp(x).$$

Example 1		Relative Errors	
ϵ	δ	$\lvert\Delta A\rvert_2$	$\lvert\Delta U^0\rvert_2$
0.000	0.020	0.0018	0.0054
0.003	0.100	0.0354	0.0042
0.005	0.140	0.0447	0.0037

Table 1. Error Norms as functions of ϵ

Example 2		Relative Errors	
ϵ	δ	$\lvert\Delta A\rvert_2$	$\lvert\Delta U^0\rvert_2$
0.000	0.010	0.0170	0.0031
0.003	0.050	0.1369	0.0262
0.005	0.050	0.1487	0.0283

Table 2. Error Norms as functions of ϵ

Table 1 shows the discrete relative errors as functions of the amount of noise in the data ϵ for $M = N = 50$. The qualitative behavior of the reconstructed functions are shown in Figures 1 and 2.

Example 2. Identify $a(x, t)$ and $u(x, 0)$ in

$$\begin{aligned} &u_t = au_{xx} + f(x, t), && 0 < x < 1, 0 < t, \\ &a(0, t) = 1 && 0 < t, \\ &u(0, t) = \exp(1 - t), && 0 < t, \\ &u_x(0, t) = \exp(1 - t), && 0 < t, \end{aligned}$$

where $f(x, t) = -(2 + 0.1xt)\exp(1 + x - t)$.

The exact solutions are

$$a(x, t) = 1 + 0.1xt \quad \text{and} \quad u(x, 0) = \exp(1 + x).$$

We set $M = N = 100$. Table 2 illustrates the stability of the method by showing the relative errors as functions of the amount of noise in the data ϵ. The quality of the reconstructions can be observed in Figures 3 and 4.

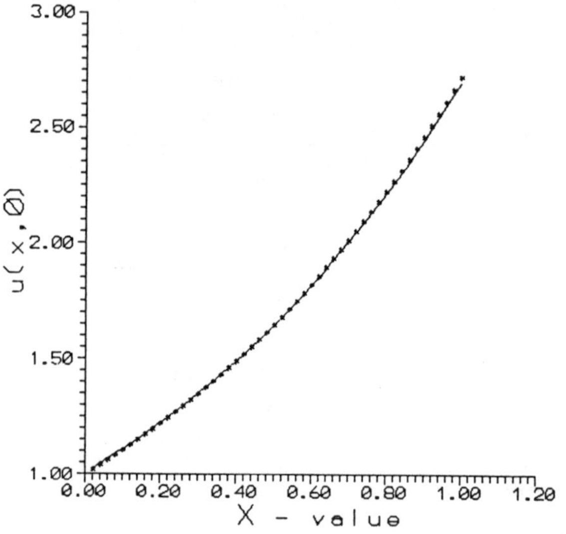

Figure 1. Initial Condition in Example 1.
$\epsilon = 0.005$, $\delta = 0.12$.
Exact: ($\star\star\star$); Computed: (___).

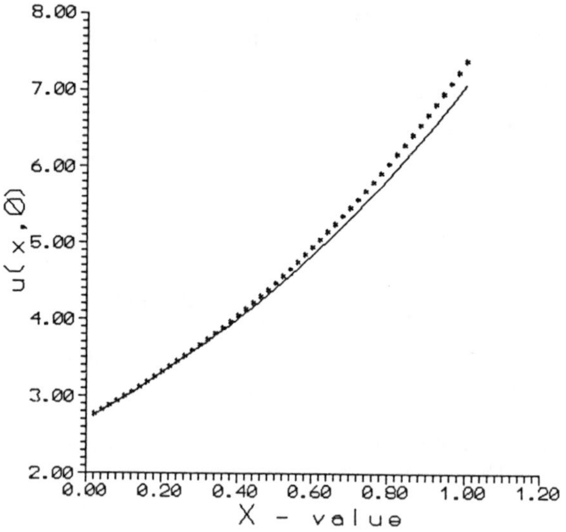

Figure 3. Initial Condition in Example 2.
$\epsilon = 0.005$, $\delta = 0.05$.
Exact: ($\star\star\star$); Computed:(___).

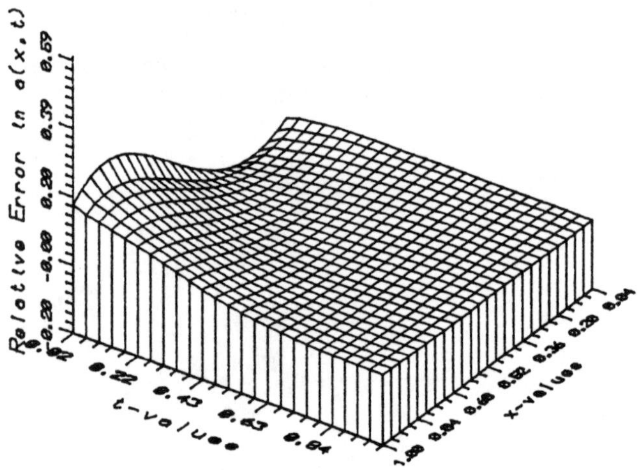

Figure 2. Relative Error in $a(x,t)$. Example 1.
$\epsilon = 0.005$, $\delta = 0.12$.

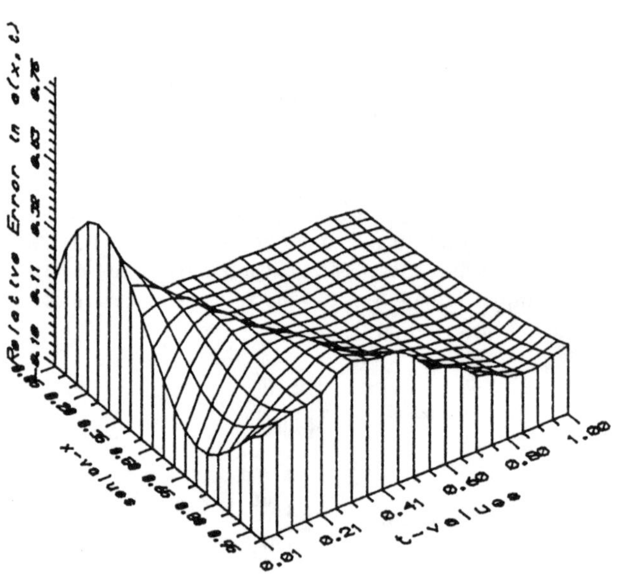

Figure 4. Relative Error in $a(x,t)$. Example 2.
$\epsilon = 0.005$, $\delta = 0.05$.

Acknowledgment. This research was partially supported by a C. Taft Fellowship and a University Research Council Summer Fellowship.

References

Chavent, G. and Jaffré, J., *Mathematical Models and Finite Elements for Reservoir Simulation,* North Holland, Amsterdam, 1986.

Ewing, R. and Lin, T., 1989, "Parameter identification problems in single-phase and two-phase flow," *International Series of Numerical Mathematics*, Vol. 91, Birkhäuser Verlag, Basel, pp. 85-108.

Hinestroza, D. and Murio, D.A., 1993, "Identification of Transmissivity Coefficients by Mollification Techniques. Part I: One-Dimensional Elliptic and Parabolic Problems," *Computers Math. Applic.,* Vol. 25, No. 8, pp. 59-79.

Mejía, C.E., 1993, "Numerical Solution of some Inverse Problems by the Mollification Method," Ph.D. Dissertation, University of Cincinnati, Cincinnati, Ohio.

Mejía, C.E. and Murio D.A., 1993, "Mollified Hyperbolic Method for Coefficient Identification Problems," to appear.

Murio, D.A., *The Mollification Method and the Numerical Solution of Ill-Posed Problems,* John Wiley and Sons, New York, 1993 .

Murio, D.A. and Hinestroza, D., 1993, "The Space Marching Solution of the Inverse Heat Conduction Problem and the Identification of the Initial Temperature Distribution," *Computers Math. Applic.,* Vol. 25, No. 4, pp. 55-63.

Wheeler, M.F., ed., *Numerical Simulation in Oil Recovery,* Springer-Verlag, New York, 1988.

DETERMINATION OF HEAT TRANSFER COEFFICIENTS FOR SPHERICAL OBJECTS IN IMMERSING EXPERIMENTS USING TEMPERATURE MEASUREMENTS

Ibrahim Dincer
Department of Energy Systems
TUBITAK-Marmara Research Center
Gebze, Kocaeli, Turkey

ABSTRACT

A simple method was proposed for determining the heat transfer coefficients of the individual spherical products during water-cooling. An experimental investigation was carried out the temperature measurements at the centers of the individual spherical products in the different batch-weights cooled in the different environments. These temperature data were used in the present model in order to determine the heat transfer coefficients of the products. The obtained results showed that the heat transfer coefficients of the individual products decreased with increasing batch-weight in the different environments.

NOMENCLATURE

a = thermal diffusivity, m^2/s
a_w = thermal diffusivity of water at the initial product temperature, m^2/s (0.148×10^{-6} m^2/s)
A,B,D = constants
Bi = Biot number
C = cooling coefficient, 1/s
Fo = Fourier number
h = heat transfer coefficient, W/m^2K
J = lag factor
k = thermal conductivity, W/mK
L = weight per crate, kg
r = radial coordinate
R = radius, m
t = time, s
T = temperature, °C or K
W = moisture content, in decimal units

Greek Symbols

Γ = dimensionless position ratio
θ = dimensionless temperature
ϕ = temperature difference, °C or K
μ = root of the characteristic equation

Subscripts

e = final
i = initial
w = water, medium condition
n = refers to the nth characteristic value
1 = refers to the 1st characteristic value

INTRODUCTION

Cooling is a very common and important preserving process and is used to maintain the quality and to prevent the spoilage of the food products.

Unsteady-state heat transfer from a solid object to any fluid medium is an important subject in many engineering-related fields and plays a role in practical processes ranging from cooling of a hot steel ball to heating and cooling of food products. As a first step to a better understanding complex heat transfer environment, one needs to carefully evaluate and understand fluid flow, and thermophysical properties of the product. Classical explanations of the cooling phenomena are largely based on the temperature distributions and heat transfer rates. However, very limited information is available in the literature on the heat transfer coefficients in the food processing applications. Most theoretical or semi-theoretical and experimental investigations of the heat transfer coefficients for the geometrical shaped objects can be classified as the Nusselt-Reynolds correlations. The surface heat transfer coefficients of the food products subjected to cooling applications are dependent upon several parameters, e.g. product's thermal and physical properties and environmental conditions. The well-known Nu-Re correlations may not show a realistic behaviour for the specific applications.

Several methods have been suggested to estimate the heat transfer coefficients during the cooling of the food products (Arce and Sweat, 1980; Ranade and Narayankhedkar, 1982; Ansari, 1987; Dincer, 1991). In particular, Dincer (1991) obtained the time-dependent heat transfer coefficients for the spherical products subjected to immersion cooling. No similar study for estimating the heat transfer coefficients of an individual spherical product using the present model appear in the literature. A new approach to the determination of the heat transfer coefficients of spherically shaped objects for the investigation of cooling process (e.g., water-cooling) is presented. This approach allows the identification and evaluation of cooling process, and the analysis of the unsteady-state heat transfer. Also, a better understanding will lead to better cooling effectiveness, and better operating conditions. This study involved both the mathematical model developed here and the temperature data.

The aim of this study is to develop a simple model to accurately determine the heat transfer coefficients of the spherical food products by means of the temperature data.

EXPERIMENTAL

In order to provide realistic results and validation of the mathematical model employed here, an experimental study program was carried out. In this program, batches of 5, 10, 15, and 20 kg from three food commodities, namely, tomatoes and pears were weighted and prepared, and each batch was located into a polyethylene crate for each trial. The trials were repeated for varying amounts of the crate load at the different environments of 0.5, 1, and 1.5° C, respectively. Experimental investigation was carried out at the Department of Refrigeration Technology of the TUBITAK-Marmara Research Center, using an experimental apparatus originally designed for water-cooling. The apparatus, illustrated in Figure 1, consisted of a conventional vapor-compression refrigeration unit including a compressor, a condenser, an expansion valve and an evaporator, and a cold water pool (test section) where the water is circulating by a centrifugal pump.

The temperature and flow velocity of the cooling water in the pool (tank) remained constant as 0.5, 1, and 1.5° C and 0.05 m/s, during the duration of an experiment. The temperatures were measured and monitored using fifteen Cu/Cu-Ni type thermocouples connected to a multichannel microprocessor device (Ellab Instruments, Denmark) with an accuracy of 0.1° C. Each thermocouple probe was inserted at the center of the product. Water-cooling tests continued until their storage temperatures. Each batch was located into a polyethylene crate, and this was cooled in the pool. During the each test, the temperatures of the individual products and water were monitored and recorded at the intervals of 30 seconds. A total of two food commodities in the different batches were tested in the different environments.

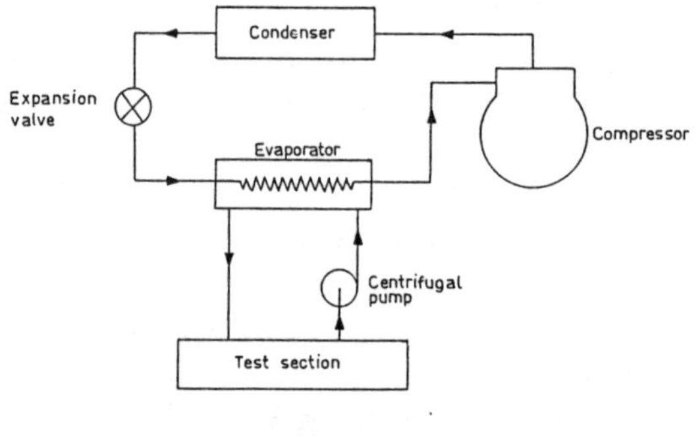

Fig.1 Schematic of the experimental set-up

The experimental facilities used in this investigation, along with the details of the facility construction, operation and accuracy have been previously reported by Dincer (1991, 1992) and Dincer et al. (1992).

ANALYSIS

It may be noted that the problem under study is the same as considering a solid spherical product of radius R, which is simultaneously subjected to the water flow. The temperature distribution at any point of the solid product is a function of time and radius.

To obtain the mathematical model of the process, the following assumptions are made:
(i) constant thermal properties of the product and water,
(ii) constant heat transfer coefficient,
(iii) homogeneous and isotropic spherical body,
(iv) uniform initial temperature of the product,
(v) no moisture transfer,
(vi) no internal heat generation,
(vii) unsteady-state conditions.

The governing heat conduction equation in the absence of the internal heat generation is

$$(\partial^2 T/\partial r^2) + (2/r)(\partial T/\partial r) = (1/a)(\partial T/\partial t) \quad (1)$$

Under these conditions the heat flow will be one-dimensional and the temperature and temperature function $\phi = T - T_w$ must satisfy.

$$(\partial^2 \phi/\partial r^2) + (2/r)(\partial \phi/\partial r) = (1/a)(\partial \phi/\partial t) \quad (2)$$

with the initial and boundary conditions

$$\phi(r,0) = \phi_i = (T_i - T_w) \quad (3)$$

$$\phi(0,t) = \text{finite} \quad (4)$$

$$-k[\partial \phi(R,t)/\partial r] = h\phi(R,t) \quad (5)$$

A detailed solution of the above equations is reported in the literature (Carslaw and Jaeger, 1959; Arpaci, 1966; Luikov, 1968). In this perspective, the dimensionless temperature distribution at any point of a spherical body is

$$\theta = \sum_{n=1}^{\infty} A_n B_n D_n \quad (6)$$

where

$$A_n = [(2\text{Bi}\sin\mu_n)/(\mu_n - \sin\mu_n\cos\mu_n)] \quad (7)$$

$$B_n = \exp(-\mu_n^2 \text{Fo}) \quad (8)$$

$$D_n = (\sin\mu_n\Gamma/\mu_n\Gamma) \quad (9)$$

For the center case ($\Gamma=0$), $D_n=1$ and hence, Eq.(6) reduces to

$$\theta = \sum_{n=1}^{\infty} A_n B_n \quad (10)$$

Consider Fo>0.2 and therefore, the first term of the series in Eq.(10) is taken and the remaining terms are neglected, and it is resulted as

$$\theta = A_1 B_1 \quad (11)$$

where

$$A_1 = [(2\text{Bi}\sin\mu_1)/(\mu_1 - \sin\mu_1\cos\mu_1)] \quad (12)$$

$$B_1 = \exp(-\mu_1^2 \text{Fo}) \quad (13)$$

$$\theta = (T - T_w)/(T_i - T_w) \quad (14)$$

A regression analysis in the exponential form is employed and there results

$$\theta = J_1 \exp(-Ct) \quad (15)$$

Consider $A_1=J_1$, and by combining Eqs.(11) and (15), the following equation is derived.

$$\mu_1^2 \text{Fo} = Ct \quad (16)$$

where

$$\mu_1^2 = (10.3\text{Bi})/(3.2 + \text{Bi}) \quad (17)$$

$$\text{Bi} = hR/k \quad (18)$$

$$Fo = at/R^2 \quad (19)$$

After making these substitutions, and some manipulations, Eq.(20) gives the heat transfer coefficients of the spherical products easily.

$$h = (3.2kRC)/(10.3a - CR^2) \quad (20)$$

The thermal conductivities and thermal diffusivities of the food products above freezing condition can be estimated using the following correlations (Sweat, 1986; ASHRAE Handbook of Fundamentals, 1981).

$$k = 0.148 + 0.493W \quad (21)$$

$$a = 0.088x10^{-6} + (a_w - 0088x10^{-6})W \quad (22)$$

RESULTS AND DISCUSSION

The center temperatures of an individual spherical object in each batch subjected to cooling in the different environments (T_w=0.5, 1, and 1.5°C) were converted to the dimensionless form and a regression analysis in the exponential form was applied to the dimensionless center temperature distribution for an individual object in each batch. Then, the effective cooling parameters expressed in terms of cooling coefficient and lag factor were determined using the regression analysis. Using the values of the thermal conductivity, thermal diffusivity, cooling coefficient, and radius of the object in the present model, the heat transfer coefficients for the individual spherical objects were obtained.

Some measured dimensions and thermophysical properties for tomatoes and pears as spherical products were determined as follows: R=0.035±0.0010 and 0.030±0.0003 m, T_i=21.0±0.5 and 22.5±0.5°C, T_e=4 and 2°C, W=94 and 83 %, k=0.61142 and 0.55719 W/mK, a=$1.444 \cdot 10^{-7}$ and $1.378 \cdot 10^{-7}$ m^2/s.

The obtained heat transfer coefficients for the individual spherical products are given in Tables 1 and 2. The regression equations described the data well and had very high regression coefficients over 0.98. It may be seen from Tables 1 and 2 that the heat transfer coefficients increase approximately linearly while the cooling coefficients and lag factors for each product vary randomly with crate load, probably due to changes in cooling water flow around individual fruits and thus to the heat-transfer environment, depending on the loading. Increasing the crate loading from 5 to 20 kg, in loads of 5 kg, decreased the effective heat transfer coefficients by 71, 61, and 48 % for tomatoes, by 75, 73, and 52 % for pears at the water temperatures of 0.5, 1, and 1.5°C, respectively.

Table 1. Heat transfer coefficients of the individual tomatoes

T_w(°C)	L(kg)	J	C(1/sx10^3)	h(W/m²K)
0.5	5	1.2087	1.020	293.70±140
	10	1.3098	0.907	165.07±40
	15	1.3295	0.800	107.98±17
	20	1.3220	0.728	83.71±10
1.0	5	1.2658	0.953	204.00±62
	10	1.2932	0.874	143.64±30
	15	1.2911	0.743	88.15±11
	20	1.3346	0.710	78.72±9
1.5	5	1.3580	0.840	125.50±23
	10	1.2849	0.829	120.32±21
	15	1.2511	0.690	73.59±8
	20	1.2948	0.649	64.19±6

Table 2. Heat transfer coefficients of the individual pears

T_w(°C)	L(kg)	J	C(1/sx10^3)	h(W/m²K)
0.5	5	1.0263	1.454	702.32±227
	10	1.0778	1.449	672.57±206
	15	1.0310	1.305	285.10±33
	20	1.2843	1.175	173.69±12
1.0	5	1.1186	1.434	595.81±157
	10	1.1567	1.419	533.62±124
	15	1.0780	1.296	274.07±31
	20	1.3658	1.151	160.56±10
1.5	5	1.1627	1.394	452.62±87
	10	1.1544	1.349	351.58±51
	15	1.1107	1.259	235.27±22
	20	1.3101	1.060	121.84±6

This investigation indicates that the present model can easily be used in order to determine the effective heat transfer coefficients as constants for the individual products, in a simple and accurate manner.

CONCLUSIONS

A simple analytical formula was developed in order to determine the effective heat transfer coefficients of the spherical shaped bodies subjected to cooling. In this study, an experimental investigation was conducted to obtain the center temperature data of the spherical food products during cooling at the temperatures of 0.5, 1, and 1.5° C and flow velocity of 0.05 m/s. Using the temperature distributions in the regression analysis, the effective process parameters, namely cooling coefficients and lag factors were obtained and hence, the effective heat transfer coefficients were determined for the individual spherical products in the batches of 5, 10, 15, and 20 kg of product by means of the present model. Based upon the results obtained, the cooling coefficients and lag factors for the individual products varied little with the batch weight. However, the effective heat transfer coefficients decreased with increasing the crate load. Finally, the present model is capable of determining the effective heat transfer coefficients for the individual spherical bodies cooled in any fluid flow.

REFERENCES

Ansari, F.A., 1987, "A Simple and Accurate Method of Measuring Surface Film Conductance for Spherical Bodies", *International Communications in Heat and Mass Transfer*, Vol.14, pp.229-236.

Arce, J. and Sweat, V.E, 1980, "Survey of Published Heat Transfer Coefficient Encountered in Food Refrigeration Processes", *ASHRAE Transactions*, Vol.86, pp.235-260.

Arpaci, V.S., 1966, *Conduction Heat Transfer*, Addison-Wesley Reading, Mass.

ASHRAE Handbook of Fundamentals, 1981, American Society of Heating, Refrigerating, and Air Conditioning Engineers, Inc., Atlanta, G.A.

Carslaw, H.S. and Jaeger, J.C., 1959, *Conduction of Heat in Solids*, 2d ed., Oxford University Press, London.

Dincer, I., 1991, "A Simple Model for Estimation of the Film Coefficients During Cooling of Certain Spherical Foodstuffs with Water", *International Communications in Heat and Mass Transfer*, Vol.18, pp.431-443.

Dincer, I., 1992, "Methodology to Determine Temperature Distributions in Cylindrical Products Exposed to Hydrocooling", *International Communications in Heat and Mass Transfer*, Vol.19, pp.359-371.

Dincer, I.; Yildiz, M.; Loker, M. and Gun, H., 1992, "Process Parameters for Hydrocooling Apricots, Plums and Peaches", *International Journal of Food Science and Technology*, Vol.27, pp.347-352.

Luikov, A.V., 1968, *Analytical Heat Diffusion Theory*, Academic Press, New York.

Ranade, M.S. and Narayankhedkar, K.G., 1982, "Thermal Characteristics of Fruits and Vegetables as Applied to Hydrocooling", Edited by International Institute of Refrigeration, pp.455-461.

Stewart, W.E.; Becker, B.R.; Greer, M.E. and Stickler, L.A., 1990, "An Experimental Method of Approximating Effective Heat Transfer Coefficients for Food Products", *ASHRAE Transactions*, Vol.96, pp.142-147.

Sweat, V.E., 1986, "Thermal Properties of Foods", In: M.A. Rao and S.S.H. Rizvi(ed.) *Engineering Properties of Foods*, Marcel Dekker, Inc., New York, pp.49-87.

Inverse Problems in Engineering: Theory and Practice
ASME 1993

INVERSE HEAT GENERATION PROBLEM IN A HOLLOW CYLINDER

N. M. Al-Najem
Department of Mechanical Engineering
Kuwait University
Safat, Kuwait

ABSTRACT

The solution of inverse heat generation problem is investigated based on Nonlinear Regression Function Specifications -NRFS- method over the whole time domain. The present algorithm is computationally efficient because it is expressed in a rapidly converging form by employing the split-up procedure. Several numerical experiments showed the capability of the current approach to treat both continuous and discontinuous variations in the behavior of internal heat generations. Furthermore, the NRFS algorithm remains stable and capable to reproduce relatively accurate results from inaccurate measured data.

INTRODUCTION

In contrast to the direct heat conduction theory in which the interior conditions are found when the surface conditions are prescribed. The Inverse heat conduction analysis deals with the attempt to estimate surface heat flux and temperature histories from inaccurate transient temperature measurements within the solid material. This concept can be extended and employed to recover time varying internal heat source utilizing measured temperature. Such situation has numerous practical applications including, among others, the estimation of waste decay heat in nuclear technology and evaluation an appropriate functional form of source strength in solar ponds.

A recent literature review shows that most workers focused mainly on estimating surface conditions by inverse techniques. It seems that the only reported work in literature is advanced by Borukhov and Kolesnikov (1988) for estimating internal heat generation rates using inverse heat conduction analysis. They were successfully recover the input time varying internal heat sources from exact measured temperature by structural factorization method. Unfortunately, the stability and sensitivity of the method to measurement errors is not verified.

The purpose of this work is to develop a whole domain estimation procedure to study qualitatively certain class of inverse heat conduction problems such as inverse heat generation problem (IHGP). This solution is based on a rapidly-converging scheme obtained by split-up approach and then employing the nonlinear regression function specification (NRFS) algorithm to recover the time variation of heat source having an abrupt change in time in a hollow cylinder.

STATEMENT OF THE PROBLEM

The inverse heat generation problem (IHGP) considered in this study is applied in an infinite long hollow cylinder, $a \le r \le b$, initially at temperature $f(r)$. For times $t > 0$ heat is generated within the solid at unknown rates of $g(t)$. The boundary surface at $r = a$ and $r = b$ dissipate heat by convection with known heat transfer coefficients h_1 and h_2, respectively, into an environment at uniform temperature T_∞. The transient temperature history $y_n(t)$ is measured at interior point r^*, $a < r^* < b$, at N discrete time intervals $[0, t_f]$.

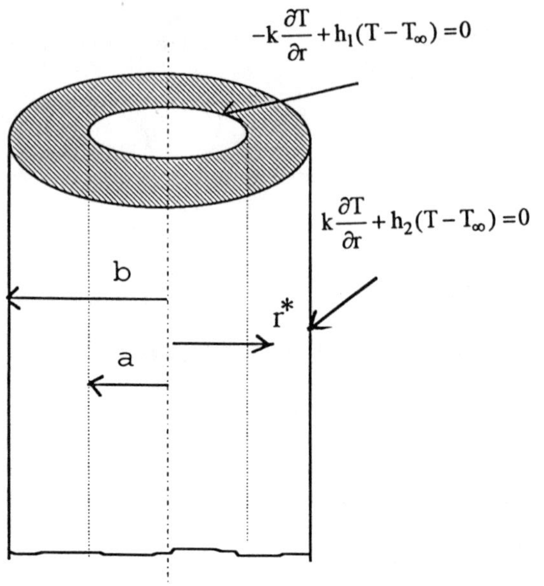

Fig. 1 Schematic of the problem

The objective of IHGP is to reproduce or recover g(t) over the time interval $[0,t_f]$ from inaccurate measured temperature $y_n(t)$ over the same time interval.

MODEL AND FORMULATION

The schematic diagram shown in Fig. 1 describes a general case of inverse generation problem in cylindrical coordinate system. The pertinent equations in this study to solve the IHGP in a hollow cylinder having constant physical properties are given below in nondimensional form.

$$\frac{1}{R}\frac{\partial}{\partial R}\left(R\frac{\partial\theta}{\partial R}\right)+G(\tau)=\frac{\partial\theta}{\partial\tau} \qquad \text{in } A<R<1,\ \tau>0, \quad (1)$$

where $G(\tau)$ is the unknown dimensionless heat generation function to be estimated by the application of nonlinear regression function specification -NRFS- method (Gallant 1977). The above various dimensionless variables are defined in the nomenclature.

The convection boundary conditions at the inner and outer surfaces with initial condition are

$$-\frac{\partial\theta}{\partial R}+B_1\theta=0 \qquad \text{at } R=A,\ \tau>0, \quad (2a)$$

$$\frac{\partial\theta}{\partial R}+B_2\theta=0 \qquad \text{at } R=1,\ \tau>0, \quad (2b)$$

$$\theta(R,\tau)=0 \qquad \text{at } \tau=0, \text{ in the region.} \quad (2c)$$

With recorded temperature at N discrete time intervals as

$$\theta(R,\tau)=Y_n \qquad \text{at } R=R^*,\ A<R^*<1,\ n=1,2,\dots N. \quad (3)$$

The function specification procedure implies that the variation of the unknown internal heat source $G(\tau)$ is expressed in an appropriate functional form. In this work, a segmented polynomial in time with unknown coefficients is assumed in the form given below to accommodate both continuous and discontinuous variations in $G(\tau)$.

$$G(\tau)=\sum_{j=o}^{2}\Delta_j^{(1)}\tau^j+\delta\left[\sum_{j=o}^{2}\Delta_j^{(1)}\tau_1^j+\sum_{j=o}^{2}\Delta_j^{(2)}(\tau-\Delta_*)^j\right], \quad (4a)$$

where

$$\delta=\begin{cases}0 & \text{for } \tau\le\Delta_*\\ 1 & \text{for } \tau>\Delta_*.\end{cases} \quad (4b)$$

The coefficients Δ's and the joint point Δ_* are unknown to be determined by the proposed inverse algorithm explained later. Here it should be pointed out that the above representation is assumed irrespective to the history of temperature profiles within the solid.

THEORETICAL ANALYSIS

In order to obtain a fast-converging algorithm for solving the above IHGP, the splitting-up procedure (Mikhailov and Ozisik 1986; Al-Najem and Ozisik 1985) is employed as now described. Since the nonhomogeneous term $G(\tau)$ is expressed in polynomial form given by equation (4), the solution $\theta(R,\tau)$ is obtained in two stages as

$$\theta(R,\tau)=\phi(R,\tau)+\delta\Phi(R,\tau-\Delta_*) \quad (5)$$

Where $\phi(R,\tau)$ is the solution of the first-stage (i.e., $\tau\le\Delta_*$) defined by the following system

$$\frac{1}{R}\frac{\partial}{\partial R}\left(R\frac{\partial\phi}{\partial R}\right)+\sum_{j=0}^{2}\Delta_j^{(1)}\tau^j=\frac{\partial\phi}{\partial\tau} \qquad \text{in } A<R<1,\ \tau>0, \quad (6a)$$

$$-\frac{\partial \phi}{\partial R}+B_1\phi=0 \qquad \text{at } R=A,\ \tau>0 \quad (6b)$$

$$\frac{\partial \phi}{\partial R}+B_2\phi=0 \qquad \text{at } R=1,\ \tau>0, \quad (6c)$$

$$\phi(R,\tau)=0 \qquad \text{at } \tau=0, \text{ in the region} \quad (6d)$$

The second-stage problem $\Phi(R,\lambda)$ is defined by

$$\frac{1}{R}\frac{\partial}{\partial R}\left(R\frac{\partial \Phi}{\partial R}\right)+\sum_{j=0}^{2}\Delta_j^{(1)}\Delta_*^j+\sum_{j=0}^{2}\Delta_j^{(2)}\lambda^j=\frac{\partial \Phi}{\partial \lambda}$$

in $A<R<1,\ \tau>0,$ (7a)

$$-\frac{\partial \Phi}{\partial R}+B_1\Phi=0 \qquad \text{at } R=A,\ \lambda>0 \quad (7b)$$

$$\frac{\partial \Phi}{\partial R}+B_2\Phi=0 \qquad \text{at } R=1,\ \lambda>0, \quad (7c)$$

$$\Phi(R,\tau)=0 \qquad \text{for } \lambda=0, \text{ in the region.} \quad (7d)$$

Where $\lambda=\tau-\Delta_*$. The solutions $\phi(R,\tau)$ and $\Phi(R,\lambda)$ are now splitted-up into several simpler problems in the form

$$\phi(R,\tau)=\sum_{j=0}^{2}\phi_j^{ss}(R)\tau^j+\phi_t(R,\tau) \qquad \text{for } \tau\le\Delta_*. \quad (8a)$$

$$\Phi(R,\lambda)=\sum_{j=0}^{2}\Phi_j^{ss}(R)\lambda^j+\Phi_t(R,\lambda) \qquad \text{for } \tau>\Delta_*. \quad (8b)$$

Where $\phi_j^{ss}(R)$ and $\Phi_j^{ss}(R)$ are the steady-state solutions and $\phi_t(R,\tau)$ and $\Phi_t(R,\lambda)$ represent transient homogeneous solutions. Due to the similarity between solutions $\phi(R,\tau)$ and $\Phi(R,\lambda)$ and for the interest of brevity, the split-up procedure is only given for $\phi(R,\tau)$ as now explained. Substituting Eq. (8a) into problem (6) yields the following two sets of simpler problems.

I. Steady-State Problem $\phi_j^{ss}(R)$

The steady-state solution satisfies the following system

$$\frac{1}{R}\frac{d}{dR}\left(R\frac{d\phi_j^{ss}}{\partial R}\right)+\Delta_j^{(1)}=(j+1)\phi_{j+1}^{ss} \quad \text{in } A<R<1 \qquad (9a)$$

$$-\frac{d\phi_j^{ss}}{dR}+B_1\phi_j^{ss}=0 \qquad \text{at } R=A \quad (9b)$$

$$\frac{d\phi_j^{ss}}{dR}+B_2\phi_j^{ss}=0 \qquad \text{at } R=1, \quad (9c)$$

with $j=2,1,0$ and $\phi_3^{ss}=0$.

The solution of the above steady-state problem can be found readily by direct integration and the solution is given as

$$\phi_2^{ss}=\Delta_2^{(1)}\alpha_0 \quad (10a)$$

$$\phi_1^{ss}=\Delta_1^{(1)}\alpha_0+2\Delta_2^{(1)}\alpha_1 \quad (10b)$$

$$\phi_0^{ss}=\Delta_0^{(1)}\alpha_0+\Delta_1^{(1)}\alpha_1+\Delta_2^{(1)}\alpha_2 \quad (10c)$$

Where α_0, α_1 and α_2 are defined in the Appendix (A).

II. Transient Solution $\phi_t(R,\tau)$

The homogeneous transient problem is defined by

$$\frac{1}{R}\frac{\partial}{\partial R}\left(R\frac{\partial \phi_t}{\partial R}\right)=\frac{\partial \phi_t}{\partial \tau} \qquad \text{in } A<R<1,\ \tau>0, \quad (11a)$$

$$-\frac{\partial \phi_t}{\partial R}+B_1\phi_t=0 \qquad \text{at } R=A,\ \tau>0 \quad (11b)$$

$$\frac{\partial \phi_t}{\partial R}+B_2\phi_t=0 \qquad \text{at } R=1,\ \tau>0, \quad (11c)$$

$$\phi_t(R,\tau)=-\phi_0^{ss} \qquad \text{at } \tau=0, \text{ in the region.} \quad (11d)$$

The integral transform technique (Ozisik 1980) is used to develop the following solution

$$\phi_t(R,\tau)=-\Delta_0^{(1)}\Theta_1+\Delta_1^{(1)}\Theta_3-2\Delta_2^{(1)}\Theta_5\ . \quad (12a)$$

Where

$$\Theta_j=\sum_{i=1}^{\infty}\frac{1}{N_i}e^{-\mu_i^2\tau}\psi_i(R)\frac{F_i}{\mu_i^{(j+1)}}, \qquad j=1,3,5 \quad (12b)$$

Substituting equations (10 & 12) into (8a), leads to the following solution for first-stage $\tau \le \Delta_*$

$$\phi(R,\tau) = \Delta_0^{(1)} W_0^{(1)} + \Delta_1^{(1)} W_1^{(1)} + \Delta_2^{(1)} W_2^{(1)} \quad , \tag{13a}$$

where

$$W_0^{(1)} = [\alpha_o - \Theta_1] \tag{13b}$$

$$W_1^{(1)} = [\alpha_o \tau + \alpha_1 + \Theta_3] \tag{13c}$$

$$W_2^{(1)} = [\alpha_o \tau^2 + 2\alpha_1 \tau + \alpha_2 - 2\Theta_5] \tag{13d}$$

The second-stage $\tau > \Delta_*$ solution is found in similar manner described previously and given by

$$\Phi(R,\lambda) = \Omega W_0^{(2)} + \Delta_1^{(2)} W_1^{(2)} + \Delta_2^{(2)} W_2^{(2)} + \tilde{W}(R,\tau,\Delta_j^{(1)},\Delta_*) \tag{14a}$$

Where

$$\tilde{W}(R,\tau,\Delta_j^{(1)},\Delta_*) = \Delta_0^{(1)} \left[\sum_{i=1}^{\infty} S_i(\beta_{oi} - \beta_{oi} e^{-\mu_i^2 \Delta_*}) \right]$$

$$+ \Delta_1^{(1)} \left[\sum_{i=1}^{\infty} S_i(\beta_{oi}\tau_1 - \beta_{1i} + \beta_{1i} e^{-\mu_i^2 \Delta_*}) \right]$$

$$+ \Delta_2^{(1)} \left[\sum_{i=1}^{\infty} S_i(\beta_{oi}\tau_1^2 - 2\beta_{1i}\tau_1 + 2\beta_{2i} - \beta_{2i} e^{-\mu_i^2 \Delta_*}) \right]. \tag{14b}$$

$$W_0^{(2)} = \left[\alpha_o - \sum_{i=1}^{\infty} S_i \beta_{o\,j} \right] \tag{14c}$$

$$W_1^{(2)} = \left[\alpha_o \lambda + \alpha_1 + \sum_{i=1}^{\infty} S_i \beta_{1i} \right] \tag{14d}$$

$$W_2^{(2)} = \left[\alpha_o \lambda^2 + 2\alpha_1 \lambda + \alpha_2 \sum_{i=1}^{\infty} S_i \beta_{2i} \right] \tag{14e}$$

$$\frac{1}{N_i} = \frac{\pi^2}{2} \frac{\mu_i^2 U_o^2}{(B_2^2 + \mu_i^2) U_o^2 - (B_1^2 + \mu_i^2) V_o^2} \tag{14f}$$

$$\psi_i(R) = S_o J_o(\mu_i R) - V_o Y_o(\mu_i R) \tag{14g}$$

$$F_i = [-\mu_i Y_i(\mu_i) + B_2 Y_o(\mu_i)] \left[\frac{1}{\mu_i} J_1(\mu_i) - \frac{A}{\mu_i} J_i(\mu_i A) \right] +$$

$$[u_i J_1(\mu_i) - B_2 J_o(\mu_i)] \left[\frac{1}{\mu_i} Y_1(\mu_i) - \frac{a}{\mu_i} Y_1(\mu_i A) \right] \tag{14h}$$

$$\Omega = \Delta_0^{(1)} + \Delta_0^{(2)} + \Delta_1^{(1)} \Delta_* + \Delta_2^{(1)} \Delta_*^2 \tag{14i}$$

$$S_i = \frac{1}{N_i} e^{-\mu_i^2 \lambda} \tag{14j}$$

$$\beta_{ji} = \frac{F_i}{\mu_i^{(2+2j)}} \, , \qquad j = 0,1,2 \tag{14k}$$

and μ_i's are the eigen-values obtained as the positive roots of

$$S_o U_o - V_o - W_o = 0 \tag{14l}$$

$$S_o = -\mu_i y_1(\mu_i) + B_2 Y_o(\mu_i) \tag{14m}$$

$$U_o = -\mu_i J_1(\mu_i A) + B_2 J_o(\mu_i A) \tag{14n}$$

$$V_o = -\mu_i J_1(\mu_i) + B_2 J_o(\mu_i) \tag{14o}$$

$$W_o = \mu_i Y_1(\mu_i A) + B_2 Y_o(\mu_i A) \tag{14p}$$

Once the solutions for the two stages $\phi(R,\tau)$ and $\Phi(R,\lambda)$ are furnished, the complete solution $\theta(R,\tau)$ over the whole time domain is determined from Eq. (5).

METHOD OF SOLUTION

The unknown parameters Δ's to be estimated so that the computed temperatures $\theta(R,\tau)$ from Eq. (5) based on these parameters agree with the inaccurate measurements Y_n. This agreement is achieved by the application of the nonlinear regression technique over the whole time domain as now described.

The least-squares criterion $S(\Delta)$ is defined by

$$S(\Delta) = \sum_{i=1}^{N} [Y_n - \theta_i]^2 \, , \tag{15a}$$

where $\theta_i = \theta(R^*, \tau_i, \Delta_*)$ is determined from Eq. (5). An important observation can be made here that this norm is

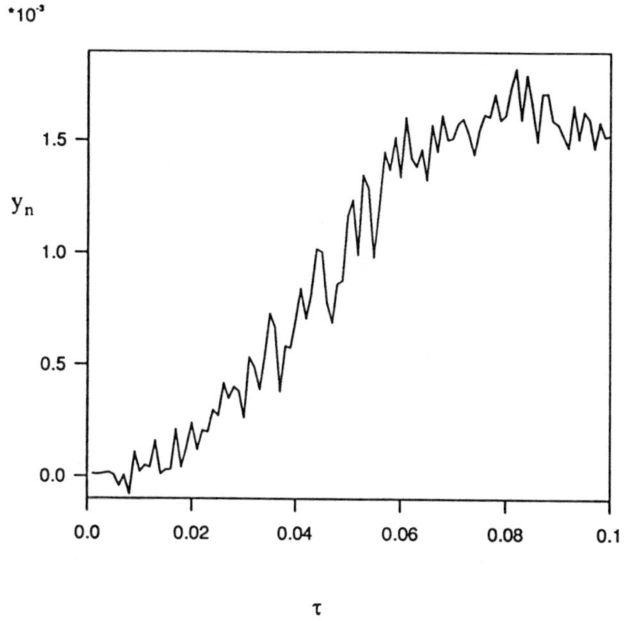

Fig. 2 Simulated measured temperature for triangular case, σ = 0.001

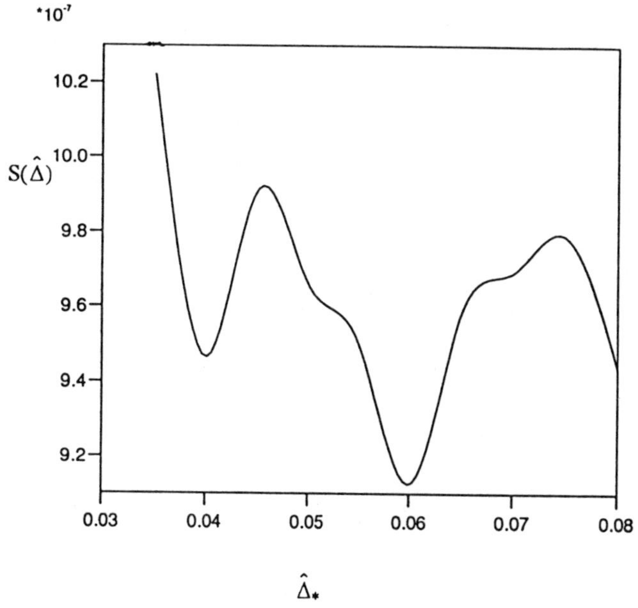

Fig. 3 Variations of least-squares function

nonlinear due to the presence of the unknown joint point Δ_*. The above nom can be written in matrix form as

$$S(\Delta) = (Y - \theta_i)^T (Y - \theta_i) \tag{15b}$$

where

$$Y = [Y_1 \; Y_2 \; Y_3 \quad Y_N]^T . \tag{15c}$$

$$\theta = [\theta_1 \; \theta_2 \; \theta_3 \quad \theta_N]^T . \tag{15d}$$

If the joint point Δ_* is specified, then the norm is linear in the other unknown parameters $\Delta_j^{(1)}$ and $\Delta_j^{(2)}$. Upon minimizing the least-squares function $S(\Delta)$ with respect to these unknown parameters

$$\frac{\partial S(\Delta)}{\partial \Delta} = \sum_{i=1}^{N} [Y_n - \theta_i] \left(\frac{\partial \theta_i}{\partial \Delta} \right) = 0 \quad , \tag{16a}$$

where

$$\frac{\partial S(\Delta)}{\partial \Delta} = \left[\frac{\partial S}{\partial \Delta_j^{(1)}} \quad \frac{\partial S}{\partial \Delta_j^{(2)}} \right]^T = 0, \qquad j = 0,1,2. \tag{16b}$$

Substituting Eq.(15) into Eq. (16) leads to the following normal equations to estimate the unknown coefficients for first and second phases, respectively.

$$[W^{(s)}]\{\hat{\Delta}^{(s)}\} = \{Y^{(s)}\} \quad , \qquad s = 1,2 \tag{17c}$$

where

$$[W^{(s)}] = \begin{bmatrix} W_{00} & W_{10} & W_{20} \\ W_{10} & W_{11} & W_{21} \\ W_{20} & W_{21} & W_{22} \end{bmatrix}^{(s)} \tag{17d}$$

$$\{\hat{\Delta}^{(s)}\} = [\hat{\Delta}_0 \; \hat{\Delta}_1 \; \hat{\Delta}_2]^{(s)} \tag{17e}$$

$$\{Y^{(s)}\} = [Y_{i0} \; Y_{i1} \; Y_{i2}]^{(s)} \tag{17f}$$

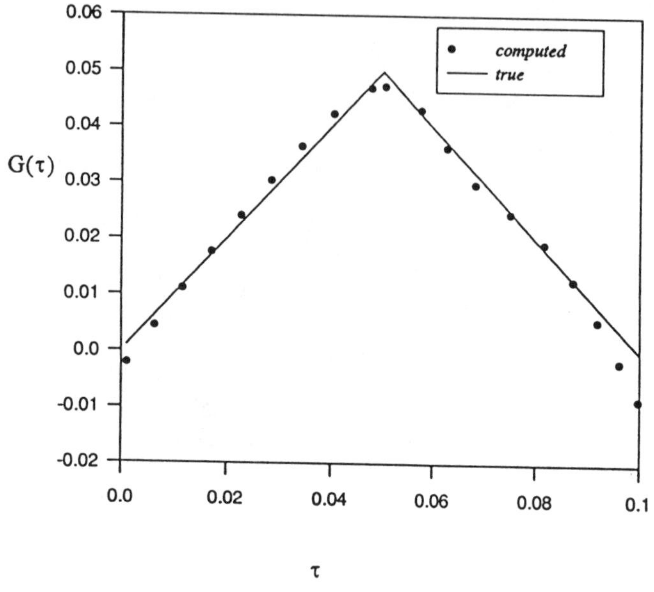

Fig. 4 Estimated triangular heat generation pulse from inexact measurements, σ = 0.001

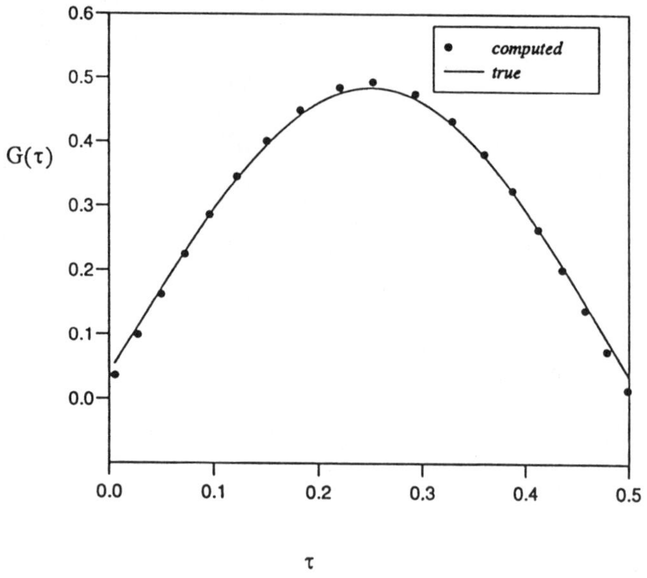

Fig. 5 Estimated half-sine heat generation wave from inexact measurements, σ = 0.001

Here we used the notation $\hat{\Delta}$ to denote the estimated values of the true ones Δ. The definition of the variables appearing in above equations are given in Appendix (B).

The NRFS algorithm begins by assigning a small starting value for the joint point $\hat{\Delta}_*$ and the corresponding values of $\hat{\Delta}$'s are estimated from Eq. (17). Then, the least-squares function $S(\hat{\Delta})$ associated with this joint point is determined from Eq. (15). The joint point is successively incremented by a small value and the procedure is repeated over the entire range of interest. This process is often known as exhaustive search and this allows one to plot of $S(\hat{\Delta})$ versus $\hat{\Delta}_*$. Obviously the estimated values of $\hat{\Delta}$'s and the corresponding $S(\hat{\Delta})$ will vary with $\hat{\Delta}_*$. Now the desired coefficients $\hat{\Delta}$ are the ones at the global minimum of $S(\hat{\Delta})$. Once the coefficients $\hat{\Delta}$'s and the joint point $\hat{\Delta}_*$ are known the estimated heat source is determined from Eq. (4).

RESULTS AND DISCUSSION

Three different benchmark tests are solved to demonstrate the ability of NFRS algorithm in estimating heat source function. These tests include triangular, half-sine wave and exponential variations of internal heat generation. The results of the present study are depicted in Figs. 2-6. The exact temperature measurements are generated at the outer surface at $R^*=1$ from Eq. (5) by assigning proper coefficients. For instant in triangular pulse case we assigned the following values for the coefficients to generate exact temperatures $\Delta_0^{(1)} = 0$, $\Delta_1^{(1)} = 1$, $\Delta_2^{(1)} = 0$, $\Delta_0^{(2)} = 0$, $\Delta_1^{(2)} = -1$, $\Delta_2^{(2)} = 0$, and with joint $\Delta_* = .05$.

In general, inverse algorithms are very sensitive to measurement errors and this can be attributed to the diffusive nature of heat conduction process. The important factor is that the error should not overwhelm the solution. Therefore, studying the effect of measurement errors is mandatory in any inverse method. In this analysis in order to simulate experimental measurements, the exact temperatures are disturbed by additive normally distributed random errors having zero mean with 0.001 standard deviatin and given in Fig. 2 for triangular case.

The variation of least-squares function $S(\hat{\Delta})$ versus $\hat{\Delta}_*$ is obtained by the previous exhaustive search process and shown in Fig. 3. shown in Fig. 3. Clearly, the global minimum of $S(\hat{\Delta})$ is approximately at $\hat{\Delta}_* \approx 0.06$ where $S(\hat{\Delta}) = 9.1\text{x}10^{-5}$. The

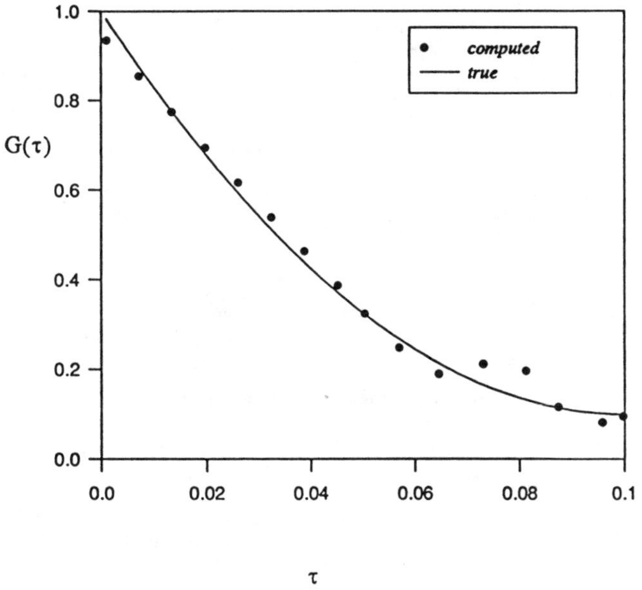

Fig. 6 Estimated exponential heat generation decay from inexact measurements, σ = 0.001

optimal estimated coefficients associated with $\hat{\Delta}_* \approx 0.06$ are

$\hat{\Delta}_0^{(1)} = -3.90\text{x}10^{-3}$, $\hat{\Delta}_1^{(1)} = 1.38$, $\hat{\Delta}_2^{(1)} = -6.62$,

$\hat{\Delta}_0^{(2)} = -3.43\text{x}10^{-2}$, $\hat{\Delta}_1^{(2)} = 9.83\text{x}10^{-1}$, and $\hat{\Delta}_2^{(2)} = -45.2$

hen, the estimated triangular pulse is calculated from Eq (4) utilizing the estimated coefficients $\hat{\Delta}$'s Figure 4 compares true and estimated triangular pulse case Obviously, the predictions are generally satisfactory except around final time because the value of heat source strength is extremely small It has been noted that most of the analytical methods advanced in inverse literature experienced difficulty in reproducing sharp variations in the un nown functions he present results showed that the ability of NRFS algorithm to accommodate, with relatively high degree of accuracy, abrupt change in heat generation function

Similarly, Figs 5 and 6 show the results for continuous variations in the form of half-sine wave and exponential decay heat source, respectively Again, the current method is accurately reproduced these variations from inaccurate measurements Here it should be pointed out that a successful inverse algorithm is the one that poses the ability to estimate the un nown parameters with error of the same order of magnitude as the error involved in the measured data

CONCLUSIONS

A fast-converging inverse algorithm is developed to estimate the time variations of heat generation source in a hollow cylinder subjected tis represented by a segmented second-order polynomial in time having un nown joint point he proposed NRFS scheme successfully estimated the un nown coefficients associated with the assumed form for the generation function Furthermore, this representation allows versatile functional forms of heat generation to be employed including functions with abrupt change

he current methodology based on nonlinear regression function specification approach over the whole time domain Several numerical experiments showed the ability of NRFS algorithm in estimating the heat generation in the presence of significant measurement errors A comprehensive statistical analysis is required to study the effect of various parameters on the inverse solution Such parameters include, among others, measurement errors, type of measurement errors, location of measurements, number of data points, step size, number of joint points, pulse duration time, and the variation of the un nown function

NOMENCLATURE

A	=	dimensionless inner radius = a/b
B_1	=	inner Biot number = bh_1/k
B_2	=	outer Biot number = bh_2/k
F_i	=	function, Eq (14h)
$G(\tau)$	=	dimensionless heat generation = $g(t)b^2/(kT_\infty)$
N	=	total number of measurements
N_i	=	normalization of integral, Eq (14f)
R	=	dimensionless radial coordinates = r/b
S_i	=	function, Eq (14j)
S_o	=	function, Eq (14m)
$S(\Delta)$	=	least-squares criterion, Eq (15a)
Y_n	=	measured temperature = $(y_n - T\infty)/T\infty$
U_o	=	function, Eq (14n)
V_o	=	function, Eq (14o)
W_o	=	function, Eq (14p)
$\theta(R,\tau)$	=	dimensionless temperature = $(T - T_\infty)/T_\infty$

$\psi_i(R)$ = eigen-functions, Eq. (14g)

μ_i = eigen-values, Eq. (14l)

τ = Fourier number = $\alpha t / b^2$

λ = time variable = $\tau - \Delta_*$

Δ_* = unknown joint point

$\Delta's$ = unknown coefficients, Eq. (4a)

$\phi(R,\tau)$ = solution of the first-stage Eq. (6)

$\Phi(R,\lambda)$ = solution of the second-stage Eq. (7)

δ = constant, Eq. (4b)

$\phi_j^{ss}(R)$ = first-stage steady-state solution, Eq. (9)

$\Phi_j^{ss}(R)$ = second-stage steady-state solution

$\phi_t(R,\tau)$ = first-stage transient solution, Eq. (11)

$\Phi_t(R,\lambda)$ = second-stage transient solution

Θ_j = function, Eq. (12b)

Ω = function, Eq. (14i)

β_{ij} = function, Eq. (14k)

REFERENCES

Gallant, A., R, 1977, " Testing a Nonlinear Regression Specification : a Nonregular Case ", J. Am. Stat. Assoc. 72, 523-530.

Borukhov, V.T., and Kolesnikov, P.M., 1988, "Method of Inverse Dynamic Systems and its Application for Recovering Internal Heat Sources" , Int. J. Heat Mass Transfer, Vol. 31, No. 8, pp. 1549 - 1556.

N.M. Al-Najem and M.N. Ozisik, A Direct Analytic Approach for Solving Linear Inverse Heat Conduction Problem", J. Heat Transfer (Transactions of the ASME) 107, 700-703, 1985.

Mikhailov, M.D., Ozisik, M.N, 1985, Unified analysis and solution of heat and mass diffusion, John Wily and Sons.

Ozisik, M.N, 1980, Heat conduction, John Wily and Sons.

APPENDIX A

$$A0 = C53*Log(R)-R^2/4+C63. \quad (A1)$$

$$A1=C32*R^2*Log(R)/2-C32*R^2/2+C42*R^2/2-R^4/64+C52*Log(R)+C62. \quad (A2)$$

$$A2=-R^6/1152+C11*R^4*Log(R)/32-3*C11*R^4/64+ C21 *R^4/32+C31*R^2*Log(R)/2-C31*R^2/2+C41*R^2/2+C11*R^2*Log(R)/4-C11*R^2/4+C21*R^2/4+C51*Log(R)+C11*Log(R) +C21+ C61. \quad (A3)$$

$$C11=B_2*B_1*A+2*B_1*A-B_1*B_2*A^3+2*A^2*B_2+B_2-B_1*B_2*A* Log(A). \quad (A4)$$

$$C21=B_2-B_1*B_2*A*B^3*Log(A)+2-2*B_1*A*Log(A)+B_1*A^3-2*A^2. \quad (A5)$$

$$C31=B_2/64-1/16-A^3*B_2/(16*B_1)+A^4*B_2/64+2*A*Log(A)*B_2/B_1-A*B_2/B_1+2*B_2/(A*H_1)-A^2*Log(A)*B_2+A^2*B_2-2*Log(A)*B_2)+C21/(4*(B_2+2*B_2+2+2*A*B_2/B_1-A^2*B^2-2* B_2))-B_2/(A*B1)+Log(A)*B_2. \quad (A6)$$

$$C32=(-B_2/8-1/4-A* B_2/(4* B_1)+A^2* B_2/8)/(-1-B_2/ (A* B_1)+Log(A)* B_2). \quad (A7)$$

$$C41=-A^3/(16*B_1)+A^4/64+(C11/4)*(2*A*Log(A)/B_1-A/B_1+2/(A*B_1)-A^2*Log(A)+A^2-2*Log(A))+(C21/4)*(2*A/B_1-A^2-2)+ C31*(1/(A*B_1)-Log(A)). \quad (A8)$$

$$C42=C32/(A*B_1)-A/(4*B_1)-C32*Log(A)+A^2/8. \quad (A9)$$

$$C51=(-B_1*B_2*A-54*C11*B_1*B_2*A+36*C21*B_1*B_2*A-576*C31*B_1*B_2*A + 576* C41 * B_1*B_2*A-288*C11*B_1*B_2*A+ 288* C21*B_1*B_2*A+ 1152 *C21*B_1*B_2*A+A^7*B_1*B_2-36*C11*A^5*Log(A)*B_1*B_2+54*C11*A^5*B_1*B_2-36*C21*A^5*B_1*B_2-576*C31*A^3*Log(A)*B_1 *B_2+576*C31*A^3*B_1*B2-576*C41*A^3*B_1*B_2-288*C11*A^3*Log(A)*B_1*B_2+288*C11*A^3*B_1*B_2-288*C21*A^3*B_1*B_2-1152*C11*Log(A)*B_1*B_2*A-1152*C21*B_1*B2*A-6*A^6*B_2+144* C11*A^4* Log(A)*B_2-180* C11* A^4*B_2+144* C21*A^4* B_2+ 1152* C31*A^2*Log(A)*B_2-576*C31*A^2*B_2+1152*C41*A^2*B_2+576*C11*A^2*log(A)*B_2-288*C11*A^2*B_2* +576* C21*A^2*B_2+1152*C11*B_2*b-6* B_1*A_1-180*C11* B_1*A+144*C21*B_1*A-576*C31*B_1*A+1152*C41*B_1*A-288*C11*B_1*A+576*C21*B_1*A+1152*C11*B_1*A)/((1152 *B_1 *A)(-1-B_2/(A*B_1)+Log(A)*B_2)) \quad (A10)$$

$$C52=(-32C32*B_1*B_2+32*C42*B_1*B_2-B_1*B_2-32*C32*A^2*Log(A)*B_1*B_2+ 32*C32*A^2*B_1*B_2- 32*C42*A^2*B_1*B_2+ A^4*B_1*B_2+ 64*C32*A*Log(A)*B_2- 32*C32*A*B_2+ 64* C42 *A*B_2- 4*A^3* B_2- 32* C32* B_1+64*C42* B_1- 4* B_1) / ((32* B_1)*(-1-B_2 / (A* B_1) +Log (A) *B_2)). \quad (A11)$$

$$C53=(-B_1*B_2+A^2* B_1* B_2-2*A*B_2-2* B_1)/((4* B_1)*(-1-B_2/(A* B_1)+Log(A)* B_2)). \quad (A12)$$

$$C61=(((A^6*B_1*A-36*C11*A^5* Log(A)*B_1+54* C11*A^5*B_1-36*C21*A^5*B_1-576*C31*A^3 *Log(A)*B_1 +576 *C31*A^3*B_1-576*C41*A^3*B_1-288*C11*A^3*Log(A)*B_1+288*C11*A^3*B_1-288*C21*A^3*B_1-1152*C11*Log(A)*$$

B1*A-1152*C21*B1*A-6*A6+ 144*C11* A4* Log(A)-180* C11* A4+144* C21*A4+1152* C31* A2* Log(A)-576*C31*A2+1152*C41*A2+576*C11*A2* Log(A)-288*C11*A2+576*C21*A2+1152C11)]/(1152 *B1*A)) -C51* Log(A))+C51/(A* B1). (A13)

C62=(((-32*C32*A2*Log(A)*B1+32*C32*A2*B1-32*C42 *A2*B1+A4*B1+64*C32*A*Log(A)-32*C32*A+64* C42 *A-4*A3)/(64*B1))-C52*Log(A))+C52/ (A*B1) (A14)

C63=(((A2*B1-2*A)/(4* B1))-C53* Log(A))+*C53/(A* B1) (A15)

APPENDIX B

$$W_{00}^{(1)} = \sum_{i=1}^{n_1} W_0^{(1)} W_0^{(1)} \tag{B1}$$

$$W_{10}^{(1)} = \sum_{i=1}^{n_1} W_1^{(1)} W_0^{(1)} \tag{B2}$$

$$W_{20}^{(1)} = \sum_{i=1}^{n_1} W_2^{(1)} W_0^{(1)} \tag{B3}$$

$$W_{11}^{(1)} = \sum_{i=1}^{n_1} W_1^{(1)} W_1^{(1)} \tag{B4}$$

$$W_{21}^{(1)} = \sum_{i=1}^{n_1} W_2^{(1)} W_1^{(1)} \tag{B5}$$

$$W_{22}^{(1)} = \sum_{i=1}^{n_1} W_2^{(1)} W_2^{(1)} \tag{B6}$$

$$Y_{i0}^{(1)} = \sum Y_i^{(1)} W_0^{(1)} \tag{B7}$$

$$Y_{i1}^{(1)} = \sum Y_i^{(1)} W_1^{(1)} \tag{B8}$$

$$Y_{i2}^{(1)} = \sum Y_i^{(1)} W_2^{(1)} \tag{B9}$$

$$W_{00}^{(2)} = \sum_{i=n_1+1}^{N} W_0^{(2)} W_0^{(2)} \tag{B10}$$

$$W_{10}^{(2)} = \sum_{i=n_1+1}^{N} W_1^{(2)} W_0^{(2)} \tag{B11}$$

$$W_{20}^{(2)} = \sum_{i=n_1+1}^{N} W_2^{(2)} W_0^{(2)} \tag{B12}$$

$$W_{11}^{(2)} = \sum_{i=n_1+1}^{N} W_1^{(2)} W_1^{(2)} \tag{B13}$$

$$W_{21}^{(2)} = \sum_{i=n_1+1}^{N} W_2^{(2)} W_1^{(2)} \tag{B14}$$

$$W_{22}^{(2)} = \sum_{i=n_1+1}^{N} W_2^{(2)} W_2^{(2)} \tag{B15}$$

$$Y_{i0}^{(2)} = \sum_{i=n_1+1}^{N} \left[Y_i^{(2)} - \Omega^*\right] W_0^{(2)} \tag{B16}$$

$$Y_{i1}^{(2)} = \sum_{i=n_1+1}^{N} \left[Y_i^{(2)} - \Omega^*\right] W_1^{(2)} \tag{B17}$$

$$Y_{i2}^{(2)} = \sum_{i=n_1+1}^{N} \left[Y_i^{(2)} - \Omega^*\right] W_2^{(2)} \tag{B18}$$

$$\Omega^* = \tilde{W} + \Omega W_0^{(2)} + \hat{\Delta}_0^{(1)} + \hat{\Delta}_1^{(1)} \Delta_* + \hat{\Delta}_2^{(1)} \Delta_*^2 \tag{B19}$$

Inverse Problems in Engineering: Theory and Practice
ASME 1993

INVERSE PHOTON TRANSPORT METHODS FOR BIOMEDICAL APPLICATIONS

N. J. McCormick
Department of Mechanical Engineering
University of Washington
Seattle, Washington

ABSTRACT

Recent advances have been made in the imaging of tumors in breast tissue with near-infrared radiation, a problem that is complicated by the multiple scattering of the radiation. Inverse methods for solving the tissue parameter estimation and imaging problems are briefly summarized in this review.

INTRODUCTION

The imaging of tumors in breast tissue is a standard procedure when done by x-ray mammography: two-dimensional images can be achieved with one image, or three dimensional images can be achieved with multiple images. The advantage of the method is that the radiation propagates in essentially straight lines, which makes the inverse problem relatively easy to solve, but the method has the disadvantage that ionizing radiation must be used. Near-infrared radiation has the advantage that the radiation is non-ionizing and that tissue is relatively transparent at these wavelengths, but the image quality is significantly degraded by multiple scattering of photons.

There are two biomedical inverse problems for near-infrared radiation of interest here: tissue parameter estimation and imaging of tumors. Both problems are complicated by the fact that in clinical applications the optical properties tend to change in a spatially-continuous manner. The purpose of this paper is to provide a summary of recent work in this field.

Explicit (non-iterative) photon diffusion methods to solve the tissue parameter estimation problem have been developed for time-dependent and frequency-dependent sources. The time-dependent methods use the long-time decay following a pulse, while frequency-modulation methods typically utilize the low frequency response. These methods generally are valid when the sample tissue volumes can be treated as a continuous, homogeneous medium, as discussed in the following section.

Implicit (iterative) methods are used to solve the inherently ill-posed imaging problem. This requires a rapid, accurate method for solving the forward (direct) problem using transport theory methods since the diffusion approximation does not apply to localization studies. Imaging is considered subsequent to the following section.

OPTICAL PARAMETER ESTIMATION

Optical Parameter Characteristics

To characterize the properties of healthy and tumor tissue so that the latter can be imaged, it is necessary to obtain the absorption coefficient Σ_a in units of mm^{-1} and coefficients that describe the scattering. Often one seeks only the reduced scattering coefficient $(1-g)\Sigma_s$, i.e., the scattering coefficient as reduced by $(1-g)$, where g is the mean cosine of the angle of an individual scattering event. These properties must be determined as a function of the radiation wavelength.

In the 600-1300 nm range the optical coefficients for soft tissues (e.g., brain, liver, lung, skin) are $\Sigma_a \approx 0.01$–$1\ mm^{-1}$, $\Sigma_s \approx 10$–$100\ mm^{-1}$, and $g \approx 0.8$–0.95 (Wilson and Jacques, 1990). The fact that the single scattering albedo is so nearly unity, i.e., $\Sigma_s/\Sigma_t \approx 1$ for $\Sigma_t = \Sigma_a + \Sigma_s$, helps justify the use of diffusion theory when simulating the migration of photons in tissue.

The highest percentage of light in biological tissue is transmitted in the 600-1000 nm transmission window. The region above ~950 nm is dominated by broad water absorption peaks while the region below 600 nm is dominated by hemoglobin absorption resonances. A traditional hope in the imaging of breast tissue has been to distinguish the absorption profile of the deoxygenated homoglobin (Hb) associated with malignant tissue from that of the oxygenated homoglobin (HbO_2) for benign tissue. Unfortunately the spectral characteristics of fibrocystic disease, a benign breast process, are quite similar to those for deoxygenated hemoglobin tissue, which can lead to false positive results in imaging applications when imaging (Marks, 1992).

Time-Resolved Estimation Methods

To estimate optical parameters with the simplest algorithm, the time-dependent diffusion equation for a homogeneous, semi-infinite medium is used. For a perpendicularly-incident beam of light incident at radial position $r = 0$ on the surface $z = 0$ at time $t = t_0$, the boundary condition can be replaced by an internal point isotropic source of strength

$$s(z_s, 0, t_0) = s(0, 0, t_0)\Sigma_s \exp(-\Sigma_t z_s) \quad (1)$$

to account for photons that undergo their first collision at point $z = z_s$. Then the Green's function $G(0 \to z_s, 0 \to r, t_0 \to t)$ for the reflected radiation at distance r away from the point of incident illumination is (Patterson et al., 1989)

$$G(0 \to z_s, 0 \to r, t_0 \to t) = \frac{z_s \exp(-\Sigma_a c\tilde{t})}{(4\pi Dc)^{3/2}} \frac{\exp\left(-\frac{r^2 + z_s^2}{4Dc\tilde{t}}\right)}{\tilde{t}^{5/2}} \quad (2)$$

where c is the speed of light, D is the diffusion coefficient, and

$$\begin{aligned} \tilde{t} &= (t - t_0 + z_s/c) \\ D &= \{3[\Sigma_a + (1-g)\Sigma_s]\}^{-1}. \end{aligned}$$

To obtain the emerging reflected flux at the surface, one combines Eqs. (1) and (2),

$$\begin{aligned} R(0, r, t) &= \int_0^\infty \int_0^\infty s(z_s, 0, t_0) \times \\ & G(0 \to z_s, 0 \to r, t_0 \to t) dz_s dt_0. \quad (3) \end{aligned}$$

Convolved forms of R from Eq. (3) have been used to estimate Σ_a. For example, Elliott et al. (1988, 1989) showed that $c\Sigma_a$ could be accurately estimated if the incident beam was large enough that the r-dependence could be ignored. Madsen et al. (1991, 1992) fit Eq. (3) to experimental data for different tissue-like phantoms by varying Σ_a, $(1-g)\Sigma_s$, and an overall scaling factor, and found that the time-resolved estimate of Σ_a was consistently underestimated and that $(1-g)\Sigma_s$ was consistently overestimated.

For the case of a source strength $s(0, 0, t_0) = \delta(t_0)$ with t long compared to z_s/c, then

$$\begin{aligned} R(0, r, t) &\approx \frac{\Sigma_s \exp\left(-\frac{r^2}{4Dct} - \Sigma_a ct\right)}{(4\pi Dc)^{3/2} t^{5/2}} \times \\ & \int_0^\infty z_s \exp\left(-\frac{z_s^2}{4Dct} - \Sigma_t z_s\right) dz_s, \quad (4) \end{aligned}$$

which can be evaluated analytically to give (Graber et al., 1992)

$$R(0, r, t) \approx \frac{\Sigma_s \exp\left(-\frac{r^2}{4Dct} - \Sigma_a ct\right)}{4(Dc)^{1/2}(\pi t)^{3/2}} F[\Sigma_t (Dct)^{1/2}] \quad (5)$$

for

$$F(\tau) = 1 - \pi^{1/2}\tau \exp(\tau^2)\mathrm{erfc}(\tau), \quad (6)$$

where $\mathrm{erfc}(\tau)$ is the complementary error function for dimensionless variable τ. In nonidealized cases, e.g., where the duration of the laser pulse is not short enough to be approximated by $\delta(t)$, it is necessary to evaluate Eq. (3) numerically as done by Graber et al. (1992).

A simple explicit inversion algorithm for estimating $\Sigma_a c$ follows by differentiating Eq. (5) and letting $t \to \infty$ to obtain

$$\begin{aligned} \frac{d \ln R(0, r, t)}{dt} &\approx -\Sigma_a c + \frac{r^2}{4Dct^2} - \frac{3}{2t} \\ & + \frac{\Sigma_t}{2}\left(\frac{Dc}{t}\right)^{1/2} \frac{d \ln F(\tau)}{d\tau}. \quad (7) \end{aligned}$$

Thus the exponential-type decay of the reflectance depends on both the properties of the medium and the separation of the source and detector, as further investigated by Haselgrove et al. (1992).

A key difficulty of a time-resolved method is the short times involved. For an index of refraction of 1.4, a typical value for tissue, the photon speed is $c = 0.214$ mm ps^{-1}, so to sample in a volume with an effective radius of only a few mm one needs a streak camera with a time resolution of ps and a laser capable of generating ps pulses. It is for this reason that frequency-dependent algorithms have recently been investigated.

Frequency-Resolved Estimation Methods

Frequency-resolved methods for estimating optical parameters with oscillatory sources are closely related to the time-dependent methods for pulsed sources: to develop an algorithm with a frequency-dependent source, one can Fourier transform results derived for a time-dependent pulse. For example, the transform $F[G(r,t)]$ of Eq. (2) leads to an equation for the phase ϕ of the measured signal at r as a function of frequency f as (Patterson et al., 1990, 1991)

$$\begin{aligned} \phi(r,f) &= \tan^{-1}\left(\frac{\text{Im } F[G(r,t)]}{\text{Re } F[G(r,t)]}\right), \\ &= \Psi_r - \tan^{-1}\left(\frac{\Psi_r}{1+\Psi_i}\right), \end{aligned} \quad (8)$$

where

$$\Psi_r = -\Psi_o \sin(\theta/2), \quad (9)$$

$$\Psi_i = \Psi_o \cos(\theta/2), \quad (10)$$

$$\theta = \tan^{-1}\left(\frac{2\pi f}{\Sigma_a c}\right), \quad (11)$$

$$\Psi_o = \left[D^{-1}(r^2+z_s^2)[(\Sigma_a c)^2 + (2\pi f)^2]^{1/2}c^{-1}\right]^{1/2} \quad (12)$$

Patterson et al. (1990) have shown that the phase shift in the detected signal for frequencies in the range 0 to 250 MHz is nearly linear, and this has been observed experimentally (Lakowicz et al., 1990; Chance et al., 1990; Sevick et al., 1991a, 1991b; Tsay et al., 1992; Tromberg et al., 1993). By expanding Eq. (8) for $f \ll (c\Sigma_a)^{-1}$ and using the fact that $(1-g)\Sigma_s$ and r are large, it follows that (Patterson et al., 1990; Sevick et al., 1991a)

$$\phi(r,f) \to -2\pi f\,\langle t \rangle, \quad (13)$$

where the mean time for photons to diffuse through the tissue from the source to the detector is

$$\begin{aligned} \langle t \rangle &= \frac{\int_0^\infty tR(r,t)dt}{\int_0^\infty R(r,t)dt}, \\ &= \frac{r[3(1-g)\Sigma_s]^{1/2}}{2c\Sigma_a}. \end{aligned} \quad (14)$$

Hence the relation between the phase and the frequency gives an estimate of the optical property defined by $\langle t \rangle$.

On the other hand, for high frequencies (Sevick et al., 1991a),

$$\phi(r,f) \propto [(1-g)\Sigma_s f]^{1/2}\left[1 - \frac{\Sigma_a c}{4\pi f}\right]. \quad (15)$$

A detailed analysis of the correlation between the frequency and attenuation coefficient, or alternatively the penetration depth, also has been carried out by Svaasand and Tromberg (1991).

ALGORITHMS FOR IMAGING

Experimental and Computational Data

An imaging experiment using multiply-scattered radiation is "diffuse tomography." The imaging of tumors (or any other type of imbedded object, for that matter) can be done with transmitted radiation or reflected radiation or a combination of the two. Because of the highly anisotropic nature of the scattering, it follows that the transmission experiment provides the best opportunity of obtaining a sharper image since the few marginally deviated photons can be better distinguished from all other scattered radiation. Then the intensity of light detected with the shortest travel times will be dependent on the absorption properties of the tissue within a narrow volume surrounding the direction of incidence. Such a technique works best if both the source and detector are moved, or alternately the specimen, so that point-by-point scanning is performed. Otherwise more multiply-scattered radiation tends to blur the contrast.

The position of the local region to be imaged is also important in a transmission experiment. If a strong absorber region is near the input surface it can effectively block the beam before too many photons can scatter around it, while if such a region is near the output surface then much of the diffuse light directly behind it does not reach the detector.

Data illustrating the use of a transmitted pulse of radiation for imaging have been given by Hebden et al. (1991). Also, Graber et al. (1992) performed both computations and experiments to assess the the change in the decay of a backscattered time-dependent pulse for an imbedded sphere in a homogeneous medium. Computed results for a medium with an imbedded plane layer have been given by Duracz and McCormick (1989) and Graber et al. (1992) and for an imbedded sphere by Duracz and McCormick (1990). A deeper object is more difficult to detect, especially if the medium scatters in a mildly anisotropic manner so the radiation becomes diffuse with only a few scatterings.

In a frequency-dependent experiment, oscillations in radiation are rapidly damped out as the photons diffuse outwards from a localized source, and there is a maximum depth below which the radiation intensity will essentially be in a steady state (Fishkin et al., 1991).

Methods For Solving The Inverse Problem

At least three different methods have been devised in an attempt to image obscured objects. An *ad hoc* explicit approach consists of attempting to look at modifications to the algorithms developed for the parameter estimation problem (Sevick and Chance, 1991). The idea is that if photons diffuse in two media, an *o*uter one and an *i*nner one, then it may be possible to decompose the total mean transit time $\langle t \rangle$ for a time-domain transmission experiment into two parts as

$$\langle t \rangle = \langle t \rangle_o + \langle t \rangle_i \,, \tag{16}$$

or, using Eq. (13), the total phase of a frequency-domain experiment as

$$\phi = \phi_o + \phi_i \,. \tag{17}$$

A more interesting implicit approach for identifying imbedded objects has been developed by Grünbaum (1990, 1991) and coworkers (Singer et al., 1990) who considered particles undergoing discrete random walks in an optically-thick body that was partitioned into little volume elements (voxels). Particles that enter voxel n through one surface can undergo absorption within the voxel with probability a_n, or they are forward scattered out through the opposite surface with probability $(1-a_n)f_n$ or a side surface with probability $(1-a_n)s_n$, or are backscattered through the entrant surface with probability $(1-a_n)b_n$. The f_n, s_n, and b_n are three conditional scattering transition probabilities that obey the constraint $f_n + 4s_n + b_n = 1$. The inverse problem consists of estimating the a_n and two of the three conditional transition probabilities for each voxel in the body from measurements at different locations on its external surface. The spatial resolution of the imaging is governed by the number of voxels in the body; measurements at a large number of locations are necessary if the voxel size is small.

The inverse problem is solved by using a set of intensity measurements M_{jk} obtained for an array of sources at positions j, $j = 1$ to J, and detectors at positions k, $k = 1$ to K. Using the values of a_n, b_n, f, s, and b from the previous iteration (or assumed initially), computed values C_{jk} of the measurements are obtained by solving the direct problem. Minimization of a least-squares error is then done. The imaging has not worked too well because of the small number of scattering directions, so models with a larger number of directions are under development to overcome this limitation (Grünbaum et al., 1991).

In a third approach, the "importance" or weight of a voxel's contribution to the reflected radiation intensity is computed (Barbour et al., 1991, 1992). The idea is to express the first-order perturbation in the photon flux, ΔI_{jkn}, in time interval Δt_n that is measured at surface element (pixel) k following illumination of pixel j by a time-dependent source. The perturbation is a composite of all small changes in the absorption cross section $\Delta\Sigma_{a,i}$ at each voxel i. If the weight w_{ijkn} is defined as the negative of the partial derivative of the detected photon flux with respect to $\Sigma_{a,i}$, then

$$\Delta I_{jkn} = \sum_i w_{ijkn} \Delta\Sigma_{a,i} \,. \tag{18}$$

This equation can be written in the form of a standard inverse problem,

$$\mathbf{y} = \mathbf{W}\mathbf{x} \tag{19}$$

where the vector of absorption cross section changes, $\mathbf{x} = \{\Delta\Sigma_{a,i}\}$ can be estimated from the measured values of vector $\mathbf{y} = \{\Delta I_{jkn}\}$ once matrix $\mathbf{W} = \{w_{ijkn}\}$ has been computed.

For a source at pixel location j and a detector at pixel k, the perturbation in the photon flux in time interval Δt_n due to absorption in location i has been reported as (Barbour et al., 1991, 1992)

$$w_{ijkn} = \frac{S_j}{4\pi V_i \Sigma_t^2} \int_{\Delta t_n} \int_0^t F_{ij}(\tau) F_{ik}(t-\tau)\, d\tau\, dt \,, \tag{20}$$

for the unperturbed total cross section $\Sigma_t = (\Sigma_a + \Sigma_s)$. Here S_j is the source strength in photons/unit time, V_i is the volume of voxel i, and F_{ij} is the rate of collisions in voxel i per photon injected into the medium from a source at pixel j. Equation (20) is based on the use of the radiation transport reciprocity relation to interchange source and detector positions.

Different standard methods have been used to solve the inverse problem of Eq. (19), including "iterative backprojection" and "sequential iterative backprojection" (Barbour et al., 1991, 1992) and "iterative perturbation" methods (Wang et al., 1992).

REFERENCES

Barbour, R.L., Graber, H.L., Aronson, R., and Lubowsky, J., 1991, "Imaging of subsurface regions of random media by remote sensing," in *Time-Resolved Spectroscopy and Imaging of Tissues*, Proc. Soc. Photo-Opt. Instrum. Eng. **1431**, 192-203.

Barbour, R.L., Graber, H.L., Lubowsky, J., Aronson, R., Das, B.B., Yoo, K.M., and Alfano, R.R., 1992, "Imaging of diffusing media by a progressive iterative backprojection method using time-domain data," in *Physiological Monitoring and Early Detection Diagnostic Methods,* Proc. Soc. Photo-Opt. Instrum. Eng. **1641**, 21-34.

Chance, B., Maris, M. Sorge, J. and Zhang, M.Z., 1990, "A Phase Modulation System for Dual Wavelength Difference Spectroscopy of Homoglobin Deoxygenation in Tissues," in *Time-Resolved Laser Spectroscopy in Biochemistry II,* Proc. Soc. Photo-Opt. Instrum. Eng. **1204**, 481-491.

Duracz, T. and McCormick, N.J., 1989, "Radiative transfer calculations for characterizing obscured surfaces using time-dependent backscattered pulses," *Appl. Opt.* **28**, 544-552.

Duracz, T. and McCormick, N.J., 1990, "Multiple scattering corrections for lidar detection of obscured objects," *Appl. Opt.* **29**, 4170-4175.

Elliott, R.A., Duracz, T., McCormick, N.J., and Emmons, D.R., 1988, "Experimental test of a time-dependent inverse radiative transfer algorithm for estimating scattering parameters," *J. Opt. Soc. Am. A* **5**, 366-373.

Elliott, R.A., Duracz, T., McCormick, N.J., and Bossert, D.J., 1989, "Experimental test of a time-dependent inverse radiative transfer algorithm for estimating scattering parameters: addendum," *J. Opt. Soc. Am. A* **6**, 603-606.

Fishkin, J., Gratton, E. van de Ven, M.J., and Mantulin, W.W., 1991, "Diffusion of intensity modulated near-infrared light in turbid media," in *Time-Resolved Spectroscopy and Imaging of Tissues,* Proc. Soc. Photo-Opt. Instrum. Eng. **1431**, 122-135.

Graber, H.L., Barbour, R.L., Lubowsky, J., Aronson, R., Das, B.B., Yoo, K.M., and Alfano, R.R., 1992, "Evaluation of steady-state, time- and frequency-domain data for the problem of optical diffusion tomography," in *Physiological Monitoring and Early Detection Diagnostic Methods,* Proc. Soc. Photo-Opt. Instrum. Eng. **1641**, 6-20.

Grünbaum, F.A., 1990, "Relating microscopic and macroscopic parameters for a 3-dimensional random walk," *Commun. Math. Phys.*, **129**, 95-102.

Grünbaum, F.A., Kohn, P., Latham, G.A., Singer, J.R., and Zubelli, J.P., 1991, "Diffuse tomography," in *Time-Resolved Spectroscopy and Imaging of Tissues,* Proc. Soc. Photo-Opt. Instrum. Eng. **1431**, 232-238.

Haselgrove, J.C., Schotland, J.C., and Leigh, J.S., 1992, "Long-time behavior of photon diffusion in an absorbing medium: application to time-resolved spectroscopy," *Appl. Opt.* **31**, 2678-2683.

Hebden, J.C., Kruger, R.A., and Wong, K.S., 1991, "Time resolved imaging through a highly scattering medium," *Appl. Opt.* **30**, 788-794.

Lakowicz, J.R., Berndt, K.W., and Johnson, M.L., 1990, "Photon Migration in Scattering Media and Tissue," in *Time-Resolved Laser Spectroscopy in Biochemistry II,* Proc. Soc. Photo-Opt. Instrum. Eng. **1204**, 468-479.

Madsen, S.J., Patterson, M.S., Wilson, B.C., Young, D.P., Moulton, J.D., Jacques, S.L., and Hefetz, Y., 1991, "Time-resolved diffuse reflectance and transmittance studies in tissue simulating phantoms: a comparison between theory and experiment," in *Time-Resolved Spectroscopy and Imaging of Tissues,* Proc. Soc. Photo-Opt. Instrum. Eng. **1431**, 42-51.

Madsen, S.J., Wilson, B.C., Patterson, M.S., Park, Y.D., Jacques, S.L., and Hefetz, Y., 1992, "Experimental tests of a simple diffusion model for the estimation of scattering and absorption coefficients of turbid media from time-resolved diffuse reflectance measurements," *Appl. Opt.* **31**, 3509-3517.

Marks, F.A., 1992. "Optical determineation of the hemoglobin oxygenation state of breast biopsies and human breast cancer xenografts in nude mice," in *Physiological Monitoring and Early Detection Diagnostic Methods,* Proc. Soc. Photo-Opt. Instrum. Eng. **1641**, 227-237.

Patterson, M.S., Chance, B., and Wilson, B.C., 1989, "Time resolved reflectance and transmittance for the non-invasive measurement of tissue optical properties," *Appl. Opt.* **28**, 2331-2336.

Patterson, M.S., Moulton, J.D., Wilson, B.C. and Chance, B., 1990, "Applications of Time-Resolved Light Scattering Measurements to Photodynamic Therapy Dosimetry," in *Photodynamic Therapy: Mechanisms II,* Proc. Soc. Photo-Opt. Instrum. Eng. **1203**, 62-75.

Patterson, M.S., Moulton, J.D., Wilson, B.C., Berndt, K.W., and Lakowicz, J.R., 1991, "Frequency-domain reflectance for the determination of the scattering and absorption properties of tissue," *Appl. Opt.* **30**, 4474-4476.

Sevick, E.M., and Chance, B., 1991, "Photon migration in a model of the head measured using time- and frequency-domain techniques: potentials of spectroscopy and imaging,"in *Time-Resolved Spectroscopy and Imaging of Tissues,* Proc. Soc. Photo-Opt. Instrum. Eng. **1431**, 84-96.

Sevick, E.M., Chance, B., Leigh, J., Nioka, S., and Maris, M., 1991a, "Quantitation of time- and frequency-resolved optical spectra for the determination of tissue oxygenation," *Anal. Biochem.* **195**, 330-351.

Sevick, E.M., Weng, J., Maris, M., and Chance, B., 1991b, "Analysis of absorption, scattering, and hemoglobin saturation using phase modulation spectroscopy," in *Time-Resolved Spectroscopy and Imaging of Tissues,* Proc. Soc. Photo-Opt. Instrum. Eng. **1431**,

264-275.

Singer, J.R., Grünbaum, F.A., Kohn, P. and Zubelli, J.P., 1990, "Image reconstruction of the interior of bodies that diffuse radiation," *Science*, **248**, 990-993.

Svaasand, L.O. and Tromberg, B.J., "On the properties of optical waves in turbid media," in *Future Trends in Biomedical Applications of Lasers*, Proc. Soc. Photo-Opt. Instrum. Eng. **1525**, 41-51.

Tromberg, B.J., Svaasand, L.O., Tsay, T.-T., and Haskell, R.C., 1993, "Properties of photon density waves in multiple-scattering media," *Appl. Opt.* **32**, 607-616.

Tsay, T.-T., Tromberg, B.J., Cho, E.H., Vu, K.T., and Svaasand, L.O., 1992, "Monitoring photochemistry in tissue using frequency domain photon migration," in *Laser-Tissue Interaction III*, Proc. Soc. Photo-Opt. Instrum. Eng. **1646**, 213-218.

Wang, Y., Chang, J-H., Aronson, R., Barbour, R.L., Graber, H.L., and Lubowsky, J., 1992, "Imaging of scattering media by diffusion tomography: an iterative perturbation approach," in *Physiological Monitoring and Early Detection Diagnostic Methods,* Proc. Soc. Photo-Opt. Instrum. Eng. **1641**, 58-71.

Wilson, B.C., and Jacques, S.L., 1990, "Optical reflectance and transmittance of tissues: principles and applications," *IEEE J. Quantum Electron.* **26**, 2186-2199.

RETRIEVAL OF ABSORPTION AND TEMPERATURE PROFILES IN AXISYMMETRIC AND NON-AXISYMMETRIC EMITTING-ABSORBING MEDIA BY INVERSE RADIATIVE METHODS

Mohamed Sakami and Michel Lallemand
Laboratoire d'Etudes Thermiques
ENSMA
Poitiers, France

ABSTRACT

The radiative inverse problem is studied in order to reconstruct the absorption coefficient and temperature fields in bi or monodimensional semi-transparent thermal structures. Several inversion methods are exposed and tested.

i) The gas absorption tomography is proceeded by an inverse Radon transform of monochromatic transmission data collected from angular and lateral scanning.

In the case of an axisymmetric flow, inversions are carried out by four different methods, an Abel equation after the preconditionning of the data, the Fourier-Bessel transform, the Regularised-Adjoint-Conjugate-Gradient algorithm and finally the Mollification method.

ii) The inversion of spectral angular emission data is used for the reconstruction of the thermal image when the absorption coefficient field is still recovered from the previous step. For an axisymmetric flow, the solution of the corresponding Volterra equation is obtained by means of the Regularised-Adjoint-Conjugate-Gradient method.

The above methods are applied to retrieve at the optically thin medium approximation both the absorption and the temperature profiles, in a premixture flame of propane-air, the data coming from either the outgoing thermal emission or transmission measurements collected from an Infrared Fourier Transform spectrometer.

NOMENCLATURE

A, A*	Operators
e_λ	relative emission
h (p)	Filter function
J_α	Tikhonov function
K_λ (r,θ)	Local Absorption Coefficient
$L^\circ_\lambda(r,\theta)$	Planck function
$L_\lambda(p,\phi)$	outgoing spectral intensity
p	Impact parameter (lateral scanning)
r,θ	polar coordinates
STM	Semi-Transparent Medium
t_λ (p)	transmissivity
z	z axis
α	regularization parameter
ε	error measure
δ	radius of mollification
ρ_A	regulatrizing family
ϕ	angle(ox,ω)
σ	spatial frequency
ξ	line of sight
ω	unit vector

INTRODUCTION

This paper is devoted to the restitution and imaging processes of thermal structures settled in semi-transparent objects from remoted measurements of their outgoing angular thermal emission or transmission. It is divided in three parts.

In section II we examine the possibility of computerized tomography of the absorption coefficient and of the thermal field by an inverse Radon transform from transmission and emission collected data. In section III the case of an axisymmetric semi-transparent object is considered. Four kinds of inversion are performed, an Abel inversion (with prepared data and spline interpolation), the Fourier-Bessel inversion method, a regularization method associated to the conjugated gradient method, and the mollification method. In section IV, as an application reconstruction of the absorption coefficient and of the temperature profiles for a premixture flame is carried out at the optically thin approximation from experimental transmission and emission data coming from FTIR measurements.

TOMOGRAPHY OF SEMI-TRANSPARENT MEDIA

Density and thermal tomographies of semi-transparent media may be processed from projections of the infrared transmission and emission data. The first one is simply an extension of the widespread image reconstruction technique used in medicine; as for the second one we have previously presented some results in the *Eurotherm 21* held at *Lyon* (Sakami et al.,1992) for data provided from light beam deflection due to inhomogenous refracive index field and have find quite a good agreement with the original thermal object (four thermal gaussian peaks scattered on a plane). In this section a thermal mapping of an nonhomogeneous object is reconstructed from its outgoing infrared emission or transmission data .

Data Collected from Transmission Measurements

Let us considere a bi-dimensional semi-transparent non-homogenous media of absorption coefficient $K_\lambda(r)$ (see Fig.1) For a given cross-section located at the altitude z the spectral transmission along a line of sight ξ perpendicular to the z axis and characterized by the impact parameter p, is

$$t_\lambda(p) = \frac{L_\lambda(p)}{Linc_\lambda(p)}$$

it is such as

$$\text{Log}\, t_\lambda(p) = -\Big[\int_{\xi.\omega=p} K_\lambda(\xi)\, d\xi\Big] \qquad (1)$$

Thus the transform

$$R\, K_\lambda(\xi.\omega) = \int_{\xi.\omega=p} K_\lambda(\xi)\, d\xi$$

is the Radon transform of K_λ and $T_\lambda(p) = -\text{Log}\,[t_\lambda(p)]$ are the projection data associated with the parameter p (Barrett,1984). R insures a correspondence between the function $K_\lambda(\xi)$ and the set of line integrals.

For a tomography of the absorption coefficient we proceed in a way completely similar to those of Herman (1980). By taking the inverse 2D Radon Transform of Eq.(1) it can be shown that

$$K_\lambda(r,\theta) = \int_0^\pi T_\lambda(p,\phi) * h(p)\, d\phi \qquad (2)$$

where * denotes the convolution product and the function h(p) is defined as

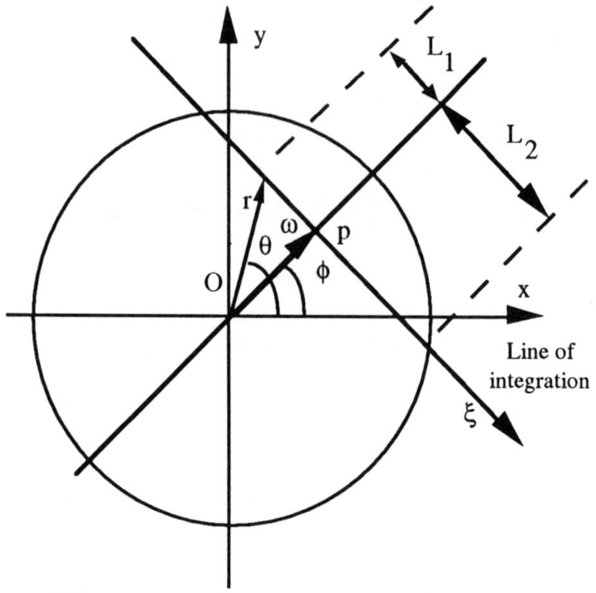

Fig.1 Geometry for the 2-D Radon Transform

$$h(p) = \int_{-\infty}^{\infty} |\sigma| \exp(2i\pi\sigma p)\, d\sigma = \frac{-1}{2\pi p^2} \qquad (3)$$

As a consequence $h * T_\lambda$ is the Hilbert Transform of T'

$$H\,[T_\lambda'] = [T'_\lambda * \frac{1}{p}] \qquad (4)$$

Due to the existence of a singularity this convolution product is evaluated by a regularization process. One defines a set of function ρ_A such as

$$\lim_{A\to\infty} T' * \rho_A = H\,[T'_\lambda] \qquad (5)$$

which can be calculated for any fixed value of A. According to Herman one takes

$$\rho_A(u) = -2 \int_0^{A/2} F_A(u) \sin(2\pi up)\, du \qquad (6)$$

F_A is the window of bandwidth A. It is a monotonically nonincreasing function of u such as $0<F_A<1$, $F_A(u)=0$ if $u \geq A/2$ and $\lim_{A\to\infty} F_A = 1$.

We shall examine two cases,

$$F_A(u)=1$$

and

$$F_A(u)= \alpha + (1-\alpha)\cos 2\pi u/A \quad \text{(Hamming Window).}$$

In view of Eq.(4), we can integrate by part and obtain ,

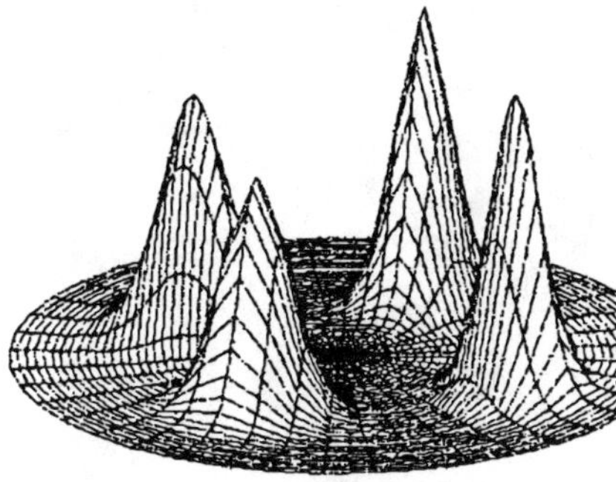

Fig.2-a Original absorption field

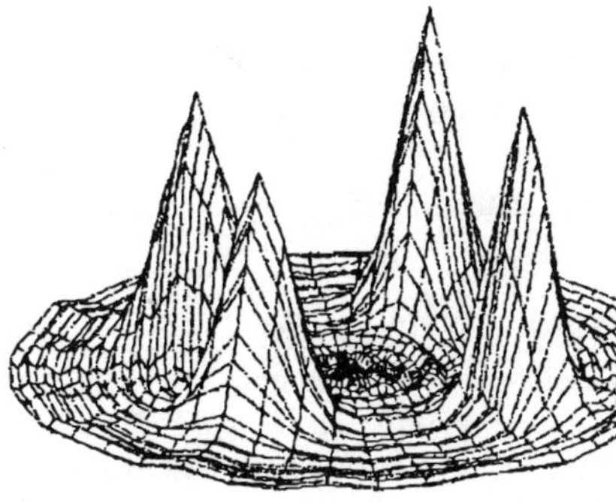

Fig.2-b Retrieved absorption field

$$T * \rho_A = T * \rho'_A$$

According to Herman (1981) , we take $A=1/\Delta p$,where Δp is sampling interval.

The final step consists of taking equally spaced projection data and to replace the integral in Eq.(2) by a Riemann sum which gives

$$K_\lambda(k\Delta r, n\Delta\theta) = \frac{-1}{2\pi} \Delta p \Delta\phi \sum_{j=0}^{M-1} \sum_{m=-N}^{N} T(m,j) \rho'_A(m\Delta p - m'\Delta p) \quad (8)$$

where $\Delta\phi = \pi / M$ and $T(m,j) = -\log[t_\lambda(m\Delta p, j\Delta\phi)]$

and according to Fig.1 the correspondence between indices (k,n) and (m,j) is given by $m\Delta p = k\Delta r \cos(n\Delta\theta - j\Delta\phi)$. In Eq.(8) ρ'_A is the derivative of ρ_A.

Data Collected from Thermal Emission

The same Semi-Transparent Medium as above is considered but this time a temperature field is established. It is easy to show that the outgoing spectral intensity, for a particular direction of observation, can be expressed as

$$L_\lambda(p,\phi) = \int_{\xi.\omega=p} K_\lambda(r,\theta) L^\circ_\lambda(r,\theta) \exp[-K_\lambda(L_1+L_2)]\, d\xi \quad (9)$$

where $L^\circ_\lambda(r,\theta)$ is the local Planck function, L_1 and L_2 the two distances shown in the Fig.1. In Eq.(4) reflections of the emited beam onto the interface are neglected. Equation (9) is the so-called attenuated Radon transform of the product $K_\lambda(r,\theta) L^\circ_\lambda(r,\theta)$. For a constant value of K_λ the Tretiak and Metz's (1980) theorem permit the inversion of this equation. We obtain (Sakami et al.,1992),

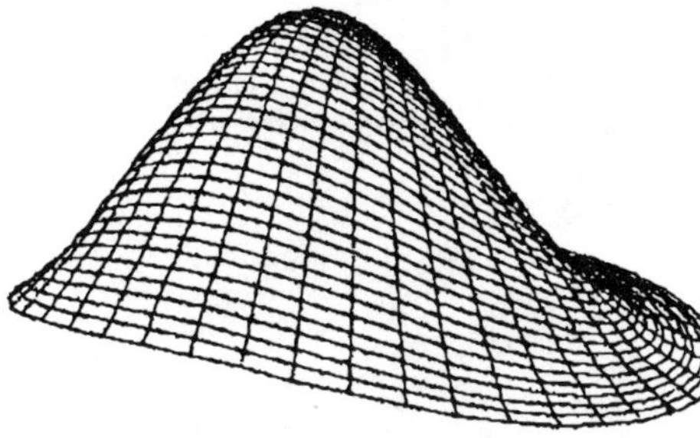

Fig.3-a Original temperature field

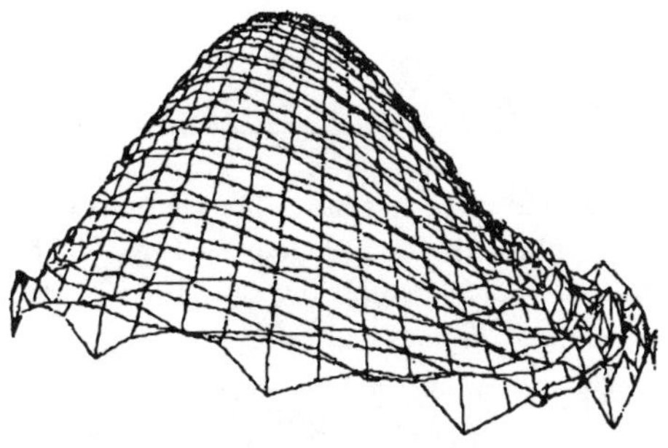

Fig.3-b Retrieved temperature field

$$L^\circ_\lambda(k\Delta r, n\Delta\theta) = \frac{1}{M\pi} \sum_{j=1}^{M-1} C(n,j) \sum_{m=-N}^{N} L(m,j)\, q(m-m') \quad (10)$$

where

$$C(n,j) = \exp K_\lambda[\, k\Delta r \ \sin(n\Delta\theta - j\Delta\phi)] ,$$

$$q(m) = \begin{cases} 0 & \text{if } m = 0 \\ [\cos(\kappa_v m\ \Delta p) + (-1)^{m+1}]/2m & \text{otherwise .} \end{cases}$$

and $L(m,j) = [\, L_\lambda(m\ \Delta p, j\ \Delta\phi) \exp(K_\lambda L_1)\,]'$

If the condition of a constant K_λ is not fulfilled , we could make the additional assumption that the medium is optically thin. Under this condition

$$L_\lambda(p,\phi) = \int_{\xi.\omega=p} K_\lambda(r,\theta) L^\circ_\lambda(r,\theta)\, (r,\theta)\, d\xi \quad (11)$$

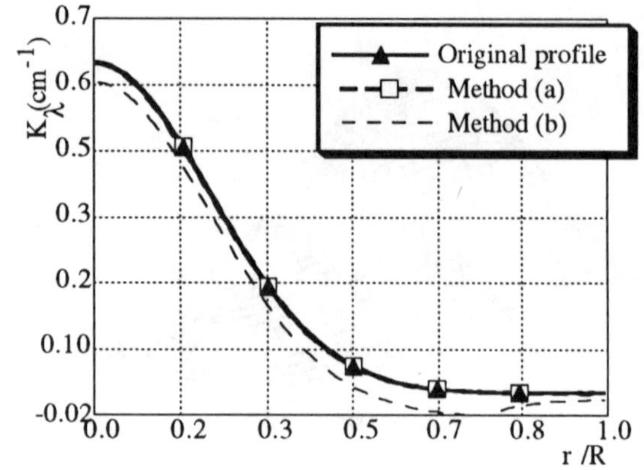

Fig. 4 Absorption profiles retrieved by methods (a) and (b)

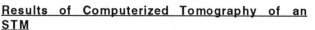

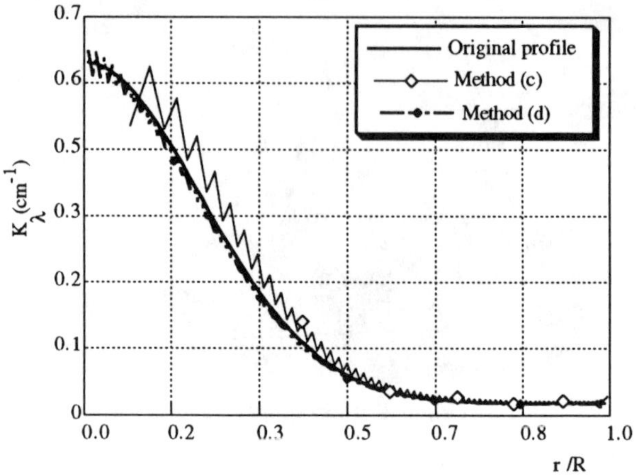

Fig. 5 Absorption profile retrieved by method (c) (α=0.07) and method (d) (δ=0.009)

Results of Computerized Tomography of an STM

A computerized tomography of four absorbing jets scattered on a plane is reproduced in Figs. 2a-b. Data transmissions were calculated by the resolution of the direct problem from 20 regular scanning orientation angles (upon π) and for 161 projections and an added artificial noise of $\pm 1\%$. Tomography of the absoption field was computed by taking the bandlimited function F_A either equal to 1 or being the generalized Hamming window. For noise free data the first one gives accurate results, but for more smeared data the second one is recommanded. It is seen by comparison with the original mapping, in Figs.2a-b, that positions of the maxima and the general shape were retrieved with a good accuracy.

Figure. 3a is represents the original temperature field set in an absorbing-emitting media of constant absorption coefficient and Fig.3 b shows the mapping of that field reconstructed by means of the inverse attenuated Radon transform. Differences between the original bi-dimensional representation and the recovered tomography are apparent mainly in the wings of the mapping; the maximum error on the z axis being about $\pm 4\%$.

RETRIEVAL OF TEMPERATURE AND ABSORPTION PROFILS IN AN AXISYMMETRIC STM

Absorption Coefficient Profiles (Problem I)

For an axisymmetric STM body Eq.(1) reduces merely to the Abel 's equation

$$-\text{Log}\,[t_\lambda(p)] = 2\int_p^R \frac{K_\lambda(r)\, r\, dr}{(r^2-p^2)^{1/2}} \qquad (12)$$

and the product $K_\lambda(r,\theta)L^\circ_\lambda(r,\theta)$ will be recovered by the same technique as in

where R is the radius of the object and p is the lateral impact parameter.

Temperature Profiles (Problem II)

Similarly, in presence of a radial temperature field T(r) the outgoing spectral intensity initiated in a STM body cross section (perpendicular to the z axis) issued from a direction repered by the lateral parameter p, is

$$L_\lambda(p) = 2\sqrt{t(p)} \int_p^R K_\lambda(r)\, L^0_\lambda\,[T(r)]\; ch\Big[\int_p^r \frac{K_\lambda(r')\, r'\, dr'}{\sqrt{r'^2-p^2}}\Big] \frac{r\, dr}{\sqrt{r^2-p^2}} \qquad (13)$$

with L°_λ [T(r)] representing the Planck function evaluated for the local temperature T(r). It is apparent that the resolution of problem II needs resolution of problem I.

Resolution Methods

a)Precondionned Data and Abel's Equation.

The absorption profile, expressed by Eq.(12), may be recovered by an Abel's inversion, which can be expressed as

$$K_\lambda(r) = \frac{1}{\pi}\int \frac{[\text{Log}\, t_\lambda(p)]'}{\sqrt{r^2-p^2}}\, dp \qquad (14)$$

where [Log $t_\lambda(p)$]' denoting the derivative of [Log $t_\lambda(p)$] with respect to p.

But application of Eq.(14) requires some precautions and data have to be prepared. In fact, for a real object the experimental response is not truly symmetric due to faults in experimental conditions and presence of noise. Thus a

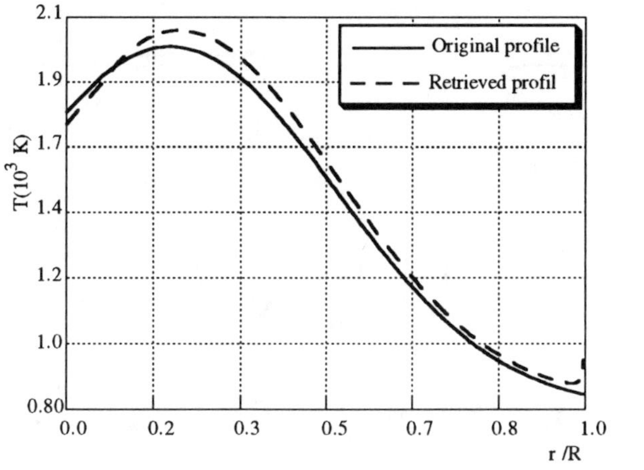

Fig. 6 Temperature profile retrieved by method c (α=0.06)

preconditionning of data is needed; it involves the following processing proposed by Dong and Kearney (1991),

* taking the discrete Fourier Transform of the data,
* converting the complex data in the Fourier space to real values. In fact, the exixtence of non symmetric data yields to an imaginary antisymmetric Fourier Transform, then the known dephazage can be used to obtain symmetrized data,
* filtering the data by using a low pass-filter which eliminated high frequency noise but preserving the information contained in the low frequency part
* taking inverse Fourier Transform of the symmetrized data,
* interpolation by cubic splines,
* taking the derivative of Log of the resulting data.

The final step consists of calculating numerically the Abel's integral by a Gauss quadrature with adaptative abscissa.

b)Fourier-Bessel Inversion Method. Equation (12) may also be be written as

$$-\mathrm{Log}\, t_\lambda(p) = 2 \int_{\xi.\omega=p} K_\lambda(\sqrt{r^2 + p^2})\, d\xi$$

which may be expressed as

$$I(\sigma)=\mathbf{F}\,[-\mathrm{Log}\, t_\lambda(p)\,] = 2\pi\int K_\lambda(r)J_0(2\pi r\sigma)r dr \qquad (15)$$

$J_0(2\pi r\sigma)$ being the zero-order Bessel function. Equation (15) is the so-called Hankel or Fourier-Bessel Transform and may be inversed as

$$K_\lambda(r) = 2\pi\int \sigma\, \mathbf{I}(\sigma)\, J_0(2\pi r\sigma)d\sigma$$

or

$$K_\lambda(r) = 2\int_0^{\pi/2}\int_{-\infty}^{\infty} \sigma\, \mathbf{I}(\sigma)\, \exp(2\pi i\sigma r\cos\theta)d\sigma\, d\theta \qquad (16)$$

this latter integral being calculated by the Candel (1981) algorithm for the assessment of which we use the preprocessing data method outlined above.

c)Regularization and conjugated gradient method. Equations.(6 and 8) are Volterra equations of first kind types. When discretized they can be put in the form

$$\mathbf{A}\ u = f \qquad (17)$$

with $\mathbf{A}$ denoting a linear operator depending on the discretization and on the kernel form.The matrix element A_{ij} corresponding to Eqs.(12 and 13) being respectively

$$Aij = \frac{h^{1/2}}{(j+1-i)^{1/2}} \qquad \text{if } j > i-1 \text{ else } Aij = 0$$

$$Aij = \frac{h^{1/2}}{(j+1-i)^{1/2}}\, ch\left[\int_{\sqrt{ih}}^{\sqrt{jh}} \frac{K_\lambda(r')\, r'\, dr'}{(r'^2-ih)^{1/2}}\right] \quad \text{if } j> i-1 \text{ else } Aij = 0$$

Under these conditions,the determination of the unknown profile **u** from the f data's is a ill-posed problem.

Conside

$$\mathbf{A}^*h = 0 \qquad (17')$$

we search for a regularized solution u_α which renders minimal the Tikhonov function

$$J(u_\alpha) = \frac{1}{2}\,[\|\, \mathbf{A}u_\alpha - f\, \|^2 + \alpha^2\|u_\alpha\|^2]$$

Murio and Mejia (1991) have shown that the regularized solution can be carried out from the following equivalent iterative coupled equation system

$$\begin{cases} \alpha h_n = \mathbf{A}\, u_n - f \\ u_{n+1} = u_n - \beta_n[\alpha^2 u_n + \alpha \mathbf{A}^* h_n] \end{cases} \qquad (18)$$

But as it can be noticed that the bracketted part of Eq.(18) is just the grad $J(u_n)$; thus the second equation of the system is the one we can find when using the congugated gradient method.

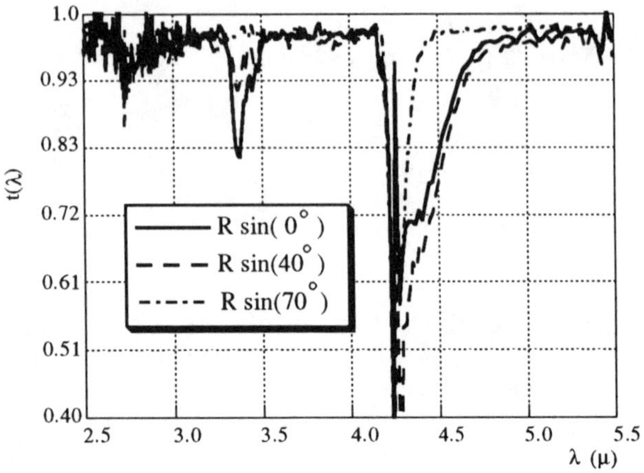

Fig.7 The infrared transmission spectra of a propane-air flame for various p parameters

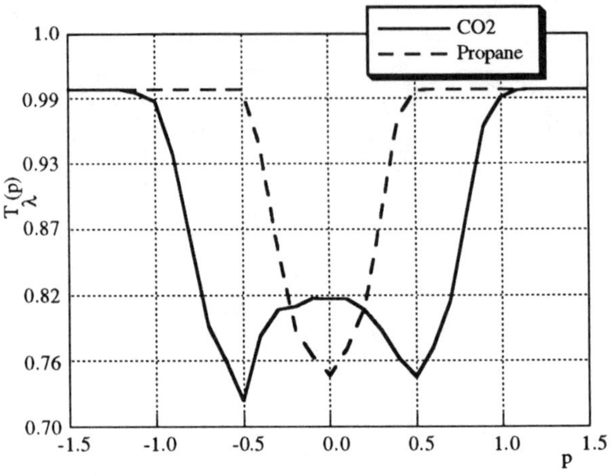

Fig.9 Angular transmission of propane and CO2

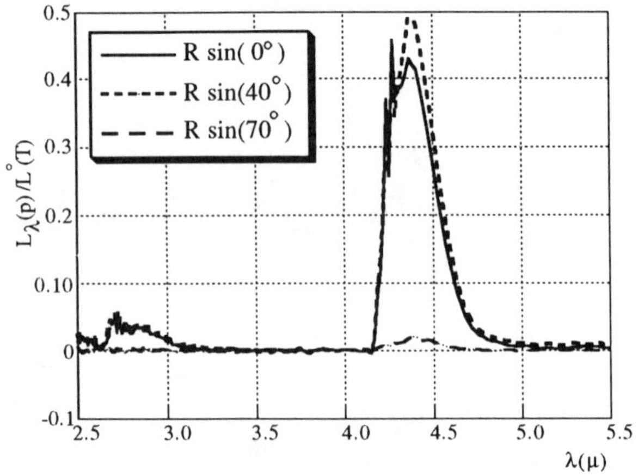

Fig.8 The emission spectra of spectra of a propane-air flame for various p parameters

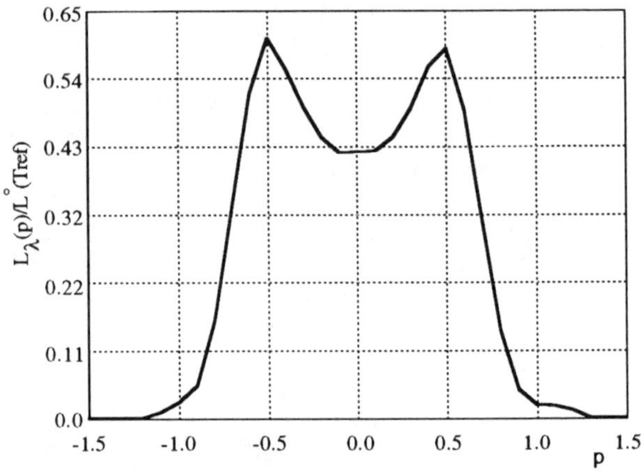

Fig.10 Angular relative emission of CO2 (Tref = 1275 K)

d)Mollification method. This method proposed by Murio (1992) replaces in Eq.(12) the term $[-\text{Log } t_\lambda(u)]'$ by $[-\text{Log } t_\lambda * p_\delta]'$ where

$$p_\delta(x) = \frac{1}{\delta} \pi^{-1/2} \exp(-x^2/\delta^2)$$

with δ the radius of mollification determined by the condition (Murio,1987)

$$\| \text{Log } t_\lambda * p_\delta - \text{Log } t \|^2 = \varepsilon \quad (19)$$

Simulations results

The presented inversion methods have been tested for simulation transmission and emission data coming from numerical resolution of the direct problems for given absorption and temperature profiles setting in an axisymmetric semi-transparent medium ; afterward the data are smeared by an added noise of ± 3%. In Figs.4 and 5 the absorption profile has been retrieved by methods *a,b,c* and *d*, and the results can be compared with the original absorption profile. The comparison is very favorable for inversion treated by methods *a* and *d*. The reconstruction of the temperature profile resolving the complete Eq. 13 through the regularization method (with noise on t_λ but without noise on $L_\lambda(p)$) is shown in Fig.6; again a good agreement is obtained.

APPLICATION TO A FLAME OF PREMIXTURE

Experimental Setup

The inversion methods have been applied to reconstruct absorption and temperature profiles in a premixture flame of propane-air. The experimental setup is composed of

* a Fourier Transform Infrared Spectrometer (Perkin Elmer 1760X),
* a black body,
* a burner (Bunzen type) of diameter 1.2 cm ,

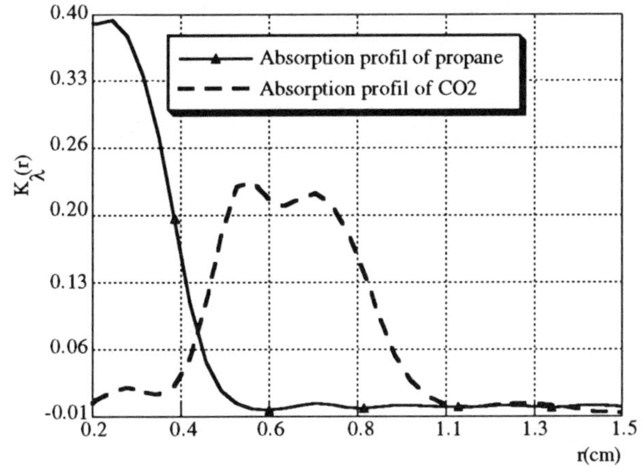

Fig.11 Retrieved absorption profil of propane and CO2

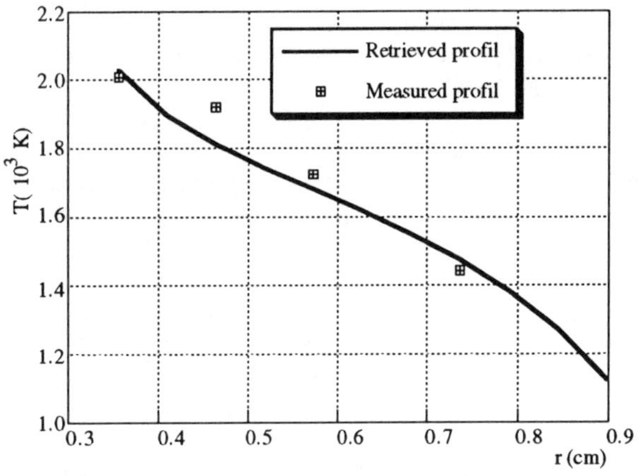

Fig.12 Retrieved temperature profil of CO2

* a set of mirrors (off axis and on axis) and a set of pin holes,
* a micro-displacement device supporting the burner,

The generated flame is fed with 99% pure propane and air with regularized flow rates.

Some advantages of a FTIR Spectrometer over conventional dispersive devices or filter-detectors systems, are

* multiplexing,
* constant resolution,
* possibility of improvement of Signal/Noise ratio by scanning accumulation,
* use of software and routines.

Lateral Scanning of Infrared Absorption and Emission Spectra

The absorption and emission spectra of the flame were recorded at a resolution of 4 cm^{-1} at a fixed altitude position for several angular positions p according to the following experimental steps,

* recording of the emission spectrum of the black body source at a known temperature (in general 1275 K), ($S_{\lambda 1}$),
* recording of the emission spectrum of the flame ($S_{\lambda 2}$),
* recording of the transmission across the flame ($S_{\lambda 3}$).

The corresponding transmittance and relative emittance functions are

$$t_\lambda(p) = \frac{S_{\lambda 3}(p)-S_{\lambda 2}(p)}{S_{\lambda 1}(p)} \quad (20)$$

$$e_\lambda(p) = \frac{S_{\lambda 2}(p)}{S_{\lambda 1}(p)} \quad (21)$$

Since the spectra have been recorded at a low resolution the measured transmission and emission are in fact spectral averages over a finite spectral interval and in general due to spectral correlations they cannot yield to a correct mean absorption coefficient or $K_\lambda L^0_\lambda$ function, but when the medium is optically thin ($\tau_\lambda = K_\lambda d << 1$) the influene of these correlations can be neglected.

In Figs.7 and 8 are displayed a transmission and an emission spectra of the propane-air flame for three choosen angular positions taken at altitude position z=0 (bottom part); one can observe in Fig.7 for projections in the vicinity of the z axis both the propane band in the range of 2960 cm-1 and the CO2 band at 2340 cm-1, whereas for the external part only the CO2 band subsists. The CO2 absorption band of the flame oberved by transmission at the bottom of the flame is represented in Fig.8, it was taken in the same experimental conditions as those for the emission spectra. Utilization of these spectra at a prescribed frequency (2980 cm-1 for propane, 2280 cm-1 for CO2) for each of the 32 selected channel yields the spectral transmittance and relative emittance input data needed for inversion. For the lower part of the flame they are represented as a function of the angular parameters in Figs.9 and 10. The non-symmetric shape of the raw experimental results should be noticed .

Inversion results

Retrieval of the absorption profiles of the flame at the base of the burner is shown in Fig.9. It has been determined by inversions *a*, i.e by the Abel's integral inversion after symmetrization and filtering of the data. One observes in the central part of the graph the propane distribution alone, then the annular distribution of CO2.

Finally in Fig.10 is represented the reconstructed radial temperature profile at the flame base obtained by Abel's inversion for an optically thin medium. The result is compared with the temperature measurements obtained by a Pt/PtRh thermocouple.

REFERENCES

Barrett, H.H.,1984, "The Radon Transform and its Applications," *Progress in Optics* Wolf E. Ed. Elsevier Sciences Publishers.

Candel, S.,1981, " An Algorithm for the Fourier-Bessel Transform," *Comput.Phys.Commun.* Vol.19,pp 343-353.

Dond J. and Kearney R.J., 1991, "Symmetrizing, filtering and Abel inversion using Fourier transform techniques",*J.Quant.Spectrosc.Radiat.Transfer*,Vol.46,pp141 -149.

Herman G.T, 1980, "Image reconstruction from projections," Academic Press, N.Y.

Murio D.A.,1987, "Automatic numerical differentiation by discrete mollification," *Comput.Math.Applic.* Vol.13,pp381-386, .

Murio D.A and Mejia C.E.,1991,"Comparison of four stable numerical Methods for Abel's Integral Equation," *Proceedings of the Third International Conference on Inverse Design Concept and Optimization in Engineering Sciences* ; G.S.Dulikavich, Washington D.C

Murio D.A, 1992," Stable Numerical Inversion of Abel's Integral Equation by Discrete Mollification," Theoretical Aspect of Industrial Design. Field D.A and Komkov V., SIAM, pp.92-104.

Sakami M., Sandhita E.,Saulnier J.B. and Lallemand M., 1992, "Computerized tomography of asymmetric thermal structure in transparent and semi-transparent media,". *Eurotherm Seminar* 21, Lyon .

Tretiak, O. and Metz C.,1980 "The exponential Radon Transform,". *SIAM J.Appl.Math.* Vol.39, pp.341.

Inverse Problems in Engineering: Theory and Practice
ASME 1993

AN INVERSE PROBLEM OF ESTIMATING THERMAL CONDUCTIVITY, OPTICAL THICKNESS, AND SINGLE SCATTERING ALBEDO OF A SEMI-TRANSPARENT MEDIUM

A. J. Silva Neto† and M. N. Özişik
Department of Mechanical and Aerospace Engineering
North Carolina State University
Raleigh, North Carolina

ABSTRACT

An inverse problem of simultaneous conduction and radiation is used for estimating the optical thickness, single scattering albedo, and the thermal conductivity of a plane-parallel semi-transparent slab. The input data used in the inverse analysis consists of simulated measured angular distribution of the transmitted radiation intensity and the temperature inside the medium.

The Levenberg-Marquardt method is used to solve the nonlinear system of algebraic equations resulting from the optimization procedure. The method works well so long as the optical thickness of the medium is sufficiently thin in order to provide a measurable transmitted radiation intensity.

The confidence bounds are established for the estimated values of the parameters.

NOMENCLATURE

e	normally distributed random number in Eq.(32)
E_n	exponential integral functions
G^*	dimensionless incident radiation
I	number of radiation intensity measurements
J	number of temperature measurements
k	thermal conductivity
L	highest order for the expansion of G^*, Eq.(7)
N	conduction-to-radiation parameter, Eq.(2b)
P	vector of unknowns, Eq.(16)
R	dimensionless source term for the radiation problem, Eq.(4)
S	least squares norm, Eq.(17)
X_i	dimensionless measured exit radiation intensity
Z_j	dimensionless measured temperature

Greek symbols

ε	tolerance for the stopping criteria, Eq.(29)
Θ	dimensionless temperature
λ	damping parameter
μ	direction cosine
σ	standard deviation of measurement errors
$\bar{\sigma}$	Stefan-Boltzmann constant
τ	optical variable
τ_0	optical thickness
φ	dimensionless radiation intensity
ω	single scattering albedo

I. INTRODUCTION

The inverse analysis of radiation in a participating medium has been studied extensively for a variety of practical applications, including, among others, remote sensing for the atmospheric or aerosol properties (Ustinov,1977;Freund,1983), temperature measurement in semitransparent materials at high temperature such as encountered in glass manufacturing (Van Laethem et al.,1961), combustion systems and estimation of radiative properties of materials (Roux and Smith,1981; Sacadura et al.,1986;Dobkin and Son,1991;Silva Neto and Özişik,1992). On the other hand, a recent review on the subject (McCormick,1992) reveals that, only a limited amount of work is available on the inverse problem of simultaneous conduction and radiation.

In the present work, an inverse analysis of simultaneous steady-state conduction and radiation is applied to estimate the optical thickness, the single scattering albedo and the thermal conductivity of a semi-transparent medium from the knowledge of transmitted exit radiation intensity and interior temperature of the medium.

† Permanent address : Promon Engenharia, Av. Pres. Juscelino Kubitschek 1830, São Paulo, Brazil 04543

II. THE FORMULATION OF THE DIRECT PROBLEM

Consider one-dimensional, steady-state combined conduction and radiation in an absorbing, emitting, isotropically scattering slab of optical thickness τ_0, with transparent boundaries and subjected to an external isotropic irradiation at the boundary $\tau = 0$, as illustrated in Fig.1. The boundaries $\tau = 0$ and $\tau = \tau_0$ are kept at prescribed temperatures T_1 and T_2, respectively.

The mathematical formulation of this problem is given, in dimensionless form, as

$$\frac{d^2\Theta}{d\tau^2} - \frac{(1-\omega)}{N}\left[\Theta^4(\tau) - G^*(\tau)\right] = 0, \quad \text{in } 0 < \tau < \tau_0 \tag{1a}$$

$$\Theta = 1 \qquad \text{at } \tau = 0 \tag{1b}$$

$$\Theta = \Theta_2 \qquad \text{at } \tau = \tau_0 \tag{1c}$$

where the dimensionless variables are defined as

$$G^*(\tau) = \frac{1}{2}\int_{-1}^{1} \varphi(\tau,\mu)\, d\mu = \text{dimensionless incident radiation} \tag{2a}$$

$$N = \frac{k\beta}{4n^2\bar{\sigma}T_1^3} = \text{conduction} - \text{to} - \text{radiation parameter} \tag{2b}$$

$$\Theta = \frac{T}{T_1} = \text{dimensionless temperature} \tag{2c}$$

Here, τ is the optical variable, μ is the direction cosine of the radiation beam with the positive τ axis, ω is the single scattering albedo, k is the thermal conductivity, β is the extinction coefficient, n is the refractive index, $\bar{\sigma}$ is the Stefan-Boltzmann constant and $\Theta_2 < 1$. The dimensionless radiation intensity, $\varphi(\tau,\mu)$, is determined from the solution of the radiation problem

$$\mu\frac{\partial\varphi(\tau,\mu)}{\partial\tau} + \varphi(\tau,\mu) = R(\Theta) + \frac{\omega}{2}\int_{-1}^{1}\varphi(\tau,\mu')\, d\mu'$$

$$\text{in } 0 < \tau < \tau_0\,,\ -1 \le \mu \le 1\,, \tag{3a}$$

$$\varphi(0,\mu) = 1\,, \qquad \mu > 0\,, \tag{3b}$$

$$\varphi(\tau_0,-\mu) = 0\,, \qquad \mu > 0\,, \tag{3c}$$

where the dimensionless temperature dependent source term $R(\Theta)$ is defined as

$$R(\Theta) = (1-\omega)\,\Theta^4(\tau)\,. \tag{4}$$

Equations (1) and (3) give a complete mathematical formulation for the steady-state simultaneous conduction and radiation problem in a plane-parallel medium. They are to be solved simultaneously by an iterative process, because the heat conduction problem (1) requires the knowledge of the radiation intensity

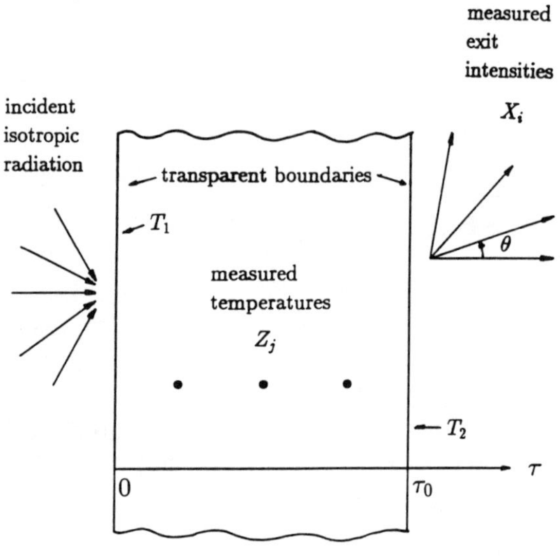

Figure 1. Schematic representation of the physical system and coordinates.

through the incident radiation $G^*(\tau)$, and the radiative transfer problem (3) requires the knowledge of the temperature through the source term $R(\Theta)$.

The integral form of the radiation problem (3) is given as (Özişik,1973)

$$G^*(\tau) = Y(\tau) + \frac{\omega}{2}\int_{\tau'=0}^{\tau_0} E_1(|\tau-\tau'|)\, G^*(\tau')\, d\tau' \tag{5}$$

where

$$Y(\tau) = \frac{1}{2}\Big[E_2(\tau) + \Theta_2^4(\tau)\, E_2(\tau_0-\tau) + \int_{\tau'=0}^{\tau_0} R(\tau')\, E_1(|\tau-\tau'|)\, d\tau'\Big] \tag{6}$$

where $E_n(\tau)$ are the exponential integral functions (Le Caine,1947).

In this work, the Galerkin method (Özişik and Yener,1982; Cengel,Özişik and Yener,1984) is used to solve the integral form of the radiation problem given by Eqs.(5) and (6). In this approach $G^*(\tau)$ is represented as a power series in optical variable as

$$G^*(\tau) = \sum_{n=0}^{L} c_n \tau^n \tag{7}$$

where the unknown expansion coefficients c_n are determined in the following manner. Let

$$\mathcal{L}\{G^*(\tau)\} \equiv G^*(\tau) - Y(\tau) - \frac{\omega}{2}\int_{\tau'=0}^{\tau_0} E_1(|\tau-\tau'|)\, G^*(\tau')\, d\tau' = 0 \tag{8}$$

Applying the Galerkin method (Özişik and Yener,1982; Cengel, Özişik and Yener,1984)

$$\int_0^{\tau_0} \mathcal{L}\{G^*(\tau)\}\, \tau^m \, d\tau \;=\; \int_0^{\tau_0} \mathcal{L}\left\{\sum_{n=0}^{L} c_n \tau^n\right\} \tau^m \, d\tau \;=\; 0\,, \qquad m = 0,1,2,\ldots,L \tag{9}$$

we obtain the following $L+1$ algebraic equations for the determination of $L+1$ unknown expansion coefficients c_n:

$$\sum_{n=0}^{L} c_n\, b_{mn} \;=\; d_m\,, \qquad m = 0,1,2,\ldots,L \tag{10}$$

where

$$b_{mn} \;=\; \frac{\tau_0^{n+m+1}}{n+m+1} - \frac{\omega}{2} T_{mn} \tag{11}$$

$$d_m \;=\; \int_{\tau=0}^{\tau_0} Y(\tau)\,\tau^m\, d\tau \;=\; \frac{1}{2}\Big\{T_m + \Theta_2^4 T_m^* + \int_{\tau=0}^{\tau_0}\int_{\tau'=0}^{\tau_0} R(\tau')\, E_1(|\,\tau-\tau'\,|)\,\tau^m\, d\tau'\, d\tau\Big\} \tag{12}$$

$$T_{mn} \;=\; \int_{\tau=0}^{\tau_0}\int_{\tau'=0}^{\tau_0} E_1(|\,\tau-\tau'\,|)\,\tau'^{\,n}\,\tau^{\,m}\, d\tau'\, d\tau \tag{13}$$

$$T_m \;=\; \int_{\tau=0}^{\tau_0} E_2(\tau)\,\tau^m\, d\tau \tag{14}$$

$$T_m^* \;=\; \int_{\tau=0}^{\tau_0} E_2(\tau_0-\tau)\,\tau^m\, d\tau \tag{15}$$

The analytical expressions for various integrals in equations (11-15) are available (Özişik and Yener,1982; LeCaine,1947).

Once the expansion coefficients c_n are determined from the solution of system (10) for a given temperature distribution, the dimensionless incident radiation $G^*(\tau)$ become known and a new temperature distribution is determined from the solution of the heat conduction problem (1). A finite difference scheme with an SOR is used to solve problem (1). For the conduction-to-radiation parameter less than about $N = 0.05$, some convergence difficulty was experienced in the numerical computations. For such cases the difficulty could be alleviated by performing the calculations with a larger value for N and using the results thus obtained as an initial value for the next iteration and gradually reducing the value of N in the subsequent calculations. These iterations are continued until N reaches the desired value.

III. THE FORMULATION OF THE INVERSE PROBLEM

The inverse problem is similar to the above direct problem except the optical thickness τ_0, the single scattering albedo ω and the thermal conductivity k (or N), are considered unknown. Instead, measured transmitted exit intensities, X_i, at boundary $\tau = \tau_0$, at different polar angles, $i = 1,2,\ldots,I$, and measured temperatures, Z_j, at different locations, $j = 1,2,\ldots,J$, are available, as shown in Fig.1. Then, the inverse problem can be stated as: By utilizing measured intensity data $\{X_i\}$, $i = 1,2,\ldots,I$, and the temperature data $\{Z_j\}$, $j = 1,2,\ldots,J$, determine τ_0, ω, and N, denoted by

$$\underset{\sim}{P} \;=\; \begin{Bmatrix} \tau_0 \\ \omega \\ N \end{Bmatrix} \tag{16}$$

As the number of measured data $I+J$ is larger than the number of parameters to be estimated, the problem is overdetermined. One way to solve such problems is to require that the unknown quantities minimize a norm. Here we consider a least squares norm

$$S \;=\; \sum_{i=1}^{I} [X_i - \varphi(\tau_0,\mu_i)]^2 + \sum_{j=1}^{J} [Z_j - \Theta(\tau_j)]^2 \;=\; \underset{\sim}{F}^T \underset{\sim}{F} + \underset{\sim}{H}^T \underset{\sim}{H} \tag{17}$$

where the elements of $\underset{\sim}{F}$ and $\underset{\sim}{H}$ are defined as

$$F_i \;=\; X_i - \varphi(\tau_0,\mu_i)\,, \qquad i = 1,2,\ldots,I \tag{18}$$

$$H_j \;=\; Z_j - \Theta(\tau_j)\,, \qquad j = 1,2,\ldots,J \tag{19}$$

X_i and $\varphi(\tau_0,\mu_i)$ are the measured and calculated exit intensities, respectively, and Z_j and $\Theta(\tau_j)$ are the measured and calculated temperatures, respectively.

To minimize S, Eq.(17) is differentiated with respect to each of the unknown parameters, $\underset{\sim}{P} \equiv (\tau_0,\omega,N)^T$ as

$$\frac{\partial S}{\partial \underset{\sim}{P}} \;=\; 0 \tag{20}$$

Equation (20) is linearized by expanding $\varphi(\tau_0,\mu_i)$ and $\Theta(\tau_j)$ in Taylor series and retaining the first order terms. Then a damping parameter λ is added to the resulting expression to improve convergence, leading to the Levenberg-Marquardt method (Marquardt,1963) given by

$$(C + \mathcal{I}\lambda)\,\Delta \underset{\sim}{P} \;=\; \underset{\sim}{D} \tag{21}$$

where $\mathcal{I}$ is the identity matrix,

$$C \;=\; A^T A + B^T B \tag{22}$$

$$\underset{\sim}{D} \;=\; A^T \underset{\sim}{F} + B^T \underset{\sim}{H} \tag{23}$$

the elements of matrices A and B are given by

$$A_{ml} \;=\; \frac{\partial \varphi(\tau_0,\mu_m)}{\partial P_l} \tag{24}$$

$$B_{ml} \;=\; \frac{\partial \Theta(\tau_m)}{\partial P_l} \tag{25}$$

and

$$\Delta \underset{\sim}{P}^k = \underset{\sim}{P}^{k+1} - \underset{\sim}{P}^k . \tag{26}$$

Equation (21) is now written in a form suitable for iterative calculations as

$$\Delta \underset{\sim}{P}^k = \left[C^k + \mathcal{I}\lambda^k \right]^{-1} \underset{\sim}{D}^k \tag{27}$$

and $\underset{\sim}{P}^{k+1}$ is determined from Eq.(26),

$$\underset{\sim}{P}^{k+1} = \underset{\sim}{P}^k + \Delta \underset{\sim}{P}^k \tag{28}$$

The solution algorithm starts with a suitable initial guess $\underset{\sim}{P}^0$. Then, $\Delta \underset{\sim}{P}^k$ and $\underset{\sim}{P}^k$ are calculated iteratively using Eqs.(27) and (28), until the following convergence criterion is satisfied

$$\Delta P_l^k < \varepsilon \qquad \text{for } l = 1, 2 \text{ or } 3 \tag{29}$$

where ε is a small number (i.e. 10^{-5}).

IV. CONFIDENCE BOUNDS

The confidence bounds for the estimated parameters P can be established by following a procedure discussed by Huang and Özişik (1990). For normally distributed measurement errors with zero mean and constant variance, the standard deviation of the estimated parameters, $\sigma_{\underset{\sim}{P}}$, is determined as

$$\sigma_{\underset{\sim}{P}} = \sigma \left\{ \mathrm{diag} \left[\left(\frac{\partial \varphi^T}{\partial P} \right) \left(\frac{\partial \varphi}{\partial P^T} \right) + \left(\frac{\partial \Theta^T}{\partial P} \right) \left(\frac{\partial \Theta}{\partial P^T} \right) \right]^{-1} \right\}^{\frac{1}{2}} \tag{30}$$

If we assume a normal (Gaussian) distribution for measurement errors and the 99 % confidence, then the resulting bounds for the estimated quantities, P_l, are determined as (Flach and Özişik, 1989)

$$(P_l - 2.576\, \sigma_{P_l}) < P_l < (P_l + 2.576\, \sigma_{P_l}) \tag{31}$$

Equation (31) defines the approximate confidence bounds for the estimated parameters. That is we expect the estimated values of the parameters to lie between this two bounds with a 99 % confidence level.

V. RESULTS AND DISCUSSION

In order to examine the accuracy of the inverse method of analysis considered here for the estimation of the optical thickness, τ_0, the single scattering albedo, ω, and the conduction-to-radiation parameter, N, several test cases have been studied.

To simulate the measured data, Q_{meas} (i.e. transmitted intensities and temperatures), containing measurement errors, random errors of normal distribution and of standard deviation σ are added to the exact quantities, Q_{exact}, as

$$Q_{meas} = Q_{exact} + \sigma\, e \tag{32}$$

where Q stands for temperature or exit intensity. For normally distributed random errors, there is a 99 % probability of e lying in the range

$$-2.576 < e < 2.576 \tag{33}$$

The IMSL subroutine RNNOR (IMSL,1987) is used to generate the e values randomly.

Several test cases were run using input data with and without measurement errors. For the test cases with no measurement errors (i.e. $\sigma = 0$), the exact values for the parameters were recovered.

Figure 2 shows the estimated values of τ, ω and N for the test case of $\tau_0 = 1.0$, $\omega = 0.1$ and $N = 0.1$, using initial guesses $\tau_0^0 = 0.1$, $\omega^0 = 0.7$ and $N^0 = 0.7$, and input data with a standard deviation $\sigma = 0.01$ which represents a 5 % error. Convergence could be achieved within 15 iterations of the Levenberg-Marquardt algorithm, using an initial value for the damping parameter $\lambda^0 = 1$. This initial value of λ worked well with all the test cases studied here. Once the calculations are started, the values of λ are adjusted from iteration to iteration according to the criteria originally suggested by Marquardt (1963).

The several runs presented in Fig. 2 were obtained using different sets of random numbers, e, in Eq.(32). A CPU time of about three minutes was required for each run on the CRAY Y-MP computer.

The sensitivity coefficients, $\frac{\partial \varphi}{\partial \underset{\sim}{P}}$ and $\frac{\partial \Theta}{\partial \underset{\sim}{P}}$, for τ_0 and N were found to be larger than those for ω. This fact is reflected in Fig.2, since the confidence bounds for τ_0 and N are narrower than those for ω.

Figure 3 shows the estimation of τ_0, ω and N for the test case of $\tau_0 = 1.0$, $\omega = 0.1$ and $N = 0.5$, using initial guesses $\tau_0^0 = 0.1$, $\omega^0 = 0.7$ and $N^0 = 0.2$, and the standard deviation of measurement errors $\sigma = 0.01$, which represents an error of 5 % on the input data. Convergence was achieved within 20 iterations with the Levenberg-Marquardt algorithm.

The confidence bounds for τ_0 are narrower than those for ω and N, which implies a stronger dependence of the problem on τ_0 than on ω or N. This is a consequence of larger values of sensitivity coefficients with respect to τ_0.

Figure 4 presents the estimations for the test case $\tau_0 = 1.0$, $\omega = 0.9$ and $N = 0.05$, using as initial guesses $\tau_0^0 = 0.1$, $\omega^0 = 0.7$ and $N^0 = 0.2$. The standard deviation of measurement errors, taken as $\sigma = 0.01$ for this test case, represents an error of 4 % on the input data. Convergence was achieved within 15 iterations. Each run presented in Fig.4 required about 4 minutes of CPU time on the CRAY Y-MP computer.

The strong dependence of the problem on the single scattering albedo, ω, is reflected by its narrower confidence bounds.

Figures 5 and 6 show the effects of standard deviation of measurement errors on the estimations of τ_0, ω and N, for the test case of $\tau_0 = 1.0$, $\omega = 0.9$ and $N = 0.5$, using initial guesses $\tau_0^0 = 0.1$, $\omega^0 = 0.2$ and $N^0 = 0.15$. The standard deviations of measurement errors $\sigma = 0.01$ and $\sigma = 0.0025$ used in Figs.5 and 6, respectively,

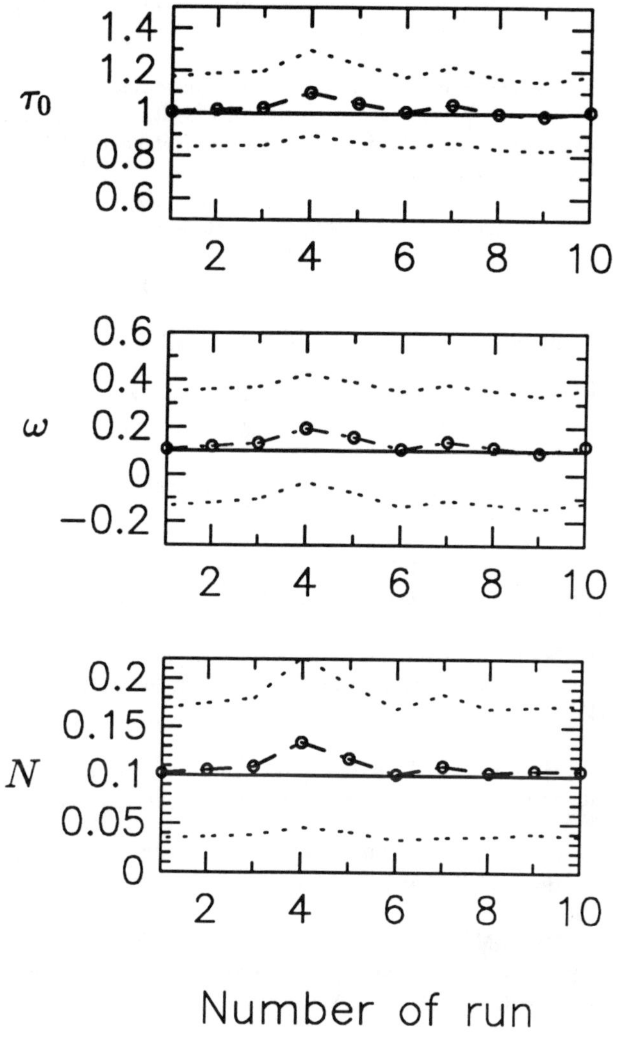

Figure 2. Estimations and confidence bounds for case $\tau_0 = 1.0$, $\omega = 0.1$ and $N = 0.1$. $\sigma = 0.01$.

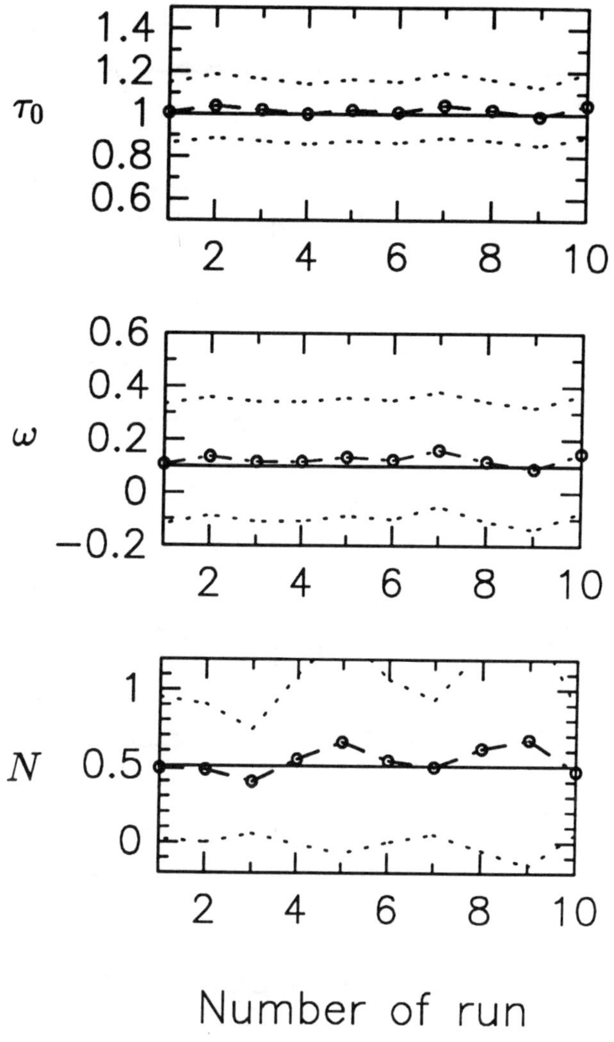

Figure 3. Estimations and confidence bounds for case $\tau_0 = 1.0$, $\omega = 0.1$ and $N = 0.5$. $\sigma = 0.01$.

correspond to errors of 4 % and 1 % respectively. As expected the estimations improve when the input data is more accurate. These figures also show that the confidence bounds for τ_0 and ω are narrower than those for N. The test cases shown in Figs. 5 and 6 require 30 and 20 iterations, respectively, to achieve convergence.

With measurement errors in the input data, convergence difficulty was experienced for the test cases with small values of N (i.e. $N < 0.05$) and for large optical thicknesses leading to total average transmitance of the plate less than about 0.2.

For all the test cases considered in this work, seven exit intensities and one temperature measurements were considered, i.e. $I = 7$ and $J = 1$ in Eq.(17). The exit intensity measurements were taken evenly spaced between the polar angles $\theta = 0$ and $\theta = 90^0$. The temperature measurement was taken at the location $\frac{\tau}{\tau_0} = 0.4$. The use of more exit intensity and temperature measurements did not seem to improve the estimations.

VI. CONCLUSIONS

An inverse analysis of simultaneous conduction and radiation is applied to estimate the optical thickness, the single scattering albedo and the thermal conductivity of a semi-transparent plane-parallel medium. The methodology works well so long as the optical thickness of the medium is sufficiently small to allow enough energy to be transmitted through the plate, in order to provide a measurable transmitted radiation intensity.

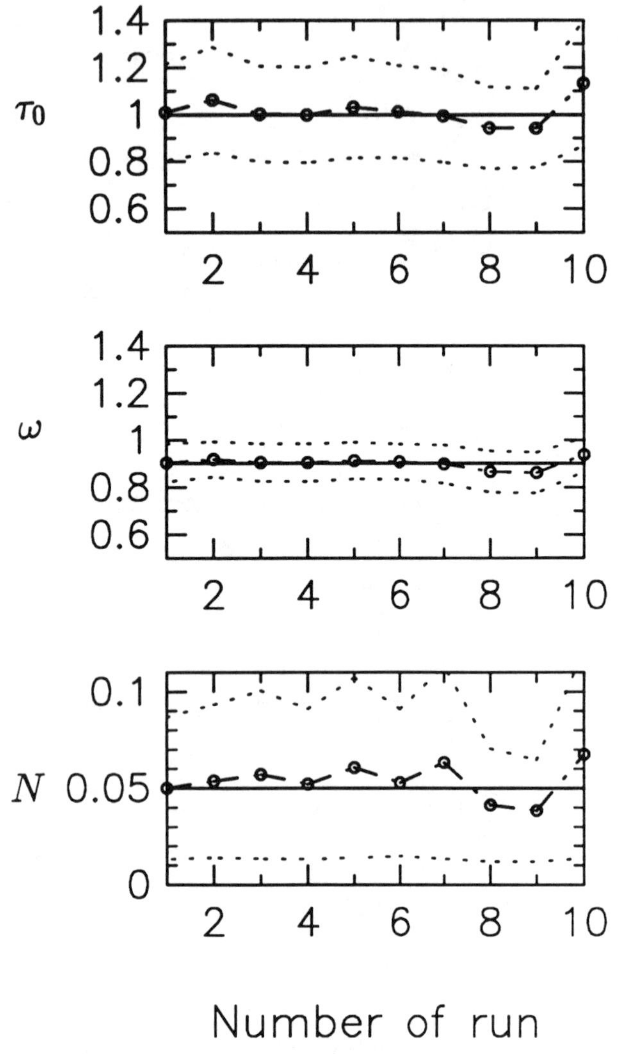

Figure 4. Estimations and confidence bounds for case $\tau_0 = 1.0$, $\omega = 0.9$ and $N = 0.05$. $\sigma = 0.01$.

ACKNOWLEDGEMENTS

The authors wish to acknowledge the North Carolina Supercomputing Center for providing time on the CRAY Y-MP computer.

One of the authors (AJSN) acknowledges the financial support given by CNPq - Conselho Nacional de Desenvolvimento Científico e Tecnológico and Promon Engenharia.

REFERENCES

Cengel,Y.A., Özişik,M.N. and Yener,Y.,1984,"Determination of Angular Distribution of Radiation in an Isotropically Scattering Slab",*Journal of Heat Transfer*,Vol.106, pp.248-252.

Dobkin,S.V., and Son,E.E.,1991,"Determining the Radiative Thermal Conductivity of Uranium Hexafluoride by Reactor Heating", *J.High Temper.*, Vol.29,No3,pp.468-473.

Flach,G.P. and Özişik,M.N.,1989,"Inverse Heat Conduction Problem of Simultaneously Estimating Spatially Varying Thermal Conductivity and Heat Capacity per Unit Volume",*Numerical Heat Transfer*,Vol.16,pp.249-266.

Freund,J.,1983,"Aerosol Single Scattering Albedo in the Arctic Determined from Ground-based Nonspecial Solar Irradiance Measurements," *J.Atmos.Sci.*, Vol.40,pp.2724-2731.

Huang,C.H.and Özişik,M.N.,1990,"A Direct Integration Approach for Simultaneously Estimating Spatially Varying Thermal Conductivity and Heat Capacity," *Int.J.Heat and Fluid Flow*, Vol.11,pp 262-268.

IMSL Library,Edition 10.,1987,NBC Building, 7500 Ballaire Blvd., Houston, TX 77036-5085.

LeCaine,J.,1947,"A Table of Integrals Involving $E_n(x)$", *National Research Council of Canada*,NRC No1553.

Marquardt,D.W.,1963,"An Algorithm for Least-Squares Estimation of Nonlinear Parameters," *J.Soc.Indust.Appl.Math.*, Vol.11,pp.431-441.

McCormick,N.J.,1992,"Inverse Radiative Transfer Problems: A Review," *Nuclear Science and Engineering*,Vol.112, pp.185-198.

Özişik,M.N.,1973,"Radiative Transfer and Interaction with Conduction and Convection," Wiley, New York.

Özişik,M.N. and Yener,Y.,1982,"The Galerkin Method for Solving Radiation Transfer in Plane-parallel Participating Media," *Journal of Heat Transfer*,Vol.104,pp.351-354.

Roux,J.A.and Smith A M 1981,"Determination of Rad[illegible] Properties from Transport Theory and Experimental Data," *AIAA 16th Thermophysics Conf.*, Palo Alto, California, 23-25 June.

Sacadura,J.F.,Uny,G. and Venet,A.,1986,"Models and Experiments for Radiation Parameter Estimation of Absorbing, Emitting, and Anisotropically Scattering Media," *Proc. 8th Heat Transfer Conf.*, Vol.2,pp.565-570.

Silva Neto,A.J. and Özişik,M.N.,1992,"An Inverse Analysis of Simultaneously Estimating Phase Function, Albedo and Optical Thickness",*HTD-Vol203, Developments in Radiative Heat Transfer, ASME.*

Ustinov,E.A.,1977,"The Inverse Problem of Multi-scattering Theory and the Interpolation of Measurements of Scattered Radiation in the Cloud Layer of Venus," *Cosmic Res.*,Vol.15,pp.667-672.

Van Laethem,R.,Leger,M.and Plumat,E.,1961,"Temperature Measurement of Glass by Radiation Analysis," *J.Am.Ceram.Soc.*, Vol.44,
pp.321-332.

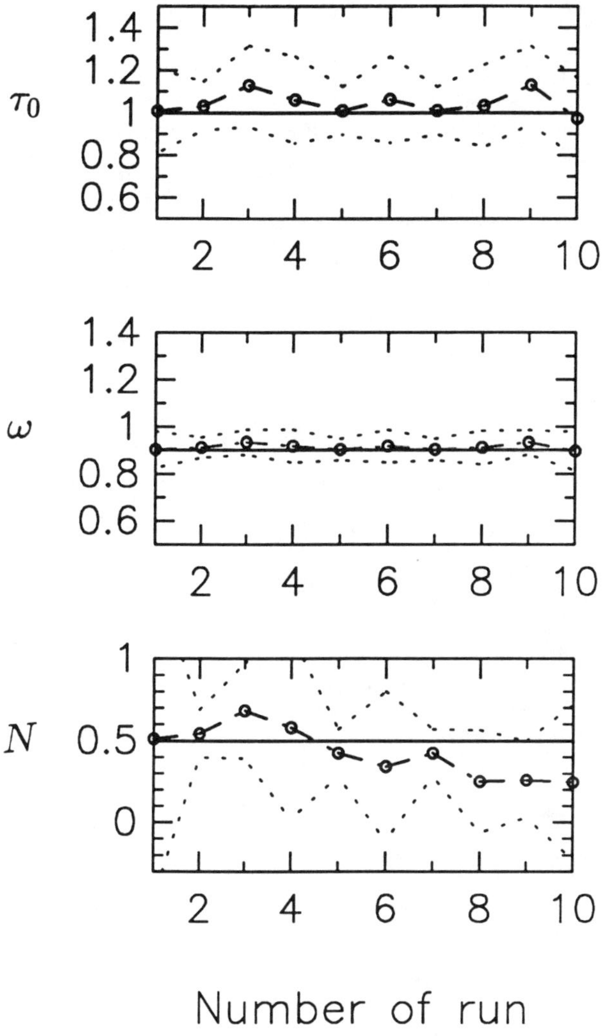

Figure 5. Estimations and confidence bounds for case $\tau_0 = 1.0$, $\omega = 0.9$ and $N = 0.5$. $\sigma = 0.01$.

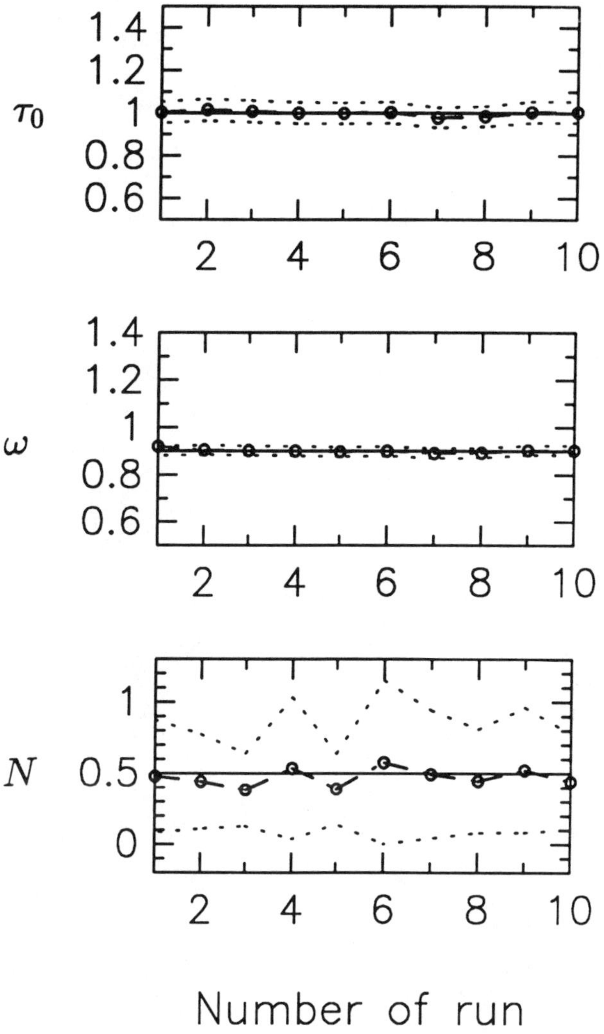

Figure 6. Estimations and confidence bounds for case $\tau_0 = 1.0$, $\omega = 0.9$ and $N = 0.5$. $\sigma = 0.0025$.

Inverse Problems in Engineering: Theory and Practice
ASME 1993

NOVEL APPROACHES TO THE IHCP: NEURAL NETWORKS AND EXPERT SYSTEMS

V. Dumek, M. Druckmüller, and M. Raudenský
Technical University of Brno
Brno, Czechoslovakia

Keith A. Woodbury
Department of Mechanical Engineering
University of Alabama
Tuscaloosa, Alabama

ABSTRACT

Inverse problems are problems of determining causes on the basis of the knowledge of their effects. The object of the inverse heat conduction problem is to determine the external heat transfer (the cause) given observation of the temperature history at one or more interior points (the effect).

This paper demonstrates two novel approachs to inverse problems. These approaches exploit two artificial intelligence mechanisms: a neural network and an expert system. The examples shown in this paper were computed using back propagation software (neural network) and a system based on Lukaszewicz's many valued logic (expert system). The numerical technique of neural networks evolved from the effort to model the function of the human brain and expert systems replace expert knowledge in several areas.

The solution to an inverse problem by a neural network and also by an expert system can be divided into two parts:

For neural networks:

- training (when the weights of the connections among the processing elements are modified)
- evaluation (when the trained network responds to input data).

For expert systems:

- creation of a knowledge base (pieces of knowledge are put into the base).
- answer of the system (when the system gives answers to questions).

THE INVERSE TASK

The inverse heat conduction problem is a problem of determining the external heat transfer (the cause) given observation of the temperature history at one or more interior points (the effects). This contrasts with the usual forward problem in heat conduction, which is to determine the internal temperature field for a given set of boundary conditions.

The one dimensional direct problem is the solution to the partial differential equation:

$$\frac{\partial T}{\partial t} = \frac{1}{\alpha}\frac{\partial^2 T}{\partial x^2} \tag{1}$$

where T are the unknown temperatures, x the spatial coordinate, α is the thermal diffusivity and t is time. Boundary conditions must be known for the solution of the equation (1). These boundary conditions in the present case are described by ambient temperature and the heat transfer coefficient at the boundary of the body (boundary condition of the third kind). The calculation of temperature fields from equation (1) is a routine problem; finite difference methods, finite element methods, or suitable linear superposition techniques may be used.

The nature of the problem changes when an internal temperature history is known (usually from an experiment) and boundary conditions are to be found. This is the inverse problem. This problem is from a mathematical point of view is incorrect and from a numerical point of view is ill-posed. The incorrectness of the problem is associated with the lack of uniqueness proofs and the ill-posedness with the fact that small changes in the input data (the measured temperature) can cause large deviations in the output (the predicted surface condition).

Beck's inverse method of sequential function specification is well-established for one-dimensional problems (Beck, et al. (1985) and Raudenský and Dumek (1992)). The essence of this method is that a functional form for the unknown variation in the external heat transfer is presumed; typically a piecewise constant or piecewise linear variation is assumed. The remaining task in the method is to determine the unknown parameters in the assumed functional form which will minimize the error between the observed and computed values of temperature at the measurement points. Usually, a sum-squared-error performance measure is used, hence a least-squares minimization routine is generally used.

NEURAL NETWORKS

It is not the aim of this paper to describe the inherent problems of the neural network technique. Thus, the neural network concept will only be briefly outlined. Further infomation can be found in Zeidenberg (1990) or Hecht-Nielson (1989).

A neural network is a structure for parallel information processing. It may be considered as a "black box" or a transfer function, which converts input signals to their corresponding output values. The network is loaded by the input signal

$$\mathbf{X} = (x_1, x_2, ..., x_n)$$

and the network responds with the output signal

$$\mathbf{Y} = (y_1, y_2, ..., y_m).$$

The transformation of the signal can be written in the form:

$$\mathbf{Y} = \mathbf{f}(\mathbf{X}, \mathbf{W}),$$

where $\mathbf{W}$ is the vector of weights. Weights are the dynamic structure of the neural network that are adjusted to perform a required task. Each element in the network receives as its input the output of the previous element to which it is attached, multiplied by the weight value on the connection between the two elements.

The neural network structure consists of a number of processing elements which are referred to as neurons. Each neuron can receive in parallel a finite amount of input information. The processing element transforms the input data with an activation function. The output signal can be received by an arbitrary number of processing elements. The schematic structure of a neuron is shown in Fig. 1.

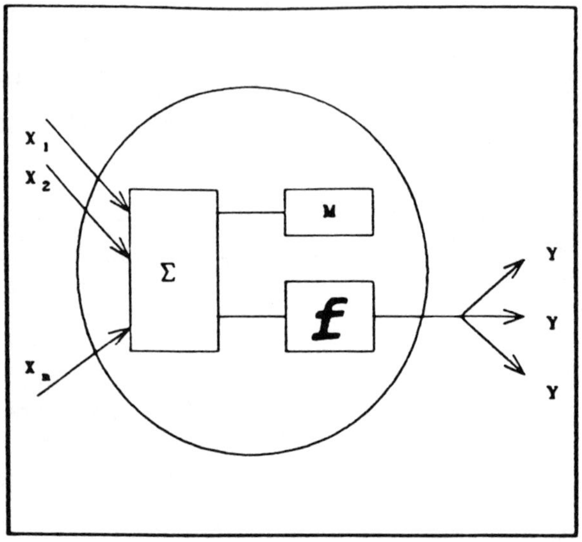

FIGURE 1 – Processing element

The input to this element is the sum of the activations of the preceding elements to which it is connected, multiplied by the corresponding weight value of each connection. This can be written in the form:

$$I_j = \sum_{i=1}^{n} W_{i,j} X_i$$

where I_j is the input to the element j, $W_{i,j}$ is the weight value of the connection from the element i to element j and X_i is the output from the previous element i. The output of the element j is given by transforming the input I_j by the activation function.

Any non-linear activation function can be used with the back-propagation network. The only restrictions are that the function exhibit some non-linearity and that its first derivative exist. In the following examples a sigmoidal function was used:

$$f = \frac{1}{(1 + e^{-I})}$$

Organization

The study of neural network simulation originates from the effort to understand the process of thinking in the human brain. The structure and ability of the human brain and neural network which is used for the present computations can hardly be compared.

Even if the human brain is extremely complicated it can be stated that the overwhelming structure is layered. The elements and connections which form a back-propagation neural network are organized into layers.

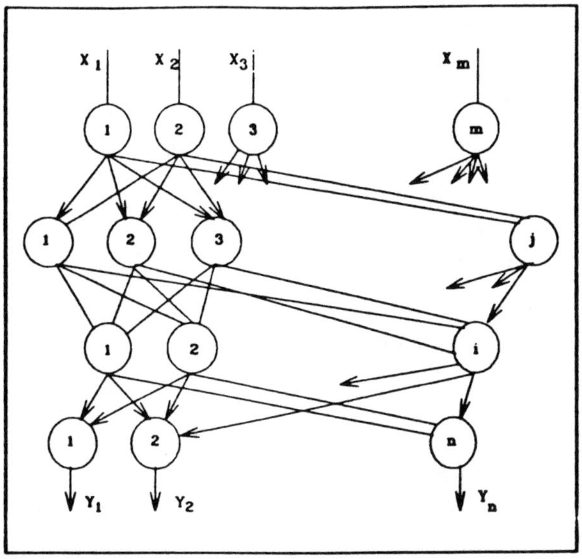

FIGURE 2 – Neural network structure

Fig. 2 shows the general structure. The structure consists of one input layer, one output layer and one or more intermediate (hidden) layers. The input layer has one element for each element of the input data pattern. The input layer processing elements only distribute the input signal to the elements of the first hidden layer. The output layer has one element for each of the desired outputs. The exact topology of a network for a given application is subject to the whim or intuition of the network designer. There need not be the same number of neurons in each layer of the net.

In dealing with any neural network problem there are two distinct phases in the use of the back propagation network: training and application. In the first phase, the network is shown input/output pairs from the system which the user wishes to simulate. The input patterns are represented by a set of **X** vectors, and the output by a set of **Y** vectors. A process involving a *learning algorithm* takes place, in which the weights of the network connections are adjusted with each pair of **X**, **Y** vectors so as to better reproduce the desired behavior. In the second phase, the trained network is used to predict unknown outputs for specified inputs.

In the training phase, the network initially begins with random weight values connecting the various elements. A set of values from the input data file is assigned to the input units. These values are propagated forward through the network by summing element inputs and calculating element outputs from the first layer to the last. At the output layer the calculated values are compared with the desired values. The difference is then propagated back toward the input units as the error signal. As this signal reaches each individual connection in the network the weight is modified so as to reduce the overall modelling error. The process of fixing input/output values and changing weights is repeated for a number of data sets. With each iteration the weights are modified so that the network more closely produces the correct output from the given input. The learning process is continued until the prescribed accuracy of response or a prescribed number or learning runs is reached.

Neural Networks Usage in Inverse Problems

An inverse problem can be formulated as searching for the input data knowing the output. All inverse problems are based on deterministic mathematical models which describe the response of a system to its input signal; this is merely a forward solver. The availablility of this deterministic mathematical model which can produce large numbers of sets of training cases make the IHCP an ideal candidate for neural nets. The possibility of including the effects of random noise in the computed temperatures presented to the net for training also exists.

The forward solver is used to generate training data in the following manner. Different types of and values for the surface heating conditions are used in turn to obtain solutions to the Eq. (1). These different heat transfer coefficients are considered inputs to the forward problem, but are classified as outputs from the neural network, $\mathbf{O_t}$. Correspondingly, the resulting temperature history at a particular point in the domain is an output from the forward solver, but is considered the input to the neural network, $\mathbf{I_t}$. In this way pairs of $\mathbf{I_t}, \mathbf{O_t}$ are set up for use in the training phase of the neural network. A well trained net can respond to a vector **X** (a vector of an observed temperature history at the specified location) with the proper **Y** (a vector of corresponding heat transfer coefficients). The vector **X** need not, naturally, come from the training data set.

Example - Inverse Heat Conduction

The surface $x^+ = 0$ of a one-dimensional body is subjected to time dependent boundary conditions. Boundary conditions are described in this case by the constant ambient temperature $T^+_\infty = 0$ and the time dependent heat transfer coefficient (HTC). The starting temperature is constant and equal to 1 (dimensionless). The surface at $x^+ = 1$ is adiabatic.

Various types of heat transfer coefficient histories were tested: constant ($h = h_c$), linearly increasing or

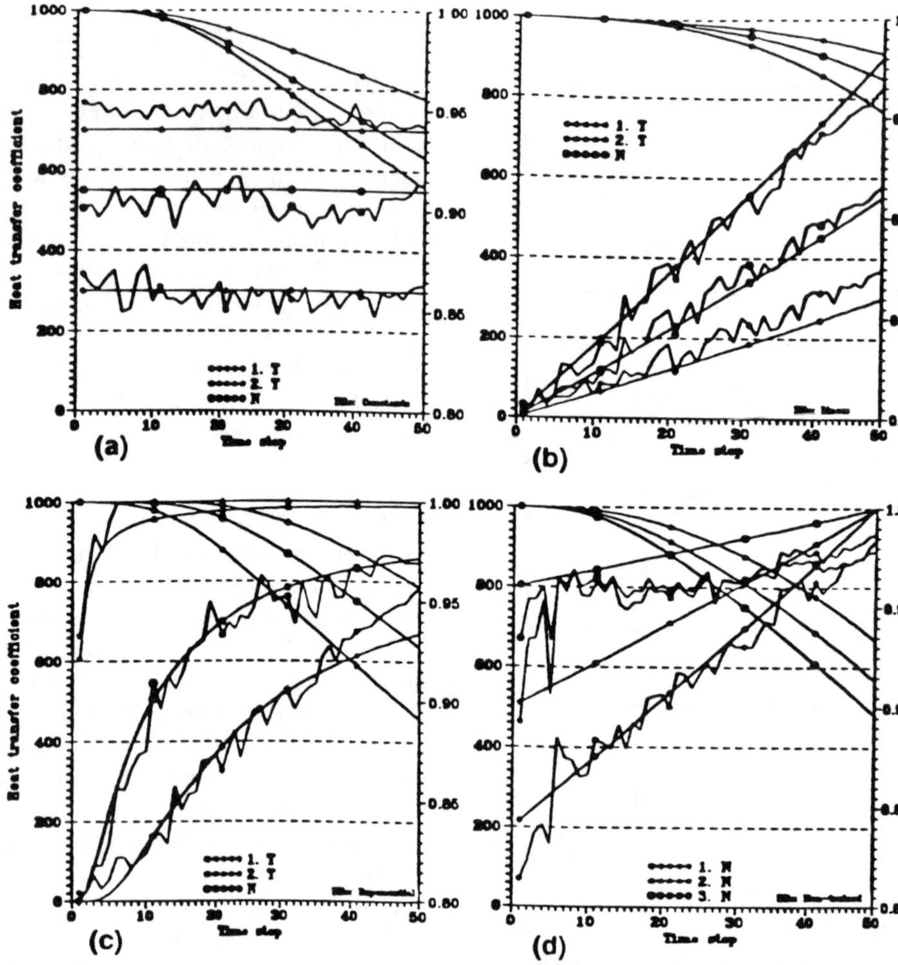

FIGURE 4 – The response of the neural network to the input temperature history

decreasing ($h = h_c t$ or $h = h_c(1-t)$), and exponential ($h = h_c(1-e^{-t})$).

The observed time interval was divided into 50 time steps. The length of the time steps does not directly play a role in the described approach to the problem. However, the approach used herein will require retraining of the net for different observation time intervals.

For a given type of heat transfer coefficient history, the training data was formed by a set of ten pairs of input and output vectors. For example, for the constant heat transfer coefficient case, ten values of the constant h_c =100, 200, 300, 400, 500, 600, 700, 800, 900 and 1000 W/m^2/C were used in subsequent solutions to the forward problem, as described earlier. The input vector contains 50 values of computed temperatures at the point $x^+ = 0.5$ and the output vector contains 50 values of the corresponding heat transfer coefficients. Typical pairs of input/output data are plotted in Fig. 3.

The net used in the tests had an input layer with 50 elements, one hidden layer with 50 elements and an output layer with 50 elements. This topology was intuitive and perfectly arbitrary.

The results for the network trained with 3x10 pairs of input/output data (HTC: constant, linear, exponential) are shown in Fig. 4. The prescribed training accuracy was 10 %. Each of these four figures shows the input temperature histories, the exact heat transfer coefficient history and the network's response to the input data.

Figures 4 a, b, c show the response to the type of data which was used for training. The curves marked 'T' belong to training data. The curve 'N' indicates intermediate data that were not used for training. For example, an intermediate value of h_c = 550 W/m^2/C was used to generate a temperature history to use as input to the network. Fig. 4d reports the case when the trained network is used for input data slightly different from the trained patterns. Specifically, a linear variation of $h(t)$ was used, but one that does not pass through the origin. Note that the net response becomes worse as the character of the input data differs from the character of the training data.

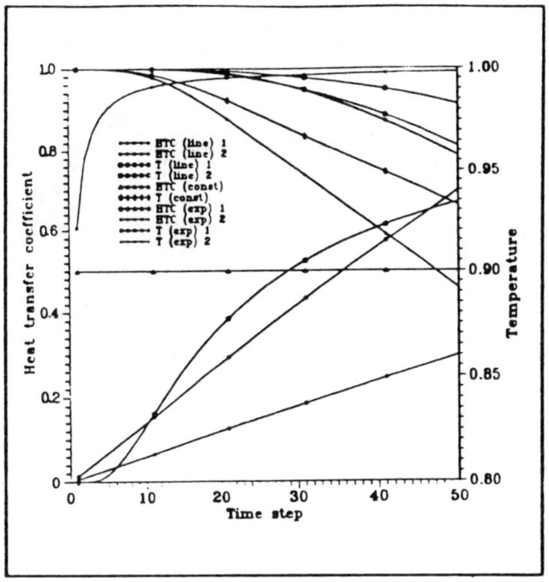

FIGURE 3 – Input and output data to the model, T - temperature, HTC - heat transfer coefficient

Conclusions of Neural Network Concept

The major advantages of the neural network approach are:

- Applicable to a wide class of inverse problems.
- Fast computation time.
- Stability.
- Suitable for multi-dimensional cases.
- Independence on the length of the time step.

The major disadvantages of the neural network approach are:

- Limited accuracy.
- Expectation of the results must be known in advance (function specification).
- Computational time required for training.

EXPERT SYSTEMS

People use natural language to give descriptions of the behavior of complex systems. But it is necessary to create mathematical models for technical areas where it is impossible to use linguistic expressions. When the creation of mathematical models is impossible or not expedient expert systems may be used which facilitate use of linguistic expressions.

In technical practice a linguistic description of the investigated phenomenon is often used. This is true especially in cases where it is not possible or there is no reason to devise a mathematical model. Introducing L. A. Zadah's term *linguistic variable* permits even the processing of the above-mentioned linguistic models by mathematical methods. The linguistic description of the following function is of interest:

$$y = f(x_1, x_2, ..., x_n).$$

This function is given by a table of linguistic values. The linguistic variables employed may be considered as a fuzzy set with a real universum (set of variable values) as they can be quantified (specifiically, temperature and heat transfer coefficient).

In order to use the expert system the linguistic values are reduced to triangles where only a single value exists for which the given variable gains the given linguistic value with high certainty .

A *rule* is defined as a sentence composed of the names of the linguistic variables, their linguistic values and logical conjunctions. Joining the rules with the help of the logical conjunctions gives a *linguistic model.*

The software system LMPS (Linguistic Model Processing System) was used for the solution of the inverse heat conduction problem. The system was developed by the firm JANES (Brno, Czech Republic). The main advantage of this system is that it uses very simple but highly effective mathematical methods (see below). The application of fuzzy sets, multi-valued logic, linguistic model compilation and rule formulation are described in the LMPS manual (1992).

Mathematical base of LMPS

The rule-based expert system LMPS was used for solving the inverse task. This expert system uses rules of the IF-THEN type with variable semantic interpretation in Lukaszewic's many-valued logic (LMPS Manual (1992)).

A knowledge base of the following form (2) was created, where x_i, y are linguistic variables and L_{ji}, K_j their corresponding linguistic values. Every linguistic value is supposed to be a convex normalized fuzzy set with a real universum.

$$\text{IF} \quad \left\{ \begin{array}{l} x_1 = L_{11} \\ x_2 = L_{12} \\ \dots \\ x_n = L_{1n} \end{array} \right\} \quad \text{THEN} \quad y = K_1$$

$$\text{IF} \left\{ \begin{array}{l} x_1 = L_{21} \\ x_2 = L_{22} \\ \dots \\ x_n = L_{2n} \end{array} \right\} \quad \text{THEN} \quad y = K_2 \qquad (2)$$

$$\vdots$$

$$\text{IF} \left\{ \begin{array}{l} x_1 = L_{m1} \\ x_2 = L_{m2} \\ \dots \\ x_n = L_{mn} \end{array} \right\} \quad \text{THEN} \quad y = K_m$$

The following forms of semantic interpretation of the knowledge base (2) were used:

1. CIC-model (contradiction sensitive, redundance insensitive). This model accounts for contradictions implied by different rules, but does not reinforce redundnat information supplied by the rules:

   $((\mu_{11}\text{and } \mu_{12} \text{ and } \dots \text{ and } \mu_{1n}) => \mu_1)$ and
   $((\mu_{21}\text{and } \mu_{22} \text{ and } \dots \text{ and } \mu_{2n}) => \mu_2)$ and

   $\vdots$

   $((\mu_{m1}\text{and } \mu_{m2} \text{ and } \dots \text{ and } \mu_{mn}) => \mu_m)$

2. CI&-model (contradiction sensitive, redundance sensitive). This model enforces both contraditions in the rules and effectively gives more weight to repeated rules.

   $((\mu_{11}\text{and } \mu_{12} \text{ and } \dots \text{ and } \mu_{1n}) => \mu_1)$ &
   $((\mu_{21}\text{and } \mu_{22} \text{ and } \dots \text{ and } \mu_{2n}) => \mu_2)$ &

   $\vdots$

   $((\mu_{m1}\text{and } \mu_{m2} \text{ and } \dots \text{ and } \mu_{mn}) => \mu_m)$

3. CCD-model (contradiction and redundance insensitive). This model effectively ignores contradictions and redundance in the supplied rules.

   $((\mu_{11}\text{and } \mu_{12} \text{ and } \dots \text{ and } \mu_{1n}) => \mu_1)$ or
   $((\mu_{21}\text{and } \mu_{22} \text{ and } \dots \text{ and } \mu_{2n}) => \mu_2)$ or

   $\vdots$

   $((\mu_{m1}\text{and } \mu_{m2} \text{ and } \dots \text{ and } \mu_{mn}) => \mu_m)$

Expert System Usage in Inverse Problem

Before questioning the expert system, it is necessary to fill its knowledge base with some quantity of high-quality information. The knowledge base is a data file holding linguistic variable definitions, linguistic values, rule declarations, author information and further auxilliary information. The knowledge base was filled with rules generated from data calculated from the cooling produced by different values of the heat transfer coefficient. This calculation gave corresponding heat transfer coefficients, h_c, and temperature history pairs. Again, the availability of a deterministic mathematical model for generation of the input information makes the IHCP a good candidate for use in an expert system.

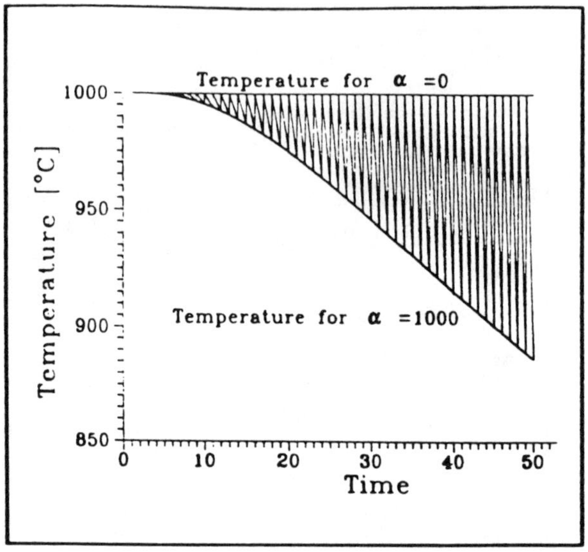

FIGURE 5 – The working area for solutions

The minimum value of h_c was taken to be 0 W/m^2/K and the maximum 1000 W/m^2/K. These bounding values together with corresponding temperature histories define the working area for all the solutions. This is shown in Fig. 5.

The temperatures were transformed into the interval [0,1] before fuzzification. The temperature history before and after this transformation is shown in Fig. 6 and Fig. 7.

Time was not used as an independent variable, but the values of h_c and temperature are bound to certain points whose position in the history carry information about time. Only values of temperature prior to the point for which h_c must be found are seen in the rules. This is because values of temperature after this point cannot (for physical reasons) influence the value of h_c. The fuzzified values are shown in Fig. 8. The arrangement of the rule is described in the next section.

Only after the knowledge base has been filled can the expert system be questioned to find a solution. To ask a question, a number of variables are set to known values. In the present case, these variables represent a temperature history. From these variables the value of h_c is determined. The question must be repeated for every point of h_c's history, because LMPS is only capable of answering a question at one point. The values of h_c at other points is not exploited by the system.

The answer of the system is a fuzzy set in a graphi-

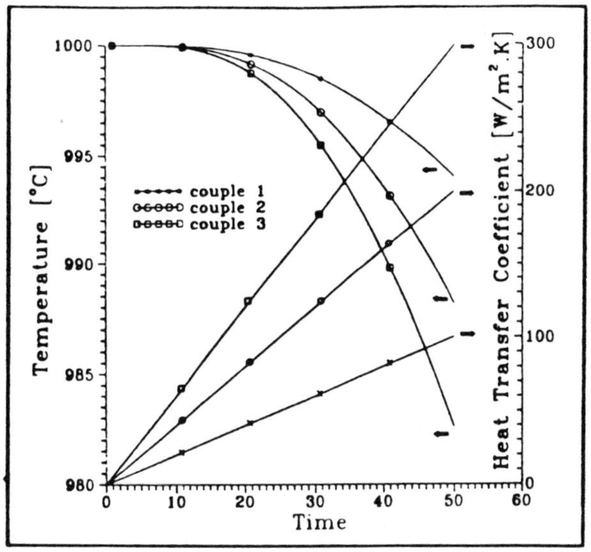

FIGURE 6 – The temperature history before transformation

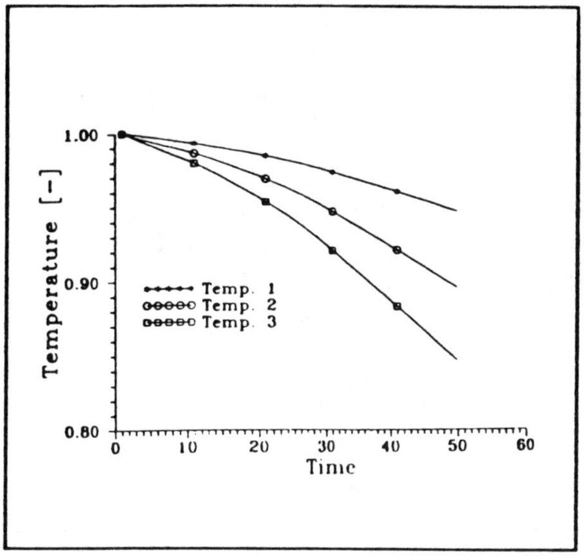

FIGURE 7 – The temperature history after transformation

cal form, as shown in Fig.9. Because only a single value of h_c must be determined, defuzzification must be performed. The defuzzification procedure recommended is the 'centre of mass' method.

The arrangement of the rule

The illustration of the rule for one value of h_c is shown below.

$$\text{if}\left(\begin{array}{c}\text{temperature0 is } T_{10} \text{ and}\\ \text{temperature1 is } T_{10} \text{ and}\\ \text{temperature2 is } T_9 \text{ and}\\ \vdots \\ \text{temperature11 is } T_4 \text{ and}\\ \text{temperature12 is } T_3\end{array}\right) \text{then } h_c \text{ is } A_8$$

The same rule must be compiled for all points at which temperature history is observed. Temperature histories which perfectly cover the whole working space are used. Remember that the values required to fill the rules base are generated from solutions of the forward problem.

CONCLUSIONS

The availability of a deterministic mathematical model for generation of training sets and/or expert rules make the IHCP a good candidate for solution by artificial intelligence. The feasibility of using both neural networks and expert systems for the solution of the IHCP has been demonstrated in this paper.

The neural network gives a response even if the net is not trained for the input data. The system is nonlinear and does not yield an error message when a given question has not been defined in the training data.

The expert system is able to give very precise information if the working space is well covered by the rules. The system informs the user that the answer has high or low certainty. This information can be used for suplementation of the knowledge base.

REFERENCES

Beck, J. V., Blackwell, B., and St. Clair, C., 1985, "Inverse Heat Conduction: Ill-posed Problems". New York: Wiley-Interscience.

Hecht-Nielson, R., 1989, "Neurocomputing", Addison-Wesley.

Raudenský, M., and Dumek, V., 1992, "Inverse Task for Determining Heat Transfer Coefficient," *Journal of Czechoslovak Academy of Science*, No. 1.

Druckmüller, M., 1991, "Many-valued logic system for linguistic model processing,". *Int.J.of General Systems*, Vol. 20, No. 1, pp. 31-38.

LMPS - Reference Guide 1992, , Brno

Zeidenberg, M., 1990, "Neural Networks in Artificial Intelligence," Ellis Horwood Ltd.

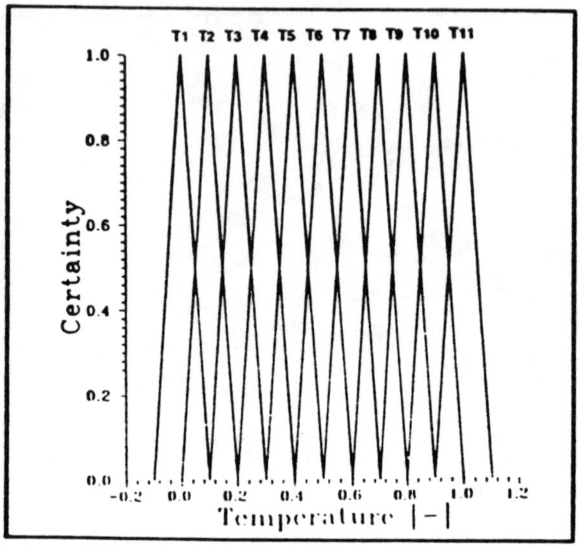

FIGURE 8 – The fuzzified value

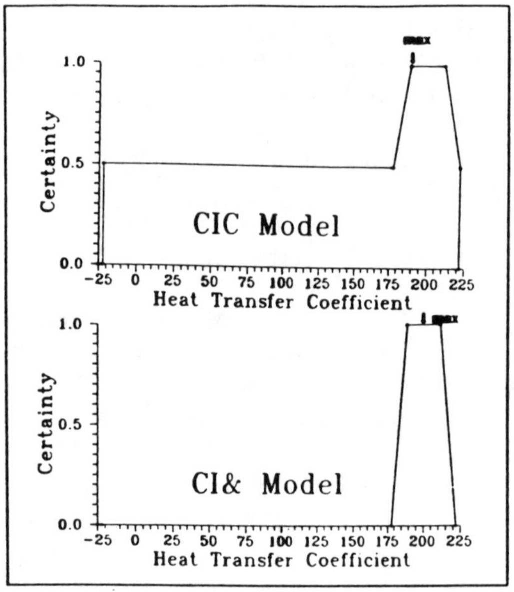

FIGURE 9 – Results from the expert system

Inverse Problems in Engineering: Theory and Practice
ASME 1993

AN INVERSION APPROACH FOR SOLVING GOVERNING EQUATION INVERSE PROBLEMS

Shiro Kubo, Kiyotsugu Ohji, and Akio Shiojiri
Department of Mechanical Engineering
Faculty of Engineering
Osaka University
Osaka, Japan

ABSTRACT

The governing equation inverse problems deal with an inference of a differential equation governing variation of field quantity ϕ from observations. In the present paper an inversion analysis scheme is examined. In this scheme derivatives of field quantity ϕ are evaluated numerically from observed field quantities using finite difference approximation. The identification of the governing differential equation is then reduced to the estimation of the order and coefficients of the differential equation, which can be made by using simultaneously the values of field quantity ϕ observed under several conditions. For cases where the order of the differential equation is not known in advance, the successive elimination method is proposed to determine the order as well as the coefficients. The applicability of the principle of parsimony for determining the order is also discussed. Numerical simulations are made of the application of the inversion scheme to the identification of a second-order ordinary differential equation and a second-order partial differential equation. It is found that the proposed inversion scheme with the successive elimination method and the principle of parsimony is useful for estimating the ordinary differential equation and the partial differential equation, even when the order of the differential equations is not known in advance.

INTRODUCTION

Much attention has been devoted recently to inverse problems in science and engineering. Many reviews, books and monographs on inverse problems are available (Tikhonov and Arsenin, 1977, Amer. Math. Soc., 1984, Groetch, 1984, Gladwell, 1986, Lavrent'ev, et al., 1986, Kubo and Ohji, 1987, Romanov, 1987, Kubo, 1988, 1992, 1993a, 1993b, Yamaguti, et al., 1991, Japan Soc. Mech. Engrs., 1991, Musha and Okamoto, 1992, Tanaka and Bui, 1993). It has been recognized that there are various kinds of inverse problems in science and engineering. The present authors discussed the classification of inverse problems related to analyses of the variation of a field quantity (Kubo and Ohji, 1987, Kubo, 1988). They pointed out that the inverse problems related to field problems can be classified into (1) domain/boundary inverse problems, (2) boundary value/initial value inverse problems, (3) material properties inverse problems, (4) force/source inverse problems, and (5) governing equation inverse problems. Although inverse problems of categories (1) to (4) are investigated extensively, inverse analyses dealing with the governing equation inverse problems of category (5) are very limited.

The governing equation inverse problems deal with an inference of a differential equation governing the variation of a field quantity from observations. We may be faced with these governing equation inverse problems in understanding an undeveloped phenomenon. Related to the governing equation inverse problems system identification deals with estimation of coefficients of ordinary differential equations. Estimation of an operator in a second-order hyperbolic equation is also made (Romanov, 1987). No general schemes, however, are available for estimating governing equations.

In the present paper a general inversion analysis scheme for governing equation inverse problems is proposed and examined. In this scheme approximate evaluations of derivatives of field quantities are made and then the identification of the governing differential equation is reduced to the estimation of the order and coefficients involved in the differential equation. Extension of the scheme is made to make the identification possible even when the order of the governing equation is not known in advance. Applicability of the inversion scheme is examined by numerical simulations.

BASIC IDEA OF ESTIMATION OF GOVERNING EQUATION

This study is concerned with a method for finding a differential equation which represents consistently the variation of a field quantity. Let ϕ denote certain quantity representing physical states of the present concern. If the variation of ϕ with coordinate x is governed by an ordinary differential equation of the n-th order, we have the following equation.

$$a_n d^n\phi/dx^n + a_{n-1} d^{n-1}\phi/dx^{n-1} + \cdots + a_1 d\phi/dx + a_o\phi + b = 0 \qquad (1)$$

where a_j and b denote a coefficient and a force/source term. Then the estimation of governing equation may be reduced to the estimation of coefficient a_j and order n of the differential equation. When the governing equation is nonlinear we can add nonlinear terms in Eq. (1), and the inference of the nonlinear differential equation is again reduced to the estimation of coefficients in the assumed nonlinear equation.

When ϕ varies with coordinates x and y and its variation is expressed by the n-th partial differential equation, we have the following equation.

$$a_{n.o}\partial^n\phi/\partial x^n + a_{n-1.1}\partial^n\phi/\partial x^{n-1}\partial y + \cdots + a_{o.1}\partial\phi/\partial y + a_{o.o}\phi + b = 0 \qquad (2)$$

where $a_{i.j}$ denotes a coefficient. In this case the estimation of the governing partial differential equation can be reduced to the estimation of coefficient $a_{i.j}$ and order n. When the governing equation is nonlinear, additional nonlinear terms should be involved in Eq. (2). The inference of the governing equation is hence equivalent to the estimation of the coefficients involved in the equation.

Even when ϕ varies with three or more coordinates the inference of the governing equation can be reduced to the estimation of finite number of coefficients. In many cases, which we are faced with in science and engineering, order of the governing equation is not high. Typical order is 2 and 4. Hence the number of coefficients required in the estimation is not so large. When no force/source term is involved, the governing equation also holds even if all the coefficients are multiplied by some number. One coefficient, therefore, has to be designated to be a finite non-zero number.

The value of ϕ and its derivatives can be evaluated approximately from its value observed at an observation point and its neighbourhood. By substituting the approximate values thus obtained into Eq. (1) or Eq. (2), a linear equation for coefficients is given, whether the equation is linear or nonlinear. This kind of equations obtained under different conditions or at different observation points are summarized in a matrix equation as,

$$[\mathbf{D}]\{\mathbf{a}\} = \{\mathbf{f}\} \qquad (3)$$

where, $[\mathbf{D}]$ is a matrix defined by approximate values of ϕ and its derivatives, and $\{\mathbf{a}\}$ is a vector composed of unknown coefficients, and $\{\mathbf{f}\}$ is a vector which can be calculated from known or designated coefficients and a derivative.

When coefficients a_j are independent of location, Eq. (3) can be constructed based on observation at several observation points. For example, for the identification of the ordinary differential equation expressed by Eq. (1) the following expressions for $[\mathbf{D}]$, $\{\mathbf{a}\}$ and $\{\mathbf{f}\}$ can be obtained when no force/source term is involved and a_n is taken to be unity.

$$[\mathbf{D}] = \begin{bmatrix} d^{n-1}\phi/dx^{n-1(1)} & \cdots & d\phi/dx^{(1)} & \phi^{(1)} \\ d^{n-1}\phi/dx^{n-1(2)} & \cdots & d\phi/dx^{(2)} & \phi^{(2)} \\ \vdots & & \vdots & \vdots \\ d^{n-1}\phi/dx^{n-1(m)} & \cdots & d\phi/dx^{(m)} & \phi^{(m)} \end{bmatrix}$$

$$\{\mathbf{a}\} = \begin{Bmatrix} a_{n-1} \\ \vdots \\ a_1 \\ a_o \end{Bmatrix} \qquad \{\mathbf{f}\} = \begin{Bmatrix} -d^n\phi/dx^{n(1)} \\ -d^n\phi/dx^{n(2)} \\ \vdots \\ -d^n\phi/dx^{n(m)} \end{Bmatrix} \qquad (4)$$

where suffix (i) of $d^k\phi/dx^{k(i)}$ and $\phi^{(i)}$ in parentheses designates that the values are evaluated at the i-th observation points $x = x_i$.

Similarly in the identification of the partial differential equation (2), $[\mathbf{D}]$, $\{\mathbf{a}\}$ and $\{\mathbf{f}\}$ can be given by the following equations, when it is assumed that no force/source terms are involved in the equation and that coefficient $a_{n.o}$ for $\partial^n\phi/\partial x^n$ term can be set to be unity.

$$[\mathbf{D}] = \begin{bmatrix} \partial^n\phi/\partial x^{n-1}\partial y^{(1)} & \cdots & \partial\phi/\partial y^{(1)} & \phi^{(1)} \\ \partial^n\phi/\partial x^{n-1}\partial y^{(2)} & \cdots & \partial\phi/\partial y^{(2)} & \phi^{(2)} \\ \vdots & & \vdots & \vdots \\ \partial^n\phi/\partial x^{n-1}\partial y^{(m)} & \cdots & \partial\phi/\partial y^{(m)} & \phi^{(m)} \end{bmatrix}$$

$$\{\mathbf{a}\} = \begin{Bmatrix} a_{n-1.1} \\ \vdots \\ a_{o.1} \\ a_{o.o} \end{Bmatrix} \qquad \{\mathbf{f}\} = \begin{Bmatrix} -\partial^n\phi/\partial x^{n(1)} \\ -\partial^n\phi/\partial x^{n(2)} \\ \vdots \\ -\partial^n\phi/\partial x^{n(m)} \end{Bmatrix} \qquad (4)'$$

When the number of equations obtained are equal to the number of unknown parameters, matrix $[\mathbf{D}]$ in Eq. (3) is square. In this case Eq. (3) can be solved for unknown coefficients $\{\mathbf{a}\}$ as,

$$\{\mathbf{a}\} = [\mathbf{D}]^{-1}\{\mathbf{f}\} \qquad (5)$$

When the number of equations available exceeds the number of unknown coefficients, $\{\mathbf{a}\}$ is obtained by the the least residual criterion as,

$$\{\mathbf{a}\} = ([\mathbf{D}]^T[\mathbf{D}])^{-1}[\mathbf{D}]^T\{\mathbf{f}\} \qquad (5)'$$

where T denotes a matrix transpose.

In cases where coefficients a_j or $a_{i.j}$ are dependent on locations, observations are repeatedly made at the same observation points under different situations. By solving Eq. (3) constructed based on the observations thus obtained for $\{\mathbf{a}\}$, the coefficients at the observation point is estimated.

When the order of the differential equation is not known in advance, the order has to be assumed to estimate coefficients. When the assumed order N is smaller than the actual one, estimated results are expected to depend on conditions employed in the estimation, and consistent results are not obtained. When the assumed order is coincident with the actual one, consistent results can be obtained. When the assumed order is larger than the actual one, inconsistent results are obtained again. Therefore, the principle of parsimony can be applied to identify the real governing equation: when the assumed order N is increased incrementally, the smallest assumed order N giving consistent estimation of coefficients is coincident with the actual one.

It may be also possible to deduce the differential equation from the coefficients estimated with the assumed order N larger than the actual one. In this case estimated governing equation can be expressed as a superposition of the actual governing equation and its differentiated forms. To decompose the original governing equation successive elimination method can be formulated, in which high order terms are eliminated from estimated governing equations obtained under several conditions. The elimination is repeated until a consistent estimation of the coefficients is obtained, which gives the solution.

NUMERICAL SIMULATION OF ESTIMATION OF ORDINARY DIFFERENTIAL EQUATION

Numerical simulations were made of the applications of the scheme to the identification of a second-order ordinary differential equation. Two cases are considered in the simulations: (1) when order n of the differential equation is known in advance and (2) when order n is not known in advance.

When Order is Known in Advance

Since order n of the ordinary differential equation is known to be 2, the differential equation can be written as

$$a_2 d^2\phi/dx^2 + a_1 d\phi/dx + a_0\phi + b = 0$$

When it is assumed that there is no force/source term, and therefore the governing equation can be expressed as follows taking a_2 to be unity.

$$d^2\phi/dx^2 + a_1 d\phi/dx + a_0\phi = 0 \quad (6)$$

As an example of the governing equation to be identified, the following differential equation was treated.

$$d^2\phi/dx^2 + 4d\phi/dx + \phi = 0 \quad (7)$$

The general solution of Eq. (7) is given by

$$\phi(x) = C_1 e^{(-2+\sqrt{3})x} + C_2 e^{(-2-\sqrt{3})x} \quad (8)$$

where C_1 and C_2 are constants.

It was assumed that only quantity ϕ can be observed and its derivatives can not be observed directly. The derivatives were estimated approximately by using finite difference evaluation from observations at observation point $x = x_i$ and its neighborhood:

$$d^2\phi/dx^{2(i)} = [\phi(x_i+h) + \phi(x_i-h) - 2\phi(x_i)]/h^2$$
$$d\phi/dx^{(i)} = [\phi(x_i+h) - \phi(x_i-h)]/(2h)$$
$$\phi^{(i)} = \phi(x_i) \quad (9)$$

where h denotes the step of finite difference.

Distributions of field quantity ϕ was given by designating the values of C_1 and C_2. Two observation points were selected and ϕ was evaluated from Eq. (8). Measurement error was not considered in this numerical simulation. By substituting values of ϕ at observation points and their neighborhood into Eq. (9), the values of ϕ and derivatives were evaluated. These values were introduced in Eq. (6) to construct linear equations to be solved for a_1 and a_0.

Table 1 shows the distribution of ϕ and combination of observation points and estimated values of coefficients. In the table the estimated values of a_j deviate from the real ones for

$$\phi = \exp\{(-2+\sqrt{3})x\}$$

which corresponds to $C_1 = 1$ and $C_2 = 0$. This is due to the fact that identical relation

$$d^2\phi/dx^2 = (-2 + \sqrt{3})d\phi/dx = (-2 + \sqrt{3})^2\phi \quad (10)$$

holds for this distribution of ϕ, and independence between $d^2\phi/dx^2$, $d\phi/dx$ and ϕ is lost. It can be seen from the table that the estimation of coefficients, and therefore, the estimation of differential equation can be made for other distributions functions.

When Order is not Known in Advance

The estimation of the governing differential equation treated in the foregoing section was made for the case where the order is not known in advance.

Table 1 Estimation of ordinary differential equation ($N=2=n$, $h=0.01$)

Distribution $\phi(x)$	x_i	a_1	a_0
$\phi(x)=e^{(-2+\sqrt{3})x}$	1, 3	-2.59146	-0.767531
$\phi(x)=e^{(-2+\sqrt{3})x}+2e^{(-2-\sqrt{3})x}$	1, 2	+3.99835	+0.998803
$\phi(x)=e^{(-2+\sqrt{3})x}+2e^{(-2-\sqrt{3})x}$	1, 3	+3.99917	+0.999193
$\phi(x)=e^{(-2+\sqrt{3})x}+2e^{(-2-\sqrt{3})x}$	2, 3	+4.02402	+1.00586
(Actual)	-	+4.0	+1.0

(x_i: Observation points)

Table 2 Estimation of ordinary differential equation ($N=1<n$, $h=0.01$)

Distribution $\phi(x)$	x_i	a_0
$\phi(x)=e^{(-2+\sqrt{3})x}+2e^{(-2-\sqrt{3})x}$	1	+0.472084
$\phi(x)=e^{(-2+\sqrt{3})x}+2e^{(-2-\sqrt{3})x}$	2	+0.274722
$\phi(x)=e^{(-2+\sqrt{3})x}+2e^{(-2-\sqrt{3})x}$	3	+0.268162

(x_i: Observation points)

Estimation of coefficients a_j was made for assumed order N = 1, 2, 3 and 4. The value of a_N was set to be unity. The procedure of the evaluation is the same as that described in the foregoing section. The distribution function used in the estimation is that given by setting $C_1 = 1$ and $C_2 = 2$ in Eq. (8). Derivatives were evaluated by substituting ϕ at observation points and their neighborhood into Eq. (9) and similar finite difference formulae for $d^4\phi/dx^4$ and $d^3\phi/dx^3$. Matrix **[D]** and vector **{f}** were evaluated from these values, and vector **{a}** was estimated using Eq. (5) or Eq. (5)'.

Estimated results for the assumed order N = 1, 3 and 4 are shown in Tables 2, 3 and 4, respectively. The estimated coefficients for N = 2 are already shown in Table 1. As can be seen from Table 2, when the assumed order is smaller than the actual one, estimated values depend on observation points, i.e. consistent results can not be obtained. When the assumed order is coincident with the actual one, consistent results are obtained for usual distribution functions of ϕ, as is shown in Table 1. Tables 3 and 4 show that when the assumed order is higher than the actual one, consistent results can not be obtained again. Therefore, the principle of parsimony works well: when the assumed order was increased incrementally from one the smallest assumed order which gives consistent estimation of coefficients is the actual one.

The successive elimination method may be useful for deducing the differential equation from the values of coefficients estimated with high assumed order shown in Tables 3 and 4. The applicability of the method is examined in the followings. For simplicity of expressions a differential operator Δ defined by the following equation is introduced.

Table 3 Estimation of ordinary differential equation ($N=3>n$, $h=0.01$)

Observation points x_i	a_2	a_1	a_0
1 , 2 , 3	-29.05871	-131.22083	-33.05500
-1, -2, -3	-10.68329	-57.72893	-14.68217
0.1,0.2,0.3	39.22985	141.89701	35.22390

(Distribution $\phi(x)=e^{(-2+\sqrt{3})x}+2e^{(-2-\sqrt{3})x}$)

Table 4 Estimation of ordinary differential equation ($N=4>n$, $h=0.01$)

Observation points x_i	a_3	a_2	a_1	a_0
1 , 2 , 3 , 4	-26.4122	4027.0721	16558.3819	4147.1660
-1, -2, -3, -4	-0.9889	7082.2971	28531.2767	7136.2577
0.1, 0.2, 0.3, 0.4	19.4871	3550.6123	13964.2392	3487.1581
-0.1,-0.2,-0.3,-0.4	-59.9015	1689.2885	7710.8824	1943.6842

(Distribution $\phi(x)=e^{(-2+\sqrt{3})x}+2e^{(-2-\sqrt{3})x}$)

$$\Delta = d/dx \tag{11}$$

The second row in Table 3 means that the estimated governing equation is expressed as

$$(\Delta^3 - 29.0587\Delta^2 - 131.2208\Delta - 33.0550)\phi = 0 \tag{12}$$

This equation is almost equivalent to

$$\{\Delta - 33.055\}\underline{(\Delta^2 + 4\Delta + 1)\phi = 0} \tag{13}$$

The underlined part of this equation is identical with the actual governing equation (7). Thus, as was expected in the foregoing, the original governing equation superposed with its differentiated forms is identified. The third and fourth rows in Table 3 were also found to be expressed in the form similar to Eq. (13).

The successive elimination method was applied to the results shown in Table 4 to decompose the original governing equation from the superposed forms. The second to fifth rows of Table 4 mean that the estimated governing equations are written as,

$$(\Delta^4 - 26.4122\Delta^3 + 4027.0721\Delta^2 + 16558.3819\Delta + 4147.1660)\phi = 0 \tag{14}$$

$$(\Delta^4 - 0.9889\Delta^3 + 7082.2971\Delta^2 + 28531.2767\Delta + 7136.2577)\phi = 0 \tag{15}$$

$$(\Delta^4 + 19.4871\Delta^3 + 3550.6123\Delta^2 + 13964.2392\Delta + 3487.1581)\phi = 0 \tag{16}$$

$$(\Delta^4 - 59.9015\Delta^3 + 1689.2885\Delta^2 + 7710.8824\Delta + 1943.6842)\phi = 0 \tag{17}$$

Some arrangements after substitution of Eq. (14) into Eqs. (15) to (17) for eliminating Δ^4 terms lead to

$$(\Delta^3 + 120.1742\Delta^2 + 470.9418\Delta + 117.5729)\phi = 0 \tag{18}$$

$$(\Delta^3 - 10.3805\Delta^2 - 56.5181\Delta - 14.3795)\phi = 0 \tag{19}$$

$$(\Delta^3 + 69.8069\Delta^2 + 264.1888\Delta + 65.7966)\phi = 0 \tag{20}$$

The estimated equations are still inconsistent with each other. So the order of the differential equation is less than 3. Elimination of Δ^3 terms in Eqs. (19) and (20) using Eq. (18) gives

$$(\Delta^2 + 4.0401\Delta + 1.0107)\phi = 0 \tag{21}$$

$$(\Delta^2 + 4.1049\Delta + 1.0280)\phi = 0 \tag{22}$$

which are almost identical with each other and close to the actual differential equation (7). Thus the successive elimination method is useful for the estimation of the governing equation when the order of the governing equation is unknown.

NUMERICAL SIMULATION OF ESTIMATION OF PARTIAL DIFFERENTIAL EQUATION

Numerical simulations were made of the applications of the scheme to the identification of a second-order partial differential equation when its order is known and when its order is unknown in advance.

When Order is Known in Advance

A general form of second-order partial differential equation defined as a function of Cartesian coordinates x, y is expressed by the following equation.

$$a_{2,0}\partial^2\phi/\partial x^2 + a_{0,2}\partial^2\phi/\partial y^2 + a_{1,1}\partial^2\phi/\partial x\partial y + \cdots + a_{1,0}\partial\phi/\partial x + a_{0,1}\partial\phi/\partial y + a_{0,0}\phi + b = 0 \tag{23}$$

When force/source term b is not involved and $a_{2,0}$ is set to be unity, Eq. (23) becomes

Table 5 Estimation of partial differential equation
($N=2=n$, $h=k=0.01$)

Distribution $\phi(x,y)$	$a_{0.2}$	$a_{1.1}$	$a_{1.0}$	$a_{0.1}$	$a_{0.0}$
sin(2x)·cosh(2y)	+0.309249	0.000000	0.000000	0.000000	+0.690739
sin(x)·cosh(y)+sin(2x)·cosh(2y)	+0.999917	0.000000	0.000000	0.000000	+0.000067
sin(x)·cosh(y)+sin(3x)·cosh(3y)	+0.999833	0.000000	0.000000	0.000000	+0.000150
sin(x)·cosh(y)+sin(4x)·cosh(4y)	+0.999717	0.000000	0.000000	0.000000	+0.000267
sin(2x)·cosh(2y)+sin(3x)·cosh(3y)	+0.999783	0.000000	0.000000	0.000000	+0.000600
(Actual)	+1.0	0.0	0.0	0.0	0.0

Observation points: $(x_i, y_i)=(0.1,0.1),(0.2,-0.1),(0.3,0.1),(0.4,-0.1),(0.5,0.1)$

Table 6 Estimation of partial differential equation
($N=1<n$, $h=k=0.01$)

Distribution $\phi(x,y)$	$a_{0.1}$	$a_{0.0}$
sin(2x)·cosh(2y)	-25.2512	-7.44978
sin(x)·cosh(y)+sin(2x)·cosh(2y)	-8.60025	-7.34428
sin(x)·cosh(y)+sin(3x)·cosh(3y)	-3.84556	-7.13166
sin(x)·cosh(y)+sin(4x)·cosh(4y)	-2.20401	-6.80817
sin(2x)·cosh(2y)+sin(3x)·cosh(3y)	-3.83761	-7.13793

Observation points: $(x_i, y_i)=(0.1,0.1),(0.2,-0.1)$

$$\partial^2\phi/\partial x^2 + a_{0.2}\partial^2\phi/\partial y^2 + a_{1.1}\partial^2\phi/\partial x\partial y + \cdots + a_{1.0}\partial\phi/\partial x + a_{0.1}\partial\phi/\partial y + a_{0.0}\phi = 0 \quad (24)$$

As an example of the governing equation to be identified, the Laplace equation which is expressed by the following equation was examined.

$$\partial^2\phi/\partial x^2 + \partial^2\phi/\partial y^2 = 0 \quad (25)$$

A solution of the Laplace equation is given by

$$\phi(x,y) = C_3\sin(C_4x)\cdot\cosh(C_4y) \quad (26)$$

where C_3 and C_4 are constants.

It was assumed again that only quantity ϕ can be observed. The derivatives were estimated approximately by the following finite difference formulae from observations at observation point $(x,y) = (x_i, y_i)$ and its neighborhood.

$$\begin{aligned}
\partial^2\phi/\partial x^{2(i)} &= [\phi(x_i+h,y_i) + \phi(x_i-h,y_i) - 2\phi(x_i,y_i)]/h^2\\
\partial^2\phi/\partial y^{2(i)} &= [\phi(x_i,y_i+k) + \phi(x_i,y_i-k) - 2\phi(x_i,y_i)]/k^2\\
\partial^2\phi/\partial x\partial y^{(i)} &= [\phi(x_i+h,y_i+k) - \phi(x_i+h,y_i-k) - \phi(x_i-h,y_i+k) + \phi(x_i-h,y_i-k)]/(4hk)\\
\partial\phi/\partial x^{(i)} &= [\phi(x_i+h,y_i) - \phi(x_i-h,y_i)]/(2h)\\
\partial\phi/\partial y^{(i)} &= [\phi(x_i,y_i+k) - \phi(x_i,y_i-k)]/(2k)\\
\phi^{(i)} &= \phi(x_i,y_i)
\end{aligned} \quad (27)$$

where h and k are steps of finite difference.

Distributions of field quantity ϕ was given by a linear combination of the solutions with different values of constant C_4. Five observation points are selected as shown in Table 5. Measurement error was not considered in this numerical simulation. Values of ϕ were substituted into Eq. (27) to evaluate ϕ and its derivatives. These values were introduced in Eq. (24) for the five observation points to construct linear equations, which were solved for unknown coefficients in the governing equation.

Table 5 shows the the estimated values of coefficients. In the table the estimated values of a_i deviate from the real ones when distribution function is given by

$$\phi(x,y) = \sin(2x)\cdot\cosh(2y)$$

This is due to the fact that for this distribution of ϕ, the following identity holds.

$$\partial^2\phi/\partial x^2 = -\partial^2\phi/\partial y^2 = -4\phi \quad (28)$$

It can be seen from the table that the partial differential equation can be estimated when the distribution of ϕ is given as a linear combinations of two terms with different C_4 values.

When Order is not Known in Advance

In this section estimation of the partial differential equation is made when the order is not known in advance, as was treated in the foregoing section for the estimation of ordinary differential

Table 7 Estimation of partial differential equation ($N=3>n$, $h=k=0.01$)

Distribution $\phi(x,y)$	$a_{0,3}$	$a_{2,1}$	$a_{1,2}$	$a_{1,1}$	$a_{2,0}$	$a_{0,2}$	$a_{1,0}$	$a_{0,1}$	$a_{0,0}$
sin(x)cosh(y) +sin(2x)cosh(2y)	8.7979	8.7497	0.9998	0.0000	4.5386	4.5382	0.0002	0.0015	0.0003
sin(x)cosh(y) +sin(3x)cosh(3y)	26.1106	26.1215	0.9996	0.0000	-1.3523	-1.3520	0.0004	0.0098	-0.0002
sin(x)cosh(y) +sin(4x)cosh(4y)	56.9178	56.9581	0.9993	0.0000	-39.0412	-39.0310	0.0007	0.0380	-0.0104
sin(2x)cosh(2y) +sin(3x)cosh(3y)	32.8706	32.8884	0.9995	0.0000	1.4750	1.4747	0.0015	0.0493	0.0009
sin(2x)cosh(2y) +sin(4x)cosh(4y)	63.6743	63.7274	0.9992	0.0000	-36.2144	-36.2023	0.0027	0.1699	-0.0386
sin(3x)cosh(3y) +sin(4x)cosh(4y)	81.0234	81.1079	0.9990	0.0000	-42.1072	-42.0897	0.0060	0.4864	-0.1010

Observation points (x_i,y_i)=(0.1,0.1),(0.2,-0.1),(0.3,0.1),(0.4,-0.1),(0.5,0.1),(0.6,-0.1),(0.7,0.1),(0.8,-0.1),(0.9,0.1)

equation.

The Laplace equation expressed by Eq. (25) was estimated again. The estimation of coefficients in the governing equation was made for assumed order N = 1, 2 and 3. The value of $a_{N,0}$ was set to be unity. The procedure of the estimation is the same as that described in the foregoing section. Evaluation of ϕ and its derivatives was made at several observation points for given distributions of ϕ. In the evaluation of derivatives finite difference equations (27) was used. Similar finite difference formulae were employed for evaluating $\partial^3\phi/\partial x^3$, $\partial^3\phi/\partial y^3$, $\partial^3\phi/\partial x^2\partial y$ and $\partial^3\phi/\partial x\partial y^2$. These values were used to calculate matrix **[D]** and vector **{f}**, which were substituted into Eq. (5) or (5)' to estimate vector **{a}**.

Estimated results for the assumed order N = 1 and 3 are shown in Tables 6 and 7, respectively. Table 6 shows that when the assumed order is smaller than the actual one, estimated values of the coefficients depend on distribution of ϕ used in the estimation. As can be seen from Table 5 consistent results can be obtained for most distribution functions when the assumed order is coincident with the actual one. From Table 7, when the assumed order is higher than the actual one, consistent results can not be obtained again. The principle of parsimony, therefore, works well again and the governing equation can be estimated when the assumed order was increased incrementally from one until consistent estimation is obtained.

The successive elimination method was applied to the results for N = 3 given in Table 7. Differential operators defined by the following equation are introduced.

$$p = \partial/\partial x,\quad q = \partial/\partial y \tag{29}$$

Then, the third row in Table 7 reads as follows.

$$(p^3 + 26.1106q^3 + 26.1215p^2q + 0.9996pq^2 - 1.3523p^2 - 1.3520q^2 + 0.0004p + 0.0098q - 0.0002)\phi = 0 \tag{30}$$

This equation is approximated by

$$\{0.998p + 26.116q - 1.352\}\underline{(p^2+q^2)\phi = 0} \tag{31}$$

whose underlined part is identical with the actual governing equation (25). Thus the original governing equation superposed with its differentiated forms is estimated when the assumed order N is higher than the actual one for the estimation of partial differential equation also. When the equation corresponding to the second row of Table 7 is used to eliminate p^3 terms, the equations corresponding to the third to seventh rows are reduced to,

$$(q^3 + 1.0034p^2q - 0.3403p^2 - 0.3402q^2 + 0.0005q)\phi = 0 \tag{32}$$

$$(q^3 + 1.0018p^2q - 0.9057p^2 - 0.9054q^2 + 0.0001p + 0.0008q - 0.0002)\phi = 0 \tag{33}$$

$$(q^3 + 1.0027p^2q - 0.1273p^2 - 0.1273q^2 + 0.0020q)\phi = 0 \tag{34}$$

$$(q^3 + 1.0018p^2q - 0.7426p^2 - 0.7424q^2 + 0.0031q - 0.0007)\phi = 0 \tag{35}$$

$$(q^3 + 1.0018p^2q - 0.6458p^2 - 0.6456q^2 + 0.0001p + 0.0067q - 0.0014)\phi = 0 \tag{36}$$

Since these equation are not consistent with each other, q^3 terms are eliminated using Eq.(32).

$$(p^2q + 353.3750p^2 + 353.2500q^2 - 0.0625p - 1.1875q + 0.1250)\phi = 0 \tag{37}$$

$$(p^2q - 304.2857p^2 - 304.1429q^2 - 2.1429q)\phi = 0 \tag{38}$$

$$(p^2q + 251.4375p^2 + 251.3750q^2 - 1.6250q + 0.4375)\phi = 0 \tag{39}$$

$$(p^2q + 190.9375p^2 + 190.8750q^2 - 0.0625p - 3.8750q + 0.8750)\phi = 0 \tag{40}$$

Further elimination of p^2q terms yields

$$(p^2 + 0.9959q^2 - 0.0001p + 0.0014q + 0.0002)\phi = 0 \tag{41}$$

$$(p^2 + 0.9994q^2 - 0.0006p + 0.0043q - 0.0031)\phi = 0 \tag{42}$$

$$(p^2 + 0.9996q^2 - 0.0000p + 0.0165q - 0.0046)\phi = 0 \tag{43}$$

Theses results are almost identical with each other

and close to the actual one.

An extension of the inversion scheme was made, which enabled us to estimate the governing equation from noisy observations (Shiojiri, et al., 1993). A method was proposed for determining the optimum value of the step in finite difference evaluation of the derivatives, which minimized the effect of errors in the observed field quantities on the estimated derivatives. The least distance method was proposed taking account of the amount of error involved in the estimated derivatives.

Numerical simulations showed that the proposed scheme was useful for estimating the governing differential equation from noisy observations.

CONCLUSIONS

An inversion scheme for governing equation inverse problems was proposed. The identification of the governing differential equation was reduced to the estimation of the order and the coefficients of the differential equation. The values of the field quantity observed under several conditions were simultaneously used in the estimation.

Numerical simulations were made on the application of the proposed scheme to the identification of an ordinary differential equation and a partial differential equation. The proposed scheme was found useful for estimating the governing differential equation, even when the order of the differential equation was not known in advance.

ACKNOWLEDGEMENTS

The work is partly supported by the Ministry of Education, Culture and Science under Grant-in-Aid for Co-operative Research (principal investigator: S. Kubo, No. 02302038).

REFERENCES

Amer. Math. Soc., 1984, SIAM AMS Proc., "Inverse Problems", Amer. Math. Soc., Providence, RI.

Gladwell, G.M.L., 1986, "Inverse Problems in Vibration", Martinus Nijhoff Publ., Dordrecht.

Groetch, C.W., 1984, "The Theory of Tikhonov Regularization for Fredholm Equations of the First Kind", Martinus Nijhoff, London.

Japan Soc. Mech. Engrs. (Ed.), 1991, Computer Analyses of Inverse Problems, Corona Publ., Tokyo.

Kubo, S., and Ohji, K., 1987, "Applications of the Boundary Element Method to Inverse Problems", Chapter 10, Applications of the Boundary Element Method, (Ed. Japan Soc. for Computational Methods in Eng.), Corona Publ., Tokyo, pp.181-198.

Kubo, S., 1988, "Inverse Problems Related to the Mechanics and Fracture of Solids and Structures", JSME Int. J., Ser.I, Japan Soc. Mech. Engrs., Vol.31, pp. 157-166.

Kubo, S., 1992, "Inverse Problems", Baifukan, Tokyo.

Kubo, S., 1993a, "Classification of Inverse Problems Arising in Field Problems and Their Treatments", in Proc. IUTAM Symposium on Inverse Problems in Engineering Mechanics, Tokyo (Tanaka, M., and Bui, H.D. (Ed.), Springer-Verlag, Tokyo, pp. 51-60.

Kubo, S. (Ed.), 1993b (to be published), "Inverse Problems", Specialized Monograph of Int. Conf. on Computational Engineering Science, Hong Kong, Technology Publ., Atlanta, GA.

Lavrent'ev, M.M., Romonov, V.G., and Shishatshii, S.P., 1986, "Ill-Posed Problems of Mathematical Physics and Analysis", Amer. Math. Soc., Providence.

Musha, T., and Okamoto, Y., 1992, "Inverse Problems and Their Solvers", Ohm Publ., Tokyo.

Romanov, 1987, "Inverse Problems of Mathematical Physics", VNU Sci. Press, Utrecht.

Shiojiri, A., Kubo, S., and Ohji, K., 1993, "An Analysis Scheme for Governing Equation Inverse Prolems", Preprint of Japan Soc. Mech. Engrs., No. 934-3, pp. 43-45.

Tanaka, M., and Bui, H.D. (Ed.), 1993, "Proc. IUTAM Symposium on Inverse Problems in Engineering Mechanics", Springer-Verlag, Tokyo.

Tikhonov, A.N., and Arsenin, V.Y., 1977, "Solutions of Ill-Posed Problems", John Willy & Sons, New York, NY.

Yamaguti, M., Hayakawa, K., Iso, Y., Mori, M., Nishida, T., Tomoeda, K., and Yamamoto, M. (Ed.), 1991, "Inverse Problems in Engineering Science", ICM-90 Satellite Conf. Proc., Springer-Verlag, Tokyo.

INVERSE ANALYSIS OF THE ELASTOMER CURE CONTROL OF THE VULCANIZATION DEGREE

Y. Jarny and D. Delaunay
Laboratoire de Thermocinétique
ISITEM

J. S. Le Brizaut
Laboratoire d'Automatique
Ecole Centrale de Nantes

Université de Nantes
Nantes, France

ABSTRACT

An inverse analysis is developed to design the heating cycle at the boundaries of thick pieces of elastomer in order to control the state of cure at the end of the cycle. The problem is formulated as a functional optimization problem and it is solved for a one-dimensional geometry. Energy equation in the material is coupled with a kinetic equation. Kinetic model is built from the standard isothermal calorimetry analysis and experimental validation is presented. Results obtained with isothermal and optimal heating cycles are compared.

INTRODUCTION

Optimization of heating cycles in thermal processing of such materials as elastomer or polymer is a key point actively being sought in rubber and plastics industries. The goals are to reduce capital and energy requirements and to improve the quality of products.

This problem cannot be solved without extensive and time consuming experimental works. In this paper it is shown how experiments can be combined with an inverse analysis in order to formulate the problem of the determination of the heating cycle during the cure and then to solve it by the use of general functional optimization tools. The methodology is presented here for the determination of optimal heating for the cure of a thick piece of rubber and could be used for other materials.

The first step consists in building a mathematical model which includes heat conduction equation coupled with a kinetic equation. The kinetic equation describes the exothermic reaction occuring in the material during the cure. Determination of the kinetic parameters is the important phase which has been done by standard isothermal calorimetry analysis. Experimental validation must involves two distinct parts: comparison of temperature measurements during the cure with calculated values and comparison of calculated state of cure with the actual degree of cure in the piece. Mechanical properties of the material are related to the final state of cure , they are not discussed here.

The second step is to predict a time-dependant thermal boundary conditions for reaching a desired state of cure. This determination is achieved by inverse analysis and consists in finding the minimum of a functional criterion.

In this paper we describe how experiments leads to the kinetic model building and to the choice of a minimization criterion. Results obtained by the cure of an industrial rubber with usual isothermal cycles and with optimal heating cycles are compared .

THE VULCANIZATION MODEL

To simulate coupling between heat transfer and the rate of vulcanization in a thick piece of rubber, we consider [1] a one-dimensional distributed parameter system which describes the temperature T(x,t) and the state of cure α(x,t) on the spatial domain (0,L), over the time interval (0,tf)

$$\rho C\frac{\partial T}{\partial t} = \frac{\partial}{\partial x}(\lambda\frac{\partial T}{\partial x}) + \rho Q_\infty\frac{\partial \alpha}{\partial t} \quad (1)$$

$$\frac{\partial \alpha}{\partial t} = F(\alpha,T) \quad (2)$$

$$T(0,t) = T(L,t) = u(t) \quad (3)$$

$$T(x,0) = T0(x) \quad (4)$$

$$\alpha(x,0) = 0. \quad (5)$$

u(t) = the boundary temperature is the control variable.

Kinetic modeling

The parameters of the kinetic equation (2) are determined from isothermal calorimetry analysis.

Samples of rubber are cured at different constant temperatures T_{iso}. The heat flux $\Phi(t)$ delivered during the cure is used to determine the rate of cure $\alpha(t)$ as follows:

$$Q_\infty = \int_0^\infty \Phi(t)\,dt$$

$$Q_\infty \frac{d\alpha}{dt} = \Phi(t)$$

then $$\alpha(t) = \frac{1}{Q_\infty}\int_0^t \Phi(t)\,dt. \quad (6)$$

Figure 1. shows the influence of the temperature on the rate of cure under isothermal conditions.These results lead [1] to the consideration of two distinct phases during the cure :

- phase #1 : induction period

$$\frac{\partial\alpha}{\partial t}(t) = 0, \qquad t < ti(T) \quad (7)$$

- phase #2 : cross-linking period

$$\frac{\partial\alpha}{\partial t}(t) = k(T)\,g(\alpha),\ t > ti(T) \quad (8)$$

Experimental determination of g(α) and k(T)

From calorimetry measurements $\alpha(t)$ and $\frac{\partial\alpha}{\partial t}$ are available according to equation (6). By plotting $\frac{\partial\alpha}{\partial t}$ with respect to α for different temperatures T_{iso} we get $k(T_{iso})g(\alpha)$ as in figure 2, maxima are observed for a unique value of α denoted $\alpha*$ which does not depend on T_{iso}. So, these functions can be normalized by a unique function $g(\alpha)$ and its maximum value is equal to one for $\alpha=\alpha*$. Since $g(\alpha*)=1$, the reaction rate $\frac{d\alpha*}{dt}$ equals to $k(T_{iso})$. Using measurements of $\frac{d\alpha*}{dt}$ at different temperatures T_{iso}, the function k(T) can be approximated in the temperature range of cure $[T_a , T_b]$, by the Arrhenius law in the form :

$$k(T) = k_{ref} \exp\left(-A\left(\frac{T_{ref}}{T} - 1\right)\right) \quad (9)$$

where T_{ref} is an arbitrary reference temperature (Kelvin) chosen in the interval $[T_a , T_b]$. Figure 3. shows a typical variation of k(T) obtained with an industrial rubber (IR). A physical interpretation of the parameter A is given by considering the variation of temperature ΔT around the reference T_{ref} for which the kinetic is twice faster, that is

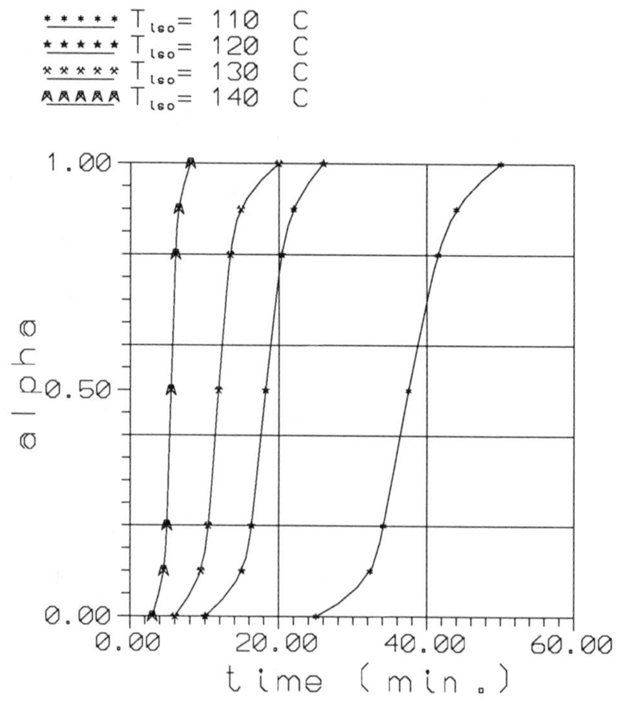

Figure 1 : Rate of cure for different temperature T_{iso}

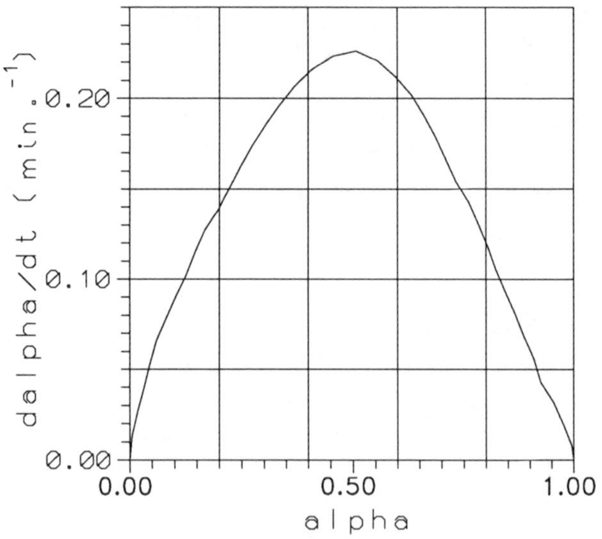

Figure 2: $k(T_{iso})*g(alpha)$
T_{iso} = 120 C

$$k(T_{ref}+\Delta T) = 2\,k_{ref}$$

then, from equation (9), it comes :

$$A = \ln 2\,\frac{T_{ref}+\Delta T}{\Delta T} \approx \ln 2\,\frac{T_{ref}}{\Delta T}. \quad (10)$$

Example : for IR, T_{ref} = 400K , A=24, then $\Delta T \approx 12°C$.

Variation of the induction time

The induction time t_i is defined by equations (7), (8). Its variation with temperature is shown in figure 1 and is described as an implicit function of T by :

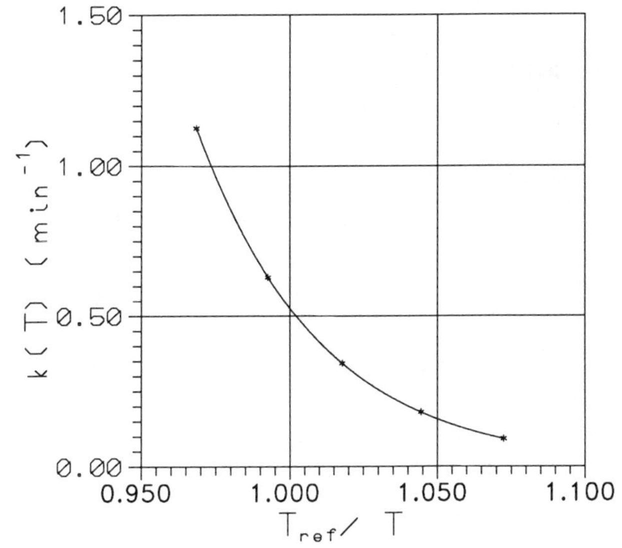

Figure 3: Variation of the rate of cure with temperature

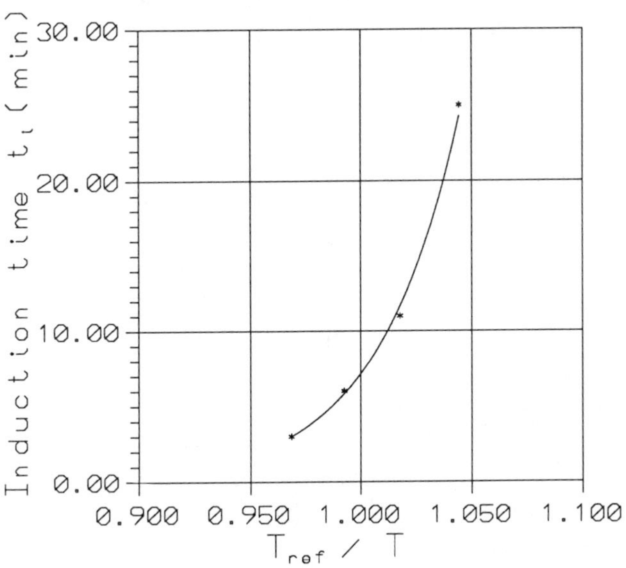

Figure 4: Variation of t_i with temperature

$$\int_0^{t_i} \exp\left(-B\left(\frac{T_{ref}}{T}-1\right)\right) dt = t_{ref} \qquad (11)$$

where t_{ref} = induction time corresponding to an isothermal cure at the temperature $T=T_{ref}$. Figure 4. shows the variation of t_i obtained by isothermal cure of the rubber (IR). As in the law k(T), the parameter B is related to the variation of temperature ΔT around the reference T_{ref}, for which induction time is equal to $0.5t_{ref}$ under isothermal condition of cure. From equation (11), we get :

$$2\, t_{ref} = \exp\left(\frac{B\Delta T}{T_{ref}+\Delta T}\right) t_{ref}$$

then $$B = \ln 2 \left(\frac{T_{ref}}{\Delta T}+1\right) \approx \ln 2\, \frac{T_{ref}}{\Delta T}. \qquad (12)$$

Example : for rubber IR, T_{ref}=400K and B= 27.5,ΔT = 10 °C

Kinetic equation

Let us introduce the functions h and w :

$$h(T)(x,t) = \int_0^t \exp\left(-B\left(\frac{T_{ref}}{T}-1\right)\right) dt - t_{ref} \qquad (13)$$

$$w(x) = 0 \qquad \text{if } x \le -\varepsilon$$

$$w(x) = 1-2\left(\frac{x}{\varepsilon}\right)^3-3\left(\frac{x}{\varepsilon}\right)^2 \qquad \text{if } -\varepsilon < x \le 0 \qquad (14)$$

$$w(x) = 1 \qquad \text{if } x>0.$$

Then the kinetic equation is written in the form of eq.(2) :

$$\frac{\partial \alpha}{\partial t} = F(T,\alpha)$$

by taking :

$$F(T,\alpha) = w(h(T))\, k(T)\, g(\alpha). \qquad (15)$$

The set of six parameters $\{Q_\infty, T_{ref}, t_{ref}, k_{ref}, A, B\}$ and the function $g(\alpha)$ which satisfies the inequalities

$$0 < g(\alpha) \le g(\alpha*) \le 1 \text{ for } 0 \le \alpha \le 1$$

are needed to determine the kinetic equation.

A Lagrangian method [2] can be used for the identification of the function $g(\alpha)$ by minimizing the residual functional based on temperature measurements during the cure in a thick piece. The function calculated by this way are identical to those derived from calorimetry measurements and shown in figure 2. This result constitutes a first step in the validation of the vulcanization model for thick pieces. But the method seems to be limited to highly exothermic reactions (Q_∞ > 20 J/g) for which the temperature field into the thick piece is significantly affected by the heat source due to the reaction. For weakly exothermic reactions, for example Q_∞ = 4 J/g, no thermal effect has been observed during the cure of thick pieces when vulcanization occurs, with isothermal boundary conditions. In that cases, sensitivity of the residual functional to the temperature measurements would be too low to identify the function $g(\alpha)$.

Numerical approximation

The discretization of time in the energy equation is done by the Crank-Nicholson scheme and in the kinetic equation by the Runge-Kutta method. Equations are non-linear, so an iterative method is needed to compute at each time step t_k the temperature $[T_i^k]_{i=1..N-1}$ and the rate of cure $[\alpha_i^k]_{i=1..N-1}$ from the boundary conditions u^k, N=number of nodes in the interval $]0,t_f[$. Numerical results are given in the last section.

CONTROL OF THE RATE OF CURE

Problem Statement

In order to optimize the heating process of cure, we consider the above mathematical model which associates to any admissible control u defined on the time interval $(0,t_f)$ at the boundary of the piece, the spatial distributions of temperature and state of cure denoted by T(u) and $\alpha(u)$ respectively and called the state of the curing process :

$$u \rightarrow \{T(u),\alpha(u)\}.$$

Let us introduced a desired state of cure $\{T_d,\alpha_d\}$, then the control problem consists to determine u* in U_{ad} which minimizes a cost criterion J(u) . The criterion will be defined by a convenient "distance" between $\{T(u),\alpha(u)\}$ and $\{T_d,\alpha_d\}$.

Set of admissible controls

The constraints on the control variable u at the boundary of the piece of rubber are given by inequalities such as :

$$u_{min} < u < u_{max}$$
$$v_{min} < \frac{du}{dt} < v_{max}.$$

To be closer to experimental conditions, the function u(t) have to satisfy other constraints. For example, the cure of the thick pieces have been done with an automatic heating apparatus which is thermaly controled by time-varying heating cycles in the forms shown on figure 6. In that case, the function u(t) is piecewise linear with respect to a given grid $[t_1, t_2, ...t_N]$ of the time interval $[0, t_f]$.

Minimization Criterion

In order to define J(u), a desired state $\{T_d,\alpha_d\}$ has to be specify by choosing:

- a desired temperature $T_d(x,t)$
- a desired rate of cure $\alpha_d(x,t)$

It can be noticed that different formulations could be considered including different set of admissible controls, different constraints, different criteria. For example, the problem could be formulated to optimize energy or time of cure .

In this paper, the study is limited to the specification of the distribution of partial rate of cure $\alpha_d(x)\leq 1$ over the spatial interval (0,L). Partial state of cure is considered because of the non reversibility of the vulcanization process. Any heating cycle with a long enough time of cure t_f leads to the final state $\alpha(x)=1$ everywhere in the piece. So, to avoid this situation which is non realistic for the validation of the method the curing process has to be stopped by an efficient cooling at the end of the heating cycle. This point will be illustrated further.

Three different criteria have been studied in order to predict heating cycle u(t) which leads to a partial state of cure $\alpha_d(x)$ at the final time t_f.

Case #1: $\alpha_d(x)$ is desired at $t = t_f$, so the criterion is defined by :

$$J_1(u) = \int_0^L (\alpha(x,t_f;u) - \alpha_d(x))^2\, dx.$$

Case #2. As in case #1., but during the last period of the time interval, the rate of cure is desired to be as little as possible :

$$J_2(u) = J_1(u) + \mu \int_0^L \int_{t_d}^{t_f} \left(\frac{\partial\alpha}{\partial t}(x,t;u)\right)^2 dt\, dx.$$

μ is a positive parameter .

Case #3. As in case #1 but $\alpha_d(x)$ is desired on the time interval (t_d,t_f) instead of t_f. So, the criterion is :

$$J_3(u) = \int_0^L \int_{t_d}^{t_f} (\alpha(x,t;u) - \alpha_d(x))^2\, dx\, dt.$$

t_d is a parameter which is adjusted to slow down the cure rate at the end of the heating cycle.

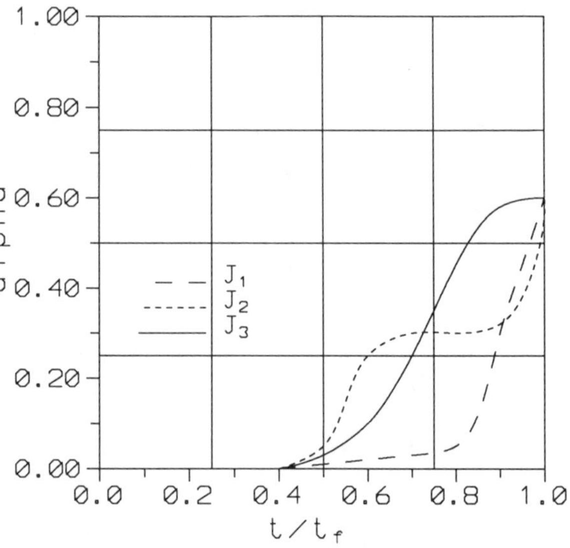

Figure 5: Calculated state of cure alpha(L/2,t;u*) with different criteria

These three criteria are compared by considering the simple case $\alpha_d(x)$ = constant . Figures 5 shows the state of cure $\alpha(L/2,t;\, u_n^*)$ at the middle of the piece calculated with the optimal heating $u_n^*(t)$ which minimizes J_n, n = 1,2,3.

- with the criterion J_1, α_d is reached at $t=t_f$, but the cure rate is maximum at this final time so the cure will continue,
- with the criterion J_2, the time derivative $\frac{d\alpha}{dt}$ goes to zero into a part of the time interval (t_d,t_f) but, at the final time, $\frac{d\alpha}{dt}$ is maximum too, then the cure rate is reduced but it is not stopped,
- with the criterion J_3, the rate of cure is slow down and stopped at the end of the time interval. It means that the calculated optimal heating cycle u* involves a cooling period at the end of the time interval.

Determination of the optimal control

The numerical determination of the optimal control u*(t) which minimizes the criterion J_3 is based on a Lagrangian procedure. This procedure leads to the computation of the gradient ∇J of the criterion by introducing an adjoint equation. Descent method using this gradient are then implemented. Details and commentaries on this procedure are given in [3][4].

The Lagrangian procedure to calculate the gradient with the adjoint state (P,Θ) consists in :

i)- introducing a Lagrangian L which depends on the criterion and on the variational formulation of state equation,

$$L(T,\alpha,P,\Theta,u) = \int_0^L \int_{t_d}^{t_f} (\rho C\frac{\partial T}{\partial t} - \frac{\partial}{\partial x}(\lambda\frac{\partial T}{\partial x}) - \rho Q_\infty F(\alpha,T))P \, dt\, dx$$

$$+ \int_0^L \int_{t_d}^{t_f} (\frac{\partial \alpha}{\partial t} - F(\alpha,T))\, \Theta \, dt\, dx + J_3(u)$$

ii)- taking the variables $\{T, \alpha, P, \Theta, u\}$ as independent variables,

iii)- calculating $\frac{\partial L}{\partial T}(T(u),\alpha(u),P,\Theta,u)$ and

$\frac{\partial L}{\partial \alpha}(T(u),\alpha(u),P,\Theta,u)$

where T(u) and $\alpha(u)$ are the solutions of (1)-(5),

iv)- solving the system :

$\frac{\partial L}{\partial T}(T(u),\alpha(u),P,\Theta,u)\,\delta T = 0$ for all δT admissible,

$\frac{\partial L}{\partial \alpha}(T(u),\alpha(u),P,\Theta,u)\,\delta\alpha = 0$ for all $\delta\alpha$ admissible,

and thus obtaining P(u) and $\Theta(u)$,

v)- writing that $\nabla J_h = \frac{\partial L}{\partial u}(T(u),\alpha(u),P(u),\Theta(u),u)$

where T, α, P and Θ satisfy ii), iii) and iv).

NUMERICAL AND EXPERIMENTAL RESULTS

Isothermal heating cycles

To analyze which kinds of partial state of cure $\alpha_d(x)$ can be reached in a thick piece, the simplest heating cycles are studied first. They have the form given in figure 6. They involve four distinct phases, two of them are for heating and two for cooling:

- phase 1, $\frac{du}{dt} = v_{max}$, $\quad o < t < t_1$,
- phase 2, $u = T_{cure} < u_{max}$, $\quad t_1 < t < t_2$,
- phase 3, $\frac{du}{dt} = v_{min}$, $\quad t_2 < t < t_3$,
- phase 4, $u = u_{min}$, $\quad t_3 < t < t_f$.

The maximum rate of heating v_{max} and of cooling v_{min} are fixed by experimental constraints of the apparatus. Initial temperature T_0 and final time t_f being fixed, the heating cycle u(t) is defined by a piecewise linear function which has only two free parameters: the curing temperature T_{cure} and the duration of heating $\Delta t_{cure} = t_2 - t_1$ at this temperature. Two particular cases which are non realistic to reach a desired partial state of cure but easy to realize are:

- $T_0 = T_{cure}$, the material to be cured is heated before the curing process, in that case the induction time of the model has to be updated,

- $t_2 = t_f$, (no cooling phase). The only way to hold a partial state of cure in the piece is to have low temperature , thus this case will give a partial cure which cannot be held or a complete cure.

Because of their forms, these simple heating cycles are called isothermal .

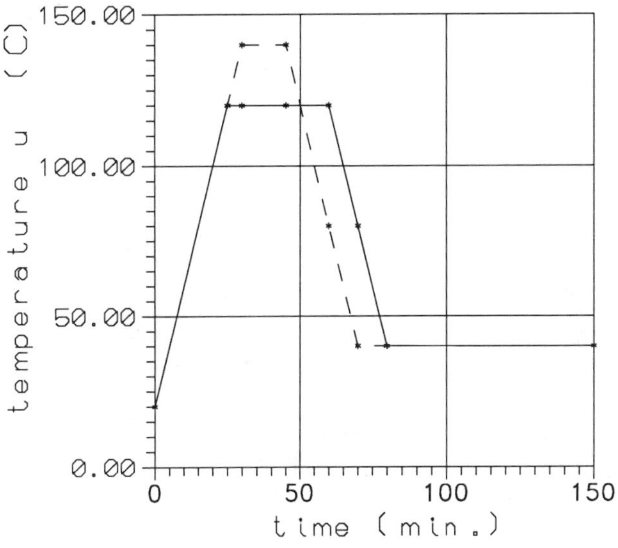

Figure 6: Isothermal heating cycles

Experiments with complete cure

The evolution of T(x,t) and $\alpha(x,t)$ calculated with the following experimental conditions of cure are shown on figures 7 and 8

- thickness of the piece, L = 62,5 mm
- Temperature of cure $T_{cure} = 160$ °C
- final time $t_f = 6800$ s, $t_2 = t_f$, (no cooling phase)
- industrial rubber (IR)

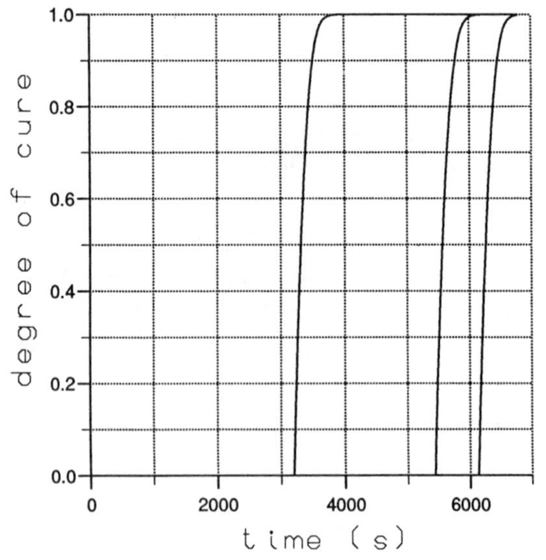

Figure 7.a: Variation of alpha(x_j,t); x_j=j*L/6, j=1,2,3

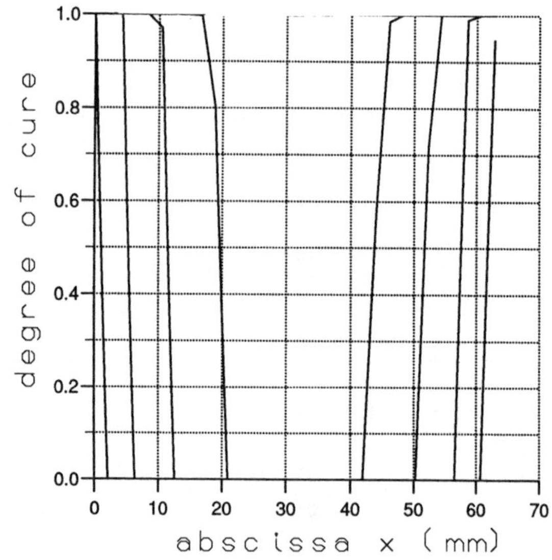

Figure 7.b: Variation of alpha(x,t_k);t_k=k*t_f/5, k=1 to 5

At the final time, the state of cure $\alpha(x,t_f)$ is equal to one everywhere in the piece. It can be observed that for this material and for this heating cycle, the spatial distribution $\alpha(x,t)$ during the cure is very narrow and can be viewed as a phase-change front moving from the boundary to the middle of the piece. Temperature measurements have been recorded during the cure into the piece, and were in good agreement with the calculated values but no measurements of the state of cure were available to validate the kinetic equation of the model.

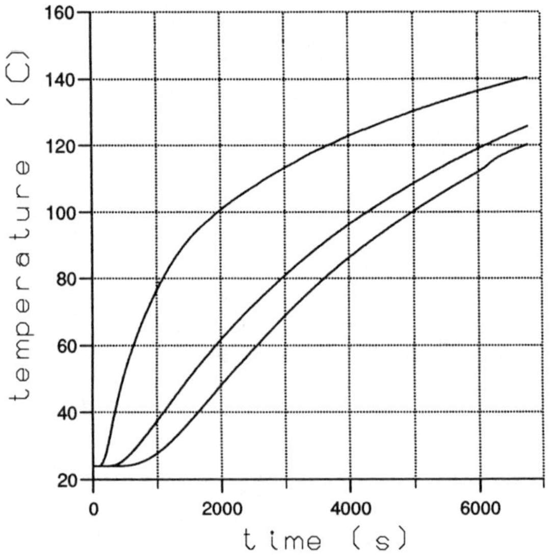

Figure 8.a: Variation of T(x_j,t); x_j=j*L/6, j=1 to 3

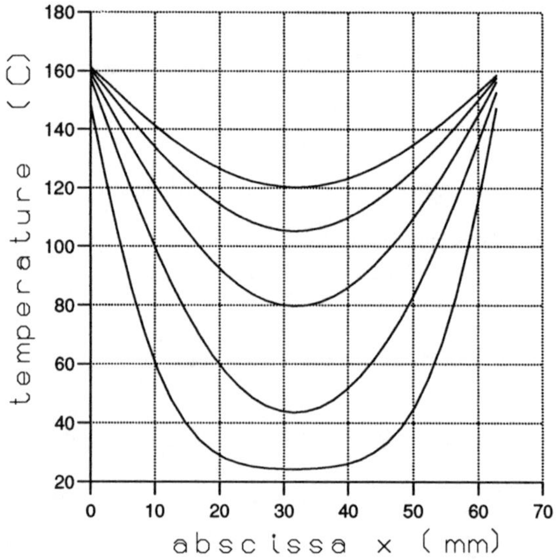

Figure 8.b: Variation of T(x,t_k); t_k=k*t_f/5, k=1 to 5

Experiments with partial cure

Partial cure have been studied with isothermal heating cycles in the form described above. To characterize the front location s(t) which separates the cured and uncured parts in the material during the cure, the mean value $\alpha(s(t),t) = 0.5$ is considered. Because of symetry in the piece, the front location s(t) determines the percentage of cured material, s=L/2 corresponds to a complete cure. Numerical experiments show that the front $s(t_f)$ can be held at a desired value $s_d < L/2$ by choosing the duration Δt_{cure}, figure 9. Due to the fast increasing variation of $s(t_f)$ with Δt_{cure}, the control is difficult or even impossible to do precisely for values of s_d close to L/2. Experiments with different duration of cure have been done to reach s_d = L/8, L/4 and 3L/4 with the following conditions

- thickness of the piece, L = 46 mm
- Temperature of cure T_{cure} = 135°C , T_0 = 20 °C
- duration of cure , Δt_{cure} = 345, 690, 1035 s
- $v_{max} = -v_{min}$, ,v_{max} = 4 °C / min
- final time t_f = 10800 s, $t_2 < t_f$,
- industrial rubber (IR)

At the end of the heating cycles, each piece has been cut to observe the actual state of cure. A front is clearly visible in the piece and the measured location of $s(t_f)$ were very closed to the desired values.

Optimal heating cycles

Optimal heating cycles u*(t) are computed by minimizing criterion J_3. The target $\alpha_d(x)$ is specified by the final front location $s(t_f)$. Figure 10 shows temperature calculated inside the material with a boundary condition u* which minimizes the criterion J_3, we have taken s_d =L/4 = 11.5 mm, and the duration of cure t_f as in the previous experiments.

To compare isothermal and optimal heating cycles, the function $t_{0.5}(x)$ of half cure which satisfies $\alpha(x,t_{0.5}(x)) = 0.5$ is drawn on figure 11. It can be observed that with both isothermal and optimal heating cycles, the desired state $\alpha_d = 0.5$ corresponding to s_d =L/4 can be reached. An advantage of the automatic determination of the heating cycle is to avoid to proceed by trial and errors as it was done for isothermal cycle.

The main difference between isothermal and optimal heating cycles can be observed by comparing the period of the cycle during which the material is completely cured. With isothermal heating this period is longer than with optimal heating in the region close to the boundary of the piece. It is interesting to reduce this period for quality requirements, because a too long heating of the piece when it is cured, can decrease the mechanical properties of the rubber. This phenomenon has not been taken into account in this study .To analyze it, the experimental characterization of the state of cure which is based here on calorimetry measurements , has to be perfected and related to physical properties of the rubber.

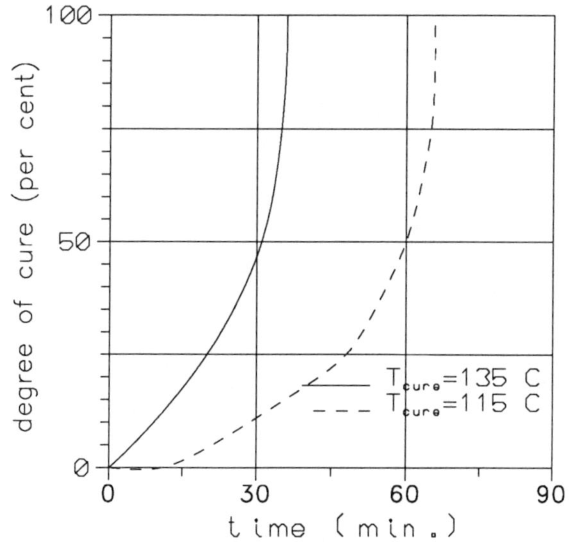

Figure 9: Isothermal cycles Influence of Temperature and duration of cure

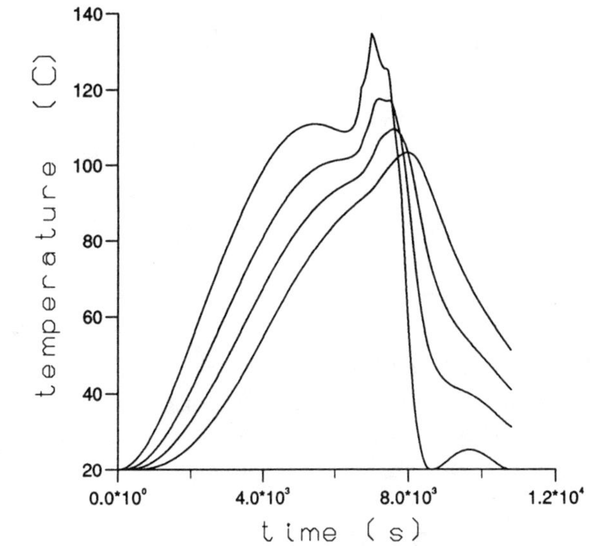

Figure 10: Optimal heating cycle — $T(x_j,t)$; $x_j=j*L/6$ j = 0 to 3

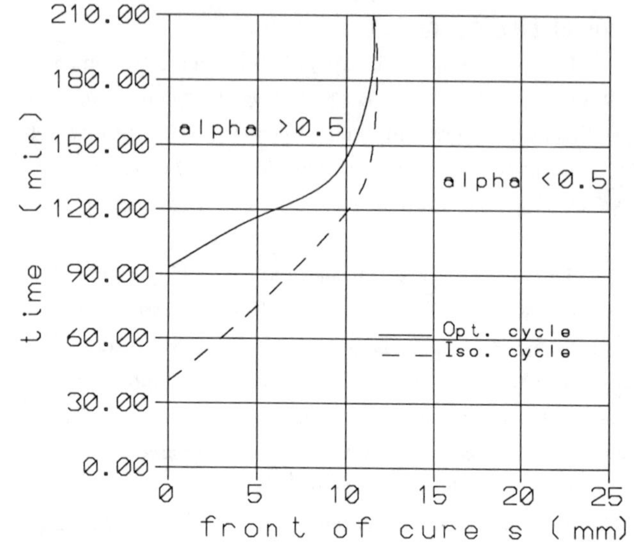

Figure 11: Isothermal and optimal heating cycles Cured and uncured parts in thick piece (L=46 mm)

CONCLUSION

The determination of optimal heating cycle for vulcanization of a thick piece of rubber has been studied in the base of a Lagrangian formulation by the consideration of a model coupling energy and kinetic equations This methodology has been associated to an experimental analysis in order to specify the desired state of cure, to choose a convenient criterion to be minimize, to adjust the duration of the cycle, and has led to satisfying results. One of the main difficulties which had been solved was to take into account the weak value of the rubber thermal diffusivity combined with a high cure rate. Thus the numerical simulation of heating for a thick piece of an industrial rubber has led to characterize the evolution of the state of cure as a moving front in the material. Experimental results have confirmed that the front location can be controled by acting on the temperature boundary condition of the piece.

ACKNOWLEDGMENTS

The authors wish to thanks VIBRACHOC Company for its collaboration and financial support.

REFERENCES

[1] Garnier B., Danes F., Delaunay D., "Coupling between heat transfer and rubber vulcanization", Proc. of IXth Int. Heat Transfer conference,1990, Jerusalem, Israel

[2] Le Brizaut J.S., Delaunay D., Garnier B.,Jarny Y.,"Implementation of an inverse method for identification of reticulation kinetics from temperature measurements in a thick sample" to appear in Int. Journal of Heat and Mass Transfer, 1993.

[3] Burger J., Pogu M., "Functional and Numerical solution of a control problem originating from Heat Transfer". J.OT.A., Vol. 68, N° 1, pp. 49-73, 1991.

[4] Le Brizaut J.S., "Identification et commande d'un système à paramètres répartis modélisant un processus de vulcanisation", Thèse de doctorat, ECN, Nantes, France, 1993

Inverse Problems in Engineering: Theory and Practice
ASME 1993

INVERSE PROBLEMS FOR DETERMINATION OF DROP BREAKAGE AND COALESCENCE FUNCTIONS

A. N. Sathyagal and H. Wright*
School of Chemical Engineering

G. Narsimhan
Department of Agricultural Engineering

D. Ramkrishna†
School of Chemical Engineering

Purdue University
West Lafayette, Indiana

ABSTRACT

Dispersed phase processes are used in a wide range of engineering operations. The processes of drop breakage and coalescence play an important role in the behavior of these operations. Accurate information on the rates of breakage and coalescence is therefore necessary in order to optimise such systems. Direct observation and physical modelling of drop breakage and coalescence processes in turbulent flow fields is extremely difficult. A promising alternative is to solve inverse problems which use experimentally measured transient drop size distributions to get quantitative information on the rates of breakage and coalescence. We have developed techniques to solve the inverse problems of breakage and coalescence. In each case, we scale the measured transient distributions by an appropriate scaling variable. The resulting self-similar distributions are used with a regularization technique to calculate the breakage and coalescence functions. The choice of regularization parameter is based on an analysis of the error in the experimental measurements. Both the inverse problem techniques have been tested out with simulated as well as real experimental data. The techniques give us excellent results in terms of the calculated functions and the predicted transient distributions.

INTRODUCTION

Dispersed phase systems occur in a broad range of engineering applications. For example, liquid-liquid dispersions are commonly used in separation processes where the increased surface area due to the dispersion facilitates mass transfer. Such dispersions are commonly achieved by agitating the two phases together. The agitation causes the drops of the dispersed phase to break up and form smaller droplets, while simultaneously causing two or more drops to join together to form a larger drop. The breakage and coalescence processes change the size distribution of drops in the system which, in turn, can affect the overall performance of the separation process. To get the best performance from the dispersed phase system, we need to control the drop size distribution near the optimum. This requires quantitative knowledge of the drop breakage and coalescence rates and the daughter drop distribution, i.e. the size distribution of drops formed from the breakage of a larger drop.

There have been various approaches used to determine drop breakage and coalescence rates in stirred liquid-liquid dispersions. Park and Blair (1975) used high-speed photography to directly observe drop coalescence events, while Konno et al. (1983) used the same technique to observe drop breakage events. The direct observation technique has some major drawbacks. It is almost impossible to measure a sufficient number of drop breakage or coalescence events to accurately determine their rates. Park and Blair measured only several hundred drop pair interactions while Konno et al. measured less than hundred breakage events.

The most common approach to determine drop breakage and coalescence rates is to model these processes occurring in a stirred environment. The turbulent environment in the stirred vessel makes it very difficult to model the processes. In each of the models in the literature, many simplifying assumptions are made to make the problem tractable. Some examples are the breakage and coalescence models of Coulaloglou and Tavlarides(1977), the coalescence models of Muralidhar and Ramkrishna (1986) and the breakage models of Narsimhan et al.(1979) and Nambiar et al.(1992).

*Present Address: Conoco Research and Engineering, Ponca City, OK
†Author to whom correspondence should be addressed.

Due to the major drawbacks in the previous two approaches, we believe that the best way to get quantitative information on breakage and coalescence is through the solution of inverse problems. We have dealt with the breakage and coalescence phenomena separately. The population balance framework is employed in each case. In the coalescence inverse problem, the population balance equation is inverted to determine the coalescence frequency, while in the breakage inverse problem, the equation is inverted to determine the breakage frequency and the daughter drop distribution.

The remainder of this paper is organized as follows. We first discuss the coalescence inverse problem. The technique is presented followed by an example which uses simulated data. We then discuss the breakage inverse problem by describing the technique and presenting examples with simulated data and experimental data.

INVERSE PROBLEM FOR COALESCENCE

The population balance framework gives us a convenient method to track the evolving drop size distributions. For a dispersed phase system evolving by pure coalescence, the population balance equation is given by

$$\frac{\partial F(v,t)}{\partial t} = -\int_0^v \partial_v F(v',t) \int_{v-v'}^{\infty} \frac{\partial_v F(v'',t)}{v''} K(v',v'') \tag{1}$$

where $F(v,t)$ is the cumulative volume fraction of drops of size less than or equal to v at time t and $K(v,v')$ is the binary coalescence frequency of drops of size v and v'. The inverse problem approach allows the extraction of the coalescence frequency from measurements of the transient drop size distributions. The approach takes advantage of self-preserving size distributions to extract the coalescence frequency.

The self-preserving size distribution is actually a similarity transformation of the population balance equation. The similarity transformation that we seek is $F(v,t) \rightarrow f(z)$ with $z = \frac{v}{S(t)}$ where $S(t)$ is a time-dependent scaling volume to be determined. Substitution of this transformation into the population balance yields

$$zf'(z) = \frac{1}{S'(t)} \int_0^z df(x) \int_{z-x}^{\infty} \frac{df(y)}{y} K(xS(t), yS(t)) \tag{2}$$

where the primes denote differentiation with respect to the argument concerned. By taking the partial derivative on both sides of the above equation, we find the general criterion for the applicability of the similarity transformation:

$$\frac{\partial}{\partial t}\left[\frac{1}{S'(t)} \int_0^z dx f'(x) \int_{z-x}^{\infty} dy \frac{f'(y)}{y} K(xS, yS)\right] = 0 \tag{3}$$

Wright and Ramkrishna(1992) have shown that the above condition constrains the form of the coalescence frequency. It implies that

$$K(xS, yS) = \sigma[S(t)] b(x,y) \tag{4}$$

and

$$\frac{S'(t)}{\sigma[S(t)]} = < b > \tag{5}$$

where σ is an arbitrary function on $S(t)$; b(x,y) is a function of only x and y and is called the scaled coalescence frequency. $< b >$ is a constant that corresponds to the average coalescence rate. We assume that $\sigma = S(t)^\lambda$ since most model frequencies possess this property. Equation (4) becomes

$$K(xS, yS) = S(t)^\lambda b(x,y) \tag{6}$$

and equation (5) becomes

$$\frac{dS(t)}{dt} = < b > S(t)^\lambda \tag{7}$$

Choice of Similarity Variable

In earlier work (Wright et al., 1990), ratios of successive moments of the drop size distribution have been considered as possibilities for $S(t)$. Wright and Ramkrishna(1992) have determined the relationship between the various moments of the size distribution during self-similarity. They have shown that certain choices of $S(t)$ do not yield a similarity solution. For example, in some cases, $\frac{M_1(t)}{M_0(t)}$, where $M_i(t)$ is the i-th moment of the size distribution, is not a good choice for $S(t)$. They also conclude that M_2/M_1 can always be used as $S(t)$. Hence, in this paper, we shall use

$$S(t) = \frac{M_2(t)}{M_1(t)} \tag{8}$$

Computational Details of the Inverse Problem

Substituting equations (6) and (7) into equation (2), we get

$$zf'(z) = \int_0^z dx f'(x) \int_{z-x}^{\infty} dy \frac{f'(y)}{y} \frac{b(x,y)}{< b >} \tag{9}$$

This equation is used to extract the scaled coalescence frequency $b(x,y)$.

Equation (9) is a Volterra-type integral equation of the first kind, and as such may be ill-posed. To solve the equation, we use the technique of Tikhonov regularization (Tikhonov and Arsenin, 1977). We recast equation (9) in operator notation as

$$\mathbf{g} = \mathbf{K}\mathbf{b} \tag{10}$$

where g is the vector representing the left-hand side of

equation (9), $\mathbf{b}$ is the unknown vector representing the function $b(x,y)$, and $\mathbf{K}$ is an integral operator which acts on $\mathbf{b}$. Note that both $\mathbf{K}$ and $\mathbf{g}$ are only known approximately within experimental error. Using Tikhonov regularization, $\mathbf{b}$ is determined by

$$\min ||\mathbf{Kb} - \mathbf{g}||_1^2 + \lambda_{reg}||\mathbf{b}||_2^2 \quad (11)$$

where $||\mathbf{f}||$ is the norm and the subscripts 1 and 2 represent the possiblity that different norms can be used.

In order to solve equation (11) we discretize the similarity coordinate into a set of $\{z_i\}$. Also, $b(x,y)$ is expanded in terms of basis functions,

$$\frac{b(x,y)}{<b>} = \sum_j a_j l_j(x,y) \quad (12)$$

This transforms the operator equation (11) into a matrix minimization problem. The choice of the discretization points and the basis functions is determined by the nature of the similarity distribution.

Wright and Ramkrishna(1992) have shown that the similarity distributions for various model coalescence frequencies show the largest variation in the region of small z. Since the small z behavior of the similarity distribution dominates the behavior of the extracted coalescence frequency, we choose the discretization points to be logarithmically spaced from z_{min} to z_{max}.

The choice of basis functions is also based on this knowledge about the similarity distributions. We define the inner product to be

$$<\mathbf{f},\mathbf{g}> = \int_0^\infty dx e^{-x} \int_0^\infty dy e^{-y} f(x,y) g(x,y) \quad (13)$$

which weights the small x and y regions. Orthogonal functions on such a space are tensor products of the Laguerre polynomials $L_i(x)$. Thus,

$$l_n(x,y) = L_i(x) L_j(y) \quad (14)$$

where $n = (i-1)n_{basis} + j$. n_{basis} is the number of basis functions along one axis.

We also need a good method to represent the similarity distribution. Fitting the distribution has the effect of limiting experimental error and guaranteeing a continuous distribution. The functional form chosen for fitting the similarity distribution is

$$f'(z) = \sum_{k=1}^{n_{term}} A_k z^{\alpha_k - 1} \exp(-\beta_k z) \quad (15)$$

This form has the ability to exhibit the different qualitative and quantitative behaviors of the similarity distribution.

Substituting the expressions for the similarity distribution and the unknown function $\mathbf{b}$, the solution of equation (11) becomes a constrained minimization problem to determine the unknown coefficients of expansion, a_j. The solution is constrained to be symmetric and positive.

Example Application

We present an example of the coalescence inverse problem procedure. The example uses the self-similar size distribution from a known model frequency where we compare the inverse problem result with the known frequency.

The known model frequency chosen in this case is $K(x,y) = 1$. The self-similar size distribution is known for this case and is given by

$$\frac{f'(z)}{z} = 4e^{-2z} \quad (16)$$

The average size evolves as

$$S(t) = 1 + t \quad (17)$$

Choosing 80 points (corresponding to the discretization points) from the analytic similarity distribution, we add normally distributed error with a standard deviation equal to 10% of the analytical value. The resulting distribution is fit to equation (15). Figure 1 shows the resulting distribution with the best fit. Similarly, 10% error is also added to the actual values of $S(t)$ and the data is fit to equation (7).

The fitted parameter values are used to extract the scaled coalescence frequency for various values of the regularization parameter. The optimum regularization parameter is chosen to be the largest possible that is consistent with the uncertainty in the self-similar distribution. Wright and Ramkrishna(1992) have analyzed the asymptotic behavior of the similarity distributions for quite general coalescence frequencies. The results of this asymptotic analysis are used to determine the uncertainty in the distribution. More details on the choice of the regularization parameter are given in Wright and Ramkrishna(1992).

Using this technique, the optimum regularization parameter is determined to be $\lambda_{reg} = 10^{-8}$. The results of the extracted frequency are shown in Figure 2.

The inverse problem procedure has also been successfully used to calculate the aggregation frequencies of fractal clusters (Wright et al., 1992) and to obtain drop coalescence frequencies from experimental data (Wright, 1991).

INVERSE PROBLEM FOR BREAKAGE

For a dispersed phase system evolving by pure breakage, the population balance equation is given by

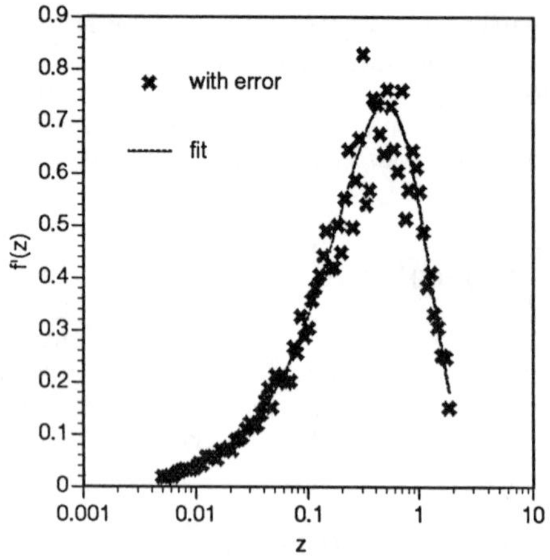

Figure 1: Similarity distribution from constant frequency with 10% added error.

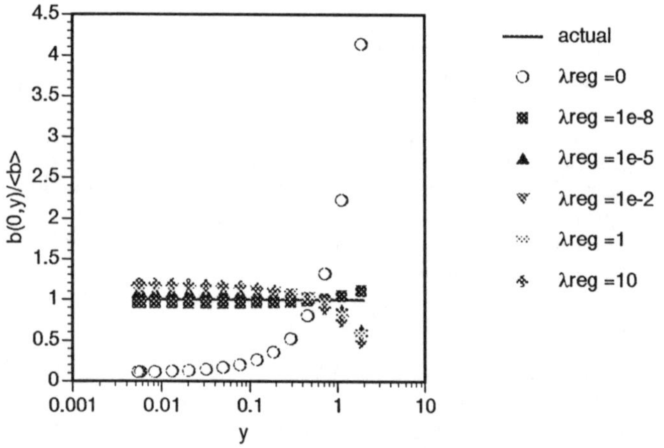

Figure 2: Inverse problem results from similarity distribution of constant frequency with 10% added error. The axes values $b(0,y)$ plotted versus y.

$$\frac{\partial F(v,t)}{\partial t} = \int_v^\infty \Gamma(v')G(v,v')\partial_v F(v',t) \qquad (18)$$

where $\Gamma(v)$ is the breakage rate of drops of size v and $G(v,v')$ is the cumulative daughter drop distribution, i.e. it is the cumulative volume fraction of drops of size less than or equal to v formed from the breakage of a drop of size v'. The inverse problem for breakage allows the extraction of the breakage rate and the daughter drop distribution from measurements of transient drop size distributions. As in the drop coalescence case, the inverse problem for breakage takes advantage of self-preserving size distributions to extract the breakage rate and daughter drop distribution.

The basis for similarity lies in assuming that $G(v,v')$ is of the form

$$G(v,v') = g\left(\frac{\Gamma(v)}{\Gamma(v')}\right) \qquad (19)$$

As explained by Narsimhan et al.(1980), the above form is based on the hypothesis that larger drops break more thoroughly than smaller drops. Equation (20) admits a similarity transformation $F(v,t) \rightarrow f(\zeta)$, where $\zeta = \Gamma(v)t$. Substitution of this transformation into equation (20) yields

$$\zeta f'(\zeta) = \int_\zeta^\infty g\left(\frac{\zeta}{\zeta'}\right)\zeta' f'(\zeta')d\zeta' \qquad (20)$$

We also considered the possibility of the population balance equation for breakage admitting a similarity transformation of the same form as that used for coalescence, i.e. a transformation with $z = \frac{v}{S(t)}$. In this case, it is assumed that $G(v,v')$ is of the form $G(v,v') = g_1(v/v')$. An analysis of this transformation shows that it is only valid when $\Gamma(v)$ is a power-law function, i.e. $\Gamma(v) = av^\lambda$. All the models for drop breakage in stirred dispersions show a breakage rate that is not a power-law function. The ζ transformation is not restricted to power-law functions. Since the ζ transformation covers a broader range of functions and since it encompasses the range of functions valid for the z transformation, we have developed the inverse problem for breakage in terms of the ζ transformation.

Note that in the transformation $\zeta = \Gamma(v)t$, the breakage rate $\Gamma(v)$ is unknown. Narsimhan et al.(1980) have shown that an experimental test of the applicability of the similarity transformation is to examine a plot of $\ln(t)$ versus $\ln(v)$ curves for various fixed values of the volume fraction F. If the $\ln(t)$ versus $\ln(v)$ curves for different F can be collapsed into a single curve by a vertical translation, then the similarity hypothesis is upheld. As shown by Narsimhan et al.(1980),

$$\frac{d\ln\Gamma(v)}{d\ln t} = -\left(\frac{\partial \ln t}{\partial \ln v}\right)_F \qquad (21)$$

The collapsed curve corresponding to the above equation also allows us to calculate the breakage rate up to a multiplicative constant. By integrating equation (21), we get

$$\Gamma(v) = \gamma \exp\left[-\int_{\ln v_0}^{\ln v}\left(\frac{\partial \ln t}{\partial \ln v}\right)_F d\ln v\right] \qquad (22)$$

where v_0 is a reference volume and γ corresponds to the breakage rate of a drop of volume v_0. By redefining the similarity variable as $\zeta = \frac{\Gamma(v)}{\gamma}t$, the population balance equation in terms of ζ becomes

$$\zeta f'(\zeta) = \gamma\int_0^1 \frac{\zeta^2}{u^3} f'(\frac{\zeta}{u})g(u)du \qquad (23)$$

This equation is used to extract the daughter drop distribution, $g(x)$.

Computational Details of the Inverse Problem

The similarity hypothesis is tested by checking if the different $\ln t$ versus $\ln v$ curves collapse under a vertical translation. This is done by calculating the curvature and arc length of the curves. If the arc length and curvature of the different curves coincide in the regions where the curves overlap, the curves are said to have collapsed into a single curve. The equation of the collapsed curve is calculated from the common arc length curve. The breakage rate, $\Gamma(v)$, (up to a multiplicative constant γ) is then determined from this collapsed curve via equation (22).

Equation (23) for the determination of the daughter drop distribution is a Fredholm equation of the first kind and hence is an ill-posed problem. Since the daughter drop distribution, $g(x)$ is a cumulative distribution function, it is a monotonic function of its argument. Since monotonic functions are convex in the space L_2 (Tikhonov and Arsenin, 1977), we can use the method of quasisolutions (Ivanov, 1962) to solve equation (23). The fact that $g(x)$ is a cumulative distribution function also allows us to determine the unknown constant γ. Since at $x = 1$, $g = 1$, the solution of equation (23) for $\gamma g(x)$ will allow us to determine γ. The equation can be recast in operator notation as

$$\boldsymbol{\Phi} = \mathbf{K}\mathbf{g} \tag{24}$$

where $\boldsymbol{\Phi}$ is the vector representing the left-hand side of equation (23), $\mathbf{g}$ is the unknown vector representing the function $\gamma g(x)$ and $\mathbf{K}$ is the integral operator which acts on $\mathbf{g}$. Using the method of quasisolutions, $\mathbf{g}$ is determined by

$$\min_{\mathbf{g} \in \mathbf{M}} \|\mathbf{K}\mathbf{g} - \boldsymbol{\Phi}\|^2 \tag{25}$$

where $\mathbf{M}$ is the set of compact functions to which $\mathbf{g}$ belongs. In this case, $\mathbf{M}$ corresponds to the set of monotonic functions.

As in the case of the coalescence inverse problem, the similarity coordinate is discretized into a set of ζ_i and $g(x)$ is expanded in terms of basis functions.

$$\gamma g(x) = \sum_{j=1}^{nb} a_j G_j(x) \tag{26}$$

where nb is the number of basis functions. The discretization intervals were chosen that there were equal areas under the $f'(\zeta)$ curve in each interval. The inner product space was defined as

$$< \mathbf{u}, \mathbf{v} >= \int_0^1 x u(x) v(x) dx \tag{27}$$

Orthogonal polynomials on such a space are the Jacobi polynomials, $G_j(x)$. We also fit the similarity distribution with a functional form similar to that used in the coalescence case.

$$\zeta f'(\zeta) = \sum_{k=1}^{n_{term}} A_k \zeta^{\alpha_k - 1} \exp(-\beta_k \zeta) \tag{28}$$

Substituting the above expressions for the similarity distribution and the unknown function g, the solution of equation (23) becomes a constrained minimization problem to determine the unknown coefficients of expansion a_j. The solution is constrained to be a monotone increasing function.

We also found that as we increased the number of basis functions used to construct the unknown function, some residual amount of regularization was required. This is due to the fact that as the number of basis functions is increased, the solution has greater freedom to act within the constraints and give spurious results.

Example Applications

We present two examples of the breakage inverse problem procedure. In the first case, we use known breakage functions and compare the inverse problem result with the known functions. In the second case, we use experimental data of transient drop size distributions in a purely breaking system of a neutrally buoyant mixture of benzene and carbon tetrachloride in water.

To test out the inverse problem procedure, we generated transient drop size distributions using a Monte Carlo simulation procedure. The initial drop size distribution was monodisperse. Since the data was generated by simulations, it inherently has some error. The breakage functions used were

$$\Gamma(v) = \exp(1.046(\ln v + 3.49) - 0.042((\ln v)^2 - 12.18)) \tag{29}$$

and

$$g(x) = \sin \frac{\pi x}{2} \tag{30}$$

From the simulated data, plots of $\ln t$ versus $\ln v$ at 14 different values of F were made. Calculating the curvature and arc length of the various curves, it was found that the curves all collapsed into one. The breakage rate was determined from the collapsed curve. Figure 3 shows the calculated breakage rate along with the actual one given by equation (29). From the figure it can be seen that the extracted breakage rate is very close to the actual. Figure 4 shows the resulting similarity distribution calculated from the extracted breakage rate along with the fit of the distribution using equation (28). This similarity distribution was then used to calculate the daughter drop distribution. This result is shown in Figure 5. The different curves are the extracted daughter drop distribution for different numbers of basis functions. As the figure shows, the result with 4 basis functions comes very close to the actual function. The figure also shows that the result with 7 basis functions shows some spurious detail. When the inverse problem was run again with 7 basis functions with some regularization, the spurious detail was eliminated. This result is shown in Figure 6.

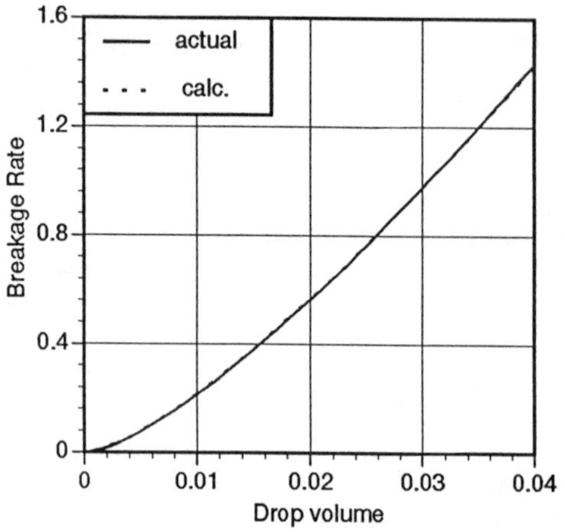

Figure 3: Comparison of calculated breakage rate from the inverse problem with the actual function.

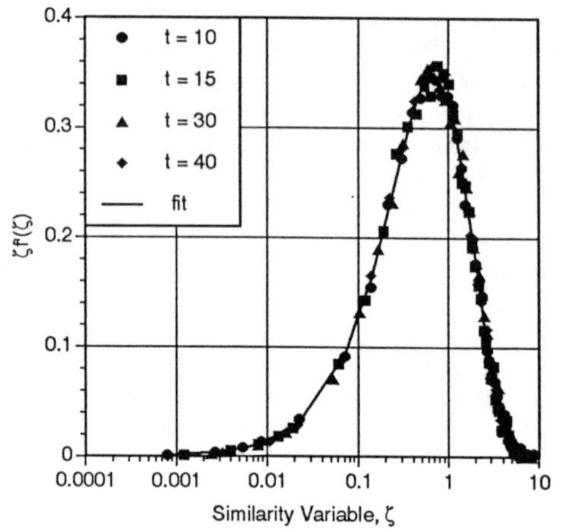

Figure 4: Similarity distribution for data simulated with known breakage functions.

Example with Experimental Data. Transient breakage experiments were carried out with a system of neutrally buoyant benzene and carbon tetrachloride in water. By using a very small amount of the organic phase, the drop size distribution in this system evolves by pure breakage for some time. The experimental data used in this example were obtained by stirring the system at 500 RPM. Samples were taken out of the stirred vessel at various times and stabilized on a slide with a small amount of surfactant. Images of the drops on the slides were taken, digitized and analyzed to get the size distribution using the Quantimet Q570 image analysis system. Up to 3000 drops were measured to obtain each transient size distribution.

From the plots of $\ln t$ versus $\ln v$ for 14 different values of F, it was found that the curves all collapsed into one, upholding the similarity hypothesis. From the collapsed curve, the breakage rate was calculated. The resulting breakage rate is shown in Figure 7. The resulting similarity distribution is then shown in Figure 8. Using the method of quasisolutions, the daughter drop distribution is calculated from this similarity distribution. The results for different numbers of basis functions are shown in Figure 9. As in the case of the simulated data, the results with larger number of basis functions seems to show some spurious detail. The results for the breakage rate and daughter drop distribution obtained from the inverse problem were used to predict the transient size distributions by the population balance equation. The initial condition for the solution of the population balance equation is taken to be the distribution at the time at which similarity is first observed. The predicted distributions in terms of the volume fraction density, $f(v,t) = \frac{\partial F(v,t)}{\partial v}$, for the daughter drop distribution determined with 5 basis functions is shown in Figure 10. From the figure it can be seen that the transient distributions are predicted very well by the extracted functions. The result with 5 basis functions predicts the data within the experimental error and is chosen to be the daughter drop distribution for the given experimental conditions.

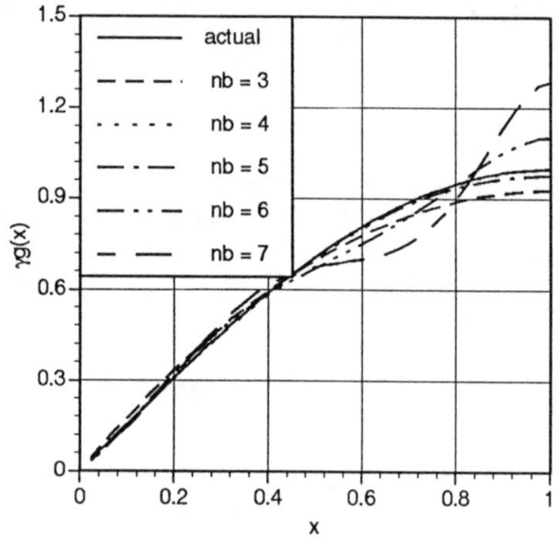

Figure 5: Daughter drop distribution calculated from the inverse problem.

SUMMARY AND CONCLUSIONS

Techniques for the extraction of the coalescence frequency, breakage rate and daughter drop distribution from transient drop size distributions have been given. The techniques make use of similarity transformations of the population balance equation. Conditions under which these transformations are valid and choices for the similarity variable are given.

Computational details of the technique for the determination of the functions from the similarity distributions are given. In the coalescence inverse problem, Tikhonov regularization is used to determine the coalescence frequency. In the breakage inverse problem, the method of quasisolutions is used to determine the daughter drop distribution.

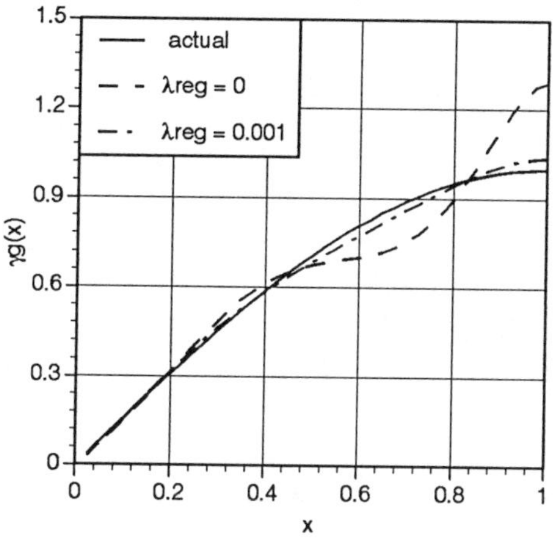

Figure 6: Effect of regularization on the daughter drop distribution extracted with 7 basis functions.

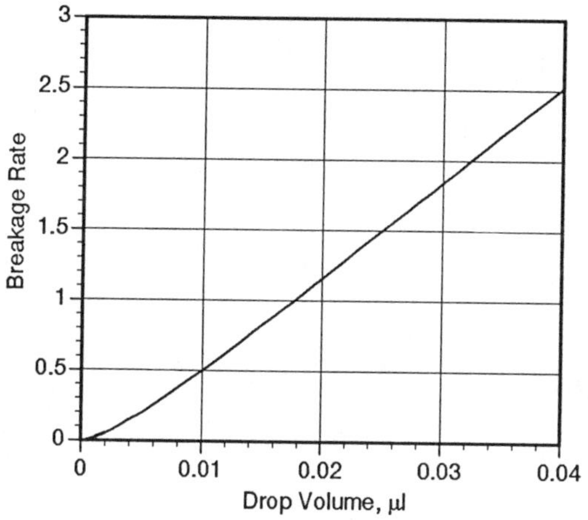

Figure 7: Breakage rate calculated from inverse problem using experimental data.

The inverse problem techniques have been successfully employed to extract the drop coalescence and breakage functions from transient size distributions. The extracted functions are very close to the actual functions in the cases where they are known, and predict the transient size distributions very well in the case of the experimental data.

REFERENCES

1. Coulaloglou, C.A., and Tavlarides, L.L., 1977, "Description of Interaction Processes in Agitated Liquid-Liquid Dispersions," *Chemical Engineering Science*, **32**, 1289-1297.

2. Ivanov, V.K., 1962, "On Linear Problems which are not Well-Posed," *Soviet Mathematics Doklady*, **3**, 981-983.

3. Konno, M., Aoki, M. and Saito, S., 1983, "Scale Effect on Breakup Process in Liquid-Liquid Agitated Tanks," *Journal of Chemical Engineering of Japan*, **16**, 312-319.

4. Muralidhar, R. and Ramkrishna, D., 1986, "Analysis of Droplet Coalescence in Turbulent Liquid-Liquid Dispersions," *Industrial and Engineering Chemistry Fundamentals*, **25**, 554-560.

5. Nambiar, D.K.R., Kumar, R., Das, T.R., and Gandhi, K.S., 1992, "A New Model for the Breakage Frequency of Drops in Turbulent Stirred Dispersions," *Chemical Engineering Science*, **47**, 2989-3002.

6. Narsimhan, G., Gupta, J.P., and Ramkrishna, D., 1979, "A Model for Transitional Breakage Probability of Droplets in Agitated Lean Liquid-Liquid Dispersions," *Chemical Engineering Science*, **34**, 257-265.

7. Narsimhan, G., Ramkrishna, D., and Gupta, J.P., 1980, "Analysis of Drop Size Distributions in Lean Liquid-Liquid Dispersions," *AIChE Journal*, **26**, 991-1000.

8. Park J.Y., and Blair, L.M., 1975, "The Effect of Coalescence on Drop Size Distribution in an Agitated Liquid-Liquid Dispersion," *Chemical Engineering Science*, **30**, 1057-1064.

9. Tikhonov, A.N., and Arsenin, V.Y., 1977, "Solutions of Ill-Posed Problems," V.H. Winston and Sons, Washington.

10. Wright, H., 1991, "Inverse Problems in Agglomeration," Ph.D. Thesis, Purdue University, West Lafayette, IN.

11. Wright, H., Muralidhar, R., and Ramkrishna, D., 1992, "Aggregation Frequencies of Fractal Aggregates," *Physical Review A*, **46**, 5072-5083.

12. Wright, H., Muralidhar, R., Tobin, T., and Ramkrishna, D., 1990, "Inverse Problems of Aggregation Processes," *Journal of Statistical Physics*, **61**, 843-863.

13. Wright, H., and Ramkrishna, D., 1992, "Solutions of Inverse Problems in Population Balances - I. Aggregation Kinetics," *Computers and Chemical Engineering*, **16**, 1019-1038.

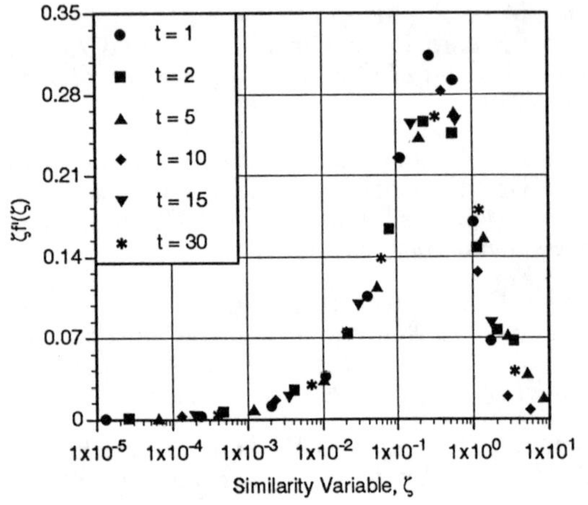

Figure 8: Similarity distribution calculated from the experimental data.

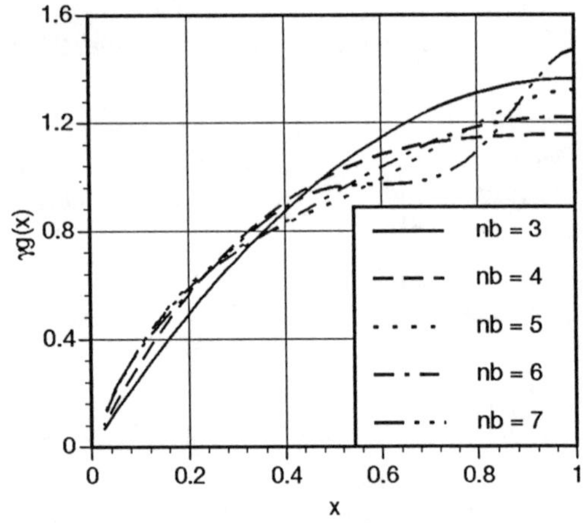

Figure 9: Inverse problem result for the daughter drop distribution using experimental data.

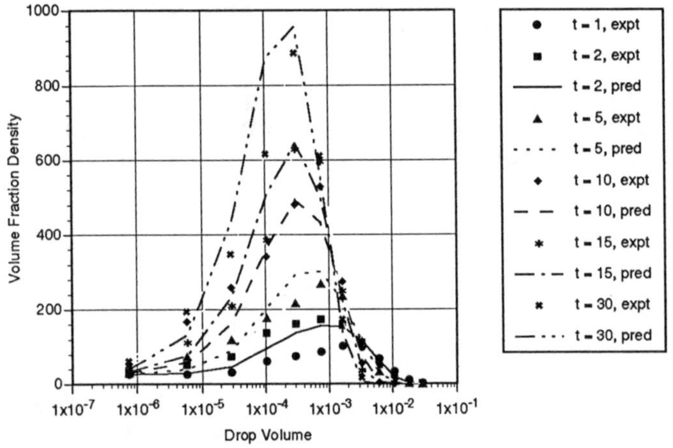

Figure 10: Prediction of the transient size distributions using the functions calculated from the inverse problem.

Inverse Problems in Engineering: Theory and Practice
ASME 1993

EXPERIMENTAL RESULT TO TEST INVERSE METHODS IN PHASE CHANGE

A. Sarda and D. Delaunay
Laboratoire de Thermocinétique
Université de Nantes
ISITEM
Nantes, France

ABSTRACT

The experimental set-up presented in this paper is used to study the physical phenomenon occurring during the vertical Bridgman configuration solidification process. The objective was to carry out experiments allowing validation of algorithms for identification and control of free boundaries. We have therefore characterized the phase change material as carefully as possible. We measured the temperature precisely by using sensors implanted in the material. We have shown the effect and origin of convection phenomenon. We present some results to test the algorithms for identification of front solidification in steady state (bidimensional case) and identification of boundary condition(unidimensional case) .

NOMENCLATURE

A_{xz}	: Form ratio in XZ
A_{xy}	: Form ratio in XZ
Cps,l	: Specific heat at constant pressure (J/kg/K)
g	: Acceleration due to gravity (ms^{-2})
H	: Height of study cavity (m)
L	: Width of study cavity (m)
m	: Mass (kg)
P	: Depth of study cavity
s	: Front location
t	: Time (s)
T	: Temperature
TCn	: Thermocouple (number)
TC1	: Temperature of vertical hot exchanger (°C)
TC2	: Temperature of horizontal hot exchanger (°C)
Tf	: Melting point (°C)
TF1	: Vertical exchanger cold zone temperature (°C)
TF2	: Horizontal cold exchanger temperature (°C)
V	: Movable partition displacement speed (m/s)
Vdsc	: Calorimeter heating and cooling rate (°K/min)
x	: Abscissa axis
z	: Ordinate axis
ΔHf	: Change of state enthalpy on melting (kJ/kg)
$\lambda_{s,l}$	:Thermal conductivity of solid, liquid (W/mK)

1 INTRODUCTION

The thermal history of a material undergoing solidification affect its structural qualities [1,2]. Solidification rate and temperature gradient are the macroscopic parameters to control. The metal and semi-conductor industry have developped systems as Bridgman [3] or Czochralsky [4]. process to control the quality of the materials. In an industrial situation, the phenomenon are very complex and movements in the liquid phase are due both to natural and thermosolutal convection . Some authors simulate this process [5, 6, 7], taking into account the more phenomenon as possible but they limit them selves to steady situations, the problem being already very complex. An alternative method consists in using an inverse method based on the phenomenom only in the solid phase. Our objective was to qualify an apparatus to test the posibility of using inverse methods in this kind of apparatus. More precisely we wanted to check the feasability of control algorithms and the possibility of identification of the interface location. It is then necessary to have a good knowledge of the heat transfer and the thermophysicals properties of the material. The mains objectives of this paper are to show the complexity of the phenomenon and to give complete experimental results allowing test of algorithms. A more complete analysis of the experimental set up and the heat transfer is to be publisched [8].

2 THERMAL PROPERTIES OF THE PHASE CHANGE MATERIAL

A material with a precisely specified melting point and change of state enthalpy was used, as close as possible to ambient temperature to minimize heat exchanges between the experimental set-up and external environment. In order to allow the observation of the solid-liquid interface and any possible convection movements in the liquid phase, it was transparent. We have chosen 99 % pure n-paraffin n-hexadecane. As this product have been commonly used in previous works, the values of the physical properties could be confirmed. The major disadvantage is that these materials are that they are thermal insulators and that the characteristic time constants of heat transfer are very different of the characteristic times corresponding to metals.

2.1 Thermal conductivity

The method used to measure the thermal conductivity is described in [11] and is a guarded hot plate thermal conductivimeter. The following conductivities was deduced of the measures (details on the values may be founded in [9]).

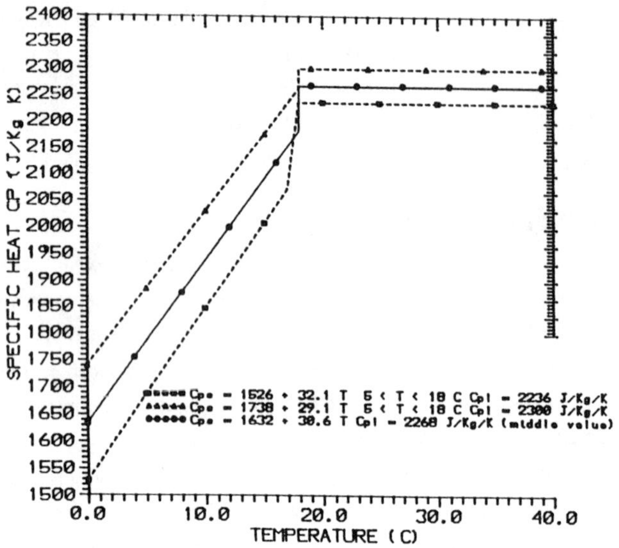

Fig 1 Specific heat of Hexadecane

n-hexadecane: λs = 0.39 W/m/K, λl = 0.18 W/m/K

2.2 Melting point

Several methods were used. The first one consisted in plunging vertically a platinum sensor, (secondary reference to 0.02 K accuracy), into a recipient with a diameter equal to 40 diameters of the sensor, and filled with phase change material. The apparatus was cooled slowly, and analysis of the solidification plateau obtained by the sensor yielded to the solidification temperature.

n-hexadecane Tf = 18.0 °C ± 0.04

To confirm these results, a second method used thermograms obtained during experimentation with the set-up described in the next section. We obtained:

n-hexadecane Tf = 18.0 °C ± 0.2.

2.3 Specific heat and Latent heat

We used a Perkin Elmer DSC 7 power compensation type differential calorimeter. The pans containing the samples were in aluminium, with a volume at 50 µl, with holes on the cover on covers. Experiments were carried out in a nitrogen atmosphere. The base line was adjusted and calibration carried out for temperature rates of Vdsc = 10 °C/min over the temperature interval 0-200 °C. The references were gallium, Tf = 30.1 °C and indium, Tf = 156.6 °C, ΔHf = 28.6 J/kg. Melting and solidification experiments were carried out with samples of different masses and at different heating and cooling rates. The analysis of the results, with a precise description of the phase change thermal kinetic is developped in [8]. As a result, the specific heat may be represented by the temperature dependence fig.1.

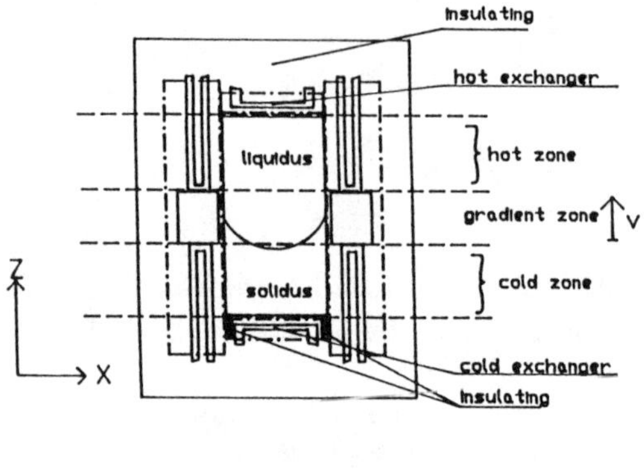

Fig 2 Boundary condition of the experiment

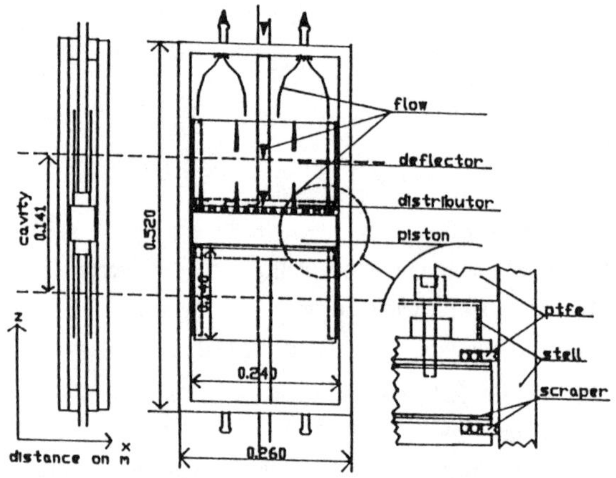

Fig 3 The moving piston system

The latent heat is : ΔHf = 231 J/kg

The specific heat of phase are :

solid : Cps = 1632 + 29,15 T J/Kg/K (T in °C)

liquid: Cpl = 2268 J/Kg/K

The paraffin is not very pure and the values are slightly different of the literature results due to the influence of the impureties.

3 EXPERIMENTAL APPARATUS

3.1 Measurement cell

The measurement cell is parallelpiped with dimensions:

$$H = 0.1401 \text{ m} \pm 1\ 10^{-4}$$
$$L = 0.070 \text{ m} \pm 1\ 10^{-4}$$
$$P = 0.242 \text{ m} \pm 1\ 10^{-3}$$

It therefore has the following aspect ratios:

$$A_{xz} = L/H = 1.9986$$
$$A_{xy} = 2L/P = 3.452$$

A gradient zone moves along the verical walls separating a hot and cold temperature zone fig.2. It is produced by vertical stainless steel exchangers equipped with a piston of low thermal conductivity.

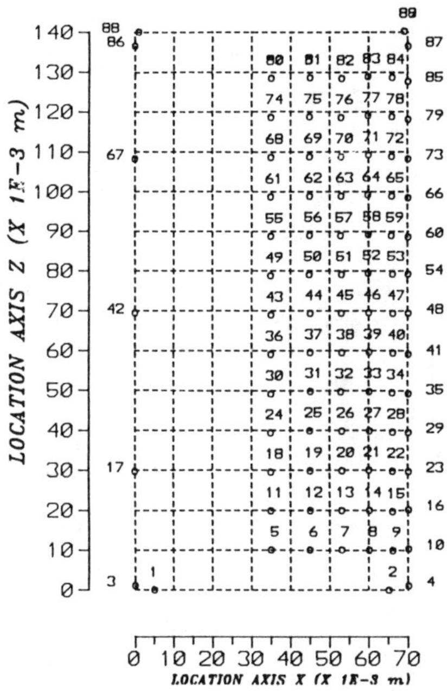

Fig 4 Thermocouples locations

This piston is deplaced by a tube used to convey the heat transfer fluid to a distributor with calibrated holes, and which sprays the fluids along the exchanger plates. Sealing between the movable partition and vertical walls is provided by PTFE scrapers Fig.3. The two other horizontal walls form brass exchangers. The end walls are in polymethacrylate to allow the phase change material to be observed.

3.2 Instrumentation systems

The measurement cell is equipped with 130 chromel-alumel thermocouples. They are distributed over the exchanger walls, at the inputs and outputs of the heat transfer fluid circuits and in the phase change material. 64 of them are stretched in the cavity between the two end walls. The wires are welded end-to-end and each thermocouple is equipped with a 2 kgf tension spring to compensate the effects of thermal expansion. Their location are given on fig 4. Only half of the cavity is equipped to define a mesh. The thermocouples are spaced by distances of 4 10^{-3}, 6 10^{-3} , 7 10^{-3} , 8 10^{-3}, 10^{-2} along -x from the exchanger wall at x = L . The step, constant along z, is 10^{-2} m. All these theoretical positions were verified by optical measurement in order to obtain the exact position of the welds as well as their largest diameter. The vertical wall situated on the half-cavity side with sensors is equipped with 15 thermocouples of diameter 8 10^{-5} m . The other only has 5, in order to check the apparatus symmetry. Horizontal exchangers are equipped with thermocouples in the corners. We implemented thermocouple in glove finger measurement systems in liquid flows in order to determine the average temperature. An optical laser tomography method allows to determine the position of the solidification front in the plane of the symmetry of the cavity. Photographs are taken at programmed time intervals with a Nikon F801 camera equipped with a Nikor 55 mm micro-objective. 1000 ASA Ektar films are used. The photographs are enlarged to 0.21 x 0.297 m and digitized with a Benson 6440 table. The data aquisition and automatic systems are controlled by a PC AT microcomputer via an IEEE 488 interface. A Keithley 181 nanovoltmeter and Keithley 706 multiplexer were used for the voltage measurements. A Keithley 195 multimeter was used for measuring the standard resistance. Several programmable power supplies were also used to control the temperature setpoints of thermostatic baths and step to step motors. The thermocouple wires were connected to copper wires in constant-temperature boxes comprising aluminium blocks immersed in a recipient filled with vermiculite. The temperatures of these blocks were determined with a special apparatus plunged in the thermostatic bath TC1. Its temperature is obtained with a platinum sensor, secondary reference. Taking account this procedure and the accuracy of the measurement apparatus, the absolute uncertainty on the temperatures is $\pm$ 0.2 K. The error on the position of the termocouple in the phase change material is from $\pm$ 5 10^{-5} m to $\pm$ 62.5 10^{-5} in the z direction. In the x direction it is $\pm$ 5 10^{-5} m. The absolute uncertainty on the solidification front position is $\pm$ 0.44 10^{-3} m in both directions x and z. The position of the movable partitions, determined with a resistive sensor, is given with a practical limit of error of $\pm$ 10^{-4} m.

4 EXPERIMENTAL RESULTS

We present two experimental cases. The first one concern a steady state in a bidimensional case. The temperature are then given at every point corresponding to a sensor. The location of solidification front is also given. The second is a case in which heat transfer in the solid phase remain unidimensional. The datas are the heat flux at the moving boundary, the interface location. The temperature on the fixed boundary has to be identified.

4.1 Steady state case

The initial state is a fully liquid cavity at temperature TC2 for the horizontal exchangers and TF2 for vertical exchangers fig.5A. The diference of TC1 - TC2 is kept to a value less than 0.2 °C. The micrometric system driving the moving piston is deplaced with a constant speed V= 6,6 10^{-6} m/s. It is stopped to a predefined position. When the piston beging to move the temperature of the horizontal exchanger TF2 is enslaved to TC4 see fig 5B (temperature in the lower corner of the vertical exchangers). We have chosen they allowing values:

TC1 = 40,4 °C, TC2 = 39,8 °C, TF1 = 7,65 °C, TF2 = 14.5 °C.

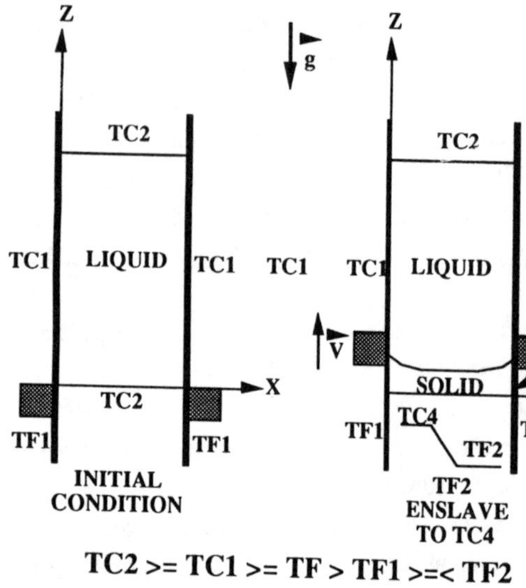

Fig 5A Initial condition experimental parameters
Fig 5B Initial condition of temperature horizontal exchanger

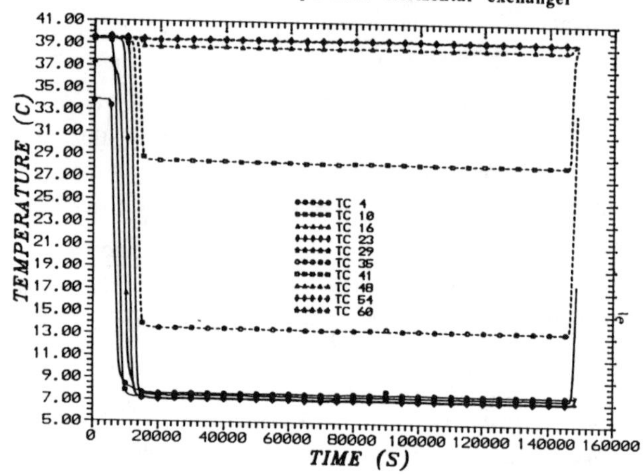

Fig 6 Time evolution of the temperature of the thermocouple situaed at x=0,0701 m in the wall

4.3 The temperature fields

Fig. 6 represents the time evolution of the temperature in the vertical wall. We note that the profile is reproducted very faithfully before the steady state reached after t = 140000 s. On the figures 7A,7B we see that solidification occurs at temperature of 18 ± 0.2 °C, corresponding to the plateau . There is no apparent subcooling effect. In the liquid phase we see clearly that heat transfer are driven by natural convection. A tentative of interpretation may be done by observing temperature profile along x axis for each thermocouple line (Fig 8). The general analysis of natural convection during the transient regim is developped by SARDA in [8,9]. In the steady state figure 9 shows the natural convection loops which allow to interpret the temperature fields. The upper loops which is present since the initial mode remains in the steady state by viscous effects. It is reavealed by the existence of domain in which temperature is constant and by a boundary layer near the

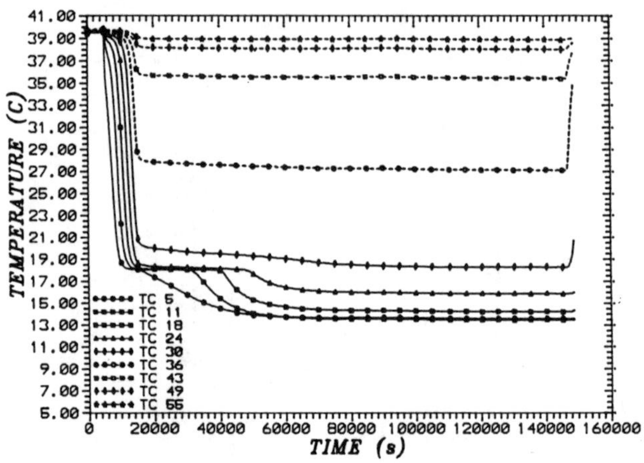

Fig 7A Time evolution of the temperature of the thermocouple situaed at x=0,035 m on the wall

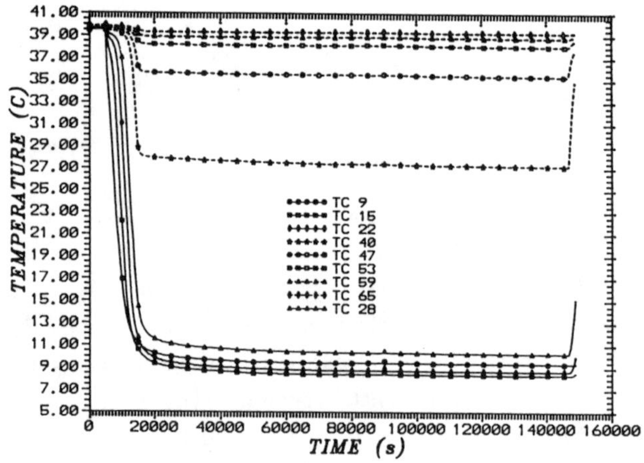

Fig 7B Time evolution of the temperature of the thermocouple situaed at x=0,066 m on the wall

vertical wall . The mecanism of the second one is detailed in Fig.10 It is due to the temperature gradient along the vertical metallic plate. In the central point of the cavity the isotherm is imposed by the interface and its deformation induces a horizontal gradient. This is the motor for the natural convection loop.

The temperatures in the solid phase at the termocouple location are given in the table (1). The location of the isotherm Tf in the vertical wall is computed by interpolation of the temperature profile on this plate. It is given by:

$$Tf = 18\ °C\ ,\ x = 70{,}1\ 10^{-3}\ m,\ \ z = 52{,}6\ 10^{-3}\ m$$

The temperature of the horizontal exchanger is measured by a surface thermocouple. Its value is T = 14.5 °C for z = 0. The solidification front determined from photographie analysis is shown on Fig.11 and the values are listed in table (2).

4.2 Results for a 1D transient case

At the begining of the solidification proces heat transfer remain unidimensional. This is illustrated on the Fig.12 on which

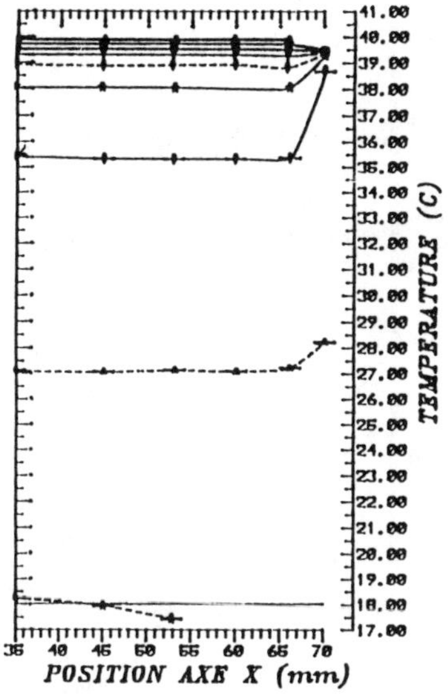

Fig 8 Temperature profile along x in the liquid phase

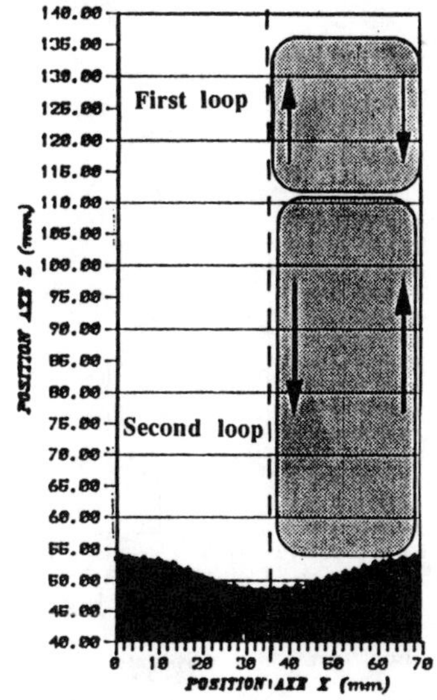

Fig 9 Lopp of natural convection

n T °C	X (10^{-3} m)	Z (10^{-3} m)	n T °C	X (10^{-3} m)	Z (10^{-3} m)
05 13,64	35,90	10,00	06 13,35	45,81	10,01
07 12,57	53,94	10,03	08 11,06	60,98	10,01
09 09,42	67,00	09,90	10 07,03	70,10	10,28
11 13,54	35,86	19,81	12 13,09	45,93	19,86
13 12,02	53,95	19,89	14 10,23	61,03	19,91
15 08,40	66,95	19,61	16 07,03	70,10	20,09
18 14,27	35,83	29,46	19 13,88	45,89	29,78
20 12,62	54,06	29,79	21 10,72	60,93	29,96
22 08,75	66,97	29,65	23 06,98	70,10	29,62
24 15,92	35,81	39,34	25 15,58	45,82	39,84
26 14,61	54,02	39,74	27 12,97	61,07	39,79
28 10,32	67,03	39,49	29 07,33	70,10	39,42
30 18,31	35,86	49,11	31 18,01	45,94	49,77
32 17,53	53,70	49,60	33 16,71	61,00	49,69
34 15,38	66,92	49,43	35 13,31	70,10	49,21

Table (1) Temperature of thermocouple

we see that the extension of the edge effects are limitated until a time t = 10000 s (at this time we can estimate a zone of 20 milimeter on each side perturbated by bidimensional effects). In this experiment temperature TC1 and TC2 are chosen very close to Tf to limit natural convection in the liquid phase. The time evolution of solidification front is given by a polynomial equation. For more easy use of the results we precise the experimental value for the moving boundary location on the table 3.

The chronological evolution of solidification front is:

$$s\,(t, x=0{,}035\text{ m}) = a0 + a1\,t + a2\,t^2 + a3\,t^3 + a4\,t^4 \qquad (1)$$

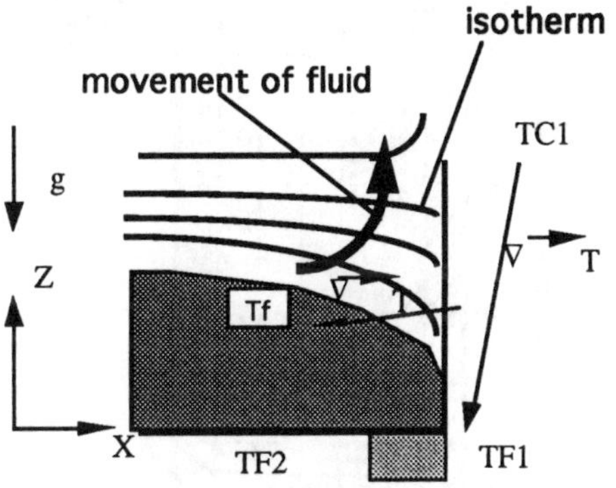

Fig 10 mechanism of the second loop

x (10^{-3} m)	z (10^{-3} m)	x (10^{-3} m)	z (10^{-3} m)	x (10^{-3} m)	z (10^{-3} m)
00,65	52,49	23,24	47,72	51,65	49,55
02,05	52,29	24,47	47,66	54,12	50,10
04,10	52,12	25,68	47,27	57,11	50,75
04,52	52,05	27,40	47,20	60,17	51,14
07,71	51,79	30,69	46,75	62,94	51,76
11,52	51,11	34,33	46,65	65,77	52,08
15,03	50,26	37,07	46,71	68,11	52,64
18,22	49,09	40,55	47,17	69,61	52,63
20,24	48,73	43,80	47,98		
23,11	48,28	48,36	48,99		

Table (2) solidification front

with:

$a0 = 1{,}05485$ $a1 = -0{,}000333$ $a2 = 6{,}0931\ 10^{-7}$
$a3 = -5{,}98993\ 10^{-11}$ $a4 = 1{,}88882\ 10^{-15}$

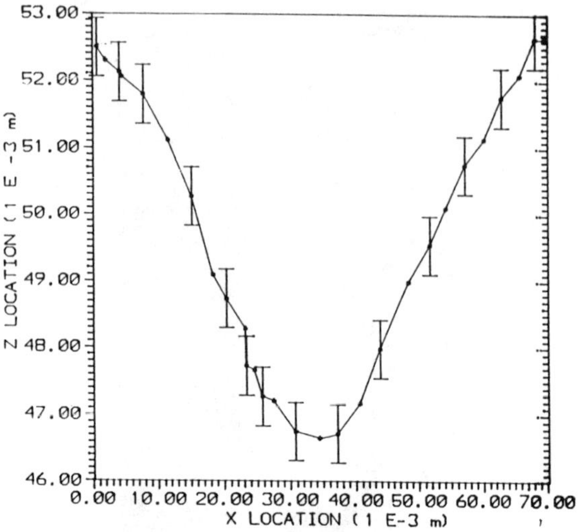

Fig 11 Solidification front in steady-state

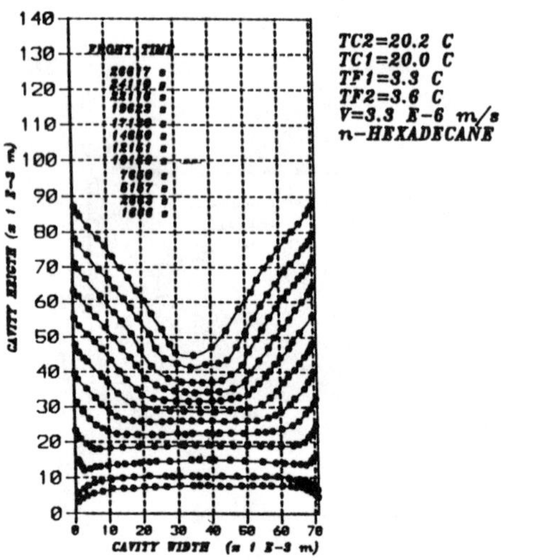

Fig 12 Chronological evolution of solidification front

t (s)	z (10^{-3} m)	t (s)	z (10^{-3} m)	t (s)	z (10^{-3} m)
1754	2,02	2845	3,84	4946	8,14
7570	13,72	9688	17,16	12317	20,94

Table (3) front solidification evolution:experimental points

The heat flux at the moving boundary may be computed with the heat balance at the interface:

$$\phi (t) = \rho \ \Delta Hf \ \frac{ds(t)}{dt} \qquad (2)$$

where $\rho = 792 - 0{,}71\ T$

The chronological evolution of the temperature on the surface of the upper horizontal exchanger is given on the figure.13. It may be computed by the relation:

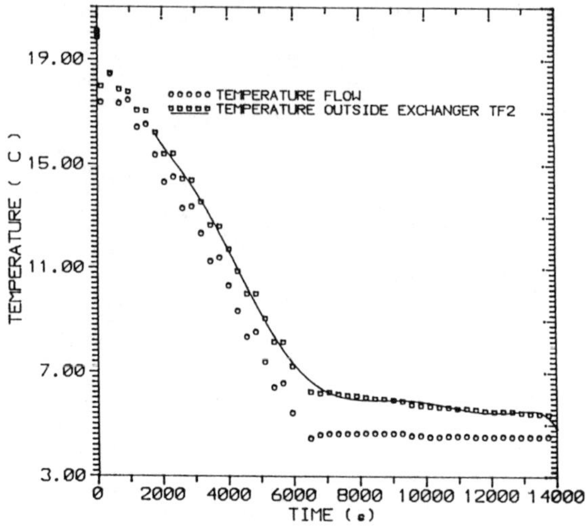

Fig 13 Temperature of the horizontal exchanger

$$Tex (t) = a0 + a1\, t + a2\, t^2 + a3\, t^3 + a4\, t^4 + a5\, t^5 + a6\, t^6 + a7\, t^7 \qquad (3)$$

a0 = 27,393   a1 = -0,015676   a2 = 9,15367 10^{-6}
a3 = -2,95256 10^{-9} a4 = 4,99581 10^{-13} a5 = -4,5262 10^{-17}
a6 = 2,08973 10^{-21} a7 = -3,86932 10^{-26}

The initial profile is:

at t0 = 1754 s T (z) = \F((Tf - Tex(to)); s(t0)) z + Tex (4)

5 CONCLUDING REMARKS

Heat transfers in a configuration representative of a simplified Bridgman vertical process were analysed. Phase change material was very carefully characterized. The results show the revelant importance of natural convection in the liquid phase, even in the absence of thermosolutal convection. Natural convection is induced by temperature gradient in the vertical conducting walls. The results presented may be used to test inverse algorithms. We tried to do it and prelemary result swere obtained by SAMAI [10]. We has show the possibility of using experimental value to identify the position of the interface. This feasability suppose very precise experimental results and a good determination of thermal properties of the phase change material. Result for identification in transient cases may be used to test 2D transient algorithms.

REFERENCES

[1] KURZ, FISHER " Fundamentals of solidification "
Third edition ,chapter four , 63-92, Trans. Tech. Publications 1989

[2] G.LAMANTHE " La solidification dirigée d'aubes de turbines "
SFT, Journée d'étude sur la Thermocinétique, 1987

[3] T.W.CLYNE " Numerical modelling of directional solidification of metallic alloys ", Metal Science, Vol 16, 441-450, 1982

[4] Y. RYCKMANS, P. NICODEME, F. DUPRET," Numerical simulation of crystal growth influence of melt convection on global heat transfer and interface shape ", Journal of crystal growth 99, 702-706, 1990

[5] M.J. CROCHET, F.T. BEYLING, J.J.VAN SCHAFTINGEN " Numerical simulation of the horizontal Bridgman growth of a gallium arsenide crystal ", Journal of crystal growth 65,166-172, 1983

[6] C.J. CHANG, R.A. BROWN ," Radial segregation unduced by natural convection and melt/solid interface shap on vertical Bridgman Growth ",Journal of crystal growth 69, 343-364, 1983

[7] P.M.ARDONATO,R.A.BROWN " Convection and segregation in directional solidification of dilute and non dilute binary alloys effect of ampoule and furnace design ", Journal of crystal growth 80, 155-190, 1987

[8] A.SARDA D.DELAUNAY, " Experimental study of vertical Bridgman configuration", submitted at experimental heat transfer.

[9] A.SARDA, Thesis," Contribution experimentale en vue de valider des algorithmes d'identification et de commande de frontières mobiles application à la solidification dirigée ", Université de Nantes, 1992

[10] M.SAMAI ," Etude de méthodes inverses pour des problèmes de conduction thermique avec changement de phase. Algorithmes numriques et validation expérimentales ", Thesis, Université de Nantes,1992

[11] P CARRE, Thesis, Université de Paris,1984

Inverse Problems in Engineering: Theory and Practice
ASME 1993

SOLUTION OF THE TWO-DIMENSIONAL INVERSE STEFAN DESIGN PROBLEM WITH THE ADJOINT METHOD

Shinill Kang and Nicholas Zabaras
Sibley School of Mechanical and Aerospace Engineering
Cornell University
Ithaca, New York

ABSTRACT

The aim of this work is to calculate the optimum history of boundary cooling conditions that results in a desired solidification morphology and grain size. The grain size together with the grain morphology is selected such that desired macroscopic mechanical properties and soundness of the final cast product are achieved. In the present work, the history of freezing front location/motion is the main control/target variable that is used to define the obtained solidification microstructures. The adjoint method is used in conjunction with the conjugate gradient method for the solution of the multidimensional inverse Stefan design problem. The gradient of the cost functional is first obtained by solving the adjoint equations backward in time and, then, the sensitivity equations are solved forward in time to compute the optimal step size for the gradient method. A two-dimensional numerical example is analyzed to demonstrate the performance of the present method.

INTRODUCTION

The crystallographic growth morphology, scale of microstructures, and grain orientation in a casting process are directly related with the macroscopic properties of the final product, and as such their control is of extreme importance. For pure metals, the freezing front motion/velocity and the interface heat flux on the solid side define the crystallographic growth morphology and scale of microstructures (Kurz and Fisher (1989)). In one-dimensional inverse Stefan problems, both interface velocity and interface thermal gradient to the solid side can be independently prescribed. Two unknown control functions (boundary conditions at the fixed boundaries) in the solid side as well as in the liquid side must then be calculated. However, in multi-dimensional problems, only the interface location can be controlled via the cooling function at the fixed boundary and, therefore, one should introduce a new independent function to control the thermal gradient at the solid side of the interface. Such a function can be a heat source in the liquid region related to the liquid feeding to the contracting freezing front or an electromagnetic force which is used to stir the liquid region (one should incorporate melt convection in this case). In this work, however, control of the interface thermal gradient is not taken into consideration and the history of freezing front location/motion is the main control/target variable.

Among a few mathematicians who approached the two-dimensional inverse Stefan problem numerically, we particularly mention the work of Colton and Reemtsen (1984). They solved (the single phase) inverse Stefan problem in two space variables by minimizing the maximal defect in the initial-boundary data with regularization. Zabaras et al (1992) analyzed two-dimensional design Stefan processes using the future information method developed by Beck et al (1985). The sequential minimization process was stabilized by using future information. To further stabilize their solution algorithm, they also introduced spatial regularization in the sense of Tikhonov (1977).

Alifanov and his colleagues (1981,1984) solved several two-dimensional inverse heat conduction problems (IHCP) without phase change using the so called iterative regularization gradient algorithms. They computed the gradient of the discrepancy functional by solving the adjoint problem.

In this work, the minimization process is performed over the whole time domain of interest and the unknown heat flux function is discretized after the minimization process is analytically performed in the L_2 space. The conjugate gradient method used in this work is different from the one presented by Luenberger (1968) and Fletcher (1987), where the cost functional was minimized in a finite vector space.

DEFINITION OF THE INVERSE STEFAN DESIGN PROBLEM

The inverse Stefan design problem is defined as follows:

Given the material properties, initial temperature, melting temperature, and the motion/location of the freezing front $\partial\Omega_I(t)$ at all times, calculate the history of the flux and temperature on the fixed boundary $\partial\Omega_o$ and the history of temperature field in $\Omega(t)$ (Fig. 1).

The governing equations defining the Stefan problem are as follows:

Governing equation:

$$C(T)\frac{\partial T(\underset{\sim}{x},t)}{\partial t} = \nabla \cdot (K(T)\, \nabla\, T(\underset{\sim}{x},t)) \qquad (1)$$

where

$$C = C_S,\quad K = K_S,\quad T = T_S \quad \text{for } \underset{\sim}{x} \in \Omega_S\,(t)$$

$$C = C_L,\quad K = K_L,\quad T = T_L \quad \text{for } \underset{\sim}{x} \in \Omega_L\,(t)$$

Initial and Boundary Conditions:

$$T\,(\underset{\sim}{x},\, 0\,) = 0, \qquad \underset{\sim}{x} \in \Omega_L(0) \qquad (2)$$

$$K_S\,(T(\underset{\sim}{x},t))\frac{\partial T(\underset{\sim}{x},t)}{\partial n_o} = q(\underset{\sim}{x},\, t), \qquad \underset{\sim}{x} \in \partial\Omega_o \qquad (3)$$

$$T\,(\underset{\sim}{x},t) = T_m, \qquad \underset{\sim}{x} \in \partial\Omega_I\,(t) \qquad (4)$$

Stefan Condition:

$$K_S\,(T_S\,(\underset{\sim}{x},t))\frac{\partial T_S(\underset{\sim}{x},t)}{\partial n} - K_L(T_L(\underset{\sim}{x},t))\frac{\partial T_L(\underset{\sim}{x},t)}{\partial n} = \rho L\, \underset{\sim}{V}\cdot\underset{\sim}{n}\,, \quad \underset{\sim}{x} \in \partial\Omega_I\,(t) \qquad (5)$$

It is noted that, unlike the conventional direct Stefan problem, the heat flux history q at the fixed boundary in eq. (3) is unknown. Instead, the history of motion of the freezing front, $\partial\Omega_I(t)$, is known at all times. The prescribed freezing front motion defines the flux discontinuity on $\partial\Omega_I(t)$ (eq. (5)). In this sense, eqs. (4) and (5) constitute a set of overspecified boundary conditions on $\partial\Omega_I(t)$.

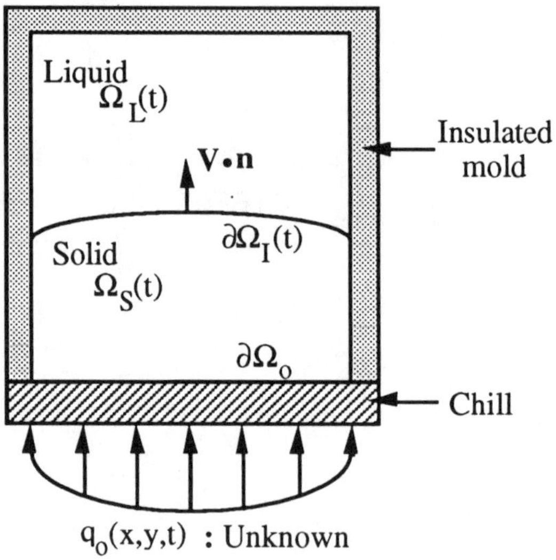

Fig. 1 Geometry of Design Solidification

Mathematically the problem is formulated as an infinite dimensional minimization problem: Given the material properties, initial temperature, melting temperature, and the motion of the freezing front at all times, find the optimal boundary heat flux $q^*(\underset{\sim}{x},t)$, $(\underset{\sim}{x},t) \in ([\partial\Omega_o]x[0,t_{max}])$ such that

$$S(q^*) \le S\,(q), \quad q \in L_2\,([\partial\Omega_o]x[0,t_{max}]\,) \qquad (6)$$

where

$$S(q) = \frac{1}{2}\,\|\, T_m - T\,(\underset{\sim}{x},t;\, q)\,\|^2_{L_2([\partial\Omega_I(t)]x[0,t_{max}])}$$

$$= \frac{1}{2}\int_0^{t_{max}}\int_{\partial\Omega_I(t)} (T_m - T(\underset{\sim}{x},t;q))^2\, d\Gamma dt \qquad (7)$$

This cost functional is the L_2 norm of the error between the calculated temperature distribution $T(\underset{\sim}{x},t;q)$ at the freezing front and the given melting temperature T_m. The minimization process will be performed over the whole time domain of interest and the unknown heat flux function is discretized after the minimization process is performed in the L_2 space.

PREPROCESSING

For the inverse design problem of concern here, certain preprocessing operations are necessary before the main solution algorithm is undertaken. At first, a time step $\Delta t = t^{n+1} - t^n$ is selected based on previous experiences with the FEM calculation of similar direct Stefan problems. Since the deforming FEM is used to analyze the problem, a finite element mesh must be introduced that at each time step conforms with the moving interface. Here, the mesh generation scheme developed by Ruan and Zabaras (1990) and McDaniel (1993) is followed. In their work, the whole region of solidification is divided into solid and liquid regions. The solid region $\Omega_S(t)$ consists of a non-deforming sub-region and a deforming sub-region. Nodes are fixed in time in the non-deforming sub-region. The elements in the solid side next to the freezing front change their sizes as $\partial\Omega_I(t)$ moves and constitute the deforming region of the solid phase. When the sizes of these interface elements reach the prescribed maximum allowable element size, new deforming elements are generated by splitting each previously deforming element into two new elements. The new group of elements next to $\partial\Omega_I(t)$ defines the new deforming region, while the other element group becomes part of the non-deforming region. The nodes in the liquid phase are assumed stationary except when remeshing is necessary. This remeshing process is performed utilizing a transfinite mapping and is considered to be induced by a continuous mesh motion. In the nearly unidirectional solidification, as new elements are added to the solid region equal number of elements are subtracted from the liquid region following McDaniel (1993).

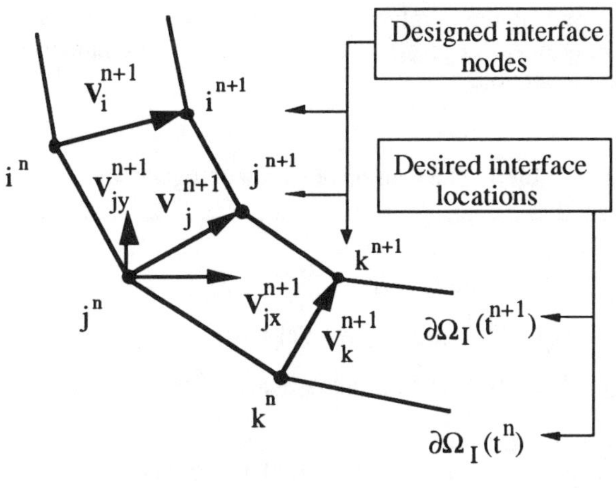

Fig. 2 Design of interface nodes

In the design problem of concern here, the location of the interface $\partial\Omega_I(t)$ is considered prescribed by the casting designer at all times. The position (x_i^{n+1}, y_i^{n+1}) of the ith interface node at time t^{n+1} is selected such that the interface elements have favourable element shapes, i.e. such that they provide a well-conditioned finite element system.

Once the interface nodal positions are designed, then the components of the interface nodal velocities are obtained using the finite difference approximation:

$$V_{ix}^{n+1} = \frac{x_i^{n+1} - x_i^n}{t^{n+1} - t^n} \quad , \quad V_{iy}^{n+1} = \frac{y_i^{n+1} - y_i^n}{t^{n+1} - t^n} \qquad (8)$$

where $(x_i^{n+1}, y_i^{n+1}) \in \partial\Omega_I\,(t^{n+1})$ and $(x_i^n, y_i^n) \in \partial\Omega_I\,(t^n)$. This approximation of interface nodal velocity components at all discrete times are saved and then used for the mesh generation in the deforming FEM simulation of direct, adjoint, and sensitivity problems. It is also used for evaluating the interface flux jump.

CALCULATION OF THE INTERFACE JUMP CONDITION

The melting temperature condition of eq. (4) is used as a pseudo-measurement at the interface in the ISDP, while the heat flux jump condition of eq. (5) is used as the interface boundary condition. In the finite element formulation context, the force vector at the interface has the form [Ruan and Zabaras (1990)]:

$$\begin{aligned} F_I &= \sum_{b=1}^{E_I} F_k^b \\ &= \sum_{b=1}^{E_I} \int_{\partial\Omega_I^b} [K_S \nabla T_S - K_L \nabla T_L] \cdot \mathbf{n} \Psi_i^b(x,y,t)\, d\Gamma \\ &= \sum_{b=1}^{E_I} \int_{\partial\Omega_I^b} [\rho\, L \mathbf{V} \cdot \mathbf{n}]\ \Psi_i^b(x,y,t)\, d\Gamma \end{aligned} \qquad (9)$$

Here, the velocity can be approximated using the interpolation functions:

$$\mathbf{V} = \mathbf{V}_i^b\, \Psi_i^b\,(x,y,t) \qquad (10)$$

where $\mathbf{V}_i^b$ is the nodal velocity at the interface defined through eq. (8) and $\Psi_i^b(x,y,t)$ are the one-dimensional shape functions defined at the interface boundary segments. Then, we have

$$f_i = \sum_{b=1}^{E_I} \int_{\partial\Omega_I^b} \rho\, L\, \mathbf{N}^b\, \Psi_i^b\, \Psi_h^b\, d\Gamma \cdot \mathbf{V}_h^b \qquad (11)$$

where f_i is the component of the force vector at the interface and $\mathbf{N}^b$ is the unit normal to the interface boundary segment b as defined through the interface nodal coordinates. It is clear that once the directional cosines of the unit normal $\mathbf{N}^b$ to the interface segment b are defined, then the term f_i can be evaluated without any difficulty because the Cartesian components V_{hx}^b and V_{hy}^b of the interface nodal velocities $\mathbf{V}_h^b$ along with the interface nodal coordinates are provided from the preprocessing procedure.

THE ADJOINT METHOD

The adjoint method is used in conjunction with the conjugate gradient method. The conjugate gradient method consists in constructing the minimizing sequence $q^0, q^1, \ldots, q^k, \ldots$:

$$q^{k+1} = q^k + \alpha^k p^k(q^k) \tag{12}$$

where q is the minimizer of the cost functional, α^k is the step size, and p^k is the conjugate search direction. The optimal step size α^k is obtained by minimizing the cost functional $S(q^{k+1})$ at each iteration step, i.e. α^k is the minimizer of the functional $S(q^k+\alpha p^k)$. Calculation of p^k requires the gradient of the cost functional defined in eq. (7). Detail expressions for α^k and p^k are shown in Table 1.

The gradient of the cost functional is obtained first by defining the directional derivative of S(q) at q in the direction of increment Δq, where q and $\Delta q \in L_2([\partial\Omega_o] \times [0,t_{max}])$. Linearization of the governing equations (1)-(5) by means of a perturbation to the unknown heat flux q gives the following **sensitivity problem**:

$$\frac{\partial}{\partial t}[C(T)\,\theta(\underset{\sim}{x},t;q,\Delta q)] = \nabla^2 [K(T)\,\theta(\underset{\sim}{x},t;q,\Delta q)] \tag{13}$$

where

$C = C_S,\ K = K_S,\ \theta = \theta_S$ for $\underset{\sim}{x} \in \Omega_S(t)$

$C = C_L,\ K = K_L,\ \theta = \theta_L$ for $\underset{\sim}{x} \in \Omega_L(t)$

$$\theta(\underset{\sim}{x},t;q,\Delta q)\,|_{t=0} = 0, \quad \underset{\sim}{x} \in \Omega_L(0) \tag{14}$$

$$\frac{\partial}{\partial n_o}[K_S(T_S)\,\theta_S] = \Delta q(\underset{\sim}{x},t), \quad \underset{\sim}{x} \in \partial\Omega_o \tag{15}$$

$$\frac{\partial}{\partial n}[K_S(T_S)\theta_S] - \frac{\partial}{\partial n}[K_L(T_L)\theta_L] = 0, \quad \underset{\sim}{x} \in \partial\Omega_I(t) \tag{16}$$

where the sensitivity function θ is the directional derivative of the temperature at q in the direction of Δq:

$$\theta(\underset{\sim}{x},t;q,\Delta q) \equiv T_q(\underset{\sim}{x},t;q)(\Delta q). \tag{17}$$

Finally, the gradient of the cost functional is the solution of the adjoint problem at the fixed boundary:

$$S'(\underset{\sim}{x},t;q) = \Psi(\underset{\sim}{x},t;q), \qquad (\underset{\sim}{x},t) \in ([\partial\Omega_o] \times [0,t_{max}]) \tag{18}$$

where the adjoint function $\Psi(\underset{\sim}{x},t;q)$ is the solution of the following **adjoint problem**:

$$C(T)\frac{\partial \Psi}{\partial t}(\underset{\sim}{x},t;q) + K(T)\,\nabla^2\Psi(\underset{\sim}{x},t;q) = 0 \tag{19}$$

where

$C = C_S,\ K = K_S,\ \Psi = \Psi_S$ for $\underset{\sim}{x} \in \Omega_S(t)$

$C = C_L,\ K = K_L,\ \Psi = \Psi_L$ for $\underset{\sim}{x} \in \Omega_L(t)$

$$\Psi(\underset{\sim}{x};t)\,|_{t=t_{max}} = 0, \qquad \underset{\sim}{x} \in \Omega(t_{max}) \tag{20}$$

$$K\frac{\partial \Psi(\underset{\sim}{x};t)}{\partial n_o}\Big|_{\partial\Omega_o} = 0, \qquad \underset{\sim}{x} \in \partial\Omega_o \tag{21}$$

$$-K\left[\left|\frac{\partial \Psi}{\partial n}\right|\right]_{\partial\Omega_I(t)} = T_m - T(\underset{\sim}{x},t;q), \quad x \in \partial\Omega_I(t) \tag{22}$$

where

$$K\left[\left|\frac{\partial \Psi}{\partial n}\right|\right]_{\partial\Omega_I(t)} = K_S\frac{\partial \Psi_S}{\partial n} - K_L\frac{\partial \Psi_L}{\partial n}$$

and the direction of the normal n is inward at the interface $\partial\Omega_I(t)$. Calculation of the optimal step size requires the information on the values of the sensitivity function at the freezing front.

Both the sensitivity and the adjoint equations are derived based on the assumption that the material properties are functions of temperature. However, only the case with constant material properties is examined in the numerical example section.

Table 1. The Conjugate gradient algorithm

<u>Step 1</u>: Pick an initial guess $q^o(\underset{\sim}{x},t)$ in $L_2([\partial\Omega_o] \times [0,t_{max}])$. Set k=0.

<u>Step 2</u>: Calculation of conjugate search direction $p^k(\underset{\sim}{x},t)$

a. Define the scalar γ^k .

a1. Solve the direct Stefan problem forward in time for $T(x,t;q^k(\underset{\sim}{x},t))$

a2. Compute the residual $T_m - T(\underset{\sim}{x},t;q^k(\underset{\sim}{x},t))$ in $L_2([\partial\Omega_I(t)] \times [0,t_{max}])$

a3. Solve the adjoint equations backward in time for $\Psi(\underset{\sim}{x}, t; q^k(\underset{\sim}{x},t))$.

Evaluate $S'(q^k(\underset{\sim}{x},t))(\underset{\sim}{x},t) = \Psi(\underset{\sim}{x},t;q^k(\underset{\sim}{x},t))$ in $L_2([\partial\Omega_o] \times [0,t_{max}])$

a4. Set $\gamma^k = 0$ if k=0, otherwise set

$$\gamma^k = \frac{\left(S'(q^k)\,,\,S'(q^k)-S'(q^{k-1})\right)_{L_2([\partial\Omega_o]x[0,t_{max}])}}{\|S'(q^{k-1})\|^2_{L_2([\partial\Omega_o]x[0,t_{max}])}}$$

b. Define the direction $p^k(\underset{\sim}{x},t)$.

If k=0, set $p^0(\underset{\sim}{x},t) = -\,S'(q^k)(\underset{\sim}{x},t)$,

otherwise, set $p^k(\underset{\sim}{x},t) = -\,S'(q^k)(\underset{\sim}{x},t) + \gamma^k\, p^{k-1}(\underset{\sim}{x},t)$

Step 3: Calculate the optimal step size α^k

a. Solve the sensitivity equations to calculate $\theta(\underset{\sim}{x},t;q^k,p^k)$

b. Set optimal step size

$$\alpha^k = \frac{-\left(S'(q^k)(\underset{\sim}{x},t),p^k(\underset{\sim}{x},t)\right)_{L_2([\partial\Omega_o]x[0,t_{max}])}}{\|\theta(\underset{\sim}{x},t;q^k,p^k)\|^2_{L_2([\partial\Omega_I(t)]x[0,t_{max}])}}$$

Step 4:

Update $q^{k+1}(\underset{\sim}{x},t) = q^k(\underset{\sim}{x},t) + \alpha^k\, p^k(\underset{\sim}{x},t)$

Step 5:

If $\|q^{k+1}(\underset{\sim}{x},t) - q^k(\underset{\sim}{x},t)\|_{L_2([\partial\Omega_o]x[0,t_{max}])} < \varepsilon$

(specified tolerance), stop

Otherwise set k = k+1 and go to step 2.

NUMERICAL INTEGRATION OF INNER PRODUCT AND NORM IN L_2

At $\partial\Omega_o$ the lengths of the boundary segments are fixed in time. As such, the inner product of two functions f and g that are defined at $\partial\Omega_o$ is computed using the following approximation:

$$(f,g)_{L_2([\partial\Omega_o]x[0,t_{max}])} = \int_0^{t_{max}}\int_{\partial\Omega_o} f\,g\,d\Gamma dt$$

$$= \sum_{j=1}^{N-1}\sum_{i=1}^{E_o} f_{i+1/2,\,j+1/2}\; g_{i+1/2,\,j+1/2}\,\Delta S_i\,\Delta t_j \qquad (23)$$

where

$$f_{i+1/2,j+1/2} = \frac{1}{4}\left(f_{i,j} + f_{i+1,j} + f_{i+1,j+1} + f_{i,j+1}\right),$$

$$g_{i+1/2,j+1/2} = \frac{1}{4}\left(g_{i,j} + g_{i+1,j} + g_{i+1,j+1} + g_{i,j+1}\right),$$

and the temporal step size $\Delta t_j = t_{j+1} - t_j$. $f_{i,j}$ is the value of the function f obtained on the ith node of $\partial\Omega_o$ (i=1,.., E_o+1) and at time t_j (j=1,..,N). N is the total number of temporal points, E_o is the total number of line segments for the discretization of the fixed boundary $\partial\Omega_o$, and ΔS_i is the length of the ith boundary segment.

However at the moving interface the spatial step size (the length of the interface line segment) changes as the interface moves. Therefore, the inner product at the moving boundary is evaluated in the same manner as that at the fixed boundary except the evaluation of the spatial step size:

$$(f,g)_{L_2([\partial\Omega_I(t)]x[0,tmax])} = \int_0^{t_{max}}\int_{\partial\Omega_I(t)} f\,g\,d\Gamma dt$$

$$= \sum_{j=1}^{N-1}\sum_{i=1}^{E_I} f_{i+1/2,j+1/2}\; g_{i+1/2,j+1/2}\,\Delta S_{i,j+1/2}\Delta t_j \qquad (24)$$

where $\Delta S_{i,j+1/2} = \frac{1}{2}\left(\Delta S_{i,j} + \Delta S_{i,j+1}\right)$, $\Delta S_{i,j}$ is the length of the ith boundary segment at time t_j, and E_I is the total number of line segments for the discretization of the moving boundary $\partial\Omega_I(t)$. The corresponding norms are computed in the same manner as inner products.

NUMERICAL EXAMPLE

A nearly unidirectional problem introduced earlier by Zabaras et. al. (1992) was solved to test the present method (Fig. 1). A sinusoidal heat flux q_o was applied to the bottom of the casting (y=0, 0≤x≤1.2 m), where

$$q(x,y,t) = 5\{1+2\sin[(\pi/1.2)\,x]\}\ \mathrm{W/m^2} \qquad (25)$$

as shown in Fig. 3. The desired interface location was obtained by solving the corresponding direct problem. The present example is ill-posed in nature in the sense that the shape of the interface becomes flatter as the freezing front moves away from the fixed boundary as shown in Fig. 4, and therefore the front motion does not realize the spatial variation of the above function q.

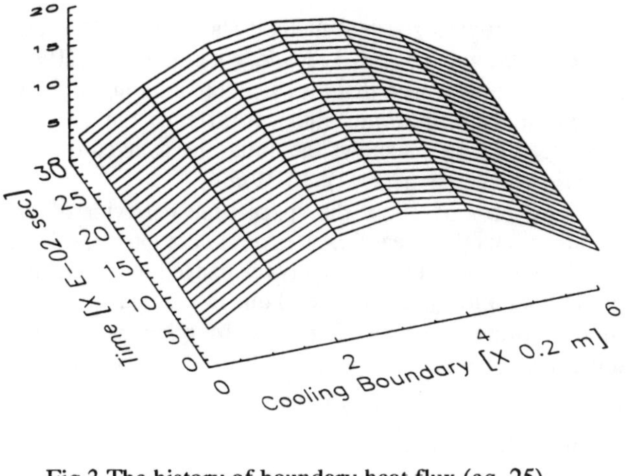

Fig.3 The history of boundary heat flux (eq. 25)

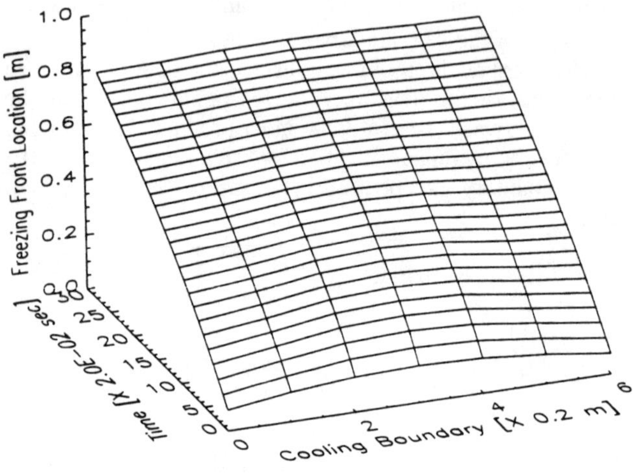

Fig. 4 Evolution of the freezing front shape

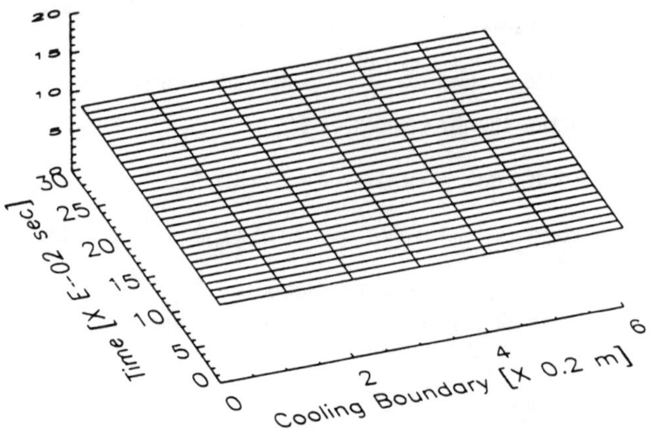

Fig. 5a Initial guess of the cooling history

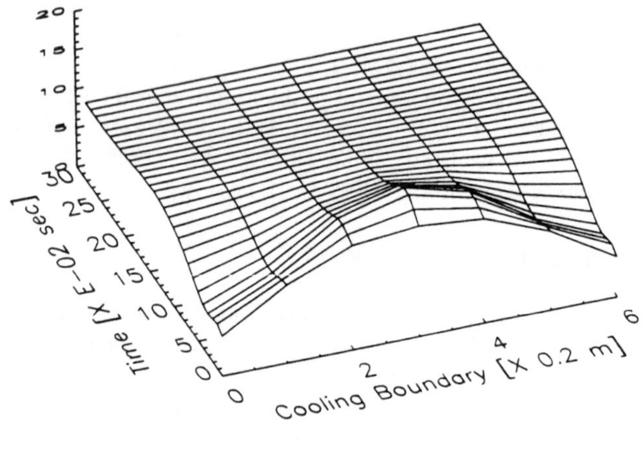

Fig. 5b Cooling history at iteration # 10

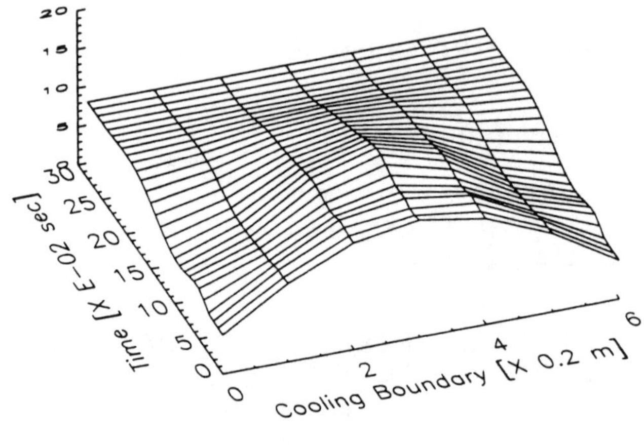

Fig. 5c Cooling history at iteration # 50

The conjugate gradient method shown in table 1 was implemented with t_{max}=0.3 secs. Zabaras and Kang (1993) previously emphasized that the selection of an initial guess is extremely important since it affects the rate of convergence. Fig. 5a shows the initial guess taken for this specific test where $q^0(x,y,t)=10$ W/m^2.

Figs. 5b - 5e show the space time behavior of the unknown boundary heat flux at several iteration steps. Note that the solution stays practically unchanged in the vicinity of the final time t_{max} (=0.3 secs) due to the final condition in the adjoint problem (eq. 20). In other words, the unknown function $q^{k+1}(\underset{\sim}{x},t)$, $\underset{\sim}{x} \in \partial\Omega_o$, was updated such that

$$q^{k+1}(\underset{\sim}{x},t_{max}) = q^k(\underset{\sim}{x},t_{max})$$

because the search direction $p^k(\underset{\sim}{x},t_{max})=0$ for all iteration steps. Some preliminary ideas for correcting the problem of the final condition are discussed by Zabaras and Kang (1993).

However, the calculated solution converged to the exact one in the time interval of [0,0.2] secs. Zabaras and co-workers (1992) solved the same test problem using the future information method and spatial regularization. Their solution converged up to 0.07 secs without future time or regularization, 0.12 secs with future time, and 0.16 secs with future time and regularization.

In this specific test run, the maximum number of iteration was specified instead of a tolerance as a stopping criterion. Table 2 indicates that the CGM converges as the iteration number increases. It should be mentioned that the

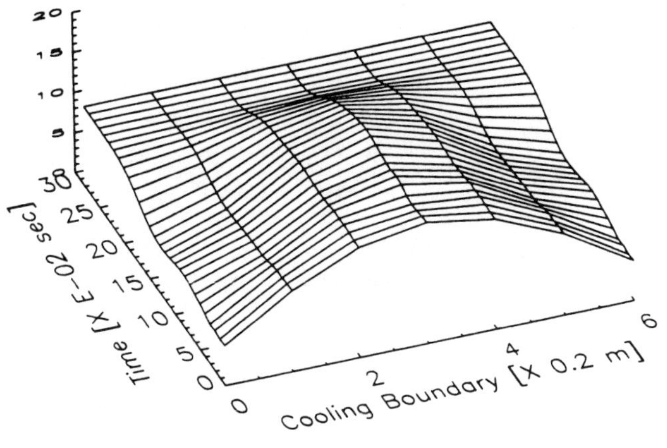

Fig. 5d Cooling history at iteration # 100

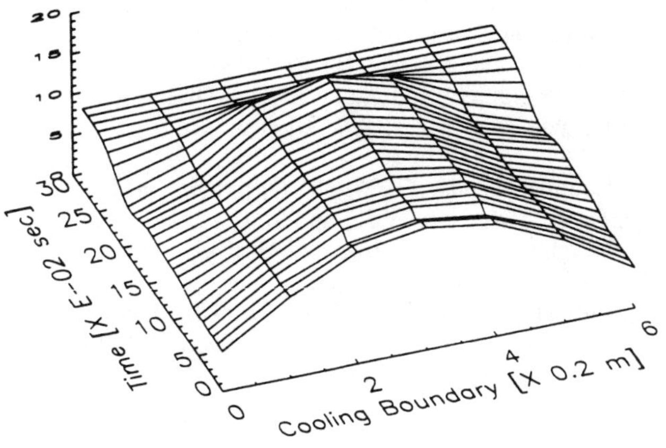

Fig. 5e Cooling history at iteration # 400

Table 2. Change of the value of the cost functional at several iteration steps

Iteration #	Cost Functional	Iteration #	Cost Functional
1	.3239E-02	100	.1593E-06
5	.7531E-04	200	.2315E-07
10	.1847E-04	300	.8801E-08
50	.7429E-06	400	.5479E-08

value of the cost functional can not often be a good stopping criterion and that, instead, the L_2 norm of the change of the unknown function during two successive iterations is a better one as shown in Table 1.

CONCLUSION

The optimum history of boundary cooling condition that results in a desired location/motion of the freezing front was obtained. The adjoint method was presented as part of the infinite dimensional minimization technique in solving the two-dimensional inverse Stefan design problem. The present method gave a converged solution in a longer time interval than the sequential method with regularization. The accuracy and rate of convergence of the solution were negatively affected by the final condition in the adjoint problem.

A nearly unidirectional solidification example was solved to demonstrate the performance of the present methodology. More complicated numerical examples are under investigation.

ACKNOWLEDGEMENTS

This work was funded by NSF grants CTS-9115438 and DDM-9157189 to Cornell University. The computing for this project was supported by the Cornell National Supercomputer facility, which receives major funding by the NSF and IBM Corporation, with additional support from the New York State.

REFERENCES

Zabaras, N., Ruan, Y., and Richmond, O., 1992, " On the Design of Two-Dimensional Stefan Processes with Desired Freezing Front Motion", *Numerical Heat Transfer*, Part B, Vol. 21, pp. 307-325.

Alifanov, O. M. and Kerov, N. V., 1981, "Determination of External Thermal Load Parameters by Solving the Two-Dimensional Inverse Heat-Conduction Problem", *J. Eng. Phys.*, Vol. 41., pp. 1049-1053.

Alifanov, O. M. and Egorov, Y. V., 1985, "Algorithms and Results of Solving the Inverse Heat Conduction Boundary Problem in a Two-Dimensional Formulation", *J. Eng. Phys.*, Vol. 48, pp. 489-1053.

Zabaras, N. and Kang, S., 1993, "On the Solution of an Ill-Posed Design Solidification Problem Using Minimization Techniques in Finite and Infinite Dimensional Function Spaces", *Int. J. Numer. Methods Eng.*, accepted for publication.

Beck, J. V., Blackwell, B. and St. Clair, C. R. Jr., 1985, *Inverse Heat Conduction, Ill Posed Problems*, Wiley-Interscience, New York.

Tikhonov, A. N., and Arsenin, V. Y., 1977, *Solution of Ill-Posed Problems*, V. H. Winston, Washington, D.C.

Colton, D. and Reemtsen, R., 1984, "The Numerical Solution of the Inverse Stefan Problem in Two Space

Variables", *SIAM J. Appl. Math.*, Vol. 44, No. 5, pp. 996-1013 (1984).

Kurz, W. and Fisher, D. J., 1989, *Fundamentals of Solidification,* Trans Tech Publications, Switzerland.

Ruan, Y. and Zabaras, N., 1990, "Moving and Deforming Finite Element Simulation of Two-Dimensional Stefan Problems", *Comm. Appl. Numer. Meth.*, Vol. 6, pp. 495-506.

Luenberger, D. G., 1968, *Optimization by Vector Space Methods,* John Wiley and Sons.

Fletcher, R., 1987, *Practical Methods of Optimization,* John Wiley & Sons, New York.

McDaniel, D. J., 1993, *Front Tracking Finite Element Analysis of Phase Change with Natural Convection,* Master's Thesis, Cornell Univ., New York.

Inverse Problems in Engineering: Theory and Practice
ASME 1993

EXPERIMENTAL INVESTIGATION OF HEAT TRANSFER IN A GUN BARREL BASED ON A SPACE MARCHING INVERSE CONDUCTION METHOD

J. J. Serra, J. M. Gineste, and S. Serror
Département Physique des Surfaces
ETCA/CREA
Arcueil, France

Y. Guilmard and M. Cantarel
Centre de Recherches et d'Etudes Technologiques
GIAT
Bourges, France

Abstract :

Frictional and convective heating knowledge is of eminent importance when attempting to understand the degradation mechanisms of gun barrel inner walls. Bore surface temperature and heat transferred to the gun tube can be experimentally determined by applying an inverse conduction method on thermograms obtained at different depths in the gun barrel.

The intrinsic thermocouples (Steel/Alumel type) are attached at the bottom of measuring holes using a capacitor discharge technique. Remaining width was 0.2 mm from the inner face for one of them and 0.5 mm for the other. These sensors are calibrated (thermoelectric power and response time) in a representative form before the tests. In addition, commercially available heat flux and temperature sensors have been fixed on the outer face of the tube.

This experimental technique has been tested on a 20 mm gun barrel during one shot and burst tests. Couples of thermograms recorded at two depths were obtained at different locations along the tube. For each one, the surface temperature and the incoming heat flux are calculated using a one-dimensional (radial) inverse conduction algorithm.

This approach allows the space and time distribution determination of the boundary conditions for the heat transfer inside the gun tube wall. It constitutes an entrance parameter for the thermo-mechanical modelling of gun barrels.

Nomenclature

- Cp specific heat of the barrel material
- $\mathbb{C}$ volumetric specific heat ($\rho x Cp$)
- F_i non-dimensional numbers (modified Fourier modulus type)
- k conductivity of the barrel material
- q heat flux in the tube wall
- r radial coordinate
- Ro inner radius of the tube
- Re outer radius of the tube
- t time coordinate
- T temperature of the tube wall
- ρ density of the barrel material

1 - Introduction

The erosion of the gun barrel interior wall is the primary factor which reduces the performances of ballistic devices. Many erosion sources contributes to the tube wear, but it is well known that a reduction of the working temperature will delay the onset of tube erosion. This can be obtain, to a certain extent, with additives in the propellant powder or with protective coatings inside the tube.

Launch tube heating is due to different phenomena, convection with hot gases, shell sliding friction, and in the breech region, radiative exchanges. The pressure, speed and thermochemical characteristics of the combustion gases and the physical properties of the tube materials are known or computable parameters. But the exchange coefficients concerning the convective heat transfers or the frictional heat distribution, the radiative characteristics of gases or particles inside the barrel are more difficult to determine. That reduces, at this time, the reliability of functioning thermomechanical models.

Nevertheless, the selection of new bulk materials or coatings to protect high performance barrels supposes the knowledge of the thermal conditions encountered by the materials. In response to that need, experimentals studies [BROSSEAU, 1974; WARD, 1979; LAWTON, 1986] have been carried out to determine temperatures, and with some assumptions heat flows, in gun barrels. It is commonly assumed that, for transient phenomena, the axial heat transfer is neglectible versus the radial one; (see, for example [BOISSON, 1992]). In these conditions, a monodirectional space marching inverse conduction algorithm can be applied to determine the flow incoming to the barrel interior wall from thermograms recorded at different depths during the internal ballistic phase.

This paper describes a method of recording transient temperature gradients inside the barrel wall and how these measurements can be processed to determine the thermal conditions at the bore surface.

2 - Temperature measurements

Internal temperatures were measured using intrinsic steel/alumel thermoelectric sensors. This measurement procedure minimizes the thermal contact resistance and presents a very short response time.

2.1 - Sensors location

The gun barrel involves twelve seatings, distributed in six sections along its length. The axial location and the barrel local external diameter are reported in fig.1a.

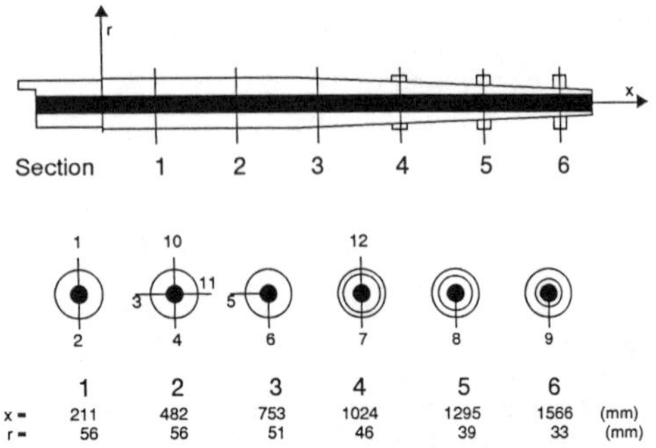

Fig.1a : Temperature sensors location.

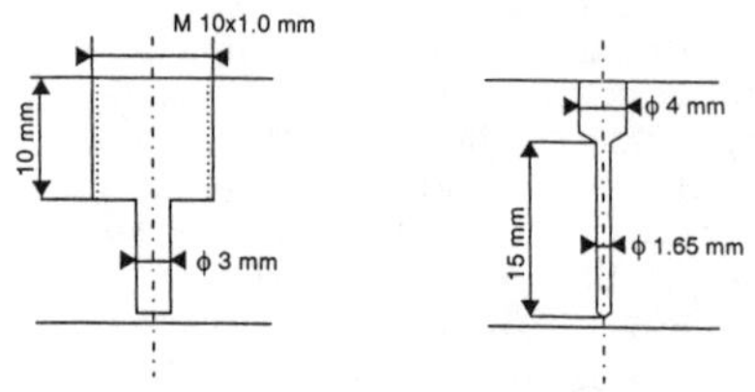

Fig.1b : Measuring holes geometries.

Due to the very steep temperature gradients existing near the inner surface of the gun barrel, the measuring holes are drilled as to set the sensors at the shortest distance from this surface. The closest distance that could be practically located without destruction due to the internal pressure, is 0.2 mm. This thickness is measured with an accuracy of 25 μm (after checking of the concentricity of the outside diameter with the bore diameter [BROSSEAU 1974]), using a depth micrometer . The gun barrel being rifled, the temperature sensors are spaced so as to be located in the center of a grove.

Two geometries (fig.1b) are used for the holes corresponding to different measurements approaches: the first kind (ISL type) involves a 3 mm diameter flat-bottomed terminal hole and the second kind (BRL type) a 1.65 mm diameter tapered hole. Sensors n° 1 to 11 belong to the first kind, the last three ones to the second kind. The remaining width is 0.2 mm from the inner face for the sensors n° 1, 3, 5, 8, 9 and 10 and 0.5 mm for the sensors n° 2, 4, 6, 7, 11, 12.

2.2 - Calibration of the thermocouples

The thermoelectric sensors are constituted by 0.127 mm diameter alumel wires welded at the bottom of each measuring hole. A precision ceramic guide is used to ensure that the measuring wires are located exactly in the center of the drilled holes. The steel/alumel junction is obtained using a capacitor discharge technique [(1)]. A common steel wire constitutes the other leg of these sensors.

The e.m.f. delivered by such thermocouple versus temperature has been determined by comparison with is a standard K type sensor set in the same calibration furnace. Calibration data are shown in figure 2.

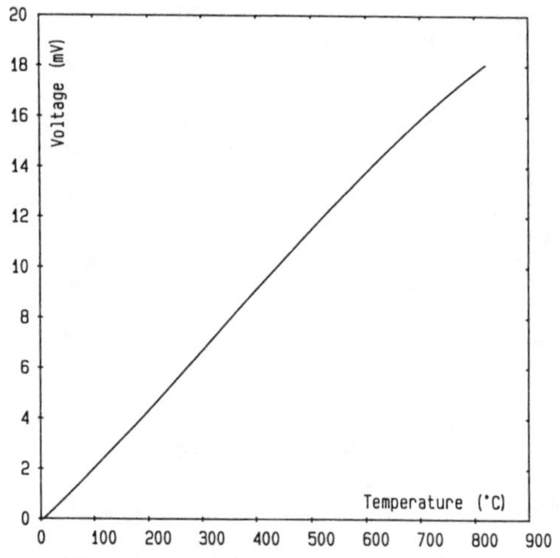

Fig.2 : Temperature response of the steel-alumel thermocouple.

Conductive heat flux and temperature sensors have been fixed on the external surface of the tube in section 2 and 3, for additional thermal monitoring.

2.3 - Data acquisition system

A schematic drawing of the instrumentation arrangement is shown in fig.3.

The thermocouple signals, referenced to a ice-point cold junction, were connected to adjustable gain (x 200) wideband (dc to 10 KHz) amplifiers [2]. The measurement of the amplified signals is performed by a multi-channel 16 bits data acquisition module [3] involving an on-board 128 K memory and programmable amplifiers. After capture, at a 1000 measurements/channel/second rate, the data are transferred through an IEEE488 interface to a PC compatible microcomputer.

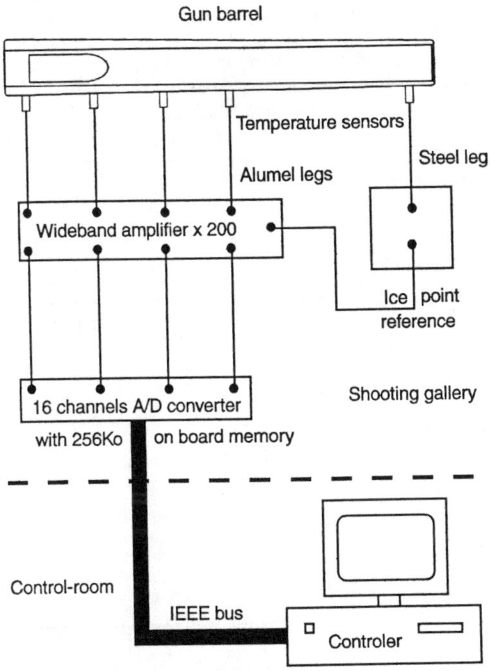

Fig.3 - Data acquisition and processing.

2.4 - Signal processing.

The gun barrel, nearly two meters long, acts like an antenna, and so low level sensors signals undergo radioelectric disturbances. To eliminate "noise spikes" produced by the 50 Hz (and harmonics) local supply circuit frequencies the recorded data are numerically filtered. First order Chebyshev bandstop filters are designed to reject 50, 150 and 250 Hz frequencies and applied two ways in forward and backward directions. The resulting sequence has precisely zero-phase distortion and double the filter order. [PARKS & BURRUS, 1987] .

3 - Tests conditions

The tests were carried out on a 20 mm calibre rifled gun barrel. The material constituting the tube is a 32CVD13 steel containing about 3% of chromium and minor quantities of other elements. Thermal data of such steel are given in [SMITHELLS, 1976]. The inner surface of this tube were erosion protected by a 12 µm thick chromization coat.

Different type of firing were investigated from single rounds to 16 shots rounds (1 ,2, 4 ,8 ,16 shots) with a nominal rate of 900 rd/mn. The complete sequence was repeated with two types of inert amunitions :

- 90 g shells propelled by 55.5 g of B7T powder (flame temperature = 2969 K; potential = 913 cal/g)
- 120 g shells propelled by 50.4 g of BTu powder (flame temperature = 2747 K; potential = 842 cal/g).

Mean velocity, measured at 25 meters of the muzzle, during the tests appears to be about 1250 m/s for the 90 g shells and 1050 m/s for the other ones.

3- Principle of thermograms analysis.

The heat transfer through the barrel wall can be solved by considering a one-dimensional radial flow in a hollow cylinder of inner and outer radii Ro and Re, initially at an uniform temperature To. The heat conduction equation can be written as :

$$C \frac{\partial T}{\partial t} = k \left(\frac{\partial^2 T}{\partial r^2} + \frac{1}{r} \frac{\partial T}{\partial r} \right) \qquad (1)$$

where C is the volumetric specific heat and k the thermal conductivity.

At an initial time t = 0, the barrel bore is subjected to a transient thermal flux. The objective is to estimate this flux q(0,t) at r = Ro (node 1), given the temperatures at $r = r_j$ and $r = r_k$. The problem is solved using a space-marching finite-difference algorithm derived from [RAYNAUD & BRANSIER, 1986] and adapted to cylindrical coordinates.

The temperature field in the direct region $r_j \leq r \leq r_k$ can be calculated with a pure implicit scheme because the boundary conditions of the first kind in r_j and r_k are known.

In the inverse region ($Ro \leq r \leq r_j$), we note T_i^n the temperature at time $n\Delta t$ and at position $Ro + i\Delta r$. The temperatures T_{i-1}^n, above the measuring points, are determined using the mathematical reasoning cited above.

First, equation (1) is written in the form :

$$C \frac{\partial T}{\partial t} = - \frac{1}{r} \frac{\partial (rq)}{\partial r} \qquad (1a)$$

$$q = - k \frac{\partial T}{\partial r} \qquad (1b)$$

Then, heat flux densities are approximated with central differences alternatively in time and space directions.

Equation (1a) gives :

$$C_i^n r_i \Delta r . \frac{T_i^{n+1} - T_i^{n-1}}{\Delta t} = 2r_{i-1/2}q_{i-1/2}^n - r_{i+1/2}(q_{i+1/2}^{n+1} - q_{i+1/2}^{n-1})$$

and the different values of q (equation 1b) are approximated as :

$$q_{i-1/2}^n = -k_{i-1/2}^n . \frac{T_i^n - T_{i-1}^n}{\Delta r} \text{ and } q_{i+1/2}^m = -k_{i+1/2}^m . \frac{T_{i+1}^m - T_i^m}{\Delta r} ;$$

in which m = n-1 and n+1

Finally, the temperature T_{i-1}^n, at point Ro + (i-1)Δr and time nΔt, can be calculated as a function of the temperatures T_i^n, T_i^{n-1} and T_{i+1}^{n-1} (past temperatures), T_i^{n+1} and T_{i+1}^{n+1} (future temperatures).

$$T_{i-1}^n = \frac{1}{2F_1} \times \qquad (2)$$

$$[(F_3-1)T_i^{n-1} + 2F_1T_i^n + (F_2+1)T_i^{n+1} - F_3T_{i+1}^{n-1} - F_2T_{i+1}^{n+1}]$$

where the quantities F_1, F_2 and F_3 (non dimensional numbers) are given by :

$$F_1 = (1 - \frac{\Delta r}{2r_i}) \frac{k_{i-1/2}^n}{C_i^n} . \frac{\Delta t}{\Delta x^2} ,$$

$$F_2 = (1 + \frac{\Delta r}{2r_i}) \frac{k_{i+1/2}^{n+1}}{C_i^n} . \frac{\Delta t}{\Delta x^2} , \qquad (3)$$

$$F_3 = (1 + \frac{\Delta r}{2r_i}) \frac{k_{i+1/2}^{n-1}}{C_i^n} . \frac{\Delta t}{\Delta x^2}$$

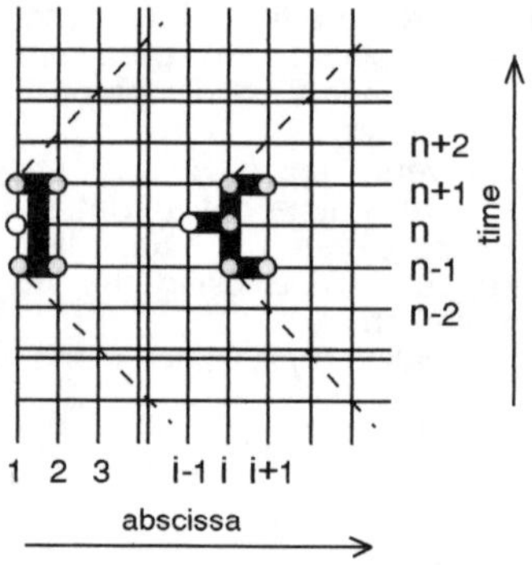

The interface heat flux q_1^n is calculated from an energy balance on the half (Δr/2 width) control volume on the right of the node 1.

$$q_1^n = \frac{C_1^n . \Delta r}{4\Delta t} \times \qquad (4)$$

$$[(2F_3-1)T_1^{n-1} + (2F_2+1)T_1^{n+1} - 2F_3T_2^{n-1} - 2F_2T_2^{n+1}]$$

4- Experimental results.

Tests were carried on a aged chromized tube (8 years from its first use). This coating involves two layers, M_7C_3 and $M_{23}C_6$ (M representing the metallic elements), but the phase ratio, cleft state, and thermal properties of these materials are unknown. For this reason, the iterative calculation described in (2) is repeated until the interface is reached. The calculated values are thus the interfacic temperature and heat flux.

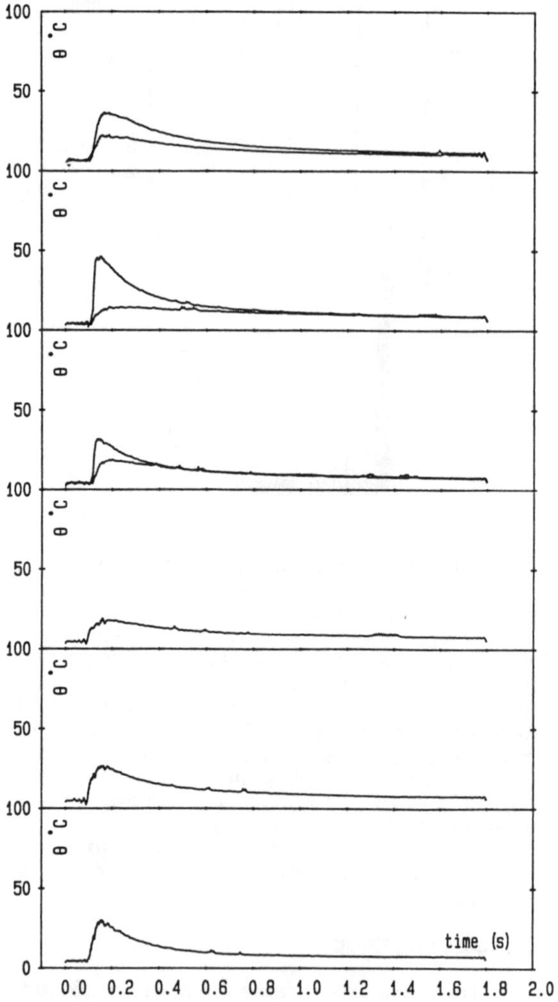

Fig.4 - Thermograms recorded during a one shot firing (90 g shell) in the six measuring sections.

An example of thermograms recorded during one shot firing in the different sections of the tube is presented in fig.4.

It have to be noted than chromizing gives a better heat protection than classical nitriding. In addition, the temperature difference observed behind the two types of coatings increases with tube ageing [SERRA, 1987]. The measured temperature levels appears to be a little lower than those observed by [BROSSEAU, 1974].

In the three sections in which temperature sensors are located at two different depths, inversion can be carried out. Interfacic fluxes are calculated and the figure 5 shows the observed differences between 90 g and 120 g shells firing heating in section 3.

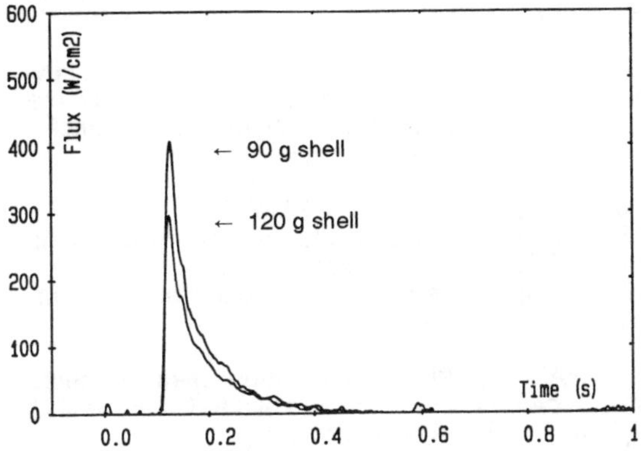

Fig.5 : Interfacic flux measured in section 3 for 90 g and 120 g shells.

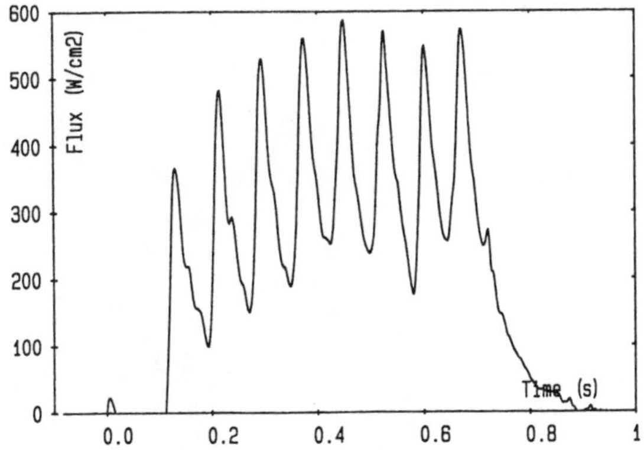

Fig.6 - Interfacic flux measured in section 3 during a 8 shots burst (90 g shells)

During the bursts of gun-fire, the delay between successive shots is about 70-80 ms. Fig.5 shows that the flux falling rate is so than a new shot is fired before a return to zero. Consequently, a cumulative thermal effect is imposed to the inner wall. Fig.6 presents an example of results showing that a steady state flux cycling is achieved after a few shots. The same result can be expressed as a function of the total heat transferred through the interface (fig.7).

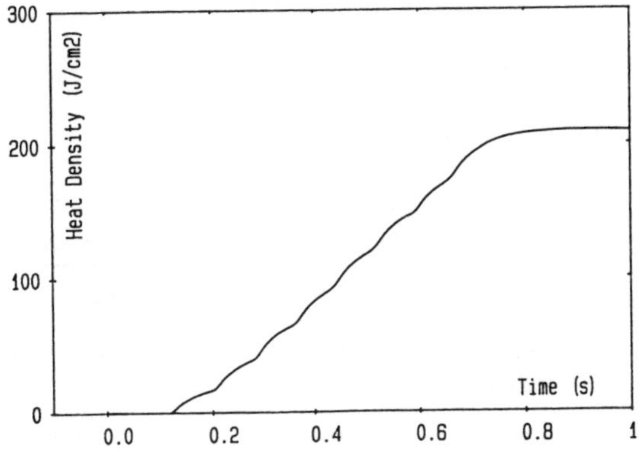

Fig.7 - Heat transferred through the interface in section 3 during a 8 shots burst (90 g shells)

5- Error analysis.

The reliability of the presented results is a function of the accuracy of different entrance data:
- thermal properties of the material in which the sensors are embedded,
- location of the sensors,
- thermometry problems (thermoelectric power determination, response time, signal disturbances)

The different tables giving the conductivity of 3%Cr steels does not differ from more than a few percent [SMITHELLS , 1976]. So, the thermal properties reliability appears to be not a major cause of errors. On the other hands, a 25 µm accuracy for sensors depths ranging from 200 to 500 µm, will cause observable differences, as it is presented in fig.8.

Concerning the disturbances caused by these sensors, and more particularly the measuring holes, it can be demonstrated that they can be neglected for short thermal pulses. The thermal field distorsion becomes important only after a few seconds [VIFDOMINE, 1993]. In this conditions, the sensor response time appears to be suitable for the temperature evolution dynamics.

The static calibration of the intrinsic thermocouples, according to §2.2 does not present difficulties. It supposes that we have at our disposal a steel rod sample of same nature than that constituting the gun barrel. It is also assumed that the steel composition remains constant and it is not polluted around the junction.

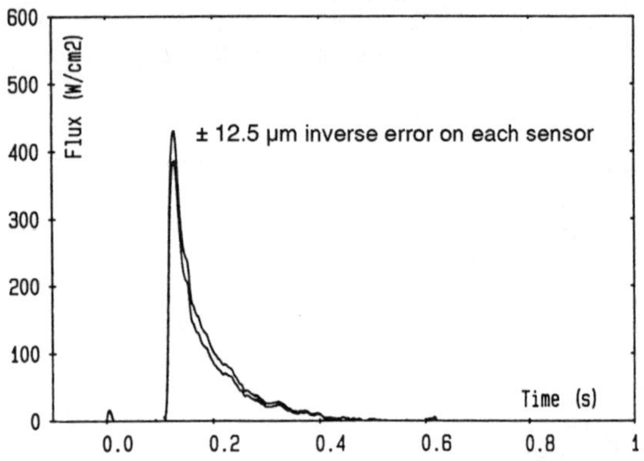

Fig.8 - Effect of depth error on heat flux determination

The major thermometric problem is due to the electrical signal noises, especially those caused by the electromechanical igniter . A manually firing appears to be preferable for the future investigations.

6 - Conclusion

The temperature behind a thermal protective coating and the heat transferred to the gun tube can be determined by measuring the temperature evolution at different depths in the inner wall and applying a monodirectional inverse conduction algorithm on these thermograms. The temperatures are measured using intrinsic thermocouples in order to reduce the response time and the thermal disturbances. The inverse algorithm used for this study is simple to code and can be run on the microcomputer controling the data acquisition system.

The feasibility of this experimental technique has been tested on a 20 mm chromized gun barrel. The lack of knowledge concerning the coating thermophysical properties prevent the determination of bore conditions, only the heat flux and temperature behind the coating have been determined. With an uncoated tube (or a thermally characterized coating), this method will allow the boundary conditions determination for the heat transfer inside the gun tube wall.

Several applications of this experimental procedure are considered in the future, such comparison of coatings efficiency and heat exchange coefficients estimations.

Notes. (Measurement devices used)
1 - microwelder TECHMETAL - France
2 - model 3B40 / ANALOG DEVICE - USA
3 - model 576 / KEITHLEY - USA

Bibliography.

BOISSON, D. and GRIGNON C., 1992, "1D and 2D thermal modelling of gun barrel heating and cooling during a burst," GIAT Rpt DIV.E/CRET/TEC/GB/BI/NT n°9/92/NC (in french)

BROSSEAU, T.L., 1974, "An experimental method for accurately determining the temperature distribution and the heat transferred in gun barrels," BRL Rpt N°1740

LAWTON, B, 1986, "Bore temperature and heat flux in a 40 mm gun barrel," AGARD Conf. Proc. N°392, pp.10.1-12

PARKS, T.W, and BURRUS, C.S, 1987, "Digital Filter Design," John Wiley & Sons, chapter 7, section 7.3.3

RAYNAUD, M., and BRANSIER, J., 1986a "A new finite-difference method for the non-linear inverse heat conduction problem," Num. Heat Transfer Vol.9, N°1, pp.27-42

RAYNAUD, M., and BRANSIER, J., 1986b "Experimental validation of a new space marching finite difference algorithm for the inverse heat conduction problem," 8th Int. Heat Transfer Conf. San Fransisco CA

SERRA, J.J., 1987, "Temperature measurements in gun barrels chromized and nitrided inner walls," ETCA Rpt 87-R-114 (in french)

SMITHELLS C.J., 1976, "Metals Reference Book," 6th Ed. BUTTERWORTH

VIFDOMINE, S., 1993, "Thermal disturbance caused by thermocouple sensors set normally to the isotherms," INSA Lyon - report in preparation (in french)

Inverse Problems in Engineering: Theory and Practice
ASME 1993

APPLICATION OF A SPACE MARCHING ALGORITHM TO THE HIGH SPEED TRIBOMETER

O. Lesquois
DGA/ETCA/CREA
Arcueil, France

J. J. Serra
DGA/ETCA/CREA/CEO
Font-Romeu, France

P. Kapsa
Laboratoire de Tribologie
et Dynamique des Systèmes
Ecole Centrale de Lyon
Ecully, France

S. Serror
DGA/ETCA/CREA
Arcueil, France

ABSTRACT:

The ETCA high speed tribometer is a facility able to achieve tests involving sliding speeds up to 350 m/s and loads up to 5000 N. The moving part is a 360 mm diameter stainless steel cylinder and the slider is a 5 mm side parallelepipedic pin. The test duration does not exceed a few seconds.

Several measurements have been carried out during the tests:

- the load and the friction forces by a piezoelectric triaxial sensor,
- the pin length loss by a capacitive sensor,
- the temperatures at different depths in the pin, using fast response intrinsic thermocouples.

In addition, a high speed video camera (1000 pictures/second) has been used to point out the different phases of the wear process.

During the high speed frictional tests, competitive process between two types of physical phenomena appears: mechanical and thermal ones. When the pressure x velocity product (homogeneous to a heat flow) is low, mechanical phenomena are governing the processes. Heat effects become more and more important when the PV product increases, and for very high values, the pin frictional surface is partially molten.

In such conditions, the incoming heat flux and surface temperature are important parameters for the understanding of pin wear process. These two parameters can be calculated from thermocouple measurements, using a space marching inverse conduction algorithm taking into account the free boundary.

The obtained results show heat flux evolution curves versus time involving instability phases. These instabilities are due to metal particles transferred from the pin to the cylinder, which form asperities pushing away the pin for a certain time. This aspect has been confirmed by the high speed film recorded during the tests.

NOMENCLATURE

- WR: wear rate,
- f : friction coefficient,
- V : sliding speed,
- P : contact pressure,
- N : normal load,
- Z : melted zone limit,
- H : heat flux,
- ρ : volumic mass,
- L : latent heat of fusion,
- T_f : melting temperature,
- T_o : room temperature,
- T : temperature,
- x : space,
- t : time,
- F : non dimentional numbers, Fourier modulus type,
- C : volumic specific heat,
- K : thermal conductivity,
- q : interface heat flux,
- Q_m : heat flux consummated by melting.

1/ INTRODUCTION

High speed friction corresponds to situations where superficial melting of at least one of the two sliding materials happens in the contact. This phenomenon generally occurs for high contact pressures (a few Mpa or more) and for high sliding speeds (several 10 m/s and more). High speed friction is, for example, observed for braking systems, electrical contacts and friction btween a projectile and a gun tube.

The first important works on high speed friction are due to Montgomery [1] during the fifties. These studies have been made for the U.S. Army aiming to reduce the wear of weapon's tubes. Several points have been pointed out:

- the amount of heat which is received by a projectile during a shot is due to three phenomena: the friction, the deformation of the belt and the creation of heat by the propellant. Then the formation of a liquid metal film at the surface of the belt is observed earlier in real cases than in laboratory tests.
- high friction coefficients and fluctuations have been recorded for low values of the PV product (pressure x speed). For values higher than 6000 MPa/s, the friction coefficient is low, stable and decreases slowly when the PV product increases.

- the wear rate is related to the speed of heat production, and then to the friction coefficient, the sliding speed and the contact pressure:

$$WR = k\ (fPV)^2 \qquad (1)$$

where k is a proportionality constant. The fPV product is used because it is homogeneous to a power per unit surface equal to a heat flux.

- the evolution of wear speed versus the inverse of the melting temperature of sliding materials, on semi-log coordinates, is a straight line with a positive slope indicating that high melting temperature materials are more resistant to wear.

- the results of these studies have shown that the wear mechanism for high speed friction is superficial melting followed by an elimination of a part of the melted layer. The surfaces are not in direct contact but are separated by a liquid layer. Then the wear speed is only a function of the melting temperature. The other material parameters, such as the thermal conductivity, the materials compatibility, the crystalline structure, the hardness, ... are of secondary importance or even negligible. However, if the sliding duration is not sufficient to produce a total melting, these factors can become important.

Sternlicht and Apkarian [2] have modified the Montgomery friction machine to make experiments with a contact subjected to an electrical current. They have then obtained a decrease of the friction force due to an increase of the energy dissipated in the contact by Joule's effect.

Williams and Giffen [3] pointed out that the increase of the material temperature produces a decrease in the shearing force of the interface. The friction coefficient is then reduced. The friction coefficient is related to the sliding speed and the normal load:

$$f = k\ (VN)^{-0,45}\text{; where k is a constant} \qquad (2)$$

High speed friction situations are characterised by melting phenomena due to the energy dissipation in the sliding interface. To have a better understanding of these situations, some authors have studied the propagation speed of the melted zone limit in the pin. One can notice in particular the works of Landau [7] used by Serra [8] for the high speed tribometer of ETCA which exhibited important thermo-mechanical parameters. The speed of the melted zone limit can be expressed by :

$$Z = \frac{H}{\rho\ (L + C(T_f - T_0))} \qquad (3)$$

The aim of the present work is to understand and modelize the phenomena governing the friction between two metals, at high speed and in a transient regime. The thermal exchanges in the contact have been considered in order to determine the therma distribution coefficient between the two sliding bodies. This study have used a high speed friction machine developed at ETCA.

2/ EXPERIMENTATION

FRICTION MACHINE

The high speed pin on cylinder tribometer is presented figure 1. The sliding speed ranges from 5 to 350 m/s, corresponding to a rotation speed up to 18 000 rpm.

An originality of this device is the use of active magnetic bearings (S2M) to support the cylinder which have a mass of around 3000 N. The axle and the cylinder are rotating around their inertia axle limiting then the vibrations.

The stainless steel (Z6CN17 4 1) cylinder is of 360 mm diameter. The axle is horizontal and the pin, with a 5 x 5 mm square section and a 22 mm length, is below. The cylinder is polished with abrasive papers to a 0.15 Ra value.

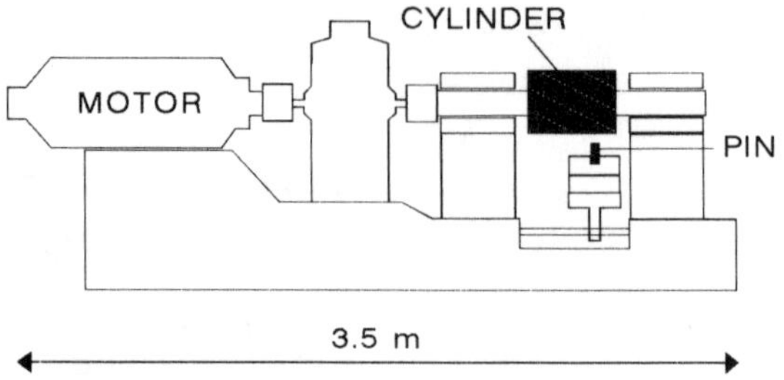

FIGURE 1: the friction machine.

Tested materials, iron, AISI 52100 steel, copper, aluminium, zinc, chromium, tungsten and lead have been chosen for their known thermo-mechanical properties. All materials, except steel, are pure. For this study, only the case of pure iron is considered.

The normal load, between 1 and 100 daN, is applied with a arm and dead load. The test duration is controlled by two pneumatic jacks which move the pin. This duration ranges from 0.2 to several minutes. For most of the tests, the wear of the pin limits the test duration to about 1 second.

The whole pin support can be moved horizontally to change the wear track on the cylinder. Around 30 various tests can be performed on the same cylinder.

MEASUREMENTS

A triaxle force piezo-electric transducer is used to measure the normal load and the friction force. A capacitive position transducer is used to follow the evolution of the pin position during the test, indicating then the pin length worn. A tachometer gives the rotating speed of the cylinder.

The pins are weighed before and after the tests. The mass loss can be then compared to the indication given by the position transducer.

From some authors [1,2], the pin surface is melted when the PV product (contact pressure x speed) is high. Then it has been interesting to perform thermal measurements during the tests to validate or not this assumption. So, iron/alumel thermocouples have been attached on the side of the pin at 1.5, 2.5 and 3.5 mm from the sliding surface.

The data are collected during the test by a PC computer, then stored and processed after the tests. Figure 2 shows evolutions of normal force, friction force and friction coefficient during the test N°161 (load 200 N, sliding speed 28,15 m/s, time of contact 1 s).

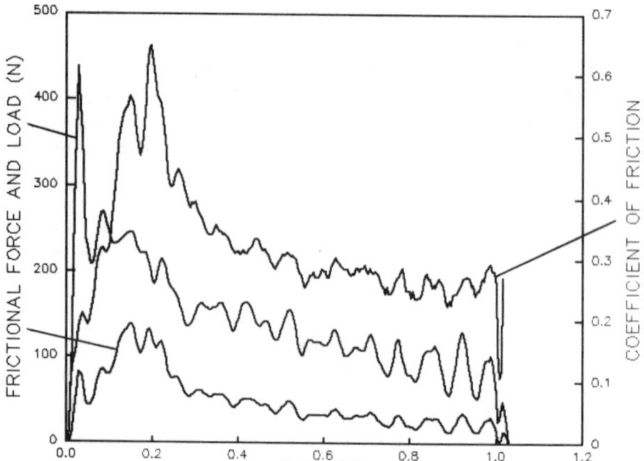

FIGURE 2: evolution during the test of load, friction force and friction coefficient.

PRINCIPLE OF THERMOGRAMS INTERPRETATION

Let us consider a one-dimensional heat conduction in the pin, initially at an uniform temperature T_0. At time t = 0, the contact surface of the sample, noted x = 0 in our problem, is subject to a frictional thermal flux. The objective is to estimate the interface heat flux q(0,t) at x = 0 (node 1), given the inner temperatures at $x = x_j$ and $x = x_k$. The problem is solved using the space-marching finite-difference algorithm developped by Raynaud & Bransier [9, 10].

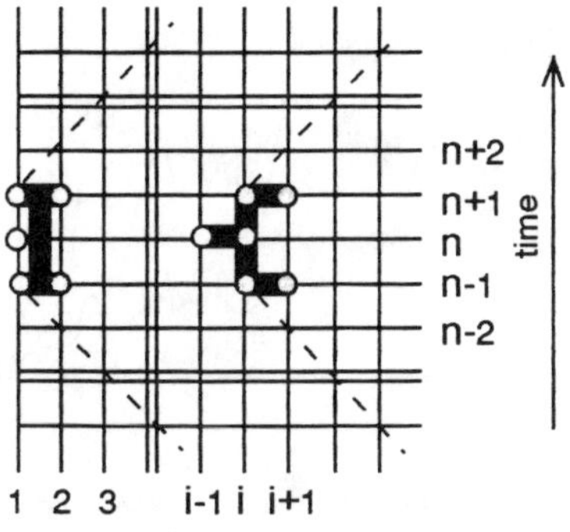

FIGURE 3: principle of the space-marching finite-difference algorithm.

The temperature field in the direct region $x_j \leq x \leq x_k$ can be calculated with a pure implicit scheme because the boundary conditions of the first kind in x_j and x_k are known. Similarly, in the inverse region $(0 \leq x \leq x_j)$, we note T_i^n the température at time nΔt and at depth iΔx. In this region, the authors demonstrate that the temperature T_{i-1}^n, at point (i-1)Δx and time nΔt, can be calculated as a function of the temperatures T_i^n, T_i^{n-1} and T_{i+1}^{n-1} (past temperatures), T_i^{n+1} and T_{i+1}^{n+1} (future temperatures).

$$T_{i-1}^n = \frac{1}{2F_1}\left[(F_3-1)T_i^{n-1} + 2F_1T_i^n + (F_2+1)T_i^{n+1} - F_3T_{i+1}^{n-1} - F_2T_{i+1}^{n+1}\right] \qquad (4)$$

Where the quantities F_1, F_2 and F_3 are given by :

$$F_1 = \frac{k_{i-1/2}^{n}}{C_i^n} \cdot \frac{\Delta t}{\Delta x^2}, \quad F_2 = \frac{k_{i+1/2}^{n+1}}{C_i^n} \cdot \frac{\Delta t}{\Delta x^2}, \quad F_3 = \frac{k_{i+1/2}^{n-1}}{C_i^n} \cdot \frac{\Delta t}{\Delta x^2} \qquad (5)$$

Where C is the volumic specific heat and k the thermal conductivity.

The interface heat flux q_1^n is calculated from an energy balance on the half (Δx/2 width) control volume on the right of the node 1.

$$q_1^n = \frac{C_1^n\,\Delta x}{4\,\Delta t}\left[(2F_3-1)T_1^{n-1}+(2F_2+1)T_1^{n+1}-2F_3T_2^{n-1}-2F_2T_2^{n+1}\right] \qquad (6)$$

When the pin abrades (by melting, wear or any other way) the contact surface moves about the temperature sensors. The surface displacement is continuously measured using a capacitive sensor, so its abscissa s(t) is known at any time. The iterative calculation - equation (4) - giving the space forward temperature at a given time is repeated until the next node be over the ablation front location. Then, the calculation cell size is reduced in order to place the forward temperature on the contact surface and the backward temperatures symmetrically compared with the last calculated position (fig.4). These temperatures are interpolated between the known values $T_i^{n\pm1}$ and $T_{i+1}^{n\pm1}$, and the T_{i-1}^n temperature is given by the relation (1) in Δx is replaced by $\Delta x' = i\Delta x - s(t)$. The interface heat flux is calculated from an energy balance on the last half (Δx'/2 width) control volume using an explicit formulation.

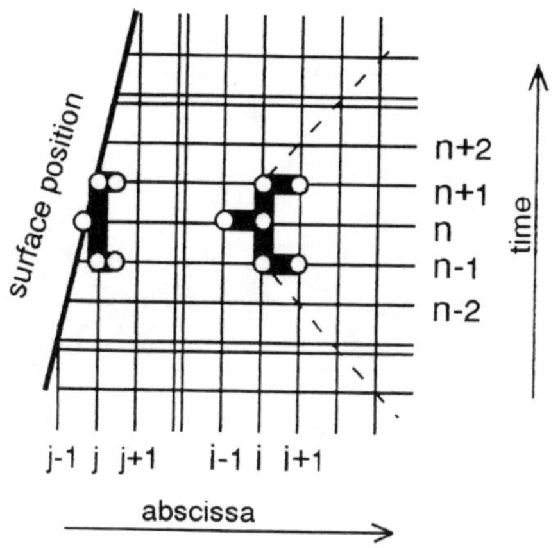

FIGURE 4: arrival at the surface when using the space-marching finite-difference algorithm.

If the pin abrades by melting, the frictional heat flux involves an additional part which is consummated by the melting front displacement :

$$Q_m = \rho\, L\, \Delta s/\Delta t \qquad (7)$$

3/ RESULTS

TRIBOLOGICAL RESULTS

At the beginning of the test, there is a competition between the linear expansion caused by the increased of the pin temperature and the wear. This causes low apparent wear which could be either positive or negative depending of the prime phenomenon. If the contact time is long enough, a phase of rapid wear of a few mm/s occurs.

Iron pins create on the cylinder an irregular transfer layer of material composed of important rough deposits which can go up to a few tenth of millimetre high. The cylinder does not appear to be deteriorated. However, an Electron Probe Microanalysis (EPMA) of the pins reveals sometimes some chromium or nickel traces coming from the cylinder in the pad.

For iron, a friction coefficient of about 0,4 is measured at the beginning of the test. Then, it grows, pass through a maximum and decreases to finish in the range of the initial values. The initial growth comes from the bad tribological behavior of iron when the temperature increases. However when the temperature is high enough, iron in plastic flowing offers a lower resistance, and the contact is partially lubrificated by melting iron. So the coefficient of friction decreases.

VISUALIZATION OF THE CONTACT

A high speed video system (60 to 1000 pictures/s) was used to observe the behavior of the pin during the test. Thus, we observe that from time to time, the pin is skipping, so the contact is periodically broken. The video film shows also the formation of the pad, the occurence of sparks and, at the back of the contact, the ejection of particles which are droplet of molten iron or oxyde particules.

THERMAL RESULTS

In tribology, three temperatures are used: ambient temperature, global temperature of the contact surface and flash temperature which are ephemeral and localized around the real contact zones which usually are microgeometrical. These flash temperatures can reach very high values even in low sliding speed ([5, 11]).

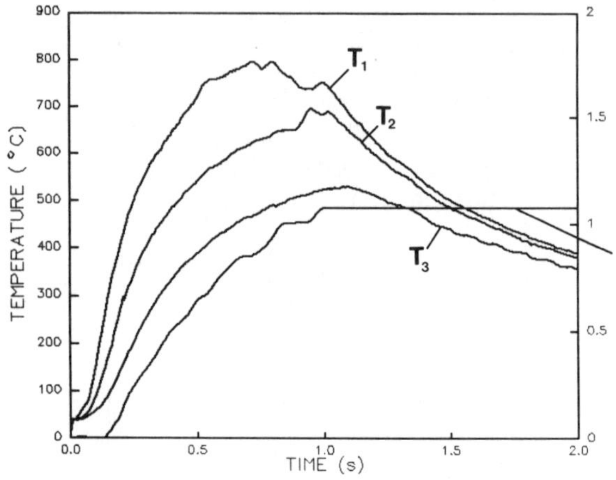

FIGURE 5: the wear and the 3 thermocouples measurements.

The temperature of contact and the thermal flux coming into the pin can be obtained in solving a one-dimensional inverse heat conduction problem with a moving boundary. This problem is not analytically solvable, and numerous authors have proposed numerical solutions. However, a method recently developed by Raynaud and Bransier (explicited in point 2) can be modified to take this conditions into account. So a code adapted to the tribometer and using this method has been made.

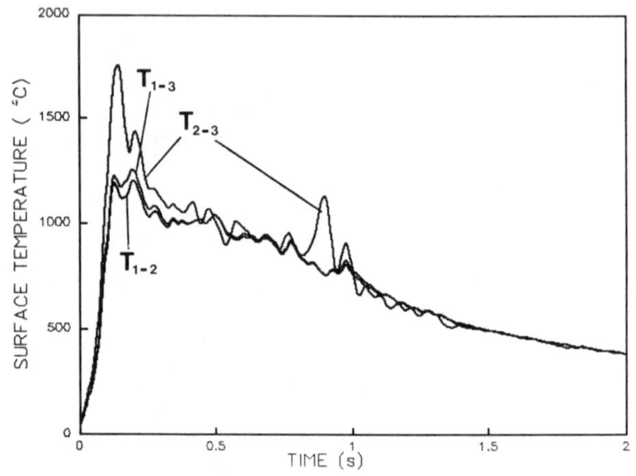

FIGURE 6: calculated surface temperatures.

In the present case, after the calculations (Δt = 0.004 s, Δx = 0.001 mm) we can see that the global temperature of the surface reaches a maximum value of 1200°C (fig.6). The three curves of the figure correspond to the calculated values of the surface temperature using respectively thermocouples 1 and 2, 1 and 3 and 2 and 3 to have boundary conditions. T_{1-2} and T_{1-3} results are in good agreement, but T_{2-3} is, mainly in the beginning of the curve, different. A numerical explaination is that the calculation by the inverse method begins at the themocouple 1 in the case 1-2 and 1-3 and at thermocouple 2 in the case 2-3. The number of iterations which is greater in the case 2-3 gives more errors mainly at the beginning of the test where the thermal gradient is high and where the thermocouple 2 which is farer from the contact than the first one is more subject to ground noises.

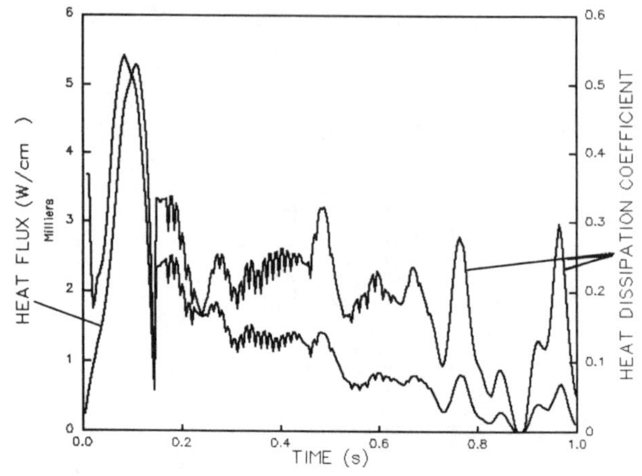

FIGURE 7: heat flux and heat dissipation coefficient.

The heat flux evolution curves versus time are presented figure 7. In this curve, we observe instabilities in the form of abrupt falls, one which reaches zero a little bit before 0.9 s. This phenomenon can be explained by the video observation. We see, figure 8, a picture coming from the video film. The pin is at the bottom of the picture and the pad can be easily seen. The cylinder is at the top of the picture, and one can see the reflexion on the pin on it. In this picture we have the contact at the left of the pin by an asperity stuck on the cylinder. This contact corresponds to the 1 cm length zone of the interface where there is a supplement of light. The figure 9 is the same, but a few thousandth of second after the contact is broken because of the push caused by the passage of the asperities of the figure 8.

The important break of the curve comes from a great rupture of the contact, but the video shows that there is also a lot of little breaks of a few thousandth of second. The existence of these little breaks can explain the little picks in some part of the curve. They do not go down to zero because the acquisition frequency is actually only 250 Hertz, so it is difficult to see completely so short phenomena.

The curve of the heat dissipation coefficient which is calculated as the ratio of the heat flux coming into the pin versus the heat flux produced by the friction follows approximatively the variations of the heat flux. The values obtained are between 0.1 and 0.5.

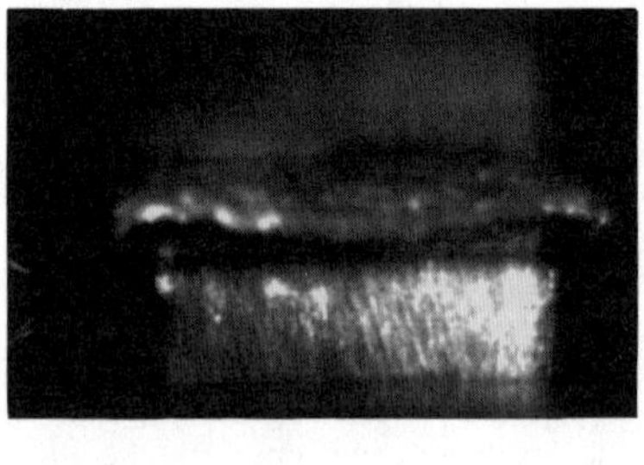

FIGURE 8: contact between the cylinder and the pin.

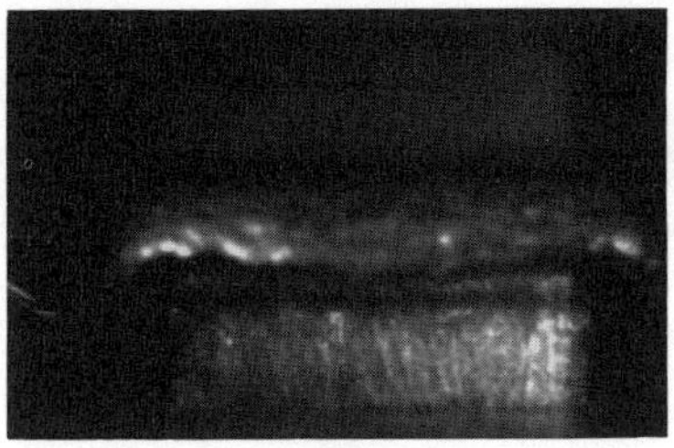

FIGURE 9: break in the contact.

4/ CONCLUSIONS

A test machine was used to characterize tribological phenomena in high load and high speed conditions. Three thermocouples give thermal measurements in addition with tribological measurements.

The thermal measurements are treated by of code using the Raynaud and Bransier's space marching inverse conduction algorithm, which takes into account the free boundary. The results are enough accurate to show the skipping of the pin which is confirmed by the observation made on a high speed video system film.

More tests with different materials and a better speed of acquisition are expected in the future.

REFERENCES

[1] R.S.MONTGOMERY; Friction and wear at high sliding speeds; Wear, vol. 36, n°3 pp 275-278, Mars 1976.

[2] B.STERNLICHT et H.APKARIAN: Investigation of melt lubrication; A.S.L.E. Trans. 2, 248 (1956).

[3] K.WILLIAMS et E.GIFFEN; Friction between unlubricated steel surface at sliding speeds up to 750 feet per second; Proc. Inst. Mech. Engr 178, 24 (1963/4).

[4] F.P.BOWDEN, F.R.S. et E.H.FREITAG; The friction of solids at very high speed 1. metal on metal; 2. metal on diamond; Proc. R. Soc., Ser. A 248, 350 (1958).

[5] S.C.LIM et M.F.ASHBY; Wear-mechanism maps; Acta metall., vol. 35, n°1, pp1-24 (1987).

[6] H.S.CARSLAW & J.C.JAEGER; Conduction of heat in solids; 2nd ed. Oxford: Clarendon Press (1959).

[7] H.G. LANDAU; Heat conduction in a melting solid.; Quart. Appl. Math. 8(1): 81-94 (1951).

[8] J.J. SERRA et P RIBERTY; Analyse thermique de l'usure par ablation de métal fondu en frottement à grande vitesse.; Rapport ETCA N°89 R 139.

[9] M.RAYNAUD et J.BRANSIER; A new finite-difference method for the nonlinear inverse heat conduction problem ? Numérical heat transfert, vol 9, N°1, p27 à 42, 1986.

[10] M.RAYNAUD et J.BRANSIER; Evaluation au moyen d'une méthode inverse, du partage du flux généré par frottement entre deux solides en mouvement relatif; Société Française des thermiciens, journée d'étude du 8 mars 1989.

[11] T.F.J. QUINN and W.O. WINER; Wear 102, 67 (1985).

Inverse Problems in Engineering: Theory and Practice
ASME 1993

IDENTIFICATION OF A TIME VARYING HEAT FLUX STARTING FROM TEMPERATURE MEASUREMENTS: IMPORTANCE OF THE DISCRETIZATION

Denis Maillet, Jean-Christophe Batsale, and Alain Degiovanni
Laboratoire d'Energétique et de
Mécanique Théorique et Appliquée
Institut National Polytechnique de Lorraine
Vandoeuvre-lès-Nancy, France

ABSTRACT

Estimation of surface heat flux variation using interior temperature measurements is a classical ill-posed inverse problem. The problem that is delt with here consists of identifying the very rapid absorbed heat flux variation of the stimulated (front) side of a sample while temperature is recorded in the other (rear) side in a heat pulse (flash) experiment. Because of the rapid variation of this pseudo-Dirac heat pulse stimulation, one can wonder whether the thermogram (temperature versus time curve) can "remember" the precise shape of this stimulation. An appropriate direct analytical solution is obtained using a N components Dirac comb for the flux excitation. The limits of the inversion of this linear estimation problems are set using a sensitivity analysis. Experimental thermograms obtained through the use of a flash diffusimeter are used to validate the preceding study on unbiased estimators, with a strong emphasis on the importance of the estimator covariance matrix.

NOMENCLATURE

a	= diffusivity
cov	= covariance matrix
$\mathbf{C}$	= matrix
e	= sample thickness
$\mathbf{G}$	= matrix
n	= number of usable temperature measurements
m	= number of temperatures skipped in inversion by derivation of the signal
N	= number of components of flux Dirac comb
q	= flux (same dimension as a temperature)
Q	= cumulated q
t	= time or Fourier number
t_p	= pulse duration
T	= temperature
T_δ	= temperature response to a Dirac Heat pulse
T_d	= dimensionless T_δ
X, $\mathbf{X}$	= sensitivity coefficient (to q) and matrix
y	= experimental temperature signal
Z, $\mathbf{Z}$	= sensitivity coefficient (to Q) and matrix
Δt	= flux time step
$\Delta t'$	= temperature time step
φ	= heat flux density
λ	= thermal conductivity
ρc	= volumetric heat capacity
σ	= noise standard deviation
σ_j	= standard deviation of $\hat{q}_j$
τ_d	= time of energetic barycenter
Subscripts	
i	= related to measurement time
j	= related to heat flux component number
max	= maximum
Superscript	
t	= transposed matrix
^	= estimator or estimate
*	= reduced quantity

INTRODUCTION

Indirect measurement of a transient heat flux, absorbed by the front face of a slab, starting from the measurement of the time variation of its rear face temperature, constitutes a classical example of an ill-posed inverse problem. One considers here the ideal conditions of an opaque homogeneous solid, of constant thermophysical properties (conductivity λ, volumetric heat capacity ρc and diffusivity $a = \lambda/\rho c$). Furthermore it is assumed that the front face absorbed flux is uniform with negligible heat losses. In that case problem of heat conduction inside the material (thickness e) is linear, one-dimensional and transient - see figure 1. The ill-posed nature of the inverse problem stems from the fact that the data are associated to rear face boundary conditions - zero heat flux and noisy temperature measurement ; These data are used to go back by "extrapolation" to the excitation heat flux, that is the front face boundary condition.

To overcome this difficulty, different techniques can be considered. The main ones are the regularization method of Tikhonov and Arsenine (1977) or the sequential estimation method using future times of Beck et al. (1985). They both have the following points in common :

- the problem of estimation of function $\varphi(t)$, the heat flux density variation with time t, is converted into a parameter estimation problem with a parameter vector of finite dimension (discretization).

- the solution of this parameter estimation problem, using a linear least squares method, provides estimators of the heat flux components that are unbiased, but their variances (that stem from the random nature of the temperature noise) can become excessibly large compared to the noise variance under certain conditions. The estimators are therefore modified - a regularization term is added to the least squares term or several adjacent heat flux components are assumed equal during one step of sequential estimation in the method of future times of Beck ; Their standard deviations then go back to an order of magnitude that can be compared to the noise 's one (numerical stability of the inversion) but a bias is generated, which means that, in the absence of measurement noise, the estimates are different from the exact values.

In this article the limits of the linear estimation of a transient flux - that is different from zero on the finite time interval [0, *tp*] - will be studied and this, in the only case where an unbiased estimator is used. Application of this type of method is interesting when one tries to measure diffusivity or other thermal characteristics of multimaterials by a transient method (heat pulse technique) where the stimulation flux is of pulse nature.

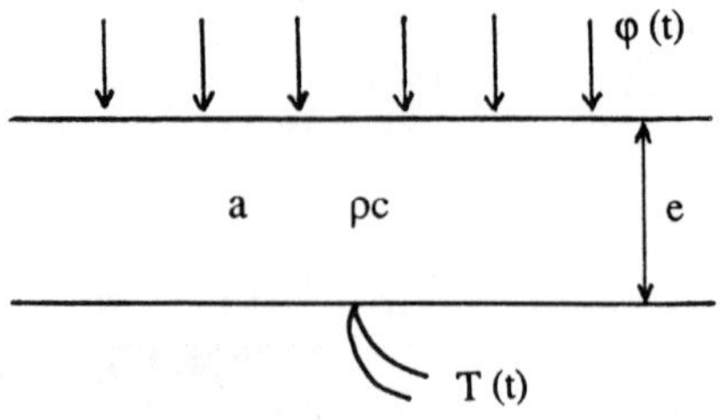

Fig. 1 Excitation of a slab with a transient heat flux

MODELING

The rear face temperature response $T(t)$ to the heat flux density $\varphi(t)$, in the configuration of figure 1, can be written under the form of a convolution product :

$$T(t) = \int_0^t \varphi(\tau)\, T_\delta(t-\tau) d t \qquad (1)$$

where T_δ is the response to a Dirac heat pulse excitation :

$$T_\delta(t) = \frac{1}{\rho c e} T_d(t^*)$$

$$\text{with} \quad T_d(t^*) = 1 + 2 \sum_{n=1}^{\infty} (-1)^n \exp\left(-n^2 \pi^2 t^*\right) \qquad (2)$$

t^* being the slab Fourier number (= at/e^2). Function T_d is plotted in figure 2. Equation (1) can be written the following way :

$$T = \int_0^{t^*} q(\tau^*)\, T_d\ (t^* - \tau^*)\ d\tau^* \qquad (3)$$

where $q(t^*) = e\varphi(t)/\lambda$ has the dimension of a temperature. In the following parts of this article the asteriks on the reduced times will be omitted. Inversion of equation (3) requires discretizing q into N components q_j. Usually the asumption of a piecewise constant function q is used but it is also possible to discretize this equation with a Dirac comb. $q(t)$ is then equal to zero everywhere, except for times t_j where pulses of area q_j are considered. Equation (3) then becomes :

$$T(t) = \sum_{j=1}^{N} q_j\ X_j \qquad (4)$$

$$\text{with} \qquad X_j(t) = T_d\ \left(t - t_j\right)$$

functions X_j are the sensitivity coefficients of temperature T to energies q_j (that are converted in temperature units using the slab resistance e/λ). The variations of these coefficients with Fourier number are plotted in figure 2 for a Fourier number relative to the excitation duration t_p equal to 0.3 and for N (=4) equidistant pulses

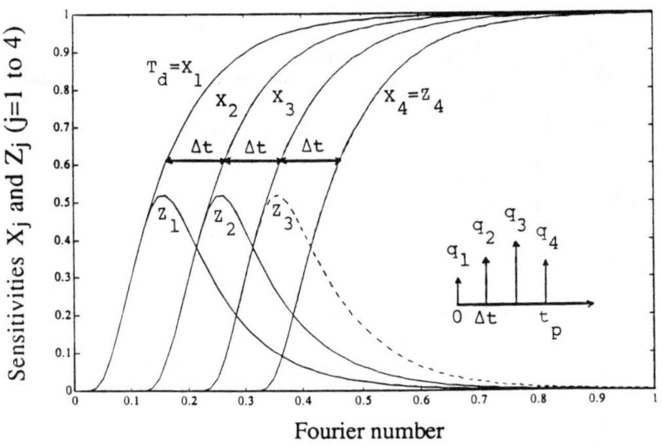

Fig. 2 Impulse temperature response and sensitivities to energies q_j (X_j) and to cumulated energies Q_j (Z_j) - t_p = 0.3 - N = 4

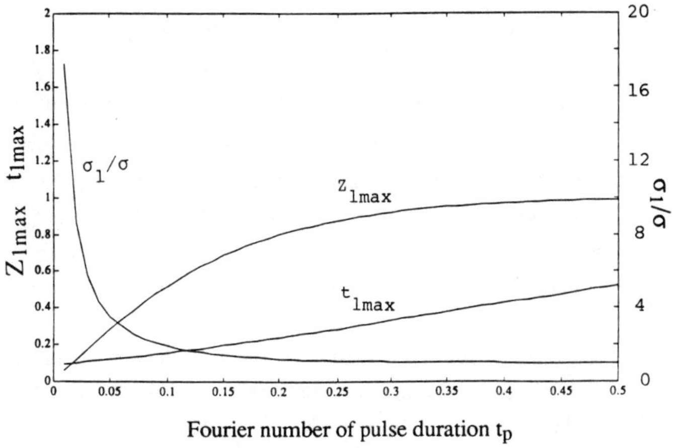

Fig. 3 Discretization using N = 2 components - Maximum sensitivity to q_1 (=Q_1), and corresponding reduced time t_{1max} and ratio of standard deviations

(step $\Delta t = t_p/3$).

It is obvious that the smaller the reduced time step Δt becomes, the closer the sensitivity coefficients, that constitute in fact the pulse response (2) with a reduced time lag t_j, will be. This reduction of the time step eventually leads to a correlation of the parameter estimators. A more direct way to study the estimation possibilities of the N flux components is to use a new parametrization :

$$Q(t) = \int_0^t q(t)\ dt \qquad (5)$$

$Q(t)$ is the reduced time integral of the flux, that is a quantity proportional to the energy absorbed by the slab since the start of excitation. It also represents here exactly the average temperature of the slab (calculated over thickness e) at reduced time t, because of conservation of energy. The sensitivity coefficients $Z_j(t)$ to cumulated energies Q_j can be deduced from the X_j's :

$$Z_1 = X_2 - X_1 \qquad Z_2 = X_3 - X_2 \qquad Z_3 = X_4 - X_3 \qquad Z_4 = X_4$$

These new sensitivity coefficients are also plotted in figure 2. Z_4 is the sensitivity to the total energy absorbed by the slab ; This one can be easily evaluated for long times where the slab has got an uniform temperature and where the other sensitivities are equal to zero. The other components of $Q(t)$ have sensitivities that have maxima of equal value $Z_{max} = Z_{jmax}$ reached for reduced times t_{jmax}. Their estimation is possible only if these Z_{max} levels are high enough and if their times of occurence t_{jmax} are spaced out sufficiently.

IDENTIFIABILITY CONDITIONS OF THE FLUX COMPONENTS

The preceding approach can be implemented in the particular case $N = 2$, which means that two Dirac pulses are looked upon at times 0 and t_p. Equation (4) can then be rewritten under its discrete form using column parameter vectors q and Q :

$$y = Xq = ZQ \qquad (6)$$

with : $X_{ij} = X_j(t_i)$ and $\mathbf{Z} = \mathbf{X}\mathbf{G}^{-1}$ (7)

for $i = 1, n$ and $j = 1, N$

where y is the measured temperature vector which, in the absence of noise is equal to the exact temperature vector $[T(t_1).....T(t_n)]^t$, n being the number of measurements at times t_i.

$\mathbf{X}$ and $\mathbf{Z}$ are therefore the sensitivity matrices for parametrizations q and Q. Matrix $\mathbf{G}$ is a $(n \times n)$ lower triangular matrix with all non-zero coefficients equal to unity.

Sensitivities Z_{1max} and related Fourier numbers t_{1max} are plotted in figure 3 for different values of the excitation duration t_p. Under the standard assumptions of a measurement noise that is additive, uncorrelated, unbiased with a constand standard deviation σ - see Beck et al. (1977) - the best unbiased estimator of Q is :

$$\hat{Q} = (\mathbf{Z}^t\mathbf{Z})^{-1}\mathbf{Z}^t y \qquad (8)$$

with a variance-covariance matrix :

$$\mathbf{cov}\ (\hat{Q}) = \sigma^2\ (\mathbf{Z}^t\mathbf{Z})^{-1} \qquad (9)$$

In the hypothesis of a "minimal" estimation using only two experimental points ($n = N = 2$), the temperature signal measured at times t_{1max}, and t_2, the ratio σ_1/σ of the standard deviations of $\hat{q}_l = \hat{Q}_l$ and noise, can be calculated. It is plotted on the same figure 3. One notices that this ratio varies in the opposite direction to the maximum sensitivity Z_{1max} (let us note that if the number of measurements n is equal to 100, the σ_1/σ ratio can be divided by a coefficient close to 4). Two extreme cases can be therefore considered :

- Fourier number corresponding to the stimulation end t_p, that is its duration divided by the diffusion characteristic times, is lower than 0.05 : it is impossible to discriminate even two heat flux components and only the total absorbed energy can be reached.
- This Fourier number is quite large compared to unity : the whole slab can be considered as isothermal at a given time and convolution product (1), after discretization, becomes equal to $Q(t)$. This is correct only if the reduced discretization step Δt of $Q(t)$ is not too small - $T_d\,(t - \tau) = T_d\,(\Delta t) \approx 1$ in equation (1) - and unbiased sequential estimation methods such as Stolz's one -see Beck et all (1985) - can be applied.

In between, only a study of the covariance matrix of $\hat{q}$ or $\hat{Q}$ can allow the determination of the precision of the least square estimators on the whole time domain.

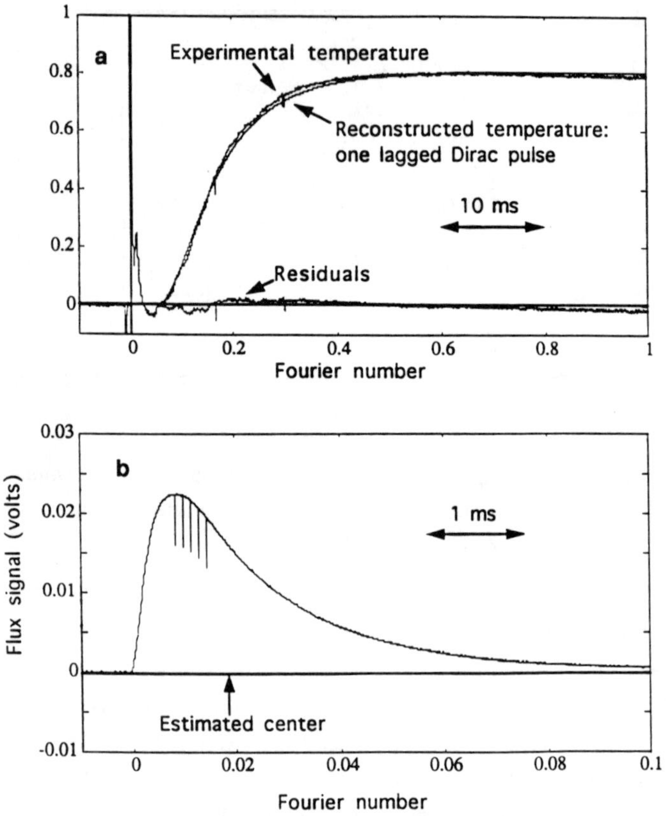

Fig. 4 Diffusion limited experiment
a- experimental and reconstructed temperatures
b- Heat flux variation recorded by a photovoltaïc cell

EXPERIMENTAL VALIDATION

Two flash experiments have been realized to test the preceding theoretical conclusions for the two extreme cases. In both cases a diffusimeter was used : heat pulse was generated by flash tubes discharge and the sample rear side temperature was measured through a semi-conductive Bi_2Te_3 thermocouple with separated contacts and a massive metal block as cold junction. This type of thermocouple is characterized both by a high thermoelectric power ($\approx$ 360 μv/k) and a rapid response time (lower than 100 μs). The temperature signal was amplified (gain = 3000) and fed into a Nicolet 310 digital oscilloscope for data acquisition. In both cases the true shape of the heat flux generated by the flash tubes was measured by a photovoltaïc cell connected to the same oscilloscope.

Diffusion limited experiment

This experiment was done on a thin duralumin sample (e = 1.80 mm) with 10 kHz low pass analog filtering (measurement time step = 20 μs). Thermal diffusivity was measured with a thicker sample of the same material and was found to be equal to 62.5 mm^2/s (with a raw precision). This gives a characteristic diffusion time e^2/a of

51.8 ms, large compared to the stimulation duration -see time scale of figure 4b for the photovoltaïc cell signal. The dimensionless pulse duration has been taken equal to 0.1. Figure 4a gives the experimental temperature recording. The number of useful points (after the electrical induction interference) is n = 2463. The estimation results for a Dirac comb heat flux with N = 2 or 3 components at times τ_i are given in Table 1.

Table 1 - Heat flux estimation using a N components Dirac comb - t_p = 0.1

N = 2						
i	1		2	ρ_{12}		
τ_i	0		0.1	- 0.90		
$\hat{q}_i$	0.67		0.13			
σ_i/σ	0.11		0.12			
N = 3						
i	1	2	3	ρ_{12}	ρ_{23}	ρ_{13}
τ_i	0	0.05	0.1	- 0.96	- 0.95	0.83
$\hat{q}_i$	0.31	0.74	- 0.23			
σ_i/σ	0.37	0.72	0.39			

For estimation of fluxes with a two-component model the estimators of q_1 and q_2 are highly correlated (ρ_{ij} is the correlation coefficient between $\hat{q}_i$ and $\hat{q}_j$: it is equal to $C_{ij}\ (C_{ii}C_{jj})^{-1/2}$ with $C = (X^tX)^{-1}$), but, since the ratios of their standard deviation to noise $\sigma_i/\sigma\ (= \sigma_{ii}^{1/2})$ remain small ($\hat{\sigma}$ = 0.07, measured before excitation), it has no consequence on the precision of the estimated energies q_i. When a three component model is used, the levels of the energies q_i decrease since their sum must remain constant, and σ_i/σ increases due to the decrease of sensitivities. A consequence of these two effects is that the relative "error" of q_i, σ_i/q_i becomes important, especially where the exact value of q_i is low, which explains the absurd negative value of $\hat{q}_3$. All these effects are still worsened by high negative correlations between adjacent fluxes that completely blurs the flux curve shape. This confirms the limitations described in the preceding section for low values of t_p.

In this case where excitation time is a lot smaller than diffusion time, Degiovanni (1987) has shown, starting from a Taylor series expansion in τ of function $T_d\,(t - \tau)$ -equation (2)- that has been explicited in convolution product (3), and from a corresponding expansion in τ_d of the delayed response T_d to a unique Dirac heat pulse occuring at a time lag τ_d, that the two responses (real one and lagged Dirac response) can be found equal, provided that only the first order terms are considered. This time lag is simply the time of the energetic barycenter of function $q(t)$:

$$\tau_d = \frac{1}{Q(\infty)}\int_0^\infty \tau\, q(\tau)d\tau \qquad (10)$$

An estimator $\hat{\tau}_d$ of this barycenter has been looked upon, by equating, according to the least squares sense, model temperatures $T_d\,(t_i - \tau_d)$ and measurements y_i (at Fourier numbers t_i). This problem is non-linear but, since only one parameter is looked upon, the solution is straigthforward. One finds $\hat{\tau}_d$ = 0.0192, compared with a real barycenter τ_d, calculated from the cell's signal of figure 5b, equal to 0.0262. This difference may stem from the lack of precision in the value of the material diffusivity. The corresponding reconstructed temperature curve, with lag $\hat{\tau}_d$, is plotted in figure 4a with the corresponding residuals curve. The level of these residuals explains why the estimation of more than one or two components of the flux curve is impossible.

"ISOTHERMAL" BODY EXPERIMENT

The second experiment concerned a brass pipe of 8 mm inside diameter and 0.4 mm thickness (length : 42 mm). A unique flash pipe generated the inside excitation while temperature of the outside surface was measured by the same type of thermocouple as before and recorded through the same oscilloscope (Gain 1000, no filtering, measurement time step $\Delta t'$ = 10 µs). Because of the small thickness

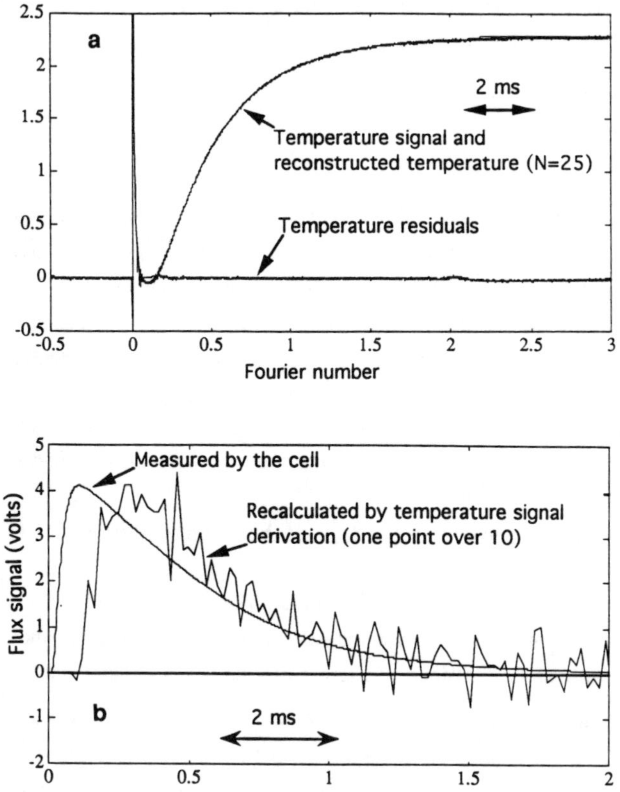

Fig. 5 "Isothermal" body experiment

a experimental and reconstructed temperatures using the 25 component flux (method of sensitivity matrix)

b - Heat flux variation recorded by a photovoltaïc cell and estimated heat flux (inversion through signal derivation)

over diameter ratio, model (2) was still used as a transfer function for the system .

With a characteristic diffusion time e^2/a of 4.76 ms (for a nominal diffusivity equal to 33.6 mm²/s) small compared to the stimulation duration - see time scale of figure 5b for the cell signal - the dimensionless pulse duration t_p has been taken equal to 2. This signifies that, to a certain extent, the sample can be considered as isothermal and the flux time step Δt is now the controling factor.

Inversion Through Signal Derivation

Figure 5a gives the experimental temperature variation with Fourier number (n = 3381 points could be used after the interference). For the isothermal model one has :

$$q(t) = \frac{dT}{dt} \qquad (11)$$

It is therefore necessary to make a numerical derivation of the experimental temperature signal $y(t)$:

$$\hat{q}_i = \frac{y_{i+m} - y_i}{m\Delt'} \qquad (12)$$

where $\Delta t'$ is the acquisition time step. For m smaller than 5 numerical derivation (12) produced unstable results. With m larger, it gave a more regularized shape for heat flux q. The estimated flux is plotted with the cell's signal in figure 5b for the case m = 10 (the cell signal have been normalized to have the same maximum as the estimated flux curve). One notices that general agreement is good, except for short times where there is a time lag equal to about 0.3 : it corresponds to a diffusion time that is not taken into account in this isothermal model.

Inversion Through Inversion of Sensitivity Matrix

Equation (8) written for the pair *(X, q)* instead of *(Z, Q)* has been implemented starting from the same experimental points as above. Result of this inversion in terms of energies $\hat{q}_i$ (that have been normalized with respect to $t_p\hat{Q}_N / N$ in order to have a unit aera under the curve is shown. In figures 5c (N = 10 - Δt = 0.2) and 5d (N = 25 - Δt = 0.08). in the same figures, points corresponding to plus or minus one standard deviation with respect to the $\hat{q}_i$'s are also shown. The cell signal (normalized by its area) is also plotted. One notices that, compared to the previous inversion, the short times steep variation of the flux is better approximated for N = 10, even if the standard deviation of the q_i estimates seem undersestimated (it may be due to the increase of the standard deviation of the noise during the flash duration). When a larger number (N = 25) of flux components is looked upon, the standard deviation of the q_i estimators increase more than proportionally. In our case all the "exact" points of the cell's curve fall within plus or minus one standard deviation around the estimated points, with the exception of the curve maximum that is overestimated. With larger standard deviations the correlation coefficients, that are large and negative between adjacent fluxes, play a significant role, which mean that the flux shape becomes degraded in the steepest region of the flux curve. Inversion of the model with the lower number of flux components restores therefore more faithfully the excitation shape.

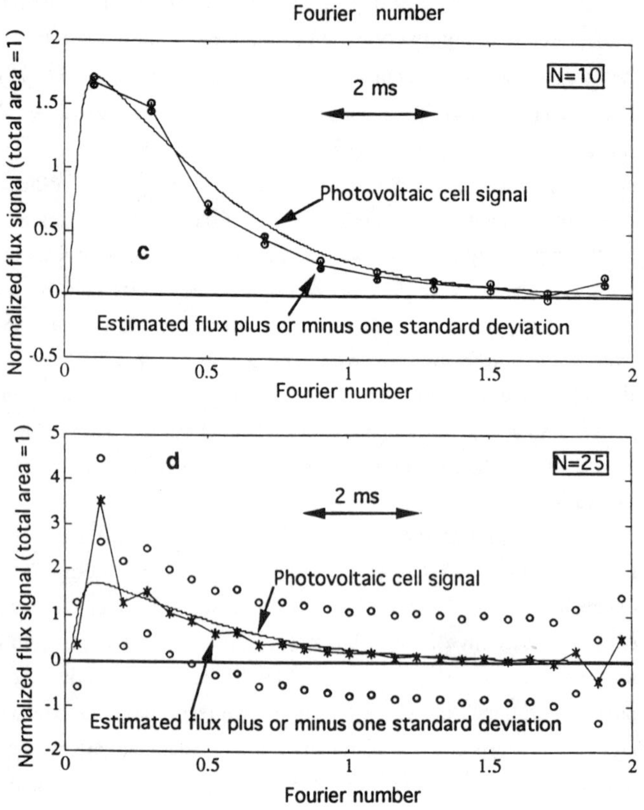

Fig. 5 "Isothermal" body experiment

c - Heat flux variation recorded by a photovoltaïc cell and estimated heat flux (inversion using the sensitivity matrix with N = 10 components)

d - Heat flux variation recorded by a photovoltaïc cell and estimated heat flux (inversion using the sensitivity matrix with N = 25 components

CONCLUSION

The theoretical study that has been done in the first part of this article has shown that the most important parameter in a problem of inversion of temperature data measured on the rear side of a sample is the ratio of the front side flux duration by the diffusion characteristic time. A criterion has been proposed for the lowest value of this ratio. Experimental validation has been accomplished using a flash diffusimeter with recording of both rear side temperature and radiative heat flux by a photovoltaïc cell. Two extreme cases have been considered. They have shown that , when identification is possible, an unbiased least square estimator, based on a not too important number of flux components, was the most efficient to restore the shape of the real flux curve.

REFERENCES

Beck, J.V. and Arnold, K.J., 1977, *Parameter Estimation in Engineering and Science*, John Wiley and Sons, New-York.

Beck, J.V., Blackwell B. and St. Clair Jr., 1985, *Inverse Heat Conduction*, Wiley Interscience, New-York.

Degiovanni A., 1987, Correction de Longueur d'Impulsion pour la Mesure de la Diffusivité Thermique par Méthode Flash, *Int. J. Heat and Mass Transfer*, Technical note, Vol 30, pp. 2199-2200.

Tikhonov, A. and Arsenine, V., 1977, *Solution of Ill-posed Problems*, V.H. Winston and Sons, Washington D.C.

Inverse Problems in Engineering: Theory and Practice
ASME 1993

AN ESTIMATION OF THE THICKNESS OF METAL SKULL FORMED ON A WALL OF AN ELECTRIC ARC FURNACE

Sui Lin
Department of Mechanical Engineering
Concordia University
Montreal, Quebec, Canada

ABSTRACT

A metal skull formed on the inside wall of an electric arc furnace is needed to protect overheating of the furnace wall. The thickness of the skull varies with variation of the load inside the furnace. A minimum thickness of the skull should be warranted in order to have a safety operation of the furnace.

Two thermocouples are installed in the furnace wall to measure the local temperatures. From the result of the measured temperatures, the thickness of the skull can be estimated.

NOMENCLATURE

a	thickness of the steel shell, m
b	thickness of the steel shell and the graphite layer, m
F	heat flux on a surface of a slab, W/m^2
h	heat transfer coefficient, W/m^2, °C
k	thermal conductivity, W/m, °C
L	thickness of a slab, m
q	heat flux, W/m^2
s	thickness of the skull, m
t	time, s
T	temperature, °C
x	position coordinate, m
α	thermal diffusivity, m^2/s
β_i	coefficients of a polynomial
δ	penetration distance, m

SUBSCRIPTS

s	inside surface of the skull
∞	ambient
o	outside surface of the furnace
1	steel shell
2	graphite layer
3	skull

INTRODUCTION

An inverse problem of heat conduction can be distinguished from a direct problem by the following statement presented by Beck et al. (1985): If the heat flux or temperature histories at the surface of a solid are known as functions of time, then the temperature distribution can be found. This is termed a direct problem. If the surface heat flux and temperature histories of a solid must be determined from transient temperature measurements at one or more interior locations, this is an inverse problem. The inverse heat conduction problem is much more difficult to solve analytically than the direct problem, because it is extremely sensitive to the measurement errors. Beck et al. (1985) presented general techniques for solving the inverse heat conduction problem.

In the present paper, a different heat conduction problem is presented. It concerns a safety operation of an electric arc furnace. We consider the wall of the furnace which consists of three layers: the steel shell, the graphite layer and the skull as shown schematically in Fig. 1. The metal skull formed on the inside wall of the furnace is to protect overheating of the furnace wall. The thickness of the skull varies with variation of the load inside the furnace. A minimum thickness of the skull should be warranted in order to have a safety operation of the furnace.

The purpose of the investigation is to find the

relations between the temperatures measured at the inside and outside surfaces of the outer steel shell of the furnace and the thickness of the skull formed on the inside wall of the furnace.

During the normal operation of an electric arc furnace, the skull is usually much thicker than the minimum thickness required for the safety operation. A small variation of the skull thickness will not affect the furnace operation. Therefore the main purpose of the analysis is to determine the condition when the skull approaches its required minimum thickness. To find the exact thickness of the skull is not the purpose of the present analysis. Hence a coarse estimation of the thickness of the skull would satisfy the requirement of the operation of the furnace.

FORMULATION OF THE PROBLEM

To estimate the thickness of the skull formed on the inside wall of the furnace, two thermocouples are installed on the inside and outside surfaces of the steel shell, respectively, to measure their local temperatures.

The skull thickness will be estimated from the measured local temperatures.

For analysis, the following assumptions are made;

1. The heat conduction process is one-dimensional.
2. Material properties of the skull, graphite and the steel shell are respectively constant.
3. The temperature at the inside skull surface is at the melting temperature of the molton metal existing inside the furnace and is constant.
4. The temperature of the ambient fluid cooling the outside surface of the steel shell is constant. The convective heat transfer coefficient is constant.

Systems 1, 2 and 3 used in the analysis refer to the steel shell, the graphite and the skull, respectively, as shown in Fig. 1. Subscripts o, a, b and s refer to the positions at x = o, a, b and s , respectively, as shown in Fig. 1. The thickness of the steel shell, the graphite and the skull are presented by a, (b - a) and s*, respectively, as shown in Fig. 1. The one-dimensional heat conduction equation can be expressed by

$$\frac{\partial T_i}{\partial t} = \alpha_i \frac{\partial^2 T_i}{\partial x^2} \quad , \quad i = 1, 2, 3 \tag{1}$$

As a first approximation, we consider that the melting or solidification of the skull is relatively slow and that the temperature distribution in the furnace wall is not far from the steady state temperature distribution. Under this consideration the melting or solidification of the skull may be considered as a quasi-steady state process. The heat flux across the furnace wall per unit area for a stedy-state condition can be expressed by

$$q = (T_s - T_\infty) / \left(\frac{s^*}{k_3} + \frac{b-a}{k_2} + \frac{a}{k_1} + \frac{1}{h}\right) = (T_s - T_\infty) / \left(\frac{s^*}{k_3} + A\right) \tag{2}$$

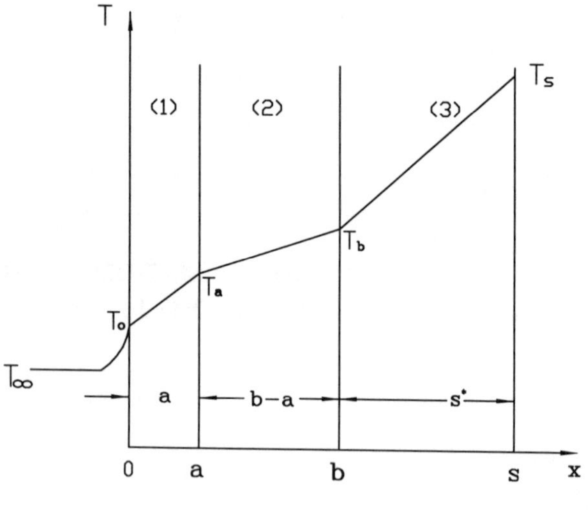

Figure 1 Schematic diagram of the wall of an electric arc furnace with its steady-state temperature distributions.
(1) Steel Shell
(2) Graphite Layer
(3) Skull

$$q = (T_s - T_a) / \left(\frac{s^*}{k_3} + \frac{b-a}{k_2}\right) = (T_s - T_a) / \left(\frac{s^*}{k_3} + B\right) \tag{3}$$

and

$$q = (T_s - T_o) / \left(\frac{s^*}{k_3} + \frac{b-a}{k_2} + \frac{a}{k_1}\right) = (T_s - T_o) / \left(\frac{s^*}{k_3} + C\right) \tag{4}$$

where

$$A = \frac{b-a}{k_2} + \frac{a}{k_1} + \frac{1}{h} \tag{5}$$

$$A = \frac{b-a}{k_2} \tag{6}$$

$$C = \frac{b-a}{k_2} + \frac{a}{k_1} \tag{7}$$

From Eqs. (2) and (3) and from Eqs. (2) and (4), we obtain the temperatures at x = a and x = 0, and the temperature difference between x = a and x = 0 as follows:

$$T_a = T_s - (T_s - T_\infty) \left(\frac{s^*}{k_3} + B\right) / \left(\frac{s^*}{k_3} + A\right) \tag{8}$$

$$T_o = T_s - (T_s - T_\infty) \left(\frac{s^*}{k_3} + C\right) / \left(\frac{s^*}{k_3} + A\right) \tag{9}$$

and

$$T_a - T_o = -(T_s - T_\infty)(B - C) / \left(\frac{s^*}{k_3} + A\right) \tag{10}$$

The ratios of the rate of change in the temperatures at x = a and x = 0 to the rate of change in the skull thickness and the ratio of the rate of change in the temperature difference between x = a and x = 0 to the rate of change in the skull thickness can be obtained by differentiating Eqs. (9), (8) and (10) with respect to time t as follows:

$$\frac{dT_a/dt}{ds^*/dt} = -(T_s - T_\infty)\left(\frac{a}{k_1} + \frac{1}{h}\right) / \left[k_3\left(\frac{s^*}{k_3} + A\right)^2\right] \quad (11)$$

$$\frac{dT_o/dt}{ds^*/dt} = -(T_s - T_\infty)\left(\frac{1}{h}\right) / \left[k_3\left(\frac{s^*}{k_3} + A\right)^2\right] \quad (12)$$

and

$$\frac{d(T_a - T_o)/dt}{ds^*/dt} = -(T_s - T_\infty)\left(\frac{a}{k_1}\right) / \left[k_3\left(\frac{s^*}{k_o} + A\right)^2\right] \quad (13)$$

Compare Eq. (13) with Eqs. (11) and (12), the numerator of Eq. (13) does not content the convective heat transfer coefficient, h. Therefore the ratio of the rate of change in the temperature difference, $d(T_a - T_o)/dt$, to the rate of change in the skull thickness, ds^*/dt, is less affected by the surrounding cooling condition which involves a great uncertainty for the analysis. Therefore the estimation of the rate of change in the skull thickness by using Eq. (13) is more reliable than by using Eq. (11) or Eq. (12).

Figure 2 shows the steady state temperatures at x = 0 and x = a, and the steady state temperature difference between x = 0 and x = a as functions of the skull thickness. The results presented in Fig. 2, are calculated from Eqs. (8), (9) and (10) with the following data:

$$T_s = 1650°C \qquad T_\infty = 31°C$$

$$k_1 = 46.73\ \frac{W}{m°C} \qquad k_2 = 141.07\ \frac{W}{m°C} \quad (14)$$

$$k_3 = 26.93\ \frac{W}{m°C} \qquad h = 3500\ \frac{W}{m^2°C}$$

$$a = 4\ cm \qquad (b - a) = 15\ cm$$

The slopes of curves T_a, T_o and $(T_a - T_o)$ appearing in Fig. 2 can be written as $\frac{dT_a}{ds^*} = \frac{dT_a/dt}{ds^*/dt}$, $\frac{dT_o}{ds^*} = \frac{dT_o/dt}{ds^*/dt}$ and $\frac{d(T_a - T_o)/dt}{ds^*/dt}$ which are shown in Eqs. (11), (12) and (13) respectively.

It can be seen that the ratios of the rate of change in the temperatures at x = 0 and x = a to the rate of change in the skull thickness and the ratio of the rate of change in the temperature difference between x = a and x = 0 to the rate of change in the skull thickness, increase with a decrease of the skull thickness. This is a good indication to use the measured temperatures at x = a or x = 0, or the temperature difference between x = a and x = 0, to estimate the skull thickness when it moves to the direction towards the minimum thickness required for the safety operation.

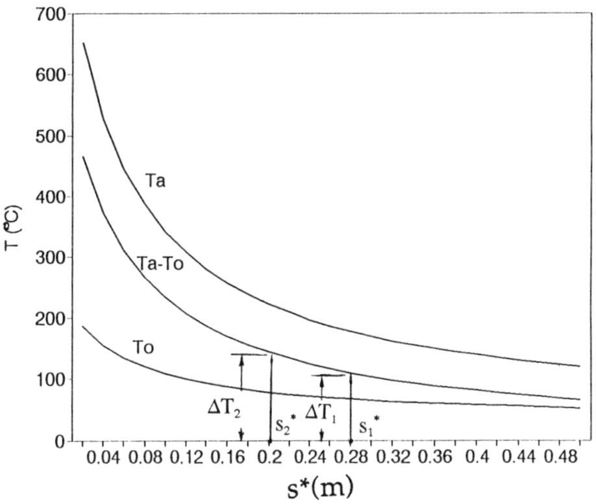

Figure 2 The steady-state temperatures at x = a and x = 0, and the steady-state temperature difference between x = a and x = 0, as functions of the skull thickness.

ESTIMATION OF THE TIME DELAY FOR THE TEMPERATURE RESPONSE TO THE CHANGE IN THE SKULL THICKNESS

The change in the skull thickness can not be detected instantaneously by the temperature measured at x = a or x = 0. There occurs a time delay. To estimate the time delay, the heat balance integral method will be used with the following consideration (Goodman, 1964):

Let us assume there is a slab having a thickness L. Initially, the temperature T is equal to zero, and at the surface x = 0 a constant heat flux, F, is given for time t > 0. The other surface at x = L is insulated. The heat conduction equation is

$$\alpha\frac{\partial^2 T}{\partial x^2} = \frac{\partial T}{\partial t}, \qquad 0 < x < L,\ t > 0 \quad (15)$$

The boundary conditions are

$$k\frac{\partial T}{\partial x} = -F \qquad x = 0,\ t > 0 \quad (16)$$

and

$$\frac{\partial T}{\partial x} = 0 \qquad x = L,\ t > 0 \quad (17)$$

We define a quantity $\delta(t)$ called the penetration distance. Its properties are such that for $x > \delta(t)$ the slab is at an equilibrium temperature and there is no heat transferred beyond this point. Integrating Eq. (15) we obtain

$$\frac{d\theta}{dt} \Rightarrow = \alpha\left[\frac{\partial T}{\partial x}(\delta,t) - \frac{\partial T}{\partial x}(0,t)\right] \qquad (18)$$

where

$$\theta = \int_0^{\delta(t)} T\,dx \qquad (19)$$

Since there is no heat transferred beyond $x = \delta$,

$$T(\delta,t) = 0 \qquad (20)$$

$$\frac{\partial T}{\partial x}(\delta,t) = 0 \qquad (21)$$

and

$$\frac{\partial^2 T}{\partial x^2}(\delta,t) = 0 \qquad (22)$$

Let us assume that T can be represented by a third-degree polynomial in x of the form

$$T = \beta_0 + \beta_1 x + \beta_2 x^2 + \beta_3 x^3 \qquad (23)$$

where the coefficients β_i may depend on t. Applying Eqs. (16,20,21,22), the temperature profile must take the form

$$T = \frac{F}{3k\delta^2}(\delta - x)^3 \qquad (24)$$

Substituting into Eq. (19), it is seen that

$$\theta = \frac{\delta^2 F}{12k} \qquad (25)$$

Introducing Eqs. (16, 21,25) into the heat-balance integral, Eq. (18), gives

$$t = \frac{\delta^2}{12\alpha} \qquad (26)$$

We consider the penetration distance δ reaching the surface of the slab, $\delta = L$, and obtain

$$t_L = \frac{L^2}{12\alpha} \qquad (27)$$

This time required for the penetration distance to be equal to L may be considered as an approximation of the time delay for the temperature response to the change in the skull thickness across the skull as

$$t_3 = \frac{s^{*2}}{12\alpha_3} \qquad (28)$$

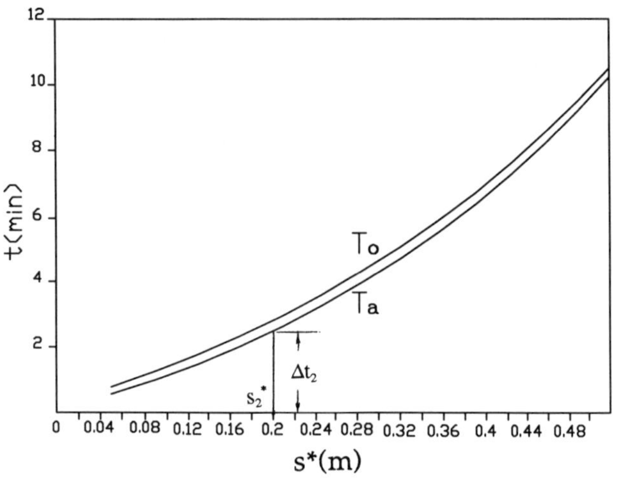

Figure 3 Time delay for the temperature responses at x = 0 and x = a to the change in the skull thickness.

the time delay across the graphite,

$$t_2 = \frac{(b - a)^2}{12\alpha_2} \qquad (29)$$

and the time delay across the steel shell,

$$t_1 = \frac{a}{12\alpha_1} \qquad (30)$$

Therefore, the time delayer for the temperature response at x = a and x = 0 to the change in the skull thickness can be expressed by

$$t_{x=a} = t_3 + t_2 = \frac{1}{12}\left[\frac{s^{*2}}{\alpha_3} + \frac{(b - a)^2}{\alpha_2}\right] \qquad (31)$$

and

$$t_{x=0} = t_3 + t_2 + t_1 = \frac{1}{12}\left[\frac{s^{*2}}{\alpha_3} + \frac{(b - a)^2}{\alpha_2} + \frac{a^2}{\alpha_1}\right] \qquad (32)$$

Figure 3 shows the time delay at x = 0 and x = a as functions of the skull thickness with

$$\alpha_1 = 1.267 \times 10^{-5}\ m^2/s,$$
$$\alpha_2 = 19.78 \times 10^{-5}\ m^2/s,$$
$$\text{and } \alpha_3 = 0.867 \times 10^{-5}\ m^2/s \qquad (33)$$

ESTIMATION OF THE SKULL THICKNESS

The skull thickness can be estimated by using combination of Figs. 2 and 3. Because curve T_o in Fig. 2 is too flat, it is not a proper one used for the estimation of the skull thickness. Either curve T_a or curve (T_a - T_o) may be used for the estimation. As an example, we consider the change of the measured temperature difference (T_a -

T_o) from a value of ΔT_1 at time t_1 to a value of ΔT_2 at time t_2. As shown in Fig. 2, the corresponding change in the skull thickness is from s_1 to s_2. The rate of change in the skull thickness for this case is $(s_1^* - s_2^*)/(t_2 - t_1)$. From Fig. 3, it can be seen that the time delay at $s^* = s_2^*$ is Δt_2. Using a linear interpretation, a decrease of the skull thickness during the time delay, Δt_2, is

$$\Delta s^* = \left(\frac{s_1^* - s_2^*}{t_2 - t_1}\right)(\Delta t_2) \tag{34}$$

Therefore a better estimation of the skull thickness at time t_2 can be obtained by

$$s^* = s_2^* - \Delta s^*. \tag{35}$$

CONCLUSIONS

The metal skull formed on the inside wall of an electric arc furnace is to protect overheating of the furnace wall. A minimum thickness of the skull should be warranted in order to have a safety operation of the furnace.

Applying the quasi-steady-state temperature distribution in the furnace wall, the local temperatures measured at the inside and outside surfaces of the steel shell can be used to estimate the skull thickness as a first approximation.

The concept of the penetration distance used in the heat balance integral method provides an estimation of the time delay for the temperature responses at the surfaces of the steel shell to the change in the skull thickness located inside the furnace. During this estimated time delay, a small change in the skull thickness can be determined, which can serve as a correction term to obtain a better estimation of the skull thickness.

A computer program for solving the present problem is in progress which will provide more accurate results.

ACKNOWLEDGEMENT

The present work is being supported by the Natural Science and Engineering Research Council of Canada under Grant No. 0GP0007929.

REFERENCES

Beck, J.V., Blackwell, B., and St. Clair, Jr., C.R., 1985, "Inverse Heat Conduction", John Wiley & Sons Inc., New York.

Goodman, T.R., 1964, "Application of Integral Methods to Transient Nonlinear Heat Transfer", Advances in Heat Transfer, Ed. by Irvine, T.F. and Hartnett, J.P., Vol. 1, pp. 51-122.

Inverse Problems in Engineering: Theory and Practice
ASME 1993

APPLICATION OF A SIMPLE PARAMETER ESTIMATION METHOD TO PREDICT EFFLUENT TRANSPORT IN THE SAVANNAH RIVER

S. J. Hensel and D. W. Hayes
Westinghouse Savannah River Company
Aiken, South Carolina

ABSTRACT

A simple parameter estimation method has been developed to determine the dispersion and velocity parameters associated with stream/river transport. The unsteady one dimensional Burgers' equation was chosen as the model equation, and the method has been applied to recent Savannah River dye tracer studies. The computed Savannah River transport coefficients compare favorably with documented values, and the time/concentration curves calculated from these coefficients compare well with the actual tracer data. The coefficients were used as a predictive capability and applied to Savannah River tritium concentration data obtained during the December 1991 accidental tritium discharge from the Savannah River Site. The peak tritium concentration at the intersection of Highway 301 and the Savannah River was underpredicted by only 5% using the coefficients computed from the dye data.

INTRODUCTION

Many large industrial facilities, such as the Savannah River Site, make use of potentially hazardous chemicals and materials within site processes. A portion of the site emergency response requirements is to accurately predict the transport of accidental releases of hazardous liquid materials within the stream and river systems on the Savannah River Site. The inherent complexities associated with the modeling of surface water systems coupled with the number of them on the site virtually eliminates the possibility of creating detailed hydrodynamic models for each one. A more realistic approach is to apply a relatively simple model that simulates the physics associated with hydrodynamic transport coupled with a good parameter estimation technique. The accuracy of virtually all simple models used to simulate natural phenomena is largely dependent on the accuracy of the parameters or coefficients within the model. In this case the predictive capability desired for the simulation of an accidental release will depend on the accuracy of the parameters used in the simulation.

Parameter estimation and inverse methods have been successfully applied to a wide variety of hydrologic problems. Some of the more recent applications include groundwater transport by Xiang (1992), estimation of aquifer parameters by Ajayi (1989), modelling of oceans by Copeland (1991), hydrologic data analysis Schilling (1986), and surface water hydrology by Awwad (1992), Singh (1988), O' Donnell (1988), and Aldama (1990). Surface water transport analyses at the Savannah River Site have been reported by Buckner (1975), and Hayes (1984). A computer code has been recently developed to determine the surface water transport parameters for the streams and rivers on the Savannah River Site.

Hydrodynamic transport is mostly an advection and diffusion process. Thus, the unsteady 1-D advection/diffusion equation was chosen as the model equation. A dead zone model has been added to simulate the sorption/desorption process within the system. The use of a dead zone increases the number of governing equations to two and the number of parameters from two to four. A finite difference method is used to discretize the governing equations, and the error between the numerical predictions and the experimental measurements is minimized through a Gauss-Newton iterative procedure. Thus, the parameters are directly related to experimental measurements from the field.

The experimental data used in the parameter estimation method are in the form of time vs. concentration. Dye is released into the stream/river at some upstream location, and samples are taken periodically at various points downstream. At each point where samples are taken a collection of time vs. concentration data exists. A unique set of parameters is determined for each stretch or reach in the stream/river. A reach is defined as the segment bounded by the two locations where the dye data was taken. The parameters of the governing transport equations are assumed constant within each reach. Clearly, the stream/river may be modeled more accurately with a greater number of reaches. The parameters can be stored in a data base for future use as a predictive tool during accidental releases.

DISCUSSION

The unsteady 1-D advection/diffusion equation can be written as

$$\frac{\partial \phi}{\partial t} + U\frac{\partial \phi}{\partial x} = D\frac{\partial^2 \phi}{\partial x^2} \tag{1}$$

where ϕ is the concentration of material, t is time, and x is the spatial coordinate. The parameters U and D are are assumed constant over the domain, and they represent the velocity and dispersion coefficient of the reach under consideration in the stream/river.

A dead zone model developed by Reichert (1991) may be added to account for sorption/desorption processes through an additional source term of the form

$$(1-\alpha)\frac{\partial \phi}{\partial t} + U\frac{\partial \phi}{\partial x} = D\frac{\partial^2 \phi}{\partial x^2} + S \tag{2}$$

$$S=H(\psi-\phi) \tag{3}$$

where a is a dead zone "fraction" parameter, H is an exchange coefficient and, ψ is the concentration of the material in the dead zone. A balance equation on ψ is also needed.

$$\alpha\frac{\partial \psi}{\partial t} = H(\phi - \psi) \tag{4}$$

Both ϕ and ψ are functions of the independent variables t and x.

Using the Briley-McDonald (1973) finite difference method both equations can be discretized and written as,

$$(1-\alpha)\frac{\phi_i^{n+1}-\phi_i^n}{\Delta t} + \frac{U}{2\Delta x}(\phi_{i+1}^n-\phi_{i-1}^n) + \frac{U(\phi_{i+1}^{n+1}-\phi_{i+1}^n) - U(\phi_{i-1}^{n+1}-\phi_{i-1}^n)}{4\Delta x} = \frac{D}{2(\Delta x)^2}(\phi_{i+1}^n-2\phi_i^n+\phi_{i-1}^n+\phi_{i+1}^{n+1}-2\phi_i^{n+1}+\phi_{i-1}^{n+1}) - \frac{H}{1+\Delta tH}\phi_i^{n+1} + \frac{H}{1+\Delta tH}\psi_i^n \tag{5}$$

$$\alpha\frac{\psi_i^{n+1}-\psi_i^n}{\Delta t} = H(\phi_i^{n+1}-\psi_i^{n+1}) \tag{6}$$

where the superscript refers to the time step level and the subscript i refers to a node in the domain. The spacing between nodes i and i+1 is assumed uniform throughout the domain. If there are N nodes in the domain, the above equations are applied from nodes 2 to N-1.

The values of ϕ and ψ at node 1 are set via Dirichlet boundary conditions. At node N the concentration gradients are set to zero, and the computational domain is twice the length of the modeled reach. Thus, the time/concentration data are known at nodes 1 and N/2. The purpose of extending the domain through a second reach is to eliminate any upstream propagation to node N/2 of downstream boundary effects. The boundary conditions at the upstream node, node 1, are calculated using quadratic interpolation within the dye data. Whenever possible, the time step is altered such that actual experimental values can be used without interpolation.

Both the transport equation and the dead zone balance equation are solved during each time step. The advection/diffusion equation may be solved first and independently of the dead zone equation because the discretization uncouples the two equations. The resulting ϕ is used to update ψ in the dead zone balance equation. This procedure is repeated for each time step.

Although the discretized form of the transport equation is stable in time, the time step size should be small enough to ensure accuracy. The maximum time step allowed is one ten thousandth of the time difference between the first data point at node 1 and the last data point at node N/2.

The determination of the best fit parameters requires the numerical predictions of the concentrations to be evaluated at the same time as the dye tracer measurements. As already discussed, the upstream boundary condition imposes requirements on the time steps early in the transient. Occasionally the dye data curves at the upstream location and at the end of the reach overlap. The code has been written to account for these situations.

After the parameters have been optimized for a reach, the dye data that was previously used in the iterative process at the downstream end of the reach becomes the boundary condition at node 1. Thus, given M locations of concentration vs. time dye data in one stream/river, the code computes M-1 sets of coefficients. The coefficients will be piecewise constant across the M-1 reaches.

A Gauss-Newton iteration scheme is used to minimize the least squares error between the numerical predictions and the experimental data in each reach. This process requires an initial guess of the parameters, and the dye data at the extremes of the reach. The initial guess of the dead zone parameters are specified by the user. The initial guess of the velocity and dispersion coefficients in each reach is computed from the original tracer dye data using the centroidal method.

The Gauss-Newton method requires computation of the derivatives of the computed concentrations with respect to each of the parameters. These derivatives are computed by computing the difference between two computed concentrations when the parameters are each varied slightly, and dividing by the difference between the two values of the parameters. The parameters were changed by 0.01% to calculate the derivatives.

The dead zone model parameters were iterated in a separate step from the velocity and dispersion coefficients in order to accelerate convergence and stabilize the iterative process. Thus, during one iteration the velocity and dispersion parameters were updated, and then using these updated values the dead zone parameters were updated. The splitting of the parameters into two sub iterations proved to be very beneficial.

The iterative process continues until a preset convergence level of normalized least squares error is reached. In some cases this level may never be reached due to the inability of the model to accurately reproduce the measured data, and the process is terminated after a maximum number of iterations. This same procedure is repeated in each reach.

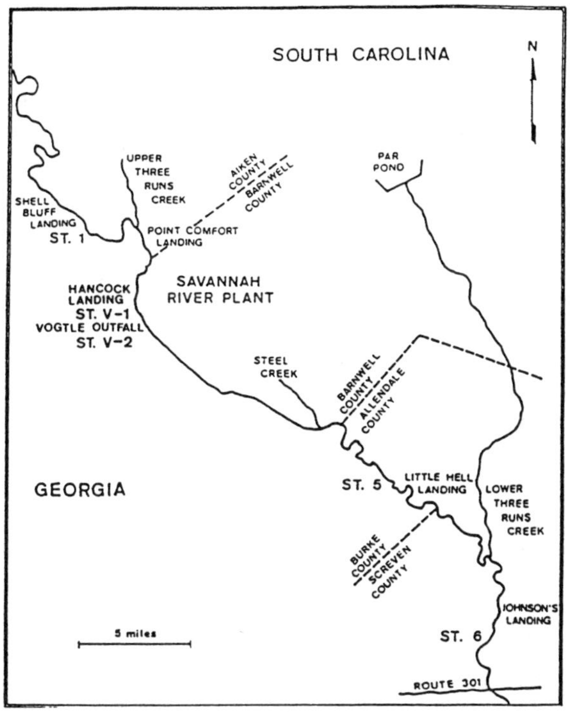

FIGURE 1: GEORGIA-SOUTH CAROLINA BORDER

The parameter estimation method was used to compute transport parameters for the Savannah River. The Savannah River Site and its proximity to the Savannah River are shown in Figure 1. The dye tracer data used in this analysis, presented in Figure 2, was taken at four locations in the Savannah River. The present model assumes mass flow in the river under consideration is constant. The areas under the time concentration dye curves are equal when the mass flow is constant from reach to reach. The normalized dye data in Figure 3 was used to compute the transport parameters. As previously mentioned, N reaches of data can produce N-1 parameters. The calculated best fit parameters were used to reproduce the tracer dye data. The comparison in Figure 4 of the original data and the "predictions" from the best fit parameters clearly illustrates the adequacy of this simple numerical method. The real applicability of the parameter estimation method is in forecasting travel times and peak concentrations during an accidental release of hazardous liquid materials. In December 1991 an accidental release of tritiated water from the Savannah River Site occurred and substantial time concentration data was taken in tracking the release. The parameters computed using the tracer dye data were applied to the tritium data between Four Mile Creek and Highway 301.

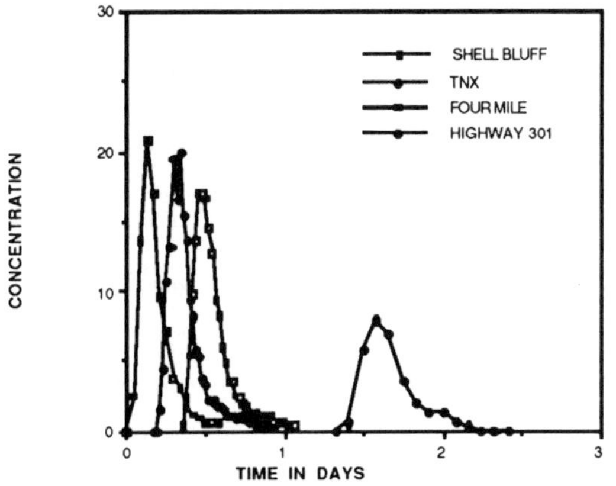

FIGURE 2: RAW DYE DATA

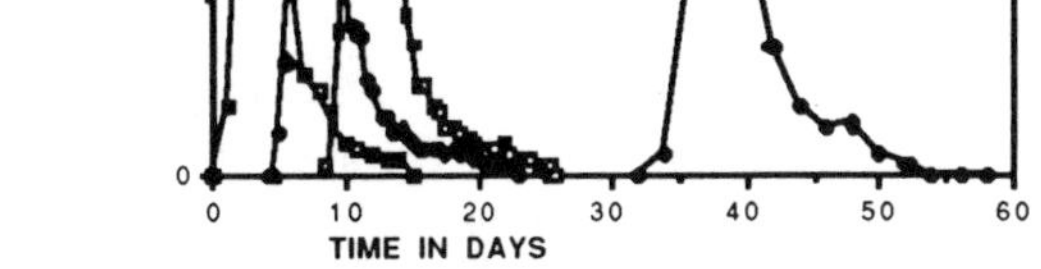

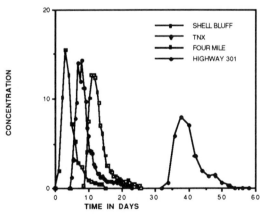

FIGURE 3: NORMALIZED DYE DATA

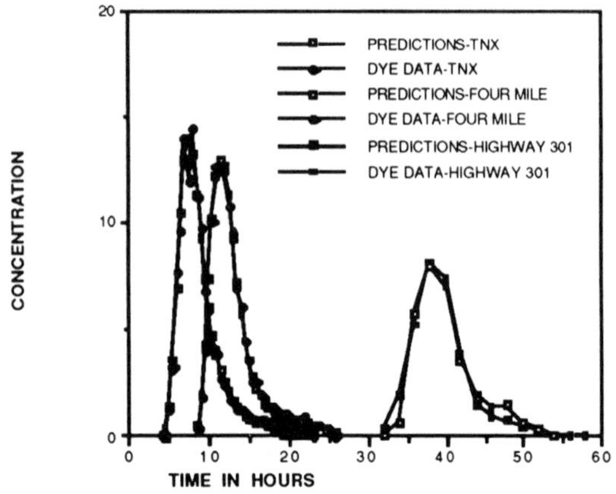

FIGURE 4: DYE DATA AND PREDICTIONS

The computed and measured time concentration data in Figure 5 illustrates the applicability of the present method to forecast peak concentrations and travel times during an accidental release.

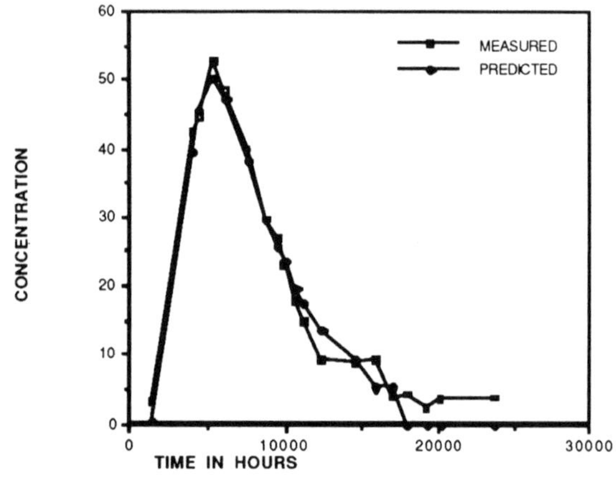

FIGURE 5: TRITIUM DATA AND PREDICTIONS

CONCLUSIONS

The transport parameters for the Savannah River appear to be valid during much of the year. The tracer dye data was not taken in December 1991 when the tritium spill occurred, but the best fit parameters from the dye data proved to be sufficient in computing tritium concentrations. This may not be valid if the best fit transport parameters were applied during a severe drought or flood. The simple 1-D model equation is sufficient to simulate transport in a river. The forecasting of hazardous material transport can not commence until time concentration data is taken at an upstream location where best fit parameters already exist. Lastly, the present model assumes constant flow, thus dilution due to increased flow in the river should be taken into account when forecasting. Conservative predictions will result if dilution effects are ignored.

SUMMARY

A simple parameter estimation method for computing best fit surface water transport parameters has been implemented and applied to the Savannah River. Comparisons of time/concentration curves reveals the sufficiency of the simple model equation and the robustness of the transport parameters. Application of the parameters to an accidental release can be performed provided release data is known (either as a source or as measured time/concentration data) and dilution factors are known.

ACKNOWLEDGMENTS

The information contained in this article was developed during the course of work under Contract No. DE-AC09-89SR18035 with the U.S. Department of Energy.

REFERENCES

O. Ajayi and T. O. Obilade, "Numerical Estimation of Aquifer Parameters Using Two Observational Wells", *Journal of Hydraulic Engineering*, Vol. 115, No. 7, pp. 982-988, 1989.

A. A. Aldama, "Least-Squares Parameter Estimation For Muskingum Flood Routing", *Journal of Hydraulic Engineering*, Vol. 116, No. 4, pp. 580-586, 1990.

H. M. Awwad and J. B. Valdes, "Adaptive Parameter Estimation For Multisite Hydrologic Forecasting", *Journal of Hydraulic Engineering*, Vol. 118, No. 9, pp. 1201-1221, 1992.

W. R. Briley and H. McDonald, "Solution of the Three-Dimensional Compressible Navier-Stokes Equations by an Implicit Technique", *Proceedings Fourth Intl. Conf. Num. Methods Fluid Dynamics*, Boulder, CO, 1973.

M R. Buckner and D. W. Hayes, "Pollutant Transport In Natural Streams", *Proceedings of Computational Methods In Nuclear Engineering*, Charleston, SC, April 15-17, 1975.

A. H. Copeland, R. S. Segall, C. D. Ringo, and B. Moore, "Mathematical Modelling of Inverse Problems For Oceans", *Appl. Math. Modelling*, Vol. 15, pp. 586-595, 1991.

D. W. Hayes, "Aquatic Emergency Response Model At The Savannah River Plant", *ANS Topical Meeting on Radiological Accidents- Perspectives and Emergency Planning*, pp. 141-143, 1984.

T. O'Donnell, C. P. Pearson, and R. A. Woods, "Improved Fitting For Three-Parameter Muskingum Procedure", *Journal of Hydraulic Engineering*, Vol. 114, No. 5, pp. 516-528, 1988.

P. Reichert and O. Wanner, "Enhanced One-Dimensional Modeling Of Transport In Rivers", *Journal of Hydraulic Engineering*, Vol. 117, No. 9, pp. 1165-1183, 1991.

W. Schilling and J. Martens, "Recursive State and Parameter Estimation With Applications in Water Resources", *Appl. Math. Modelling*, Vol. 10, pp. 433-437, 1986.

V. P. Singh and K. Singh, " Parameter Estimation For Log-Pearson Type III Distribution By POME", *Journal of Hydraulic Engineering*, Vol. 114, No. 1, pp. 112-122, 1988.

J. Xiang and D. Elsworth, "Direct and Integration Methods of Parameter Estimation in Groundwater Transport Systems", *Appl. Math. Modelling*, Vol. 16, pp. 404-413, 1992.

Inverse Problems in Engineering: Theory and Practice
ASME 1993

ANALYSIS OF DEEP PENETRATION LASER WELDING USING BEM SENSITIVITY SCHEME

Junghwan Lim and Cho Lik Chan
Department of Aerospace and Mechanical Engineering
University of Arizona
Tucson, Arizona

ABSTRACT

This paper presents a Boundary Element Method (BEM) formulation for the three-dimensional steady-state convection-conduction problems and a BEM sensitivity formulation for the determination of sensitivities of temperature and heat flux to a boundary shape parameter. The BEM sensitivity formulation is based on the direct differentiation approach (ADD). These formulations are applied to simulate deep penetration laser welding. It is found that the BEM and sensitivity formulations are very efficient and robust in the determination of the solid-liquid interface in a workpiece.

INTRODUCTION

The heat transfer phenomenon in various manufacturing processes may be modeled as steady-state convection-conduction problems. In many cases, the convective velocity may be considered to be uniform. This is also true for the heat transfer in the solid region of the workpiece during laser welding.

Laser welding has shown many advantages over the conventional gas and arc welding. One of those is that the laser can deliver a focused and high power beam to a very small surface area. It makes the laser very effective tool in many difficult welding situations. Owing to the power deposition of high density, the laser beam can create a deep narrow cavity which is referred to as the keyhole on the workpiece. This keyhole allows a deep penetration welding of the workpiece.

There have been various analytical and numerical studies to predict depth of penetration or in a more sophisticated model shape of the keyhole given welding parameters (Hashimoto and Matsuda, 1965; Swift-Hook and Gick 1973; Tong and Giedt, 1971; Miyazaki and Giedt, 1982; Giedt and Tollerico, 1990; Wei *et al*, 1990). However, all these models to some extent simplified the mechanism of heat flow from the laser heat source to the solid part of a workpiece by employing a line or a cylindrical heat source, neglecting heats of phase change or convective flow of molten material, or assuming shape of the keyhole.

In light of inappropriate simplifications in the previous approaches, it is deemed necessary to develop a more detailed model which accounts for all the important physics around the keyhole. Thus, a computational model was developed for deep penetration laser welding which is capable of predicting shape of the solid-liquid interface when welding parameters, such as distribution of laser power density and welding speed, and welding environments such as atmospheric pressure and convective cooling on the surface of the workpiece are provided (Lim, 1992).

In this model, an explicit analysis of the keyhole was performed. In laser welding, there exist three phases; solid, liquid, and vapor, around the keyhole. Each of these phases is analyzed individually by a most suitable method. For example, the heat transfer in the solid phase is solved by the BEM. In the liquid phase, the thin liquid layer approximation is made which simplifies the analysis of the heat and fluid flow. Finally the analysis for the vapor phase adopts an existing one-dimensional gasdynamic model (Knight, 1979). Then the solutions of these phases are matched by a matching scheme. In the matching scheme, the solid-liquid interface is represented in a functional form with several unknown coefficients. These unknown coefficients are determined by a non-linear least squares method (Marquardt, 1963) as an energy conservation at the interface becomes satisfied. The matching scheme requires the heat flux into the solid material at the solid-liquid interface and the sensitivity of the heat flux to each of the unknown coefficients. This paper will focus on the BEM formulations for the heat flux at the solid-liquid interface and for its sensitivities to each unknown coefficient in the solid-liquid interface shape function, and discuss briefly the solution methods for the three phases and the matching scheme with a few results of the solid-liquid interface shape predicted by the present model.

FORMULATION

Governing Equation and Boundary Conditions

The geometrical configuration of the three-dimensional (3-D) solid domain with the keyhole its boundary and the cartesian coordinate system is shown in Fig. 1. The governing equation and appropriate boundary conditions may be introduced in dimensionless forms as follows:

$$Pe\frac{\partial\theta}{\partial x} = \frac{\partial^2\theta}{\partial x^2} + \frac{\partial^2\theta}{\partial y^2} + \frac{\partial^2\theta}{\partial z^2}\ , \qquad (1)$$

where Pe is a Peclet number defined as $U_o r_o/\kappa$. U_o is the laser scanning velocity, r_o is the laser beam radius, and κ is the thermal diffusivity. The boundary conditions are given as

$$-\frac{\partial\theta}{\partial n} = Bi\ (\theta+1) \text{ at surfaces parallel to } U_o\ , \qquad (2)$$

$$\theta = -1 \text{ at upstream}\ , \qquad (3)$$

$$\frac{\partial\theta}{\partial n} = 0 \text{ at downstream}\ , \qquad (4)$$

$$\theta = 0 \text{ at solid-liquid interface}\ , \qquad (5)$$

where Bi is a Biot number defined as hr_o/k. h is the heat transfer coefficient, and k is the heat conductivity. Above are used the following dimensionless variables.

$$x = \frac{X}{r_o},\ y = \frac{Y}{r_o},\ z = \frac{Z}{r_o}\ ,$$

$$\text{and } \theta = \frac{T - T_m}{T_m - T_o}\ , \qquad (6)$$

where T is the temperature, T_o is the ambient temperature, and T_m are the melting temperature. The BEM formulations are done based on the dimensionless governing equation Eq.(1).

Boundary Element Formulation

The BEM formulation for the two-dimensional (2-D) convection-conduction problems has been conducted by Chan and Chandra (1991). In principle, the formulation for the 3-D convection-conduction problems will follow the exactly same path as that for the 2-D problems. However, In the 3-D formulation, the resulted boundary integrals contain the surface integrations as opposed to the line integrations in the 2-D. Thus, more difficulties are faced in the discretization of the boundary, the connectivity of the boundary elements, and the computation of the boundary integrals.

A fundamental solution of Eq.(1) (Carslaw and Jaeger, 1959) is given as

$$G(p,q) = \frac{1}{4\pi r}\exp\left[\frac{Pe\{r-(x(q)-x(p))\}}{2}\right]\ , \qquad (7)$$

where p and q represent a source point and a field point, respectively, in the domain, and r is the distance between the source and field point.

In order to obtain the boundary integral equation, a residual form of Eq.(1) should be multiplied by the principal solution as a weight function and integrated over the entire domain. The domain integration can then be transformed into a boundary integration by applying the divergence theorem. Thus, a resulting boundary integral equation can be found to determine an internal temperature as

$$T(p) = \int_\Gamma \left(G(p,Q)\frac{\partial T(Q)}{\partial n} - \frac{\partial G(p,Q)}{\partial n}T(Q)\right) d\Gamma(Q)$$

$$- Pe\int_\Gamma G(p,Q)\ T(Q)\ n_x(Q)\ d\Gamma(Q)\ , \qquad (8)$$

where n_x is a x component of an outward normal vector, and P and Q are the source and field point at the boundary. Finally, the boundary integral equation for the steady convection-conduction problem can be obtained by taking a limit as p tends to P as follows:

$$c(P)T(P) = \int_\Gamma \left(G(P,Q)\frac{\partial T(Q)}{\partial n} - \frac{\partial G(P,Q)}{\partial n}T(Q)\right) d\Gamma(Q)$$

$$- Pe\int_\Gamma G(P,Q)\ T(Q)\ n_x(Q)\ d\Gamma(Q)\ , \qquad (9)$$

where the coefficient c(P) takes care of the singularity in the boundary integrations occurring when the source point p in the domain approaches to a source point P in the boundary. In general, the coefficient c(P) depends only on the local geometry at P. When the boundary is locally smooth at P, c = 1/2. Thus, the boundary integrations on the right side of Eq.(9) are regular.

The numerical implementation of Eq.(9) requires discretization of the boundary and also interpolation of the primary unknowns, i.e., potential and flux. In this work, a linear interpolation is used for both. So, each node at the boundary has two unknowns, and yet one or a relationship of which should be provided as a boundary condition so as to render the problem well-posed. In this way, a matrix equation is established to be solved through a proper numerical method.

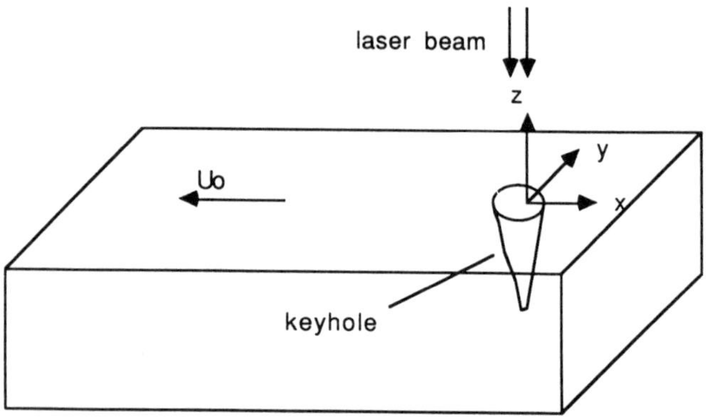

FIG. 1 GEOMETRY OF THE 3-D SOLID DOMAIN AND CARTESIAN COORDINATE SYSTEM

Boundary Element Sensitivity Formulation

In this section, the boundary element sensitivity formulation for temperature and heat flux at the boundary will be discussed. First, the boundary integral equation for the sensitivities to arbitrary shape parameters a_k which determine the shape of the solid-liquid interface is derived through the direct differentiation approach (DDA) (Chandra and Chan, 1992). Later, these shape parameters will be specified so that necessary derivatives with respect to the parameters can be computed.

A direct differentiation of the boundary integral equation Eq.(9) yields

$$\int_\Gamma \left[\left(\frac{\partial G}{\partial n} + G\ Pe\ n_x \right) \left(T^*(Q) - T^*(P) \right) - G \frac{\partial T^*}{\partial n} \right] d\Gamma(Q)$$

$$+ \int_\Gamma \left[\left(\frac{\partial G^*}{\partial n} + G^*\ Pe\ n_x^* \right) \left(T(Q) - T(P) \right) - G^* \frac{\partial T}{\partial n} \right] d\Gamma(Q)$$

$$+ \int_\Gamma \left[\left(\frac{\partial G}{\partial n} + G\ Pe\ n_x \right) \left(T(Q) - T(P) \right) - G \frac{\partial T}{\partial n} \right] d\Gamma^*(Q) = 0\ , \quad (10)$$

where a superposed asterisk (*) denotes a derivative with respect to a_k. It should be noted from the above equation that the coefficient terms for T^* and $\partial T^*/\partial n$ are identical to those for T and $\partial T/\partial n$ in Eq.(9). Thus the system matrix for the sensitivities is not necessary to be computed additionally, and thus only the right-hand-side vector needs to be calculated.

It is necessary to calculate some derivatives with respect to a_k in computation of the right-hand-side vector. Shape of the solid-liquid interface is approximated to be

$$z = F(x,\ y;\ a_k)\ . \quad (11)$$

Thus, the derivatives of the kernels $G(P,Q;a_k)$ and $\partial G(P,Q;a_k)/\partial n$ to the boundary shape parameters may be expressed as

$$\frac{\partial G}{\partial a_k} = \frac{\partial G}{\partial z} \left(\frac{\partial z(Q)}{\partial a_k} - \frac{\partial z(P)}{\partial a_k} \right), \quad (12)$$

$$\frac{\partial}{\partial a_k} \left(\frac{\partial G}{\partial n} \right) = \frac{\partial}{\partial a_k} \left(\frac{\partial G}{\partial x_i} \right) n_i + \frac{\partial G}{\partial x_i} \frac{\partial n_i}{\partial a_k}\ . \quad (13)$$

Here the subscript i varying from 1, 2, to 3, denotes x_i component of a given quantity with $\{x_1, x_2, x_3\} = \{x, y, z\}$. Also, the summation rule of repeating indices is applied. In Eq.(13), the first term in the right-hand-side term can be rewritten in a readily computable form,

$$\frac{\partial}{\partial a_k} \left(\frac{\partial G}{\partial x_i} \right) n_i = \frac{\partial}{\partial z} \left(\frac{\partial G}{\partial x_i} \right) n_i \left(\frac{\partial z(Q)}{\partial a_k} - \frac{\partial z(P)}{\partial a_k} \right)\ . \quad (14)$$

However, the second term contains $\partial n_i/\partial a_k$ so that the computation of this term is discussed in what follows.

n_i is a x_i component of the outward unit normal vector which is

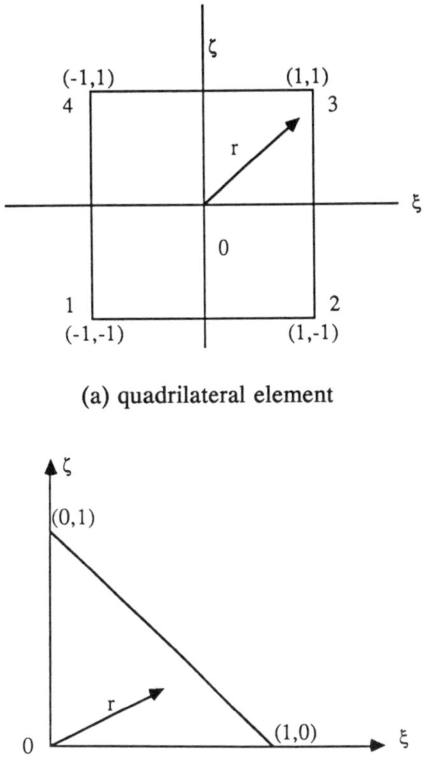

(b) triangular element

FIG. 2 THE PARAMETRIC COORDINATE SYSTEMS FOR (A) A QUADRILATERAL (B) A TRIANGULAR ELEMENT

perpendicular to a surface boundary element. The coordinates of the boundary surface element may be represented in a parametric form using a linear interpolation function as shown in Fig. 2. In the present work, a quadrilateral and a triangular boundary surface are employed. The parametric representation of the surface elements is carried out by transforming the global coordinates into the local ξ and ζ coordinates. Thus, the global coordinates on the surface element may be described in terms of the nodal global coordinates and linear interpolation functions which are expressed in the local coordinates, as follows:

$$x_i = N_j\, x_i^j\ , \quad (15)$$

where i = 1, 2, 3, and j = 1, 2, 3, 4 for the quadrilateral element and j = 1, 2, 3 for the triangular element, denoting the nodal points. The interpolation functions N_j are given for the quadrilateral element,

$$N_j = \frac{1}{4}(1+\xi\xi_j)(1+\zeta\zeta_j)\ , \quad (16)$$

and for the triangular element,

$$N_1 = \xi,\ N_2 = \zeta,\ N_3 = 1-\xi-\zeta\ . \quad (17)$$

Discussion will be focused on the quadrilateral element from now on, for the same argument can be applied to the triangular element except for replacing a superscript 4 by 3 in the following equations. An outward normal vector can be determined only with coordinates of the nodal points,

$$C_i = (-1)^{i-1}\{(x_j^2-X_j^1)(x_k^4-x_k^1)-(x_k^2-x_k^1)(x_j^4-x_j^1)\}\ , \qquad (18)$$

where i, j, and k are cycling indices such as (1,2,3), (2,3,1), or (3,1,2). Thus, a unit normal vector can easily be obtained from Eq.(18),

$$n_i = \frac{C_i}{|C|}\ , \quad \text{where } |C| = \sqrt{C_i C_i}\ . \qquad (19)$$

Now the derivative of the unit normal vector may be found as follows:

$$\frac{\partial n_i}{\partial a_k} = \frac{\partial C_i}{\partial a_k}\frac{1}{|C|} - \frac{C_i}{|C|^2}\frac{\partial |C|}{\partial a_k}\ , \qquad (20)$$

where

$$\frac{\partial C_1}{\partial a_k} = (y^2-y^1)\left(\frac{\partial z^4}{\partial a_k}-\frac{\partial z^1}{\partial a_k}\right)-\left(\frac{\partial z^2}{\partial a_k}-\frac{\partial z^1}{\partial a_k}\right)(y^4-y^1)\ ,$$

$$\frac{\partial C_2}{\partial a_k} = -(x^2-x^1)\left(\frac{\partial z^4}{\partial a_k}-\frac{\partial z^1}{\partial a_k}\right)+\left(\frac{\partial z^2}{\partial a_k}-\frac{\partial z^1}{\partial a_k}\right)(x^4-x^1)\ ,$$

$$\frac{\partial C_3}{\partial a_k} = 0, \text{ and } \frac{\partial |C|}{\partial a_k} = n_i\frac{\partial C_i}{\partial a_k}\ .$$

Finally, the derivative of a differential surface element is discussed. A differential surface element is expressed in a parametric description,

$$d\Gamma = \sqrt{E\ H - F^2}\ d\xi d\zeta\ , \qquad (21)$$

$$\text{where } E = \frac{\partial x_i}{\partial \xi}\frac{\partial x_i}{\partial \xi},\ F = \frac{\partial x_i}{\partial \xi}\frac{\partial x_i}{\partial \zeta},\ H = \frac{\partial x_i}{\partial \zeta}\frac{\partial x_i}{\partial \zeta}\ .$$

Here $\partial x_i/\partial\xi$ and $\partial x_i/\partial\zeta$ can be rewritten using Eq.(15),

$$\frac{\partial x_i}{\partial \xi} = \frac{\partial N_j}{\partial \xi}x_i^j,\ \frac{\partial x_i}{\partial \zeta} = \frac{\partial N_j}{\partial \zeta}x_i^j\ .$$

Thus, the derivative of the differential surface element may be obtained as follows:

$$\frac{\partial(d\Gamma)}{\partial a_k} = \frac{\frac{\partial E}{\partial a_k}H+E\frac{\partial H}{\partial a_k}-2F\frac{\partial F}{\partial a_k}}{2(E\ H - F^2)}d\Gamma\ , \qquad (22)$$

$$\text{where } \frac{\partial E}{\partial a_k} = 2\frac{\partial z}{\partial \xi}\frac{\partial}{\partial a_k}\left(\frac{\partial z}{\partial \xi}\right),\ \frac{\partial H}{\partial a_k} = 2\frac{\partial z}{\partial \zeta}\frac{\partial}{\partial a_k}\left(\frac{\partial z}{\partial \zeta}\right),$$

$$\frac{\partial F}{\partial a_k} = \frac{\partial}{\partial a_k}\left(\frac{\partial z}{\partial \xi}\right)\frac{\partial z}{\partial \zeta}+\frac{\partial z}{\partial \xi}\frac{\partial}{\partial a_k}\left(\frac{\partial z}{\partial \zeta}\right)\ .$$

Solution Methods in Each Phase

Around the keyhole are three different phases of a workpiece. The keyhole cavity filled with vapor is surrounded by a liquid layer which in turn encompassed by a solid phase. Due to the complexities involved in the fluid and heat flow in laser welding, an entire solution domain is separated into three phases, and then each region is handled independently with a most suitable method. Once each phase is solved for an assumed solid-liquid interface, they are matched together through the matching which will find a new interface. The matching scheme makes use of the energy balance at the solid-liquid interface. Thus, it is necessary to calculate the heat fluxes in the solid phase and in the liquid phase. Details of the matching scheme will be explained later. First, the solution methods for three phases will be described.

Solid Phase. In order to calculate the heat flux in the solid side, the BEM formulation is used. The BEM has a great advantage in the computation of the boundary heat flux, compared to other numerical methods such as a finite difference method (FDM) and a finite element method (FEM), for the BEM is able to calculate the heat flux directly rather than interpolate temperatures near the boundary as do the FDM and FEM. Accuracy of this interpolation inherently depends on a grid size.

Liquid Phase. As to the calculation of the heat flux in the liquid region, the heat and fluid flow in the liquid layer should be solved with appropriate boundary conditions at both the solid-liquid interface and the liquid-vapor interface. However, we assume that the thickness of the liquid layer is very small compared to a representative radius of curvature of the keyhole. The assumption is deduced from an experimental observation (Arata and Miyamoto, 1978) and results of theoretical approaches (Wei and Ho, 1990; Wei and Giedt, 1985). The thin layer approximation enables a scaling analysis to determine the importance of the convection term in the energy equation. Two different velocity scales can be used; one is a velocity scale from an assumption that the flow is driven by a pressure gradient force in a tangential direction to a liquid layer, and the other is from that the flow is driven by a thermocapillary force. For both velocity scales, it is found that the convective heat transfer is negligible and the conduction perpendicular to the layer is the dominant heat transfer when the liquid layer is thin.

This analysis shows that the heat flux is constant in the direction normal to the layer. Thus, the heat flux in the liquid side at the solid-liquid interface is equal to the heat flux going into the liquid layer at the liquid-vapor interface, which can be obtained by subtracting a vaporization energy from an absorbed laser energy at the liquid-vapor interface.

Vapor Phase. When a high-density laser power irradiates on the surface of a metal, intensive heat is generated, melts a solid surface, and invokes a strong vaporization on the free surface of the liquid layer. The surface temperature of the liquid layer is usually higher than the boiling temperature of a molten metal since the recoil pressure due to a strong vaporization increases an effective saturated vapor pressure. Therefore, the vaporization of this kind can not be explained by the diffusion-dominated models. Thus, the convection-dominated model should be considered in the case of laser welding. In principle, three-dimensional gasdynamic equations should be solved with appropriate boundary conditions at the surface of a condensed phase to calculate the properties of the gas phase such as velocities, density, pressure, and temperature. however, the present study requires the gasdynamic properties just above the liquid-vapor interface where one-dimensional behavior of the vapor flow may prevail due to a violent vaporization. Thus, a one-dimensional vaporization model is adopted which was developed by Knight (1979).

Matching Scheme

The solid-liquid interface shape is approximated by a function of unknown parameters,

$$z(x,y;a_k) = \sum_{l=1}^{3} A_l \exp\left\{-\frac{(x-B_l)^2}{D_l^2} - \frac{y^2}{P_l^2}\right\} + Q \ . \tag{23}$$

The function is expressed as a combination of three different gaussian distributions, guided by a fact that a profile of laser power density is in a gaussian shape. The matching scheme is to determine the values of the unknown parameters. It is convenient to define the following vector for the parameters,

$$a_k = \{A_l, B_l, D_l, P_l, Q\} \ . \tag{24}$$

The energy conservation at the solid-liquid interface may be rewritten in a residual form,

$$\Delta q(a_k) = -k_l \frac{\partial T_l}{\partial n}(a_k) + k_s \frac{\partial T_s}{\partial n}(a_k) + \rho_s u_{s,n}(a_k) L_f \ , \tag{25}$$

where subscripts l and s denotes liquid and solid respectively, ρ_s is the solid density, $u_{s,n}$ is the normal velocity of the solid to the interface, and L_f is the heat of fusion. When N number of node points on the solid-liquid interface is used, the following least squares sum which is also called a merit function is defined,

$$\Xi^2 = \sum_{j=1}^{N} \left(\frac{\Delta q_j(a_k)}{\sigma_j}\right)^2 , \tag{26}$$

where σ_j are standard deviations. Hence the matching scheme will determine the parameters a_k which minimize the Ξ^2 merit function. $\Delta q(a_k)$ behaves non-linearly with a_k so that the fitting should rely on a non-linear least squares method. In the present study, the Marquardt method is used (Marquardt, 1963).

Marquardt Method. Most method for the least squares estimation of non-linear coefficients have centered on either of the following two approaches; the Taylor series expansion assuming local linearity, and the steepest descent method. The Marquardt method, a maximum neighborhood method, is developed to optimally perform the interpolation between the Taylor series method and the gradient method, the interpolation being based on the maximum neighborhood in which the truncated Taylor series gives an adequate representation of the non-linear model.

The mathematical implementation of the Marquardt method is introduced below. The following matrix equation is solved at each iteration in order to compute the corrections to the parameters,

$$\sum_{l=1}^{M} \alpha'_{kl}\, \delta a_l = \beta_k, \quad \text{where } M = 13 \ ,$$

$$\alpha'_{kl} = \alpha_{kl}\,(1+\delta_{kl}\,\lambda) \ , \tag{27}$$

where δ_{kl} is the Kronecker delta, and λ is the parameter supplied by the algorithm. When λ is much larger than 1, then the method works as the steepest-descent method, the matrix α_{kl}' becoming diagonally dominant. Meanwhile, when λ gets smaller than 1, it performs like the Taylor series expansion. The matrix α_{kl} and the right-hand-side vector β_k are given as

$$\alpha_{kl} = \frac{1}{2}\frac{\partial^2 \Xi^2}{\partial a_k \partial a_l} \ , \quad \beta_k = -\frac{1}{2}\frac{\partial \Xi^2}{\partial a_k} \ . \tag{28}$$

Consecutive substitution of Eqs.(26) and (25) into the above expressions produces a term for the sensitivity of the heat flux to a interface shape parameter, which may be calculated by the BEM sensitivity formulation. Once δa_k are obtained, these corrections are added to old a_k, leading to new a_k. The iteration continues until a_k converges.

RESULTS AND DISCUSSION

The present model is applied to find shape of the solid-liquid interface in the case of pure iron at various welding conditions. Fig. 3 shows the interface shapes at various welding speeds with a constant laser power where Q_o is the maximum laser power density of the gaussian power density distribution. According to the figure, obviously a lower welding speed generates a deeper penetration, which is qualitatively correct, as shown by the previous studies. There is a little change in a keyhole diameter compared to penetration depth as the welding velocity varies. This property of laser welding, that most laser energy is transferred in a laser beam direction rather than conducts radially, enables deep penetration welding.

Also, the predicted values of penetration depth are in a reasonable range, compared to the empirical formula created by Hashimoto and Matsuda (1965), as shown in Fig. 4. Although they have performed the experiment using a electron beam welder, the effects on the workpiece are considered to be the same as far as heat transfer is concerned. In this comparison, we assume that the radiative absorptivity of the laser energy at the keyhole surface is 100 %, the beam profile is gaussian, the heat transfer coefficient on the workpiece surface is 100 $W/m^2{\cdot}K$, and the ambient pressure 1 atm.

CONCLUSIONS

A boundary element method (BEM) formulation for the three-dimensional steady convection-conduction problems is developed,

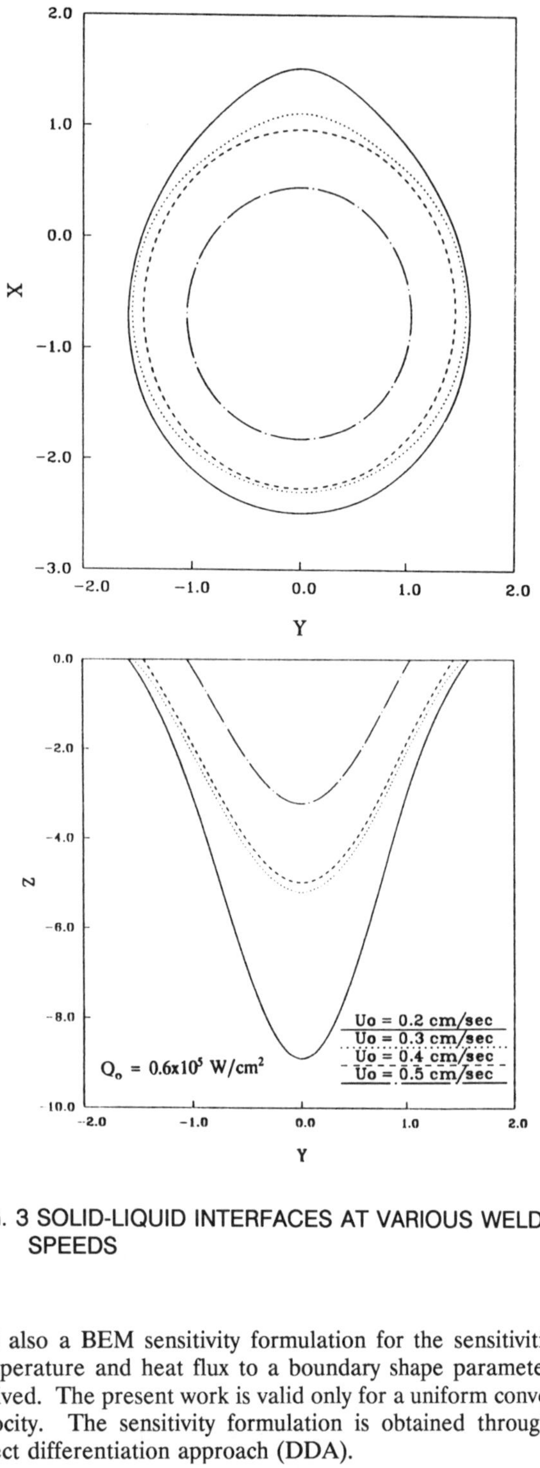

FIG. 3 SOLID-LIQUID INTERFACES AT VARIOUS WELDING SPEEDS

and also a BEM sensitivity formulation for the sensitivities of temperature and heat flux to a boundary shape parameter are derived. The present work is valid only for a uniform convective velocity. The sensitivity formulation is obtained through the direct differentiation approach (DDA).

These formulations are developed in conjunction with laser welding application. That is, shape of the solid-liquid interface needs to be determined under given welding parameters such as distribution of laser power density, laser scanning velocity, and so on. A computational model was developed for this purpose and requires the determination of the heat flux at the solid-liquid interface and its sensitivity to a shape parameter of the interface. Thus, these BEM formulations serve as parts of the whole strategy determining the solid-liquid interface.

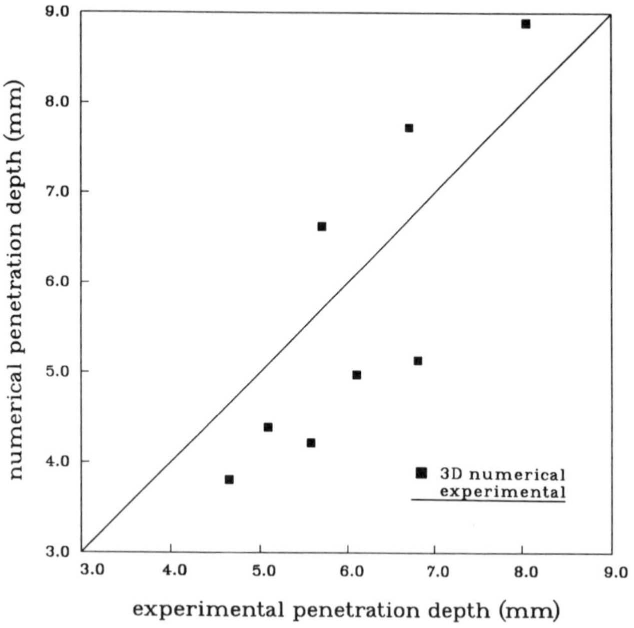

FIG. 4 CONPARISON OF PENETRATION DEPTH

The solid-liquid interface is successfully predicted by the present model for an example of pure iron.

ACKNOWLEDGMENTS

This research has been supported by Grant No. CTS-8909101 of the U.S. National Science Foundation.

REFERENCES

Arata, Y., and Miyamoto, I., 1978, "Laser Welding," *Technocrat*, Vol. 11, No. 5, pp. 33-42.

Carslaw H. S., and Jaeger, J. C., 1959, *Conduction of Heat in Solids*, 2nd ed., Oxford Science Publications, pp. 266-270.

Chandra, A., and Chan, C. L., 1992, "A Boundary Element Method Formulation for Design Sensitivities in Steady-State Conduction-Convection Problems, ASME *J. Appl. Mech.*, Vol. 59, pp. 182-190.

Giedt, W. H., and Tollerico, L., 1988, "Prediction of Electron Beam Depth of Penetration," *Welding Journal*, Vol. 67, pp. 299s-305s.

Hashimoto, T., and Matsuda, F., 1965, "Effects of Welding Variables and Material upon Bead Shape in Electron Beam Welding," *Trans. National Research Institute for Metals*, Vol. 7, pp. 22-25.

Knight, C. J., 1979, "Theoretical Modelling of Rapid Surface Vaporization with Back Pressure," *AIAA Journal*, Vol. 17, No. 5, pp. 519-523.

Lim, J. H., 1992, "A computational Analysis of Deep Penetration Laser Welding," Ph. D. Dissertation, The University of Arizona, Tucson, AZ.

Marquardt, D. W., 1963, "An Algorithm for Least Squares Estimation of Nonlinear Parameters," *J. Soc. Indust. Appl. Math.*, Vol. 11, No. 2, pp. 431-441.

Miyazaki, T., and Giedt, W. H., 1982, "Heat Transfer from an

Elliptical Cylinder Moving through an Infinite Plate Applied to Electron Beam Welding," *Int. J. Heat Mass Transfer*, Vol. 25, pp. 807-814.

Swift-Hook, D., and Gick, A., 1973, "Penetration Welding with Lasers," *Welding Journal*, Vol. 52, pp. 492s-499s.

Tong, H., and Giedt, W. H., 1971, "Depth of Penetration during Electron Beam Welding," ASME *J. Heat Transfer*, pp. 155-163.

Wei, P. S., and Giedt, W. H., 1985, "Surface Tension Gradient-Driven Flow around an Electron Beam Welding," *Welding Journal*, pp. 251s-259s.

Wei, P. S., and Ho, J. Y., 1990, "Energy Considerations in High-Energy Beam Drilling," *Int. J. Heat Mass Transfer*, Vol. 33, No. 10, pp. 2207-2217.

Wei, P. S. et al, 1990, "Investigation of High-Intensity Beam Characteristics on Welding Cavity Shape and Temperature Distribution," ASME *J. Heat Transfer*, Vol. 112, pp. 163-169.

Inverse Problems in Engineering: Theory and Practice
ASME 1993

OPTIMAL SENSOR LOCATION CONSIDERING PARAMETER VARIABILITY

A. F. Emery and T. D. Fadale
University of Washington
Seattle, Washington

A. V. Nenarokomov
Moscow Aviation Institute
Moscow, Russia

Abstract

Accurate modelling of thermal system depends upon the determination of physical properties (ρ, c_p, k) and surface transfer coefficients (h, ϵ). These parameters are frequently estimated from temperatures measured within the system or on the surface of from measured surface heat fluxes. Because of sensor errors or lack of sensitivity, the measurements may lead to erroneous estimates of the parameters. The errors can be ameliorated of the sensors are placed at points of maximum sensitivity. This paper describes two methods to optimize sensor locations, one to account for signal error; the other to consider interacting parameters. The methods are based upon variants of the normalized Fisher information matrix and are shown to be equivalent in some cases, but to predict differing sensor locations under other conditions, usually transient. Examples of correlated paramters are give.

NOMENCLATURE

Roman

B_l	Specific outcome of **B**
$B(\vec{x})$	Continuous random field
B	Random variable vector
c	specific heat capacity (J/m^3/K)
c	Capacitance matrix
$Cov[\,]$	Covariance operator
D	Domain
D_i	ith elemental domain
$E[\,]$	Expected value operator
F_o	Fourier Number
h	Heat transfer coefficient (W/m^2-K)
k	Thermal conductivity (W/m-K)
k	Conductance matrix
L	Maximum Likelihood
M	Fisher information matrix
P_1, p_1	Initial conditions
P_2, p_2	Boundary conditions
q	Heat flux (W/m^2)
q	Source matrix
s	length parameter (m)
S	covariance matrix related to measurement
t	Time (s)
T_∞	Ambient Temperature (C)
$Var[\,]$	Variance operator
V	covariance matrix of temperatures related to parameters
x, y	Position (m)
$\bar{\mathbf{X}}$	Vector of sensor positions (m)
z	Measured temperatures (C)

Greek

α	Thermal diffusivity
γ	Variance function
κ	Thermal diffusivity (m^2/s)
$\sigma_{\mathbf{B}}$	Standard deviation of **B**
ϕ	Temperature field
$\mathbf{\Phi}$	Temperature vector
$\mathbf{\Phi}_k$	Vector of sensor temperature at time k
ϕ_{B_l}	Derivative of ϕ w.r.t. B_l

Subscripts

i, j,	sensors
k,	time
l, m,	parameters

Introduction

Mathematical modeling of systems or processes relies upon accurate estimates of parameters or properties. Frequently these parameters are determined by an inverse method in which measured responses are compared with direct simulations. In the sequence of simulation the parameter values are altered until a predicted response matches a measured response. The premise is that if agreement can be attained for a large number of responses the parameters are uniquely and accurately determined. It is not uncommon that errors or non repeatable measured temperatures may lead to conflicting conclusions about inferred property values.

This approach yields approximate values of the parameters (the degree of approximation is related to the heterogeneity of the material, the accuracy of the measurements, the spatial gradients existing in the neighborhood of the sensor, and the refinement of the discretization of the model). In addition, the results are subject to uncertainty in that the response upon which the estimation is based may not be sensitive to the parameter sought. One solution to a possible lack of sensitivity is to locate the sensors at the points of maximal sensitivity. Unfortunately, in most transient problems the point of maximal sensitivity is a function of time and the ideal location of a single sensor is not unequivocally defined. An example of this is given by Fadale (1990) in the analysis of the temperature response of a high temperature test of a thermal protection system (Gorton, 1988). Temperatures were measured at several different points in the structure and conductivities were to be deduced from simulations of the tests. The temperatures were affected by radiation from an internal surface whose emissivity was unknown. Consequently it was important to determine the effect of the radiation (as characterized by the surface emissivity) on the measured temperatures. It was shown that the different locations had different sensitivity histories. Clearly, one cannot in general move sensors during an experiment. Thus the choice is twofold: a) employ a large number of sensors and use the response from a given sensor while it possesses the greatest sensitivity; b) employ only a few sensors, locating them at positions which will give the best integrated sensitivity over the entire experiment.

Mathematical modeling data also show that the error of determination of the parameters of the mathematical model from the solution of the inverse problem may depend significantly on the spatial placement of the temperature sensors (Beck, 1992). The proper setup of the thermophysical experiment requires the solution of an optimal experimental design problem, viz.; to place a fixed number of sensors in the sample in such a way as to minimize the error of identification of the required characteristics. The stated problem can be solved on the basis of the fundamental principles of experimental design theory for distributed-parameter system, although the precise criterion is not unique. It is also clear that in multi-parameter systems, the response of a sensor will be affected by more than one parameter and the question naturally arises as to where to place a sensor to best determine the parameters sought without other parameters confounding the issue.

The general problem is to locate the sensors such that the parameter (PAR) estimation is not adversely affected by error in the responses, that is, $\partial \text{PAR}/\partial\phi$ is small. Another way of expressing this is that the sensitivity of the temperature, $\partial\phi/\partial \text{PAR}$, should be as large as possible. At the same time it must be recognized that all measurements are subject to noise and the sensors should be located to minimize the noise effects. It is not obvious that maximizing sensitivity to parameter variations and minimizing effects due to noise lead to the same sensor locations. This paper compares the use of sensitivity analysis and of the Fisher information matrix to define optimal sensor locations to gather the most statistical information about a parameter. Although the methods are applicable to any system, we restrict our study to transient and steady state two dimensional thermal systems.

Analysis

Fisher Information Matrix (FIM)

The Fisher Information Matrix approach (Goodwin and Payne, 1977) is one of the experimental design techniques used to minimize the effects of noise. If the noise is taken to be *white*, the approach can be used to estimate parameters, as in the theory of Inverse Problems. In this approach it is sought to estimate one or more parameters by using the measurements of the temperature at one or several points in the interior or on the boundary of the system. Let the measured temperatures be denoted by z_i where

$$z_i(t_k) = \phi(x_i, y_i, t_k), \qquad i = 1, 2, ...N, \quad k = 1, 2, ...K \quad (1)$$

and N is the number of sensors and K the number of readings taken over time.

The usual method for estimating the unknown parameters is to minimize the time summed mean square deviation (Γ) between the temperature values at the sensor placement sites calculated by means of a mathematical model and the experimentally measured values:

$$\Gamma = \sum_{i=1}^{N}\sum_{k=1}^{K}(\phi(x_i, y_i, t_k) - z_i(t_k))^2 \quad (2)$$

It must be emphasized that the minimum required number of temperature sensors and the limits on their placement are completely determined by an analysis of the conditions for the existence and uniqueness of the solution of the corresponding inverse problem. Fedorov (1992) has shown that although the maximum number of sensors is equal to the number of parameters, there are circumstances in which a smaller number can be employed. For example, in order to determine either the conductivity k or the specific heat capacity c, Musylev (1980) has shown that it is necessary to measure the transient temperature at at least one point. This point must be situated in the interior if Dirichlet boundary conditions are specified or on the boundary for a Neumann-type boundary condition. However, if the problem is to identify both k and c, the unique solution of such an inverse problem requires measuring the temperature at two different points and specifying a Neumann boundary condition on at least one of the boundaries, in which case the heat flux must be nonvanishing. On the other hand, if both k and c are known to be constants, then a unique solution can be obtained with only one measurement point. Requirements of this type on the measurements dictate the minimum possible experimental information that is required in principle for the solution of a specific inverse problem. If a greater number of measurements is performed, the problem becomes overdetermined.

Let the unknown parameters be expressed as a vector, B, and the location of the temperature sensors be defined in the form of a vector of space coordinates, $\bar{\mathbf{X}}$:

$$\bar{\mathbf{X}} = \{x_i, y_i, \; i = 1, 2, ..., N \;\; \bar{\mathbf{X}} \in \mathbf{D}\} \quad (3)$$

Almost all methods to define optimal sensor locations are based upon maximizing some scalar function of the Fisher information matrix, **M**, based upon the principle of maximum likelihood, **L** where **M** is given by

$$\mathbf{M} = E_{z|B}\left\{\left[\frac{\partial log\ \mathbf{L}(z|B)}{\partial B}\right]\left[\frac{\partial log\ \mathbf{L}(z|B)}{\partial B}\right]^T\right\} \qquad (4)$$

which can be written as

$$M_{lm} = \sum_{k=1}^{K}(\frac{\partial \mathbf{\Phi}_k}{\partial B_l})^T\ \mathbf{S}_k^{-1}\ (\frac{\partial \mathbf{\Phi}_k}{\partial B_m})$$
$$+\frac{1}{2}\sum_{k=1}^{K} trace\{\mathbf{S}_k^{-1}\frac{\partial \mathbf{S}_k}{\partial B_l}\} trace\{\mathbf{S}_k^{-1}\frac{\partial \mathbf{S}_k}{\partial B_m}\} \qquad (5)$$

where $\mathbf{\Phi}_k$ is a vector of the sensors readings at time **k**, K is the total number of readings, and $\mathbf{S}_k^{-1}$ is the inverse of the covariance of the readings. It is usual to simplify the form of **M** by assuming that

$$\mathbf{S}_k^{-1} = \sigma_{noise}^{-2}\mathbf{I} \text{ and } \frac{\partial \mathbf{S}_k}{\partial B_l} = 0 \qquad (6)$$

That is: all sensors are exposed to the same noise; the sensor noise is uncorrelated; and the noise is independent of the parameters. Thus **M** reduces to

$$M_{lm} = \sigma_{noise}^{-2}\sum_{k=1}^{K}(\frac{\partial \mathbf{\Phi}_k}{\partial B_l})^T\ (\frac{\partial \mathbf{\Phi}_k}{\partial B_m}) \qquad (7)$$

The partial differential equations for the sensitivity functions, $\partial\mathbf{\Phi}/\partial B_l$, are found by appropriately differentiating the energy equation, see Fadale (1992).

The information matrix, **M**, characterizes the total sensitivity of the entire set of measurements to the variation of all of the components of the vector of unknown parameters but without requiring an estimate of the variation of the parameters, only of the noise. The optimal measurement design problem entails finding a design $\bar{\mathbf{X}}$ for which the total sensitivity of the system in some sense will be a maximum.

Various criteria are used for the optimization of the experimental conditions (Kubrusly and Malebranche, 1982, Polis, 1982). The so-called D- optimal design is widely used to ensure the minimum error of estimation of the unknown parameters. In this case, the measurement design can be determined by requiring that some measure of the information matrix is a maximum, i.e.

$$\max_{\bar{\mathbf{X}}}\{Measure\ \mathbf{M}(\bar{\mathbf{X}})\} \Rightarrow \bar{\mathbf{X}},\ \ \bar{\mathbf{X}} \in D \qquad (8)$$

Artyukhin (1985, 1989) has shown that the best criterion is found by maximizing *cond*(M), but that maximizing *det*(M) is a nearly equivalent criterion. It is important to note that the use of M requires knowledge about the noise of the readings, but not a knowledge about the range of values of the parameters.

It must be emphasized that the elements of the information matrix **M** and, hence, the optimal sensor positions, $\bar{\mathbf{X}}$, depend on the estimate of the vector of unknown parameters $\bar{B}$. This is because of the strongly nonlinear dependence of the temperature on the unknown parameters and is typical of measurement designs for the solution of inverse problem in thermal - structural problems. Accordingly, it is only meaningful to speak of locally ptimal designs, which are formulated with the use of a-priori information about the unknown parameters.

Sensitivity Analysis

For the sensitivity analysis, the parameter under consideration is treated as a random variable which is characterized by a probability distribution function. An analysis can be carried out by using either a statistical approach or a non- statistical approach. The statistical analysis uses either experiments or Monte Carlo simulations. In both, repetetive tests are conducted in which one or more parameters are varied and the resulting response of interest is analyzed statistically. The non-statistical approach, unlike sampling, is based upon an analytical treatment of the uncertainty. This approach involves procedures like numerical integrations, finite differencing, and perturbation methods (Lin, 1967). The non-statistical approach is not only an elegant way to handle uncertainties in systems but also has an advantage over the statistical approach in terms of computer time and in ease of interpretation.

Perturbation methods have been commonly used for nonlinear analysis, e.g. turbulence, nonlinear vibrations. One such perturbation method, which can be used for uncertainty modelling, is the First order/Second order - second moment analysis (Haftka and Kamat, 1985, Ditlevesen, 1981). This technique consists of expressing the response in terms of a Taylor series expanded about the mean value of the random parameter and truncating the series at the first or second order terms. The system is then solved for the mean value and the standard deviation of the response. The first order and the second order derivatives of the responses with respect to the random parameters are in effect the sensitivities of the responses with respect to the random parameters.

The uncertainty parameters whose effects on the response are being studied are modeled as either random variables or as random fields defined in the interior or on the boundary of the system. For the finite element implementation, the continuous random field $B(\vec{x})$, which represents any of the variable parameters (e.g. material properties, heat transfer coefficient), is discretized over the domain of application. Thus if we consider the discretized random field as an elemental quantity, then the average of the continuous random field over a finite element can be treated as an elemental quantity (Vanmarke 1977). The local integral over a finite element of the random field is itself a random variable, referred to as B_i for the ith element.

$$B_i = \frac{1}{D_i}\int_{D_i} B(\vec{x})d\vec{x} \qquad (9)$$

Due to the discretization of the random field over the finite elements it is easier to account for the correlation between the discretized random variables and it is also convenient to evaluate their stochastic properties. It also makes it easier to develop the matrices of the first and second order equations. These advantages of using the finite element to define a localized random variable are lost for a Finite difference method. Since a two dimensional finite element code is being used, the random field, if present on the boundary (surface) of the system, will be defined in a 1/D space, i.e. $\vec{x} = (s)$ where s is a length parameter along the surface, and if present in the interior (volume) of the system, will be defined in a 2/D space, i.e. $\vec{x} = (x_1, x_2)$. One stochastic property for a 2/D space localized random variables is given by $Var[\mathbf{B}] = \sigma^2_{\mathbf{B}}\gamma(D)$ where $\gamma(D)$ is a variance function of **B** which measures the reduction of the point variance $\sigma^2_{\mathbf{B}}$ under local averaging over the element D and can be precomputed for the mesh. The stochastic properties for the 1/D space localized random variables are analogous to those of 2/D space.

The finite element formulation of the first order, second moment technique for a transient heat conduction problem is given in the paper by Fadale (1992). The formulation leads to the evaluation of derivatives of nodal temperature ϕ with respect to the elemental random variables B_i. These derivatives indicate the sensitivities of nodal temperatures to the random variables **B** at various locations. This large amount of data on temperature sensitivities can be concisely represented by the statistical property of covariance of the temperatures at the various sensor locations. This being a first order method the first order estimate of the covariance matrix is given as,

$$E[(\phi - \overline{\phi})^2] = \sum_{l=1}^{M}\sum_{m=1}^{M} \overline{\mathbf{\Phi}_{B_l}}\ \overline{\mathbf{\Phi}_{B_m}}^T Cov\,[B_l, B_m] \qquad (10)$$

where B_l represents the random variable associated with the lth parameter and $Cov[B_l, B_m]$ represents the interaction between the parameters. For nonlinear cases the second order estimates of the covariance of the temperature do a better job of predicting the covariance of the temperature. The finite element formulation and the covariance estimates for the second order calculations can be found in Fadale and Emery (1992).

Following the development of **M** a tentative criterion for optimal sensor location is the maximizing of det(**V**) where **V** is given by

$$V_{ij} = \sum_{k}^{K}\sum_{l}^{P}\sum_{m}^{P} \frac{\partial \phi_{ki}}{\partial B_l}\frac{\partial \phi_{kj}}{\partial B_m} Cov[B_l, B_m] \quad i,j = 1, 2..., N \quad (11)$$

and K denotes the number of readings and P the number of parameters. If $cov(B_l, B_m) = \delta_{lm}\sigma_l\sigma_m$, then it can be shown that if M=P,

$$det\ \mathbf{V} = \sigma^{-2}_{noise}\ det\ \mathbf{M} \prod_{l}^{P} \sigma_l^2 \qquad (12)$$

and the optimizations based on the sensitivity and Fisher information matrix are equivalent. If the off diagonal terms of the covariance are not zero, then the methods are not equivalent.

Note that there is an intrinsic difference between the two methods. The Fisher information matrix is a measure of the sensitivity of the signals to noise, but with no prescription of the statistical properties of the parameters. Optimizing **M** leads to the determination of the optimal sensor location but reveals nothing about the individual parameters. The sensitivity approach assumes that the signals are well defined (i.e. not noisy) but that they vary because of variations in the parameters. Thus sensitivity analysis can lead to conclusions about the effect of parameters, but not about sensor location. The first method requires a knowledge of $\mathbf{S}_k^{-1}$ while the second requires a knowledge of $Cov[B_l, B_l]$. Naturally there are intrinsic difficulties in specifying either and neither are easy to come by unless much is already known about the system being studied.

Examples and Discussions

A.) Transient Slab

Consider a slab of thickness L with an initial temperature of 0C and with a boundary condition of the third kind on both the back (x=0) and the front (x=L) surfaces. The plate is convectively heated with an ambient temperature of 100C on both the sides but with different heat transfer coefficients (h_b on the back surface and h_f on the front surface). At steady state, the temperature will be uniform through out the plate, equal to the ambient temperature. Therefore an experiment carried out at steady state will not furnish any information which can lead to the estimation of the thermal diffusivity of the plate or the heat transfer coefficients on the boundaries. On the other hand, the transient temperatures of the plate depends upon the thermal properties of the plate and the nature of the convective boundary conditions and so can be used to for estimation of these parameters. A transient analysis was thus carried out using the Fisher Information matrix and the Sensitivity analysis to determine the optimal locations of sensors for the estimation of (a) thermal conductivity only and (b) the thermal conductivity along with the heat transfer coefficients h_b and h_f from a single experiment. Since, for the optimal placement of sensor locations *a priori* information on the parameters to be estimated is necessary, the effect of gross changes in the expected value of one of the parameters on the sensor location for both the above mentioned cases was investigated. The heat transfer coefficient on the front surface (h_f) was considered to be badly misjudged while the thermal conductivity (k) and the heat transfer coefficient on the back surface (h_b) were considered to be close to the expected estimates, viz, $k = 1.0W/(m-K)$ and $h_b = 10.0W/(m^2-K)$. The volumetric heat capacity was taken to be $1.0x10^6 J/(m^3-K)$.

One of the fundamental differences in the two approaches i.e. using the Fisher Information Matrix (Eqn. 7) or the Covariance of the temperature matrix from Sensitivity analysis (Eqn. 11), for optimal sensor locations is the statistical condition on the parameters to be estimated. The Fisher Information matrix cannot take into account any correlation which might exist between two desired parameters, for example correlation between the heat transfer coefficient of the front and the back surface. The Sensitivity analysis can account for such correlation through the covariance $Cov[h_b, h_f]$, used in defining the covariance of the temperature matrix. The influence of such a correlation on the optimal sensor locations was also investigated. The correlation between the two heat transfer coefficients was defined by a correlation coefficient given as,

$$\rho_{h_b h_f} = \frac{Cov[h_b, h_f]}{\sigma_{h_b}\sigma_{h_f}} \qquad (13)$$

Finally the effect of the length (L) of the slab on the sensor locations was also studied.

Figures 1, 2 and 3 present the variation of det(**M**) over the thickness of the slab for h_f = 50.0, 100.0, and $500.0W/(m^2-K)$ respectively. For the case (a) where only the thermal conductivity is to be estimated, the optimal sensor location when

$h_f = 50.0W/(m^2 - K)$ is at x/L = 1.0 but a nearly equivalent sensitivity exists near x/L = 0.5, while for h_f = 100.0 and $500.0W/(m^2 - K)$ the optimal sensor location is in the neighbourhood of x/L = 3/8. This implies that for high values of h_f the sensor location is going to be very close to the optimal location if placed in the region x/L = 3/8 while for low values of h_f ($h_f < 62.0W/(m^2 - K)$) the sensor location should be at x = L.

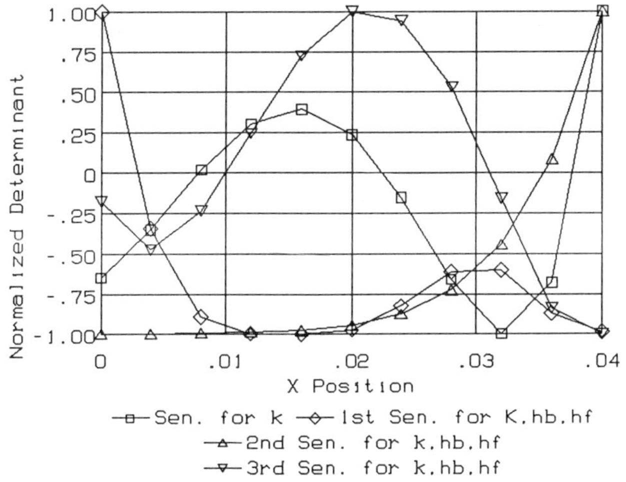

Figure 1: FIM Determinant for k (Case a) and k,h_b,h_f (Case b) (Tmax=500, h_f=50)

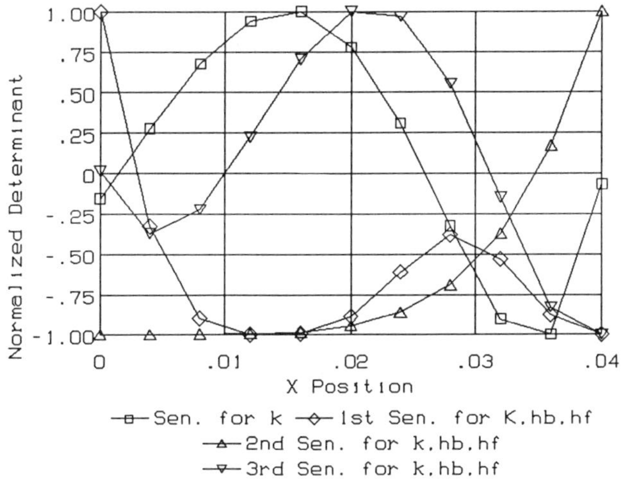

Figure 2: FIM Determinant for k (Case a) and k,h_b,h_f (Case b) (Tmax=500, h_f=100)

In case (b) when estimating three parameters, viz, k, h_b and h_f, one can use either one sensor, two sensors or three sensors in the experiment. If four sensors are used to estimate these three parameters, the optimal location for the fourth sensor is seen to coincide with one of the first three sensor location implying that nothing more will be gained by using the forth sensor. When only one sensor is used for estimation of all the three parameters the optimal location for $h_f < 500.0W/(m^2 - K)$ will always be at x=0. With the use of two sensors Figures 1, 2 and 3 show the variation of det(**M**) when the first sensor is fixed at x=0.0. It is seen that for all values of h_f the second sensor should be placed at x=L. For the use of three sensors, Figures 1, 2 and 3 represent the det(**M**) with the first two sensors fixed at x=0 and x=L respectively. Again the optimal location of the third sensor is not affected by the magnitude of h_f and remains fixed near x/L= 0.5.

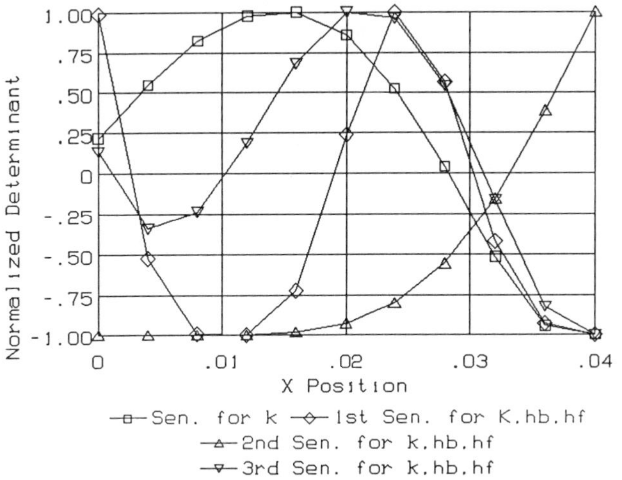

Figure 3: FIM Determinant for k (Case a) and k,h_b,h_f (Case b) (Tmax=500, h_f=500)

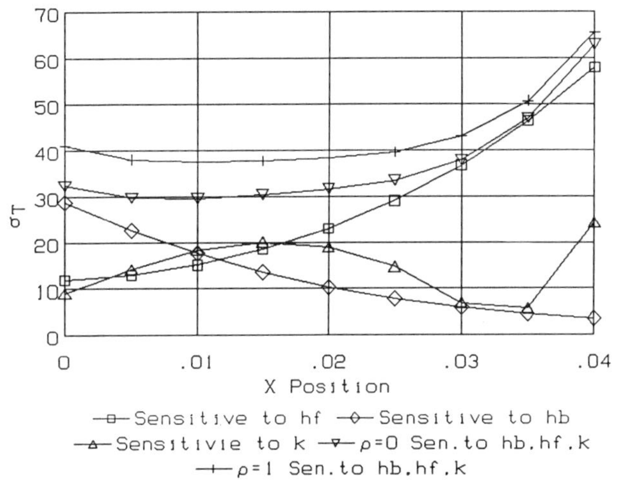

Figure 4: Sensitivity to k (Case a) and to k,h_b,h_f (Case b) (Tmax=500, h_f=50)

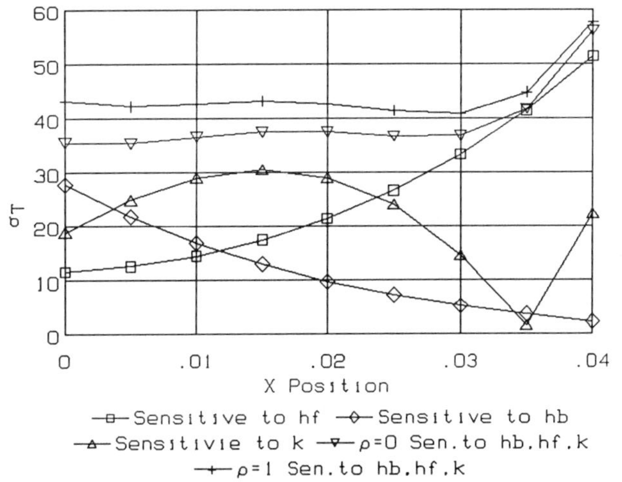

Figure 5: Sensitivity to k (Case a) and to k,h_b,h_f (Case b) (Tmax=500, h_f=100)

Figures 4, 5 and 6 show the results of the time integral sensitivity of temperature (Sensitivity analysis approach) to the desired parameters corresponding to results of det(M) from Figures 1, 2 and 3 respectively (FIM approach). For the case of a single desired parameter k (case(a)), the results obtained from Sensitivity for the optimal locations are identical to the results obtained from FIM.

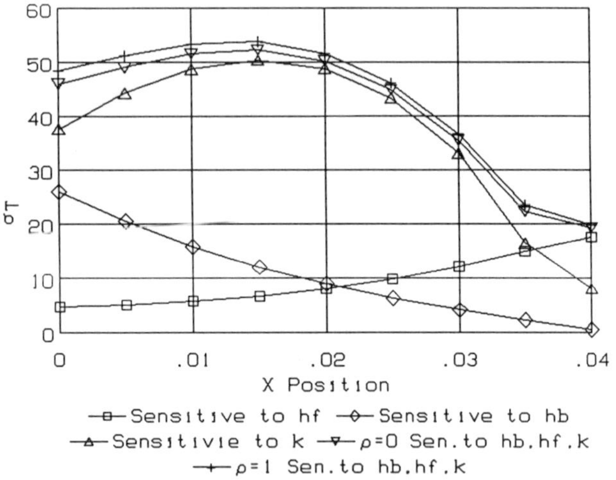

Figure 6: Sensitivity to k (Case a) and to k,h_b,h_f (Case b) (Tmax=500, h_f=500)

Figures 4, 5 and 6 also show results of temperature sensitivity to h_b and h_f indicating that the optimal sensor locations are at x=0 and x=L for estimation of h_b and h_f respectively. These sensor locations are not affected by the magnitude of h_f and so the locations are optimal for a wide range of h_f values. For the case (b) the sensitivity of temperature shows a summing effect of the sensitivity of the individual parameters (k, h_b and h_f respectively), thus the maximum sensitivity indicates the effect of the most influential parameter on temperature compared to the rest of the paramters, for example when $h_f = 50$ and $100W/(m^2-K)$, the temperature in the region with highest sensitivity (x/L = 1.0) is most influenced by h_f and only slightly affected by k or h_b, while when $h_f = 500W/(m^2-K)$ the temperature in the region with highest sensitivity (x/L = 3/8) is most influenced by k, not h_b or h_f.

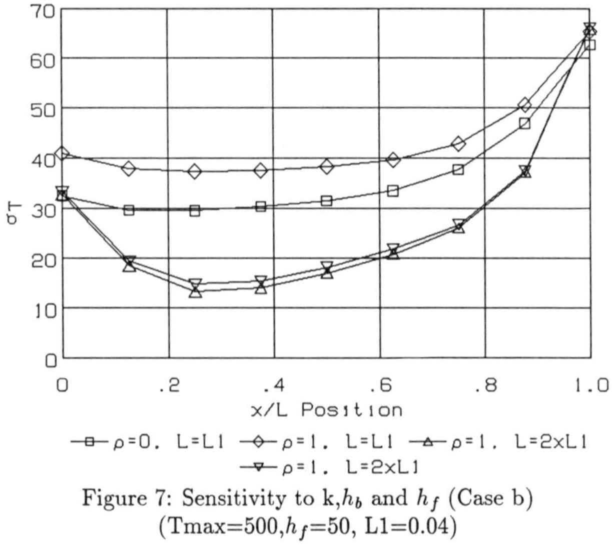

Figure 7: Sensitivity to k,h_b and h_f (Case b) (Tmax=500,h_f=50, L1=0.04)

The influence of the slab thickness (L) is presented in Figure 7. There is a significant change in the magnitude of the sensitivity but the thickness has no influence on the optimal locations of sensors.

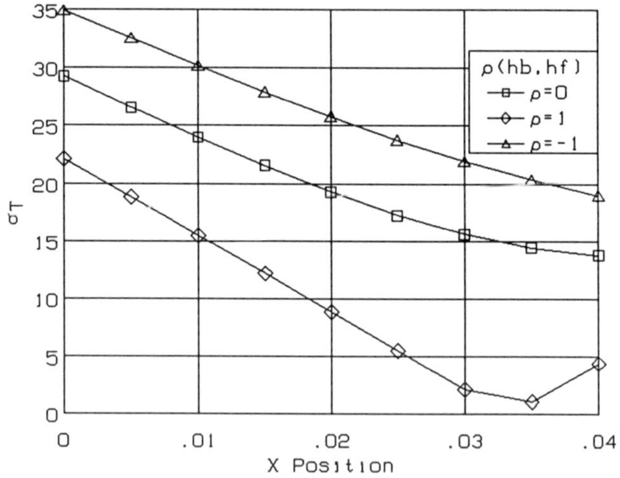

Figure 8a: Effect of Coefficient of Correlation between h_b and h_f the Sensitivity to k,h_b,h_f (with h_f=1) for steady state temperature

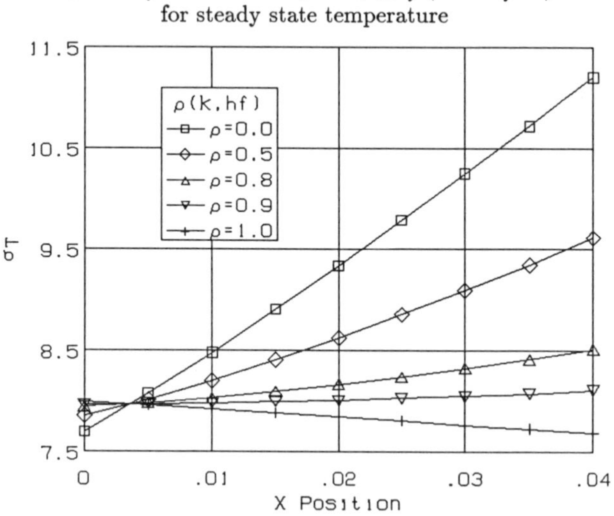

Figure 8b: Effect of Coefficient of Correlation between k and h_f on the Sensitivity to k,h_f (with h_f=1) for steady state temperature

The effect of the correlation coefficient on the sensitivity of temperature was studied by varying $\rho_{h_b h_f}$ from 0.0 (no correlation) to 1.0 (complete correlation). Only the results for the extreme correlations have been shown in Figure 4, 5 and 6, since the sensitivity exhibited a monotonic change between these two extreme correlations. The correlation did not influence the sensitivity distribution over the slab and hence the optimal location of the sensors, although there was as much as a 30% increase of sensitivity for $\rho_{h_b h_f} = 1$ compared to $\rho_{h_b h_f} = 0$. Although, the correlation coefficient did not seem to have a strong influence on the optimal location of sensors for the transient case when both the slab surfaces were exposed to the same ambient temperature of 100 C, the effect was quite substantial for the steady state case with the slab surfaces exposed to different ambient temperatures, when the conductivity and convective heat transfer coefficient on the front surface (h_f) are correlated. Figure 8a shows the effect when $\rho_{h_b h_f}$ varies from -1 to +1, the most important effect is that of $\rho_{h_b h_f} = 1$ for which a local minimum (i.e. two maximum) occurs. Figure 8b shows the effect for a steady state analysis with the front and the back surface exposed to an ambient temperature of 100 C and 200 C respectively. It can be seen that the optimal location is at x=L for ρ_{kh_b} of less than 0.9, but as ρ_{kh_f} increases the temperature sensitivity at the optimal location drops. In fact, for $\rho_{kh_f} = 1.0$ the optimal sensor location moves from x=L (for $\rho_{kh_f} = 0.0$) to x=0. Thus, the correlation between parameters can influence the optimal sensor locations and hence should be acconted for during experiment design. It is also interesting to note that there is a point near x/L=0.2 at which the sensitivity is independent of ρ_{kh_f}. It is easy to realize a correlation between h_b and h_f if both surfaces are exposed to similar flow conditions. It is unlikely to have a correlation between k and h_f on physical grounds, but if both quantities were estimated from a common experiment, such a correlation could exist.

B.) Two D Problem

Consider the problem of Figure 9. At steady state the temperature distribution is a function only of h/k and the sensitivities $\partial\phi/\partial k$ and $\partial\phi/\partial h$ are related and have the same spatial distribution. In this case the optimal location for the sensor is at x=0, y=H, and locating the sensor at any point along the convective boundary near x=0 will suffice. Because of steady state, Vand Mhave identical contours and both give the same optimal sensor location.

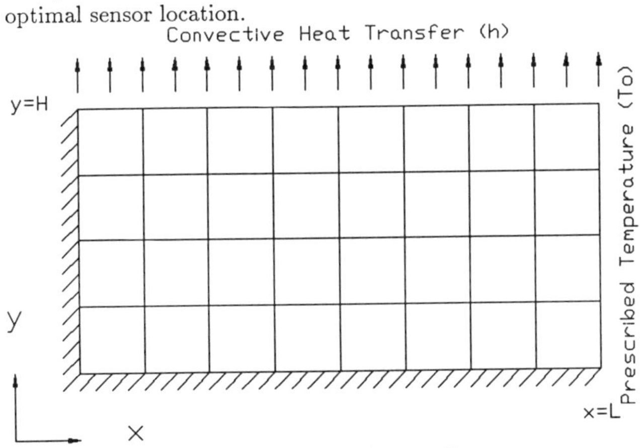

Figure 9: Rectangular Test Case

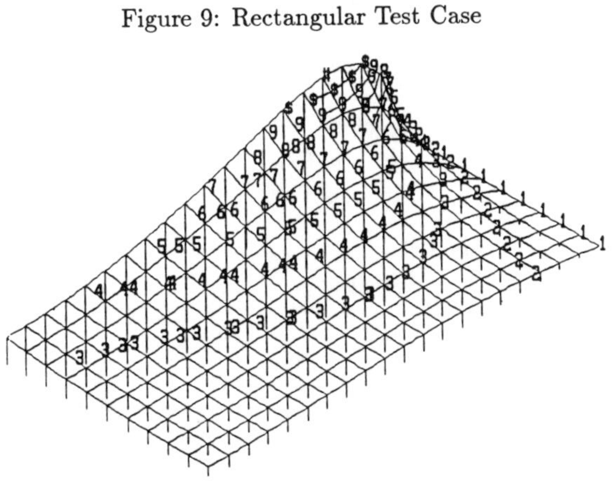

Figure 10a: Temperature Sensitivity to h at t=200sec.

The transient problem yields much different results as indicated in Figure 10a and b. The optimal location for h is on the convective boundary near x/L = 0.8, while for k it is about x/L = 0.65 and quite insensitive to the value of y – at lest for a total measurement time of 200 sec. For a time of 400 sec., the contours for sensitivity to h do not change much, while the field for k becomes even flatter near x=0. This relative flatness of the k surface is fortunate because it means that while the h sensor must be accurately placed, that for k is free to be located somewhat arbitrarily.

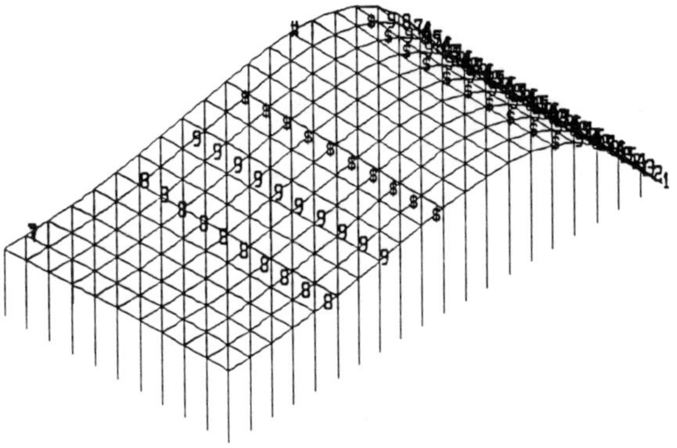

Figure 10b: Temperature Sensitivity to k at t=200sec.

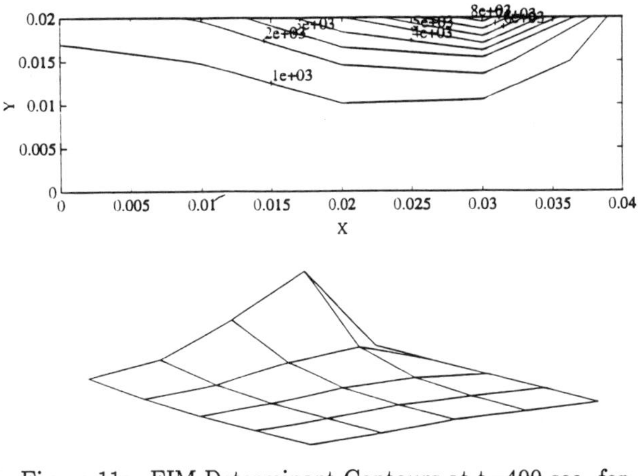

Figure 11a: FIM Determinant Contours at t=400 sec. for Estimation of k and h with one sensor fixed at x=0 and y=0

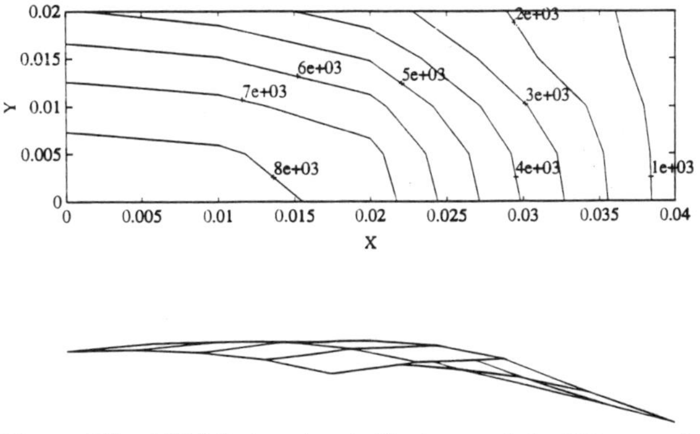

Figure 11b: FIM Determinant Contours at t=400 sec. for Estimation of k and h with one sensor fixed at x=0.03 and y=0.02

If both h and k are to be determined, two sensors can be used. Figure 11 compares the sensitivities associated with the second sensor. In Figure 11a, one sensor is located at x = y = 0 and used to estimate k. The second sensor should be located on the convective boundary and should be accurately placed – as shown by the steepness of the surface.

On the other hand, if the sensor is located at x/L = 0.75 and y/H = 1, the second sensor can be located very approximately near x=y=0. Thus some sensor locations are much more critical than others.

Conclusions

Two different approaches have been described to determine the optimal location of sensors for thermal problems. The first, based upon the Fisher information matrix, requires a knowledge of the statistical properties of the sensor signal, but not of the parameters of the system. The second, based upon sensitivity analysis, requires information about the parameters and their interactions. Although the methods are intrinsically different, if all signals have the same error characteristics, if the parameters are uncorrelated, and if the number of sensors equals the number of parameters, the results of the two approaches are almost identical. If these conditions are not met, the methods yield different results. The Fisher information matrix approach is better because it can predict optimal locations of multiple sensors, even without statistical information about the parameters. On the other hand, the sensitivity approach allows one to understand how the different parameters interact with each other and influence the temperature and to choose locations such that the parameters can be separated. This is particularly important for steady state problems in which several independent parameters may affect the temperature in the form of a combined property (e.g. thermal diffusivity). In this case, the Fisher information matrix approach may be ineffective. Because of the differing nature of the two approaches, both should be used and both may be included in a finite element method of solution.

Acknowledgements

A portion of this work was done while one of the authors (A. Nenarokomov) held a CAST grant from the National Academy of Sciences.

REFERENCES

Artyukhin, E. A., A. V. Nenarokomov, "Optimal experimental design for determining the total emissivity of materials", *High Temperatures*, **26**, (5), pp. 761-767, 1988

Artyukhin, E. A., "Experimental design of measurement of the solution of coefficient-type inverse heat conduction problem", *Journal of Engineering Physics*, **48**, (3), pp. 372-376, 1985

Beck, J. V. et al., "Joint American-Russian NSF Workshop on Inverse Problems in Heat Transfer," Michigan State University, June, 1992

Ditlevsen, O., 1981, *Uncertainty Modeling - with Application to Multidimensional Civil Engineering Systems*, McGraw - Hill.

Emery, A. and Fadale, T. D., 1990, "The Sensitivity of the Temperature Histories in the Shuttle Thermal Protection System to Radiation Heat Transfer," ASME HTD-Vol 137, pp. 81-88, AIAA/ASME Thermophysics and Heat Transfer Conference, Seattle, Wa.

Fadale, T. and Emery, A. F., "Transient Effects of Uncertainties on the Sensitivities of Temperatures and Heat Fluxes using Stochastic Finite Elements," Submitted to the *Journal of Heat Transfer*, Paper No. 92-V-775, 1992

Fedorov, V. V., *Theory of Optimal Experiment,* Academic Press, 1972

Gill, P. E. et al., "User's Guide for NPSOL - A Fortran Package for Nonlinear Programming," Department of Operations Research, Stanford University, Palo Alto, Calif., 1986

Goodwin, G. E. and Payne, R. L., *Dynamic System Identification. Experiment Design and Data Analysis,* Academic Press, 1977

Gorton, M. P., and Shideler, J. L., "Measured and Calculated Temperatures of a Superalloy Honeycomb Thermal Protection System Panel," Proc. Workshop on Correlation of Hot Structures Test Data with Analysis, NASA Ames Dryden Flight Research Center, 1988

Haftka, R. T. and Kamat, M. P., 1985, *Elements of Structural Optimization*, Martinus Nijhoff Publishers.

Kubrusly, C. S. and Malebranche, H., "A Survey on Optimal Sensors and Controllers Locations in DPS," *IFAC 3rd Symposium, Control of Distributed Parameter Systems*, pp 59-73, 1982

Lin, Y. K., 1967, *Probabilistic Theory of Structural Dynamics*, McGraw - Hill.

Musylev, N. V., "Uniqueness Theorems for Certain Inverse Heat Conduction Problems," *Zh. Vychisl. Mat. Fiz.*, vol 20., Num 2, pp. 388 - 400, 1980

Polis, M. P., "The Distributed System Parameter Identification Problem: A Survey of Recent Results," *IFAC 3rd Symposium, Control of Distributed Parameter Systems*, pp 45-58, 1982

Vanmarcke, E., 1977, "Probabilistic Modeling of Soil Profiles", *Journal of the Geotechnical Engineering Division*, **103**, GT11.

AUTHOR INDEX

Inverse Problems in Engineering: Theory and Practice

Book Number: I00357